Advances in Delivery Science and Technology

Series Editor
Michael J. Rathbone

For further volumes:
http://www.springer.com/series/8875

Abraham J. Domb • Wahid Khan
Editors

Focal Controlled Drug Delivery

Springer

Editors
Abraham J. Domb
Institute of Drug Research
School of Pharmacy-Faculty of Medicine
The Hebrew University of Jerusalem
Jerusalem, Israel

Wahid Khan
Department of Pharmaceutics
National Institute of Pharmaceutical Education and Research (NIPER)
Balanagar, Hyderabad, India

ISSN 2192-6204 ISSN 2192-6212 (electronic)
ISBN 978-1-4614-9433-1 ISBN 978-1-4614-9434-8 (eBook)
DOI 10.1007/978-1-4614-9434-8
Springer New York Heidelberg Dordrecht London

Library of Congress Control Number: 2013957870

Printed on acid-free paper

Springer is part of Springer Science+Business Media (www.springer.com)

Preface

The concept of focal drug delivery has been applied for treating illnesses that are localized to a certain tissue or organ. These delivery systems are applied directly to the diseased site and deliver a desired dose for an extended time period while minimizing systemic distribution of toxic drug. Overall, this book contains two sections: first section includes fundamental introductory chapters for focal drug delivery, whereas second section includes chapters describing drug delivery to body sites/system.

Biodegradable polymers have been playing a very important role in delivery aspects of therapeutic molecules because of their biocompatibility and biodegradability. Chapter 1 discusses the importance of these polymeric carriers in focal controlled drug delivery and explores a wide range of polymers, including those from natural and synthetic sources. Chapter 2 presents the role of implantable medical devices such as stents in controlled local drug delivery. Various classes and requirements of implantable devices such as mechanical properties, biocompatibility, and sterilization have been listed in this chapter. A detail about focal drug delivery applications of cardiovascular and orthopaedic implantable devices has been provided.

Tumor targeting comes first in discussion in emphasizing the role of systemic targeting. Chapter 3 provides various means of drug targeting like polymeric nanoparticles, liposomes, polymersomes, and solid lipid nanoparticles, the role and endogenous factors making an impact on EPR effect in tumor targeting along with the ligand-based targeting based on carbohydrates, proteins, antibodies, aptamers, and small molecules for various malignancies. Chapter 4 discusses the emerging role of liposomes in focal drug delivery for cancer and noncancer therapies. Chapter 5 emphasizes the expanding role of polymer drug conjugates and suggests their potential in drug delivery platforms. An understanding of in vivo pharmacokinetics and pharmacodynamics of a delivery system is important in determining the clinical effectiveness of these systems in site-specific targeting. Chapter 6 focuses on the anticancer drug delivery systems for treatment of solid tumors and on the quantitative assessment of the analyzed factors/parameters.

Chapter 7 covers various treatment modalities in brain tumor, describing a brief about locally delivered chemotherapy following surgical resection. Strategies like drug delivery microchips, gene-targeted drugs, and nanocarriers have been explored in their role in effective delivery in this aspect. Chapter 8 describes intranasal administration of neuropeptide-loaded nanoparticles as a feasible means to noninvasively deliver neuropeptides for CNS therapeutics. Chapter 9 examines the scope of focal drug delivery in inner ear therapy, comparing advantages and disadvantages of different techniques like intratympanic perfusion, organ-targeted delivery, and direct cochlear drug delivery. Chapters 10 and 11 focus on the emerging role of nanotechnology in ocular drug delivery that presents a big challenge to current therapies because of its complex anatomical and physiological barriers and the major types of nucleic acids and various strategies that have been used to achieve site-specific delivery of nucleic acids to the eye for the treatment of ocular diseases. Chapter 12 gives a brief about basic mechanisms of transport in iontophoretic drug delivery systems in localized delivery of drugs in addition to its growing investigating potential in treatment of various diseases related to eye and skin. A drug designated to act locally in the oral cavity has to remain in the site for a measured period of time and withstand the dynamic conditions in the mouth such as changing pH levels, masticatory abrasion, slippery mucosa, and smooth teeth surfaces. Chapter 13 focuses on potential of polymeric carriers in treating various oral cavity diseases of bacteriological, viral, and fungal origin. The major reason for the resistance of the oral microbes is their inherent organization into characteristic biofilms. Chapter 14 begins with the introduction of these biofilms and advantages of focal drug delivery in this aspect and concludes with the recent development of various novel technologies for the prevention, control, and treatment of oral infections including controlled focal delivery modalities. Chapter 15 discuses focal drug delivery to stomach that includes gastro-retentive drug delivery systems. A brief for gastro-retentive dosage form (GRDF), different GRDF technologies, and their unique application has been discussed. Chapter 16 discusses focal drug delivery to intestine emphasizing variable intestinal conditions like environmental pH values, transporter expression levels, and CYP3A4 expression necessary for a successful targeted drug delivery.

Chapter 17 gives a review of the challenges and opportunities in systemic and local drug delivery to the arterial tissue, advances in systemic (e.g., targeted nanotechnology-based formulations) and local (e.g., drug-eluting stent (DES) implantation) delivery technologies, and their future perspective in the development of multifunctional nano-systems resulting in localized intracellular drug delivery with improved efficacy. Chapters 18 and 19 focus on the role of stents and vascular grafts in localized drug delivery to vascular tissues. The first gives the readers an introduction about bare-metal stents (BMS) that have been used in coronary artery diseases, long-term result problems of in-stent restenosis (ISR), and stent thrombosis resulting in introduction of DES to overcome the problems of ISR along with the advancement to development of a more user-friendly bioabsorbable and polymer-free stents. The second one discusses the advantages of drug-eluting vascular graft compared to coronary artery bypass grafting as well as trends in their development with a prime focus on electrospinning as a promising platform technology for creating a new generation of vascular grafts. Chapter 20 introduces the importance of and major hurdles for lymphatic targeting as well as the potential of new delivery platforms for nanocarriers such as liposomes, solid lipid nanoparticles, etc. for better penetration into diseased areas. Osteomyelitis, a disease of bone and bone marrow, presents a major challenge in therapy. The challenges along with the potential of newer forms of sustained-release antibiotic delivery systems in delivering antibiotics at constant rates over a prolonged period of time, eliminating the need for multiple dosing, are indicated in Chapter 21. Chapter 22 reviews an up-to-date overview of the acellular biomaterial-based strategies, aimed at simultaneous regeneration of bone and cartilage by the controlled focal delivery of the appropriate factors. Development and clinical testing of various delivery systems (microspheres, hydrogels, and macroporous scaffolds) are also discussed.

Chapter 23 presents locally and systemically delivered polymeric drug carrier systems in treatment of solid tumors. Different types of in situ forming injectable hydrogels, various micro- and nanoparticulate systems, and the role of polymeric drug delivery systems in addressing the multidrug resistance are discussed. Chapter 24 describes the advantages offered by multifaceted nanocarrier systems over current conventional formulations for skin ailments and their research and market potential along with factors vital for selection of appropriate nano delivery system. Chapter 25 deals with anatomy of the nail unit, related diseases, and challenges presented by topical and oral anti-infective. It also explores the potential of focal delivery approaches to nails that are currently being investigated. In Chap. 26, the reader is presented with an overview of different wound management dressings and advanced technologies for achieving improved healing for wounds, burns, and diabetes-related ulcers and related pathogenesis. Also pharmacological agents like gene therapy and cytokine, growth factors, stem cells, etc. are also discussed. Chapter 27 previews the potential of vaginal drug delivery systems focusing on the barriers presented by vaginal mucosa. The various strategies discussed include nanomaterials like nanoparticles, electrospun fibers, and HIV microbicide applications for effective and successful vaginal drug delivery. Lastly, discussed in Chap. 28 are the selected developments in the extensive field of prolonged duration of local anesthesia for the goal of enhanced anesthetic duration following administration.

Jerusalem, Israel — Abraham J. Domb
Hyderabad, India — Wahid Khan

Contents

Contributors

Mansoor Amiji Department of Pharmaceutical Sciences, School of Pharmacy, Northeastern University, Boston, MA, USA

Nurit Beyth Faculty of Dentistry, The Hebrew University—Hadassah, Jerusalem, Israel

Department of Prosthodontics, The Hebrew University-Hadassah School of Dental Medicine, Jerusalem, Israel

Hansen Bow Department of Neurosurgery, School of Medicine, The Johns Hopkins University, Baltimore, MD, USA

Henry Brem Department of Neurosurgery, Oncology and BioMedical Engineering, School of Medicine, The Johns Hopkins University, Baltimore, MD, USA

Yaron S. Brin Department of Orthopaedic Surgery, Meir Medical Center, Tel Aviv University, Kfar-Saba, Israel

Venu Gopala Swami Challa Department of Pharmaceutics, National Institute of Pharmaceutical Education and Research (NIPER), Hyderabad, India

Kun Cheng Division of Pharmaceutical Sciences, School of Pharmacy, University of Missouri-Kansas City, Kansas City, MO, USA

Smadar Cohen Avram and Stella Goldstein-Goren Department of Biotechnology Engineering, Ben-Gurion University of the Negev, Beer-Sheva, Israel

The Ilse Katz Institute for Nanoscale Science and Technology, Ben-Gurion University of the Negev, Beer-Sheva, Israel

The Regenerative Medicine and Stem Cell (RMSC) Research Center, Ben-Gurion University of the Negev, Beer-Sheva, Israel

Arik Dahan Department of Clinical Pharmacology, School of Pharmacy, Faculty of Health Sciences, Ben-Gurion University of the Negev, Beer-Sheva, Israel

Preshita P. Desai Department of Pharmaceutical Sciences and Technology, Institute of Chemical Technology, Matunga, Mumbai, India

Dipti Deshpande Department of Pharmaceutical Sciences, School of Pharmacy, Northeastern University, Boston, MA, USA

Fahima Dilnawaz Nanomedicine Laboratory, Institute of Life Sciences, Bhubaneswar, Orissa, India

Abraham J. Domb Faculty of Medicine, Institute of Drug Research, School of Pharmacy, The Hebrew University of Jerusalem, Jerusalem, Israel

Sean Essex Department of Pharmaceutical Sciences, Center for Pharmaceutical Biotechnology and Nanomedicine, Northeastern University, Boston, MA, USA

Shady Farah Faculty of Medicine, Institute of Drug Research, School of Pharmacy, The Hebrew University of Jerusalem, Jerusalem, Israel

Taís Gratieri Faculdade de Ciências da Saúde, Universidade de Brasília, Brasília, DF, Brazil

School of Pharmaceutical Sciences, University of Geneva & University of Lausanne, Geneva, Switzerland

Uri Hadelsberg Technion-Israel Institute of Technology, Haifa, Israel

Jingjia Han Department of Pediatrics, School of Medicine, Emory University and Parker H. Petit Institute of Biotechnology and Biosciences, Georgia Institute of Technology, Atlanta, GA, USA

Orith Haramaty Department of Prosthodontics, The Hebrew University-Hadassah School of Dental Medicine, Jerusalem, Israel

Michael Jaffe Medical Device Concept Laboratory, New Jersey Institute of Technology, Newark, NJ, USA

Sonam Jain Department of Pharmaceutics, National Institute of Pharmaceutical Education & Research (NIPER), Mohali, Punjab, India

Aziza Jamal-Allial Department of Pharmaceutical Sciences, School of Pharmacy, Northeastern University, Boston, MA, USA

Yogeshvar N. Kalia School of Pharmaceutical Sciences, University of Geneva & University of Lausanne, Geneva, Switzerland

Wahid Khan Department of Pharmaceutics, National Institute of Pharmaceutical Education and Research (NIPER), Balanagar, Hyderabad, India

Daniel S. Kohane Department of Anesthesiology, Laboratory for Biomaterials and Drug Delivery, Division of Critical Care Medicine, Boston Children's Hospital, Harvard Medical School, Boston, MA, USA

Division of Health Sciences and Technology, Harvard-Massachusetts Institute of Technology, Boston, MA, USA

Emily A. Krogstad Department of Bioengineering, University of Washington, Seattle, WA, USA

Daniel J. Kubek Department of Anatomy and Cell Biology, Indiana University School of Medicine, Indianapolis, IN, USA

School of Dentistry, University of Louisville, Louisville, KY, USA

Michael J. Kubek Department of Anatomy and Cell Biology, Indiana University School of Medicine, Indianapolis, IN, USA

Department of Psychiatry, Indiana University School of Medicine, Indianapolis, IN, USA

Program in Medical Neuroscience, Indiana University School of Medicine, Indianapolis, IN, USA

Dinesh Kumar Department of Pharmaceutics, National Institute of Pharmaceutical Education and Research (NIPER), Hyderabad, India

Neeraj Kumar Department of Pharmaceutics, National Institute of Pharmaceutical Education & Research (NIPER), Mohali, Punjab, India

Robert Langer Department of Chemical Engineering, Massachusetts Institute of Technology, Cambridge, MA, USA

Peter I. Lelkes Department of Bioengineering, College of Engineering, Temple University, Philadelphia, PA, USA

J. Brian McAlvin Department of Medicine, Medicine Critical Care Program, Boston Children's Hospital, Harvard Medical School, Boston, MA, USA

Laboratory for Biomaterials and Drug Delivery, Department of Anesthesiology, Division of Critical Care Medicine, Boston Children's Hospital, Harvard Medical School, Boston, MA, USA

Division of Health Sciences and Technology, Harvard-Massachusetts Institute of Technology, Cambridge, MA, USA

Eameema Muntimadugu Department of Pharmaceutics, National Institute of Pharmaceutical Education and Research (NIPER), Hyderabad, India

Sudaxshina Murdan Department of Pharmaceutics, UCL School of Pharmacy, London, UK

Meir Nyska Department of Orthopaedic Surgery, Meir Medical Center, Tel Aviv University, Kfar-Saba, Israel

Vandana B. Patravale Department of Pharmaceutical Sciences and Technology, Institute of Chemical Technology, Matunga, Mumbai, India

Prasad V. Pawar Department of Pharmaceutics, National Institute of Pharmaceutical Education & Research (NIPER), Mohali, Punjab, India

Rajendra P. Pawar Department of Chemistry, Deogiri College, Aurangabad, India

Michael Perez-Davidi Department of Prosthodontics, Hadassah School of Dental Medicine, Hebrew University, Jerusalem, Israel

David Polak Department of Prosthodontics, The Hebrew University-Hadassah School of Dental Medicine, Jerusalem, Israel

Gustavo Pradilla Department of Neurosurgery, School of Medicine, The Johns Hopkins University, Baltimore, MD, USA

Sheeba Qureshi Braun School of Public Health, The Hebrew University of Jerusalem, Jerusalem, Israel

Sistla Ramakrishna Department of Pharmaceutics, National Institute of Pharmaceutical Education and Research (NIPER), Hyderabad, India

Department of Pharmacology, Indian Institute of Chemical Technology, Hyderabad, India

Michael J. Rathbone School of Pharmacy, International Medical University, Kuala Lumpur, Malaysia

Emil Ruvinov Avram and Stella Goldstein-Goren Department of Biotechnology Engineering, Ben-Gurion University of the Negev, Beer-Sheva, Israel

Sanjeeb Kumar Sahoo Nanomedicine Laboratory, Institute of Life Sciences, Bhubaneswar, Orissa, India

Kinjal Sankhe Department of Pharmaceutical Sciences, School of Pharmacy, Northeastern University, Boston, MA, USA

Nalini Shastri Department of Pharmaceutics, National Institute of Pharmaceutical Education and Research (NIPER), Hyderabad, India

Ariella Shikanov Lurie Biomedical Engineering, University of Michigan, Ann Arbor, MI, USA

Ravi S. Shukla Division of Pharmaceutical Sciences, School of Pharmacy, University of Missouri-Kansas City, Kansas City, MO, USA

Indu Singh Department of Pharmaceutics, National Institute of Pharmaceutical Education and Research (NIPER), Hyderabad, India

David Stepensky Department of Clinical Biochemistry and Pharmacology, Faculty of Health Sciences, Ben-Gurion University of the Negev, Beer-Sheva, Israel

Ian Suk Department of Neurosurgery, The Johns Hopkins University School of Medicine, Baltimore, MD, USA

Rajan Swami Department of Pharmaceutics, National Institute of Pharmaceutical Education and Research (NIPER), Hyderabad, India

Rajesh Thipparaboina Department of Pharmaceutics, National Institute of Pharmaceutical Education and Research (NIPER), Hyderabad, India

Vladimir Torchilin Department of Pharmaceutical Sciences, Center for Pharmaceutical Biotechnology and Nanomedicine, Northeastern University, Boston, MA, USA

Betty M. Tyler Department of Neurosurgery, School of Medicine, The Johns Hopkins University, Baltimore, MD, USA

Michael C. Veronesi Department of Radiology, University of Chicago School of Medicine, Chicago, IL, USA

Judah Z. Weinberger Cardiovascular Division, Department of Medicine, Columbia University, New York, NY, USA

Ervin I. Weiss Department of Prosthodontics, The Hebrew University-Hadassah School of Dental Medicine, Jerusalem, Israel

Omri Wolk Department of Clinical Pharmacology, School of Pharmacy, Faculty of Health Sciences, Ben-Gurion University of the Negev, Beer-Sheva, Israel

Kim A. Woodrow Department of Bioengineering, University of Washington, Seattle, WA, USA

Jing Zou Hearing and Balance Research Unit, field of Otolaryngology, School of Medicine, University of Tampere, Tampere, Finland

Department of Otolaryngology-Head & Neck Surgery, Center for Otolaryngology-Head & Neck Surgery of Chinese PLA, Changhai Hospital, Second Military Medical University, Shganghai, China

Part I
Introductory Chapters

Chapter 1
Biodegradable Polymers for Focal Delivery Systems

Wahid Khan, Venu Gopala Swami Challa, Robert Langer, and Abraham J. Domb

1.1 Introduction

Polymer science has been undergoing tremendous advancements during the past three decades in terms of chemical modifications with desirable characters and synthesis of copolymers in the development of new delivery systems. Several strategies have been explored to deliver drugs to a specific site or body compartment, but delivery via polymeric carriers is one of the simplest and most successful approaches due to its physicochemical properties. Biodegradable polymers for focal drug delivery application can play structural and functional roles. They are classified in two categories, viz., natural and synthetic based on their origin [1] (Fig. 1.1).

Natural polymers are the first biodegradable biomaterials used clinically. They possess several inherent advantages such as bioactivity, the ability to present receptor-binding ligands to cells, susceptibility to cell-triggered proteolytic

W. Khan • V.G.S. Challa
Department of Pharmaceutics, National Institute of Pharmaceutical Education and Research (NIPER), Hyderabad 500037, India

R. Langer
Department of Chemical Engineering, Massachusetts Institute of Technology, Cambridge, MA 02139, USA

A.J. Domb (✉)
Faculty of Medicine, Institute of Drug Research, School of Pharmacy, The Hebrew University of Jerusalem, Jerusalem 91120, Israel
e-mail: avid@ekmd.huji.ac.il

A.J. Domb and W. Khan (eds.), *Focal Controlled Drug Delivery*, Advances in Delivery Science and Technology, DOI 10.1007/978-1-4614-9434-8_1,

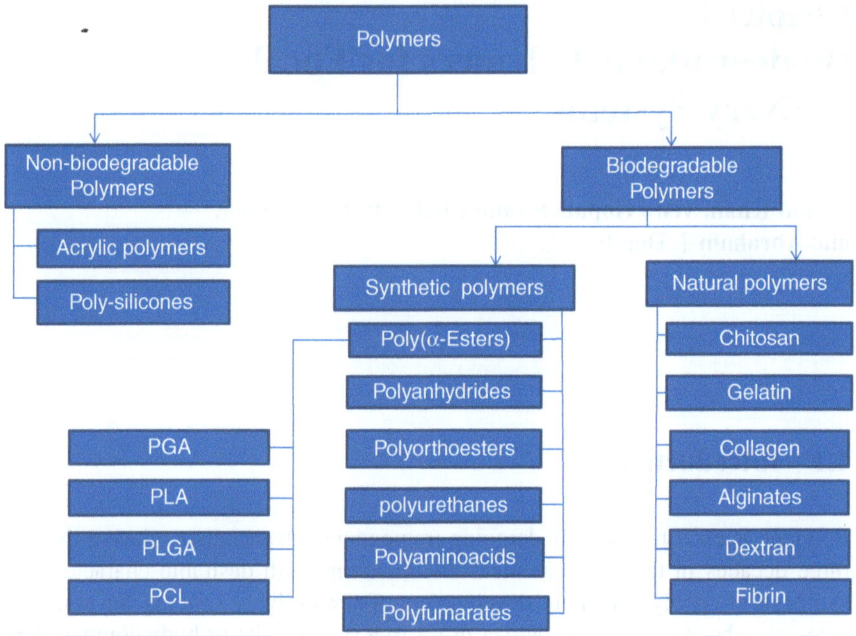

Fig. 1.1 Polymers used in focal drug delivery systems

degradation, and natural remodeling. However, they have drawbacks that include a strong immunogenic response associated with most of the polymers, complexities associated with purification, and the possibility of disease transmission. Synthetic biomaterials on the other hand are generally biologically inert; they possess batch-to-batch uniformity and have an advantage of possessing tailored property profiles for specific applications. Thus, they are devoid of many of the disadvantages of natural polymers. Extensive research has examined custom-designed biodegradable polymer systems with predictable erosion kinetics as drug/gene delivery vehicles or as scaffolds for tissue engineering.

There are numerous advantages of focal drug therapy [2]:

1. Dose-dependent activity can be regulated.
2. Therapeutically effective dose required at the site is reduced, thus diminishing or completely eliminating adverse effects.
3. Drugs with a relatively short half-life can also be delivered locally with minimal loss in therapeutic activity.
4. Drugs with low bioavailability can be targeted directly to the required site.
5. Patient-to-patient variability in drug pharmacokinetics is reduced, which is of significance in the case of drugs with a narrow therapeutic index.

6. The need of premedication for drugs that show adverse effects when given systemically is obviated.

This chapter provides a review of both introductory and state-of-the-art information about the polymeric carriers used in focal drug therapy for diseased states that are most exploited for localized treatment. Major emphasis is placed on the biodegradable polymers of both natural and synthetic origins, focusing on their sources, designs, physicochemical properties, and applications.

1.2 Natural Polymers

Natural polymers have the advantage of possessing an intrinsic property of environmental responsiveness via degradation, bioactivity, the ability to present receptor-binding ligand to cells, and remodeling by cell-secreted enzymes. They are generally nontoxic, even at high concentrations. Therefore, they can be fabricated, as a tissue-engineering scaffold, or used as a growth factor delivery system, e.g., chitosan. They have been successfully used in oral and nasal delivery due to their mucoadhesive properties [3]. However, they suffer from some drawbacks such as immunogenic response, batch-to-batch variation, restrictions with the versatility of designing devices having specific biomechanical properties, variable rate of in vivo degradation, and the possibility of disease transmission [3].

1.2.1 Collagen

Collagen is the most abundant protein in the animal kingdom and is a major component of the extracellular matrix and connective tissues. It renders strength and flexibility to connective tissues such as tendons, bones, cartilage, blood vessels, skin, and other musculoskeletal tissues. The main applications of collagen as a drug delivery system are collagen shields in ophthalmology, sponges for burns/wounds, mini-pellets and tablets for protein delivery, gel formulation in combination with liposomes for sustained drug delivery, as controlling material for transdermal delivery, nanoparticles for gene delivery, and basic matrices for cell culture systems [4]. Collagen has broad utility as gene-activated matrices capable of delivering large quantities of DNA in a direct, localized manner [3]. A modulated type II collagen-glycosaminoglycan (CG) scaffold serves as a nonviral gene delivery vehicle for transferring gene-seeded adult articular chondrocytes that produces a prolonged and local expression of insulin-like growth factor (IGF)-1 for enhancing cartilage regeneration [5]. Type 1 collagen-based viscous bulking material can serve as a retrievable implant for systemic delivery of erythropoietin [6]. Today collagen is widely used as a coating material for synthetic biodegradable polymers like poly-ε-caprolactone and poly(D,L-lactide-co-glycolide) (PLGA) in controlled delivery of drugs and in tissue engineering [7, 8].

Collagen matrices sponges are available for tissue engineering of skeletal muscle; they are helpful in the treatment of soft tissue defects in reconstructive surgery [9]. Collagen scaffolds for corneal tissue engineering have been reported. A modification of collagen is a biofunctional polymer to enhance its properties in tissue engineering. The effects of the peptide modified collagen gels on corneal epithelial cell behavior were examined to improve the potential of these materials as tissue-engineering scaffolds [10]. The use of UroMaix scaffolds was reported in tissue engineering of the urinary tract. Generally, in the urinary tract reconstruction of bladder and ureter tissue is indicated in cases of injury, stenosis, infection, or tumor. Substitution by ileum, colon, or pure synthetic polymers generates a variety of complications. Collagen has promising advantages over synthetic polymers which support the attachment and proliferation of urinary tract cells in tissue engineering of the urinary tract with its high mechanical demands [11].

Collagen has been used for tissue engineering including skin replacement, bone substitutes, and artificial blood vessels and valves. There has been extensive collagen research into an array of other tissue-engineering and biomedical applications owing to its mechanical, hemostatic, and cell-binding properties. Collagen already has a wide range of applications as hemostats, implants, device coatings, and stabilizers for biologics. Examples are collagen sponges, collagen membranes, collagen stents, and vascular graft coatings. Revitix™, Vctoi™, Forta-Derm™, Infuse®, Collagraft®, Healos®, Biomend, Integra® Dermal Regeneration Template, Orcel®, and Apligraf® are a few FDA-approved products of collagen [12].

1.2.2 *Gelatin*

Gelatin is derived by denaturing collagen and is hence free of antigenicity associated with collagen. Due to its promising properties such as biodegradability and biocompatibility, gelatin is widely used in drug delivery. It also has potential properties as a controlled local drug delivery system, e.g., cancellous bone applications [13]. Recent studies have explored gelatin usage in the treatment of tuberculosis through controlled release of drugs in the gastrointestinal tract. Rifampicin-containing microspheres were designed to treat and prevent tuberculosis infection in the entire intestine [14]. Gelatin shows excellent results in regional drug delivery systems due to its inherent properties like biodegradability and biocompatibility, being absorbed into the human body. There has been a report of colloidal drug delivery of gelatin in the form of microspheres loaded with curcumin for targeting lung diseases [15]. Gelatin drug conjugates release a drug molecule in a specified site for treatment. Cisplatin-conjugated gelatin microspheres are a potential antitumor agent; these microspheres confirm the release of cisplatin to prevent solid tumors but not metastasis tumors [16]. Cross-linked gelatin microspheres have reported as a drug delivery system for doxycycline to treat chronic wounds. These microspheres have also been used for tissue-engineering applications and as hemostats, implants, coating device, and stabilizers for biologics. Gelfoam® is an absorbable gelatin sponge available in powder form by milling the gelatin sponges. Some other gelatin-based products in the market are Surgifoam®, CultiSpher-G®, and H.P. Acthar® gel, all approved by the FDA.

1.2.3 Fibrin

Fibrin (derived from fibrinogen) is one of the earliest biopolymers used as a biomaterial, due to its excellent biocompatibility, biodegradability, and injectability. Fibrin sealants (also known as tissue adhesives or glues) are designed as hemostasis agents by mimicking the final steps of the blood coagulation cascade, forming a stable, physiological fibrin clot that assists hemostasis and wound healing; they are also used in the delivery of many anticoagulant agents in hemorrhagic conditions [17]. Fibrin seal compounds are also used to carry out antibiotic formulation for the prolonged drug therapy of localized bacterial infections [18]. Fibrin sealants can be useful for transscleral and the uniform drug delivery of fluorescein-labeled dexamethasone, and methotrexate for the posterior segment diseases such as rheumatoid arthritis [19]. Fibrin films are widely used in ophthalmic preparations for coating of antiglaucoma drugs like pilocarpine for their long-term release profile [20].

For drug delivery systems fibrin can be used either as microparticles or fibrin-coated particulates for sustained release or as fibrin sheets suitable as implant material [21]. Fibrin as a hydrogel carrier is widely used in cartilage tissue engineering. Fibrin is rapidly degraded with plasmin, produced by the cells. Thus, ε-aminocaproic acid can be used to inhibit this enzyme and thus save the fibrin carrier [22]. Sugitachi et al. studied drug-entrapped fibrin clots for regional cancer chemotherapy and showed a sustained release of oncolytic agents and favorable antineoplastic effects [23]. Fibrin is a versatile biopolymer used in tissue regeneration and wound healing applications. It is a crucial blood component responsible for hemostasis and used extensively as a biopolymer scaffold in tissue engineering in the form of a fibrin hydrogel by fibrinogen and thrombin. The incorporation of bioactive peptides and growth factors via a heparin-binding delivery system improves the functionality of fibrin as a scaffold for stem or primary cells to regenerate adipose tissue, bone, cardiac tissue, cartilage, liver, and nerve tissue [24]. Breen et al. reported fibrin as a potent carrier for gene transfer and controlled vector delivery. Fibrin scaffold is used to enhance delivery of the adenovirus to a wound site to regenerate the damaged tissue [25]. In cardiovascular tissue engineering, fibrin-based scaffolds are widely used in encapsulated gel form as a cell carrier by providing structural integrity to the developing tissue [26]. Recent studies report that fibrin gel can inhibit potential immunogenic, toxic degradation, and inflammatory reactions of three-dimensional biodegradable scaffolds, which are generally used as a basic structure for cell anchorage, cell proliferation, and cell differentiation in tissue engineering [27]. Tisseel®, Evicel®, Crosseal®, and Bioseed® are a few of the fibrin-based products available for clinical use.

1.2.4 Silk Fibroin

Silk is a protein polymer spun into fibers by silkworms, spiders, scorpions, mites, and flies. Degradable silk is a mechanically robust biomaterial that offers a wide range of mechanical and functional properties for biomedical applications, including drug delivery. Silk proteins are a promising material for controlled drug delivery due to their aqueous processability, biocompatibility, and biodegradability [28]. Silk fibroin is an interesting polymer for drug delivery due to its controllable level of crystallinity and the ability to process the biomaterial in biocompatible fashion under ambient conditions to avoid damage to labile compounds to be delivered [29].

Regenerated silk fibroin is another promising biomaterial for localized drug delivery systems [30]. Wenk et al. studied the fabrication of drug-loaded silk fibroin spheres as a platform for the controlled release of a sensitive biological-like, insulin-like growth factor in diabetic disorders [31]. Silk fibroin microparticles offer a promising approach for the sustained delivery of different bone morphogenetic proteins (BMPs) in tissue-engineering applications. BMPs like BMP-2, BMP-9, or BMP-14 are cytokines with the strong ability to promote new bone formations [32]. Silk fibroin-coated liposomes are used in long-term and targeted drug delivery of many oncolytic agents. Silk fibroin-coated, emodin-loaded liposomes show prominent diffusion kinetics and specific cell targeting in tumor sites [33].

1.2.5 Alginates

Alginates are naturally derived polysaccharides. A block copolymer of alginate contains two uronic acid units. Due to the presence of high acid content in alginates, alginate undergo reversible gelation in aqueous solution under mild conditions through interaction with divalent cations such as Ca^{2+}. High acid reactivity allows their wide use as cell transplantation vehicles to grow new tissues, as wound dressings and also in the three-dimensional culture of chondrocytes. Alginates have several applications as scaffolding material for bone regeneration, as wound dressings and as drug and cell delivery vehicles, chemically immobilizing the drug to the polymer back bone.

Alginate is a biodegradable hydropolymer widely used in the local drug delivery of steroid hormones. In a study, Bhowmik et al. slowed testosterone-loaded nanocapsules prepared and evaluated, showing sustained-release pattern [34]. Alginates showed significant effects on deoxycholic-acid-induced cancer in esophageal mucosal. Reflux of bile acids into the esophagus is implicated in the progression to Barrett's oesophagus (a metaplastic condition and precursor to oesophageal adenocarcinoma), since bile acids at pH 4 have been shown to induce c-myc expression. Alginates were able to prevent the induction of c-myc by acidified deoxycholic acid in vitro [35]. Several alginate-based products are commercially available, that includes Nu-Derm®, Curasorb®, and AlgiSite®.

1.2.6 Dextran

These are polysaccharides available in a wide range of molecular weights. They contain a high density of hydroxyl groups that make the polymers highly hydrophilic and capable of being further functionalized chemically. They are also used as the hydrophilic part of amphiphilic block copolymers. Dextran is widely used in focal delivery of antitumor agents like daunorubicin, doxorubicin as gold, and coated nanoparticles. Cai et al. prepared biocompatible gold nanoparticle-loaded lysozyme-dextran (Au@Lys-Dex) nanogels. The antitumor agent, doxorubicin-loaded Au@Lys-Dex nanogels, has antitumor activity similar to free doxorubicin. The nanogels can also be used as a contrasting agent in optical cell imaging and for simultaneous usage in biomedical applications and drug delivery systems [13]. Varshosaz et al. studied dextran polymeric carriers in novel drug delivery systems; dextran-antitumor drugs conjugate, enhancing effectiveness and improving cytotoxic effects of chemotherapeutic agents in localized drug delivery [36]. Liu et al. studied the block copolymers of dextran with poly(D,L-lactide) to form mucoadhesive nanoparticles as a potential treatment for anterior eye diseases and hence can be used to improve the bioavailability of topical formulations [37]. Dextran-drug conjugates are widely used in focal drug delivery of active agents, e.g., budesonide-dextran conjugates, using glutarate spacer as a colon-targeted drug delivery system for the treatment of ulcerative colitis [36, 38, 39].

Dextran complexes with nonbiodegradable polymers (methacrylates) and, partially biodegradable polymers (poly(*N*-isopropylacrylamide)/dextran-maleic acid), to form hydrogels has great potential as a drug carrier because of its combined stimuli-response capability, as well as partial biodegradability in controlled drug delivery of doxorubicin in tumor site [40]. A hydrogel formulation of dextran and polyaspartamide derivatives (poly amino acid) has been reported for colon-specific drug delivery. This formulation undergoes an enzymatic hydrolysis in the colon and is potentially useful for treating inflammatory bowel diseases. Abdullah et al. studied novel methods of dextran polymer delivered genes to specific cell lines, thereby treating genetic disorders. A novel cationic polymer, dextran-spermine, has been found to mediate gene expression in a wide variety of cell lines and in vivo through systemic delivery of plasmid DNA, partially protected from degradation by nuclease. This exhibited optimal gene transfer efficiency [41].

1.2.7 Chitosan

Chitin is a linear polymer found in the shells of crabs, lobsters, shrimps, and insects. It is the fully or partially deacetylated form of chitin. The degree of deacetylation of typical commercial chitosan is usually between 70 and 95 % and the molecular weight between 10 and 1,000 kDa. Chitosan is biologically renewable, biodegradable, biocompatible, nonantigenic, nontoxic, and biofunctional. Due to its properties it has attracted much attention in tissue engineering with a wide variety of applications ranging from skin, bone, cartilage, and vascular grafts to substrates for mammalian cell culture. Chitosan has been found to enhance blood coagulation, accelerating wound healing; thus, it can act as an ideal wound dressing, since it exhibits a positive charge, film-forming capacity, mild gelation characteristics, and strong tissue adhesion. Chemically modified chitosan polymers contain reactive sites amenable for ligand conjugation mucoadhesives [3].

Currently, chemically modified chitosan is widely used in tissue engineering and drug delivery systems. For example, *N*,*N*-dicarboxymethyl chitosan has been found to be useful as a molecular carrier or drug delivery agent for cartilage related problems. BMP is used to induce or facilitate the repair of articular cartilage lesions [42]. Trapani et al. studied chitosan as a potential carrier for drugs to treat serious brain disorders, such as Parkinson's disease, with nanoparticles formulation of chitosan-dopamine [43]. Chitosan acts as a scaffold material in hydrogels that has been used due to the polymer's biocompatibility, low toxicity, and biodegradability [44]. Chitosan is widely used as mucoadhesive polymer that can increase cellular permeability and improve the bioavailability of orally administered protein drugs. It can also be readily formed into nanoparticles able to entrap drugs or condense plasmid DNA [45]. Chitosan-based micro- and nanoparticles are effective in local drug delivery systems due to their inherent properties and drug loading, encapsulation efficiency, and release characteristics [46]. HemCon® is a commercially available chitosan-based bandage.

Chitosan exhibits favorable biological behavior, with bioadhesion- and permeability-enhancing properties, which makes it a unique material for the design of ocular drug delivery vehicles. Chitosan-based colloidal drug delivery systems have shown its potential action, either facilitating the transport of drugs to the inner eye (chitosan-coated colloidal systems containing indometacin) or their accumulation into the corneal/conjunctival epithelia (chitosan nanoparticles containing cyclosporine) [47]. Chitosan nanoparticles for brain targeting are another promising application of chitosan. Nano-sized particles have been widely used to treat brain infection caused by AIDS viruses. Didanosine-loaded chitosan nanoparticles delivery systems are used to enhance systemic and brain targeting efficiency to CSF and brain [48].

Chitosan and its carboxymethyl derivatives nanoparticles (curcumin loaded) have been reported to treat tumor cells, where curcumin acts as a chemotherapeutic agent with antioxidant, anti-inflammatory, anti-proliferative, anticancer, and antimicrobial effects [49]. Chitosan-gold hybrid hydrogel has exhibited an excellent water absorbing property and can be applied as a drug delivery system for anticancer drug. Doxorubicin, due to its high equilibrium water swelling content, is an ideal local drug delivery system [50]. Poor water solubility and low transfection efficiency of chitosan are major drawbacks for its use as a gene delivery carrier, but now modified chitosan polymers with polyethylene glycol (PEG) to enhance its solubility and folate conjugation may improve gene transfection efficiency due to promoted uptake of folate receptor-bearing tumor cells. Cross-linked folate-poly(ethylene glycol)-grafted chitosan has been used to target plasmid DNA to tumor cells to induce remarkable cytotoxicity against potential tumor HEK 293 cells [51].

1.2.8 Polyhydroxyalkanoates

Polyhydroxyalkanoates (PHAs) are biological polyesters produced by microorganisms under unbalanced growth conditions.The main biopolymer of the PHA family is the poly hydroxybutyrate (PHB), but many copolymers were synthesized based on PHB [poly (hydroxybutyrate-co-hydroxyvalerate) (PHBV), poly (hydroxybutyrate-co-hydroxyhexanoate) (PHBHx), poly (hydroxybutyrate-co-hydroxyoctanoate) (PHBO) etc [52].

Ye et al. studied copolymer of 3-hydroxybutyrate and 3-hydroxyhexanoate (PHBHHx) to produce neocartilage upon seeding with differentiated human adipose-derived stem cells (hASCs). hASCs were grown on a three-dimensional PHB/PHBHHx scaffold in vitro with or without chondrogenic media for 14 days. The differentiated cells/(PHB/PHBHHx) implants formed cartilage-like tissue after 24 weeks of implantation [53]. A cross-linked polymer of PHB, 3-hydroxybutyrate, and 3-hydroxyvalerate (PHBV) microspheres-PLGA matrix composite has been reported as a scaffold for bone tissue engineering and for applications in drug delivery systems [52]. Biodegradable poly-beta-hydroxybutyrate scaffold seeded with Schwann cells to promote spinal cord repair. Modification of PHB surfaces with fibronectin, laminin, or collagen significantly increases Schwann cell attachment and proliferation in vitro. The results demonstrate that a PHB scaffold promotes attachment, proliferation, and survival of adult Schwann cells and supports marked axonal regeneration within the graft [54].

1.2.9 Arabinogalactan

Arabinogalactan is a highly branched water-soluble natural polysaccharide extracted from the larix tree (*Larix laricina*). It is available in a 99.9 % pure form with reproducible molecular weight and physicochemical properties. Its high water solubility, biocompatibility, biodegradability, and ease of drug conjugation in an aqueous medium make arabinogalactan an attractive and potential drug carrier. Arabinogalactan is a highly branched polysaccharide consisting of a galactan backbone with side chains of galactose and arabinose with unusual water solubility (70 % in water) [55–57]. Elgart et al. studied conjugation drug delivery of arabinogalactan with potent antifungal agent amphotericin B (AMB). AMB shows variability in solubility, pharmacokinetics, and pharmacodynamics. Arabinogalactan conjugation with the poorly soluble drug AMB affects its physicochemical properties with improved solubility and reduced toxicity [58]. Arabinogalactan has reportedly been used for hepatic drug delivery as a natural polymer due to its properties (biocompatible, biodegradable, ease of availability). It is mainly suitable as a carrier for the delivery of diagnostic or therapeutic agents to hepatocytes via the asialoglycoprotein receptor [59].

1.3 Synthetic Polymers

Synthetic polymers provide flexibility to tailor drug release, mechanical properties, and degradation kinetics to suit different applications. Most synthetic polymers undergo hydrolytic degradation that binds hydrolytically labile chemical bond in their backbone. Functional groups susceptible to hydrolysis include esters, orthoesters, anhydrides, carbonates, amides, and urethanes. Furthermore, synthetic polymers can be designed on a need basis and are available in a wide variety of compositions with readily adjustable properties [60]. Similar to natural polymers, synthetic polymers are also being developed and used by industry and academia. Table 1.1 summarizes the technological capacity overview of synthetic biodegradable polymers.

Table 1.1 Technological capacity overview of synthetic biodegradable polymers

Polymer	Product	Drug	Application	Reference
PLA	Atridox	Doxycycline hyclate	Periodontal disease	[61, 62]
PEG-PLA	Trenantone	Leuprorelin acetate	Prostate cancer	
	Genexol-PM	Paclitaxel	Cancer (breast, lung, prostate, pancreatic)	[63]
PLGA	Zoladex	Goserelin acetate	Prostate cancer, endometrioses	[64–66]
	Lupron Depot	Leuprolide acetate	Prostate cancer	[67]
	Eligard	Leuprolide acetate	Prostate and breast cancer	[68, 69]
	Sandostatin LAR	Octreotide acetate	Growth hormone suppression	[70, 71]
	Nutropin Depot	Somatotropin (hGH)	Growth hormone deficiency	[72, 73]
	Risperdal Consta	Risperidone	Antipsychotic	[74, 75]
	Arestin	Minocycline	Adult periodontitis	[76]
	Decapeptyl	Triptorelin acetate	Advanced and metastatic prostate cancer	[77, 78]
	Vivitrol	Naltrexone	Alcohol dependence	
	Profact Depot	Buserelin	Prostate cancer	[79, 80]
PCL	Capronor	Levonorgestrel	Contraceptive	[81]
Polyanhydride	Gliadel	Carmustin	Brain cancer	[82–84]
	Septacin	Gentamicin sulfate	Osteomyelitis	[72]
Gelatin	H.P. Acthar gel	Adrenocorticotropic hormone	Endocrine disorders	
Collagen	Apligraf®	Collagen from human skin	Wound healing	[85, 86]
Fibrin	Tisseel®	Fibrin	Fibrin sealant for hemostasis	[87]

PLA poly(lactic acid), *PLGA* poly(lactide-co-glycolide), *PCL* poly-ε-caprolactone, *PEG-PLA* PEGylated-PLA

1.3.1 Polyesters

Poly(α-ester)s are thermoplastic aliphatic polyesters having hydrolytically labile aliphatic ester linkages in their backbone. Hence, they are biodegradable. These polymers are usually synthesized by ring-opening polymerization (ROP) and polycondensation. Bacterial bioprocess routes can also be used to develop some poly(α-ester)s. These are bulk-degrading polymer degrades by the nonspecific scission of the ester backbone. In the class the most extensively investigated polymers are the poly(α-hydroxy acid)s, which include poly(glycolic acid) (PGA), the stereoisomeric forms of poly (lactic acid) (PLA) and their copolymers (PLGA), and polycaprolactone (PCL) along with there copolymers.

Both, homopolymers and copolymers of poly(α-ester)s have been investigated as potential biomaterials for a variety of biomedical applications due to their good biocompatibility and controllable degradation rates tailored to specific demand of each tissue type. Polyester polymers have also shown to be very attractive polymers in many biomedical fields as conducting polymers (CPs) that have both electrical and optical properties similar to those of metals and inorganic semiconductors. The fact that several tissues are responsive to electrical fields and stimuli has made CPs attractive for a number of biological and medical applications such as biosensors, tissue engineering, and neural probes [88].

1.3.1.1 Poly(Glycolic Acid)

PGA is a crystalline polymer (45–55 % crystallinity) and, therefore, exhibits a high-tensile modulus with very low solubility in organic solvents. Glass transition temperatures range from 35 to 40 °C with melting point nearly 200 °C. Polyglycolide shows excellent mechanical properties due to its high crystallinity. In spite of its low solubility, this polymer has been fabricated into a variety of forms and structures. Extrusion, injection, and compression molding as well as particulate leaching and solvent casting are some of the techniques used to develop polyglycolide-based structures for biomedical applications. It is the first synthetic polyester polymer to show diverse applications in tissue engineering and controlled, localized drug delivery of medicaments. Dexon® and Biofix® are PGA-based and FDA-approved products available on the market. PGA is also used for bioresorbable sutures (e.g., Dexon®) due to its excellent fiber-forming ability.

High-rate degradation, acidic degradation, and low-solubility products, however, limit biomedical applications for polyglycolic acid. Therefore, several copolymers containing glycolide units are being developed to overcome the inherent disadvantages of polyglycolide. One example is poly(lactide-co-glycolide). Among the few other copolymers, cross-linked structures of PGA-chitosan [P/C] hybrid matrices have been used for tissue-engineering applications with novel, porous, biocompatible, degradable, and modifiable hybrid matrices for biomedical applications. The presence of chitosan in the P/C matrices provides many amino groups for further modifications such as biomolecule conjugation and thus enhances the application potential [89].

1.3.1.2 Poly(Lactic Acid)

Poly(lactic acid) is a chiral molecule and exists in two optically active forms, L- and D-lactide. L-lactide is a naturally occurring monomer. Polymerization of racemic (D,L)-lactide and mesolactide results in the formation of amorphous polymers, whereas PLLA is semicrystalline. The degree of crystallinity depends on molecular weight and polymer processing parameters. It has a glass transition temperature of 60–65 °C and a melting temperature of approximately 175 °C. PLLA is a more hydrophobic and slow-degrading polymer than PGA and has good tensile strength; thus, it is used for sutures. High molecular weight PLLA can take between 2 and 5 years for total resorption in vivo. It is more hydrophobic than PGA but less than PCL. Several copolymers of L-lactides with glycolides or DL-lactides are currently being researched for the development of polymers with better property modulation for controlled drug delivery. PDLLA is an amorphous polymer due to the random distribution of L- and D-lactide units. It has a glass transition temperature of 55–60 °C due to its amorphous nature. In addition the polymer shows much lower strength compared to poly(L-lactic acid). Polylactides undergo hydrolytic degradation via the bulk erosion mechanism by the random scission of the ester backbone. It degrades into lactic acid, a normal human metabolic by-product, which is broken down into water and carbon dioxide via the citric acid cycle.

Polylactics have a wide range of applications in bone fixtures, sutures, drug delivery, tissue engineering, and carrier for gene delivery. Some PLA-based products that are FDA approved and available in the market are Phantom Soft Thread Soft Tissue Fixation, Screw®, Phantom Suture Anchor®, Full Thread Bio Interference Screw®, BioScrew®, Bio-Anchor®, Meniscal Stingers®, Clearfix®, Meniscal Dart®, Sculptra®, Atridox®, Trenantone®, and Genexol®.

New advances in the polymerization process of PLA have shown prominent results in controlled drug release. Bishara et al. describe the controlled drug release of peptide/protein molecules. Reversible stereoselective complexes are spontaneously formed by mixing acetonitrile solutions of enantiomeric D-poly(lactide) (D-PLA), L-poly(lactide) (L-PLA), and octreotide (an octapeptide, somatostatin analogue). This results in (a) increasing peptide concentration in the stereocomplex, increases the release rate of the peptide, and (b) increasing the polymer degradation rate as monitored by lactic acid release form stereocomplexes [90]. A new application of biodegradable polyester polymer has shown a controlled release sirolimus-eluting biodegradable poly-L-lactide (PLLA) stent for peripheral vascular application [91]. PLA has been explored as a potential vehicle in tissue engineering due to its versatility. Montjovent et al. describe the potential usage of porous PLA obtained by supercritical gas foaming and reinforced with 5 % beta-tricalcium phosphate in bone cell tissue engineering [92].

1.3.1.3 Poly(D,L-Lactide-co-Glycolide)

Among the co-polyesters investigated, extensive research has been performed to develop a full range of PLGA. In this connection, both L- and DL-lactides have been used for copolymerization. PLGA forms amorphous polymers in the composition range of 25–75 %. However, 50/50 PLGA is very hydrolytically unstable and is resistant to hydrolytic degradation. PLGA has been explored as the most widely used polymer for biomedical applications, especially in drug delivery. Properties of the copolymer can vary depending on the ratio of lactide and gylcolide. Undergoing bulk erosion through hydrolysis of the ester bonds and the rate of degradation depend on various parameters including the LA/GA ratio, molecular weight, and the shape and structure of the matrix. Different ratios of PLGA have been commercially developed and are being investigated for a wide range of biomedical applications. PuraSorbs® PLG is a semicrystalline bioresorbable copolymer of L-lactide and glycolide with a monomer ratio of 80L:20G. A copolymer containing 90 % glycolic acid (GA) and 10 % L-lactic acid (LA) was initially used for the development of the multifilament suture Vicryls®. A modified version of the suture, Vicryl Rapid® is currently on the market; it is an irradiated version of the suture to increase the rate of degradation. Panacryl® is another commercially developed suture from the copolymer with a higher LA/GA ratio to decrease the rate of degradation.

PLGA shows excellent biocompatibility in the tissue engineering of cardiac valves as a scaffold for the development of a new generation of heart valves. Previously, prosthetic heart valves were used, but they suffer from the risks of infection and thromboembolic complications. Biological devices have limited durability. The problem of lack of growth potential remains a serious issue particularly for pediatric cardiac patient. The multidisciplinary field of tissue engineering potentially offers an attractive pathway to overcome these disadvantages. Recently biodegradable, biocompatible polymers have shown attractive results by providing structural integrity and biomechanical profile for the newly developed tissue structure [93]. Karp et al. studied the potential activity of bone formation on two-dimensional PLGA films and three-dimensional PLGA tissue-engineering scaffolds in vitro. PLGA supports appositional bone growth on both two-dimensional films and three-dimensional scaffolds, including the formation of a mineralized cement line matrix [94]. Osteogenesis of mesenchymal stem cells for bone tissue engineering is another application of modified porous PLGA scaffolds from which ascorbate-2-phosphate and dexamethasone were continuously released for a month [95]. Tian et al. studied the biocompatibility of spongy PLGA-collagen membrane as a dermal scaffold. The PLGA-collagen membrane possesses good biocompatibility and cell affinity with an appropriate degradation speed to meet the requirements of a tissue-engineering dermal scaffold [96]. PLGA has shown prominent applications in tissue fixation, sutures, tissue engineering, and drug delivery. Examples of marketed products include PuraSorb®, Vicryl®, Panacryl®, Dermagraft®, Cytoplast Resorb® Risperdal Consta®, Arestin®, Decapeptyl®, Pamorelin®, Vivitrol®, Proffact® Depot, and Plenaxis®. There are also FDA-approved drug delivery products like Lupron Depot®, Zoladex®, Trelstar®, Enatone®, Procin®, Eligard®, Sandostatin LAR®, and Nutropin Depot®.

Biocompatible, biodegradable PLGA has been used as a potential carrier for gene delivery. PLGA Nanoparticle bearing polyethyleneimine (PEI) on the surface are reported for the potential in serving as nonviral gene carriers to the pulmonary epithelium in treating genetically related diseases. NPs may induce immunity towards pathogens entering the body via the airways for pulmonary gene delivery [97]. Csaba et al. conducted in vitro experiments on gene delivery of plasmid DNA with nanoparticulate carrier systems and with different PLGA:poloxamer and PLGA:poloxamine ratios [98].

PLGA has also been widely investigated as a potential carrier in ophthalmic drug delivery system due to its potential penetration properties. Efforts in ophthalmic drug delivery have been devoted to increasing the corneal penetration of drugs with the goal of improving the efficiency of treatments of different ocular diseases such as viral infections. Attempts include the use of colloidal drug delivery systems like liposomes, microparticles, biodegradable nanoparticles, and nanocapsules. A microsphere formulation for intravitreal delivery by dispersing ganciclovir-loaded PLGA microspheres in thermogelling PLGA-PEG-PLGA gel has reported for focal delivery to the retina in the treatment of cytomegalovirus retinitis [99].

Porous PLGA microsphere has been a useful tool in delivering therapeutic drugs and biologically active proteins, utilizing hydrogen peroxide as a novel porogen [100]. Chan et al. reported PLGA-lecithin-PEG core-shell nanoparticles for controlled drug delivery in clinically approved therapeutics such as Doxil/Caelyx and Genexol-PM, respectively [101]. A triblock copolymer of PLGA (PLGA-PEG-PLGA) has shown thermosensitive and controlled release of peptide/protein formulations like growth hormone in vitro and in vivo for longer duration. The delivery system is biocompatible [102]. RGD-grafted PLGA-nanoparticles have the ability to target tumor sites with loading of antitumor agents like paclitaxel and daunorubicin to treat solid tumors [103].

1.3.1.4 Polycaprolactone

PCL is a semicrystalline polyester of great interest, as it can be obtained by the ROP of a relatively inexpensive monomeric unit, "ε-caprolactone." PCL can be easily processed, since it is soluble in a wide range of organic solvents. It has a low-glass-transition temperature and low melting point (55–60 °C). It undergoes bulk hydrolytic degradation of hydrolytically labile aliphatic ester linkages, and it is autocatalyzed by carboxylic end groups; the rate of degradation, however, is rather slow (2–3 years). Polycaprolactones are semicrystalline polyesters, due to their slow degradation and biocompatibility. They offer an attractive material for long-term biomedical and drug delivery applications [104]. PCL has been widely used in sutures, tissue engineering, and drug delivery systems such as Capronor®, Monocryl®, and SynBiosys®, which are FDA-approved products available in the market.

PCL has the ability to form compatible blends with other polymers, which provides opportunities to manipulate the drug release rate from matrix. Recent advances in colloidal drug delivery of drugs with biodegradable polymers like PCL show significant progress in the sustained release of drugs. Microsphere formulation of PCL with prednisolone acetate for orbital administration shows prolonged-release features in the treatment of inflammatory orbitopathy diseases [105]. Kyun et al. studied a local drug delivery of medicaments with biocompatible PCL and release kinetics of minocycline from monolithic film prepared from PCL and PEG to carry out its antimicrobial activity in vitro. PCL films delivered minocycline to show its antimicrobial activity for elimination of pathogenic microflora from periodontal pockets or reducing inflammation in periodontal disease in a sustained manner [106].

PCL also shows immense versatility in biomedical applications like tissue engineering as a scaffold to generate new tissue. Hoque et al. studied a novel triblock copolymer, poly(ethylene glycol)-block-poly(ε-caprolactone)-block-poly(DL-lactide). (PEG-PCL-P(DL)LA) has shown good cell biocompatibility of the polymer and efficient tissue-engineering scaffolding [107]. Williamson et al. prepared vascular scaffolds with a new compliant scaffold. PCL fibers form the luminal surface, and then porous polyurethane (PU) is placed onto the back of the PCL fibers to form suitable substitute vessel walls for small-diameter vascular grafts. This promotes strong attachment of endothelial cells for vascular tissue engineering [108].

PCL has been widely used as a gene delivery vehicle because of its immense properties. Zhao et al. studied the gene delivery applications of PCL. Cationic PCL-pluronic-PCL copolymer nanoparticles were employed to condense and adsorb DNA onto its surface. This has great potential application in DNA delivery [109]. A novel approach to deliver genes into cancerous cells with PCL has shown potential. Another novel approach is the use of folate-conjugated ternary copolymer based on polyethylenimine-graft-PCL-block-poly(ethylene glycol) (PEI-g-PCL-b-PEG-Fol) as targeted gene delivery system into folate receptor overexpressing tumor cells [37].

1.3.2 Polyorthoesters

Polyorthoesters (POEs) are amorphous hydrophobic polymers containing hydrolytically labile, acid-sensitive backbone linkages. ALZA Corporation (Alzamer®) developed these as hydrophobic surface-eroding polymers particularly for drug delivery applications. Although the orthoester linkages are hydrolytically labile, the polymer is hydrophobic enough that its erosion in aqueous environments is very slow. The unique feature of POEs is that, in addition to its surface erosion mechanism, the rate of degradation for these polymers, pH sensitivity, and glass transition temperature can be controlled. POEs were invented during attempts to develop a bioerodible polymer and subdermal implantable that would release contraceptive steroids by close to zero-order kinetics for at least 6 months [110]. An additional objective was that the polymer erosion and drug release should be concomitant so that no polymer remnants are present in the tissue after release of the entire drug. The above objectives can be achieved only if the polymer is truly surface eroding. For a surface-eroding polymer, the erosion process at the surface of the polymer should be much faster than that in the interior of the device. To have such a phenomenon, the polymer has to be extremely hydrophobic with very labile linkages. Hence, it was envisioned that polymeric devices with an orthoester linkage in the backbone, which is an acid-sensitive linkage, could provide a surface-eroding polymer if the interior of the matrix is buffered with basic salts.

The rate of degradation for these polymers, pH sensitivity, and glass transition temperatures can be controlled by using diols with varying levels of chain flexibility. To date, four different classes of POEs have been developed (POE I, II, III, and IV). Few orthopedic applications of this class of polymers have been explored and major use of POEs has been limited to drug delivery systems [111]. Among all POEs, POE III and IV are of most interest due to their unique properties. POE III is a semisolid material that has been shown to be highly biocompatible and is currently being investigated as an adjunct to glaucoma filtering surgery and other ocular applications. POE IV can be easily prepared in a highly reproducible manner, is very stable if moisture excluded, and has shown to be highly biocompatible. It is currently under development for a variety of applications, such as ocular delivery, protein release, postoperative pain treatment, and postoperative cancer treatment [112].

The biocompatibility of a viscous, hydrophobic, bioerodible POE intended for intraocular applications is a subject of interest. Einmahl et al. conducted a study on rabbits to prove POE applications in ophthalmic local drug delivery. POE was tolerated well, and no inflammatory reaction developed during the observation period. The polymer degraded slowly, appearing as a round whitish bubble in the vitreous cavity. The presence of modulators of degradation both improved POE biocompatibility and prolonged polymer lifetime in the eye [113]. POE has investigated as a sustained drug release system for an ophthalmic application in intraocular proliferative disorders. The combination of wound healing modulators such as 5-fluorouracil and dexamethasone has advantage, since these drugs act at different stages of these diseases [114].

PEO has shown tremendous results in the delivery of drugs to localized bone tissue with inorganic phosphate fiber composites, which are prone to rapid degradation due to water sensitivity of the interface between the degradable polymer and the degradable fiber. Novel polymeric coating solutions on the inorganic composite show better results in orthopedic drug delivery due to the hydrophilicity of POE [115]. POE is also widely used in postsurgical pain management, as these can be prepared as solid materials or as viscous, injectable polymers; in this regard, POE IV is getting more attention for sustained drug delivery. POE IV in particular undergoes surface erosion and is able to moderate drug release over periods from days to many months. The local anesthetic agent, mepivacaine, has been incorporated into a viscous, injectable POE IV with potential to provide longer-acting anesthesia in postsurgical pain management [116].

1.3.3 *Polyanhydrides*

The distinctive properties of polyanhydrides make them a unique class of biodegradable polymers. Constant drug release rates are expected from these hydrophobic, surface-eroding polymers. Degradation of the polymer can be altered easily by changing various parameters such as composition, crystallinity, molecular weight, and pH of the surrounding medium. Intensive research in this class of polymers resulted in clinical use of several marketed products. Polyanhydrides are a class of hydrolytically unstable surface-eroding polymers that are either aliphatic, aromatic, or a combination of the two. Almost all polyanhydrides show some degree of crystallinity as manifested by their crystalline melting points. The melting point of these aromatic polyanhydrides, as determined by differential scanning calorimetry, is much higher than aliphatic polyanhydrides. Polyanhydrides are surface-eroding polymers usually synthesized by melt polycondensation. Drug release usually follows zero-order kinetics. Polyanhydrides have been considered to be useful biomaterials as carriers of drugs to various organs of the human body such as the brain, bone, blood vessels, and eyes. They can be prepared easily from available, low-cost resources and can be manipulated to meet desired characteristics.

Over these years intensive research has been conducted, which has yielded hundreds of publications and patents describing excellent film- and fiber-forming properties and structures of new polymer. There have been studies on the chemical and physical characterization of these polymers, degradation and stability properties, toxicity studies, and applications of polymers for mainly controlled bioactive agents. Due to their rapid degradation and limited mechanical properties, however, their main use has been limited to short-term controlled delivery of bioactive agents [117].

They find application in both tissue engineering and drug/growth factor delivery. Various drugs and proteins (including insulin, bovine groth factors, angiogenesis inhibitors, enzymes, and anesthetics) were incorporated in polyanhydride matrices, and evaluated for in vitro and in vivo release characteristics. Polyanhydrides derived from bis-*p*-(carboxyphenoxypropane), and sebacic acid is marketed under the name of Gliadel® for the delivery of BCNU (bis-chloroethylnitrosourea), a chemotherapeutic agent, to the brain for the treatment of glioblastoma multiforme, a universally fatal brain cancer.

To obtain polyanhydrides with high mechanical strength for load-bearing applications, osteocompatible poly(anhydride-co-imides) have also been designed and shown to support endosteal bone growth [118]. A copolymer of 1:1 sebacic acid and erucic acid dimer has been found useful as a potential delivery vehicle for gentamicin (Septacin®) in the treatment of osteomyelitis [119–121]. Polyanhydrides have been showing promising results in localized drug delivery. The limitations of currently available drug therapies, particularly for the treatment of diseases localized in a specific organ or tissue, have encouraged scientists to consider alternative methods of drug administration to increase specificity. This is particularly relevant in the treatment of solid tumors and localized infections, where a high rate of recurrence remains a major clinical problem associated with inadequate drug supply to the diseased site. An alternative treatment is the local administration of the agent from a degradable polymeric delivery system implanted at the site of the disease. In this way, the drug deposit can maintain a high local concentration at the site for extended periods with minimal systemic distribution of the drug. The polymer carrier is degraded and eliminated from the body shortly after the drug has been released [122]. Biocompatible, biodegradable, hydrolytically unstable polyanhydride polymers play a significance role in localized drug delivery. An injectable degradable polyanhydride polymer-poly (sebacic- co- ricinoleic-ester-anhydride) releasing gentamicin has proven to be efficient histologically in the treatment of osteomyelitis [72]. Manoharan et al. studied the potential of poly 1,3-bis-(p-carboxyphenoxy) propane-co-sebacic acid (p(CPP:SA)) microspheres for controlled delivery of basal insulin. Subcutaneous administration of CPP:SA (50:50) microspheres in diabetic rats controlled insulin release over a month; the released insulin was bioactive as determined by lowered blood glucose levels. The results indicate that CPP:SA microspheres controlled insulin release in vitro and in vivo over a month, and the released insulin is conformationally and chemically stable and bioactive [123]. An intratumoral injectable polymeric formulation, based on poly(sebacic acid-co-ricinoleic acid) of paclitaxel loaded in the polymer, was found to be an effective treatment for localized tumors like orthotopic prostate cancer [124, 125].

Polyanhydrides are also helpful in localized analgesia with local anesthetic because prolonged postoperative analgesia cannot be achieved using single injections of local anesthetic solutions. Polymer-local anesthetic conjugates have been used to prevent postoperative analgesia to overcome this problem. This is an injectable fatty acid-based biodegradable polymer-poly(sebacic-co-ricinoleic acid) with long-acting anesthetic bupivacaine for producing motor and sensory block when injected near the sciatic nerve [126]. Similarly, many orthopedic injuries lack a high-strength and degradable material with good tissue compatibility. There is also a great clinical need for materials, which are easily contoured or placed into complex-shaped defects. To overcome this problem, a new class of photocross-linkable polyanhydride monomers which in situ form high-strength and surface eroding networks of complex geometries was reported [127].

Unique features of polyanhydrides polymers are prone to be used in pulsatile drug delivery of medicaments and macromolecules like proteins and peptides. A laminated device comprised of polyanhydrides as isolating layers and pH-sensitive complexes as protein-loaded layers was designed to deliver proteins like myoglobin and bovine serum albumin in a pulsatile manner. Poly(sebacic anhydride)-b-polyethylene glycol (PSA-b-PEG) and poly(trimellitylimidoglycine-co-sebacic anhydride)-b-polyethylene glycol (P(TMA-gly-co-SA)-b-PEG) were synthesized as isolating layers for their positive processing properties at room temperature and suitable erosion duration [128].

1.3.4 Polyurethanes

PUs are prepared by the polycondensation reaction of diisocyanates with alcohols and/amines. Conventional polyols are polyethers or polyesters. The resulting polymers are segmented block copolymers with the polyol segment providing a low-glass-transition temperature (i.e., <25 °C) soft segment and the diisocyanate component, often combined with a hydrocarbon chain extender, providing the hard segment. A wide range of physical and mechanical properties are found in commercial PUs. Bio-stable PUs have been extensively investigated as materials for the preparation of long-term medical implants, especially cardiac pacemakers and vascular grafts, due to their excellent biocompatibility, mechanical properties, and their synthetic versatility [129]. Attempts have been made to develop biodegradable PUs. Biodegradable PUs have recently been investigated as candidate materials for bone regenerative medicine. Poly α-hydroxy acids, including PLA, PGA, and PCL, have been used as soft segments for biodegradable PUs [130]. PUs for biomedical engineering have been reviewed by Gunatillake and Meijs [131].

PUs are primarily composed of block copolymers with hard and soft segments. Glassy, semicrystallines of 22–70 % hard segments were developed and evaluated for bone tissue-engineering applications. As the hard segment contents increase, the polyurethane surface exhibits more phase separation, a higher content of urethane moieties, and a higher hydrophilicity. Biocompatibility results indicate that proliferation of human bone-derived cells (HBDC) cultured in vitro improves with increasing the hard segment content, while the osteogenic potential of HBDC decreases with an increasingly hard segment content [132]. Novel polyurethane-based injectable in situ curable polymer platform to determine its potential uses as a tissue-engineered implant is under investigation. PUs from pentaerythritol-based prepolymers may have potential orthopedic applications, ranging from bone glues to scaffolds for bone regeneration [133]. Jiang et al. described phenomenon in many biomedical applications. Gly-Arg-Gly-Asp-Ser-Pro (GRGDSP) peptide has frequently been used to enhance adhesion and proliferation of cells. In this study, modified, nontoxic biodegradable waterborne polyurethanes (WBPU) with GRGDSP peptide were investigated. Fabricated 3-D porous scaffold with the modified WBPU was used to investigate the effect of the retention of GRGDSP peptide on human umbilical vein endothelial cells (HUVECs). Cell adhesion and proliferation were found to exert effective control on the GRGDSP content in the modified WBPU and 3-D porous modified WBPU scaffolds. This was helpful to cell adhesion, proliferation, and tissue regeneration in soft tissue engineering [134].

1.3.5 Polyethylene Glycol-Based Polymers

PEG is a biocompatible and nontoxic polymer, which has been used to modify other polymers, since it is known to reduce protein adsorption and modify polymer conformation. PEG-based hydrogels have been widely studied for drug delivery as well as for other biomedical applications in tissue engineering and the delivery of growth factors. Blends of electrospun PEG-PLA have shown positive response towards the biological activities of seeded human dermal fibroblasts and enhanced cell growth within fibrous mats [135]. A pH-sensitive chondroitin sulfate-PEG tissue adhesive and hydrogel with several potential biomedical applications specifically in wound healing and regenerative medicine has been reported [136]. The hydrogel supported cell viability and produced minimal inflammatory response when implanted subcutaneously in a rat model. Pluronic F-127 is a commercial name for the copolymer of polyethylene oxide and polypropylene oxide. It is a thermosensitive, biocompatible hydrogel and is FDA approved for use in humans [137]. There have also been studies about its use in cartilage and lung tissue engineering [138–140]. A biomimetic hydrogel has also been developed using Pluronic F-127 and fibrinogen. It is reported to be biocompatible, aiding cell signaling through fibrinogen backbone [141].

1.3.6 Poly(Amino Acids)

Proteins are composed of amino acids, with this many researchers have tried to develop synthetic polymers derived from amino acids, i.e., poly(amino acids), to serve as models for structural, biological, and immunological studies. Polymers from amino acids were most intensively synthesized from α-hydroxyl amino acids. Polymerization of α-hydroxy acids can take place either by condensation via direct esterification of the α-hydroxy acid or ester or via lactone formation [142].

Poly(amino acids) are advantageous as biomaterials due to their diversity and the availability of side chains which offer sites for the attachment of small peptides, drugs, cross-linking agents, or pendent groups that can be used to modify the physicomechanical properties of the polymer. Since these polymers release naturally occurring amino acids as the primary products of polymer backbone cleavage, their degradation products may be expected to show a low level of systemic toxicity. Poly (aspartic acid), poly(L-lysine), and poly(L-glutamic acid) are a few examples of this class. Kohn et al. have replaced the peptide bonds in the backbone of synthetic poly(amino acids) with a variety of such "nonamide" linkages such as ester, iminocarbonate, urethane, and carbonate bonds [143]. The term, "pseudo-poly(amino acid)," is used to denote this new family of polymers, in which naturally occurring amino acids are linked together by nonamide bonds. Hydroxyproline-derived polyesters, serine-derived polyesters, tyrosine-derived polyiminocarbonates, and polycarbonates represent specific embodiments of these synthetic concepts. Recent, advances achieved in drug delivery have provided novel and versatile possibilities for the treatment of various diseases. Among the biomaterials applied in this field, it is worth highlighting the increasing importance of poly amino acids and polypeptides due to their appealing properties. They are very promising for the design of new compositions in a variety of drug delivery applications. Cancer therapy has been benefited the most from these advances, although other fields such as vaccine delivery and gene therapy have been benefited as well [144].

Polymeric drug delivery systems are used not only to improve the aqueous solubility of drug molecules but also to achieve desirable pharmacokinetics and an enhanced therapeutic index for drugs. Synthetic polyamino acids are used to deliver macromolecular prodrugs and drug targeting in antitumor therapy. This includes polyamino acids like alpha,beta-poly(*N*-2-hydroxyethyl)-DL-aspartamide (PHEA), alpha,beta-polyaspartylhydrazide (PAHy), poly(glutamic acid) (PGA), poly(aspartic acid) (PAA), and poly(L-lysine) (PLL). PLL has been extensively studied in this field [145]. Recent advances in polymer chemistry make use of polyamino acid polymers in gene delivery. The grafting of poly(L-histidine) and poly(L-lysine) makes the formation of *N*-Ac-poly(L-histidine)-graft-poly(L-lysine)(PLH-g-PLL). Its shaped (comb shape) polymer has shown more pronounced action compared with homo polymer PLL. The inclusion of chloroquine as an endosomolytic agent enhances transfection for both PLL and PLH-g-PLL gene carriers. PLH-g-PLL is an enhanced beta-galactosidase gene expression compared with PLL [146]. Davis et al. studied formulation of PEGylated poly-L-lysine DNA nanoparticles for gene delivery composed of plasmid DNA. They are soluble and stable in saline and tissue fluids, transfect nondividing cells, display minimal toxicity, and are effective in vivo and in humans. Moreover, they are easy to prepare in a reliable and reproducible fashion. These properties represent a substantial advance for nonviral gene transfer [147].

1.3.7 Polyfumarates

Polyfumarates are a class of novel, injectable, biodegradable, and biocompatible materials, which have applications for the delivery of bioactive molecules/drugs in various pharmaceutical and biomedical fields. Oligo(poly(ethylene glycol) fumarate) (OPF) is one such water-soluble synthetic polyfumarate which can be injected into a defect site and cross-linked in situ at physiological conditions, thereby eliminating the need for invasive implantation and retrieval surgeries [148]. Poly(propylene)fumarate (PPF), another polymer from this class, is a linear polyester whose repeating units contain two ester bonds and one unsaturated carbon–carbon double bond. It undergoes bulk erosion through hydrolysis of its ester bonds. The degradation products formed are primarily fumaric acid and propylene glycol. The double bonds in PPF allow the polymer to be cross-linked thermally or by photoinitiator into a solid, polymeric network. These networks are mechanically strong, biocompatible, and biodegradable [149, 150]. Polyfumarates suffer a limitation particularly with respect to bone tissue engineering, i.e., lack of mechanical strength due to flexible C-O-C region in its backbone. Hence, several strategies have been devised which include incorporation of ceramics or nanoparticles [149, 151]. Other polyfumarate-based polymers, namely, poly(caprolactone fumarate) and poly(ethylene glycol fumarate), have also been developed into injectable systems and termed as "self-cross-link," since no cross-linking agent is required. A photoinitiator and accelerator, however, are required [152].

A copolymer of polyfumarate is widely used in focal drug delivery. Poly(propylene fumarate) (PPF) oligomers were synthesized by step polymerization using bis(2-hydroxypropyl fumarate) or propylene bis(hydrogen maleate) as starting materials. Oligomers possessing identical degrees of polymerization, but varying in their end group character (either hydroxyl or carboxyl), were first prepared and characterized and then used as part of a bone cement preparation consisting of oligomer, tricalcium phosphate, calcium carbonate, and methyl methacrylate. Compressive strength of the resulting composite appears to be dependent on both the degree of polymerization of the PPF and the nature of the oligomers' end groups [153].

1.4 Summary and Prospective

Biodegradable polymers have increasingly been used as medical implants and invasive delivery systems. A comprehensive review of the polymers in clinical use was published in our recently published book (Biodegradable Polymers in Clinical Use and Clinical Development, John Wiley & Sons, Inc., 2011). The most commonly used biodegradable materials which are clinically tested for focal drug therapy are the natural polymers, gelatin and collagen, and the synthetic polymers prepared from lactic acid, glycolic acid, and caprolactone. The polymer molecular weight, copolymer composition, crystallinity, branching, surface area, drug loading and manufacturing process of the delivery device and drug water solubility and size, and the release conditions determine the release rate and biodegradation of the incorporated drug. Injectable extended release formulations are desirable for localized extended release of drugs.

References

1. Jain JP, Yenet Ayen W, Domb AJ, Kumar N (2011) Biodegradable polymers in clinical use and clinical development. Wiley, Hoboken, NJ, pp 1–58
2. Mittal A, Chitkara D, Kumar N, Pawar R, Domb A, Corn B (2007) Polymeric Carriers for Regional Drug Therapy, Smart polymers. CRC, Boca Raton, FL, pp 359–400
3. Dang JM, Leong KW (2006) Natural polymers for gene delivery and tissue engineering. Adv Drug Deliv Rev 58(4):487–499
4. Lee CH, Singla A, Lee Y (2001) Biomedical applications of collagen. Int J Pharm 221(1–2):1–22
5. Capito RM, Spector M (2007) Collagen scaffolds for nonviral IGF-1 gene delivery in articular cartilage tissue engineering. Gene Ther 14(9):721–732

6. Eliopoulos N, Lejeune L, Martineau D, Galipeau J (2004) Human-compatible collagen matrix for prolonged and reversible systemic delivery of erythropoietin in mice from gene-modified marrow stromal cells. Mol Ther 10(4):741–748
7. Aishwarya S, Mahalakshmi S, Sehgal PK (2008) Collagen-coated polycaprolactone microparticles as a controlled drug delivery system. J Microencapsul 25(5):298–306
8. Ananta M, Aulin CE, Hilborn J, Aibibu D, Houis S, Brown RA, Mudera V (2009) A poly(lactic acid-co-caprolactone)-collagen hybrid for tissue engineering applications. Tissue Eng Part A 15(7):1667–1675
9. Beier JP, Klumpp D, Rudisile M, Dersch R, Wendorff JH, Bleiziffer O, Arkudas A, Polykandriotis E, Horch RE, Kneser U (2009) Collagen matrices from sponge to nano: new perspectives for tissue engineering of skeletal muscle. BMC Biotechnol 9:34
10. Duan X, McLaughlin C, Griffith M, Sheardown H (2007) Biofunctionalization of collagen for improved biological response: scaffolds for corneal tissue engineering. Biomaterials 28:78–88
11. Becker C, Olde Damink L, Laeufer T, Brehmer B, Heschel I, Jakse G (2006) 'UroMaix' scaffolds: novel collagen matrices for application in tissue engineering of the urinary tract. Int J Artif Organs 29(8):764–771
12. Khan W, Yadav D, Domb AJ, Kumar N (2011) Biodegradable polymers in clinical use and clinical development. Wiley, Hoboken, NJ, pp 59–89
13. Cai S, Zhai Y, Xu G, Lu S, Zhou W, Ye X (2011) Preparation and properties of calcium phosphate cements incorporated gelatin microspheres and calcium sulfate dihydrate as controlled local drug delivery system. J Mater Sci Mater Med 22(11):2487–2496
14. Samad A, Sultana Y, Khar RK, Chuttani K, Mishra AK (2009) Gelatin microspheres of rifampicin cross-linked with sucrose using thermal gelation method for the treatment of tuberculosis. J Microencapsul 26(1):83–89
15. Cao FL, Xi YW, Tang L, Yu AH, Zhai GX (2009) Preparation and characterization of curcumin loaded gelatin microspheres for lung targeting. Zhong Yao Cai 32(3):423–426
16. Ohta S, Nitta N, Sonoda A, Seko A, Tanaka T, Takahashi M, Kimura Y, Tabata Y, Murata K (2009) Cisplatin-conjugated degradable gelatin microspheres: fundamental study in vitro. Br J Radiol 82(977):380–385
17. Featherstone C (1997) Fibrin sealants for haemostasis and drug delivery. Lancet 349(9048):334
18. Greco F, de Palma L, Spagnolo N, Rossi A, Specchia N, Gigante A (1991) Fibrin-antibiotic mixtures: an in vitro study assessing the possibility of using a biologic carrier for local drug delivery. J Biomed Mater Res 25(1):39–51
19. Cruysberg LP, Nuijts RM, Gilbert JA, Geroski DH, Hendrikse F, Edelhauser HF (2005) In vitro sustained human transscleral drug delivery of fluorescein-labeled dexamethasone and methotrexate with fibrin sealant. Curr Eye Res 30(8):653–660
20. Miyazaki S, Ishii K, Takada M (1982) Use of fibrin film as a carrier for drug delivery: a long-acting delivery system for pilocarpine into the eye. Chem Pharm Bull (Tokyo) 30(9):3405–3407
21. Senderoff RI, Sheu MT, Sokoloski TD (1991) Fibrin based drug delivery systems. J Parenter Sci Technol 45(1):2–6
22. Kupcsik L, Alini M, Stoddart MJ (2009) Epsilon-aminocaproic acid is a useful fibrin degradation inhibitor for cartilage tissue engineering. Tissue Eng Part A 15(8):2309–2313
23. Sugitachi A, Takatsuka Y, Numata N, Kawahara T, Hirata M, Kido T, Sakamoto I (1989) Loco-regional cancer chemotherapy with a new drug delivery system, "anticancer drug-fibrin clot". Gan To Kagaku Ryoho 16(8 Pt 2):2814–2817
24. Ahmed TA, Dare EV, Hincke M (2008) Fibrin: a versatile scaffold for tissue engineering applications. Tissue Eng Part B Rev 14(2):199–215
25. Breen A, Dockery P, O'Brien T, Pandit A (2009) Fibrin scaffold promotes adenoviral gene transfer and controlled vector delivery. J Biomed Mater Res A 89(4):876–884
26. Mol A, van Lieshout MI, Dam-de Veen CG, Neuenschwander S, Hoerstrup SP, Baaijens FP, Bouten CV (2005) Fibrin as a cell carrier in cardiovascular tissue engineering applications. Biomaterials 26(16):3113–3121

27. Ye Q, Zund G, Benedikt P, Jockenhoevel S, Hoerstrup SP, Sakyama S, Hubbell JA, Turina M (2000) Fibrin gel as a three dimensional matrix in cardiovascular tissue engineering. Eur J Cardiothorac Surg 17(5):587–591
28. Lammel AS, Hu X, Park SH, Kaplan DL, Scheibel TR (2010) Controlling silk fibroin particle features for drug delivery. Biomaterials 31(16):4583–4591
29. Hofmann S, Foo CT, Rossetti F, Textor M, Vunjak-Novakovic G, Kaplan DL, Merkle HP, Meinel L (2006) Silk fibroin as an organic polymer for controlled drug delivery. J Control Release 111(1–2):219–227
30. Cilurzo F, Gennari CG, Selmin F, Marotta LA, Minghetti P, Montanari L (2011) An investigation into silk fibroin conformation in composite materials intended for drug delivery. Int J Pharm 414(1–2):218–224
31. Wenk E, Wandrey AJ, Merkle HP, Meinel L (2008) Silk fibroin spheres as a platform for controlled drug delivery. J Control Release 132(1):26–34
32. Bessa PC, Balmayor ER, Azevedo HS, Nurnberger S, Casal M, van Griensven M, Reis RL, Redl H (2010) Silk fibroin microparticles as carriers for delivery of human recombinant BMPs. Physical characterization and drug release. J Tissue Eng Regen Med 4(5):349–355
33. Gobin AS, Rhea R, Newman RA, Mathur AB (2006) Silk-fibroin-coated liposomes for long-term and targeted drug delivery. Int J Nanomedicine 1(1):81–87
34. Bhowmik BB, Sa B, Mukherjee A (2006) Preparation and in vitro characterization of slow release testosterone nanocapsules in alginates. Acta Pharm 56(4):417–429
35. Dettmar PW, Strugala V, Tselepis C (2007) The effect of alginates on deoxycholic-acid-induced changes in oesophageal mucosal biology at pH 4. J Biomater Sci Polym Ed 18(3):317–333
36. Varshosaz J (2012) Dextran conjugates in drug delivery. Expert Opin Drug Deliv 9(5):509–523
37. Liu L, Zheng M, Renette T, Kissel T (2012) Modular synthesis of folate conjugated ternary copolymers: polyethylenimine-graft-polycaprolactone-block-poly(ethylene glycol)-folate for targeted gene delivery. Bioconjug Chem 23:1211–1220
38. Varshosaz J, Emami J, Ahmadi F, Tavakoli N, Minaiyan M, Fassihi A, Mahzouni P, Dorkoosh F (2010) Preparation of budesonide-dextran conjugates using glutarate spacer as a colon-targeted drug delivery system: in vitro/in vivo evaluation in induced ulcerative colitis. J Drug Target 19(2):140–153
39. Varshosaz J, Emami J, Fassihi A, Tavakoli N, Minaiyan M, Ahmadi F, Mahzouni P, Dorkoosh F (2010) Effectiveness of budesonide-succinate-dextran conjugate as a novel prodrug of budesonide against acetic acid-induced colitis in rats. Int J Colorectal Dis 25(10):1159–1165
40. Namkung S, Chu CC (2007) Partially biodegradable temperature- and pH-responsive poly(N-isopropylacrylamide)/dextran-maleic acid hydrogels: formulation and controlled drug delivery of doxorubicin. J Biomater Sci Polym Ed 18(7):901–924
41. Abdullah S, Wendy-Yeo WY, Hosseinkhani H, Hosseinkhani M, Masrawa E, Ramasamy R, Rosli R, Rahman SA, Domb AJ (2010) Gene transfer into the lung by nanoparticle dextran-spermine/plasmid DNA complexes. J Biomed Biotechnol 2010:284840
42. Mattioli-Belmonte M, Gigante A, Muzzarelli RA, Politano R, De Benedittis A, Specchia N, Buffa A, Biagini G, Greco F (1999) N,N-dicarboxymethyl chitosan as delivery agent for bone morphogenetic protein in the repair of articular cartilage. Med Biol Eng Comput 37(1):130–134
43. Trapani A, De Giglio E, Cafagna D, Denora N, Agrimi G, Cassano T, Gaetani S, Cuomo V, Trapani G (2011) Characterization and evaluation of chitosan nanoparticles for dopamine brain delivery. Int J Pharm 419(1–2):296–307
44. Bhattarai N, Gunn J, Zhang M (2010) Chitosan-based hydrogels for controlled, localized drug delivery. Adv Drug Deliv Rev 62(1):83–99
45. Bowman K, Leong KW (2006) Chitosan nanoparticles for oral drug and gene delivery. Int J Nanomedicine 1(2):117–128

46. Agnihotri SA, Mallikarjuna NN, Aminabhavi TM (2004) Recent advances on chitosan-based micro- and nanoparticles in drug delivery. J Control Release 100(1):5–28
47. Alonso MJ, Sanchez A (2003) The potential of chitosan in ocular drug delivery. J Pharm Pharmacol 55(11):1451–1463
48. Al-Ghananeem AM, Saeed H, Florence R, Yokel RA, Malkawi AH (2010) Intranasal drug delivery of didanosine-loaded chitosan nanoparticles for brain targeting; an attractive route against infections caused by AIDS viruses. J Drug Target 18(5):381–388
49. Anitha A, Maya S, Deepa N, Chennazhi KP, Nair SV, Jayakumar R (2012) Curcumin-loaded N, O-carboxymethyl chitosan nanoparticles for cancer drug delivery. J Biomater Sci Polym Ed 23:1381–1400
50. Chen R, Chen Q, Huo D, Ding Y, Hu Y, Jiang X (2012) In situ formation of chitosan-gold hybrid hydrogel and its application for drug delivery. Colloids Surf B Biointerfaces 97:132–137
51. Chan P, Kurisawa M, Chung JE, Yang YY (2007) Synthesis and characterization of chitosan-g-poly(ethylene glycol)-folate as a non-viral carrier for tumor-targeted gene delivery. Biomaterials 28(3):540–549
52. Huang W, Shi X, Ren L, Du C, Wang Y (2010) PHBV microspheres–PLGA matrix composite scaffold for bone tissue engineering. Biomaterials 31(15):4278–4285
53. Ye C, Hu P, Ma MX, Xiang Y, Liu RG, Shang XW (2009) PHB/PHBHHx scaffolds and human adipose-derived stem cells for cartilage tissue engineering. Biomaterials 30(26):4401–4406
54. Novikova LN, Pettersson J, Brohlin M, Wiberg M, Novikov LN (2008) Biodegradable poly-beta-hydroxybutyrate scaffold seeded with Schwann cells to promote spinal cord repair. Biomaterials 29(9):1198–1206
55. Falk R, Domb AJ, Polacheck I (1999) A novel injectable water-soluble amphotericin B-arabinogalactan conjugate. Antimicrob Agents Chemother 43(8):1975–1981
56. Golenser J, Frankenburg S, Ehrenfreund T, Domb AJ (1999) Efficacious treatment of experimental leishmaniasis with amphotericin B-arabinogalactan water-soluble derivatives. Antimicrob Agents Chemother 43(9):2209–2214
57. Avramoff A, Khan W, Mizrahi B, Domb AJ (2012) Preparation and characterization of a novel once-daily formulation of diltiazem using arabinogalactan as a channeling agent. J Appl Polym Sci 126(S1):E197–E203
58. Elgart A, Farber S, Domb AJ, Polacheck I, Hoffman A (2010) Polysaccharide pharmacokinetics: amphotericin B arabinogalactan conjugate-a drug delivery system or a new pharmaceutical entity? Biomacromolecules 11(8):1972–1977
59. Groman EV, Enriquez PM, Jung C, Josephson L (1994) Arabinogalactan for hepatic drug delivery. Bioconjug Chem 5(6):547–556
60. Gunatillake PA, Adhikari R (2003) Biodegradable synthetic polymers for tissue engineering. Eur Cell Mater 5:1–16, discussion 16
61. Buchter A, Kleinheinz J, Meyer U, Joos U (2004) Treatment of severe peri-implant bone loss using autogenous bone and a bioabsorbable polymer that delivered doxycycline (Atridox). Br J Oral Maxillofac Surg 42(5):454–456
62. Sinclair G (2000) The A-X of atridox. Ann R Australas Coll Dent Surg 15:157–158
63. Wessely R, Kastrati A, Mehilli J, Dibra A, Pache J, Schomig A (2007) Randomized trial of rapamycin- and paclitaxel-eluting stents with identical biodegradable polymeric coating and design. Eur Heart J 28(22):2720–2725
64. Asbury RF, Brunetto VL, Lee RB, Reid G, Rocereto TF (2002) Goserelin acetate as treatment for recurrent endometrial carcinoma: a Gynecologic Oncology Group study. Am J Clin Oncol 25(6):557–560
65. Matsuura M, Fujiwara T, Kataoka K, Itoh M, Ohtani S, Higaki K, Senoo N (2009) Catamenial pneumothorax with breast cancer treated successfully by goserelin acetate. Kyobu Geka 62(11):1015–1018
66. Roach M III, Izaguirre A (2007) Goserelin acetate in combination with radiotherapy for prostate cancer. Expert Opin Pharmacother 8(2):257–264

67. Litjens TT, Fernandez del Moral P, van Laarhoven JP, Weil EH, Debruyne FM (1986) Treatment of advanced prostate carcinoma with a depot form of an LH-RH analog: preliminary endocrinological and clinical results. Ned Tijdschr Geneeskd 130(17):787–790
68. Sartor O (2003) Eligard: leuprolide acetate in a novel sustained-release delivery system. Urology 61(2 Suppl 1):25–31
69. Ruiz-Hornillos J, Henriquez-Santana A, Moreno-Fernandez A, Gonzalez IG, Sanchez SR (2009) Systemic allergic dermatitis caused by the solvent of Eligard. Contact Dermatitis 61(6):355–356
70. Woltering EA, Mamikunian PM, Zietz S, Krutzik SR, Go VL, Vinik AI, Vinik E, O'Dorisio TM, Mamikunian G (2005) Effect of octreotide LAR dose and weight on octreotide blood levels in patients with neuroendocrine tumors. Pancreas 31(4):392–400
71. Fasano CJ, O'Malley G, Dominici P, Aguilera E, Latta DR (2008) Comparison of octreotide and standard therapy versus standard therapy alone for the treatment of sulfonylurea-induced hypoglycemia. Ann Emerg Med 51(4):400–406
72. Brin YS, Golenser J, Mizrahi B, Maoz G, Domb AJ, Peddada S, Tuvia S, Nyska A, Nyska M (2008) Treatment of osteomyelitis in rats by injection of degradable polymer releasing gentamicin. J Control Release 131(2):121–127
73. Kotler DP, Muurahainen N, Grunfeld C, Wanke C, Thompson M, Saag M, Bock D, Simons G, Gertner JM (2004) Effects of growth hormone on abnormal visceral adipose tissue accumulation and dyslipidemia in HIV-infected patients. J Acquir Immune Defic Syndr 35(3):239–252
74. Gefvert O, Eriksson B, Persson P, Helldin L, Bjorner A, Mannaert E, Remmerie B, Eerdekens M, Nyberg S (2005) Pharmacokinetics and D2 receptor occupancy of long-acting injectable risperidone (Risperdal Consta) in patients with schizophrenia. Int J Neuropsychopharmacol 8(1):27–36
75. Bobo WV, Shelton RC (2010) Risperidone long-acting injectable (Risperdal Consta(R)) for maintenance treatment in patients with bipolar disorder. Expert Rev Neurother 10(11):1637–1658
76. Ballabio E, Armesto M, Breeze CE, Manterola L, Arestin M, Tramonti D, Hatton CS, Lawrie CH (2012) Bortezomib action in multiple myeloma: microRNA-mediated synergy (and miR-27a/CDK5 driven sensitivity)? Blood Cancer J 2:e83
77. Al-Qashi S, Al-Qaoud KM, Ja'fer M, Khali AM (2006) Immunocytogenetic effects of gonadotropin releasing hormone analogue: Triptorelin Pamoate (Decapeptyl) during in vitro fertilization treatment. Hum Exp Toxicol 25(10):593–597
78. Palatynski A, Gruszczynska J (2001) Decapeptyl (triptorelin) in the treatment of endometriosis genitalis externa. Ginekol Pol 72(5):290–295
79. Choktanasiri W, Rojanasakul A (2001) Buserelin acetate implants in the treatment of pain in endometriosis. J Med Assoc Thai 84(5):656–660
80. Harada T, Momoeda M, Taketani Y, Aso T, Fukunaga M, Hagino H, Terakawa N (2009) Dienogest is as effective as intranasal buserelin acetate for the relief of pain symptoms associated with endometriosis–a randomized, double-blind, multicenter, controlled trial. Fertil Steril 91(3):675–681
81. Ory SJ, Hammond CB, Yancy SG, Hendren RW, Pitt CG (1983) The effect of a biodegradable contraceptive capsule (Capronor) containing levonorgestrel on gonadotropin, estrogen, and progesterone levels. Am J Obstet Gynecol 145(5):600–605
82. Attenello FJ, Mukherjee D, Datoo G, McGirt MJ, Bohan E, Weingart JD, Olivi A, Quinones-Hinojosa A, Brem H (2008) Use of Gliadel (BCNU) wafer in the surgical treatment of malignant glioma: a 10-year institutional experience. Ann Surg Oncol 15(10):2887–2893
83. Bota DA, Desjardins A, Quinn JA, Affronti ML, Friedman HS (2007) Interstitial chemotherapy with biodegradable BCNU (Gliadel) wafers in the treatment of malignant gliomas. Ther Clin Risk Manag 3(5):707–715
84. McGirt MJ, Than KD, Weingart JD, Chaichana KL, Attenello FJ, Olivi A, Laterra J, Kleinberg LR, Grossman SA, Brem H, Quinones-Hinojosa A (2009) Septacin. J Neurosurg 110(3):583–588

85. Karr J (2008) Utilization of living bilayered cell therapy (Apligraf) for heel ulcers. Adv Skin Wound Care 21(6):270–274
86. Edmonds M (2009) Apligraf in the treatment of neuropathic diabetic foot ulcers. Int J Low Extrem Wounds 8(1):11–18
87. Saffarzadeh A, Gauthier O, Bilban M, Bagot D'Arc M, Daculsi G (2009) Comparison of two bone substitute biomaterials consisting of a mixture of fibrin sealant (Tisseel) and MBCP (TricOs) with an autograft in sinus lift surgery in sheep. Clin Oral Implants Res 20(10):1133–1139
88. Guimard NK, Gomez N, Schmidt CE (2007) Conducting polymers in biomedical engineering. Prog Polym Sci 32(8–9):876–921
89. Wang YC, Lin MC, Wang DM, Hsieh HJ (2003) Fabrication of a novel porous PGA-chitosan hybrid matrix for tissue engineering. Biomaterials 24(6):1047–1057
90. Bishara A, Domb AJ (2005) PLA stereocomplexes for controlled release of somatostatin analogue. J Control Release 107(3):474–483
91. Bunger CM, Grabow N, Sternberg K, Kroger C, Ketner L, Schmitz KP, Kreutzer HJ, Ince H, Nienaber CA, Klar E, Schareck W (2007) Sirolimus-eluting biodegradable poly-L-lactide stent for peripheral vascular application: a preliminary study in porcine carotid arteries. J Surg Res 139(1):77–82
92. Montjovent MO, Mark S, Mathieu L, Scaletta C, Scherberich A, Delabarde C, Zambelli PY, Bourban PE, Applegate LA, Pioletti DP (2008) Human fetal bone cells associated with ceramic reinforced PLA scaffolds for tissue engineering. Bone 42(3):554–564
93. Stock UA, Mayer JE Jr (2001) Tissue engineering of cardiac valves on the basis of PGA/PLA Co-polymers. J Long Term Eff Med Implants 11(3–4):249–260
94. Karp JM, Shoichet MS, Davies JE (2003) Bone formation on two-dimensional poly(DL-lactide-co-glycolide) (PLGA) films and three-dimensional PLGA tissue engineering scaffolds in vitro. J Biomed Mater Res A 64(2):388–396
95. Kim H, Kim HW, Suh H (2003) Sustained release of ascorbate-2-phosphate and dexamethasone from porous PLGA scaffolds for bone tissue engineering using mesenchymal stem cells. Biomaterials 24(25):4671–4679
96. Tian JG, Bai DH, Liu ZG, Tang HT, Xia ZF (2003) Experimental studies on the biocompatibility of spongy PLGA-collagen membrane as a tissue-engineering dermal scaffold. Zhonghua Shao Shang Za Zhi 19(Suppl):1–4
97. Bivas-Benita M, Romeijn S, Junginger HE, Borchard G (2004) PLGA-PEI nanoparticles for gene delivery to pulmonary epithelium. Eur J Pharm Biopharm 58(1):1–6
98. Csaba N, Caamano P, Sanchez A, Dominguez F, Alonso MJ (2005) PLGA:poloxamer and PLGA:poloxamine blend nanoparticles: new carriers for gene delivery. Biomacromolecules 6(1):271–278
99. Duvvuri S, Janoria KG, Pal D, Mitra AK (2007) Controlled delivery of ganciclovir to the retina with drug-loaded Poly(d, L-lactide-co-glycolide) (PLGA) microspheres dispersed in PLGA-PEG-PLGA Gel: a novel intravitreal delivery system for the treatment of cytomegalovirus retinitis. J Ocul Pharmacol Ther 23(3):264–274
100. Bae SE, Son JS, Park K, Han DK (2009) Fabrication of covered porous PLGA microspheres using hydrogen peroxide for controlled drug delivery and regenerative medicine. J Control Release 133(1):37–43
101. Chan JM, Zhang L, Yuet KP, Liao G, Rhee JW, Langer R, Farokhzad OC (2009) PLGA-lecithin-PEG core-shell nanoparticles for controlled drug delivery. Biomaterials 30(8):1627–1634
102. Chen S, Singh J (2008) Controlled release of growth hormone from thermosensitive triblock copolymer systems: in vitro and in vivo evaluation. Int J Pharm 352(1–2):58–65
103. Danhier F, Vroman B, Lecouturier N, Crokart N, Pourcelle V, Freichels H, Jerome C, Marchand-Brynaert J, Feron O, Preat V (2009) Targeting of tumor endothelium by RGD-grafted PLGA-nanoparticles loaded with paclitaxel. J Control Release 140(2):166–173
104. Sinha VR, Bansal K, Kaushik R, Kumria R, Trehan A (2004) Poly-ε-caprolactone microspheres and nanospheres: an overview. Int J Pharm 278(1):1–23

105. Byrro RM, Miyashita D, Albuquerque VB, Velasco e Cruz AA, Cunha Junior Ada S (2009) Biodegradable systems containing prednisolone acetate for orbital administration. Arq Bras Oftalmol 72(4):444–450
106. Kyun KD, Yun KS, Young JS, Pyoung CC, Heui SS (1990) Development of minocycline containing polycaprolactone film as a local drug delivery. Taehan Chikkwa Uisa Hyophoe Chi 28(3):279–290
107. Hoque ME, Hutmacher DW, Feng W, Li S, Huang MH, Vert M, Wong YS (2005) Fabrication using a rapid prototyping system and in vitro characterization of PEG-PCL-PLA scaffolds for tissue engineering. J Biomater Sci Polym Ed 16(12):1595–1610
108. Williamson MR, Black R, Kielty C (2006) PCL-PU composite vascular scaffold production for vascular tissue engineering: attachment, proliferation and bioactivity of human vascular endothelial cells. Biomaterials 27(19):3608–3616
109. Zhao J, Gou M, Dai M, Li X, Cao M, Huang M, Wen Y, Kan B, Qian Z, Wei Y (2009) Preparation, characterization, and in vitro cytotoxicity study of cationic PCL-pluronic-PCL (PCFC) nanoparticles for gene delivery. J Biomed Mater Res A 90(2):506–513
110. Benagiano G, Gabelnick HL (1979) Biodegradable systems for the sustained release of fertility-regulating agents. J Steroid Biochem 11(1B):449–455
111. Sokolsky-Papkov M, Agashi K, Olaye A, Shakesheff K, Domb AJ (2007) Polymer carriers for drug delivery in tissue engineering. Adv Drug Deliv Rev 59(4–5):187–206
112. Heller J, Barr J, Ng SY, Shen HR, Schwach-Abdellaoui K, Einmahl S, Rothen-Weinhold A, Gurny R (2000) Poly(ortho esters) – their development and some recent applications. Eur J Pharm Biopharm 50(1):121–128
113. Einmahl S, Behar-Cohen F, Tabatabay C, Savoldelli M, D'Hermies F, Chauvaud D, Heller J, Gurny R (2000) A viscous bioerodible poly(ortho ester) as a new biomaterial for intraocular application. J Biomed Mater Res 50(4):566–573
114. Einmahl S, Zignani M, Varesio E, Heller J, Veuthey JL, Tabatabay C, Gurny R (1999) Concomitant and controlled release of dexamethasone and 5-fluorouracil from poly(ortho ester). Int J Pharm 185(2):189–198
115. Andriano KP, Daniels AU, Heller J (1992) Biocompatibility and mechanical properties of a totally absorbable composite material for orthopaedic fixation devices. J Appl Biomater 3(3):197–206
116. Barr J, Woodburn KW, Ng SY, Shen HR, Heller J (2002) Post surgical pain management with poly(ortho esters). Adv Drug Deliv Rev 54(7):1041–1048
117. Chasin M, Lewis D, Langer R (1988) Polyanhydrides for controlled drug delivery. Biopharm Manuf 1:33–46
118. Ibim SE, Uhrich KE, Attawia M, Shastri VR, El-Amin SF, Bronson R (1998) Preliminary in vivo report on the osteocompatibility of poly(anhydride-co-imides) evaluated in a tibial model. J Biomed Mater Res 43:374–379
119. Jain JP, Chitkara D, Kumar N (2008) Polyanhydrides as localized drug delivery carrier: an update. Expert Opin Drug Deliv 5(8):889–907
120. Jain JP, Modi S, Domb AJ, Kumar N (2005) Role of polyanhydrides as localized drug carriers. J Control Release 103:541–563
121. Jain JP, Modi S, Kumar N (2008) Hydroxy fatty acid based polyanhydride as drug delivery system: synthesis, characterization, in vitro degradation, drug release, and biocompatibility. J Biomed Mater Res A 84:740–752
122. Domb AJ (1995) Polymeric carriers for regional drug therapy. Mol Med Today 1(3):134–139
123. Manoharan C, Singh J (2009) Evaluation of polyanhydride microspheres for basal insulin delivery: effect of copolymer composition and zinc salt on encapsulation, in vitro release, stability, in vivo absorption and bioactivity in diabetic rats. J Pharm Sci 98(11):4237–4250
124. Shikanov A, Vaisman B, Shikanov S, Domb AJ (2010) Efficacy of poly(sebacic acid-co-ricinoleic acid) biodegradable delivery system for intratumoral delivery of paclitaxel. J Biomed Mater Res A 92(4):1283–1291

125. Shikanov S, Shikanov A, Gofrit O, Nyska A, Corn B, Domb AJ (2009) Intratumoral delivery of paclitaxel for treatment of orthotopic prostate cancer. J Pharm Sci 98(3):1005–1014
126. Shikanov A, Domb AJ, Weiniger CF (2007) Long acting local anesthetic-polymer formulation to prolong the effect of analgesia. J Control Release 117(1):97–103
127. Burkoth AK, Anseth KS (2000) A review of photocrosslinked polyanhydrides: in situ forming degradable networks. Biomaterials 21(23):2395–2404
128. Jiang HL, Zhu KJ (2000) Pulsatile protein release from a laminated device comprising polyanhydrides and pH-sensitive complexes. Int J Pharm 194(1):51–60
129. Nair LS, Laurencin CT (2007) Biodegradable polymers as biomaterials. Prog Polym Sci 32(8–9):762–798
130. Gorna K, Gogolewski S (2002) Biodegradable polyurethanes for implants. II. In vitro degradation and calcification of materials from poly(epsilon-caprolactone)-poly(ethylene oxide) diols and various chain extenders. J Biomed Mater Res 60:592–606
131. Gunatillake PA, Meijs GF (2008) In: Buschow KHJ, Cahn RW, Flemings MC, Ilschner BI, Kramer EJ, Mahajan S, Veyssière P (eds) Encyclopedia of materials: science and technology. Amsterdam, Elsevier, pp 7746–7752
132. Bil M, Ryszkowska J, Wozniak P, Kurzydlowski KJ, Lewandowska-Szumiel M (2009) Optimization of the structure of polyurethanes for bone tissue engineering applications. Acta Biomater 6(7):2501–2510
133. Bonzani IC, Adhikari R, Houshyar S, Mayadunne R, Gunatillake P, Stevens MM (2007) Synthesis of two-component injectable polyurethanes for bone tissue engineering. Biomaterials 28(3):423–433
134. Jiang X, Wang K, Ding M, Li J, Tan H, Wang Z, Fu Q (2011) Quantitative grafting of peptide onto the nontoxic biodegradable waterborne polyurethanes to fabricate peptide modified scaffold for soft tissue engineering. J Mater Sci Mater Med 22(4):819–827
135. Cui W, Zhu X, Yang Y, Li X, Jin Y (2009) Evaluation of electrospun fibrous scaffolds of poly(dl-lactide) and poly(ethylene glycol) for skin tissue engineering. Mater Sci Eng C 29:1869–1876
136. Strehin I, Nahas Z, Arora K, Nguyen T, Elisseeff J (2010) A versatile pH sensitive chondroitin sulfate–PEG tissue adhesive and hydrogel. Biomaterials 31(10):2788–2797
137. Vashi AV, Keramidaris E, Abberton KM, Morrison WA, Wilson JL, O'Connor AJ, Cooper-White JJ, Thompson EW (2008) Adipose differentiation of bone marrow-derived mesenchymal stem cells using Pluronic F-127 hydrogel in vitro. Biomaterials 29:573–579
138. Ruszymah BH, Chua K, Latif MA, Hussein FN, Saim AB (2005) Formation of in vivo tissue engineered human hyaline cartilage in the shape of a trachea with internal support. Int J Pediatr Otorhinolaryngol 69:1489–1495
139. Liu Y, Chen F, Liu W, Cui L, Shang Q, Xia W (2002) Repairing large porcine full-thickness defects of articular cartilage using autologous chondrocyte-engineered cartilage. Tissue Eng 8:709–721
140. Cortiella J, Nichols JE, Kojima K, Bonassar LJ, Dargon P, Roy AK (2006) Tissue-engineered lung: an in vivo and in vitro comparison of polyglycolic acid and Pluronic F-127 hydrogel/somatic lung progenitor cell constructs to support tissue growth. Tissue Eng 12:1213–1225
141. Shachaf Y, Gonen-Wadmany M, Seliktar D (2010) The biocompatibility of Pluronic®F127 fibrinogen-based hydrogels. Biomaterials 31(10):2836–2847
142. Khan W, Muthupandian S, Farah S, Kumar N, Domb AJ (2011) Biodegradable polymers derived from amino acids. Macromol Biosci 11(12):1625–1636
143. Bourke SL, Kohn J (2003) Polymers derived from the amino acidL-tyrosine:polycarbonates, polyarylates and copolymers with poly(ethylene glycol). Adv Drug Deliv Rev 55:447–466
144. Gonzalez-Aramundiz JV, Lozano MV, Sousa-Herves A, Fernandez-Megia E, Csaba N (2012) Polypeptides and polyaminoacids in drug delivery. Expert Opin Drug Deliv 9(2):183–201
145. Cavallaro G, Pitarresi G, Giammona G (2011) Macromolecular prodrugs based on synthetic polyaminoacids: drug delivery and drug targeting in antitumor therapy. Curr Top Med Chem 11(18):2382–2389

146. Benns JM, Choi JS, Mahato RI, Park JS, Kim SW (2000) pH-sensitive cationic polymer gene delivery vehicle: N-Ac-poly(L-histidine)-graft-poly(L-lysine) comb shaped polymer. Bioconjug Chem 11(5):637–645
147. Davis PB, Kowalczyk TH (2010) Preparation and analysis of PEGylated poly-L-lysine DNA nanoparticles for gene delivery. Cold Spring Harb Protoc 2010(5):pdb.prot5419
148. Holland TA, Tabata Y, Mikos AG (2005) Dual growth factor delivery from degradable oligo(poly(ethylene glycol) fumarate) hydrogel scaffolds for cartilage tissue engineering. J Control Release 101(1–3):111–125
149. Peter SJ, Lu L, Kim DJ, Mikos AG (2000) Marrow stromal osteoblast function on a poly(propylene fumarate)/β-tricalcium phosphate biodegradable orthopaedic composite. Biomaterials 21:1207–1213
150. Yaszemski MJ, Payne RG, Hayes WC, Langer R, Mikos AG (1996) In vitro degradation of a poly(propylene fumarate)-based composite materials. Biomaterials 17:2120–2130
151. Horch RA, Shahid N, Mistry AS, Timmer MD, Mikos AG, Barron AR (2004) Nanoreinforcement of poly(propylene fumarate)-based networks with surface modified alumoxane nanoparticles for bone tissue engineering. Biomacromolecules 5(5):1990–1998
152. Jabbari E, Wang SF, Lu LC, Gruetzmacher JA, Ameenuddin S, Hefferan TE, Currier BL, Windebank AJ, Yaszemski MJ (2005) Synthesis, material properties, and biocompatibility of a novel self-cross-linkable poly(caprolactone fumarate) as an injectable tissue engineering scaffold. Biomaterials 6:2503–2511
153. Domb AJ, Laurencin CT, Israeli O, Gerhart TN, Langer R (1990) The formation of propylene fumarate oligomers for use in bioerodible bone cement composites. J Polym Sci A Polym Chem 28:973–985

Chapter 2
Implantable Medical Devices

Wahid Khan, Eameema Muntimadugu, Michael Jaffe, and Abraham J. Domb

Abbreviations

MEMS	Microelectromechanical systems
NEMS	Nano-electromechanical systems
PCL	Polycaprolactone
PET	Polyethylene terephthalate
PGA	Polyglycolic acid
PLA	Polylactic acid
PU	Polyurethane
PTFE	Poly(tetrafluoroethylene)
PMMA	Polymethylmethacrylate
UHMWPE	Ultrahigh molecular weight polyethylene

W. Khan • E. Muntimadugu
Department of Pharmaceutics, National Institute of Pharmaceutical Education and Research (NIPER), Hyderabad 500037, India

M. Jaffe
Medical Device Concept Laboratory, New Jersey Institute of Technology, Newark, NJ, USA

A.J. Domb (✉)
Faculty of Medicine, Institute of Drug Research, School of Pharmacy, The Hebrew University of Jerusalem, Jerusalem 91120, Israel
e-mail: avid@ekmd.huji.ac.il

A.J. Domb and W. Khan (eds.), *Focal Controlled Drug Delivery*, Advances in Delivery Science and Technology, DOI 10.1007/978-1-4614-9434-8_2,

2.1 Introduction

Each year millions of patients improve their quality of life through surgical procedures that involve implanted medical devices. The term implant is used for devices that replace or act as a fraction of or the whole biological structure. Currently, implants are being used in many different parts of the body for various applications such as orthopaedics, pacemakers, cardiovascular stents, defibrillators, neural prosthetics or drug delivery system [1]. Concurrent with the increased life span in today's world, the number of age-related diseases has also increased. Hence, the need for new treatments, implants, prostheses and long-term pharmaceutical usage as well as the need for prolonging the life span of the current techniques has increased [2]. Joint diseases represent one of the examples of changing needs in the medical treatment: Today's estimates show that 90 % of the population over the age of 40 suffers from a degenerative joint disease [3]. In 2000, the number of total hip replacements operation was about 152,000 which represents a 33 % increase from the number of operations in 1990 and also represents about half of the estimated number of operations by 2030 [4]. Cardiovascular diseases are another example. Over the last two decades, coronary stents have become a new standard in angioplasty procedure. In 2004, the number of implanted drug-eluting stents alone exceeded two million [5]. In-stent restenosis is a consequence almost entirely of tissue hyperplasia, occurring principally around the points where the stent struts impinge upon the artery wall. Less common, but troublesome when it occurs, is subacute thrombosis, a complication not quite eliminated by modern stent deployment techniques and antiplatelet agents. By carrying a coating or drug targeted at the thrombotic or hyperplastic responses occurring locally, drug-eluting stents present a solution to the above problems [6].

2.2 Classes of Implantable Medical Devices

The regulatory authorities recognise different classes of medical devices, based on their design complexity, their use characteristics and their potential for harm if misused. Each country or region defines these categories in different ways. The authorities also recognise that some devices are provided in combination with drugs, and regulation of these combination products takes this factor into consideration. The FDA classification of medical devices is based upon the level of control required to assure safety and effectiveness of the device [7]. The classification procedures are described in the Code of Federal Regulations, Title 21, part 860 (usually known as 21 CFR 860).

Class I devices are subject to the least regulatory control "general controls". Class I devices are not intended for use in supporting or sustaining life or to be of substantial importance in preventing impairment to human health, and they may not present a potential unreasonable risk of illness or injury. Examples of class I devices include elastic bandages, examination gloves and hand-held surgical instruments. *Class II* devices are those for which general controls alone are insufficient to assure safety and effectiveness, and existing methods are available to provide such assurances. In addition to complying with general controls, class II devices are also subject to special controls. A few class II devices are exempt from the premarket notification. Special controls may include special labelling requirements, mandatory performance standards and postmarket surveillance. Examples of class II devices include powered wheelchairs, infusion pumps and surgical drapes. *Class III* device is one for which insufficient information exists to assure safety and effectiveness solely through the general or special controls sufficient for class I or class II devices. Such a device needs premarket approval, a scientific review to ensure the device's safety and effectiveness, in addition to the general controls of class I. Class III devices are usually those that support or sustain human life and are of substantial importance in preventing impairment of human health or those which present a potential, unreasonable risk of illness or injury. Examples of class III devices include implantable pacemaker, pulse generators, HIV diagnostic tests, automated external defibrillators and endosseous implants. Most of the devices discussed in this chapter will belong to class III. These are mostly implants in the orthopaedic, dental, ophthalmic and cardiovascular fields as well as soft tissue implants such as implants used in plastic surgery. Implants without bioactive coatings intended to secure teeth or prostheses to the maxillary or mandibular bones, and implants intended to be placed in teeth such as bridges, crowns, dental filling materials and dental alloys belong to class II [8].

Based on the purpose for which they are using, these implantable devices are broadly divided into three classes: cardiovascular implants, orthopaedic implants and implants for other use. Various subcategories under these classes are given in Table 2.1.

2.2.1 Cardiovascular Implants

Cardiovascular implants have strong potential to reduce the overall treatment cost for heart disease and at the same time contribute significantly to improved quality of life. Pacing devices will realise the greatest sales gains, largely due to growth in cardiac resynchronisation therapy. A focus on developing new generations of pacing devices that reduce mortality and improve patient outcomes has resulted in greater pricing flexibility in an increasingly cost-conscious health-care environment. Demand for cardiovascular stents and related devices will be similar to that of demand for pacing devices. The fastest growth will be in structural implants, as technological advances in heart valves, ventricular assist devices and implantable monitors will encourage greater use.

Cardiovascular disease broadly covers a range of conditions affecting both the heart and the blood vessels. Polymer-coated and polymer-based cardiovascular implants are essential constituents of modern medicine and will proceed to gain importance with the demographic changes toward a society of increasing

Table 2.1 Classification of implantable medical devices

Orthopaedic implants	Cardiovascular implants	Other medical implants
Reconstructive joint replacements[a] • Knee replacements • Hip replacement implants • Other reconstructive joint replacements – Shoulder implants – Elbow implants – Ankle implants – Joint replacements	Pacing devices[b] • Cardiac resynchronisation therapy devices • Implantable cardioverter-defibrillators • Implantable cardiac pacemakers • Pacing accessories—pacing leads, pacing batteries	Otolaryngeal implants[b,c] • Cochlear implants • Airway and oesophageal stents • Cosmetic implants—breast implants Ophthalmic implants[b] • Intraocular lenses • Glaucoma and other lenses Neurostimulators[b]
Spinal implants[a] • Thoracolumbar implants • Intervertebral spacers • Motion preservation devices • Cervical implants • Implantable spinal stimulators	Cardiac stents and related implants[b,c] • Coronary stents—drug-eluting stents, bare-metal coronary stents • Stent-related implants – Synthetic grafts—vascular grafts, peripheral grafts – Vena cava filters	Gastroenterological implants[b] • Gastric bands • Biliary stents • Urological implants
Orthobiologics[b] • Hyaluronic acid • Bone substitutes • Bone growth factors • Bone cement Trauma implants[b] • Internal fixation devices • Craniomaxillofacial implants • Implantable trauma stimulators	Structural cardiac implants[a] • Heart valves and accessories • Tissue heart valves • Ventricular assist devices • Implantable heart monitors – Insertable loop recorders – Implantable hemodynamic monitors	Gynaecological devices[b] • Soft tissue repair • Intrauterine devices Drug implants[c] • Hormonal implants • Brachytherapy products • Implantable drug pumps

[a]Structural and mechanical support
[b]Functional support
[c]Localised drug delivery

age-related morbidity. Based on the experiences with implants such as coronary or peripheral stents, which are presently widely used in clinical medicine, several properties of the next generation of cardiovascular implants have been envisioned that could be fulfilled by multifunctional polymers. The challenge is to combine tailored mechanical properties and rapid endothelialisation with controlled drug release in order to modulate environmental cells and tissue. Additionally, degradability and sensitivity to external stimuli are useful in several applications. A critical function in terms of clinical complications is the haemocompatibility. The design of devices with improved haemocompatibility requires advanced in vitro test setups as discussed in depth in this article. Finally, degradable, multifunctional shape-memory polymers are introduced as a promising family of functional polymers that fulfil several requirements of modern implants and are of high relevance for cardiovascular application (e.g. stent technology). Such multifunctional polymers are a technology platform for future cardiovascular implants enabling induced auto-regeneration in regenerative therapies [9]. Shape-memory materials have been proposed for cardiovascular stents due to their self-expansion ability. The most ideal way to anchor a stent is using self-expansion in the range of body temperature. Ajili et al., for the first time, report the use of polyurethane/polycaprolactone (PU/PCL) blend as a proposed material for shape-memory stents. The results showed that the blend supported cell adhesion and proliferation, which indicated good biocompatibility, and the results suggested that this blend might be a potential material as a stent implant [10]. Multifunctional shape-memory polymers are highlighted as a class of materials that combine biocompatibility and the capability for stimuli-induced active movements for anchoring of implants with a controlled degradation and drug release profile to enable a functional regeneration of the tissue at the application site. The challenge is to combine tailored mechanical properties and rapid endothelialisation with controlled drug release in order to modulate environmental cells and tissue. Additionally, degradability and sensitivity to external stimuli are useful in several applications [11]. Current cardiovascular therapies are limited by the loss of endothelium, restenosis and thrombosis. Andukuri et al. reported biomimetic hybrid nanomatrix that combined the unique properties of electrospun PCL nanofibres with self-assembled peptide amphiphiles. Results indicate that this hybrid nanomatrix has great potential application in cardiovascular implants [12].

2.2.2 Orthopaedic Implants

Orthopaedic implants will remain the largest implantable device segment in market value. Gains will also reflect the growing prevalence of degenerative musculoskeletal disorders and lifestyle changes that place people at risk for sports and exercise injuries. At the same time, as products become more durable and long-lived, demand will increasingly come from an enlarged patient base for new surgeries rather than for replacements. Also challenging this segment over the long term will be advances in pharmaceutical alternatives to treat arthritic conditions. However, the segment will benefit from a strong base of insurance approvals for orthopaedic implants, as well as a stable and well-funded medical delivery system and product designs that allow for less invasive surgeries.

One of the most prominent application areas for biomaterials is for orthopaedic implant devices. Both osteoarthritis and rheumatoid arthritis affect the structure of freely movable (synovial) joints, such as the hip, knee, shoulder, ankle and elbow. The pain in such joints, particularly weight-bearing joints such as the hip and knee, can be considerable, and the effects on ambulatory function quite devastating. It has been possible to replace these joints with prostheses since the advent of anaesthesia, antisepsis and antibiotics and the relief of pain and restoration of mobility is well known to hundreds of thousands of patients [13].

Total joint replacement is widely regarded as the major achievement in orthopaedic surgery in the twentieth century. Arthroplasty, or the creation of a new joint, is the name given to the surgical treatment of degenerate joints aimed at the relief of pain and the restoration of movement. This has been achieved by excision, interposition and replacement arthroplasty and by techniques that have been developed over approximately 180 years. In a total knee arthroplasty, the diseased cartilage surfaces of the lower femur (thighbone), the tibia (shinbone) and the patella (kneecap) are replaced by a prosthesis made of metal alloys and polymeric materials. Most of the other structures of the knee, such as the connecting ligaments, remain intact [3].

2.2.3 *Other Implants*

Evolution of cochlear implant technology resulted in enhanced hearing, speech and cost-effectiveness for children. Binaural cochlear implantation has been used in children. Development of perimodiolar electrodes, implantable microphones and rechargeable batteries promise fully implanted devices in future [14].

Intraocular sustained drug release using implantable devices has been investigated to treat vitreoretinal diseases. Possible targeted diseases include those in which repeated intraocular injections are effective (cytomegalovirus retinitis, uveitis), diseases requiring surgery (proliferative vitreoretinopathy) and chronic diseases (macular oedema, retinitis pigmentosa). Hydrophobic or hydrophilic polymers shaped into a sheet, disc, rod, plug or a larger device can be implanted into the subretinal space, intrascleral space, vitreous space or peribulbar space or at the pars plana [15].

Solid biocompatible implantable devices for sustained or controlled intravitreal drug delivery to the posterior segment of the eye have been developed employing diverse approaches and include osmotic mini-pumps, nonbioerodible and bioerodible drug-loaded pellets, configured capillary fibres, biodegradable scleral plugs, scleral discs, polymeric matrices and scaffolds of various geometries providing unique mechanisms of drug release for the delivery of drugs to the posterior segment of the eye [16].

2.3 Materials Used for Medical Devices

In general, most of the materials used for implants or devices can be divided into the following categories: metals, polymers and ceramics. Metals are based on the metallic bond, ceramics are based on ionic bonds, and polymers are based on covalent bonds; and each of these categories contains many subdivisions. The metallic materials include pure metals and alloys; ceramics include glasses, glass–ceramics and carbons; and the polymers include thermosets, thermoplastics, elastomers and textiles. As biomaterials science emerged, the conventional view of materials, as being tangible pieces of substances from which useful objects were made, prevailed [17]. The best performance of the vast majority of implantable devices is achieved when the biomaterials used in their construction are chemically and biologically inert; no biological, let alone pharmacological, activity should be sought in these devices. However, at least in theory, there are some exceptions, with the intention of either promoting some biological activity such as bone regeneration or minimising undesirable activity such as infection or blood clotting. Some materials are used with the express intention of delivering some biologically or pharmacologically active agent to the patient; the concept of drug delivery devices is of course well known [18].

2.3.1 Metals

As a class of materials, metals are the most widely used for load-bearing implants. These range from simple wires and screws to fracture fixation plates and total joint prostheses (artificial joints) for the hips, knees, shoulders, ankles and so on. In addition to orthopaedics, metallic implants are used in maxillofacial surgery and cardiovascular surgery and as dental materials. Although many metals and alloys are used for medical device applications, the most commonly employed are stainless steels [19, 20], cobalt-base alloys, commercially pure titanium and titanium alloys and some other metals [21]. Various properties of these metallic implant materials are listed in Table 2.2.

Stainless steels are iron-base alloys, and stainless characteristics are achieved through the formation of an invisible and adherent chromium-rich oxide surface film. This passive film serves as a barrier to corrosion processes in alloy systems that would otherwise experience very high corrosion rates and has the ability of self-healing, when damaged, as chromium in the steel reacts with oxygen and moisture in the environment to reform the protective oxide layer. Based on the characteristic crystallographic structure/microstructures of the alloys, stainless steels are classified into four classes: martensitic, ferritic, austenitic and duplex (austenitic plus ferritic). Stainless steels are used extensively for fracture fixation devices. Compared to the other metals used in orthopaedics, stainless steels exhibit a moderate to high elastic modulus and tensile strength. Additionally, these steels possess good ductility, which allows them to be cold worked. Stainless steels are fairly biocompatible although they never appear to fully integrate with bone or soft tissue. For instance, if stainless steel is placed in close proximity of bone in the body, a thin layer of fibrous tissue will intervene between the bone and metal at the microscopic level. This phenomenon is not conducive to the use of stainless steels in applications where the success of the implant is dependent on its close integration with tissue [22].

Cobalt–chromium alloys are highly corrosion resistant. Compared to stainless steel, they exhibit higher elastic modulus, strength and hardness, but they have relatively low ductility and are difficult to machine. Commonly used cobalt alloys are Co-28Cr-6Mo casting alloy, Co-20Cr-15W-10Ni wrought alloy, Co-28Cr-6Mo thermomechanically processed alloy and Co-35Ni-20Cr-10Mo wrought alloy.

Table 2.2 Comparison of some of the characteristics and mechanical properties of metallic implant materials [22]

	Stainless steels	Cobalt-base alloys	Ti and Ti-base alloys
Grade	Austenitic stainless steel	Cobalt–chromium alloy	α–β alloy
Composition	Fe Cr (17–20) Ni (12–14) Mo (2–4)	Co Cr (19–30) Mo (0–10) Ni (0–37)	Ti Al (6) V (4) Nb (7)
Young's modulus (GPa)	200	230	106
Tensile strength (MPa)	540–1,000	900–1,540	900
Advantages	Cost, availability, processing	Wear resistance, corrosion resistance, fatigue strength	Biocompatibility, corrosion, minimum modulus, fatigue strength
Disadvantages	Long-term behaviour, high modulus	High modulus, biocompatibility	Lower wear resistance, low shear strength
Uses	Temporary devices (fracture plates, screws, hip nails); used for total hip replacement	Dentistry castings, prostheses stems, load-bearing components in total joint replacement	Used in THRs with modular femoral heads; long-term permanent devices (nails, pacemakers)

They possess adequate fatigue properties to serve as artificial joints or total joint prostheses and are used extensively for this purpose.

Commercially pure titanium is well known for its excellent corrosion resistance. Various grades of unalloyed titanium are available with oxygen and iron as primary variants. Biomedical applications for commercially pure titanium grades include pacemaker cases, housings for ventricular assist devices, implantable infusion drug pumps, dental implants, maxillofacial and craniofacial implants and screws and staples for spinal surgery. Superior biocompatibility, enhanced corrosion resistance and lower modulus are some of the attractive properties of titanium-base alloys as biomaterials. Based upon their microstructure after processing, titanium-base alloys are divided into four classes: α, near α, α–β and β. Femoral hip stems, fracture fixation plates, spinal components, fasteners, intramedullary nails and screws are some of the biomedical applications of these alloys [23].

Among the various refractory metals, niobium, molybdenum and tungsten are used as alloying elements in stainless steels, cobalt-base alloys and titanium-base alloys [24]. Tantalum has excellent resistance to corrosion and also offers intrinsic fabrication advantages. Apart from the alloying additive, commercially pure tantalum is fabricated into various medical devices such as foils and sheets for nerve anastomoses, clips for ligation of vessels and staples for abdominal surgery and as pliable sheets and plates for cranioplasty and reconstructive surgery [25].

Table 2.3 Polymers used for implantable devices [26]

Polymers	Applications
Poly(tetrafluoroethylene)	Oxygenator membrane, vascular graft, catheter coating, soft tissue augmentation, vascular prostheses
Poly(dimethylsiloxane)	Oxygenator membrane, tubing, shunt
Polypropylene	Heart valve structures, sutures
Poly(ethylene terephthalate)	Vascular grafts and prosthesis, shunt, sutures
Polyamides (nylons)	Hemodialysis membrane
Poly(ether urethane) (e.g. Pellethane)	Percutaneous leads, catheters, tubings, intra-aortic balloons
Poly(ether urethane urea) (e.g. Biomer)	Artificial heart components, heart valve
Low- and high-density polyethylene	Tubing; knee, hip, shoulder joints
Polysulfones	Artificial heart components, heart valve
Polyvinylchloride	Tubing, blood bags
Poly(2-hydroxyethylmethacrylate)	Catheter coating
Polymethylmethacrylate	Dental restorations, intraocular lenses, joint replacement, e.g. bone cements
Polyamides	Sutures
Polyesters	Vascular prostheses, drug delivery systems like drug-eluting stents, sutures
Silicone	Soft tissue replacement, ophthalmology, finger joint
Hydrogels	Ophthalmology, drug delivery systems
Acrylic, nylon	Tracheal tube

2.3.2 Polymers

Polymeric materials are rapidly replacing other material classes such as metals, alloys and ceramics for use as biomaterials because of their versatility. Their applications range from facial prostheses to tracheal tubes, from kidney and liver parts to heart components and from dentures to hip and knee joints. Various polymers used for implantable medical devices are listed in Table 2.3.

Polymeric materials are generally classified into three different classes depending on their source: natural polymers, obtained from natural sources including both plant and animal origin; synthetic polymers, based on totally synthetic sources; and bio-inspired polymers which comprise materials synthesised to mimic a naturally occurring polymer, but not necessarily identical to it. Natural polymers suffer from various disadvantages such as possibility of antigenicity, possibility of microbial contamination and source-to-source variability of properties. Hence, synthetic polymers have been the material of choice for implants because of their ease of production, availability and versatility of manipulation [27]. Bio-inspired polymers promise innovative materials that have the potential to functionally replace diseased or unavailable cell components, such as the extracellular matrix, which plays a structural role in many organs and tissues [28].

2.3.2.1 Non-biodegradable Polymers

Ultrahigh molecular weight polyethylene (UHMWPE) was the first polymeric material, used in medicine since the 1960s. UHMWPE is highly resistant to corrosive chemicals and has extremely low moisture absorption, very low coefficient of friction, characteristic of self-lubrication and high resistance to abrasion. UHMWPE emerged as a bearing material in many joint replacement devices. Tibial bearings in knee arthroplasties and acetabular bearings in hip arthroplasties are the most common uses of UHMWPE. Recently, it has been found that generation of particulate debris from the articulating surface of this polymer is associated with osteolysis and loosening of implants. Research has been carried out to address these problems and highly cross-linked UHMWPE materials are clinically introduced. Another important medical advancement for UHMWPE in the past decade has been the increase in use of fibres for sutures, where maximum strength and minimum weight are required [25].

Polymethylmethacrylate (PMMA) is used extensively as bone cement, which is primarily used to support the stems of total joint prostheses in the medullary cavity of the bone. The primary purpose of the material is to fill the space between the prostheses and bone to achieve more uniform stress distribution, and bone cements do not serve as adhesives. PMMA is also used for replacement intraocular lenses in the eye when the original lens is removed in the treatment of cataracts. PMMA microspheres injected under the skin reduce wrinkles or scars permanently. Polyetheretherketone, a thermoplastic polymer, is used as biomaterials for trauma, orthopaedic and spinal implants [21].

The use of polyethylene terephthalate (PET) in medical devices has endured for more than 50 years. Current medical applications of PET include implantable sutures, surgical mesh, vascular grafts, sewing cuffs for heart valves and components for percutaneous access devices. The notable biological characteristics of PET are biostability, promotion of tissue ingrowth, a well-characterised fibrotic response and a long history of human implantation [29].

2.3.2.2 Biodegradable Polymers

For most applications, biodegradable materials offer advantages over other materials. The degradable nature of these materials allows for temporospatial clearance of the material from the body, enabling the surrounding and/or ingrowth tissue to autonomously restore its function over time after having benefited from the implant. Synthetic biodegradable polymers offer the ability to control surface as well as mechanical properties and degradation kinetics [28].

There are four major degradation mechanisms for polymers used in biomedical devices: hydrolysis (reaction with water in tissues), oxidation (due to oxidants produced by tissues), enzymatic degradation and physical degradation (e.g. water swelling and mechanical loading and wearing). Hydrolysis has been studied extensively, especially for biodegradable polymers. Polyesters, polyorthoesters, polyanhydrides, polycarbonates and polyamides are some of the polymers that are degraded by hydrolysis. Degradation of polyurethane material through oxidation is observed in an implantable electrical insulation lead. Biological defence action is also responsible for oxidative degradation, in which inflammatory cells generate oxidative agents that diffuse into polymeric implants and degrade them. Enzymatic degradation is also due to a defensive action against implanted foreign materials. Collagens, polysaccharides (hyaluronic acids), some polyesters (e.g. polyhydroxy-alkanoate, PHA), synthetic polycarbonates and proteins are mainly degraded due to this type of reactions. Physical degradation is mostly due to mechanical friction associated with motion under pressure, for example, the wearing of acetabular cups of total hip replacements. Water swelling is another mechanism. This can be a problem if swollen polymers have significant changes in glass transition temperature, geometry and mechanical properties such that the normal functions of the materials are affected [30].

Potentially, devices made from bioresorbable polymers can overcome problems associated with metal implants like stress protection, potential for corrosion, wear and debris formation as well as the necessity of implant removal. Resorbable polymers have proven to be good materials for a range of devices in trauma surgery [31]. Among resorbable polymers for implants, polyhydroxy acids occupy the main position. These are mainly poly(L-lactide), poly(glycolide) and/or copolymers based on L-lactide, L/DL-lactide, DL-lactide, glycolide, trimethylene carbonate and caprolactone. These polymers are well known for their good biocompatibility, with their degradation products being eliminated from the body by metabolic pathways. Many reports have shown that the different PLA-based substrates do not present toxicity since the cells were found to differentiate over the different polymers, as demonstrated by the production of extracellular matrix components by various cell types [32].

Polymers prepared from glycolic acid and lactic acids have found a multitude of uses in the medical industry, beginning with biodegradable sutures first approved in the 1960s. Since that time other medical devices, based on lactic and glycolic acid, as well as other materials, including poly(dioxanone), poly(trimethylene carbonate) copolymers and PCL homopolymers and copolymers, have been accepted for use as medical devices. In addition to these approved devices, a great deal of research continues on polyanhydrides, polyorthoesters and other materials [33].

Collagen has received increasing attention over the last years due to its excellent biocompatibility, degradation into physiological end products and suitable interaction with cells and other macromolecules [34]. Resorbable forms of collagen have been used for closure of grafts and extraction sites. Collagen-based membranes also have been used in periodontal and implant therapy as barriers to prevent epithelial migration and allow cells with regenerative capacity to repopulate the defect area [35]. Tissue-based collagen devices are mostly used for cardiovascular applications in the form of heart valves, vascular prostheses [36].

Polymer coatings such as silicone rubber, PTFE, Parylene and epoxy were used to encapsulate the implantable devices. However, these materials are biocompatible; they have limited abilities to protect the device from water ingress. Polyether-based polyurethane elastomers are currently used in a variety of blood- and tissue-contacting devices in biomedical applications due to their biocompatibility and stability in biological environment together with their superior processability. Despite their excellent mechanical properties and biocompatibility, the chemical structure and morphology of polyurethanes make them relatively permeable to gases and water [37].

2.3.3 Ceramics and Composites

Restorative materials in dentistry such as crowns, cements and dentures are made up of ceramic materials. The poor fracture toughness of ceramics severely limits their use for load-bearing applications. They are generally used to replace or fix hard connective tissue, such as the bone [22]. The bone itself is a composite, comprising an organic phase and a ceramic phase. This ceramic phase is predominantly calcium hydroxyapatite with a Ca/P ratio of 1.67. Thus, synthetic calcium hydroxyapatite is a good candidate for a successful biomaterial. Several dental and orthopaedic metal implants are coated with hydroxyapatite to ensure long-term fixation in bone [21]. Zirconium dioxide or zirconia ceramics (ZrO_2), a bioinert nonresorbable metal oxide, is being recognised for its high strength, toughness and surface finish. It is used to manufacture femoral heads for total hip replacements and this material is potentially suitable for the highly loaded environments found in joint replacement. A ceramic that is used in load-bearing applications is high-purity alumina. It is used as the bearing surface in total hip prostheses. The material is characterised by its excellent biocompatibility and high strength, hardness and fracture resistance. A class of glassy bioactive ceramics upon implantation undergo a modification of their surface and form a layer of a very bioactive form of hydroxyapatite. As new bone is formed in opposition to this layer, it forms a very strong bond such that the mechanical integrity of the bond can exceed that of bone. Widescale use of these materials has been limited due to their brittle nature [25].

The most successful composite biomaterials are used in the field of dentistry as restorative materials or dental cements. Although carbon–carbon and carbon-reinforced polymer composites are of great interest for bone repair and joint replacement because of their low elastic modulus levels, these materials have not displayed a combination of mechanical and biological properties appropriate to these applications. Composite materials are, however, used extensively for prosthetic limbs, where their combination of low density/weight and high strength makes them ideal materials for such applications.

2.3.4 Drugs and Other Biomolecules

The coated implant can be viewed as a drug/medical device combination product which represents an emerging new trend in therapeutics. Combination products facilitating localised drug delivery have already been used in a variety of applications from cardiovascular diseases to diabetes. Drug and device combinations can be designed to increase the performance and life time of currently used implants resulting in improved patient life quality [38]. Drug-coated implants function as a semipermeable compartment that holds the drug while permitting passage of preferred molecules in a controlled manner. Drug-eluting stents are good examples of such devices [4, 39, 40].

Various immunosuppressive drugs (sirolimus, everolimus, tacrolimus, ABT-578), antiproliferative drugs (paclitaxel, actinomycin, angiopeptin, etc.), antimigratory drugs (batimastat) and gene therapeutic reagents (antisense and siRNA, vascular endothelial growth factor, endothelial nitric oxide synthase (eNOS and related genes)) have been combined with stents and investigated for their local release and antirestenotic effects. FDA's approval of Cordis' CYPHER™ sirolimus-eluting stent (2003) opened the gate for adapting new technology combining both device and pharmaceutical designs [41].

2.4 Requirements of Implantable Devices

2.4.1 Mechanical Properties

Elasticity, yield strength, ductility, toughness, creep, ultimate strength, fatigue strength, hardness and wear resistance are some of the important mechanical properties of materials used in implantable medical devices.

2.4.1.1 Yield Strength

The yield strength determines the load-bearing capability of the implant. For example, in the case of TJR surgeries where a high load-bearing capability of the implant is essential, one ideally needs an appropriately high yield strength value of the alloy. Thus, the orthopaedic alloys should have a sufficiently high yield strength value with adequate ductility (defined by percentage elongation or percentage reduction of area in a standard tensile test) [42].

2.4.1.2 Elastic Modulus

There is always a concern for the relatively higher modulus of the implant compared to that of the bone (_10–40 GPa, or 1.5–6 _106 psi) [3]. Long-term experiences indicate that insufficient load transfer from the artificial implant to the adjacent remodelling bone may result in bone reabsorption and eventual loosening of the prosthetic device [43]. It has been seen that when the tensile/compressive load or the bending moment to which the living bone is exposed is reduced, decreased bone thickness, bone mass loss and increased osteoporosis occur. This is termed the stress shielding effect, caused by the difference in flexibility and stiffness, which is partly dependent on the elastic moduli difference between the natural bone and the implant material [44]. Any reduction in the stiffness of the implant by using a lower-modulus material would definitely enhance the stress redistribution to the adjacent bone tissues, thus minimising stress shielding and eventually prolonging the device lifetime [3].

2.4.1.3 Fatigue

Variable fatigue resistance of the metallic implants is also a cause of concern while developing an alloy. The orthopaedic implants undergo cyclic loading during body motion, resulting in alternating plastic deformation of microscopically small zones of stress concentration produced by notches and microstructural inhomogeneities. Standard fatigue tests include tension/compression, bending, torsion and rotation-bending fatigue testing [3].

2.4.2 Corrosion and Biocompatibility

Degree of success of an implant is determined by reactions at the interface between an implant and the environment of the body. In the case of orthopaedic implants, one of the potentially important aspects of this interfacial reaction concerns metallic corrosion. The metals and alloys used as surgical implants achieve passivity by the presence of a protective surface passive film. This film inhibits corrosion and keeps current flow and the release of corrosion products at a very low level, i.e. all the implantable materials undergo corrosion at some finite rate due to complex corrosive environment of the body, while in use [45].

Metallic implant degradation results from both electrochemical dissolution and wear but most frequently occurs through a synergistic combination of the two [46]. Electrochemical corrosion processes include both generalised dissolution uniformly affecting an entire surface and localised areas of a component [47]. Locally these areas tend to be at both identifiable areas relatively shielded from the environment (e.g. crevice corrosion) and at random sites on the surface (e.g. pitting corrosion). In the past, these electrochemical and other mechanical processes have interacted to cause premature structural failure and accelerated metal release (e.g. stress corrosion cracking, corrosion fatigue and fretting corrosion) [48].

The only way to reduce the corrosion of surgical implants is selecting an appropriate alloy with improved surface properties by addition of alloying elements, which improve the nature, composition and stability of the passive film; homogenisation treatment to dissolve the second-phase precipitates; production of cleaner varieties of steels using advanced steel-making processes in order to have control over inclusions; avoiding improper heat treatment; and fabrication in order to eliminate the formation of second phases especially at grain boundaries. Also, the nature, composition and chemical stability of passive films with superior corrosion resistance can be produced through surface treatment by ion beams. Many of the cases of corrosion could be avoided by improvements in materials selection, implant design, quality control, materials handling and education [45].

According to Williams in 2008, biocompatibility refers to the ability of a biomaterial to perform its desired function with respect to a medical therapy, without eliciting any undesirable local or systemic effects in the recipient or beneficiary of that therapy but generating the most appropriate beneficial cellular or tissue response in that specific situation and optimising the clinically relevant performance of that therapy [49]. Biocompatibility studies on an implantable device require complex experiments both in vitro and in vivo in order to test the local and systemic effects of the material on culture cells, tissue sections and the whole body [50].

In vitro assessment of tissue compatibility usually involves performing cell cultures for a wide variety of materials used in medical devices. Direct contact, agar diffusion and elution are the three different cell culture assays used for in vitro study. In all the tests, experimental variables such as cell type (usually L-929 mouse fibroblast), number of cells, duration of exposure and test sample size are kept constant [51]. Positive and negative controls are often used during the assay test to determine the viability of the test. In all cases, the amount of affected or dead cells in each assay provides a measure of the cytotoxicity and biocompatibility of the biomaterials. Cell adhesion, cell spreading, cell migration, cell proliferation and cell function are some of the key parameters that can be investigated individually in these assays [52]. In vivo assessment of tissue compatibility tests are performed to determine the biocompatibility of a prosthetic device and also to assess whether the device is performing according to expectations without causing harm to the patient. Some tests, such as toxicity, carcinogenicity, sensitisation and irritation, determine if the leachable products of the medical device affect the tissues near or far from the implant site. Other tests, such as implantation and biodegradation, study the post-surgery changes in the implant material itself and their ensuing effect on the body. For conducting the actual in vivo tests, animal models (sheep, pig, rat) are usually selected after weighing the advantages and disadvantages for human clinical applications [42].

2.4.3 Sterilisation

Infection is a major problem associated with implantable devices, and a further complication is that organisms of relatively low virulence are frequently found to be pathogenic when they contaminate the bioimplant. Further, little is known about the effect of a prosthetic device on either the host immune response or the inflammatory action of the tissue with which the prosthesis interfaces [53]. Composition of the device and its function, the location, the length of time the device is in place and the underlying condition of the patient are some of the factors affecting the pathogenesis of such infections and the relative rate of infection [54]. The same implant located in different body areas can be associated with different flora, potentially

Table 2.4 Different methods used for sterilisation of implantable medical devices [53]

Method	Principle	Advantages	Disadvantages
Steam sterilisation	Saturated steam in the range of 121–134 °C	Sterilises penetrable materials and exposable surfaces	Microcavitation in case of hydrophilic polymers Altered biocompatibility of heparinised surfaces
Dry heat sterilisation	Carried out in electrically heated ovens at 160–180 °C	Ability to penetrate solids Lack of corrosion in case of non-stainless steel metals	Rubbers, plastic, etc. do not withstand high temperature
Ethylene oxide	Biocidal activity is achieved at 30 % humidity, 45 °C temperature levels with >450 mg/l gas concentration	Suitable for heat-sensitive implants	EO residues after sterilisation cause: Anaphylactoid reactions in dialysis patients Serious tissue reactions in pump oxygenators Mutagenicity and carcinogenicity
Radiation sterilisation	Primary biocidal action is via aqueous free radical formation following the primary physical interaction of the ionising radiation with the biological material	High-energy irradiation sterilisation	Degradation and/or cross-linking of polymers Gas evolution and free radical formation from polymers

different host defence mechanisms and other alterations in milieu that may alter the potential development of prosthesis-associated infection. Implants can also alter the host immune response, through alterations of phagocytic capacity, inflammatory response or immunoglobulin synthesis [55]. Sterilisation of implantable devices is carried out to eliminate infecting microorganisms upon the device. Various sterilisation procedures used for implants are listed in Table 2.4 along with their advantages and disadvantages.

2.5 Applications

2.5.1 Drug Delivery and Scaffolds

One of the fastest growing areas for implant applications is for devices for controlled and targeted delivery of drugs. Combinations of drug and device are predicated on the principle of local controlled drug delivery from an implanted prosthetic device whose primary purpose is functional or structural replacement of host tissue. Optimal dual functions (i.e. drug release and prosthetic performance) are ideally coordinated and designed to work in tandem. Hence, drug release properties from the device are not simply adjunct to device implantation and must be thoroughly understood. Drug release strategies from implantable devices are frequently considered to address thrombosis, osteomyelitis, periodontitis, biomedical device-related infections and other microbial pathologies or inflammatory complications [41].

Currently, there are various methods available and being used for drug–implant combinations. While polymeric coatings are one of the best recognised techniques, depending on the application, there are other methods currently available and nanoporous coatings are one of them. The nanoporous ceramic templates can be fabricated directly on the currently available implants or stents [4]. Drug loading for the nanoporous templates generally is performed through capillary action by either immersing the templates in the concentrated drug solution or dropping the solution slowly on the template surfaces [56]. Different techniques have been used to increase and accelerate the intake of the drug, including surface enhancement, sonication or solution aids [57]. One of the unique features of these nanoporous coatings is the ability to precisely control the surface properties. By varying the pore size, distribution and density, drug loading and release can be altered. Surface charges of these pores can also be modified to hydrophobic or hydrophilic to accommodate variety of drug molecules. Additionally, through suitable surface modification the release kinetics of the drug molecules can be altered [4].

Many exciting new opportunities exist in the application of microfabricated drug delivery devices to medicine and biology. Microfabricated devices are poised to revolutionise drug delivery. They offer new methods to deliver compounds in a targeted manner, at the desired rate, and are compact to allow minimally invasive placement [58]. Micro- and nano-electromechanical systems (MEMS or NEMS)-based polymeric and electromechanical delivery devices create totally new drug delivery paradigms. MEMS technology has been used to construct microreservoirs, micropumps, nanoporous membranes, nanoparticles, valves, sensors and other structures using biocompatible materials appropriate for drug administration [59]. Using NEMS, complex mechanical nanostructures can be built with lateral dimensions as small as tens of nanometers. By incorporating transducers, control and measurement functions can be built into these systems [60]. A biodegradable polymer chip version of an implantable multireservoir drug delivery device incorporates an array of reservoirs capped with resorbable membranes that may differ from other membranes in the array by thickness or chemical composition. The interior of each reservoir contains drug formulation(s). An advantage of biodegradable polymer-based systems compared to microchip-based systems is the elimination of a requirement for a second surgery to remove the device. In addition, the lack of electronics reduces any size restrictions in terms of device manufacture. Such systems are simpler but will not deliver reservoir contents with as much precision as the analogous microelectronic devices [61].

Drug-eluting stents are among the most widely known combination products. Micromachining technology allowed bare-metal stents to be manufactured that had the physical capability of propping open occluded vessels. Coating the stent with a drug-containing polymer resulted in combination products featuring localised drug release capability in addition to the mechanical action of the stent. Next-generation drug-eluting stents incorporate reservoir-based drug containment on the stent surface, with release properties determined by polymer composition and layer thickness [62].

In case of orthopaedic device-based drug delivery, osteo-inductive molecules as well as biologically derived growth factors, anti-osteoporotic agents and osteo-synthetic genetic materials (DNA, siRNA) to bone injury sites are successfully delivered [63]. Osteo-precursor cell-based local delivery is reported for bone engineering [64]. These biotechnology approaches seek to accelerate and enhance bone defect healing and bone-implant stabilisation through local release of cells and mitogenic and morphogenic agents. One commonly used infection management method with orthopaedic implants utilises antibiotics loaded into clinically ubiquitous bone cement, polymethylmethacrylate beads. These non-biodegradable polymer cements have been employed clinically to prevent or treat osteomyelitis in various forms [65].

One of the key components in successful tissue engineering is the production of the correct scaffold using biomaterials. An ideal scaffold should provide cells not only with a structural framework but also with the appropriate mechanical and biochemical conditions so that these cells can proliferate and produce extracellular matrix to form tissue. Areas of research in tissue engineering include the repair or regeneration of the skin, blood vessels, nerves, liver, bone and articular cartilage. PLA and PGA are prime candidates for such scaffolds because they are biocompatible, provide the appropriate mechanical environment, can be easily fabricated and, moreover, are biodegradable. Collagen sponges are also under investigation [21]. Various FDA-approved implantable devices are listed in Table 2.5.

2.5.2 Structural and Mechanical Support

Currently, one of the main achievements in the field of arthroplasty is total joint replacement, where the entire load-bearing joint (mainly in the knee, hip or shoulder) is replaced surgically by ceramic, metal or polymeric artificial materials. Bone replacement, fracture fixation, dental implants, dental restorations, bone plates and orthodontic wires are some of the medical devices that provide structural and mechanical support (Table 2.6).

Table 2.5 FDA-approved implantable medical devices intended for localised drug delivery and functional support

S.No.	Device	Polymer/drug	Purpose/use	Manufacturer	References
1.	PROMUS® Element™ Plus	Everolimus-eluting platinum chromium stent	Coronary stent	Boston Scientific Corporation	[66]
2.	OVATION Abdominal Stent Graft System	The main portion of the graft is made of a plastic tube supported partially by polymer-filled rings and partially by a metallic stent	Used to treat abdominal aortic aneurysms during endovascular repair	Trivascular, Inc.	[67]
3.	LeGoo®	Thermosensitive polymer (poloxamer)	Injected into a blood vessel to temporarily plug a blood vessel and stop blood flow	Pluromed, Inc.	[68]
4.	Propel	The implant contains the active ingredient mometasone furoate, a steroid which suppresses inflammation in the area around the implant	Temporary, self-absorbing implant designed to keep the spaces within and around the sinuses open following sinus surgery	Intersect ENT	[69]
5.	ION™ Paclitaxel-Eluting Coronary Stent System	Expandable, stainless steel tube with paclitaxel contained within a thin polymer coating surface	Paclitaxel-Eluting Coronary Stent	Boston Scientific Corporation	[70]
6.	Gel-One®	Hyaluronate hydrogel produced from chicken combs, in a phosphate-buffered saline solution	Treatment of osteoarthritis	Seikagaku Corporation	[71]
7.	Synvisc-One (hylan GF-20)	Elastoviscous high molecular weight fluid containing hylan polymers	Treatment of osteoarthritis	Genzyme Biosurgery	[72]
8.	Medtronic® Attain StarFix™	Surgically implanted insulated wire that is designed to be used as a part of a biventricular pacemaker system	The other end of the lead connects to an implanted biventricular pacemaker or implantable cardioverter-defibrillator, so signals and electrical impulses can be sent between that device and the left ventricle	Medtronic, Inc.	[73]
9.	Epicel® cultured epidermal autograft (CEA)	Sheets of autologous keratinocytes (skin cells)	Cultured epidermal autografts, or skin grafts, used to replace the epidermal or top layer of skin on severely burned patients	Genzyme Biosurgery	[74]

(continued)

Table 2.5 (continued)

S.No.	Device	Polymer/drug	Purpose/use	Manufacturer	References
10.	INFUSE® Bone Graft	INFUSE® Bone Graft is used to fill space where the bone is needed in order to place endosseous dental implants	Endosseous dental implants are inserted in the jaw and have an exposed head that can be used to secure dental devices like a crown, fixed bridge or dentures	Medtronic Sofamor Danek	[75]
11.	Onyx® Liquid Embolic System (LES)	The Onyx® LES is an artificial material used to block blood flow in the treatment of malformed blood vessels in the brain	This material is used to block the flow of blood before surgical treatment of the malformed vessels	Micro Therapeutics, Inc.	[76]
12.	Endologix PowerLink® System	Y-shaped stent graft made out of a fabric tube of synthetic material called ePTFE	Endovascular graft and used to treat an abdominal aortic aneurysm during a surgical procedure called endovascular repair	Endologix, Inc.	[77]
13.	Nuflexxa™ (Sodium Hyaluronate)	1 % sodium hyaluronate solution	Injected into knee joints to treat pain from osteoarthritis of the knee	Savient Pharmaceuticals, Inc.	[78]

http://www.fda.gov/MedicalDevices/ProductsandMedicalProcedures/DeviceApprovalsandClearances/Recently-ApprovedDevices/ucm286493.htm

Table 2.6 FDA-approved implantable medical devices used for structural and mechanical support

S.No.	Device	Polymer	Purpose/use	Manufacturer	References
1.	Belotero Balance	Hyaluronic acid gel	Injected into facial tissue to smooth wrinkles and folds	Merz Pharmaceuticals	[79]
2.	Edwards SAPIEN Transcatheter Heart Valve	Cow tissue attached to a stainless steel mesh frame with a polyester wrap	Heart valve	Edwards Lifesciences	[80]
3.	Restylane® Injectable Gel	Restylane is a transparent hyaluronic acid gel	Injected into a patient's lips to increase their size	Aesthetics Holdings, Inc	[81]
4.	EUFLEXXA®	Hyaluronate hydrogel in a phosphate-buffered saline solution	Used to relieve pain in the knee due to osteoarthritis	Ferring Pharmaceuticals, Inc.	[82]
5.	St. Jude Medical® Trifecta™ Valve	The valve leaflets are manufactured using bovine pericardial tissue. A polyester-covered titanium stent supports the bovine pericardial tissue	The Trifecta valve is a three-leaflet stented pericardial valve designed for supra-annular placement in the aortic position	St. Jude Medical	[83]
6.	ProGEL™ Pleural Air Leak Sealant	Made of human serum albumin and a polyethylene glycol (PEG) cross-linker that forms a clear flexible gel on mixing	Surgical sealant, sprayed or "painted" on the lung tissue. The ProGEL™ forms on the lung tissue by chemical reaction	NeoMend, Inc.	[84]
7.	Gel-One®	Hyaluronate hydrogel produced from chicken combs, in a phosphate-buffered saline solution	Treatment of osteoarthritis	Seikagaku Corporation	[71]
8.	Sculptra Aesthetic	Sculptra Aesthetic is an implant of poly-L-lactic acid (PLLA) microparticles	It is injected into the facial tissue to correct shallow to deep smile lines (nasolabial fold), contour deficiencies and other facial wrinkles	Sanofi-Aventis U.S.	[85]

(continued)

Table 2.6 (continued)

S.No.	Device	Polymer	Purpose/use	Manufacturer	References
9.	EVOLENCE® Collagen Filler	Collagen Filler	Injected into the inner layers of facial skin (mid to deep dermis) in order to correct moderate to deep facial wrinkles and folds	ColBar LifeScience Ltd.	[86]
10.	Mitroflow Aortic Pericardial Heart Valve	Consists of a single piece of bovine pericardium that is preserved with glutaraldehyde and sewn onto a polyester covered polymer stent	The Mitroflow Aortic Pericardial Heart Valve is intended for the replacement of diseased, damaged or malfunctioning native or prosthetic aortic valves	CarboMedics, Inc.	[87]
11.	Cosmetic Tissue Augmentation Product	Cosmetic Tissue Augmentation Product (CTA) is a transparent hyaluronic acid gel with 0.3 % lidocaine that is injected into facial tissue	CTA works by temporarily adding volume to facial tissue and restoring a smoother appearance to the face	Anika Therapeutics, Inc.	[88]
12.	Radiesse	Radiesse is an injectable calcium hydroxyapatite implant in the form of a gel	Radiesse works by temporarily adding volume to facial tissue and restoring a smoother appearance to the face	BioForm Medical, Inc.	[89]
13.	Inamed® Silicone-Filled Breast Implants	Implant is a silicone shell filled with silicone gel	The breast implant is surgically implanted either under breast tissue or under the chest muscle	Allergan	[90]
14.	ArteFill®	Contains small polymethylmethacrylate beads, collagen and lidocaine	Filler that is injected into the nasolabial folds around the mouth to smooth these wrinkles	Artes Medical, Inc.	[91]
15.	Juvederm Gel Implants	Hyaluronic acid gel	Injected into the middle layer of skin (mid to deep dermis) to temporarily correct moderate to severe facial wrinkles and folds	Inamed Corporation	[92]

http://www.fda.gov/MedicalDevices/ProductsandMedicalProcedures/DeviceApprovalsandClearances/Recently-ApprovedDevices/ucm286493.htm

2.6 Conclusion

Medical devices are now a pervasive part of modern medical care. The medical development in terms of implantable devices has brought about the robust change in the life of the people (as offered by the cosmetic treatment, dentist, face and cardiology devices). Medical devices have extended the ability of physicians to diagnose and treat diseases, making great contributions to health and quality of life. The approach to quality of devices has depended largely on regulation. The critical nature of medical devices has caused them to come under stringent regulations. Clearance to market devices in the USA is granted only after the Food and Drug Administration (FDA) has determined through its classification and review procedure that there is reasonable assurance of the safety and effectiveness of the device. Such regulatory requirements are necessary and appropriate. A rigorous but responsive and responsible regulatory process helps to ensure that new medical technologies represent the state of the art, have the real potential to do good as demonstrated in scientifically grounded studies and reach patients promptly. Despite the enormous contribution medical devices have made to the public health, there is a fear of the possibility of liability exposure in the event of device malfunction or failure. Its influence is growing and is having a chilling effect on innovation. It also damages global competitiveness and increases health care costs directly and indirectly. Ironically, the shadow of product liability may actually be keeping better performing products from the market rather than being a force for improvement.

References

1. Regar E, Sianos G, Serruys PW (2001) Stent development and local drug delivery. Br Med Bull 59:227–248
2. Ginebra MP, Traykova T, Planell JA (2006) Calcium phosphate cements as bone drug delivery systems: a review. J Control Release 113:102–110
3. Long M, Rack HJ (1998) Titanium alloys in total joint replacement-a materials science perspective. Biomaterials 19:1621–1639
4. Gultepe E, Nagesha D, Sridhar S, Amiji M (2010) Nanoporous inorganic membranes or coatings for sustained drug delivery in implantable devices. Adv Drug Deliv Rev 62:305–315
5. Park GE, Webster TJ (2005) A review of nanotechnology for the development of better orthopedic implants. J Biomed Nanotechnol 1:18–29
6. Gunn J, Cumberland D (1999) Stent coatings and local drug delivery. State of the art. Eur Heart J 20:1693–1700
7. Zuckerman DM, Brown P, Nissen SE (2011) Medical device recalls and the FDA approval process. Arch Intern Med 171:1006–1018
8. Santos MH, Valerio P, Goes AM, Leite MF, Heneine LGD, Mansur HS (2007) Biocompatibility evaluation of hydroxyapatite/collagen nanocomposites doped with Zn^{+2}. Biomed Mater 2:135
9. Jung F, Wischke C, Lendlein A (2010) Degradable, multifunctional cardiovascular implants: challenges and hurdles. MRS Bull 35:607–613
10. Ajili SH, Ebrahimi NG, Soleimani M (2009) Polyurethane/polycaprolactane blend with shape memory effect as a proposed material for cardiovascular implants. Acta Biomater 5:1519–1530

11. Wischke C, Lendlein A (2011) Designing multifunctional polymers for cardiovascular implants. Clin Hemorheol Microcirc 49:347–355
12. Andukuri A, Kushwaha M, Tambralli A, Anderson JM, Dean DR, Berry JL et al (2011) A hybrid biomimetic nanomatrix composed of electrospun polycaprolactone and bioactive peptide amphiphiles for cardiovascular implants. Acta Biomater 7:225–233
13. Kelly A (1990) Concise encyclopedia of composite materials. MIT Press, Cambridge
14. Balkany TJ, Hodges AV, Eshraghi AA, Butts S, Bricker K, Lingvai J et al (2002) Cochlear implants in children—a review. Acta Otolaryngol 122:356–362
15. Yasukawa T, Ogura Y, Sakurai E, Tabata Y, Kimura H (2005) Intraocular sustained drug delivery using implantable polymeric devices. Adv Drug Deliv Rev 57:2033–2046
16. Choonara YE, Pillay V, Danckwerts MP, Carmichael TR, du Toit LC (2010) A review of implantable intravitreal drug delivery technologies for the treatment of posterior segment eye diseases. J Pharm Sci 99:2219–2239
17. Williams DF (2009) On the nature of biomaterials. Biomaterials 30:5897–5909
18. Williams D (2008) The relationship between biomaterials and nanotechnology. Biomaterials 29:1737
19. Khan W, Kapoor M, Kumar N (2007) Covalent attachment of proteins to functionalized polypyrrole-coated metallic surfaces for improved biocompatibility. Acta Biomater 3:541–549
20. Khan W, Marew T, Kumar N (2006) Immobilization of drugs and biomolecules on in situ copolymerized active ester polypyrrole coatings for biomedical applications. Biomed Mater 1:235–241
21. Agrawal CM (1998) Reconstructing the human body using biomaterials. J Miner Met Mater Soc 50:31–35
22. Davis JR (2003) Handbook of materials for medical devices. ASM International, Materials Park, OH
23. Rack HJ, Qazi JI (2006) Titanium alloys for biomedical applications. Mater Sci Eng C 26:1269–1277
24. Roach M (2007) Base metal alloys used for dental restorations and implants. Dent Clin N Am 51:603–627
25. BomBac DD, Brojan M, Fajfar P, Kosel F, Turk R (2007) Review of materials in medical applications. RMZ Mater Geoenviron 54:471–499
26. Angelova N, Hunkeler D (1999) Rationalizing the design of polymeric biomaterials. Trends Biotechnol 17:409–421
27. Venkatraman S, Boey F, Lao LL (2008) Implanted cardiovascular polymers: natural, synthetic and bio-inspired. Prog Polym Sci 33:853–874
28. Commandeur S, Van Beusekom HMM, Van Der Giessen WIMJ (2006) Polymers, drug release, and drug eluting stents. J Interv Cardiol 19:500–506
29. Metzger A (1976) Polyethylene terephthalate and the pillar™ palatal implant: its historical usage and durability in medical applications. Biomed Eng 11:301–306
30. Lyu SP, Untereker D (2009) Degradability of polymers for implantable biomedical devices. Int J Mol Sci 10:4033–4065
31. Gogolewski S (2000) Bioresorbable polymers in trauma and bone surgery. Injury 31:D28–D32
32. Santos AR Jr (2010) Bioresorbable polymers for tissue engineering. In: Eberlin D (ed) Tissue engineering. In-Teh, Olajnica, pp 235–246
33. Middleton JC, Tipton AJ (2000) Synthetic biodegradable polymers as orthopedic devices. Biomaterials 21:2335–2346
34. Geiger M, Li RH, Friess W (2003) Collagen sponges for bone regeneration with rhBMP-2. Adv Drug Deliv Rev 55:1613–1629
35. Patino MG, Neiders ME, Andreana S, Noble B, Cohen RE (2002) Collagen as an implantable material in medicine and dentistry. J Oral Implantol 28:220–225
36. Ramshaw JAM, Werkmeister JA, Glattauer V (1995) Collagen-based biomaterials. Biotechnol Genet Eng Rev 13:335–382

37. Roohpour N, Wasikiewicz JM, Paul D, Vadgama P, Rehman IU (2009) Synthesis and characterisation of enhanced barrier polyurethane for encapsulation of implantable medical devices. J Mater Sci Mater Med 20:1803–1814
38. Diener T, Hensel B, Heublein B, Barlach S, Rohde R, Sternberg K et al (2003) Biodegradable drug depots on coronary stents-local drug delivery in interventional cardiology. Prog Biomed Res 8:82–91
39. Khan W, Farah S, Domb AJ (2012) Drug eluting stents: developments and current status. J Control Release 161:703–712
40. Levy Y, Khan W, Farah S, Domb AJ (2012) Surface crystallization of rapamycin on stents using a temperature induced process. Langmuir 28:6207–6210
41. Wu P, Grainger DW (2006) Drug/device combinations for local drug therapies and infection prophylaxis. Biomaterials 27:2450–2467
42. Nag S, Banerjee R (2012) Fundamentals of Medical Implant Materials. In: Narayan R (eds) Materials for medical devices. ASM International, pp 6–16
43. Dujovne AR, Bobyn JD, Krygier JJ, Miller JE, Brooks CE (1993) Mechanical compatibility of noncemented hip prostheses with the human femur. J Arthroplasty 8:7–22
44. Sumner DR, Galante JO (1992) Determinants of stress shielding: design versus materials versus interface. Clin Orthop Relat Res 274:202–212
45. Kamachimudali U, Sridhar TM, Raj B (2003) Corrosion of bio implants. In: Sadhana: Academy proceedings in engineering sciences. Indian Academy of Sciences, pp 601–637
46. Black J (1988) Orthopaedic biomaterials in research and practice. Churchill Livingstone, New York, NY
47. Jacobs JJ, Gilbert JL, Urban RM (1998) Current concepts review-corrosion of metal orthopaedic implants. J Bone Joint Surg Am 80:268–282
48. Hallab NJ, Urban RM, Jacobs JJ (2004) Corrosion and biocompatibility of orthopedic implants. In: Yaszemski MJ, Trantolo DJ, Lewandrowski KU, Hasici V, Altobelli DE, Wise DL (eds) Biomaterials in orthopedics. Dekker, New York, NY, p 75
49. Williams DF (2008) On the mechanisms of biocompatibility. Biomaterials 29:2941–2953
50. Onuki Y, Bhardwaj U, Papadimitrakopoulos F, Burgess DJ (2008) A review of the biocompatibility of implantable devices: current challenges to overcome foreign body response. J Diabetes Sci Technol 2:1003–1015
51. Ratner BD, Hoffman AS, Schoen FJ, Lemons JE (2004) Biomaterials science: an introduction to materials in medicine. Academic, San Diego, CA
52. Kirkpatrick CJ, Peters K, Hermanns M, Bittinger F, Krump-Konvalinkova V, Fuchs S et al (2005) In vitro methodologies to evaluate biocompatibility: status quo and perspective. ITBM-RBM 26:192–199
53. Matthews IP, Gibson C, Samuel AH (1994) Sterilisation of implantable devices. Clin Mater 15:191–215
54. Band JD, Maki DG (1979) Infections caused by arterial catheters used for hemodynamic monitoring. Am J Med 67:735–741
55. Rae T (1975) A study on the effects of particulate metals of orthopaedic interest on murine macrophages in vitro. J Bone Joint Surg Br 57:444–450
56. Foraker AB, Walczak RJ, Cohen MH, Boiarski TA, Grove CF, Swaan PW (2003) Microfabricated porous silicon particles enhance paracellular delivery of insulin across intestinal Caco-2 cell monolayers. Pharm Res 20:110–116
57. Losic D, Cole MA, Dollmann B, Vasilev K, Griesser HJ (2008) Surface modification of nanoporous alumina membranes by plasma polymerization. Nanotechnology 19:245704
58. Tao SL, Desai TA (2003) Microfabricated drug delivery systems: from particles to pores. Adv Drug Deliv Rev 55:315–328
59. Grayson ACR, Shawgo RS, Johnson AM, Flynn NT, Li Y, Cima MJ et al (2004) A BioMEMS review: MEMS technology for physiologically integrated devices. Proc IEEE 92:6–21
60. Ekinci KL, Roukes ML (2005) Nanoelectromechanical systems. Rev Sci Instrum 76:061101–061112

61. Grayson ACR, Choi IS, Tyler BM, Wang PP, Brem H, Cima MJ et al (2003) Multi-pulse drug delivery from a resorbable polymeric microchip device. Nat Mater 2:767–772
62. Finkelstein A, McClean D, Kar S, Takizawa K, Varghese K, Baek N et al (2003) Local drug delivery via a coronary stent with programmable release pharmacokinetics. Circulation 107:777–784
63. Luginbuehl V, Meinel L, Merkle HP, Gander B (2004) Localized delivery of growth factors for bone repair. Eur J Pharm Biopharm 58:197–208
64. Leach JK, Mooney DJ (2004) Bone engineering by controlled delivery of osteoinductive molecules and cells. Expert Opin Biol Ther 4:1015–1027
65. Schmitt JM, Hwang K, Winn SR, Hollinger JO (2005) Bone morphogenetic proteins: an update on basic biology and clinical relevance. J Orthop Res 17:269–278
66. Stone GW, Teirstein PS, Meredith IT, Farah B, Dubois CL, Feldman RL et al (2011) A prospective, randomized evaluation of a novel everolimus-eluting coronary stent: the PLATINUM (a prospective, randomized, multicenter trial to assess an everolimus-eluting coronary stent system [PROMUS element] for the treatment of up to two de novo coronary artery lesions) trial. J Am Coll Cardiol 57:1700–1708
67. Jaff MR (2012) Advances in the management of patients with vascular disease. Expert Rev Cardiovasc Ther 10:151–153
68. Kretz B, Steinmetz E, Brenot R, Bouchot O (2012) First results of clampless distal anastomosis in peripheral vascular bypass with LeGoo, a thermoreversible polymer. J Vasc Surg 55:1821–1825
69. Kennedy DW (2012) The PROPEL™ steroid-releasing bioabsorbable implant to improve outcomes of sinus surgery. Expert Rev Respir Med 6:493–498
70. Mani G, Feldman MD, Patel D, Agrawal C (2007) Coronary stents: a materials perspective. Biomaterials 28:1689–1710
71. Geller J (2011) CDRH issues nonapproval letter to EES for SEDASYS system. J Clin Eng 36:88–91
72. Chevalier X, Jerosch J, Goupille P, van Dijk N, Luyten FP, Scott DL et al (2010) Single, intra-articular treatment with 6 ml hylan GF 20 in patients with symptomatic primary osteoarthritis of the knee: a randomised, multicentre, double-blind, placebo controlled trial. Ann Rheum Dis 69:113–119
73. Nagele H, Azizi M, Hashagen S, Castel MA, Behrens S (2007) First experience with a new active fixation coronary sinus lead. Europace 9:437–441
74. Junkins-Hopkins JM (2011) Biologic dressings. J Am Acad Dermatol 64:e5–e7
75. Glassman SD, Carreon L, Djurasovic M, Campbell MJ, Puno RM, Johnson JR et al (2007) Posterolateral lumbar spine fusion with INFUSE bone graft. Spine J 7:44–49
76. Young VL, Boswell CB, Centeno RF, Watson ME (2005) DMSO: applications in plastic surgery. Aesthet Surg J 25:201–209
77. Jordan WD Jr, Moore WM Jr, Melton JG, Brown OW, Carpenter JP (2009) Secure fixation following EVAR with the Powerlink XL System in wide aortic necks: results of a prospective, multicenter trial. J Vasc Surg 50:979
78. Pagnano M, Westrich G (2005) Successful nonoperative management of chronic osteoarthritis pain of the knee: safety and efficacy of retreatment with intra-articular hyaluronans. Osteoarthritis Cartilage 13:751–761
79. Glogau RG (2012) Fillers: from the past to the future. Semin Cutan Med Surg 31:78–87
80. Chau M, Patterson ME, Yi S, Geist S, Oba T. (2012) Prosthetic apparatus for implantation at mitral valve. US Patent 20,120,323,316
81. Kablik J, Monheit GD, Yu L, Chang G, Gershkovich J (2009) Comparative physical properties of hyaluronic acid dermal fillers. Dermatol Surg 35:302–312
82. Altman RD (2010) Non-avian-derived hyaluronan for the treatment of osteoarthritis of the knee. Expert Rev Clin Immunol 6:21–27
83. Roten L, Stortecky S, Scarcia F, Kadner A, Tanner H, Delacretaz E et al (2012) Atrioventricular conduction after transcatheter aortic valve implantation and surgical aortic valve replacement. J Cardiovasc Electrophysiol 23:1115–1122

84. Klijian A (2012) A novel approach to control air leaks in complex lung surgery: a retrospective review. J Cardiothorac Surg 7:49–55
85. Burgess CM (2008) Treatment of facial asymmetry with poly-L-lactic acid: a case study. Aesthetic Plast Surg 32:552–554
86. Shoshani D, Markovitz E, Cohen Y, Heremans A, Goldlust A (2007) Skin test hypersensitivity study of a cross linked, porcine collagen implant for aesthetic surgery. Dermatol Surg 33:S152–S158
87. Jamieson WRE, Koerfer R, Yankah CA, Zittermann A, Hayden RI, Ling H et al (2009) Mitroflow aortic pericardial bioprosthesis-clinical performance. Eur J Cardiothorac Surg 36:818–824
88. Becker LC, Bergfeld WF, Belsito DV, Klaassen CD, Marks JG Jr, Shank RC et al (2009) Final report of the safety assessment of hyaluronic acid, potassium hyaluronate, and sodium hyaluronate. Int J Toxicol 28:5–67
89. Busso M, Voigts R (2008) An investigation of changes in physical properties of injectable calcium hydroxylapatite in a carrier gel when mixed with lidocaine and with lidocaine/epinephrine. Dermatol Surg 34:S16–S24
90. Spear SL, Parikh PM, Reisin E, Menon NG (2008) Acellular dermis-assisted breast reconstruction. Aesthetic Plast Surg 32:418–425
91. Cohen SR, Berner CF, Busso M, Gleason MC, Hamilton D, Holmes RE et al (2006) ArteFill: a long-lasting injectable wrinkle filler material-summary of the US Food and Drug Administration trials and a progress report on 4-to 5-year outcomes. Plast Reconstr Surg 118:64S–76S
92. Suman T (2011) Enhancing facial esthetics by other modalities. Int J Dent 2011:1–6

Chapter 3
Systemic Targeting Systems-EPR Effect, Ligand Targeting Systems

Prasad V. Pawar, Abraham J. Domb, and Neeraj Kumar

3.1 Introduction

Traditional chemotherapy has been a hallmark for the treatment of widespread cancers as it causes even biodistribution of therapeutic agents throughout the body. It makes use of anticancer drugs which work on the principle of inhibiting rapidly dividing cells by interfering in the processes important for their division. However, these anticancer drugs cannot discriminate normal healthy cells from tumor cells, and they reach most of normal cells/tissues/organs as well as tumor tissues by free diffusion-dependent equilibrium resulting in a systemic toxicity limiting their clinical uses at high doses. To devise cancer-selective drug delivery, the current research focuses on the development of targeted drug delivery systems. Targeted drug delivery could be defined as predominant drug accumulation within a target zone independent of the method and route of drug administration [1, 2]. These systems can specifically reach to the cancer cells, thereby improving efficacy with reduced toxic effect on normal cells. This can expand the therapeutic index of the drug and also enables in reducing the minimum effective dose of drug and accompanying toxicity [3]. Targeting can be achieved in three principle ways: passive, active, and triggered targeting. Passive targeting is based on the accumulation of drug within the tumor cells by exploiting enhanced permeation and retention (EPR) effect. Active targeting involves ligand-receptor interactions which direct the drug carrier towards the target site [4–9]. This approach provides the widest opportunities as it avail various targeting ligands such as monoclonal antibodies and their fragments, aptamers,

P.V. Pawar • N. Kumar (✉)
Department of Pharmaceutics, National Institute of Pharmaceutical Education and Research (NIPER), S.A.S. Nagar, Mohali 160062, Punjab, India
e-mail: neeraj@niper.ac.in

A.J. Domb
Faculty of Medicine, Institute of Drug Research, School of Pharmacy, The Hebrew University of Jerusalem, Jerusalem 91120, Israel

A.J. Domb and W. Khan (eds.), *Focal Controlled Drug Delivery*, Advances in Delivery Science and Technology, DOI 10.1007/978-1-4614-9434-8_3,

peptides, carbohydrates, glycoproteins, and small molecules including folate and vitamins. These targeting ligands are now the important component of many of the targeted drug delivery system including polymeric nanoparticles, liposomes, and nanopolymersomes. The current chapter is divided into three parts: part one describes various drug targeting strategies which is followed by part two dedicated to understand the role of EPR effect in tumor targeting and important endogenous factors that can positively impact EPR effect in tumor tissue. The third part principally focuses on ligand targeting systems for the treatment of various malignancies, including breast, colorectal, lung, and prostate cancers.

3.2 Targeted Drug Delivery System

The concept of targeted drug delivery system was first suggested almost a century ago and refers to the increased/enhanced drug accumulation in the target zone independently of the method and route of drug administration [1]. Targeting may resolve many problems currently associated with the systemic drug administration, such as lack of drug specificity at the target site, non-specific toxicity and adverse effects, and high-dose requirements. Targeted drug delivery systems can provide better pharmacokinetic and pharmacodynamic profiles of drug, can improve the specificity and efficacy of the drug, and can increase drug internalization leading to intracellular delivery of drug [1].

3.2.1 Drug Targeting Strategies

Several strategies proposed for targeting can be classified into passive, active, and triggered targeting (Fig. 3.1). Passive targeting depends on passive diffusion of

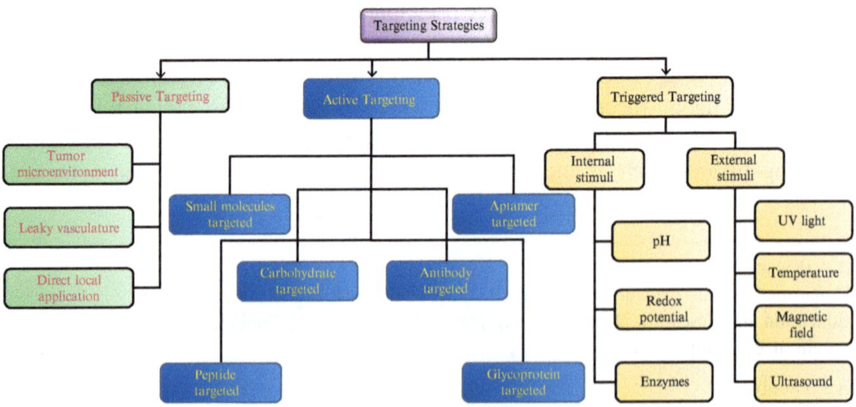

Fig. 3.1 Drug targeting strategies for enhancing the intratumoral drug payload

delivery system across the leaky vasculature exploiting enhanced permeability and retention (EPR) effect. Active targeting requires anchoring ligands/vectors onto the surface of delivery system for binding to appropriate receptors expressed at the target site. Triggered targeting depends on stimuli such as abnormal temperature or pH conditions at the disease site for releasing the therapeutic load at the diseased site.

3.2.1.1 Passive Targeting

Passive targeting takes advantage of natural properties and processes in tumors to localize a drug delivery system at a desired target site. There are different types of drug delivery system including micelles, polymeric nanoparticles [10], liposomes [11], nanopolymersomes [12–17], and dendrimers designed to evaluate drug targeting to solid tumors [18]. All these systems tend to exploit the unique difference in anatomical and pathophysiological abnormalities present in tumor vasculature in comparison to normal cells [19]. Blood vessels in tumors majorly shows two abnormalities (1) high proportion of proliferating endothelial cells leading to enhance vascular permeability for carrier system and (2) deficient or nonfunctional lymphatic vessels contributing to inefficient drainage from the tumor tissue and thereby allowing prolonged retention of carrier systems (Fig. 3.2a). Both these abnormalities lead to the phenomenon named as EPR effect.

3.2.1.2 Active Targeting

The concept of active drug targeting was suggested by Ringsdorf in the mid-1970s. It involves coupling of targeting ligand to drug-loaded nanocarrier system that binds to specific overexpressed receptors on target cell. The targeting ligands can be polysaccharide, antibodies and their fragments, hormones, proteins, lipoproteins, and low-molecular weight molecule, such as folate and vitamin. These targeting ligands bind to the cell surface receptors followed by their internalization into the cells (Fig. 3.2b). Unlike passive targeting which demonstrate relative tumor selectivity as a consequence of EPR effect, active targeting results in higher intratumoral accumulation of nanocarrier and hence higher cellular concentration of the drug [19].

3.2.1.3 Triggered Targeting

The goal of chemotherapy is to achieve the maximal drug concentration in the tumor microenvironment; thus, the drug release from a carrier system plays a vital role. The ideal targeted system is expected to release the drug at the target site rather than during the circulation. Hence, controlling the drug release rate from the delivery system could be a useful strategy for efficient drug therapy. Releasing the drug at the

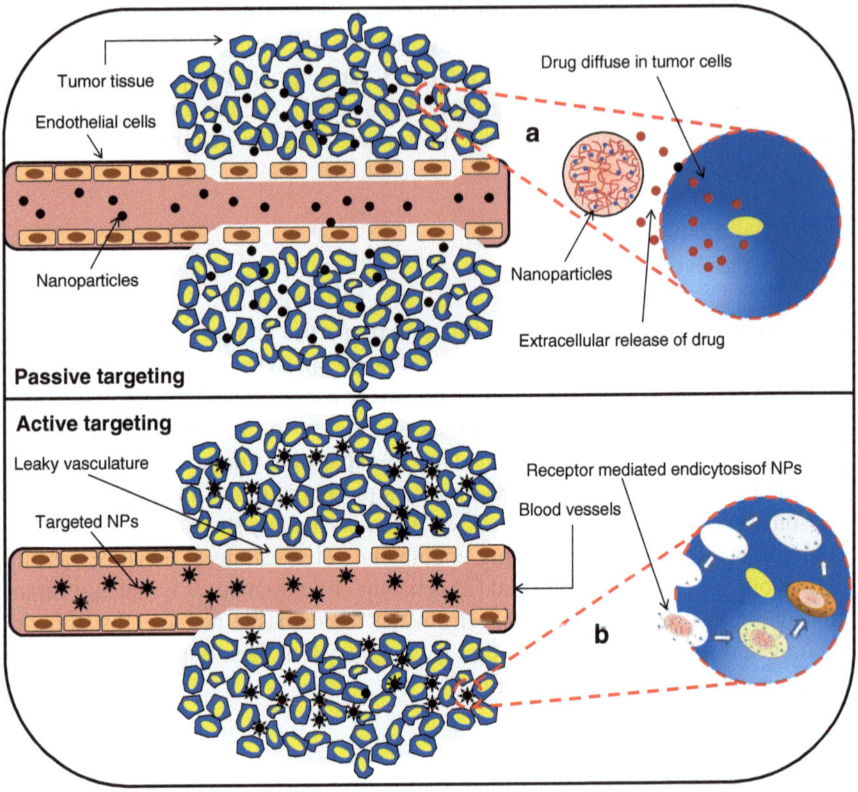

Fig. 3.2 Passive targeting versus active targeting strategies for anticancer drug delivering system. (*Top*) By the enhanced permeability and retention effect, nanoparticles (NPs) passively diffuse through the leaky vasculature and accumulate in tumor tissues. In this case, drug may be released in the extracellular matrix and then diffuse through the tissue (**a**). (*Down*) In active targeting, once particles have extravasated in the tumor tissue, the presence of targeting ligands (e.g., antibody, carbohydrate) on the NP surface facilitates their interaction with receptors that are present on tumor cells, resulting in enhanced accumulation and preferential cellular uptake through receptor-mediated endocytosis (**b**)

target site not only increases the interaction of the drug with target site but also allows more uniform diffusion of drug across the cell membrane and to the cytoplasm. Triggered targeting systems use stimuli responsive polymers which trigger the release of drug upon exposure to internal (pH, redox potential, or enzymes) or external (UV light, temperature, ultrasound, and magnetic fields) stimuli present at the target site [18, 19]. However, developing the triggered release nanocarrier systems is very difficult as the systems are not so stable to avoid the indiscriminate release of the drug. Moreover, the internal or external stimuli are not very specific to the target site [18].

3.3 Role of EPR Effect in Tumor Targeting

Enhanced permeability and retention (EPR) effect is the physiology-based principal mechanism of tumor accumulation of large molecules and small particles [20]. It is associated with the unique feature of enhanced vascular permeability and sluggish lymphatic drainage system responsible for spontaneous accumulation and retention of nanoparticles in the tumor cells. The EPR effect is applicable only to larger molecular weight (40–800 kDa) particles having particle size ranging from 10 to 500 nm but not to the low molecular weight anticancer drugs. The lower molecular weight anticancer drugs have very little tumor selectivity and therefore cannot attend higher concentrations in tumor tissue than that of blood plasma. Further, these drugs cannot be retained in tumor tissue as they undergo rapid excretion (washout) in the blood stream. In contrast to this, nanoparticles selectively accumulate and retained in tumor tissue and provide higher intratumoral concentration (10–20-fold) and release of drug in comparison to blood plasma or normal cells. This ultimately results in improved tumor targeting and improved therapeutic efficacy of drug, thereby reducing the toxic effect on the normal healthy cells [21–25].

3.3.1 Historical Background

The discovery of the EPR effect is considered to be one of the greatest breakthrough leading to the development of targeted antitumor therapy [26]. EPR effect was reported for the first time by Matsumura and Maeda in 1986 while studying the in vivo behavior for antitumor protein antibiotic and polymer conjugate SMANCS [21]. They observed a significant difference in the accumulation of low molecular weight drugs as compared to biocompatible macromolecular drug SMANCS in solid tumors [27, 28]. This difference in the uptake was also seen in macromolecules such as transferrin (90 kDa), IgG (immunoglobulin, 150 kDa), albumin (65 kDa), and lipid particles. Moreover, these macromolecules were retained in the solid tumors for the extended period up to 3 months or longer.

Accordingly several researchers also observed the same effect in various tumor models while working on series of biocompatible synthetic polymers, 4-hydroxypropylmethacrylate copolymer (HPMA) having a molecular weight ranging from 20 to 800 kDa. They found that the polymers with higher molecular weight were differentially accumulated in solid tumors at different dose levels. However, for the lower molecular weight polymer, typically less than 40 kDa, tumor uptake was lower, indicating the importance of molecular weight [21, 26–29].

3.3.2 Unique Features of Tumor Microenvironment: Angiogenesis and Abnormal Lymphatic Drainage

There is a major difference in the characteristics of normal tissue and tumor tissue. The tumor cell is characterized by a number of events including acceleration of cell cycle, invasive growth, genomic alterations, increased cell mobility, and changes in the cellular surface. Tumor cell also differs in morphology as well as functional characteristics in comparison to normal cells. Morphologically, it is characterized by large nucleus, having an irregular size and shape, and angiogenesis (fragmentation vasculature). Functionally, tumor cell causes elimination of active substances, such as growth factors and hormones, and causes lysis of important enzymes such as collagenase and plasmogen activator. Moreover, there is a vast difference in energy metabolism especially in case of glucose metabolism. Although there is a difference in both morphology and functional characteristics, drug delivery system takes the advantages of morphological dissimilarities and hence shall be discussed further [30].

3.3.2.1 Morphology of Tumor Blood Vasculature: Angiogenesis

Angiogenesis is a physiological process involving the generation of new blood vessels from preexisting capillaries to form network-like structure. In the prevascular phase, solid tumors generally have a volume of not more than 1–2 mm^3. Moreover, they are not much vascularized and have few million cells needing oxygen and nutrients that can easily cross the surface of the cells and can reach to the center of the tumor mass by simple passive diffusion [19]. However, as the tumor volume increases and goes beyond the critical volume of 2 mm^3, there is difficulty in diffusion of oxygen and nutrients across the tumor mass, leading to the state of cellular hypoxia. This state of hypoxia kicks off the onset of tumoral angiogenesis which results in the creation of a new network of blood vessels so as to supply oxygen, nutrients, growth factors, and proteolytic enzymes to tumor cells. Angiogenesis is regulated by four major steps (1) degradation of basement membrane by proteolytic enzymes, (2) migration and proliferation of endothelial cells, (3) adhesion of molecules by cell-cell and cell-matrix interaction, and (4) the construction of capillary loops [31].

The onset of hypoxia causes the release of angiogenic molecules such as platelet-derived growth factor (PDGF), vascular endothelial growth factor (VEGF), or tumor necrosis factor-α (TNF-α) resulting in the activation of endothelial cells. Activated endothelial cells reorganize their cytoskeleton and express various cell surface adhesion molecules such as integrin αvβ3, selectins, and proteolytic enzymes which interact with extracellular matrix to promote the migration and proliferation of endothelial cells. The proteolytic enzymes such as metalloproteinases are used to cause the degradation of basement membrane and extracellular matrix. Following proteolytic degradation of basement membrane, endothelial cells stimulated by a variety of growth factors starts to migrate and proliferate through the degraded extracellular matrix. During the migration and proliferation, endothelial cells interact with extracellular matrix facilitated by surface adhesion molecules such as integrin and selectins, leading to the formation of capillary buds. Angiogenic factors such as autocrine and paracrine causes migration, proliferation, elongation, orientation, and differentiation of the endothelial cells, eventually constructing the capillary loops forming microvessels [32].

3.3.2.2 Lymphatic Clearance of Tumor Tissue

The lymphatic vasculature is a system of low-pressure vessels which, in normal cells, acts to transport fluid, macromolecules, and lipid particles from the tissue interstitium back to the systemic circulation. However, in the tumor cells, there is a lack of effective lymphatic vasculature due to which all the macromolecules and lipid particles can selectively be retained in the tumor cells as compared to normal cells [22–24, 33]. Lack of lymphatic clearance was assessed by Maeda and group by using Lipiodol (iodinated poppyseed oil) as a carrier for the delivery of drug molecule (SMANCS) as it is primarily recovered via the lymphatic system. As the Lipiodol-loaded SMANCS was injected into the tumor feeding artery, they observe that the carrier was selectively retained into the tumor cells but not in the normal cells [34, 35]. Similar observation was seen when the Evans blue/albumin complex was injected intravenously to tumor-bearing mice. Evans blue/albumin complex selectively accumulated and was retained in the tumor cell for more than a week; however, in normal cells, the complex was steadily cleared by the normal lymphatic system, indicating the lack of lymphatic vasculature in tumor cells.

3.4 Factors Affecting EPR Effect

There are various endogenous factors that can impact EPR effect in the tumor tissue. These include bradykinin (BK), nitric oxide (NO), prostaglandins (PGs), vascular endothelial growth factor (VEGF), peroxynitrites, and matrix metalloproteinases (MMPs). All these factors play the major role in tumor angiogenesis process, and hence, their understanding could provide insight in designing suitable methods or drug delivery system targeting tumor tissue. For instance, utilizing nitroglycerin or other NO-releasing agents and ACE inhibitors could enhance the EPR effect and therefore could increase the therapeutic effect by delivering greater amount of drug to tumor cells. The schematic representation depicting the role/activation of various factors is shown in Fig. 3.3 and Table 3.1.

3.4.1 Bradykinin

Bradykinin (BK) is known to play a crucial role in inducing vascular permeability in common infection, inflammation, and tumor cells [36]. Generation of BK in

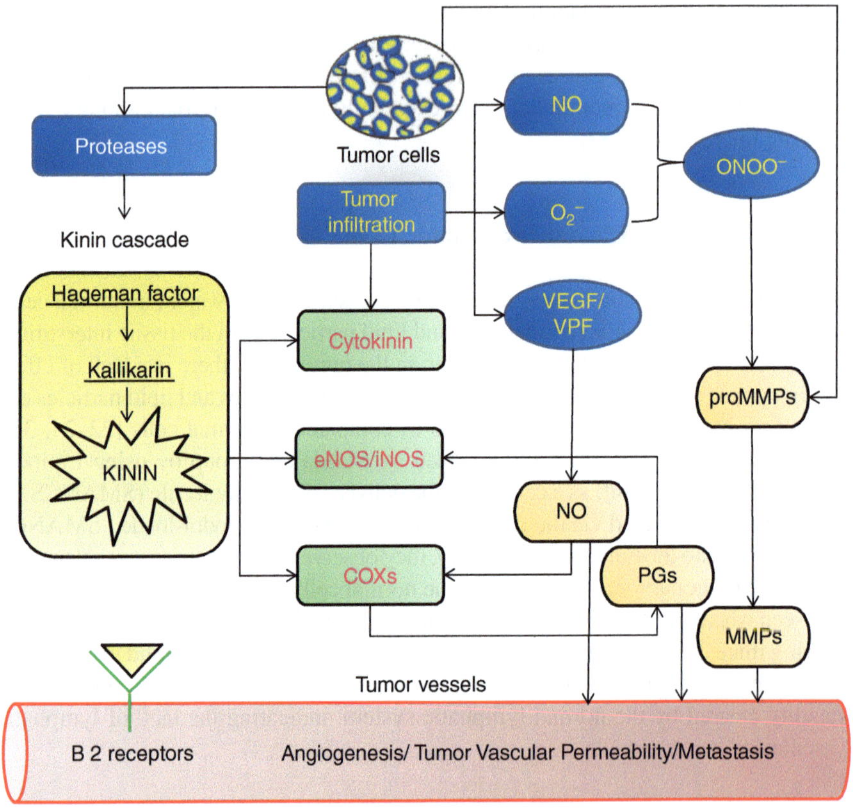

Fig. 3.3 Mechanism of various vascular mediators to induce EPR effect in the tumor tissue. *NOS* nitric oxide synthase, *COXs* cyclooxygenases, O_2^- oxygen radical, *VEGF* vascular endothelial growth factor, $ONOO^-$ peroxynitrites, *MMPs* matrix metalloproteinases

Table 3.1 Factors responsible for the enhanced permeability and retention effect of drug carrier in solid tumors

Mediators	Mechanism involved in inducing EPR effect
Bradykinin (BK)	Initiated by activation of Hageman factor (factor XII), followed by kinin generating cascade. ↑ eNOS
Nitric oxide (NO)	Synthesized from L-arginine by NOS, ↑ $ONOO^-$
Prostaglandins (PGs)	Synthesized from arachidonic acid by COXs, ↑ cytokines, ↑ tumor necrosis factor-α
Peroxynitrite	Formed by rapid reaction of NO with superoxide anion radical ($O_2^{-\bullet}$), ↑ MMPs
Metalloproteinases (MMPs)	↑ kinin generating cascade, ↑ NO
Vascular endothelial growth factor (VEGF)	↑NOS,↑NO, ↑ matrix-degrading enzymes, ↓ cell apoptosis

tumor cells involves the cascade of reaction initiated by the activation of Hageman factor (factor XII). This activation is followed by activation of prekallikrein to kallikrein and ultimately leading to the release of BK from high molecular weight kininogen. In addition to BK, its hydroxylated derivative such as 3-hydroxypropyl bradykinin (Hyp BK) also induces accumulation of body fluids in inflammatory tissues or tumor cells. Further, BK receptors are highly upregulated in the tumor cells to induce EPR effect causing accumulation and retention of the extravascular fluid [37, 38]. It has also been known that BK activates endothelial cell-derived nitric oxide synthase (eNOS) to generate nitric oxide (NO), an important vascular mediator, for inducing tumor vascular permeability [29, 39–41]. Inhibition of BK generation either by means of kallikrein inhibitors or by injecting BK antagonist such as HOE-140 significantly suppresses both fluid accumulation and tumor growth [42, 43].

3.4.2 Nitric Oxide

In addition to BK, nitric oxide (NO) has been found as an important vascular mediator involved in vasodilatation, hypotension, angiogenesis, cell proliferation, and EPR effect. It is synthesized from L-arginine by NO synthase (NOS) in presence of oxygen [39]. The enzyme NOS is extensively expressed in the tumor cells and significantly produces more NO as compared to the normal tissues [41, 44, 45]. The role of NO in the tumor cell was assessed by simultaneous injection of Evans blue/albumin complex and oily formulation of NO to guinea pigs [46]. The experiments were carried out in the presence and absence of both NO scavenger (e.g., carboxy-2-phenyl-4,4,5,5-tetramethylimidazoline-1-oxyloxide) and NO inhibitor (e.g., N_ω-monomethyl-L-arginine) [44, 47]. In the absence of both NO scavenger and inhibitor, there was a marked extravasation of Evans blue/albumin complex at the injection site, while significantly reduced extravasation was observed in the presence of both scavenger and inhibitor. Further studies using NOS inhibitor also proved the suppression of vascular permeability in the tumor cells, thereby showing significant reduction in extravasation of Evans blue/albumin complex [46, 47].

3.4.3 Prostaglandins

PGs are the proinflammatory mediators involved in inflammation and upregulation of the inflammatory cytokines such as interleukin-1 and tumor necrosis factor-α. They are enzymatically synthesized from arachidonic acid by means of cyclooxygenases (COXs). Among various PGs, PGE_1 and PGI_2 behave very similar to NO not only preventing platelet aggregation, leukocyte adhesion, and thrombosis formation but also facilitating extravasation and EPR effect [39, 41]. The stable analogue of PGI_2, Beraprost sodium, when injected either intravenously or intra-arterially was found to increase extravasation significantly in the tumor cells. Beraprost shows two- to threefold more accumulation of Evans blue/albumin complex without significantly affecting the systemic blood flow in the normal tissue or organs. However, the systemic blood flow in tumor cells was more drastically suppressed by 70–90 % [48]. In case of COXs inhibitors, there was significantly decreased accumulation of the Evans blue/albumin complex in tumor cells [49].

3.4.4 Peroxynitrites and Matrix Metalloproteinases

Peroxynitrites ($ONOO^-$) and MMPs are strong oxidizing and nitrating agents extensively formed in tumor and inflammatory tissues via rapid reaction of NO with superoxide anion radical ($O_2^{-\bullet}$). Several studies have shown their role in inducing EPR effect. There are multiple mechanisms of enhancement of vascular permeability by $ONOO^-$ radicals. Experimental studies performed by Maeda and coworkers showed that intradermal injection of $ONOO^-$ increased extravasation of Evans blue/albumin in a dose-dependent manner at the site of injection. Although the half-life of $ONOO^-$ at physiological pH is only a few seconds [50], the extravasation lasted for relatively longer time after administration of $ONOO^-$. One of the mechanism involved in the induction of EPR effect by $ONOO^-$ radicals is through the activation of matrix metalloproteinases (MMPs) [47]. MMPs are the group of endopeptidase enzymes produced by a variety of cells, such as fibroblasts, macrophages at inflammatory tissues, and also by the tumor cells. The activated MMPs are known to enhance the EPR effect and support the tumor growth. Moreover, MMPs cause disintegration and remodeling of the extracellular matrix and facilitate cancer metastasis. Another mechanism in enhancing EPR effect by $ONOO^-$ radicals is through a kinin cascade reaction via activation of MMPs [47]. Peroxynitrites are also involved in production of nitro compounds such as nitrotyrosine and nitroguanosine which serves as the major source of nitrite.

3.4.5 Vascular Endothelial Growth Factor

VEGF is an angiogenesis factor which performs dual function of angiogenesis and enhancement of vascular permeability to facilitate and sustain the rapid tumor growth. It exhibits high-binding affinity towards VEGF receptors which are overexpressed in the tumor cells. Several reports indicated that the amount of VEGF in different murine tumors was 2–30-fold higher in comparison to the normal tissues [39]. The EPR effect induced by VEGF was assessed by examining the extravasation of Evans blue/albumin complex by intradermally injecting both VEGF and BK (for comparison) in guinea pigs' skin [39]. VEGF showed significant extravasation of Evans blue/albumin complex even at picomolar concentration in comparison to the potent BK which shows activity in nanomolar concentration [39]. Further, the studies suggested that VEGF is a potential activator of NOS involved in the production of

NO and thus facilitates EPR effect [51–53]. It was also involved in inducing various cellular responses including an enhancement in the expression of matrix-degrading enzymes and inhibition of cell apoptosis [54]. Looking into the role of VEGF in EPR effect, its inhibitors such as bevacizumab and ranibizumab have been developed as a useful therapeutic tool against solid tumors.

3.5 Ligand Targeting Systems

In order to increase the targeting capability of NPs, their surfaces are coupled with targeting ligands including small molecules (folic acid and thiamine), aptamers (oligonucleotides, DNA, and RNA), carbohydrates (galactose and mannose), monoclonal antibodies (IgG), peptides, and glycoproteins. It is pertinent to mention that most of the current clinically approved nanotechnology products (e.g., Abraxane®, AmBisome®, Doxil®, Myocet®) relatively experience the passive targeting approach (i.e., lack of active targeting). Therefore, the current research focuses on developing and evaluating ligand targeted drug delivery systems for in vivo targeting to cancer cells. Figure 3.4 depicts various ligand-based targeting systems along with their examples.

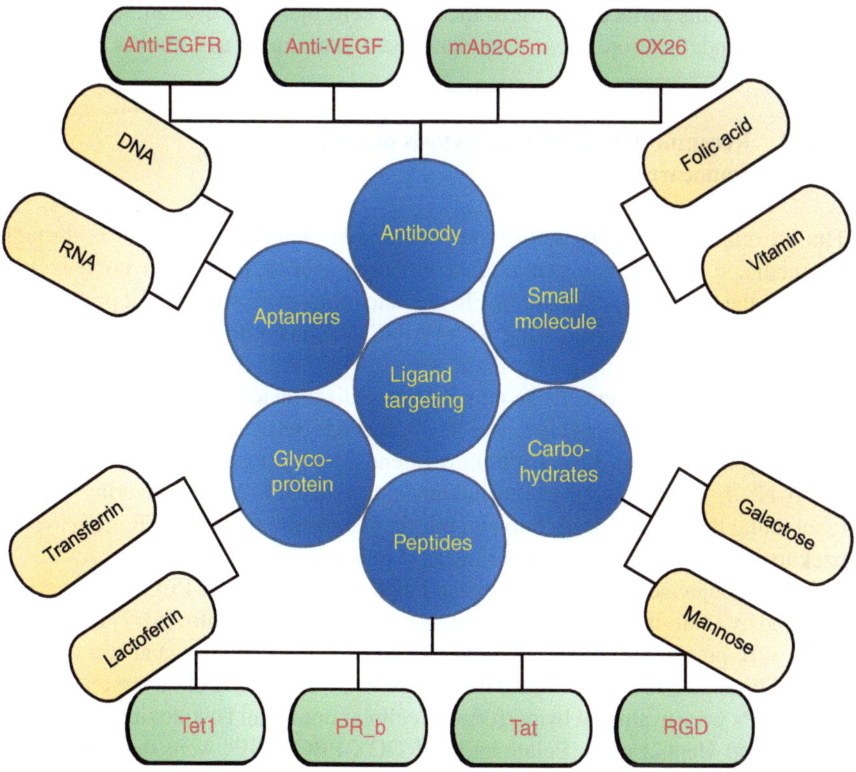

Fig. 3.4 Various types of ligand-based targeting system along with their examples

Table 3.2 provides the list of targeted drug delivery systems previously reported for the treatment of cancer.

3.5.1 *Small Molecules-Based Targeting Ligands*

Small molecules comprise an important class of targeting moieties with a diverse structures and properties making them very much suitable for functionalization of nanoparticles. Small size and low molecular weight further allows the attachment of multiple ligands onto nanoparticles.

Among all small molecule-based targeting ligand, folic acid is the most widely used ligand for tumor targeting [77]. It is an essential water soluble vitamin required by eukaryotic cells for one-carbon transfer reactions in several metabolic pathways [78]. It is involved in the biosynthesis of nucleotide bases which are required for cell proliferations. Cellular uptake of folic acid across plasma membrane is associated either with reduced form of folate carrier or with folate receptor (FR). Reduced folate carriers are virtually present in all cells; however, FR exhibits very limited distributions. Unlike reduced folate carrier, FR has the ability to transport both folic acid and folate-linked nanocarrier systems (FA-NS). Moreover, folate receptors are over expressed on the surface of most of the tumor cells including ovarian, lung, kidney, and breast and, therefore, have been identified to have tumor as a marker [79].

3.5.1.1 Receptor-Mediated Endocytosis of Folate-Linked Nanocarrier System

The movement of FA-NS across the plasma membrane to cytoplasm is through FR-mediated endocytosis. Although conflicting reports have been published to describe the precise mechanism of FR-mediated transport of FA-NS, it proceeds in a series of distant steps beginning with FA-NS attachment to cell surface FR to the release of therapeutic drug into the cytoplasm [80]. FA-NS binds to the overexpressed FR on the cancer cells and trigger internalization to form endosomes. As the endosomal compartment pH is 5, acidification of FA-NS causes its detachment from the receptors [81] (Fig. 3.5). Trafficking of endosome intracellularly causes release of FA-NS from endosome by an unknown mechanism. After delivering FA-NS (containing therapeutic drug) into cytoplasm, the membrane-bound FR are recycled back to cell surface to carry further internalization of FA-NS [78].

Folate-conjugated doxorubicin-loaded nanoparticles (DOX-PBCA NPs) were prepared by Duan et al. to selectively target cancer cell lines in vitro [5]. The nanoparticles were prepared by using poly(butyl cyanoacrylate) polymer and were coated with chitosan to effectively control the release of doxorubicin. Folate-targeted DOX-PBCA NPs were evaluated by performing cellular uptake and cytotoxicity studies in MCF-7 and HepG2 cells. Folate-targeted DOX-PBCA NPs were detected in the cytosol region within 4 h of incubation, whereas free DOX and DOX-PBCA NPs did

Table 3.2 List of some reported examples of targeted nanocarrier system for cancer therapy

Targeting moiety	Target	Carrier system	Therapeutic agent	Targeted organ/ tissue	Therapeutic outcome	References
Small molecules based targeting						
Folate	Folate receptors	PDCs	5-FU	Tumor	β-CD-PEG-FA conjugate – assembled nanoparticles exhibited a greater extent of cellular uptake against HeLa cells than A549 cells	[55]
	Folate receptors	Liposomes	Docetaxel	Tumor	Docetaxel-loaded liposomes, FA-PDCT-L demonstrated the strongest cytotoxicity against two carcinoma cell lines; FA-PDCT-L showed 6.23 times increase in AUC in comparison with docetaxel solution	[56]
	Folate receptors	NPs	Doxorubicin	Tumor	Folate-conjugated DOX-PBCA NPs showed enhanced cellular uptake, increased targeting capacity, and increased cytotoxicity against MCF-7 cells overexpressing folate receptor	[5]
	Folate receptors	Liposomes	Doxorubicin	Tumor	Significantly higher antitumor effect in mice-bearing murine lung carcinoma M109 with the folate ligand in PEGylated and masked folate-linked liposomes than with non-PEGylated liposomes independent of circulation time after i.v. injection	[57]
	Folate receptors	Liposomes	Irinotecan	Tumor	The distributions of drug in the tumor was far greater with the F-Lip group in comparison to the C-Inj and C-Lip; F-Lip exhibited a dose-dependent tumor growth inhibition, superior anticancer activity, and lower toxicity as compared to C-Lip and C-Inj after IV. administration	[58]
Aptamer-based targeting						
RNA	Epidermal growth factor receptor (EGFR)	Gold NPs	–	Targeted therapy, in vitro	Enhanced cellular uptake in vitro in A431 cells	[59]
DNA	Mucin-1	Polymeric NPs	Paclitaxel	Targeted therapy, in vitro	Enhanced cellular uptake and cytotoxicity in vitro	[60, 61]

(continued)

Table 3.2 (continued)

Targeting moiety	Target	Carrier system	Therapeutic agent	Targeted organ/ tissue	Therapeutic outcome	References
DNA	Nucleolin	Liposomes	Cisplatin	Targeted therapy, in vitro	Enhanced cellular uptake and cytotoxicity in vitro in MCF-7 cell lines	[62]
Carbohydrate-based targeting						
Galactose	ASGP Receptors	Liposomes	–	Liver	Gal-liposomes with higher molar ratio of galactose to cholesterol showed enhanced cellular uptake in vitro in HepG2 cells and in vivo, rapid liver accumulation accounting for 80 % of the initial dose	[63]
	ASGP Receptors	Liposomes	(siRNA)	Liver	The distribution of siRNA to parenchymal cells was 1.9 times higher when galactosylated cationic liposomes were used in comparison to non-galactosylated liposomes Galactosylated liposomes showed enhanced uptake of siRNA by the liver without cytotoxic effects	[64]
	ASGP Receptors	Nanogels	Oridonin	Hepatoma-targeted delivery of oridonin	ORI-loaded nanogels exhibited a higher antitumor activity. Anticancer activity increased in relation to increases in the number of galactose moieties of the nanogels in HepG2 cells	[65]
Mannose	Mannose receptor	Liposomes	–	Alveolar macrophages	The in vitro uptake of Man-liposomes took place in a concentration-dependent manner both in vitro and in vivo	[66]
Glycoprotein-based targeting						
Transferrin	Transferrin receptors	Polymersomes	Doxorubicin	Brain	Compared with PODOX and free DOX, Tf-PODOX demonstrated the strongest cytotoxicity against C6 glioma cells and the greatest intracellular delivery Significant reduction of tumor volume and a significant increase of median survival time in the group of Tf-PODOX compared with those in saline control animals, animals treated with PODOX, and free DOX solution	[67]
Transferrin and lactoferrin	Transferrin receptors	Polymersomes	–	Brain	Both Lf and Tf could increase the cell uptake of PSs at 37 °C, but the uptake of Tf-PS was significantly greater than that of Lf-PS In vivo tissue distribution and pharmacokinetics in mice revealed higher brain uptake and distribution of Tf-PS than Lf-PS	[68]

Transferrin	Transferrin receptors	Liposomes	Doxorubicin	Targeted therapy, in vitro	Enhanced intracellular uptake of entrapped doxorubicin by HepG2 cells Enhanced doxorubicin concentration in tumors and decreased doxorubicin concentration in heart and kidneys in tumor-bearing mice	[69]
Peptide-based targeting						
RGD peptide		Liposomes	Paclitaxel	Tumor	Enhanced uptake and cytotoxicity in vitro Increased antitumor activity in vivo in mice bearing SKOV-3 solid tumors	[70]
Thiolated Herceptin		Liposomes	Paclitaxel	Tumor	Enhanced cellular uptake in vitro and enhanced antitumor effect in vivo against BT-474	[71]
Antibody-based targeting						
Nucleosome-specific monoclonal antibody (mAb 2C5)	–	Liposomes	Doxorubicin	Tumor	mAb 2C5 targeted liposomes showed enhanced accumulation in tumors, and the in vivo therapeutic activity of the mAb 2C5 Doxil treatment was found to be significantly superior	[72, 73]
Anti-VEGFR2	VEGF receptors	Liposomes	Doxorubicin	Tumor	Anti-VEGFR2-targeted immune liposomes (ILs) loaded with doxorubicin (anti-VEGFR2-ILs-dox) were superior in therapeutic efficacy to empty liposomes	[74]
Anti-HER2 mAb fragments	HER2 receptors	Liposomes	Doxorubicin	Tumor	Enhanced antitumor effect over nontargeted liposomes in vivo. Similarly accumulation of targeted and nontargeted liposomes in tumors in a nude mouse model with HER2- overexpressing BT-474 cells	[75]
Anti-MT1-MMP antibody	–	Liposomes	Doxorubicin	Tumor	Increased uptake by MT1-MMP-overexpressing HT1080 fibrosarcoma cells in vitro and more effective tumor growth inhibition in vivo	[76]
Antibody trastuzumab and rituximab	HER2 receptors	Gold NPs	Trastuzumab	Tumor	HER-2 targeting via attachment of trastuzumab paradoxically decreased tumor uptake as a result of faster elimination of the targeted AuNPs from the blood while improving internalization in HER-2-positive tumor cells as compared to nontargeted AuNPs	[4]

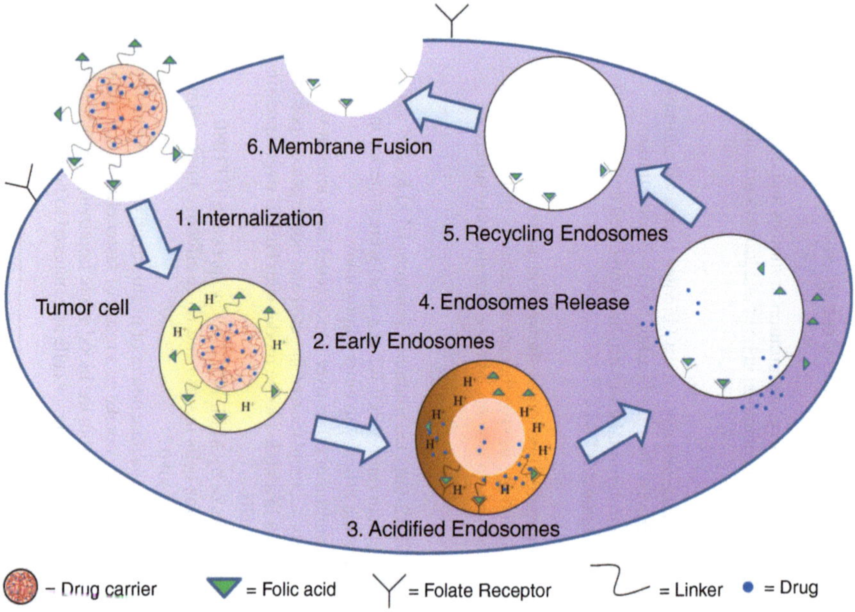

Fig. 3.5 Mechanism of receptor-mediated endocytosis for folate-linked nanocarrier system

not show detectable change in fluorescent intensity in both MCF-7 and HepG2. Further, the quantitative analysis showed that a greater fraction of folate-targeted DOX PBCA NPs were taken up within MCF-7 cells than HepG2 cells [5].

In another study, click chemistry approach was used to design and synthesize folate-conjugated β-cyclodextrin (β-CD-PEG-FA) nanoparticles using PEG as a spacer [55]. The PEG spacer used provides flexibility to folic acid molecules to interact freely with folate receptors, thereby enhancing its binding efficiency [82]. In vitro studies were carried out using 5-FU-loaded nanoparticles in two different cell lines, HeLa and A549 cells. It was observed that β-CD-PEG-FA nanoparticles showed much greater cellular uptake in HeLa cells than A549 cell. Further, in cytotoxicity studies, 5-FU-loaded β-CD-PEG-FA nanoparticles showed strong inhibitory effect on the HeLa cells in concentration-dependent manner. However, there was no significant difference in the inhibitory effect in the presence and absence of free folic acid in the culture medium [55].

Docetaxel-loaded liposomes conjugated with folic acid (FA-PDCT-L) showed strongest cytotoxicity and the most powerful apoptotic efficacy in carcinoma cell lines [56]. The pharmacokinetics studies of these NPs in rats indicated sixfold increase in AUC values in comparison to free drug solution. Tissue distribution studies reveal lower concentration of docetaxel in liver and spleen, whereas in tumor the concentration was significantly higher indicating targeted delivery of liposomes in presence of conjugated folic acid [56]. Folate-conjugated liposomes were also designed with varying length of PEG spacers to increase the level of FR targeting [83]. It was observed that shorter length PEG spacer reduced folate exposure to KB cells by interference with the ability of the liposome to recognize FR [57]. Liposomes with larger PEG spacer, F-PEG_{5000}-DSPE, showed higher association with folate receptors after 1 h incubation. In vivo efficacy studies showed higher tumor-killing effect of folate-linked PEGylated liposomes regardless of their chain length in comparison to folate-linked non-PEGylated liposomes [83].

3.5.2 Aptamer (Nucleic Acids DNA or RNA)-Based Targeting

Aptamer has been widely explored as targeting ligand in the past three decades to deliver therapeutic as well as diagnostic agents [7]. They are special class of nucleic acid molecules (DNA or RNA) which can form unique three-dimensional structures capable of specifically binding (with higher affinity) to molecular targets including small molecules, proteins, nucleic acids, and other cellular targets [84]. Aptamers are chemically equivalent to antibodies and thus serve as excellent alternative to them [60]. In comparison to other targeting agents, aptamers posses the following distinctive advantages [60, 62, 84]:

1. They are non-immunogenic and cause minimum toxicity as they are gradually degraded by nucleases.
2. They have relatively small size which allows them to easily penetrate into the tumors cells. Further, being small molecule functionalization of aptamers at specific site is easy and straightforward.
3. They are highly specific and show high-binding affinity which is almost comparable to monoclonal antibodies. However, unlike antibodies, they exhibit remarkable stability in a wide range of pH (4–9), temperature, and organic solvents without loss of activity [85].
4. Aptamers has low synthesis cost and are much easier to prepare as they do not rely on any biological system for their production and scale-up.

Ellington and coworkers prepared anti-EGFR aptamer-conjugated gold nanoparticles (GNPs) by facile hybridization method [59]. The GNPs were assessed on EGFR expressing A431 epidermoid carcinoma cells and showed specific internalization via EGFRs [59]. Chenchen et al. make use of mucin protein aptamers as targeting ligand owing to its overexpression (tenfold) in most malignant adenocarcinomas [60]. Aptamer-conjugated PLGA nanoparticles (Apt-NPs) were prepared using DNA as spacer. In comparison with nontargeted NPs, 71 % increase in fluorescent intensity was observed in MCF-7 cells treated with Apt-NPs. Also paclitaxel-loaded nanoparticles (PTX-Apt-NPs) produced more potent cytotoxicity than PTX-NPs in MCF-7 cells [60].

In another study, aptamer which specifically binds to nucleolin (NCL), mRNA-binding protein involved in cell proliferation, was conjugated on the surface of liposomes containing cisplatin [62]. The cell viability results revealed 21 % and 60 % cell viability on day 2 and day 4. However, no significant effect on cell viability was observed with non-NCL aptamer liposomes, indicating receptor-mediated cell uptake and improved therapeutic efficacy of liposomes with aptamer as a targeting ligand [62].

3.5.3 Carbohydrates-Based Targeting Ligands

Carbohydrates play a vital role in both physiological and pathological events such as cell growth and differentiation, cell–cell communication, inflammatory response, tumor metastasis, and viral infection. They show highly specific interactions with lectins [86] and asialoglycoprotein receptor (ASGP-R) which are overexpressed on the hepatic cells. Carbohydrates such as galactose and mannose readily bind with ASGP-Rs and thereby serve as effective liver targeting ligand [87].

3.5.3.1 Galactose

Managit et al. studied the effect of galactose density on asialoglycoprotein receptor-mediated uptake of galactosylated liposomes [63]. Gal-liposomes with molar ratio of 3.5, 5.0, and 7.5 % showed much higher (enhanced) uptake by HepG2 cells than that of non-galactosylated liposomes. However, this uptake was significantly inhibited in the presence of excess galactose, suggesting that uptake and internalization of Gal-liposomes is by the asialoglycoprotein receptors [63]. Further, these Gal-liposomes (with molar ratio of 3.5, 5.0, and 7.5 %) were found to be rapidly eliminated from the blood circulation and exhibited rapid liver accumulation accounting for 80 %, of the dose, within 10 min [63]. In case of Gal-liposomes with low molar ratio (1.0 and 2.5 %) of galactose to cholesterol, only slight improvement in liver accumulation was observed indicating surface density of galactose playing a crucial role in asialoglycoprotein receptor-mediated uptake of liposomes [63].

In another study, Lai et al. investigated the effect of ligand spatial orientation on the targeting specificity of magnetic nanoparticles (MNPs) [88]. The galactosylated MNPs were specifically endocytosed by HepG2 cells via asialoglycoprotein receptors and not by HeLa cells (devoid of ASGP-R) indicating targeted delivery of nanoparticles. Moreover, the results also showed that nanoparticles conjugated to triantennary galactosylated ligand provide best internalization efficiency in comparison to mono- and diantennary galactosylated ligand [88]. Galactosylated liposomes were also used for delivery of small interfering RNA (siRNA) to the liver [64]. These galactosylated liposomes showed 1.9 times higher distribution of siRNA in hepatocytes in comparison to non-galactosylated liposomes [64].

3.5.3.2 Mannose

Mannosylated nanocarrier system has been used for targeting mannose receptors which are highly expressed on alveolar macrophage cells which act as an important target for the treatment of diseases like lung inflammation and visceral leishmaniasis.

In the past decade, Hasida and colleagues developed mannosylated liposomes to target alveolar macrophage cells. Liposomes were prepared with varying ratio mannosylated to cholesterol derivatives to evaluate the targeting efficiency to alveolar macrophages [66]. In vitro uptake studies of fluorescent-labeled Man-liposomes in alveolar macrophages showed significantly higher uptake of Man-5.0- and Man-7.5-liposomes in comparisons to Man-2.5- and nude liposomes. However, this in vitro uptake of Man-liposomes was significantly inhibited in presence of excess mannose indicating mannose receptor-mediated endocytosis of liposomes. Similar results were obtained in in vivo studies in rats, where internalization of Man-liposomes was increased with increase in mannose content [66]. In another study, dexamethasone palmitate incorporated in mannosylated liposomes (DPML) were prepared to target macrophages for the treatment of lipopolysaccharide-induced lung inflammation in rats [89]. DPML showed marked reduction in inflammatory markers including cytokine and TNF-α in comparison to free dexamethasone palmitate and nontreatment groups. In addition to this, DPML significantly suppressed the activation of NF-κB and cause complete downregulation of phosphorylation of p38 MAPK an important marker in lung inflammation [89].

Mannose-coated liposomes incorporated with antibiotic MT81 was tested against visceral leishmaniasis in hamsters [90]. In comparison to control liposomes or free MT8, mannose-coated liposomes were found to be more efficient in eliminating intracellular amastigotes within splenic macrophages. The splenic parasitic load was reduced up to 79 % of the total parasite in mannose-coated liposomes, whereas at the same equivalent dose, the non-mannosylated liposomes or the free drug causes 55 and 50 % decrease in splenic parasitic load, respectively [90]. Similar in another study, mannosylated liposomes bearing amphotericin B showed maximum reduction in parasitic load up to 78 % in comparison to amphotericin B solution and drug-containing cationic liposomes which showed reduction of 42 and 61 %, respectively [91].

Mannose functionalized at terminal end of tetra (*p*-phenylene)-*block*-PEG polymer was synthesized for targeting *Escherichi coli*. The polymersomes formed showed 800-fold increase in the binding affinity towards pili of ORN 178 *E. coli* strain [92]. In the former, mannose recognized the specific lectin binding protein (concanavalin A) and showed higher inhibitory potency, whereas in the later study, galactose was unable to recognize the specific lectin binding protein. This signifies the importance of ligand specificity used for end group functionalization of polymersomes.

3.5.4 Antibody-Based Targeting

Antibodies were initially believed as "targeting missiles" which hit their specific biological targets. However, their use is much more complex for targeting and biological properties than expected by the researchers. There are various antibodies which have been found to be very much effective and well-tolerable for the treatment of different malignancies. Monoclonal antibodies such as trastuzumab, gemtuzumab, and bevacizumab have been approved by FDA and have emerged as important therapeutic agents as they have both targeting ability and anticancer activity. Other monoclonal antibodies such as anti-EGFR [93, 94], anti-VEGF [95], mAb2C5m [96], and 2C5 [72] were also used for targeting cancer cells as they specifically bind to the overexpressed receptors present on the surface of cancer cells. These monoclonal antibodies facilitate the endocytic uptake of various nanoparticulate systems to improve their antitumor efficacy.

Gold NPs bioconjugated with anti-EGFR monoclonal antibody (Ab) were prepared for targeting the epidermal growth factor receptor (EGFR) [94]. The in vitro EGFR-targeting properties of these NPs were demonstrated in $EGFR^+$ and $EGFR^-$ cells using ELISA and western blot assays. Using ELISA test, both free Ab and Ab-conjugated gold NPs showed concentration-dependent targeting in $EGFR^+$ cells but not in $EGFR^-$ cells. Similar results were observed by western blot analysis wherein Ab-conjugated gold NPs inhibit the EGFR phosphorylation in the presence of EGF indicating the EGFR-targeting property of NPs. In in vivo pharmacokinetic profile, Ab-conjugated gold NPs showed very similar results as that of uncoupled Ab indicating their suitability for EGFR targeting [94]. In another study, these anti-EGFR antibodies were coupled on the surface of polymersomes composed of mPEG-pep-PDLLA block copolymer using EDC/NHS chemistry. These antibody-coupled polymersomes showed enhanced endocytosis uptake within 3 days as compared to the polymersomes without anti-EGFR antibody [93].

Anti-VEGF was used to functionalize dextran-coated supermagnetic iron oxide nanoparticles (anti-VEGF-NPs) to achieve active targeting against human colon cancer cells [95]. Anti-VEGF-NPs were evaluated by MRI after systemic injection to tumor-bearing mice, which showed 1.34-fold change in MRI signal as compared to control group without NPs injection. However, the difference in MRI signal was not significant in mice treated with bare NPs indicating specific targetability of anti-VEGF-NPs towards colon cancer cells [95].

Monoclonal antibodies, mAb2C5m and 2C5, were conjugated on the surface doxorubicin-loaded liposomes by Torchilin and coworkers [72, 96]. Nucleosome-specific monoclonal antibodies (mAb 2C5) had increased the cytotoxicity of doxorubicin-loaded liposomes by recognizing surface-bound nucleosomes in various tumor cells. Further, the results of whole body gamma scintigraphic imaging indicate two- to threefold increase in the accumulation of doxorubicin in tumor cells when employed as immunoliposomes in comparison to nonspecific IgG-conjugated or plain liposomes [96]. Liposomes conjugated with monoclonal antibodies 2C5 also showed improved cytotoxicity of doxorubicin in human brain U-87 tumor xenograft model [72].

Brain targeting possesses one of the most complicated and challenging task in drug delivery due to the presence of blood–brain barrier (BBB). Pang et al. make use of monoclonal antibodies to develop drug carriers for brain targeting. Polymersomes composed of poly(ethyleneglycol)-poly(ε-caprolactone) (PEG-PCL) were conjugated with mouse-anti-rat monoclonal antibody OX26 (OX26-PO) for treatment of scopolamine-induced memory impairment [97]. 6-Coumarin loaded OX26-PO (IV route) reveals 2.62-fold increase in 6-coumarin concentration per gram of brain tissue as compared to the nonconjugated PEG-PCL polymersomes. The results were further confirmed using NC1900 (a peptide) loaded OX26-PO.

Concentration of NC1900 per gram of brain tissue was increased by 2.36-fold when employed as OX26-PO as compared to the PEG-PCL polymersomes. Moreover, they further showed that surface density of conjugated antibody also plays an important role in improving permeability of polymersomes across BBB [97].

Monoclonal antibodies were also used for targeting inflammatory vascular endothelium cells. Polymersomes functionalized with an anti-ICAM-1, an antibody which specifically bind to vascular cell adhesion molecules, were prepared using modular biotin-avidin chemistry [98]. The adhesiveness between anti-ICAM-1 functionalized polymersomes was measured using micropipette aspiration technique by quantifying the critical tension for detachment. It was found that the adhesion strength increases in proportion to the surface density of anti-ICAM-1 molecules, suggesting the possible application of functionalized polymersomes in targeting inflammatory vascular endothelium cells [98].

3.5.5 Peptide-Based Targeting

Although antibodies are found to be an interesting and viable option for targeting nanocarrier systems, they are highly immunogenic and produce deleterious side effects. They show poor in vivo mobility and therefore delay/reduce uptake into the desired target [99, 100]. Further, they have relatively large size and hence it is difficult to attach multiple molecules onto each nanoparticle surface. This in turn indicates that ligand molecule should have small size and peptides fulfill this requirement remarkably. Further, as compared to antibodies, peptides not only target the specific surface receptors but can also penetrate into cells or tissue to cause cell toxicity. Various peptides such as Tat, Tet 1, RGD, and PR_b have been used for targeted delivery of nanoparticles for different applications including therapy and diagnostic imaging.

Tet1 peptide functionalized polymersomes were prepared by Zhang et al. for targeting cochlear nerve in treatment of hearing loss and other neurological diseases [9]. Tet1 peptide specifically binds to the trisialoganglioside clostridial toxin receptor expressed on the cochlea. Intracochlear administration of Tet1 functionalized polymersomes showed specific localization into the neurofilaments and migration to tractus spiralis foraminous region present in cochlea. They were also observed in satellite cells of spiral ganglion as well as in the nerve fibers of CNS [9]. Another peptide which specifically binds to prestin, a unique protein in the inner ear that is solely expressed in sensorineural hearing loss, was used to functionalize polymersomes [101]. This peptide functionalized polymersomes composed of polyethylene glycol block poly-caprolactone (PEG-b-PCL) were studied for their targeting ability on outer hair cells (OHCs). Functionalized polymersomes showed abundant and specific internalization in OHCs and sparsely internalized in the other cells. In contrast, nonfunctionalized polymersomes were nonspecifically internalized by different type of cells present in cochlear explants [101].

Kokkoli and group utilized PR_b peptide for targeting prostate [102] and breast cancer cells [102]. PR_b peptide is a highly effective integrin targeting peptide that mimics the cell adhesion-binding site in fibronectin. This PR_b peptide was conjugated onto the surface of poly(ethylene oxide)-b-poly(butadiene) (PEO-PBD) polymersomes and was found to be effectively internalized by prostate cancer cells after adhering specifically to α5b1integrins expressed on their surface. Internalization of polymer vesicles was found to be dependent on the surface concentration of PR_b. Cytotoxicity studies for 0 %, 1 %, 5 %, 10 %, and 50 % PR_b-functionalized polymersomes were performed. It was found that internalization of polymersomes within the cancer cells was dependent on the surface concentration of PR_b with largest increase in cytotoxicity occurring when proceeding from 1 % PR_b to 5 % PR_b. Polymer vesicles functionalized with only 5 % (w/w) PR_b peptide achieved a 4.4-fold increase in cytotoxicity compared to nonfunctionalized polymersomes [102]. These PR_b peptide functionalized polymersomes were used for delivery of siRNA to breast cancer cells [103]. PR_b peptide functionalized polymersomes undergoes more effective internalization in MCF10A and T47D breast cancer cell lines. In cancerous T47D cells, peptide functionalized polymersomes were sevenfold more effectively delivered than nonfunctionalized polymersomes. Moreover, these peptide functionalized polymersomes showed significantly greater levels of delivery in the cancerous T47D cells as compared to the noncancerous MCF10A cells [103].

Recently Stojanov et al. demonstrated brain targeting potential of polymersomes functionalized with ganglioside (G 23) and prion (P 50)-targeting peptide. Biodistribution studies in mice show significantly higher accumulation of G 23 functionalized polymersomes in brain in comparison with P 50 functionalized polymersomes. This could be due to the strong promotion of transcytosis of polymersomes across BBB by G 23 peptide [104].

NIR emissive polymersomes functionalized with cell permeable cationic peptide Tat demonstrated significantly enhanced, time-dependent uptake by the dendritic cells [105]. The control NIR emissive polymersomes reaches half-maximal fluorescent intensity at 7 h as against 5 h required for Tat-conjugated NIR emissive polymersomes. Moreover, the Tat-conjugated NIR emissive polymersomes were found to remain localized within the cells for at least 3 days [105].

3.5.6 Glycoprotein-Based Targeting

Transferrin (Tf) and lactoferrin (Lf) are the glycoproteins which have a broad spectrum of functions particularly in the host defense mechanisms against infection and in severe inflammation. This broad spectrum of biological functions relies on their interaction with numerous cells such as platelets, endothelial cells of mesencephalic microvessels, and dopaminergic neurons [106]. Both of these molecules have been shown to cross the BBB via transferrin receptor-mediated transcytosis using in vitro model of BBB (bovine brain capillary endothelial cells and astrocytes).

Transferrin (Tf)-conjugated lipopolyplexes (LPs) containing G3139, an antisense oligonucleotide, were synthesized and evaluated in murine K562 xenograft model [107]. G3139 causes downregulation of B-cell leukemia/lymphoma 2 (Bcl-2), a protein which makes leukemia cells resistant to apoptosis. Greatest reduction in Bcl-2 protein levels was observed for Tf-LP G3139 in comparison to free G3139 and nontargeted LP-G3139. The AUC values were ten times higher for Tf-LP G3139 and the terminal half-life of 8.5 h as against the 0.63 h of free G3139, indicating rapid clearance of free drug from the blood circulation. In addition to this, the tumor growth was significantly suppressed in the mice treated with Tf-LP G3139 as compared to G3139 alone. However, contradictory results were observed in immunohistochemical staining for Bcl-2 protein, wherein free G3139 showed greater Bcl-2 downregulation as against Tf-LP G3139. The reported reason was the larger particle size of LPs (141 nm) which causes limited diffusion inside the solid tumor [107].

In another study, lysosome-targeted delivery system based on ceramide-loaded Tf-liposomes (cTf-LPs) were prepared to induce apoptosis in cancer cells by stimulating the lysosomal membrane permeabilization (LMP) [108]. The Tf-LPs showed significantly higher fluorescence as compared to plain liposomes (PL) in HeLa cells in vitro. However, in the presence of excess free Tf, the uptake of Tf-LPs was strongly decreased indicating that their internalization proceeded via TfR-mediated endocytosis. The percentage increase in number of apoptotic cells was found to be twofold higher in case of cells treated with cTf-LPs as compared to ceramic-loaded PL. LMP induction was shown by both ceramic and cTf-LPs; however, it was significantly higher in case of cTf-LPs. cTf-LPs cause strong inhibition of tumor growth leading to twofold reduction in tumor weight in comparison to PL (cer-loaded) and Tf-LP (cer-free) [108].

Pang and coworkers compared the relative superiority of Tf- and Lf-conjugated polymersomes (Tf- and Lf-PS) in brain drug targeting [68]. The results of in vitro uptake studies in mice revealed higher brain uptake of Tf-PS as compared to Lf-PS. Pharmacokinetics study showed a similar plasma concentration time profile for both conjugated ligands. However, Tf-PS showed 1.84-fold increase in AUC in comparison to Lf-PS. Further, the elimination and distribution rate constant for Lf-PS was 3.12- and 2.11-fold higher in comparison to Tf-PS [68].

In another study, Yu et al. developed lactoferrin-conjugated self-assembled poly(ethylene glycol)-poly(D,L-lactic-co-glycolic acid) (PEG-PLGA) polymersomes [8]. They investigated brain targeting ability of Lf-PS by varying the densities of lactoferrin (number of lactoferrin molecules ranging from 59 to 268) on the surface of the polymersomes. The polymersomes with lactoferrin molecules 101 were found to be optimum as demonstrated by BBB permeability surface area. Further, the plasma clearance of polymersomes was found to be inversely proportional to the lactoferrin densities conjugated on polymersomes surface. Other parameters such as AUC and percentage of injected dose per gram of brain tissue were found to increase with increased in surface lactoferrin densities. However, at higher surface density, a saturation phenomenon was observed resulting in decreased brain tissue accumulation of polymersomes. The reported reason was saturation of lactoferrin-specific receptors present in brain [8].

3.6 Challenges and Future Perspective for Systemic Drug Targeting

Although from the past few decades there is progress in the development of targeted drug delivery systems against the tumor cells, the existence of drug delivery barriers serves as a major challenge in its development. This results in insufficient and heterogeneous drug delivery to the tumor tissue and thereby limiting their clinical application. The following challenges are needed to be overcome to achieve tumor targeting.

3.6.1 Tumor Heterogeneity

The EPR effect is highly a heterogeneous phenomenon. The highly vascularized tissue that shows EPR effect is usually located at the periphery of the tumors and works well in the small tumors. However, in case of large tumors, the blood vessels are poorly perfused at the central area resulting in a diminishing EPR effect. As a result, the nanoparticles were able to deliver the drug only to the few layers of the tumor cells but not at the central area of the tumor tissue [109]. Moreover, the EPR effect is dependent on the size of nanoparticles and also on the porosity of the tumor vasculature which varies from tumor to tumor [110, 111]. In order to deliver the drug effectively to the large tumors, systemic approach needs to be designed for comprehensive characterization of tumor heterogeneity in different types of cancer cells as well as through different stages of tumor development. Further, new approaches/therapies should be developed to address unique dynamics of each patient specifically relevant for each type of malignancies.

3.6.2 Tumor Interstitial Fluid Pressure

As there is the lack of lymphatic drainage in the tumor tissue, there is a high interstitial pressure within the tumor tissue. This causes the efflux of tissue fluid against the direction of drug diffusion from blood vessel to the tumor tissue [112, 113]. Further, interstitial fibrosis in large tumors can also impede the diffusion of compounds through tumors [114]. In addition to this, increased tumor interstitial fluid pressure (TIFP) may facilitate the entrance of cancer cells into tumor blood vessels or into the surrounding normal tissue lymphatics, thereby aiding the metastatic process [115]. Due to all these limitations, the delivery system fails to deliver the drug deep into the tumor tissue. A variety of approaches have been tried to enhance the delivery of therapeutic agents to tumor tissue by reducing TIFP. For instance, US Pat. No. 5484399 to DiResta et al. discloses process and device to reduce TIFP. The patent discloses an artificial lymphatic system (ALS) which includes a plurality of aspiration tubes, each having plurality of aspiration holes connected to a vacuum source [116]. Other devices or process includes the use of aspiration probe and electroporation therapy [117]. As all these devices/processes have their own limitations and hence there exist a need in developing drug delivery system which can reduce efflux of drug associated with high interstitial pressure and could increase the drug residence time in tumor tissue.

3.6.3 Reduce Interactions

The specific interaction between the nanoparticles and cell surfaces are highly reduced within the tumor cells which affect the internalization of the nanoparticles, eventually reducing the payload of drug delivered to the tumor cells. Further, PEGylation of the nanoparticles can also reduce these specific interactions between the nanoparticles and the cells surfaces [87, 118–120].

References

1. Torchilin VP (2000) Drug targeting. Eur J Pharm Sci 11(Suppl 2):S81–S91
2. Bae YH, Park K (2012) Targeted drug delivery to tumors: myths, reality and possibility. J Control Release 153:198–205
3. Sudimack J, Lee RJ (2000) Targeted drug delivery via the folate receptor. Adv Drug Deliv Rev 41:147–162
4. Chattopadhyay N, Fonge H, Cai Z, Scollard D, Lechtman E, Done SJ et al (2012) Role of antibody-mediated tumor targeting and route of administration in nanoparticle tumor accumulation in vivo. Mol Pharm 9:2168–2179
5. Duan J, Liu M, Zhang Y, Zhao J, Pan Y, Yang X (2012) Folate-decorated chitosan/doxorubicin poly (butyl) cyanoacrylate nanoparticles for tumor-targeted drug delivery. J Nanopart Res 14:1–9
6. Wang HW, Jiang PL, Lin SF, Lin HJ, Ou KL, Deng WP et al (2012) Application of galactose-modified liposomes as a potent antigen presenting cell-targeted carrier for intranasal immunization. Acta Biomater 9:5681–5688
7. Xiao Z, Farokhzad OC (2012) Aptamer-functionalized nanoparticles for medical applications: challenges and opportunities. ACS Nano 6:3670–3676
8. Yu Y, Pang Z, Lu W, Yin Q, Gao H, Jiang X (2012) Self-assembled polymersomes conjugated with lactoferrin as novel drug carrier for brain delivery. Pharm Res 29:83–96
9. Zhang Y, Zhang W, Johnston AH, Newman TA, Pyykkö I, Zou J (2012) Targeted delivery of Tet1 peptide functionalized polymersomes to the rat cochlear nerve. Int J Nanomedicine 7:1015–1022
10. Deepa G, Ashwanikumar N, Pillai JJ, Kumar GS (2012) Polymer nanoparticles—a novel strategy for administration of Paclitaxel in cancer chemotherapy. Curr Med Chem 19:6207–6213
11. Lukyanov AN, Elbayoumi TA, Chakilam AR, Torchilin VP (2004) Tumor-targeted liposomes: doxorubicin-loaded long-circulating liposomes modified with anti-cancer antibody. J Control Release 100:135–144
12. Jain JP, Kumar N (2010) Development of amphotericin B-loaded polymersomes based on $(PEG)_3$-PLA co-polymers: factors affecting size and in vitro evaluation. Eur J Pharm Sci 40:456–465

13. Jain JP, Kumar N (2010) Self assembly of amphiphilic $(PEG)_3$-PLA copolymer as polymersomes: preparation, characterization and their evaluation as drug carrier. Biomacromolecules 11:1027–1035
14. Ayen WY, Chintankumar B, Jain JP, Kumar N (2011) Effect of PEG chain length and hydrophilic weight fraction on polymersomes prepared from branched (PEG)3-PLA co-polymers. Polymer Adv Tech 22
15. Ayen WY, Garkhal K, Kumar N (2011) Doxorubicin-loaded (PEG)3-PLA nanopolymersomes: effect of solvents and process parameters on formulation development and in vitro study. Mol Pharm 8:466–478
16. Jain JP, Ayen WY, Kumar N (2011) Self assembling polymers as polymersomes for drug delivery. Curr Pharm Des 17:65–79
17. Jain JP, Jatana M, Chakrabarti A, Kumar N (2011) Amphotericin-B-loaded polymersomes formulation (PAMBO) based on $(PEG)_3$-PLA copolymers: an in vivo evaluation in a murine model. Mol Pharm 8:204–212
18. Lammers T, Kiessling F, Hennink WE, Storm G (2012) Drug targeting to tumors: principles, pitfalls and (pre-) clinical progress. J Control Release 161:175–187
19. Danhier F, Feron O, Preat V (2010) To exploit the tumor microenvironment: passive and active tumor targeting of nanocarriers for anti-cancer drug delivery. J Control Release 148:135–146
20. Torchilin V (2011) Tumor delivery of macromolecular drugs based on the EPR effect. Adv Drug Deliv Rev 63:131–135
21. Matsumura Y, Maeda H (1986) A new concept for macromolecular therapeutics in cancer chemotherapy: mechanism of tumoritropic accumulation of proteins and the antitumor agent SMANCS. Cancer Res 46:6387–6392
22. Maeda H, Matsumura Y (1989) Tumoritropic and lymphotropic principles of macromolecular drugs. Crit Rev Ther Drug Carrier Syst 6:193–210
23. Maeda H, Sawa T, Konno T (2001) Mechanism of tumor-targeted delivery of macromolecular drugs including the EPR effect in solid tumor and clinical overview of the prototype polymeric drug SMANCS. J Control Release 74:47–61
24. Greish K, Fang J, Inutsuka T, Nagamitsu A, Maeda H (2003) Macromolecular therapeutics: advantages and prospects with special emphasis on solid tumor targeting. Clin Pharmacokinet 42:1089–1095
25. Duncan R (2006) Polymer conjugates as anticancer nanomedicines. Nat Rev Cancer 6:688–701
26. Greish K, Nagamitsu A, Fang J, Maeda H (2005) Copoly (styrene-maleic acid)-pirarubicin micelles: high tumor-targeting efficiency with little toxicity1. Bioconjug Chem 16:230–236
27. Maeda H (1991) SAMNCS and polymer-conjugated macromolecular drugs: advantages in cancer chemotherapy. Adv Drug Deliv Rev 6:181–202
28. Noguchi Y, Wu J, Duncan R, Strohalm J, Ulbrich K, Akaike T et al (1998) Early phase tumor accumulation of macromolecules: a great difference in clearance rate between tumor and normal tissues. Jpn J Cancer Res 89:307–314
29. Maeda H, Bharate GY, Daruwalla J (2009) Polymeric drugs for efficient tumor-targeted drug delivery based on EPR-effect. Eur J Pharm Biopharm 71:409–419
30. Baba A, Catoi C (2007) In comparative oncology: Chapter 3, Tumor cell morphology. The Publishing House of the Romanian Academy, Bucharest
31. Liekens S, De Clercq E, Neyts J (2001) Angiogenesis: regulators and clinical applications. Biochem Pharmacol 61:253–270
32. Gupta MK, Qin RY (2003) Mechanism and its regulation of tumor-induced angiogenesis. World J Gastroenterol 9:1144–1155
33. Konno T, Maeda H, Iwai K, Maki S, Tashiro S, Uchida M et al (2006) Selective targeting of anticancer drug and simultaneous image enhancement in solid tumors by arterially administered lipid contrast medium. Cancer 54:2367–2374

34. Iwai K, Maeda H, Konno T (1984) Use of oily contrast medium for selective drug targeting to tumor: enhanced therapeutic effect and X-ray image. Cancer Res 44:2115–2121
35. Konno T (1992) Targeting chemotherapy for hepatoma: arterial administration of anticancer drugs dissolved in lipiodol. Eur J Cancer 28:403–409
36. Maeda H (2001) The enhanced permeability and retention (EPR) effect in tumor vasculature: the key role of tumor-selective macromolecular drug targeting. Adv Enzyme Regul 41:189–207
37. Maeda H, Wu J, Okamoto T, Maruo K, Akaike T (1999) Kallikrein kinin in infection and cancer. Immunopharmacology 43:115–128
38. Wu J, Akaike T, Hayashida K, Miyamoto Y, Nakagawa T, Miyakawa K et al (2001) Identification of bradykinin receptors in clinical cancer specimens and murine tumor tissues. Int J Cancer 98:29–35
39. Maeda H, Fang J, Inutsuka T, Kitamoto Y (2003) Vascular permeability enhancement in solid tumor: various factors, mechanisms involved and its implications. Int Immunopharmacol 3:319
40. Iyer AK, Khaled G, Fang J, Maeda H (2006) Exploiting the enhanced permeability and retention effect for tumor targeting. Drug Discov Today 11:812–818
41. Fang J, Nakamura H, Maeda H (2011) The EPR effect: unique features of tumor blood vessels for drug delivery, factors involved, and limitations and augmentation of the effect. Adv Drug Deliv Rev 63:136–151
42. Matsumura Y, Kimura M, Yamamoto T, Maeda H (1988) Involvement of the Kinin-generating cascade in enhanced vascular permeability in tumor tissue. Jpn J Cancer Res 79:1327–1334
43. Matsumura Y, Maruo K, Kimura M, Yamamoto T, Konno T, Maeda H (1991) Kinin-generating cascade in advanced cancer patients and in vitro study. Jpn J Cancer Res 82:732–741
44. Akaike T, Horie H, Noguchi Y, Fujii S, Beppu T, Ogawa M et al (1998) Excessive production of nitric oxide in rat solid tumor and its implication in rapid tumor growth. Cancer 77:1598–1604
45. Doi K, Akaike T, Fujii S, Tanaka S, Ikebe N, Beppu T et al (1999) Induction of haem oxygenase-1 by nitric oxide and ischaemia in experimental solid tumours and implications for tumour growth. Br J Cancer 80:1945
46. Maeda H, Noguchi Y, Sato K, Akaike T (1994) Enhanced vascular permeability in solid tumor is mediated by nitric oxide and inhibited by both new nitric oxide scavenger and nitric oxide synthase inhibitor. Jpn J Cancer Res 85:331–334
47. Wu J, Akaike T, Hayashida K, Okamoto T, Okuyama A, Maeda H (2001) Enhanced vascular permeability in solid tumor involving peroxynitrite and matrix metalloproteinases. Jpn J Cancer Res 92:439–451
48. Tanaka S, Akaike T, Wu J, Fang J, Sawa T, Ogawa M et al (2003) Modulation of tumor-selective vascular blood flow and extravasation by the stable prostaglandin I2 analogue beraprost sodium. J Drug Target 11:45–52
49. Wu J, Akaike T, Maeda H (1998) Modulation of enhanced vascular permeability in tumors by a bradykinin antagonist, a cyclooxygenase inhibitor, and a nitric oxide scavenger. Cancer Res 58:159–165
50. Okamoto T, Akaike T, Nagano T, Miyajima S, Suga M, Ando M et al (1997) Activation of human neutrophil procollagenase by nitrogen dioxide and peroxynitrite: a novel mechanism for procollagenase activation involving nitric oxide. Arch Biochem Biophys 342:261–274
51. Leung DW, Cachianes G, Kuang WJ, Goeddel DV, Ferrara N (1989) Vascular endothelial growth factor is a secreted angiogenic mitogen. Science 246:1306–1309
52. Papapetropoulos A, Garcia-Cardena G, Madri JA, Sessa WC (1997) Nitric oxide production contributes to the angiogenic properties of vascular endothelial growth factor in human endothelial cells. J Clin Invest 100:3131
53. Murohara T, Horowitz JR, Silver M, Tsurumi Y, Chen D, Sullivan A et al (1998) Vascular endothelial growth factor/vascular permeability factor enhances vascular permeability via nitric oxide and prostacyclin. Circulation 97:99–107

54. Terman BI, Stoletov KV (2001) VEGF and tumor angiogenesis. Einstein Quart J Biol Med 18:59–66
55. Zhang H, Cai Z, Sun Y, Yu F, Chen Y, Sun B (2012) Folate-conjugated b-cyclodextrin from click chemistry strategy and for tumor-targeted drug delivery. J Biomed Mater Res A 2012:2441–2449
56. Li X, Tian X, Zhang J, Zhao X, Chen X, Jiang Y et al (2011) In vitro and in vivo evaluation of folate receptor-targeting amphiphilic copolymer-modified liposomes loaded with docetaxel. Int J Nanomedicine 6:1167
57. Gabizon A, Horowitz AT, Goren D, Tzemach D, Mandelbaum-Shavit F, Qazen MM et al (1999) Targeting folate receptor with folate linked to extremities of poly (ethylene glycol)-grafted liposomes: in vitro studies. Bioconjug Chem 10:289–298
58. Zhang Z, Yao J (2012) Preparation of irinotecan-loaded folate-targeted liposome for tumor targeting delivery and its antitumor activity. AAPS PharmSciTech 13:802–810
59. Li N, Larson T, Nguyen HH, Sokolov KV, Ellington AD (2010) Directed evolution of gold nanoparticle delivery to cells. Chem Commun 46:392–394
60. Yu C, Hu Y, Duan J, Yuan W, Wang C, Xu H et al (2011) Novel aptamer-nanoparticle bioconjugates enhances delivery of anticancer drug to MUC1-positive cancer cells in vitro. PLoS One 6:e24077
61. Kurosaki T, Higuchi N, Kawakami S, Higuchi Y, Nakamura T, Kitahara T et al (2012) Self-assemble gene delivery system for molecular targeting using nucleic acid aptamer. Gene 491:205–209
62. Cao Z, Tong R, Mishra A, Xu W, Wong GCL, Cheng J et al (2009) Reversible cell-specific drug delivery with aptamer-functionalized liposomes. Angew Chem Int Ed Engl 48:6494–6498
63. Managit C, Kawakami S, Yamashita F, Hashida M (2005) Effect of galactose density on asialoglycoprotein receptor-mediated uptake of galactosylated liposomes. J Pharm Sci 94:2266–2275
64. Sonoke S, Ueda T, Fujiwara K, Kuwabara K, Yano J (2011) Galactose-modified cationic liposomes as a liver-targeting delivery system for small interfering RNA. Biol Pharm Bull 34:1338–1342
65. Duan C, Gao J, Zhang D, Jia L, Liu Y, Zheng D et al (2012) Galactose-decorated pH-responsive nanogels for hepatoma-targeted delivery of oridonin. Biomacromolecules 12:4335–4343
66. Wijagkanalan W, Kawakami S, Takenaga M, Igarashi R, Yamashita F, Hashida M (2008) Efficient targeting to alveolar macrophages by intratracheal administration of mannosylated liposomes in rats. J Control Release 125:121–130
67. Pang Z, Gao H, Yu Y, Guo L, Chen J, Pan S et al (2011) Enhanced intracellular delivery and chemotherapy for glioma rats by transferrin-conjugated biodegradable polymersomes loaded with doxorubicin. Bioconjug Chem 22:1171–1180
68. Gao H-l, Pang Z-q, Fan L, Hu K-l, Wu B-x, Jiang X-g (2010) Effect of lactoferrin- and transferrin-conjugated polymersomes in brain targeting: in vitro and in vivo evaluations. Acta Pharmacol Sin 31:237–243
69. Li XM, Ding LY, Xu Y, Wang Y, Ping QN (2009) Targeted delivery of doxorubicin using stealth liposomes modified with transferrin. Int J Pharm 373:116–123
70. Zhao H, Wang JC, Sun QS, Luo CL, Zhang Q (2009) RGD-based strategies for improving antitumor activity of paclitaxel-loaded liposomes in nude mice xenografted with human ovarian cancer. J Drug Target 17:10–18
71. Yang T, Choi MK, Cui FD, Lee SJ, Chung SJ, Shim CK et al (2007) Antitumor effect of paclitaxel-loaded PEGylated immunoliposomes against human breast cancer cells. Pharm Res 24:2402–2411
72. Gupta B, Torchilin VP (2007) Monoclonal antibody 2C5-modified doxorubicin-loaded liposomes with significantly enhanced therapeutic activity against intracranial human brain U-87 MG tumor xenografts in nude mice. Cancer Immunol Immunother 56:1215–1223

73. ElBayoumi TA, Torchilin VP (2009) Tumor-targeted nanomedicines: enhanced antitumor efficacy in vivo of doxorubicin-loaded, long-circulating liposomes modified with cancer-specific monoclonal antibody. Clin Cancer Res 15:1973–1980
74. Wicki A, Rochlitz C, Orleth A, Ritschard R, Albrecht I, Herrmann R et al (2012) Targeting tumor-associated endothelial cells: anti-VEGFR2 immunoliposomes mediate tumor vessel disruption and inhibit tumor growth. Clin Cancer Res 18:454–464
75. Kirpotin DB, Drummond DC, Shao Y, Shalaby MR, Hong K, Nielsen UB et al (2006) Antibody targeting of long-circulating lipidic nanoparticles does not increase tumor localization but does increase internalization in animal models. Cancer Res 66:6732–6740
76. Hatakeyama H, Akita H, Ishida E, Hashimoto K, Kobayashi H, Aoki T et al (2007) Tumor targeting of doxorubicin by anti-MT1-MMP antibody-modified PEG liposomes. Int J Pharm 342:194–200
77. Ross JF, Chaudhuri PK, Ratnam M (1994) Differential regulation of folate receptor isoforms in normal and malignant tissues in vivo and in established cell lines. Physiologic and clinical implications. Cancer 73:2432–2443
78. Lu Y, Low PS (2002) Folate-mediated delivery of macromolecular anticancer therapeutic agents. Adv Drug Deliv Rev 54:675–693
79. Qiang X, Wu T, Fan J, Wang J, Song F, Sun S et al (2012) Preparation and folic acid conjugation of fluorescent polymer nanoparticles for cancer cell targeting. J Mater Chem 22:16078
80. Hilgenbrink AR, Low PS (2005) Folate receptor-mediated drug targeting: from therapeutics to diagnostics. J Pharm Sci 94:2135–2146
81. Lee RJ, Wang S, Low PS (1996) Measurement of endosome pH following folate receptor-mediated endocytosis. Biochim Biophys Acta 1312:237–242
82. Ghaghada KB, Saul J, Natarajan JV, Bellamkonda RV, Annapragada AV (2005) Folate targeting of drug carriers: a mathematical model. J Control Release 104:113–128
83. Yamada A, Taniguchi Y, Kawano K, Honda T, Hattori Y, Maitani Y (2008) Design of folate-linked liposomal doxorubicin to its antitumor effect in mice. Clin Cancer Res 14:8161–8168
84. Ni X, Castanares M, Mukherjee A, Lupold SE (2011) Nucleic acid aptamers: clinical applications and promising new horizons. Curr Med Chem 18:4206
85. Farokhzad OC, Cheng J, Teply BA, Sherifi I, Jon S, Kantoff PW et al (2006) Targeted nanoparticle-aptamer bioconjugates for cancer chemotherapy in vivo. Proc Natl Acad Sci U S A 103:6315–6320
86. Zhang H, Ma Y, Sun X-L (2010) Recent developments in carbohydrate-decorated targeted drug/gene delivery. Med Res Rev 30:270–289
87. Yu MK, Park J, Jon S (2012) Targeting strategies for multifunctional nanoparticles in cancer imaging and therapy. Theranostics 2:3–44
88. Lai CH, Lin CY, Wu HT, Chan HS, Chuang YJ, Chen CT et al (2010) Galactose encapsulated multifunctional nanoparticle for HepG2 cell internalization. Adv Funct Mater 20:3948–3958
89. Wijagkanalan W, Higuchi Y, Kawakami S, Teshima M, Sasaki H, Hashida M (2008) Enhanced anti-inflammation of inhaled dexamethasone palmitate using mannosylated liposomes in an endotoxin-induced lung inflammation model. Mol Pharmacol 74:1183–1192
90. Mitra M, Mandal AK, Chatterjee TK, Das N (2005) Targeting of mannosylated liposome incorporated benzyl derivative of Penicillium nigricans derived compound MT81 to reticulo-endothelial systems for the treatment of visceral leishmaniasis. J Drug Target 13:285–293
91. Rathore A, Jain A, Gulbake A, Shilpi S, Khare P, Jain SK (2011) Mannosylated liposomes bearing Amphotericin B for effective management of visceral Leishmaniasis. J Liposome Res 21:333–340
92. Kim B-S, Hong D-J, Bae J, Lee M (2005) Controlled self-assembly of carbohydrate conjugate rod coil amphiphiles for supramolecular multivalent ligands. J Am Chem Soc 127:16333–16337
93. Lee JS, Groothuis T, Cusan C, Mink D, Feijen J (2011) Lysosomally cleavable peptide-containing polymersomes modified with anti-EGFR antibody for systemic cancer chemotherapy. Biomaterials 32:9144–9153

94. Marega R, Karmani L, Flamant L, Nageswaran PG, Valembois V, Masereel B et al (2012) Antibody-functionalized polymer-coated gold nanoparticles targeting cancer cells: an in vitro and in vivo study. J Mater Chem 22:21305–21312
95. Hsieh WJ, Liang CJ, Chieh JJ, Wang SH, Lai IR, Chen JH et al (2012) In vivo tumor targeting and imaging with anti-vascular endothelial growth factor antibody-conjugated dextran-coated iron oxide nanoparticles. Int J Nanomedicine 7:2833–2842
96. ElBayoumi TA, Torchilin VP (2008) Tumor-specific anti-nucleosome antibody improves therapeutic efficacy of doxorubicin-loaded long-circulating liposomes against primary and metastatic tumor in mice. Mol Pharm 6:246–254
97. Pang Z, Lu W, Gao H, Hu K, Chen J, Zhang C et al (2008) Preparation and brain delivery property of biodegradable polymersomes conjugated with OX26. J Control Release 128:120–127
98. Lin JJ, Ghoroghchian PP, Zhang Y, Hammer DA (2006) Adhesion of antibody-functionalized polymersomes. Langmuir 22:3975–3979
99. Pissuwan D, Valenzuela SM, Miller CM, Cortie MB (2007) A golden bullet? Selective targeting of Toxoplasma gondii Tachyzoites using antibody-functionalized gold nanorods. Nano Lett 7:3808–3812
100. Qian X, Peng X-H, Ansari DO, Yin-Goen Q, Chen GZ, Shin DM et al (2008) In vivo tumor targeting and spectroscopic detection with surface-enhanced Raman nanoparticle tags. Nat Biotechnol 26:83–90
101. Surovtseva EV, Johnston AH, Zhang W, Zhang Y, Kim A, Murakoshi M et al (2012) Prestin binding peptides as ligands for targeted polymersome mediated drug delivery to outer hair cells in the inner ear. Int J Pharm 424:121–127
102. Demirgoz D, Pangburn TO, Davis KP, Lee S, Bates FS, Kokkoli E (2009) PR_b-targeted delivery of tumor necrosis factor-a by polymersomes for the treatment of prostate cancer. Soft Matter 5:2011–2019
103. Pangburn TO, Georgiou K, Bates FS, Kokkoli E (2012) Targeted polymersome delivery of siRNA induces cell death of breast cancer cells dependent upon Orai3 protein expression. Langmuir 28:12816–12830
104. Stojanov K, Georgieva JV, Brinkhuis RP, van Hest JC, Rutjes FP, Dierckx RAJO et al (2012) In vivo biodistribution of prion-and gm1-targeted polymersomes following intravenous administration in mice. Mol Pharm 9:1620–1627
105. Christian NA, Milone MC, Ranka SS, Li G, Frail PR, Davis KP et al (2007) Tat-functionalized near-infrared emissive polymersomes for dendritic cell labeling. Bioconjug Chem 18:31–40
106. Bin J, Maeda J, Higuchi M, Inoue K, Akita H, Harashima H et al (2006) Pharmacokinetics and brain uptake of lactoferrin in rats. Life Sci 78:851–855
107. Zhang X, Koh CG, Yu B, Liu S, Piao L, Marcucci G et al (2009) Transferrin receptor targeted lipopolyplexes for delivery of antisense oligonucleotide G3139 in a murine K562 xenograft model. Pharm Res 26:1516–1524
108. Koshkaryev A, Piroyan A, Torchilin VP (2012) Increased apoptosis in cancer cells in vitro and in vivo by ceramides in transferrin-modified liposomes. Cancer Biol Ther 13:50–60
109. Hambley TW (2009) Is anticancer drug development heading in the right direction? Cancer Res 69:1259–1262
110. Hobbs SK, Monsky WL, Yuan F, Roberts WG, Griffith L, Torchilin VP et al (1998) Regulation of transport pathways in tumor vessels: role of tumor type and microenvironment. Proc Natl Acad Sci U S A 95:4607–4612
111. Bae YH (2009) Drug targeting and tumor heterogeneity. J Control Release 133:2–3
112. Jain RK (1999) Transport of molecules, particles, and cells in solid tumors. Annu Rev Biomed Eng 1:241–263
113. Heldin CH, Rubin K, Pietras K, Ostman A (2004) High interstitial fluid pressure: an obstacle in cancer therapy. Nat Rev Cancer 4:806–813
114. Ruoslahti E, Bhatia SN, Sailor MJ (2010) Targeting of drugs and nanoparticles to tumors. J Cell Biol 188:759–768

115. Jain RK (1987) Transport of molecules in the tumor interstitium: a review. Cancer Res 47:3039–3051
116. DiResta GR, Lee J, Arbit E (1996) Process and device to reduce interstitial fluid pressure in tissue. US 5484,399 A
117. DiResta GR, Healey JH (2003) Apparatus and method for reducing interstitial fluid pressure and enhancing delivery of a therapeutic agent. US 6547,777 B2
118. Mishra S, Webster P, Davis ME (2004) PEGylation significantly affects cellular uptake and intracellular trafficking of non-viral gene delivery particles. Eur J Cell Biol 83:97–111
119. Romberg B, Hennink WE, Storm G (2008) Sheddable coatings for long-circulating nanoparticles. Pharm Res 25:55–71
120. Gullotti E, Yeo Y (2009) Extracellularly activated nanocarriers: a new paradigm of tumor targeted drug delivery. Mol Pharm 6:1041–1051

Chapter 4
Liposomal Formulations for Focal and Targeted Drug Delivery in Cancer and Other Diseases

Sean Essex and Vladimir Torchilin

4.1 FDD for Cancer Therapy

Before the concept of FDD is explored further, it is important to introduce a familiar term *"targeted drug delivery."* In the context of cancer therapy, targeted drug delivery means *"delivering therapeutic concentrations of anticancer drugs at the tumor site and to reduce/minimize deleterious effects on normal tissue"* [1]. Though there is an exhaustive list of strategies that could be employed to accomplish targeted drug delivery [1], keeping the scope of this chapter in mind, tumor targeting using systemically administered nanocarriers can be broadly classified as *passive* or *active targeting* [2, 3]. Passive targeting exploits the unique tumor microenvironment characteristics like leaky vasculature coupled with high interstitial fluid pressure and poor lymphatic drainage for accumulation of the nanocarriers at the tumor site. This is referred to as the *"enhanced permeability and retention effect"* (EPR effect) [4–6]. Active targeting, on the other hand, employs tumor-targeting ligands, such as monoclonal antibodies (mAbs) and antibody fragments or non-antibody ligands (peptides or non-peptides), grafted onto the surface of the nanocarriers. These targeting moieties, in turn, selectively bind to the receptors overexpressed on the surface of the tumor cells and/or angiogenic endothelial cells [6, 7]. Drug release from the nanocarrier at the tumor site then in turn depends on the complex interplay of tumor microenvironmental factors and the components of the drug carrier [8, 9]. The prospect of *fine tuning* this drug release to regulate drug release kinetics [10], by disrupting the nanocarriers mechanically at the tumor site, using an external energy source such as *ultrasound (US)*, is what marks the subtle difference between targeted and focal drug delivery. The concept of FDD for cancer therapy is schematically represented below (Fig. 4.1).

S. Essex • V. Torchilin (✉)
Department of Pharmaceutical Sciences, Center for Pharmaceutical Biotechnology and Nanomedicine, Northeastern University, 140 The Fenway, Room 211/214, Boston, MA 02115, USA
e-mail: v.torchilin@neu.edu

A.J. Domb and W. Khan (eds.), *Focal Controlled Drug Delivery*, Advances in Delivery Science and Technology, DOI 10.1007/978-1-4614-9434-8_4,

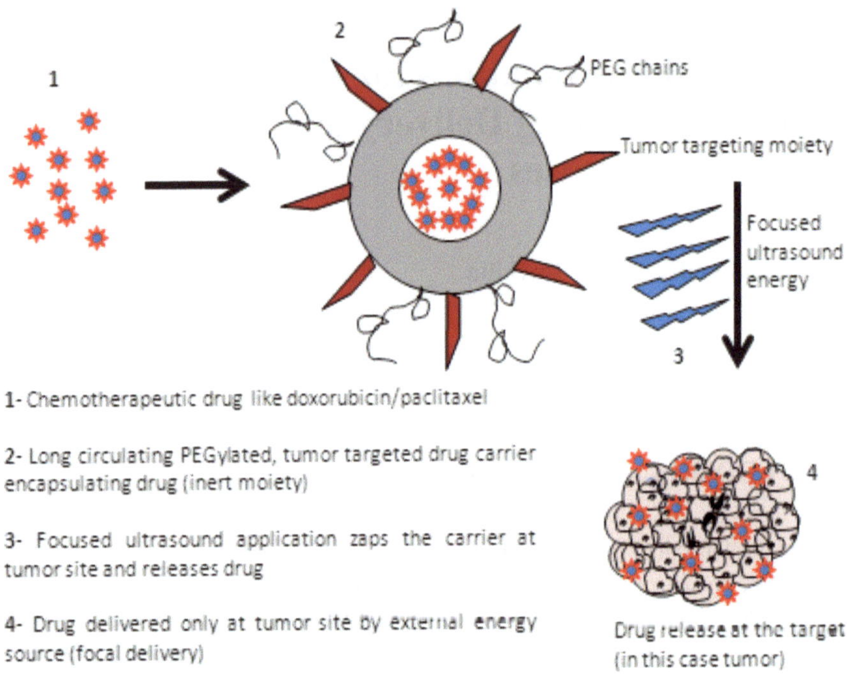

Fig. 4.1 The concept of focal drug delivery in cancer therapy

From the available literature, the two principal components that FDD employs are:

1. **Drug carrier**—This chapter will focus on "**liposomes**" as drug delivery agents.
2. **Focused ultrasound** [low (LFU) or high energy (HIFU)]—Focused ultrasound can be visualized akin to the phenomenon that occurs when sunlight is focused using a magnifying glass. The focused energy can obliterate matter and this is basically the same principle that FU employs. Briefly, HIFU has been used to obliterate tissue and it basically does it via two main physical processes: *conversion of mechanical energy into heat focally, where rate of heating >> rate of cooling,* and *acoustic cavitation*, which is the combination of mechanical stress and thermal injury. The combination of the above factors causes a rapid temperature elevation, up to 80 °C within seconds and will thus kill cells by coagulative necrosis. In the interest of FDD, HIFU can also be used to disrupt lipid (liposome)/biological barriers (cell membrane) and this could be mainly attributed to the acoustic cavitation effect, in addition to the radiation and the thermal effects (when applicable) of US [11, 12]. This can aid site-specific drug delivery, as we will see in the due course of this chapter. LFU on the other hand is less destructive and can disrupt biological/lipid membranes without causing the lesioning effects of HIFU [13, 14]. LFU may thus prove to be a desirable modality for burst release of payload for several minutes, focused to small spatial locations via transient holes formed in the cell membrane in the order of about

100 nm [15], and provide a greater degree of spatial and temporal control over biochemical and passive tumor targeting [14]. However, it should be noted that LFU is difficult to focus and often dissipates near the body surface, thus suggesting that it would only be beneficial to treat superficial tumors (skin, head, neck, and gynecological); for deeper tumors, HIFU and HIFU-responsive delivery systems may be preferred [16].

Several recent studies also employ *magnetic resonance imaging (MRI or MR)* as the imaging modality for FDD. In such studies, the MRI contrast agent like gadolinium (Gd) is encapsulated within the drug carrier. The MR imaging serves to pinpoint the area of interest, viz., the tumor area in cancer therapy that needs to be destroyed. The target region is identified by a combination of sublesioning FU exposures and monitoring local temperature rise using MR. Upon reliably identifying the focus region on the diagnostic image, higher therapeutic exposures of FU are used. MR lends the virtues of real-time imaging and temperature monitoring and thus lends more control to the procedure [17].

4.2 Liposomes

Liposomes are microscopic phospholipid bubbles with a bilayered membrane structure [18]. Over the period of several decades, appreciable advances have been achieved in the field of *liposomal drug delivery for cancer therapy*. These can be classified in several ways. One of those classifications is depicted as shown in the tree diagram below (adapted from Torchilin [19]) (Fig. 4.2).

4.2.1 Advancements in Liposomal Drug Delivery

Long-Circulating Liposomes

The rationale for developing long-circulating liposomes for tumor targeting was to exploit the EPR effect [4, 20, 21]. Briefly put, it is the ability of macromolecules and other nanoparticles to accumulate in the tumor area due to its leaky vasculature and poor lymphatic drainage. The longer the liposomes are in circulation, the greater the chances of it passing by and accumulating in the target area. The most popular method of doing this is by coating the liposomal surface with polyethylene glycol (PEG), a biocompatible, inert polymer. It has been proposed that the PEG coat prevents interaction of the liposomes with opsonins in the blood and thus prevents elimination by the mononuclear phagocyte system (MPS) [18, 22]. Doxil® is an FDA-approved PEGylated, doxorubicin-encapsulating liposomal delivery system. In an attempt to minimize the interference of PEG with cellular uptake, stimuli-detachable (low tumor pH, redox reactions) PEG conjugates have been developed [23–25]. However, recent evidence suggests against the complete inertness of PEG

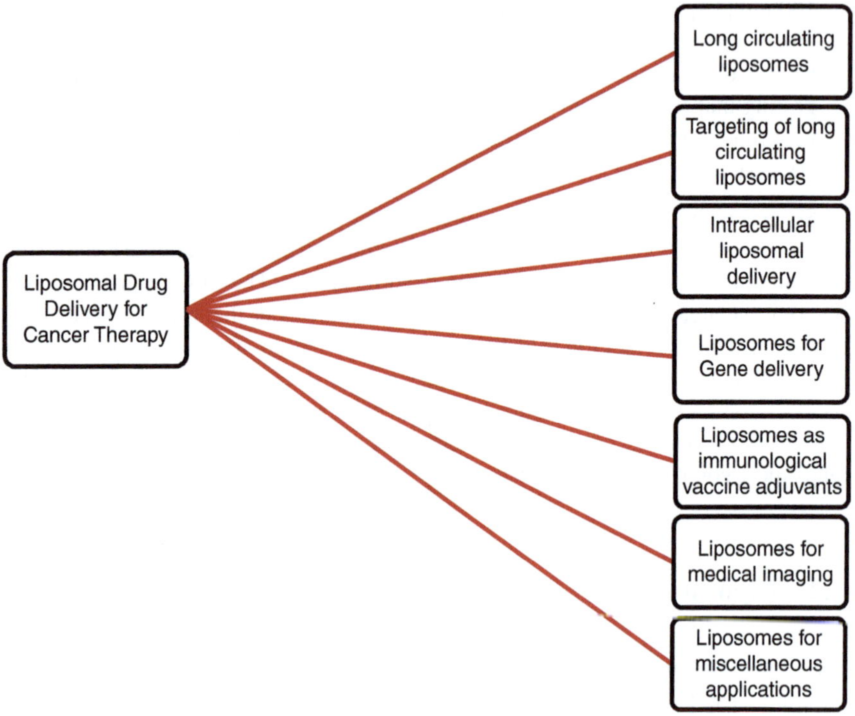

Fig. 4.2 Tree diagram depicting the advancements in liposomal drug delivery

by providing proof that it could lead to certain undesired effects such as the activation of the complement system [26, 27].

Targeting of Long-Circulating Liposomes

The most important rationale for attempts to target long-circulating, drug-loaded liposomes is to improve the tumor-targeting specificity of the formulation, thereby increasing the drug concentration at the desired site of action, and reduce the systemic levels of the drug and its toxic sequelae in healthy tissues [28]. Liposomal targeting is an exhaustive topic, and it is impossible to cover all aspects associated with it here due to the scope of this chapter. However, a table with the popular targeting moieties used in conjunction with long-circulating liposomes is listed below for perusal along with the reference manuscripts that describe the same in greater detail (Table 4.1).

Intracellular Liposomal Delivery

Liposomes have been developed with an aim of bringing the drug/payload as close as possible to its target, e.g., cytoplasm, nucleus, etc. and two popular designs have been used. *pH-sensitive liposomes* can be formulated by incorporating pH-sensitive

Table 4.1 Some examples of targeting moieties for liposomal preparations

Targeting moiety	Rationale for use in tumor targeting	References
Anti-HER2 monoclonal antibody (mAb)	HER2 receptor is overexpressed esp. in human breast cancer cells	[29]
Anti-nucleosome 2C5 mAb	Tumor cell recognition by nucleosomes bound to tumor cell surface	[30–32]
Anti-GD2 antibody	GD2 overexpressed in human melanoma and neuroblastoma	[33]
Fab fragment from humanized anti-EGFR mAb, anti-epidermal growth factor (EGFR) monoclonal antibody	EGFR is overexpressed on the surface of several tumors, e.g., colorectal tumor cells	[34, 35]
Anti-transferrin antibody, transferrin ligand	Transferrin receptors are overexpressed on the surface of many tumor cells	[36–40]
Folate	Folate receptor is frequently overexpressed in tumor cells	[41–44]
RGD peptide	Can bind to the integrins of the tumor vasculature	[45]
Phage coat fusion proteins	Specific against target cells, cheap prospective substitute for expensive and unstable mAbs	[46]
Fibroblast growth factor-binding peptide	Selective for tumors overexpressing fibroblast growth factor receptor (FGFR)	[47]
OX26 monoclonal antibody	Binds to transferrin receptor overexpressed on surface of tumor cells	[48]
Targeting antibody + endosome disrupting peptide	Improves cytosolic delivery and cytotoxicity of drug	[49]
Anti-P selecting antibody	Selective for areas of vascular inflammation and application can be extrapolated to deliver pro-angiogenic drugs at that site	[50]
Ascorbate	Antitumor activity improves cytotoxicity of other chemotherapeutics when used in combination	[51]

components in long-circulating liposomes. The liposome will interact with the endosomal membrane after being endocytosed and, under the effect of low pH in the endosomes, release their drug payload into the cytoplasm [52]. Antisense oligonucleotides and siRNA have also been delivered inside the cell cytoplasm using pH-sensitive liposomes [53, 54]. pH-sensitive liposomes in combination with targeting moieties like folate and transferrin have also been developed [55, 56]. The other approach is to employ *liposomes containing cell-penetrating peptides/ proteins (CPPs)* such as TAT peptide, penetratin, and oligoarginine. CPPs can be grafted onto the liposomal surface and have been shown to effectively deliver liposomes inside cells [57–59]. In addition to these methods, liposomes grafted with mitochondriotropic amphiphilic cations in their membrane have been shown to specifically target mitochondria within intact cells [60–62].

Liposomes for Gene Delivery

The inherent in vivo instability and susceptibility to degradation of molecules like DNA siRNA, etc. is a problem in delivery and has attracted attention in recent times.

Liposomes have been used to deliver DNA, siRNA, and oligonucleotides inside cells. PEGylated cationic liposomes have been used to condense and efficiently deliver DNA for cancer therapy, both in vitro and in vivo [63–65]. Liposomes have been used for oligo delivery against neuroblastoma and consist of a cationic core with the oligo coated with neutral lipid and grafted with monoclonal antibody [66]. A combination of a proapoptotic peptide and Bcl-2 antisense oligo $(KLAKLAK)_2$ in the same liposomal preparation has displayed synergistic effects in mice for cancer therapy [67]. Various liposomes (cationic, neutral, and anionic) have recently been used to deliver siRNA into cells for cancer therapy, and this seems to be a promising venture. Recently, novel SNALP liposomes have achieved great success in siRNA delivery [68–72].

In addition to the above, liposomes have also been used as *immunological vaccine adjuvants* [73], in *medical imaging* [74–76], *antibody-directed enzyme prodrug therapy* (ADEPT) [77, 78], and *photodynamic therapy* (PDT) [79–81].

4.2.2 Why Are Liposomes an Attractive FDD Tool? (Adapted from Allen et al. [82])

One of the important considerations for focal delivery is that the carrier should not leak out the drug en route the target (tumor). Regulated drug release from liposomes can overcome/eliminate this and in turn prevent accidental off-target tissue damage due to drug extravasation. As discussed before, LFU and HIFU (components of focal delivery) can disrupt the lipid membranes (components of liposome) to facilitate release of the payload at the tumor site. Liposomes can be designed to be thermally sensitive for exploiting the thermal/indirect effects of HIFU for focal delivery. Similarly, liposomes can be designed to be responsive to LFU (nonthermal/mechanical/direct) for focal delivery (to be described later in the chapter). Liposomes can passively (via the EPR effect) or actively (using targeting ligands) accumulate at the tumor site which makes application of focused ultrasound easier by providing a narrow foci for the treatment. Liposomes provide hydrophilic as well as hydrophobic environments allowing for the encapsulation of water-soluble and water-insoluble drugs, preventing premature degradation and rapid kidney elimination of the small drug molecules and thus effectively altering drug PK and lowering the drug dose. Most importantly, several studies elucidated below have successfully shown the benefits of using focused ultrasound with liposomes for tumor targeting.

4.2.3 Recent Advancements in LFDD

Headways are now being made to deliver chemotherapeutic drugs encapsulated in liposomes at the tumor site using US. A few examples of recent published literature on this issue are being discussed below.

4.2.3.1 HIFU-Mediated Liposomal FDD (Thermal Ultrasonic Effects Associated with Liposomal Drug Release)

The use of thermosensitive liposomes in conjunction with focused ultrasound for the delivery of anticancer drugs at the tumor site is surprisingly *not* a new concept. There are studies that go as far back as 30 years ago that successfully report this [83]. The successful delivery of methotrexate encapsulated in phase transition liposomes to bladder cancer cells in combination with hyperthermia has been shown almost 2 decades ago [83]. A study comparing the benefits of the combination treatment of the different *types* of liposomes and hyperthermia for antitumor activity has been reported, and it was shown that low temperature-sensitive liposomes (LTSL, trigger temperature ~39–40 °C) encapsulating doxorubicin, when combined with hyperthermia, exhibited best antitumor activity, over non-thermosensitive (no trigger at physiological temperature range) and thermosensitive liposomes (trigger temperature ~42–45 °C) [84]. The feasibility of using temperature-sensitive liposomes (TSLs) in conjunction with MR-HIFU for drug delivery was also shown [85]. The combination of thermosensitive liposomes and ultrasound phase arrays for thermodynamic therapy (TDT) has potential for tumor-specific drug release [86]. Using a combination of Doxil and MR-guided HIFU, the delivery of therapeutic levels of doxorubicin to the rat brain has been reported. A good correlation between the MRI enhancement and the doxorubicin delivery to the brain parenchyma was seen, and thus this study successfully shows the development of a noninvasive method for the targeted delivery of doxorubicin to the brain at therapeutic levels using an MR-HIFU platform [87]. A combination of LTSLs with noninvasive, nondestructive pulsed HIFU exposures has demonstrated enhanced delivery of doxorubicin and thus its cytotoxic effects in tumors. In mice bearing adenocarcinoma tumors, liposomes encapsulating doxorubicin followed by pulsed HIFU, 0 h and 24 h after injection, showed better tumor growth inhibition than free doxorubicin [88].

Frenkel has published a review where he has touched upon aforementioned studies describing HIFU-liposomal combination strategy for drug delivery to solid tumors [12]. In a proof-of-concept study, a combination of TSLs co-encapsulating doxorubicin and an MRI contrast agent ProHance® [Gd(HPDO3A)(H2O)] was employed to deliver the drug at the tumor site under the influence of HIFU. The use of TSLs in combination with HIFU-mediated hyperthermia led to higher uptake of the drug and the MR agent in HIFU-treated tumors as compared to control. Also, there was a linear correlation between the drug concentration, T_1 relaxivity (R1), and Gd concentration in the tumor [89]. A similar study reports an image-guided drug delivery using a combination of LTSLs (ThermoDox®, Celsion corp. USA; currently in clinical phase III trials) and MR-HIFU in a Vx2 rabbit tumor model. It was shown that it was feasible to combine a clinical grade LTSL and MR-HIFU in a relevant rabbit Vx2 tumor model. Stable and spatially accurate mild hyperthermia with MR-HIFU-applied post-LTSL administration significantly improved delivery of doxorubicin to the tumor tissue and spared the adjacent normal tissue [90]. A similar study using combination of LTSL and MR-guided HIFU was published by Negussie et al. [91] (Fig. 4.3).

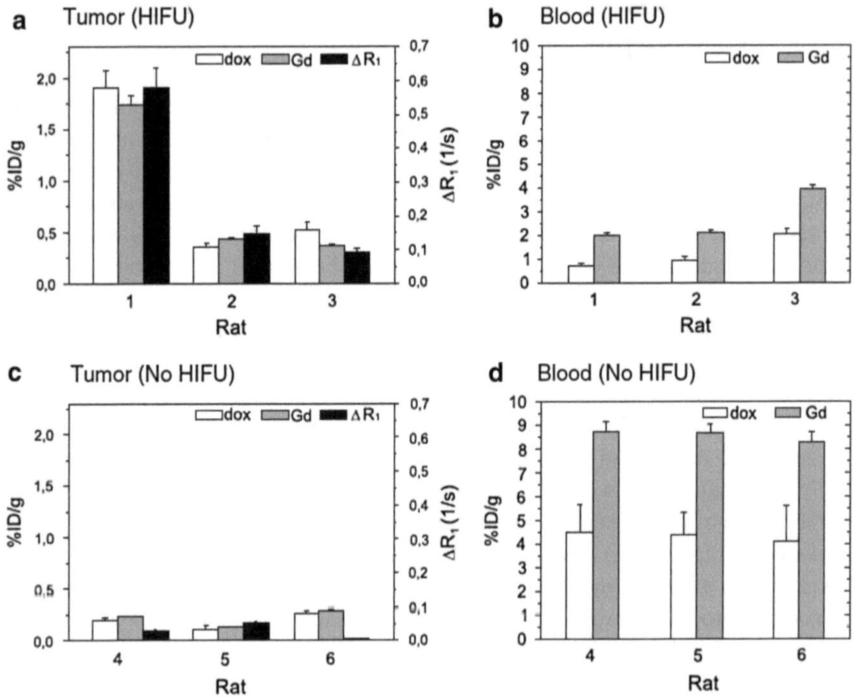

Fig. 4.3 Doxorubicin and gadolinium concentrations (expressed in % injected dose/g) and ΔR1 (1/s) in the tumor of HIFU-treated (**a**) and control (**c**) rats 7 h after time of injection. Doxorubicin and gadolinium concentrations (expressed in % injected dose/g) in blood of HIFU-treated (**b**) and control (**d**) rats. The error *bars* show the standard deviation of the doxorubicin extraction ($n=3$ per sample), ICP-MS analysis (±5 % error), and the standard deviation of the ΔR1 within the region of interest (reproduced with permission)

It is clear from the above studies that the combination of TSLs used in conjunction with HIFU-mediated hyperthermia is a viable platform for focal drug delivery to the tumor. The MR added to this platform lends the virtues of real-time drug visualization and monitoring. These features will not only preclude any off-target effects but also potentially decrease the amount of the therapeutic drug needed for a similar antitumor effect. This augurs well for a translation into a clinical setting.

It should be noted that all the above studies employ HIFU for site-specific drug delivery by exploiting its thermal effects. However, high-frequency ultrasound or HFUS (wave frequency is in the order of MHz) has also been employed with acoustically active/echogenic liposomes (AAL)/(ELIP) to release their content at the desired site. Generally, these studies exploit the acoustic reception of specifically designed liposomes towards ultrasound for releasing their contents and not the thermosensitivity to ultrasound as the previous studies described. It has been proposed that the *rarefaction* phase of the sound wave is responsible for causing stress that overcomes the elastic limit of the weakest surface in the liposome and causes the content release. The rate of release then in turn depends upon the intensity of the ultrasound and the resealing of the liposome. An echogenic liposomal system was developed with calcein as the model drug entrapped in the presence of moderate concentration of mannitol during the initial freezing of the formulation. It was shown that the bulk of the liposomes thus, formulated, entrapped air and calcein in the same formulation and released it in response to HFUS. Also, the entrapment of the drug had no deleterious effect on the echogenecity of the liposome [92]. The interesting thing to note here is that since these liposomes are echogenic, they could, in theory, be visualized using US and thus this system has potential for real-time drug delivery monitoring. Recently, doxorubicin-containing liposomes were loaded onto lipidic microbubbles encapsulating air via biotin-avidin bonds to form Dox-liposome-loaded microbubbles. In the presence of HFUS, this system exhibited greater melanoma cell killing in vitro as compared to plain doxorubicin liposomes. It was also shown that the greater cytotoxic effect seen with this system was due to the greater drug release from the formulation and increased cellular uptake of the drug in the presence of US [93]. Klibanov et al. have reported a similar drug-liposome-loaded microbubble system recently [94]. At this point an important point about the safety considerations of using HIFU or HFUS should be mentioned. It is known that focused ultrasound is safe as long as it doesn't exceed the 720 mW/cm^2 safety threshold. Recent studies show that HIFU causes acoustic cavitation (mentioned earlier in this chapter), i.e., an asymmetric violent implosion of microbubbles (gas-filled bubbles) associated with an aqueous environment, and can cause rupture of phospholipid structures in close proximity [14]. The effect of this on normal mammalian cells needs to be investigated to completely rule out potential clinical issues associated with this platform especially when liposome-microbubble conjugate systems are used for drug delivery.

4.2.3.2 LFU-Mediated Liposomal Focal Drug Delivery Systems (Nonthermal Ultrasonic Effects Associated with Liposomal Drug Release)

LFU is one of the most promising artificial triggers that could be employed for drug delivery [14], and a nice review has been published that describes this in detail [10]. As mentioned before in the introductory part of this chapter, LFU can achieve the needed disruption of the biological/lipid membranes (prerequisite for focal delivery), without causing the lesioning effects of HIFU, and it has been successfully shown to exhibit better spatial and temporal drug delivery control over passive/biochemical targeting. It has been demonstrated that LFU-mediated *microstreaming* leads to formation of transient pore-like disruptions in the plasma membrane (mainly composed of phospholipids; an essential component of liposomes) and no cavitation effects are seen [14, 95]. There is currently limited literature reporting the use of LFU in combination with liposomes for FDD in cancer therapy. This is in stark contrast to HIFU-mediated FDD, and this is surprising since initial developments in both were made almost in parallel [96]. Hence, via means of this book chapter, a conscious effort has been made to explain how some prospective systems could be employed in the future using LFU-mediated LFDD. Before moving onto the *"prospective"* systems, a brief account of a proof-of-concept study published by Ibsen this year has been briefly described [14] to reiterate the fact that LFU can indeed be used in conjunction with liposomes for FDD.

Recently, development of a hybrid liposome-microbubble drug delivery vehicle has been reported that demonstrated the potential to incorporate and deliver hydrophilic and hydrophobic therapeutics for drug delivery, in response to low-intensity pulsed ultrasound [97]. Building on this, Ibsen et al. in 2012 formulated SHERPAs or (shockwave-ruptured nanopayload carriers) encapsulated in a liposome. They showed that the SHERPA performed dual roles of delivering a drug to the tumor vicinity, and since it is associated with microbubbles, the application of LFU could rupture the system locally due to microstreaming and the drug then burst diffused along the concentration gradient into the tumor. They also proposed that drug uptake by the tumor tissue could be increased due to the added effect of sonoporation [14]. This approach seems promising for in vivo studies. To the best of our knowledge, this is currently the only study published with LFU for LFDD.

Moving on, a brief account is being presented below of the *current systems* employing a combination of *liposomal delivery* and *low-frequency ultrasound or LFUS*. PEG lipids and detergents based on ethylene oxide polymers/oligomers grafted onto the liposomal surface increase the susceptibility of liposomes (unilamellar ca. 100 nm) to LFUS (20 kHz frequency) as compared to non-PEG liposomes of the same composition. Small <10 nm particles coexist with the structurally and chemically unaltered PEGylated liposomes, whereas this is not observed in non-PEGylated liposomal dispersions. This suggests that it is possible that under ultrasonic stress, the PEGylated liposomes undergo transient structural deformities to force a fraction of lipids out, forming smaller assemblies. The responsiveness of the PEG liposomes to LFUS correlated well with the maximum amount of PEG that could be incorporated [98]. In vitro studies with PEGylated phospholipid vesicles showed that liposomal leakage was higher with exposure to 20 kHz (LFUS) as compared to 1 or 1.6 MHz [99]. The same group also proved a similar effect in the release profile of doxorubicin from Doxil®. It was also found that the release of the liposomal payload could be enhanced by increasing the PEG concentration to 5 mol% [100]. Schroeder et al. tested the mechanism and feasibility of using LFUS in combination with sterically stabilized liposomes (SSLs) for targeted delivery. They showed that the ultrasonic amplitude influenced drug release from liposomes, and this in turn was due to transient cavitation (extra- and intra-liposomal media) that occurs above the energetic threshold near the liposomal membrane, causing rupture. Pulsed/continuous mode of LFUS could be used to release similar amounts of drugs from the liposomes, and drug release is via "dumping" through transient pores formed in the liposomal membrane during the time of LFUS application. LFUS does not alter the chemical integrity of the encapsulated drugs which was confirmed via HPLC and a cytotoxicity assay in murine adenocarcinoma C26 cell lines [101]. It was also suggested that the enhanced drug release in PEG-DSPE:phospholipid liposomes (1:10 mol ratio) was probably due to high ultrasonic energy absorption by the PEG moieties and focusing it near the liposomal membrane. Also, the contribution of the PEG part

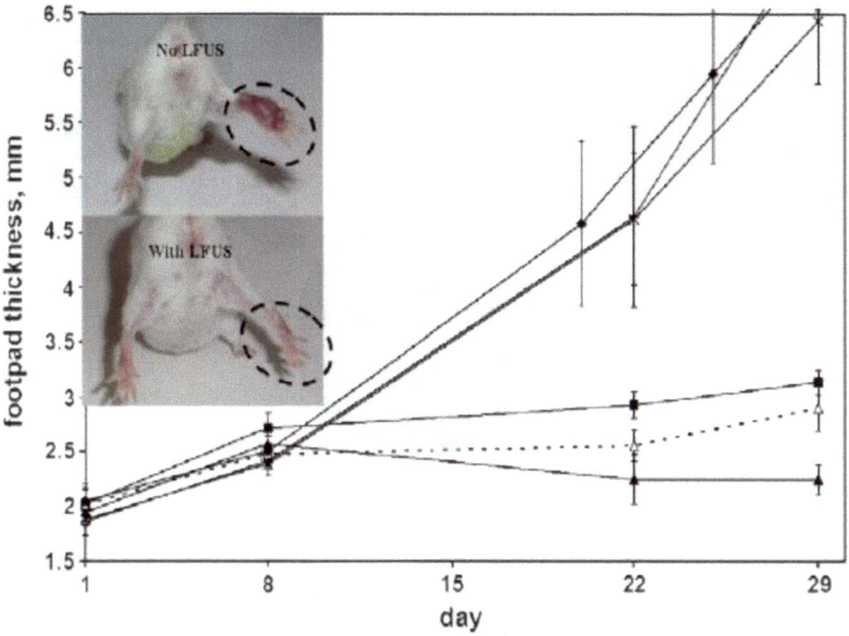

Fig. 4.4 C26 tumors in the footpad of BALB/c mice were exposed to different treatments: (1) control, no drug, no LFUS (*upper image* in graph) (*filled diamond*); (2) control, saline (placebo) plus LFUS (5.9 W/cm2, 60 s, *broken lines*); (3) nSSL cisplatin injected intravenously without LFUS (*cross*); (4) nSSL cisplatin injected intravenously plus LFUS (*upper image* in graph) (*filled triangle*); (5) free (non-liposomal) cisplatin plus LFUS (*open triangle in between lines*); (6) free cisplatin without LFUS (*filled square*). *Insert*—footpad of mice treated with nSSL cisplatin i.v., without (*top*) or with (*bottom*) LFUS. At day 29, using Student's *t*-test, a statistically significant difference ($p<0.05$) was demonstrated between animals in groups (4) and (5) and ($p<0.006$) between group (4) and all other groups. Data points indicate mean footpad thickness of six mice, in two experiments, ±SD (reproduced with permission)

closest to the lipid bilayer is important for the above effect and this requires further investigation [10]. Nano-sterically stabilized liposome (nSSL) encapsulating cisplatin has been developed and in conjunction with LFUS has proven to be a useful modality to deliver drugs like cisplatin which have low in vivo release rates, at the tumor site. This platform may, however, not add any advantage to deliver drugs like doxorubicin which exhibit good in vivo release rates when encapsulated in liposomes [16]. It is important to note that in the systems described above, real-time imaging is not possible as compared to the MR-HIFU platform and this is a potential clinical limitation (Fig. 4.4).

To summarize, the findings from the above literature and a study recently published [10] can be combined to define the parameters that influence US-mediated drug release from liposomes. These parameters should be optimized through further investigative work to develop potential LFDD systems (Fig. 4.5).

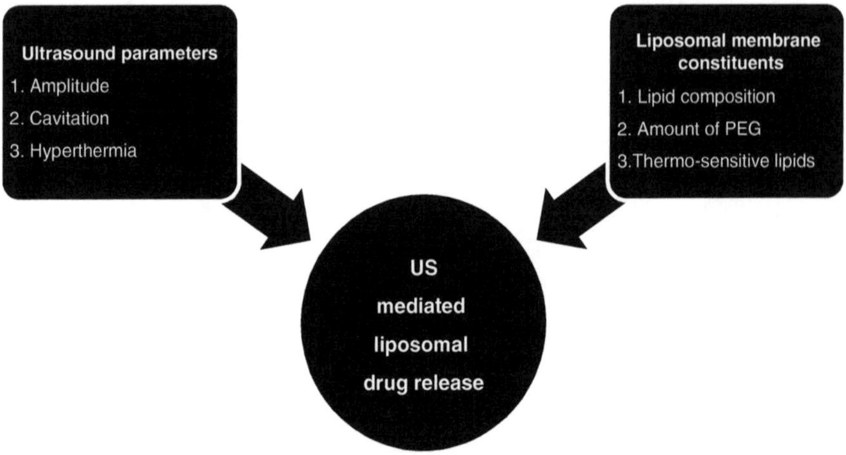

Fig. 4.5 Parameters influencing US-mediated liposomal drug delivery

4.2.4 Prospective US-Mediated Liposomal FDD

In the studies described up to this point, only temperature-sensitive PEGylated or non-PEGylated and acoustically active ELIP have been employed. There is a plethora of untapped potential for US-mediated focal delivery using other types of liposomes. It is reasonable to claim that using the other liposomes with US will share all the same aspects described by the studies above, and thus the idea of employing US in combination with say magnetic or light-sensitive liposomes for focal tumor delivery seems a promising extrapolation of the modality used above. Whether these liposomes can be employed with HFUS or LFUS or in combination with microbubbles for enhancement of ultrasound responsiveness and delivery remains to be investigated in the future. A flowchart elucidating the concept is shown below (Fig. 4.6).

1. Actively targeted liposomes
2. pH-sensitive liposomes
3. Redox potential-sensitive liposomes
4. Magnetic liposomes
5. Light-responsive liposomes

\+ US

Actively Targeted Liposomes + LFU

Long-circulating liposomes that have been grafted with tumor-targeting moieties (as described in Table 4.1 in this chapter before) will accumulate at the tumor site,

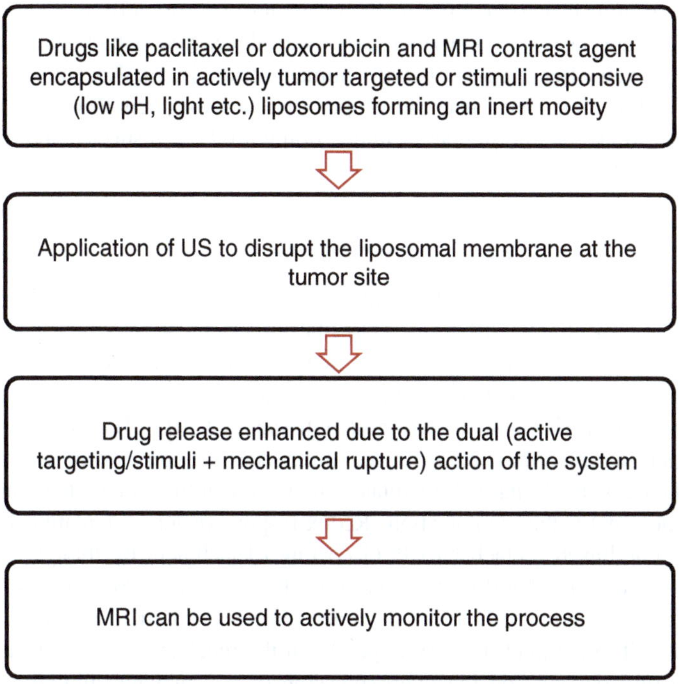

Fig. 4.6 A flowchart scheme explaining the concept of MRI-guided LFU-mediated liposomal focal drug delivery

and subsequent application of US will disrupt the liposomal membrane and release the drug at the tumor site.

pH-Sensitive Liposomes + US

These liposomes are designed to take advantage of the acidic tumor microenvironment for accumulation and the release of its contents in response to the low pH. The proposed combination of the low pH-stimuli responsive liposomes with US seems to have a sound rationale due to the dual benefits of the tumor-specific drug release, enhanced by both the inherent low pH-responsive characteristics of the formulation and mechanical intervention by US. Though described briefly before in this chapter, it is imperative to discuss how pH-sensitive liposomes could be used in conjunction with US for focal delivery. One of the most well-studied examples of developing a low pH-sensitive liposome is by using fusogenic lipids like DOPE in the liposomal composition [102]. Long- circulating (PEGylated) liposomes containing DOPE have demonstrated their ability to deliver their payload into the cell cytoplasm [52].

Low pH-stimuli-responsive liposomes consisting of terminally alkylated NIPAM (*N*-isopropylacrylamide) copolymer have been developed, which maintained pH sensitivity even after incubation with human serum [103]. Recently, promising "SMART" liposomes have been developed that can lose their PEG coating under acidic conditions, facilitating better interaction with the cell surface and thus internalization [24].

Redox Potential-Sensitive Liposomes + US

Exploiting the differences between cytosolic and extracellular GSH levels (cytosolic GSH >>> extracellular GSH, i.e., cytosolic environment is reducing relative to extracellular environment) marks the primary mechanism of these liposomes, and this fact can be extrapolated for tumor targeting since tumor tissue usually has higher levels of GSH as compared to normal tissue [104, 105]. Tumor targeting using liposomes that exploit these physiological differences between normal and tumor tissue combined with US that will facilitate drug release at the tumor site seems to be a sound rationale for this combination. Redox-responsive long-circulating liposomes with disulfide-linked (detachable) PEG coating (detachment by thiolytic cleavage) have been investigated for drug delivery [106]. Recently, as a proof-of-concept study, redox-responsive quinine-DOPE liposomes have been used to deliver a dye site specifically. The payload delivery depends on the mechanism of electron transfer which in turn depends on the overexpression of the enzyme quinine reductases (overexpressed in tumors) and can thus be used for tumor targeting [107]. Other redox-responsive liposomes have also been described [108].

Magnetic Liposomes + US

iposomes encapsulating magnetic nanoparticles so as to be capable of targeting under the influence of a magnetic field have been used for tumor targeting, and this system in combination with US augurs well for site-specific localized drug release due to the dual mechanical forces of magnetism and US-mediated drug release. Magnetic adriamycin liposomes injected i.v. have been used to deliver the drug under the influence of an implanted magnet in an osteosarcoma model [109, 110]. Recently, a folate-targeted liposomal system has been developed that delivers the payload on the basis of magnetism-induced hyperthermia [111].

Light-Responsive Liposomes + US

Liposomes have been used to deliver photosensitizers to the tumor tissue, and then the tumor foci is irradiated with nonthermal light, which then is suggested to destroy tumor tissue without affecting the healthy tissue, by production of reactive oxygen species locally. If used in combination with US, this will facilitate the release of the photosensitizer payload from the liposomes at the tumor site and may increase the benefit of the PDT for tumoricidal action. Several examples of liposomes for PDT (liposomes carrying the photosensitizer) and light-sensitive liposomes (release payload in response to light) have been mentioned in [80]. There is mention of a very interesting study where in a mouse osteosarcoma model, glucuronide-modified long-circulating liposomes have been used to deliver a photosensitizer (BPD-MA) at the tumor site (systemic i.v. injection, passive targeting) and, upon subsequent illumination, exhibited significant tumor regression as compared to the control group (BPD-MA solution or BPD-MA in conventional DPPG liposomes).

4.3 Weighing the Benefits vs. the Limitations of Liposomal Focal Drug Delivery

The LFDD platform comprises of two essential components, i.e., the liposomes and the focused ultrasound. These platforms may incorporate an imaging modality especially MRI. It is only logical that any benefit or limitation (current or prospective) associated with the platform will be a resultant of the benefits/limitations of the individual components or an amalgamation of the above. The benefits vs. limitations listed below should be read as separate. They only have been listed as a comparative table for the reader to appreciate the pros vs. cons simultaneously. Some of these aspects are reviewed in (Table 4.2) [112, 113].

4.4 Non-Cancer-Related Liposomal FDD

This section of the chapter will focus on the applications of liposomal focal drug delivery in a non-cancer therapy setting. There are several studies that elucidate the simultaneous application of focused ultrasound and a liposomal delivery system to deliver drug specifically at the target site, i.e., by definition FDD. Some of these current systems will be described below.

LFDD in Cardiovascular Disorders

Most of these studies employ focused ultrasound in combination with ELIP. As described previously in this chapter, these are liposomes filled with gas like air or some other agent that makes the liposome acoustically reflective for ultrasound imaging, and also application of ultrasound can stress and cause burst of the liposomal bilayer to release its content at the target site [114]. ELIP are shown to retain properties exhibited by conventional liposomes and hence the rationale for co-encapsulating drug/gene and a gas for site-specific delivery seems sound. For greater details on ELIP, the reader is referred to the recent review by Huang [115]. ELIP encapsulating tissue plasminogen activator (tPA-ELIP) combined with focused ultrasound effectively enhanced the thrombolytic effect of tPA in an in vivo rabbit aorta thrombus model. This treatment displayed earlier aortic recanalization

Table 4.2 A comparative reading of benefits vs. limitations of focal delivery

Benefits	Limitations
This platform requires a drug delivery system that forms an inert moiety with the toxic drug; so there is no drug leakage; this augurs well for prevention of off-target side effects	Currently, no such liposomal delivery system exists that can boast of zero drug leakage en route the target, for focal delivery. Further advancements are required in this regard
Drug release kinetics at target site can be better regulated [95]	PEG which is employed widely to make liposomes long circulating has been recently shown to elicit immune response [26, 27]
Since the drug release is limited to the narrow target site, focal delivery necessitates a smaller dose for similar therapeutic effect	There is a limitation associated with focused ultrasound penetration. A clear acoustic path is needed from the skin to the target. Air, bone, and soft tissue may interfere
Real-time imaging of the therapy is possible using MRI guidance. MR thermometry allows localization of focus and temperature monitoring. Feedback control allows for precise control of hyperthermia	A lot of preparatory work is required before employing this technique, e.g., (a) MR thermometry has to be corrected for periodic respiratory motion for imaging abdominal organs (b) Temperature elevations may vary for a given ultrasound power and duration and have to be corrected for uniformity
Personalized therapy is an exciting possibility, since the various parameters can be titrated depending on the type of tumor being treated	The MR-HIFU platform essential for focal delivery is expensive
One of the most important advantages of this platform is being either minimally/noninvasive	Though HIFU has been the preferred modality for focal delivery with liposomes until now, it can cause severe internal burns if reflected onto nontarget tissues
This method opens up a possibility of a new in vivo method of drug delivery (drug delivery to tumor down the concentration gradient due to burst release at tumor site), bypassing the need for endocytosis and endosomal escape [14]	LFU is safer, but it is difficult to focus and usually dissipates at the surface, limiting its use for deep tumors
The feasibility of this method has been established in several in vitro and a few in vivo models in cancer therapy	The use of image guidance for focal delivery, despite several years of efforts, is still in its infancy, and further improvements are needed for it to be indispensable for personalized cancer therapy [90]

rates over control and empty ELIP liposomal treatments as demonstrated by the improved vessel blood flow. The authors propose that the use of ultrasound in this study to selectively visualize the clot will prove to be beneficial in a clinical setting by avoiding off-target effects. However, the animal model used was an acute thrombosis model and so the findings cannot be extrapolated to a chronic thrombosis model [116]. ELIP have been employed in gene therapy for atherosclerosis. A proof-of-concept study using fluorescent-labeled oligodinucleotide (FITC-NF-κB ODN) encapsulated in an ELIP has shown that the ODN can be released efficiently at the target site by application of HFUS, in vitro. NF-κB-signaling pathway regulates inflammation and plays a key role in the pathogenesis of atherosclerosis and hence the authors propose that this system can be useful for atherosclerotic gene therapy [114, 117]. Another study has reported a novel thrombin encapsulating liposome-coated microbubble system that is responsive to pulsed HFUS. US-induced thrombin release from the microbubble-liposome system led to an accelerated EDTA containing canine blood clotting, in vitro. The authors propose that this system possesses the advantage of greater drug loading inside the aqueous liposomal core over the common drug-on-shell approach [94]. A similar microbubble-liposome system in conjunction with ultrasound has been successfully developed for gene-mediated transfection of cardiomyocytes in vivo and has the potential for application in primary myocardial disease [118]. Recently, an ELIP system that releases its contents in response to an ultrasound trigger followed by exposure to clinically relevant matrix metalloproteinase (MMP-9) enzyme levels has been developed. MMP-9 is overexpressed in atherosclerotic plaques, and this system has been designed to exploit this enzyme as an external trigger, in addition to HFUS, to destabilize the liposome and release its contents at that site. This system could also be employed in a metastatic cancer setting since these types of cancer cells, too, overexpress MMP-9. The fact that this system can also be imaged using ultrasound adds to its benefits [119]. A liposomal system encapsulating nitric oxide for the inhibition of intimal hyperplasia (inflammation of a vascular vessel due to injury) as a potential drug delivery platform in conjunction with ultrasound has been developed [120].

LFDD in Ocular Disorders

Pigment epithelium-derived factor (PEDF)-loaded immunoliposomes (INL) have been developed for the treatment of ocular neovascularization (CNV) which is a commonly seen pathological change in several intraocular diseases. PEDF is a potent angiogenesis inhibitor, but its delivery has been a problem. The treatment of the PEDF-loaded INLs along with ultrasonic irradiation showed the highest inhibition of CNV in a rat model over other appropriate controls. This system was shown to specifically deliver PEDF efficiently into the endothelial cells of the CNV and had no off-target effects on the normal choroidal cells [121].

LFDD for Tissue-Specific siRNA Delivery

Recently, Takahashi et al. have reported an siRNA-loaded liposomal system (si-BLs) for site-specific, US-induced delivery. The liposomal system is based on the cationic lipid DOTAP and protected the siRNA from serum degradation. In vitro experiments proved that upon application of HFUS, the intact siRNA was released from the liposome and exhibited specific luciferase gene downregulation. It was proposed that this system could be administered i.v., imaged by US by entrapping gas, and could be potentially employed for site-specific siRNA delivery in various disease models [122].

LFDD for Cerebrovascular Disorders

Currently, the only treatment for refractory cerebral vasospasm following aneurysmal subarachnoid hemorrhage is the direct intra-arterial infusions of vasodilators like papaverine. However, this method is invasive and its efficacy is transient and hence needs repeated endovascular procedures. In an effort to overcome these limitations, a novel ELIP encapsulating papaverine (PELIP) was designed. The rationale was that the system that could be potentially receptive to transcutaneous US beams focused to the internal carotid or vertebral arteries (noninvasive) and thus releases the drug site specifically. It was shown that the ELIP successfully encapsulated the drug and this procedure did not alter its biological activity, though the echogenecity of the ELIP slightly decreased. The system also has the potential for real-time drug monitoring due to the fact that it could be visualized by US [123]. The response of this system to US in vitro has not been investigated yet. Also, whether this system confers any benefit over the existing protocol in vivo needs to be seen. It should also be noted that though papaverine itself is hydrophobic, the hydrophilic hydrochloride salt has been used in this study. The nature of the salt used may also play a significant role in the drug release. Recently, an interesting study has been published that proves that under identical high-frequency US conditions, hydrophilic drugs can be better released from identical ELIP than hydrophobic drugs [124].

LFDD for Miscellaneous Applications

In addition to the applications described above, various lipid composition-based PEGylated liposomes (70–300 nm) have been developed as ultrasound (US) contrast agents that could also be employed as long-circulating, site-specific drug delivery agents. These liposomes have demonstrated comparable ultrasound reflectivity as compared to the commercially available US contrast agent, SonoVue®. They have also shown to retain stable ultrasound contrast with homogeneous contrast for up to 2 weeks [125].

4.5 Conclusions and Future Prospects

The aim of this chapter is to bring to the fore the current aspects associated with the modality of the liposomal delivery system in combination with focused ultrasound for focal drug delivery and proposing a strong possibility about similar future systems. Like any other system, it is bound to have advantages and limitations. The goal should be to further investigate the possibilities scientifically and employ this modality with a definite scientific temper where its use will augur better patient treatment.

References

1. Jain KK (2005) Editorial: targeted drug delivery for cancer. Technol Cancer Res Treat 4(4):311–3
2. Maeda H (2001) The enhanced permeability and retention (EPR) effect in tumor vasculature: the key role of tumor-selective macromolecular drug targeting. Adv Enzyme Regul 41(1):189–207
3. Allen TM (2002) Ligand-targeted therapeutics in anticancer therapy. Nat Rev Cancer 2(10):750–63
4. Maeda H et al (2000) Tumor vascular permeability and the EPR effect in macromolecular therapeutics: a review. J Control Release 65(1–2):271–284
5. Peer D et al (2007) Nanocarriers as an emerging platform for cancer therapy. Nat Nanotechnol 2(12):751–760
6. Danhier F, Feron O, Préat V (2010) To exploit the tumor microenvironment: passive and active tumor targeting of nanocarriers for anti-cancer drug delivery. J Control Release 148(2):135–146
7. Byrne JD, Betancourt T, Brannon-Peppas L (2008) Active targeting schemes for nanoparticle systems in cancer therapeutics. Adv Drug Deliv Rev 60(15):1615–1626
8. Frieboes H, Sinek J et al (2006) Nanotechnology in Cancer Drug Therapy: A Biocomputational Approach. BioMEMS and Biomedical Nanotechnology. M. Ferrari, A. Lee and L. J. Lee, Springer US: 435–460
9. Huynh NT et al (2010) The rise and rise of stealth nanocarriers for cancer therapy: passive versus active targeting. Nanomedicine 5(9):1415–1433
10. Schroeder A, Kost J, Barenholz Y (2009) Ultrasound, liposomes, and drug delivery: principles for using ultrasound to control the release of drugs from liposomes. Chem Phys Lipids 162(1–2):1–16
11. Deckers R, Moonen CTW (2010) Ultrasound triggered, image guided, local drug delivery. J Control Release 148(1):25–33
12. Frenkel V (2008) Ultrasound mediated delivery of drugs and genes to solid tumors. Adv Drug Deliv Rev 60(10):1193–1208
13. Shuto J, Ichimiya I, Suzuki M (2006) Effects of low-intensity focused ultrasound on the mouse submandibular gland. Ultrasound Med Biol 32(4):587–94
14. Ibsen S et al (2012) Ultrasound mediated localized drug delivery. In: Zahavy E et al (eds) Nanobiotechnology for biomedical and diagnostic research. Springer, Netherlands, pp 145–153
15. Zarnitsyn V, Rostad CA et al (2008) Modeling transmembrane transport through cell membrane wounds created by acoustic cavitation. J Biophys 95(9):4124–4138
16. Schroeder A et al (2009) Ultrasound triggered release of cisplatin from liposomes in murine tumors. J Control Release 137(1):63–68
17. Kennedy JE, ter Haar GR, Cranston D (2003) High intensity focused ultrasound: surgery of the future? Br J Radiol 76(909):590–599
18. Torchilin VP (2005) Recent advances with liposomes as pharmaceutical carriers. Nat Rev Drug Discov 4(2):145–160
19. Torchilin V (2012) Liposomes in drug delivery. In: Siepmann J, Siegel RA, Rathbone MJ (eds) Fundamentals and applications of controlled release drug delivery. Springer, New York, NY, pp 289–328
20. Maeda H, Bharate GY, Daruwalla J (2009) Polymeric drugs for efficient tumor-targeted drug delivery based on EPR-effect. Eur J Pharm Biopharm 71(3):409–419
21. Yuan F et al (1994) Microvascular permeability and interstitial penetration of sterically stabilized (stealth) liposomes in a human tumor xenograft. Cancer Res 54(13):3352–3356
22. Blume G, Cevc G (1993) Molecular mechanism of the lipid vesicle longevity in vivo. Biochim Biophys Acta 1146(2):157–168
23. Kale AA, Torchilin VP (2007) Design, synthesis, and characterization of pH-sensitive PEG–PE conjugates for stimuli-sensitive pharmaceutical nanocarriers: the effect of substi-

tutes at the hydrazone linkage on the pH stability of PEG – PE conjugates. Bioconjug Chem 18(2):363–370
24. Sawant RM et al (2006) "SMART" drug delivery systems: double-targeted pH-responsive pharmaceutical nanocarriers. Bioconjug Chem 17(4):943–949
25. Zalipsky S et al (1999) New detachable poly(ethylene glycol) conjugates: cysteine-cleavable lipopolymers regenerating natural phospholipid. Diacyl phosphatidylethanolamine. Bioconjug Chem 10(5):703–707
26. Moghimi SM et al (2010) Complement activation cascade triggered by PEG–PL engineered nanomedicines and carbon nanotubes: The challenges ahead. J Control Release 146(2): 175–181
27. Chanan-Khan A et al (2003) Complement activation following first exposure to pegylated liposomal doxorubicin (Doxil®): possible role in hypersensitivity reactions. Ann Oncol 14(9):1430–1437
28. Malam Y, Loizidou M, Seifalian AM (2009) Liposomes and nanoparticles: nanosized vehicles for drug delivery in cancer. Trends Pharmacol Sci 30(11):592–599
29. Park JW et al (2001) Tumor targeting using anti-her2 immunoliposomes. J Control Release 74(1–3):95–113
30. Gupta B, Torchilin V (2007) Monoclonal antibody 2C5-modified doxorubicin-loaded liposomes with significantly enhanced therapeutic activity against intracranial human brain U-87 MG tumor xenografts in nude mice. Cancer Immunol Immunother 56(8):1215–1223
31. ElBayoumi TA, Torchilin VP (2008) Tumor-specific anti-nucleosome antibody improves therapeutic efficacy of doxorubicin-loaded long-circulating liposomes against primary and metastatic tumor in mice. Mol Pharm 6(1):246–254
32. ElBayoumi TA, Torchilin VP (2009) Tumor-targeted nanomedicines: enhanced antitumor efficacy in vivo of doxorubicin-loaded, long-circulating liposomes modified with cancer-specific monoclonal antibody. Clin Cancer Res 15(6):1973–1980
33. Raffaghello L et al (2003) Immunoliposomal fenretinide: a novel antitumoral drug for human neuroblastoma. Cancer Lett 197(1–2):151–155
34. Mamot C et al (2005) Epidermal growth factor receptor–targeted immunoliposomes significantly enhance the efficacy of multiple anticancer drugs in vivo. Cancer Res 65(24): 11631–11638
35. Mamot C et al (2006) EGFR-targeted immunoliposomes derived from the monoclonal antibody EMD72000 mediate specific and efficient drug delivery to a variety of colorectal cancer cells. J Drug Target 14(4):215–223
36. Iinuma H et al (2002) Intracellular targeting therapy of cisplatin-encapsulated transferrin-polyethylene glycol liposome on peritoneal dissemination of gastric cancer. Int J Cancer 99(1):130–137
37. Ishida O et al (2001) Liposomes bearing polyethyleneglycol-coupled transferrin with intracellular targeting property to the solid tumors in vivo. Pharm Res 18(7):1042–1048
38. Joshee N, Bastola DR, Cheng PW (2002) Transferrin-facilitated lipofection gene delivery strategy: characterization of the transfection complexes and intracellular trafficking. Hum Gene Ther 13(16):1991–2004
39. Tan PH et al (2003) Antibody targeted gene transfer to endothelium. J Gene Med 5(4):311–323
40. Xu L et al (2002) Systemic tumor-targeted gene delivery by anti-transferrin receptor scFv-immunoliposomes 1 this work was supported in part by National Cancer Institute Grant R01 CA45158 (to E. C.), National Cancer Institute Small Business Technology Transfer Phase I Grant R41 CA80449 (to E. C.), and a grant from SynerGene Therapeutics, Inc.1. Mol Cancer Ther 1(5):337–346
41. Gabizon A et al (2004) Tumor cell targeting of liposome-entrapped drugs with phospholipid-anchored folic acid–PEG conjugates. Adv Drug Deliv Rev 56(8):1177–1192
42. Lee RJ, Low PS (1994) Delivery of liposomes into cultured KB cells via folate receptor-mediated endocytosis. J Biol Chem 269(5):3198–3204
43. Low YLPS (2002) Folate targeting of haptens to cancer cell surfaces mediates immunotherapy of syngeneic murine tumors. Cancer Immunol Immunother 51(3):153–162

44. Reddy JA et al (2002) Folate-targeted, cationic liposome-mediated gene transfer into disseminated peritoneal tumors. Gene Ther 9(22):1542–50
45. Schiffelers RM et al (2003) Anti-tumor efficacy of tumor vasculature-targeted liposomal doxorubicin. J Control Release 91(1–2):115–122
46. Jayanna PK, Torchilin VP, Petrenko VA (2009) Liposomes targeted by fusion phage proteins. Nanomed Nanotechnol Biol Med 5(1):83–89
47. Terada T et al (2007) Optimization of tumor-selective targeting by basic fibroblast growth factor-binding peptide grafted PEGylated liposomes. J Control Release 119(3):262–70
48. Huwyler J, Wu D, Pardridge WM (1996) Brain drug delivery of small molecules using immunoliposomes. Proc Natl Acad Sci 93(24):14164–14169
49. Mastrobattista E et al (2002) Functional characterization of an endosome-disruptive peptide and its application in cytosolic delivery of immunoliposome-entrapped proteins. J Biol Chem 277(30):27135–27143
50. Scott RC et al (2007) Targeted delivery of antibody conjugated liposomal drug carriers to rat myocardial infarction. Biotechnol Bioeng 96(4):795–802
51. D'Souza G et al (2008) Surface modification of pharmaceutical nanocarriers with ascorbate residues improves their tumor-cell association and killing and the cytotoxic action of encapsulated paclitaxel in vitro. Pharm Res 25(11):2567–2572
52. Simões S et al (2004) On the formulation of pH-sensitive liposomes with long circulation times. Adv Drug Deliv Rev 56(7):947–965
53. Fattal E, Couvreur P, Dubernet C (2004) "Smart" delivery of antisense oligonucleotides by anionic pH-sensitive liposomes. Adv Drug Deliv Rev 56(7):931–946
54. Gomes-da-Silva LC et al (2012) Lipid-based nanoparticles for siRNA delivery in cancer therapy: paradigms and challenges. Acc Chem Res 45(7):1163–71
55. Kakudo T et al (2004) Transferrin-modified liposomes equipped with a pH-sensitive fusogenic peptide: an artificial viral-like delivery system†. Biochemistry 43(19):5618–5628
56. Turk MJ et al (2002) Characterization of a novel pH-sensitive peptide that enhances drug release from folate-targeted liposomes at endosomal pHs. Biochim Biophys Acta 1559(1):56–68
57. Gorodetsky R et al (2004) Liposome transduction into cells enhanced by haptotactic peptides (Haptides) homologous to fibrinogen C-termini. J Control Release 95(3):477–488
58. Torchilin VP et al (2001) TAT peptide on the surface of liposomes affords their efficient intracellular delivery even at low temperature and in the presence of metabolic inhibitors. Proc Natl Acad Sci 98(15):8786–8791
59. Tseng Y-L, Liu J-J, Hong R-L (2002) Translocation of liposomes into cancer cells by cell-penetrating peptides penetratin and Tat: a kinetic and efficacy study. Mol Pharmacol 62(4):864–872
60. Boddapati SV et al (2008) Organelle-targeted nanocarriers: specific delivery of liposomal ceramide to mitochondria enhances its cytotoxicity in vitro and in vivo. Nano Lett 8(8):2559–2563
61. Boddapati SV et al (2005) Mitochondriotropic liposomes. J Liposome Res 15(1–2):49–58
62. Patel NR et al (2010) Mitochondria-targeted liposomes improve the apoptotic and cytotoxic action of sclareol. J Liposome Res 20(3):244–249
63. Lasic DD, Vallner JJ, Working PK (1999) Sterically stabilized liposomes in cancer therapy and gene delivery. Curr Opin Mol Ther 1(2):177–85
64. Audouy SAL et al (2002) Invivo characteristics of cationic liposomes as delivery vectors for gene therapy. Pharm Res 19(11):1599–1605
65. Smyth Templeton N (2002) Cationic liposome-mediated gene delivery in vivo. Biosci Rep 22(2):283–295
66. Brignole C et al (2003) Targeted delivery system for antisense oligonucleotides: a novel experimental strategy for neuroblastoma treatment. Cancer Lett 197(1–2):231–235
67. Ko YT, Falcao C, Torchilin VP (2009) Cationic liposomes loaded with proapoptotic peptide d-(KLAKLAK)2 and Bcl-2 antisense oligodeoxynucleotide G3139 for enhanced anticancer therapy. Mol Pharm 6(3):971–977

68. Landen CN, Merritt WM et al (2006) Intraperitoneal delivery of liposomal siRNA for therapy of advanced ovarian cancer. Cancer Biology & Therapy 5(12):1708–1713
69. Zhang C et al (2006) siRNA-containing liposomes modified with polyarginine effectively silence the targeted gene. J Control Release 112(2):229–239
70. Judge AD et al (2009) Confirming the RNAi-mediated mechanism of action of siRNA-based cancer therapeutics in mice. J Clin Invest 119(3):661
71. Pirollo KF, Chang EH (2008) Targeted delivery of small interfering RNA: approaching effective cancer therapies. Cancer Res 68(5):1247–1250
72. Kapoor M, Burgess DJ (2012) Efficient and safe delivery of siRNA using anionic lipids: formulation optimization studies. Int J Pharm 432(1–2):80–90
73. Christensen D et al (2011) Cationic liposomes as vaccine adjuvants. Expert Rev Vaccines 10(4):513–21
74. Cormode DP et al (2009) Nanotechnology in medical imaging. Arterioscler Thromb Vasc Biol 29(7):992–1000
75. Oku N et al (2011) PET imaging of brain cancer with positron emitter-labeled liposomes. Int J Pharm 403(1):170–177
76. Sofou S, Sgouros G (2008) Antibody-targeted liposomes in cancer therapy and imaging. Expert Opin Drug Deliv 5(2):189–204
77. Fonseca MJ et al (2003) Liposome-mediated targeting of enzymes to cancer cells for site-specific activation of prodrugs: comparison with the corresponding antibody-enzyme conjugate. Pharm Res 20(3):423–428
78. Storm G et al (1997) Immunoliposomes bearing enzymes (immuno-enzymosomes) for site-specific activation of anticancer prodrugs. Adv Drug Deliv Rev 24(2–3):225–231
79. Oku N et al (1997) Application of long-circulating liposomes to cancer photodynamic therapy. Biol Pharm Bull 20(6):670–3
80. Derycke ASL, de Witte PAM (2004) Liposomes for photodynamic therapy. Adv Drug Deliv Rev 56(1):17–30
81. Chen B, Pogue BW, Hasan T (2005) Liposomal delivery of photosensitising agents. Expert Opin Drug Deliv 2(3):477–487
82. Allen TM, Cullis PR (2004) Drug delivery systems: entering the mainstream. Science 303(5665):1818–1822
83. Tacker J, Anderson R (1982) Delivery of antitumor drug to bladder cancer by use of phase transition liposomes and hyperthermia. J Urol 127(6):1211
84. Kong G et al (2000) Efficacy of liposomes and hyperthermia in a human tumor xenograft model: importance of triggered drug release. Cancer Res 60(24):6950
85. De Zwart J, et al (2000) On the feasibility of local drug delivery using thermo-sensitive liposomes and MR-guided focused ultrasound. Proceedings of the International Society for Magnetic Resonance in Medicine, Denver
86. Zhang C et al (2007) Pilot study on combined cancer therapy by thermo-sensitive liposomes and ultrasound phased arrays induced heating. In: Magjarevic R, Nagel JH (eds) World congress on medical physics and biomedical engineering 2006. Springer, Berlin, pp 3238–3242
87. Treat LH et al (2007) Targeted delivery of doxorubicin to the rat brain at therapeutic levels using MRI-guided focused ultrasound. Int J Cancer 121(4):901–907
88. Dromi S et al (2007) Pulsed-high intensity focused ultrasound and Low temperature–sensitive liposomes for enhanced targeted drug delivery and antitumor effect. Clin Cancer Res 13(9):2722–2727
89. de Smet M et al (2011) Magnetic resonance imaging of high intensity focused ultrasound mediated drug delivery from temperature-sensitive liposomes: an in vivo proof-of-concept study. J Control Release 150(1):102–10
90. Ranjan A et al (2012) Image-guided drug delivery with magnetic resonance guided high intensity focused ultrasound and temperature sensitive liposomes in a rabbit Vx2 tumor model. J Control Release 158(3):487–494

91. Negussie AH et al (2011) Formulation and characterisation of magnetic resonance imageable thermally sensitive liposomes for use with magnetic resonance-guided high intensity focused ultrasound. Int J Hyperthermia 27(2):140–155
92. Huang S-L, MacDonald RC (2004) Acoustically active liposomes for drug encapsulation and ultrasound-triggered release. Biochim Biophys Acta 1665(1–2):134–141
93. Lentacker I et al (2009) Design and evaluation of doxorubicin-containing microbubbles for ultrasound-triggered doxorubicin delivery: cytotoxicity and mechanisms involved. Mol Ther 18(1):101–108
94. Klibanov AL et al (2010) Ultrasound-triggered release of materials entrapped in microbubble-liposome constructs: a tool for targeted drug delivery. J Control Release 148(1):13–17
95. Schlicher RK et al (2006) Mechanism of intracellular delivery by acoustic cavitation. Ultrasound Med Biol 32(6):915–924
96. Bystritsky A et al (2011) A review of low-intensity focused ultrasound pulsation. Brain Stimul 4(3):125–136
97. Kheirolomoom A et al (2007) Acoustically-active microbubbles conjugated to liposomes: characterization of a proposed drug delivery vehicle. J Control Release 118(3):275–284
98. Lin HY, Thomas JL (2003) PEG-lipids and oligo (ethylene glycol) surfactants enhance the ultrasonic permeabilizability of liposomes. Langmuir 19(4):1098–1105
99. Cohen-Levi D, Kost J, Barenholz Y (2000) Ultrasound for targeted delivery of cytotoxic drugs from liposomes. M.Sc. Thesis (Kost, J., Barenholz, Y., supervisors), Faculty of Engineering Sciences, Ben Gurion University, Beer Sheva, Israel
100. Pong M et al (2006) In vitro ultrasound-mediated leakage from phospholipid vesicles. Ultrasonics 45(1):133–145
101. Schroeder A et al (2007) Controlling liposomal drug release with low frequency ultrasound: mechanism and feasibility. Langmuir 23(7):4019–4025
102. Shalaev EY, Steponkus PL (1999) Phase diagram of 1,2-dioleoylphosphatidylethanolamine (DOPE):water system at subzero temperatures and at low water contents. Biochim Biophys Acta 1419(2):229–247
103. Roux E et al (2002) Steric stabilization of liposomes by pH-responsive N-isopropylacrylamide copolymer. J Pharm Sci 91(8):1795–1802
104. Perry RR et al (1993) Glutathione levels and variability in breast tumors and normal tissue. Cancer 72(3):783–7
105. Russo A et al (1986) Selective modulation of glutathione levels in human normal versus tumor cells and subsequent differential response to chemotherapy drugs. Cancer Res 46(6):2845–2848
106. Kirpotin D et al (1996) Liposomes with detachable polymer coating: destabilization and fusion of dioleoylphosphatidylethanolamine vesicles triggered by cleavage of surface-grafted poly(ethylene glycol). FEBS Lett 388(2–3):115–118
107. Ong W et al (2008) Redox-triggered contents release from liposomes. J Am Chem Soc 130(44):14739–14744
108. West KR, Otto S (2005) Reversible covalent chemistry in drug delivery. Curr Drug Discov Technol 2(3):123–160
109. Kubo T et al (2000) Targeted delivery of anticancer drugs with intravenously administered magnetic liposomes in osteosarcoma-bearing hamsters. Int J Oncol 17(2):309–15
110. Kubo T et al (2001) Targeted systemic chemotherapy using magnetic liposomes with incorporated adriamycin for osteosarcoma in hamsters. Int J Oncol 18(1):121–5
111. Pradhan P et al (2010) Targeted temperature sensitive magnetic liposomes for thermo-chemotherapy. J Control Release 142(1):108–121
112. Staruch R, Chopra R, Hynynen K (2011) Localised drug release using MRI-controlled focused ultrasound hyperthermia. Int J Hyperthermia 27(2):156–171
113. Staruch R, Chopra R, Hynynen K (2011) MRI-controlled ultrasound thermal therapy. IEEE Pulse 2(5):39–47
114. Buchanan KD et al (2010) Encapsulation of NF-κB decoy oligonucleotides within echogenic liposomes and ultrasound-triggered release. J Control Release 141(2):193–198

115. Huang S-L (2008) Liposomes in ultrasonic drug and gene delivery. Adv Drug Deliv Rev 60(10):1167–1176
116. Laing ST et al (2011) Ultrasound-enhanced thrombolytic effect of tissue plasminogen activator–loaded echogenic liposomes in an in vivo rabbit aorta thrombus model—brief report. Arterioscler Thromb Vasc Biol 31(6):1357–1359
117. Gareus R et al (2008) Endothelial cell-specific NF-kappaB inhibition protects mice from atherosclerosis. Cell Metab 8(5):372–83
118. Chen Z-Y et al (2011) Ultrasound- and liposome microbubble-mediated targeted gene transfer to cardiomyocytes in vivo accompanied by polyethylenimine. J Ultrasound Med 30(9):1247–1258
119. Nahire R et al (2012) Ultrasound enhanced matrix metalloproteinase-9 triggered release of contents from echogenic liposomes. Mol Pharm 9(9):2554–2564
120. Huang S-L et al (2009) Nitric oxide-loaded echogenic liposomes for nitric oxide delivery and inhibition of intimal hyperplasia. J Am Coll Cardiol 54(7):652–659
121. Li T et al (2010) Targeting therapy of choroidal neovascularization by use of polypeptide- and PEDF-loaded immunoliposomes under ultrasound exposure. J Huazhong Univ Sci Technolog Med Sci 30(6):798–803
122. Endo-Takahashi Y et al (2012) Efficient siRNA delivery using novel siRNA-loaded bubble liposomes and ultrasound. Int J Pharm 422(1–2):504–509
123. Kee PH et al (2008) Synthesis, acoustic stability, and pharmacologic activities of papaverine-loaded echogenic liposomes for ultrasound controlled drug delivery. J Liposome Res 18(4):263–277
124. Kopechek JA et al (2008) Ultrasound-mediated release of hydrophilic and lipophilic agents from echogenic liposomes. J Ultrasound Med 27(11):1597–1606
125. Marxer EEJ et al (2011) Development and characterization of new nanoscaled ultrasound active lipid dispersions as contrast agents. Eur J Pharm Biopharm 77(3):430–437

Chapter 5
Polymer–Drug Conjugate in Focal Drug Delivery

Rajan Swami, Dinesh Kumar, Wahid Khan, Ramakrishna Sistla, and Nalini Shastri

5.1 Introduction

Evolution is a nonpareil truth of universe and polymers are no exceptions. In the midst of outcry of nanotechnology and genomic revolution, polymers are proliferating as a prominent tool for drug delivery. The exquisiteness of this bailiwick, "polymers in drug delivery," is its longevity and self-transforming capability. From decades, polymers were part and parcel of classical drug formulations, as excipients for controlling the release rate of drugs or in circumventing the clearance of drugs. Now they have emerged as an advanced and sophisticated tool to target drugs for focal drug delivery. The importance of polymer architecture property relationships has gradually been realized and emphasized with development of various nanosystems and other related technologies. Polymers have unique cooperative properties that are not found with low-molecular-weight compounds, and there lies the root of their success. Such nano-constructs were first proposed in the 1970s [1], developed preclinically in the 1980s [2–4], and started entering the clinical pipeline in the 1990s [5–8].

Historically, polymer chemists were successful in reaching the unmet needs of formulation scientists and were able to come up with biocompatible, biodegradable, and therapeutics polymers including the "modern cousins" dendrimers. Nearly all essential polymers explored widely as "polymer therapeutics" [9, 10] encompass mainly five types [11–18] of agents: (a) polymer–drug conjugate (PDC), (b) biologically active

R. Swami • D. Kumar • W. Khan • N. Shastri (✉)
Department of Pharmaceutics, National Institute of Pharmaceutical Education and Research (NIPER), Hyderabad 500037, India
e-mail: nalini@niperhyd.ac.in

R. Sistla
Department of Pharmaceutics, National Institute of Pharmaceutical Education and Research (NIPER), Hyderabad 500037, India

Department of Pharmacology, Indian Institute of Chemical Technology, Hyderabad 500607, India

A.J. Domb and W. Khan (eds.), *Focal Controlled Drug Delivery*, Advances in Delivery Science and Technology, DOI 10.1007/978-1-4614-9434-8_5,

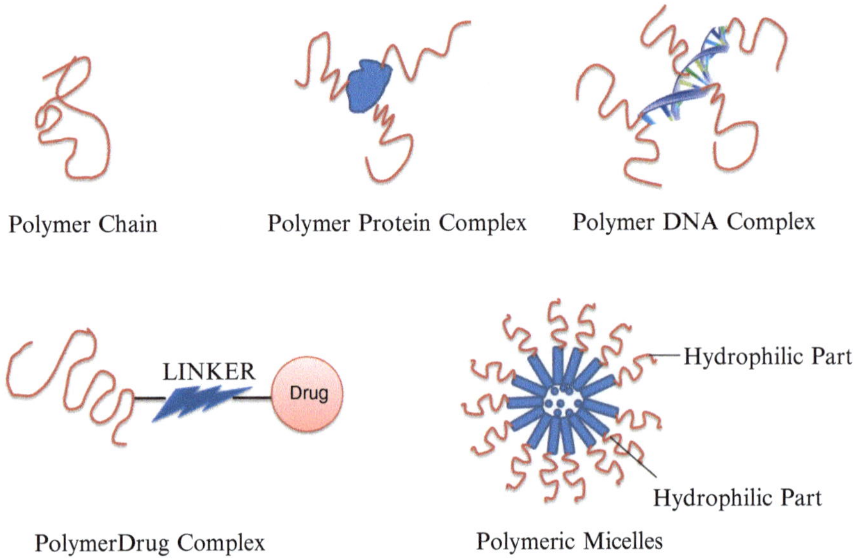

Fig. 5.1 Schematic representation of polymer therapeutics

polymers, (c) polymer–protein complex, (d) polymeric micelles, and (e) polymer–DNA complex (Polyplex). Schematic representation is shown in Fig. 5.1.

Polymer–drug conjugates came in realization by first polymer conjugate of N-vinylpyrrolidone conjugates of glycyl-L-leucine-mescaline as a drug depot formulation in 1955 [19]. However, the original idea of covalently conjugating a low-molecular-weight drug to a hydrophilic polymeric carrier to increase its therapeutic effect was proposed by Helmut Ringsdorf in 1975 [1]. His idea was revolutionized further by many authors, to espouse polymer–drug conjugate as an important area in advanced focal drug delivery. This chapter is confined to the scope of prominent role of polymers to formulate polymer–drug conjugate (PDC) for focal drug delivery. The chapter commences with historical milestones and later overlays characterization and applications followed by advantages and future prospects of polymer–drug conjugates.

5.2 Historical Milestones

Hermann Staudinger, in 1953, was awarded Nobel Prize for demonstrating the existence of macromolecules, which he characterized as "polymers" [20]. Polymers are substances having high molar masses and are composed of a large number of repeating units called as "monomers." There are both naturally occurring and synthetic polymers. Naturally occurring polymers from plant or animal origin are proteins, starches, cellulose, chitosan, alginates, glycosaminoglycans, poly-b-hydroxybutyrate, latex, etc. Synthetic polymers are produced commercially on a very large scale and have a wide range of properties and uses. They include poly(lactic-*co*-glycolic acid) (PLGA), polyamidoamine (PAMAM) dendrimer, polyacrylate, polyamides, etc. Conjugation of drugs to polymers was foremost attempted in 1955 with poly(N-vinylpyrrolidone) using a dipeptide (glycyl-L-leucine) spacer [19]. Ushakov's group synthesized numerous water-soluble polymer–drug conjugates in the 1960s and the 1970s [21]. In the 1970s, de Duve et al. discovered that macromolecules localize in the lysosome following their incubation with cells [22]. The first clinical trial for polymer–drug conjugate divinyl ether-maleic anhydride in 1960s was unsuccessful [23, 24]. However, this led to spurt of investigations on polymer–drug conjugates by incorporating various strategies to design rational formulations of polymeric drug conjugates that are stable in the bloodstream during circulation but could release the free drug specifically in the cells by the use of lysosomally degradable linkers. Nevertheless, the first clear concept for the use of polymer–drug conjugates was presented by Ringsdorf in 1975 [1]. Further pioneer work was done in the 1980s by Duncan and Kopecek, who designed the first targeted synthetic polymer–anticancer conjugates that successfully progressed to clinical trials [3, 4].

5.3 Types of Polymers Used for Polymer–Drug Conjugate

Polymers can be classified on the basis of many categories like (1) source (natural, semisynthetic, synthetic), (2) type of polymerization (addition, condensation polymers), (3) molecular forces (elastomers, fibers, thermoplastic, thermosetting, flexible, rigid), and (4) degradability (biodegradable, non-biodegradable). Although, according to drug delivery point of view, polymers can be broadly classified on the basis their architecture [25, 26] (Fig. 5.2) consisting four types: (1) linear polymers, (2) branched polymers, (3) cross-linked polymers, and (4) dendritic polymers.

5.3.1 Linear Polymer

Polyethylene is called a linear or straight-chain polymer because it consists of a long string of carbon–carbon bonds. These terms are misleading because the geometry around each carbon atom is tetrahedral and the chain is neither linear nor straight. As the polymer chain grows, it folds back on itself in a random fashion to form structures of stack of pile, e.g., polyethylene glycol (PEG), chitosan, mannose, N-(2-hydroxypropyl) methacrylamide (HPMA), hyaluronic acid, etc. However, they are further classified as homopolymer, diblock, triblock, and random polymer depending upon way of polymerization of two monomers.

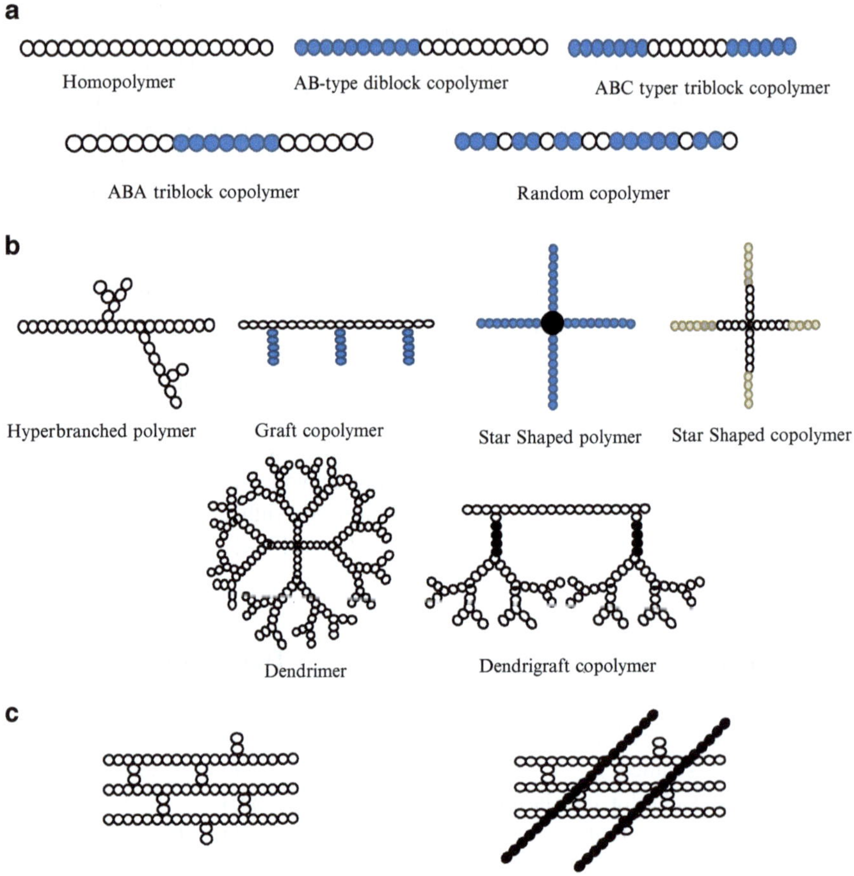

Fig. 5.2 Polymers classification according to their architecture. (**a**) Linear polymers, (**b**) branched polymers, and (**c**) crosslinked polymers

5.3.2 Branched Polymer

Polymers with branches at irregular intervals along the polymer chain are called branched. These branches make it difficult for the polymer molecules to pack in a regular array and therefore make the polymer less crystalline and less dense. The amount and type of branching also affects physical properties such as viscosity and elasticity. Branches often prevent chains from getting close enough together for intermolecular forces to work effectively. They are further classified into star-shaped polymers, star-shaped block polymers or graft polymers, dendigraft, or even dendrimer [27]. However, dendrimers are considered as fourth type of polymer because of their conformational complexity.

5.3.3 Cross-Linked Polymers

An additional type of polymer is known as the cross-linked polymer. Cross-linked polymers contain short side chains that connect different polymer chains into a "network." They can also be subdivided into polymer network, interpenetrating polymer network, or semi-penetrating polymer network. Cross-linked polymers are made firstly by adding cross-links between polymer chains that makes the polymer more elastic; later these polymer forms long chains, either branched or linear, that can form covalent bonds between the polymer molecules. Because cross-linked polymers form covalent bonds that are much stronger than the intermolecular forces that attract other polymer chains, the result is a stronger and more stable material. The higher the cross-linked polymer, the higher will be the viscosity and difficulty to conjugate them, e.g., rubber.

5.3.4 Dendritic Polymers or Dendrimer

This type of polymers can be classified under branched type of polymers, where network of branches follows dendritic architecture [28]. Research on these branched polymers started in 1970 by Vogtle and co-workers [29]. They were successful in developing dendritic arrangement of low-molecular-weight amine with large molecular cavities [29]. However, it was in 1984, when Tomalia and his team developed hyperbranched polymers with successive reaction of ethylenediamine to ammonia core, to produce a dendritic closed structure, reminiscence of a tree, whence dendrimers derive their name "Starburst Dendrimer" [30]. The word dendrimer is derived from the Greek words *dendri* (tree branch-like) and *meros* (part of) and was coined by Tomalia et al. [30]. Dendrimers are molecular (nano) architectures of well-defined size and number of terminal groups with certain symmetrical geometry, starting from a multifunctional core unit with structure branches in three dimensions from the inside outwards. These layers are designated as "generations," wherein there is an increase in the molecular weight and surface functional groups with each new generation (Fig. 5.3). A dendrimer may be based on practically any type of chemistry, the nature of which can determine its solubility, degradability, and biological activity. Some of the commonly encountered types of dendrimers in biological applications are based on polyamidoamines [31], polyamines [32], polyamides (polypeptides) [33], poly(aryl ethers) [34], polyesters [35, 36], carbohydrates [37], etc. Dendrimers can function as drug carriers either by encapsulating drugs within the dendritic structure or by conjugating drugs to terminal functional groups via electrostatic or covalent bonds [38]. This high density of functional groups on surface can be used to conjugate targeting moiety to dendrimers. Most common terminal functional groups are amine in polypropylene imine (PPI) dendrimer, polyamidoamine (PAMAM) dendrimer, and polymelamine dendrimer, hydroxyl group in polyester dendrimer, and carboxylic groups in poly(glutamic acid) dendrimer or

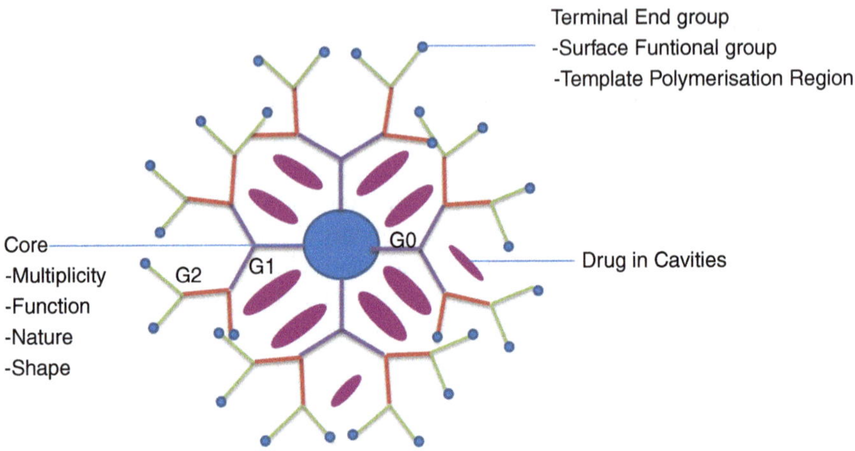

Fig. 5.3 Schematic representation of dendrimer architecture

in half generation PAMAM dendrimer. Apart from therapeutic drugs, dendrimers also serve as a unique platform for a variety of biotherapeutic and imaging agents [39]. Having a basic understanding of polymers will thus give us the opportunity to not only familiarize ourselves with the function of drug products but also possibly developing new formulations or better delivery systems.

5.4 Ringsdorf Model for Designing Polymer–Drug Conjugates

The concept of covalent polymer–drug conjugates was introduced by Helmut Ringsdorf in 1975, who called them synthetic polymeric drugs or pharmacologically active polymers [1]. Ringsdorf explained three main regions of PDC: a *hydrophilic region*, used to make the whole macromolecule soluble and nontoxic; a drug-bound region (fixed) where drug is linked through a *labile bond or linker* that is stable during conjugate transport and able to release drug at an optimum rate on arrival at the target site; and a third area that comprises a transport system or a *homing device*, whose function is to carry the whole cargo to the target cells or site of pharmacological action. Hence, one can consider the possibility of fashioning a polymeric drug with which one can separate different requisite requirements by using distinct areas along the polymer chain to accomplish specific effects. All the phases of drug action affected by polymer characteristics like (a) the pharmaceutical phase, e.g., by influencing disintegration time; (b) the pharmacokinetic phase, e.g., change in ADME of drug; and (c) the pharmacodynamic phase, e.g., by variations in drug–receptor interactions and possible cell-specific effects (Fig. 5.4).

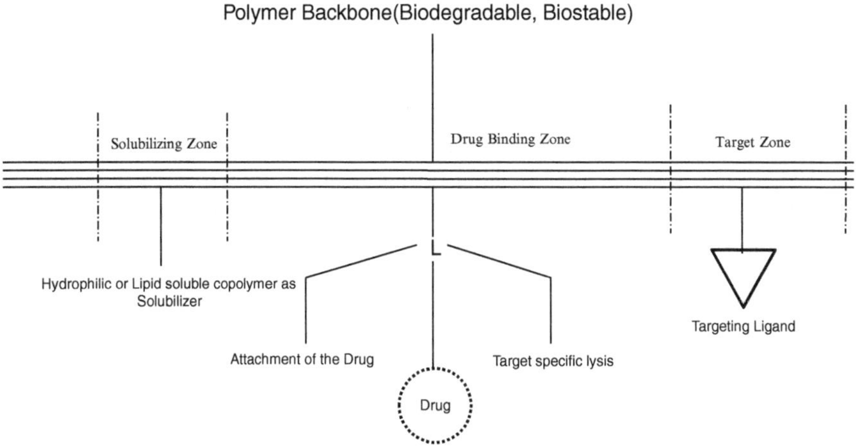

Fig. 5.4 Ringsdorf's proposed model for polymer–drug conjugate

5.5 Classification of Polymer–Drug Conjugates

Polymer–drug conjugate can be classified in two major ways. One of that is inspired from Ringsdorf's model for polymeric drugs. Another classification is based on intrinsic activity of polymer used in designing PDC.

5.5.1 Classification of PDC Based on Ringsdorf's Model

5.5.1.1 End Group System

Simple linear polymers are explored to their maximum for end group conjugation. When attachment of drug is done on the terminal groups of polymers, then they are termed as end group PDC system. Polyethylene glycol (PEG) [40, 41] has been most widely used in end group synthetic PDC. All modifications on PEG are based on the replacement of the end hydroxyl group. Molecular weight of PEG chains and site of conjugation influence the final properties of the conjugates like increasing elimination half-life, escaping reticuloendothelial system [42, 43], etc.

Dendrimers are also included in this category, where homing device or ligands and drugs are attached to terminal functional groups on the surface of dendrimers [44].

5.5.1.2 Pendent Group System

This type of polymer–drug conjugate is purely based on the Ringsdorf model. Simple free radical copolymerization reactions lead to a range of different compositions of polymer. These polymers may present a single or multiple reactive pendant groups. The number of reactive functional groups can be generated by reacting copolymer with reactive groups like benzene ring in polystyrene, methyl group in polypropylene, and hydroxyl group in polyvinyl alcohol. Copolymers of HPMA are the best-studied pendant group system, and their anticancer drug conjugates have already entered early clinical trial pipeline [45]. HPMA monomers are copolymerized with drug or methacrylic monomer bonded through a pH sensitive or a stable, biocompatible peptide [46–49]. Ulbrich et al. [50] studied the effect of polymer chain modification on HPMA polymer. The presence of carboxylic groups in the copolymer structure resulted in increase in doxorubicin (DOX) release rate to 15–20 %, while no such effect with introduction of positively charged groups was observed, when compared to the unmodified conjugate. They also reported that the molecular weight and molecular architecture significantly influenced the biodistribution and antitumor activities of their DOX conjugates in tumor-bearing mice [51].

5.5.2 PDC Classification on the Basis of Intrinsic Activity of Polymer

Some of the polymers attached to drugs are themselves biologically active. Hyaluronic acid is considered as a polymer that exhibits molecular weight-dependent biological activity. Low-molecular-weight hyaluronic acid elicits proinflammatory responses by modulating the toll-like receptor-4 or by activating the nuclear factor kappa B (NF-kB). In contrast, high-molecular-weight HA manifests an anti-inflammatory effect via CD receptors and by inhibiting NF-kB activation and is also known to induce receptor-mediated endocytosis [52–54]. Chitosan is another polymer that shows antifungal activity [55]. This activity was also explored by Uher et al. [56]. On the other hand, there are other polymers with no biological activity but known to affect pharmacokinetic properties of drugs, e.g., HPMA, PLGA, polylactic acid, polyglycolic acid, and polyethylene glycol (PEG) [42, 57].

5.6 Rationale Behind Design

PDCs are often termed as "polymeric prodrugs," as "prodrugs" are chemical entity of an active parent drug with altered physicochemical properties like increased aqueous solubility and enhanced biodistribution while retaining the inherent pharmacological properties of the drug. The utilization of prodrugs, to a certain extent, allows for the preservation of specific activity of a drug and targets its release to certain cells or their organelles. Plethora of rationale behind PDC designing makes this area charismatic. Every structural part of a PDC has a prominent role, but the most significant parts are its polymeric backbone and spacer or linker. In general, an ideal polymeric prodrug model consists mainly a combination of one or more components:

- A polymeric backbone as a vehicle
- Linker(s) for hydrolysis of the biomolecule and versatility for conjugation
- One or more drugs of the biological active components
- An imaging agent
- Targeting moiety

5.6.1 Polymeric Backbone

Polymeric backbone is selected on the basis of chemical nature (vinylic or acrylic polymers, polysaccharides, poly(α-amino acids), etc.), biodegradability, origin (either natural polymers or synthetic polymers), and molecular weight and lastly on the basis of ease of conjugation as polymers of high molecular weight may have high viscosity or can show high crowding effect [58]. Charge on polymer is an important criterion for selection, e.g., negatively charged polymers can stimulate coagulation in blood circulation, whereas positively charged polymers, e.g., polyethylenimine, chitosan, and PAMAM dendrimer known for effective cellular internalization can cause hemolysis and induce reticuloendothelial system (RES). Further, there are other requirements that should be kept in mind before selecting a polymer for PDC [59]:

- Well-characterized structure.
- Hydrophilicity for intravenous drug delivery.
- Biocompatible (nontoxic and nonimmunogenic). Metabolic products generated from the parent polymers must also be biocompatible.
- Functionality to allow conjugation of the drug in sufficient amounts.
- Allow for drug linkage, which is stable in circulation.
- Result in a construct that allows convenient administration.
- Eliminated from the body once they have performed their function or metabolized to smaller fragments that are below the renal threshold and are subsequently eliminated.

5.6.2 Linker

PDC have progressed to targeted type of drug delivery systems. They can selectively target lysosomal component, tumor microenvironment, etc. to release their cargo specifically to the site where it was meant to be released by certain biological thresholds like acids, enzymes and temperature. The linkers have emerged as a key to open the ways to focal drug delivery. Linker is an important part of PDC. It should be stable in the blood circulation but needs to be degraded by any of the above-mentioned biological triggers. There are many linkers available and studied. Linkers can be classified on the basis of their length [60] and according to cleavable conditions (Table 5.1). These linkers are intensively presented by Wagner et al. [61].

Table 5.1 Examples of different cleavable groups and their applications

Cleavage conditions	Cleavable group	Applications
Enzymes	TEV, trypsin, thrombin, cathepsin B, cathepsin D, cathepsin K, caspase matrix metalloproteinase sequences, phosphodiester, phospholipid, ester, β-galactose	Protein purification, imaging enzyme activity and tumor, drug delivery, DNA sequencing, metabolite enrichment
Nucleophilic	Dialkyl dialkoxysilane, cyanoethyl group, sulfone, ethylene glycolyl disuccinate, 2N-acyl nitrobenzene sulfonamide, α-thiophenylester, unsaturated vinyl sulfide, sulfonamide after activation, malondialdehyde (MDA) indole derivative, levulinoyl ester, hydrazone, acylhydrazone, alkyl thioester	Protein modification and purification, structural biology, imaging, synthesis of oligonucleotides
Reducing reagents	Disulfide bridges, azo compounds	Protein modification and purification, structural biology, tumor targeting, imaging, visualization of PEG shedding, drug delivery
Photo irradiation	2-Nitrobenzyl derivatives, phenacyl ester, 8-quinolinyl benzenesulfonate, coumarin, phosphotriester, bis-arylhydrazone, bimane bi-thiopropionic acid derivative	Protein purification, imaging protein activity, structural biology, drug delivery, DNA sequencing, metabolite enrichment
Electrophilic/acidic reagents	p-Methoxybenzyl derivative, tert-butylcarbamate analogue, dialkyl or diaryl dialkoxysilane, orthoester, acetal, aconityl, hydrazone, b-thiopropionate, phosphoramidate, imine, trityl, vinyl ether, polyketal, alkyl 2-(diphenylphosphino)benzoate derivatives	Protein purification, structural biology, drug delivery
Organometallic and metal catalyst	Allyl ester, 8-hydroxyquinoline ester, picolinate ester	DNA sequencing
Oxidizing reagents	Vicinal diols, selenium compound	Structural biology

5.7 Physicochemical Characterization of Polymer–Drug Conjugates

Greco and Vicent explained the physicochemical characterization of PDC [62]. When compared to low-molecular-weight compounds, PDC are considered as relatively difficult systems for absolute characterization. These systems are considered as intrinsically heterogeneous due to several reasons. First, most of the polymeric carriers lack monodispersity. Second, covalent conjugation of a drug to the carrier is often a random process, although optimization of reaction conditions ensures a good degree of batch-to-batch reproducibility, the point of attachment of the drug within the chain remains uncontrollable in many cases. Attachment of a second drug to the same carrier complicates the matter even further. As a result, an adequate physicochemical characterization is somewhat difficult to achieve. Polymer molecular weight is another challenge to choose characterization technique. Polymers having molecular weight exceeding 100 kDa are provable by matrix-assisted laser desorption/ionization mass spectrometry (MALDI-ToF-MS). Further, molecular weight distributions are determined by MALDI-ToF [63]. Solution properties of polymer were also exploited by small-angle neutron scattering (SANS) to understand the aggregation process of pharmaceutical PDC formulation in solution form [64]. Small-angle neutron scattering (SANS) is another technique to elaborate the PDC structure, size, shape, even aggregation with other polymer chain or interaction with cell membrane [65, 66].

5.8 Applications and Current Status of Polymer–Drug Conjugates in Drug Delivery

Among the new drug delivery systems, PDC have been considered as promising carriers for anticancer agents. To promote optimal targeting of cancer cells, several strategies for conjugation of drugs into polymer carriers have been developed. In PDC approach, drugs are conjugated to polymers as a means of allowing drugs to accumulate at the tumor site by the enhanced permeability and retention effect. PDCs have a lot of potential applications in pharmaceutical industries. In recent studies, it is reported that polymer–drug conjugates are not only applicable in cancer treatment but also in other therapeutic categories. In this section, the application part was classified according to therapeutic categories and kind of targeting.

5.8.1 Application of Polymer–Drug Conjugates in Cancer

Most of the PDC that entered into clinical trials belong to antitumor therapeutic category and that also with specific established chemotherapeutic agents, like paclitaxel, doxorubicin, and platinates. Results from early clinical trials of about a dozen PDC have demonstrated several advantages over the respective parent drugs, including fewer side effects, enhanced therapeutic efficacy, ease of drug administration, and improved patient compliance. Advanced conjugates containing experimental chemotherapy and novel polymer-based combinations are undergoing clinical studies after the successful studies on established molecules. Major responses have been observed in a variety of solid tumors, including gastric cancer, colorectal cancer, lung cancer, ovarian cancer, and breast cancer [67]. Clinical trials status of different polymer–drug conjugates is discussed in next the chapter (Table 5.2).

Table 5.2 Polymer–drug conjugates undergoing clinical trials and research

Conjugates	Indication	Company
HPMA–doxorubicin (PK1; FCE28068)	Lung and breast cancers	Pfizer
HPMA–doxorubicin–galactosamine (PK2, FCE28069)	Hepatocellular carcinoma	Pfizer
HPMA–camptothecin (PNU166148)	Solid tumors	Pfizer
HPMA–paclitaxel (PNU166945)	Solid tumors	Pfizer
HPMA–platinate (AP5346, ProLindac)	Ovarian, melanoma, and colorectal cancers	Access pharmaceuticals
HPMA copolymer–doxorubicin–galactosamine (AP5280)	Cancer	–
PEG–camptothecin (pegamotecan)	Solid tumors	Enzon
PEG-SN38 (EZN-2208)	Solid tumors	Enzon
Cyclodextrin-based polymer–CPT (IT-101)	Solid tumors	Insert therapeutics
Carboxymethyldextran–exatecan (DE-310)	Solid tumors	Daiichi Sankyo
PG–TXL (CT-2103, Xyotax)	Lung, ovarian, colorectal, breast, and esophageal cancers	Cell therapeutics
PG–camptothecin (CT2106)	Colorectal, lung and ovarian cancer	Cell therapeutics
PEG–naloxone (NKTR-118)	Opioid-induced constipation	Nektar
HPMA–NPC1161	Leishmaniasis	–
k-Carrageenan–zidovudine	Antiviral	–
HPMA–prostaglandin E1	Osteoporosis and other musculoskeletal diseases	–

5.8.1.1 HPMA–Doxorubicin Conjugate Named PK1-FCE28068

The first polymer–drug conjugate-based chemotherapeutic drug to enter clinical trials in 1994 was an HPMA–doxorubicin conjugate named PK1-FCE28068 [68]. In this conjugate, doxorubicin was linked to the HPMA copolymer via a tetrapeptide chain comprised of Gly-Phe-Leu-Gly. It has a molecular weight of ~30,000 Da with doxorubicin loading of ~8.5 % (w/w). Designed peptidyl linker was to be hydrolyzed by thiol-dependent proteases after entering into lysosomes. In phase I trials, PK1 was given as a short infusion every 3 weeks, and it had a maximum tolerated dose of 320 mg/m^2 (doxorubicin equivalent) [69]. This targeted system showed nearly fourfold higher than the advised safe clinical dose of doxorubicin. Maximum tolerated dose of PK1-FCE28068 was also found higher then doxorubicin liposomes.

5.8.1.2 HMPA Copolymer–Camptothecin (MAG–CPT, PNU166148)

Two selected properties of camptothecin (CPT) compounds limit its therapeutic efficacy in humans: first one is the instability of the lactone and second is its limited aqueous solubility. Conjugation of water-soluble polymer carrier resulted in improved stability of its lactone ring and additionally higher aqueous solubility. In HPMA–CPT (MAG–CPT, PNU166148), CPT is conjugated to the copolymer at the C-20 hydroxyl group of CPT through an ester linkage. This conjugation helped in increasing the half-life of CPT up to 6 days which shows the betterment of drug as original reported half-life of CPT is only 1.3 h [10, 70].

5.8.1.3 HMPA Copolymer–Paclitaxel (PNU166945)

In HPMA copolymer–paclitaxel (PNU166945), paclitaxel was linked to the same tetrapeptide linker that was used to create PK1 discussed in Sect. 5.8.1.1. In a phase I clinical study in a small patient cohort, one patient with advanced breast cancer had a partial response. Some authors reported that dose-limiting toxicity of PNU166945 (HPMA–PTX) was not observed at the studies dose levels up to 196 mg/m^2, as paclitaxel equivalents; therefore, the maximum tolerated dose (MTD) was not reached. The plasma AUC of HPMA–TX increased from 318 to 450 h·μM corresponding to the starting dose of 80 mg/m^2 and escalating to 196 mg/m^2. The concentration C_{max} also increased from 40 to 75 μM, corresponding with the doses. The $t_{1/2}$ and total clearance was about 6.0–6.5 h and from 0.5 to 0.9 L/h, respectively, for given doses. However, the trial study was discontinued due to severe neurotoxicity observed in rat studies [71]. This formulation showed toxicity consistent with commonly observed toxicities associated with paclitaxel [72].

5.8.1.4 HMPA Copolymer–Platinate (AP5346)

Many platinum-based compounds, like cisplatin and carboplatin, are standard treatment regimens for various types of cancers. The dose-limiting toxicity of these platinates includes nephrotoxicity, neurotoxicity, and myelosuppression. 1,2-Diaminocyclohexane (DACH) has received increasing attention in recent years because of its activity against cisplatin-resistant cancer cells. In AP5346, DACH–platinum was bound to HPMA via a pH-sensitive chelate. A methacrylamide monomer substituted with a triglycine aminomalonate group provided the primary binding site for the DACH–platinum moiety [73]. While negligible in neutral solutions, release of the DACH–platinum moiety was increased at low pH (tumor pH) so that platinum release was enhanced in environments such as the extracellular space of hypoxic tumors and the intracellular lysosomal compartment. The pharmacokinetics of AP5346 indicated a prolonged half-life ($t_{1/2}$ = 72 h) and evidence of high antitumor activity. Additional human clinical trials are planned for this agent.

5.8.1.5 PEG–Camptothecin (Pegamotecan) and PEG-SN38 (EZN-2208)

Pegamotecan was developed by Enzon Pharmaceuticals, Inc. The conjugate consists of two CPT molecules conjugated to a 40-kDa PEG using an alaninate ester linkage.

Free CPT must be cleaved from the PEG to give pharmacological action. The hydroxyl group (-OH) at the 20-position of CPT is the active portion of the molecule responsible for the conformational changes between the active lactone. The 20-OH of CPT in pegamotecan, is blocked by the alaninate linker, which stabilizes the CPT molecule into its active lactone conformation [74]. The main limitation of PEG as drug carrier is the presence of only two reactive groups per polymer chain, which leads to an intrinsically low drug payload. To overcome this limitation, the construction of a dendron structure at the PEG's end chain has been proposed. Enzon is developing a conjugate of SN38, an active metabolite of CPT, with a 40-kDa PEG containing four arms. EZN-2208 has shown activity in a panel of human tumor xenografts [68].

5.8.1.6 Cyclodextrin-Based Polymer–CPT (IT-101)

IT-101 is a conjugate of CPT and a linear, cyclodextrin-based polymer (CDP) [70]. The components of CDP are beta-cyclodextrin and polyethylene-glycol (PEG). Camptothecin is covalently attached to CDP through a glycine linker, which preserves CPT in its active form and increases its water solubility. Preclinical efficacy of intravenous IT-101 (molecular weight ~90 kDa) demonstrated a significant antitumor effect with IT-101 in a range of mouse xenografts, including several tumors that were resistant to CPT [68]. Pharmacokinetic and biodistribution studies indicated that CDP–CPT conjugate gave prolonged plasma half-life, which led to enhanced distribution to tumor when compared to CPT alone. Taken together, research indicated that IT-101 has good tolerability and antitumor activity against a wide range of tumors [75].

5.8.1.7 Carboxymethyldextran–Exatecan (DE-310)

In DE-310, DX-8951 (exatecan mesylate, a CPT analogue) is linked to carboxymethyldextran (340 kDa) via a glycyl-glycyl-phenylalanyl-glycyl-peptidyl spacer. This peptide spacer is intended to provide sustained release of the active moiety DX-8951 within the tumor as a result of enzymatic cleavage of the peptide by cathepsin B and cathepsin L [76]. The results suggested that the mechanism by which DX-8951 is released from DE-310 in vivo is involved in the process of uptake of DE-310 into tumor or macrophages, digestion by intracellular lysosomal cysteine proteinase, and subsequent secretion of the drugs [77]. Neutropenia, thrombocytopenia, and grade 3 hepatotoxicity with veno-occlusive disease were dose-limiting toxicities. The apparent half-life of conjugated DX-8951, glycyl-DX-8951, and DX-8951 was 13 days. The observed long blood half-lives are not surprising, considering that DE-310 had the highest molecular weight among all polymer–drug conjugates tested so far.

5.8.1.8 PG–Gly–CPT (CT-2106)

CT-2106, a camptothecin (CPT) conjugate, is a new generation of topoisomerase I inhibitors designed to deliver higher, more effective chemotherapy to tumor tissue with less toxicity to normal tissues [78]. PG was an effective solubilizing carrier of CPT and stabilized the E-ring lactone structure in CPT. PG–CPT conjugate was obtained by directly coupling the hydroxyl group at the C20(S)-position of CPT with the carboxylic acid of PG [79]. When given intravenously in four doses every 4 days at an equivalent CPT of 40 mg/kg, PG–CPT delayed the growth of established H322 human lung tumors grown subcutaneously in nude micc. In mice that received intratracheal inoculation of H322 cells, the same treatment prolonged the median survival duration in mice by fourfold when compared with that in untreated control mice. These results showed that PG was as efficient carrier of CPT [79]. Studies have systematically investigated the structural effects of the antitumor efficacy of PG-bound CPT, including linkers between PG and CPT, the point of attachment of PG on the CPT molecule, the polymer molecular weight, and drug loading. Coupling through the 20 (S)-hydroxyl group of CPT with or without a glycine linker yielded the most active conjugates, because this site is located in close proximity to the lactone ring; thus, the E-ring lactone is better protected by the linked PG chains. However, the linkers affected the maximum drug payload; for example, only 15 % (w/w) of CPT loading could be achieved for direct conjugation of PG–CPT by the ester linkage because of steric hindrance, whereas up to 50 % CPT loading could be achieved for the case with a glycine as linker. Increasing the molecular weight of PG from 33 to 50 kDa improved the antitumor efficacy of PG–Gly–CPT, probably because of an increased plasma half-life and reduced renal clearance. CT-2106 has a more manageable toxicity profile compared with unconjugated camptothecin. The maximum tolerated dose is 25 mg/m^2 weekly given 3 of 4 weeks. This compound results in prolonged release of unconjugated camptothecin [80].

5.8.2 Active Targeting with Polymer–Drug Conjugates

The major drawback of polymer–drug conjugates is the absence of specificity for cancer cells. Although one of the well-known water-soluble paclitaxel conjugates, poly(L-glutamic acid)–paclitaxel has shown better antitumor efficacy in vivo, but this "binary nanoparticle" also binds to the proteins and enzymes of normal cells. The development of ternary biomolecules is also underdevelopment to overcome the problem of nonspecific binding and to allow a drug to reach tumor cells specifically. Ternary nanomolecules are composed of three elements: (a) drug carrier, in most cases, a polymer; (b) a drug; and (c) a ligand. Many researchers have studied several targeting ligands such as antibodies, cytokines, and homing peptides to improve the tumor selectivity of polymeric drug carriers.

Drugs conjugated to PG are also being extensively studied. Doxorubicin and daunorubicin have also been attached to PG and have shown activity in vivo against L1210 leukemia and B16 melanoma cells, although the cytotoxic effect was weaker than the free drug [81, 82]. Similar results in vivo have been reported when PG was conjugated with arabinofuranosyl cytidine, cyclophosphamide, and melphalan [82]. However, in vivo cytotoxic effect has been shown to be better than free drug for all these three drug–PG conjugates. Wosikowski et al. were able to show that methotrexate bound to human serum albumin enters cells by albumin-mediated endocytosis. They also observed that thymidylate synthesis was inhibited when KB cells expressing folate receptors were exposed to this conjugate. Polymer–drug conjugates can passively accumulate in the tumor tissue. However, it has been suggested that the addition of a targeting moiety would increase selectivity by actively targeting the tumor. A number of strategies have been adopted to achieve active targeting, including the use of peptides such as melanocyte-stimulating hormone [83], EBV peptide to promote targeting to lymphocytes [84], RGD4C peptide to promote targeting to endothelial cells ($\alpha_v\beta_3$ integrin receptor) [85], luteinizing hormone-releasing hormone (LHRH) used to target LHRH receptors [10], antibodies such as antitransferrin receptor antibody [57], or other ligands such as folate. However, to date only few clinical studies have been conducted on conjugates containing targeting agents.

5.8.2.1 HMPA Copolymer–Doxorubicin–Galactosamine (PK2)

PK2 is a 27-kDa HMPA copolymer synthesized with 6.5 % mol/wt, <2 % free doxorubicin, and 2 % mol/wt galactose, with efficient targeting of the asialoglycoprotein receptor selectively expressed on hepatocytes and in hepatomas. In preclinical murine models, 80 % of an administered dose targeted the liver, associated with high anticancer activity. PK2 remained the only targeted polymeric conjugate to be tested clinically to date [57]. In phase I/II trial, the maximum tolerated dose of PK2 was 160 mg/m^2, with neutropenia being the dose-limiting toxicity. The toxicity profile was similar to that seen for anthracyclines, but interestingly the maximum tolerated dose was half of that of the nontargeted conjugate.

5.8.2.2 HPMA Copolymer–DOX–Human Immunoglobulin Conjugate

The other targeted conjugate tested clinically is HPMA copolymer–DOX–human immunoglobulin (HuIg) conjugate [86]. Preliminary clinical experiments were carried out in four patients with this HPMA conjugate. Antitumor activity was seen in some of the patients, but conclusions are difficult to draw since this trial was not carried out according to “Good Clinical Practice” (GCP) guidelines.

5.8.3 *Polymer–Drug Conjugate Application in Other Diseases*

Polymer–drug conjugates can also be used for the treatment of diseases other than cancer [87].

5.8.3.1 HPMA–Antileishmaniasis Conjugate

In 2001, the first HPMA copolymer carrying an antileishmanial drug 8-(4-amino-1-methylbutyl) amino)-5-(3,4-dichlorophenoxy)-6-methoxy-4-methylquinoline (NPC1161) was proposed [88]. Later on, the same group improved this conjugate with the addition of N-acetylmannosamine to target the specific mannose receptor of macrophages [88]. Studies carried out in macrophages showed that the targeted conjugate had a high uptake as compared to the nontargeted conjugates. Domb et al. [89, 90] investigated a method to conjugate amphotericin B (AmB), a water-insoluble antifungal and antileishmanial agent, to arabinogalactan via tosylate or mesylate derivatives with an amine bond. The conjugates showed comparable inhibitory concentration values against the pathogenic yeast *Candida albicans* and against *Leishmania major* parasites. Conjugates were about 60 times less hemolytic against sheep erythrocytes than the free drug and less toxic when injected intravenously to BALB/c mice. Similarly in another study, the contribution of aldehyde groups to the toxicity of polymer–drug conjugates, such as dextran-AmB was evaluated. It was found that oxidized dextran was toxic against the RAW 264.7 cell line with an IC50 of 3 μmol/mL aldehydes. Modification of aldehyde groups and their reaction with ethanolamine reduced the toxicity to least 15-fold [91].

5.8.3.2 Zidovudine–Polymer Conjugates

A large number research groups are working on development of zidovudine (AZT) polymer conjugates for the treatment of HIV. The objective behind many of these has been to develop agents with higher potency along with improved toxicological profile. However, only a few studies have investigated the potential of developing controlled release conjugates. One of these studies was described by Vlieghe et al. [92]. They used kappa-carrageenan as polymer carrier, and AZT was linked through an ester bond. In vitro studies using MT-4 cells showed that the conjugates displayed increased anti-HIV activity as compared to the free drug.

5.8.3.3 Musculoskeletal Diseases

Bone-targeted HPMA copolymer conjugate with a bone anabolic agent prostaglandin E1 (PGE1) was designed for the treatment of osteoporosis and other musculoskeletal diseases [93]. The specificity of the conjugates by the skeleton was carried out by an octapeptide of D-aspartic acid (D-Asp8) or alendronate [94]. After attachment to bone, the PGE1 selectively released at the sites of higher osteoclast activity. On administration in anabolic dose range, the released PGE1 activates specific receptors on bone cell surface to attain net bone formation [95]. Remarkably, preferential deposition of D-Asp8-targeted conjugate to the bone resorption sites was observed in ovariectomized rats. In vivo experiments on ovariectomized rats have proven the concept. Following a single intravenous administration of the HPMA copolymer–D-Asp8–PGE1 conjugate to ovariectomized rats, bone formation rates were substantially greater than controls. Obviously, a similar concept can be used for targeting bone cancer metastasis.

5.8.3.4 Inflammatory and Infectious Diseases

Wang et al. have proved in their research that macromolecular therapeutics specifically accumulates in inflammatory tissues [96]. They gave the new targeting mechanism "ELVIS" (Extravasation through Leaky Vasculature and the subsequent Inflammatory cell-mediated Sequestration) [97]. This concept shows large potential of polymer–drug conjugates in the treatment of inflammatory disease like osteoarthritis. Preliminary studies with polymer–drug conjugates to prevent scar tissue formation have also been described [87]. They used a multivalent dendrimer conjugate of glucosamine or glucosamine-6-sulfate. They showed that PAMAM generation 3.5-glucosamine inhibited synthesis of proinflammatory chemokines and cytokines. Also, PAMAM generation 3.5 glucosamine-6-sulfate blocked fibroblast growth factor-2-mediated endothelial cell proliferation and angiogenesis. More importantly, combination therapy with these two conjugates prevented scar tissue formation after glaucoma filtration surgery.

5.9 Advantages of PDC

Polymer–drug conjugates offer several advantages over the free drug.

- *Enhanced Permeability and Retention (EPR) Effect*: In order to stimulate tumor cell grow quickly and to satisfy the ever-increasing nutrient demands of tumor, tumor induces angiogenesis [98]. The blood vessels in the tumor are irregular in shape, dilated, leaky, or defective, and the endothelial cells are poorly aligned or disorganized with large fenestrations. This defective tumor architecture with other combinations of deficient drainages results in a pooling of high-molecular-weight agents in the tumor tissue. This phenomenon is termed as EPR effect [99]. Passive accumulation of polymeric drug conjugate can be possible by selecting appropriate molecular weight of polymers.
- *Escaping Reticuloendothelial System (RES)*: Polymers can sterically shield high-molecular-weight compounds such as antibodies from the antigen-presenting

cells (APCs), which initiate the immune response and reduce the immunogenicity of immunogenic drugs. Increasing the molecular weight can increase the size of the drug and stave off its possibility of renal elimination hence increases the half-life and thereby prolonging the duration of action of drug. The increased plasma residence of PDC (compared to free drug) encourages passive accumulation within solid tumor tissue. PEG is linear or branched polymer, exists in different molecular weights and has been exploited to the most [100]. Many biologics secure their ways to get into the market like in Pegasys® (peginterferon alfa-2a) [101, 102], PEGINTRON® (peginterferon alfa-2b), Somavert® (pegvisomant) [103], Neulasta® (pegfilgrastim) [104], etc.

- As Ringsdorf's model predicted that hydrophilic portion of polymer carrier can aid in increasing the solubility of hydrophobic drugs thus can be easily *administered intravenously*.
- Low-molecular-weight drugs enter cell cytoplasm via diffusion and are exposed to plasma membrane efflux pumps, e.g., p-glycoproteins. Polymer assists *endosome-mediated internalization* of PDC drug conjugate assist drug entry into cell. Endosome encapsulation renders the efflux of drug ineffective. These polymers also help to buffer the effect of pH in the endosomes [105].
- *Targeting moieties* or homing devise contribute in active targeting of the drug and polymer provide easy sites for conjugating these ligands.

5.10 Future Trends, Challenges and Opportunities

Although the pharmaceutical industry prefers to develop small-molecule anticancer agents that can be administered orally (patient compliant), macromolecular drugs like polymer conjugates are establishing their niche in modern chemotherapy. With recent studies polymer–drug conjugates showed a great hope for therapeutic problems. But they are also having many challenges in front of them like polymer architecture, drug combinations, and their characterization. These challenges and opportunities can be divided according to polymers, drugs, and diseases. Early works unveiled unexpected therapeutic benefits but raised new issues, in particular in relation to "system design." A better understanding of how drug combinations impact on cellular and molecular mechanisms is needed to rationally design new therapeutics. In nutshell it is to be believed that the interdisciplinary scientific approach to the applications of PDCs will result in their translation into the clinic within this decade.

5.10.1 Novel Polymers for Polymer–drug Conjugates

The development of polymeric carriers with better properties is an ongoing challenge. There is a need to develop high-molecular-weight biodegradable polymeric

Table 5.3 Difference between dendrimer and linear polymer–drug conjugates

Main properties	Dendrimers	Polymer–drug conjugates
Structure	Compact/globular (predefined molecular weight) with stringent structure control	Not compact and heterogeneous structure
Solubility	High aqueous solubility refine pharmacokinetic properties of drug	Low water solubility or depends on polymer attached
EPR effect	Nanosize increases the penetration and uptake by tumor tissue allow passive targeting	Depends on charge and molecular weight of polymer
Route of administration	Intravenous, intraperitoneal, ocular, transdermal, oral, intranasal, pulmonary, etc.	Mainly intravenous

carriers which can enhance the EPR-mediated tumor targeting. They have considerable potential for further evaluation as a treatment different type of cancers. On the other hand, there is also a need to move on from heterogeneous random coiled polymeric carriers towards better-designed polymer architectures. All the conjugates that have been tested clinically so far are linear. However, enormous research is going on developing the novel polymeric carries. These include hyperbranched polymers, star polymers, and dendrimers as of new carriers.

5.10.1.1 Dendrimers in Polymer–Drug Conjugates

Dendrimers are novel fourth class of polymer, and hence, their comparison with well-known linear polymer–drug conjugates is of real interest (Table 5.3). Although dendrimer–drug conjugates are result of covalent bonds, but globular and 3D structure of dendrimer dendronized polymers provide monodisperse nanoscale geometry with high end group density at their surface which provides potential for high drug loading. They have been proven as attractive candidates for anticancer drug delivery and specific targeting moieties. Although dendrimer-based imaging agents have been evaluated clinically, there is still a need to establish the safety and chemical characteristics of many dendrimer structures drawing in preclinical development. VivaGel® is an example of poly-L-lysine dendrimer-based drug conjugate in clinical development by Starpharma as a topical vaginal microbicide for HIV prevention [106]. The active ingredient of VivaGel® has been shown in scientific studies to inactivate and inhibit infection with viruses, including HIV and the genital herpes virus. Starpharma has a licence agreement with condom manufacturer Ansell Limited to develop a VivaGel®-coated antiviral condom and a licence agreement with Okamoto Industries Inc. in relation to the VivaGel®-coated condom for the Japanese market. Subsequently, Starpharma-completed animal studies demonstrated that the Starpharma dendrimer–docetaxel formulation was significantly more efficacious than docetaxel in a breast cancer model. The improved efficacy is understood to be due to a longer circulating half-life, extended release of the docetaxel, and targeting of the dendrimer–docetaxel construct to the tumor tissue.

Dendrimers are further used in brain targeting due to their positive charge and high functional group density to bind ligand on the surface. Leptin is a 30-amino acid peptide having ability to cross blood–brain barrier. Dendrigraft poly-L-lysine-PEG-Leptin30 was synthesized and complexed with plasmid DNA yielding nanoparticles. Complex found to be effecting and safe to deliver gene to brain [107]. Santos et al. [108] successfully functionalized generation 5 PAMAM dendrimer with hydrophobic alkyl chains that varied in length and number results into increased pDNA encapsulation and higher in vitro efficiency. Higher density of terminal functional groups on the surface allows conjugation of more than one type of ligands. Transferring (Tf) is a peptide that belongs to the transferrin family. Overexpressed Tf receptors on blood–brain barrier in Alzheimer's and Parkinson's disease are responsible for its internalization in brain. A dual-targeting drug carrier (PAMAM–PEG–WGA–Tf) based on the PEGylated fourth generation ($G=4.0$) PAMAM dendrimer with Tf and wheat germ agglutinin (WGA) on the periphery and doxorubicin loaded in the interior was synthesized. Authors demonstrated that PAMAM–PEG–WGA–Tf delivered 13.5 % of DOX in a period of 2 h, indicative of an enhanced transport ratio as compared to the ratio of 8 % for PAMAM–PEG–WGA, 7 % for PAMAM–PEG–Tf and 5 % for free DOX in the same period of time [109]. Dendrimers also showed immense potential to change the physiochemical properties of drugs like solubility and permeability. D'Emanuele et al. [110–112] conjugated many drugs to different generations of PAMAM dendrimer and found promising results. The main successes of dendrimer resulted on their appropriate, consistent, and optimized design parameters addressing physicochemical limitations of classical drugs (e.g., solubility, specificity, stability, biodistribution, and therapeutic efficiency) and their ability to overcome biological issues to reach the right targets (e.g., first pass effect, immune clearance, cell penetration, off target clearance, etc.). Improvement of pharmacokinetic (PK) and pharmacodynamic (PD) behavior of both drug–dendrimer conjugates and drug–dendrimer encapsulated verses plain drugs demonstrates their strong potential in medicine as nano carrier. Dendrimers have been evaluated as novel nano carrier in nearly all route of administration as presented in Table 5.4. Although dendrimers are one of the promising area in polymer–drug delivery, they are still far away from successful commercialization due to their generation dependent toxicity [127, 128].

5.10.2 Drug Combination in Polymer–Drug Conjugates

The use of polymer–drug conjugates has been generally limited to the delivery of a single drug. However, the polyvalency of polymeric carriers permits their use to deliver combination of different drugs. Drug combination has remarkable potential in future as it is becoming increasingly clear that multi-agent therapy is preferable for diseases such as cancer [129]. An HPMA copolymer carrying the aromatase inhibitor aminoglutethimide and the chemotherapeutic agent doxorubicin was the first conjugate that combined endocrine therapy and chemotherapy agents on a

Table 5.4 Overview of route of administration of dendrimer for therapeutic applications

Route	Targeting cells	Dendrimer type	Surface feature	Therapeutic area	Drug/imaging moiety	Reference
IV	Epidermal growth factor receptor overexpressed tumor cells	PAMAM	Epidermal growth factor conjugation	Improved nucleic acid targeting and tumor imaging	Quantum dots	[113]
	Fibroblast growth factor receptor overexpressed tumor cells	PAMAM	Fibroblast growth factor-1 conjugation	Ameliorated breast cancer cells targeting	–	[114]
	Murine B16 melanoma cells	PAMAM	PEG conjugation and cis-aconityl ph-sensitive linkage	Enhanced tumor targeting and release	Doxorubicin	[109]
	Folate receptor overexpressing tumor cells	PAMAM	Folic acid conjugation	Better breast cancer targeting	5-Fluorouracil	[109]
	Brain tumor	PAMAM	Transferring and wheat germ agglutinin duel ligand conjugation	Brain tumor targeting	Doxorubicin	[109]
	Brain capillary endothelial cells	Poly-L-lysine	Leptin conjugation	Brain tumor targeting	Red fluorescent protein (RFP) expressing plasmid	[107]
	Overexpress biotin-specific receptors	Partially acetylated PAMAM dendrimer	Biotin conjugation	Cervical cancer targeting	Fluorescein isothiocyanate	[115]
	Integrin receptors overexpressed tumor	PAMAM	RGD conjugation and pH-sensitive linkage	Glioma (brain tumor) targeting	Doxorubicin	[116]
	Tumor targeting	PAMAM	Polyethylene glycol conjugation	Better pharmacokinetics and tumor targeting by EPR	5-Fluorouracil	[117]
	–	PAMAM-alpha cyclodextrin	Mannose conjugation	Kidney targeting	DNA	[118]
	–	PAMAM	Folic acid conjugation	Activated macrophages in cancer tissue and arthritis joints	Indomethacin	[119]

TD	Skin	PAMAM	Size and surface charge	Influence ocular residence time and enhanced response	Indomethacin	[120]
				Hyperproliferative skin disease improvement	8-Methoxypsoralene	[121]
				Improved oral bioavailability and inflammatory	Ketoprofen and Diflunisal	[122]
OR	Transepithelial transport	PAMAM	Cationic charge and drug complexation	Improved oral bioavailability	5-Aminosalisylic acid	[123]
		PAMAM	Drug conjugation	Enhanced permeability and reduced Pgp efflux	Terfenadine	[112]
OC	–	Phosphorus dendrimers	Drug conjugation	Reduced hypertension and glaucoma with better formulation	Carteolol	[124]
	Remanence time on the cornea	PAMAM		Mitotic activity and mydriatic activity	Pilocarpine nitrate and tropicamide	[125]
Nasal	Nasal mucosa	PAMAM	Fluorescein isothiocyanate–PAMAM complex	Absorption of hydrophilic drugs	Fluorescein isothiocyanate-labeled dextran	[126]

IV intravenous, *TD* transdermal, *OR* oral, *OC* ocular

individual polymeric chain [6]. This conjugate displayed markedly increased antitumor activity in vitro in breast cancer cells, compared to the conjugate carrying only DOX, whose activity has been proven clinically [7]. Another group prepared a PEG conjugate containing the combination of epirubicin and nitric oxide (NO) [130]. The objective for this approach is twofold. First, epirubicin and nitric oxide (NO) have a synergistic effect. In addition NO displayed cardioprotective action which counterbalance epirubicin-induced cardiotoxicity. Using this concept of combination therapy, together with the use of targeting residues, Minko et al. have investigated the feasibility of a two tier targeting of camptothecin (CPT)–PEG conjugates to LHRH receptors and cellular antiapoptotic defense using a synthetic analogue of Bcl-2 homology 3 (BH3) domain peptide [131]. The multicomponent hyperbranched PEG polymer bearing an equimolecular amount of CPT, BH3, and LHRH moieties was almost a hundred times more cytotoxic and displayed enhanced antitumor activity when compared with other synthetic analogues. An alternative approach is to combine a polymer conjugate carrying a single drug with standard chemotherapy (administered as a free drug). For example, a phase III clinical trials compared PGA–paclitaxel + carboplatin versus paclitaxel + carboplatin [132]. The same conjugate has also been tested in combination with radiotherapy in a phase I trial for esophageal and gastric cancer [133]. In addition, combination of two conjugates each carrying a single therapeutic agent have also been suggested. For instance, Minko and co-workers tested free CPT, CPT–PEG, CPT–PEG–BH3, or CPT–PEG–LHRH conjugates and the mixture of CPT–PEG–BH3 and CPT–PEG–LHRH conjugates in human ovarian carcinoma cell [57]. It was demonstrated that conjugation of CPT to PEG increased its proapoptotic activity and that further enhancement was achieved by using BH3 peptide in a CPT–PEG–BH3 and LHRH peptide in a CPT–PEG–LHRH conjugates and their mixture. Also, Kopececk and colleagues found increased activity in vivo following administration of a mixture of an HPMA copolymer–DOX (chemotherapy) together with an HPMA copolymer–meso-chlorin e6 monoethylene diamine disodium salt (Mce6) conjugate (photodynamic therapy) [134]. Another elegant combination therapy approach is the concept of polymer-directed enzyme prodrug therapy (PDEPT) and polymer–enzyme liposome therapy (PELT) [135]. In the case of PDEPT approach, two conjugates are administered sequentially: one carries an anticancer agent (administered first) and the other carries an enzyme capable of degrading the linker between the carrier and the anticancer agent. Although combination PDC therapy accompanied with serious challenges like difference in loading capacity of polymer, difficulty in correlation of in vitro and behavior in vivo, and altered kinetics of drug release, Linker design to ensure optimum release kinetics for each drug depending on their molecular mechanism of action and careful selection of drug combination and its ratio. In addition there are substantial challenges for characterization of such complex constructs as every component must be quantified in the presence of the others [136].

These early research findings brought out unexpected therapeutic benefits but raised new issues, in particular in relation to “system design.” A better understanding of how drug combinations impact on cellular and molecular mechanisms is needed to rationally design new therapeutics. In nutshell it is to be believed that the interdisciplinary scientific approach to the applications of polymer–drug conjugate (PDC) will result in their translation into the clinic within this decade.

References

1. Ringsdorf H (1975) Structure and properties of pharmacologically active polymers. J Polym Sci Polym Symp 51:135–153, Wiley Online Library; 2007
2. Duncan R, Seymour LCW, Scarlett L, Lloyd JB, Rejmanova P, Kopecek J (1986) Fate of N-(2-hydroxypropyl)methacrylamide copolymers with pendent galactosamine residues after intravenous administration to rats. Biochim Biophys Acta (BBA)-General Subjects 880:62–71
3. Duncan R, Lloyd JB, Kopecek J (1980) Degradation of side chains of N-(2 hydroxypropyl) methacrylamide copolymers by lysosomal enzymes. Biochem Biophys Res Commun 94:284–290
4. Duncan R, Kopecek J (1984) Soluble synthetic polymers as potential drug carriers. Adv Polym Sci 57:51–101
5. Meerum Terwogt JM, ten Bokkel Huinink WW, Schellens JHM, Schot M, Mandjes IAM, Zurlo MG et al (2001) Phase I clinical and pharmacokinetic study of PNU166945, a novel water-soluble polymer-conjugated prodrug of paclitaxel. Anticancer Drugs 12:315
6. Vicent MJ, Duncan R (2006) Polymer conjugates: nanosized medicines for treating cancer. Trends Biotechnol 24:39–47
7. Vasey PA, Kaye SB, Morrison R, Twelves C, Wilson P, Duncan R et al (1999) Phase I clinical and pharmacokinetic study of PK1 [N-(2-hydroxypropyl) methacrylamide copolymer doxorubicin]: first member of a new class of chemotherapeutic agents—drug-polymer conjugates. Clin Cancer Res 5:83–94
8. Julyan PJ, Seymour LW, Ferry DR, Daryani S, Boivin CM, Doran J et al (1999) Preliminary clinical study of the distribution of HPMA copolymers bearing doxorubicin and galactosamine. J Control Release 57:281–290
9. Duncan R (2003) The dawning era of polymer. Nat Rev Drug Discov 2:347
10. Duncan R (2006) Polymer conjugates as anticancer nanomedicines. Nat Rev Cancer 6: 688–701
11. Gros L, Ringsdorf H, Schupp H (2003) Polymeric antitumor agents on a molecular and on a cellular level? Angew Chem Int Ed Engl 20:305–325
12. Maeda H, Seymour LW, Miyamoto Y (1992) Conjugates of anticancer agents and polymers: advantages of macromolecular therapeutics in vivo. Bioconjug Chem 3:351–362
13. Pasut G, Veronese F (2007) Polymer–drug conjugation, recent achievements and general strategies. Prog Polym Sci 32:933–961
14. Pack DW, Hoffman AS, Pun S, Stayton PS (2005) Design and development of polymers for gene delivery. Nat Rev Drug Discov 4:581–593
15. Wagner E, Kloeckner J (2006) Gene delivery using polymer therapeutics. Polym Ther I:135–173
16. Niidome T, Huang L (2002) Gene therapy progress and prospects: nonviral vectors. Gene Therapy 9:1647–1652
17. Luo D, Haverstick K, Belcheva N, Han E, Saltzman WM (2002) Poly (ethylene glycol)-conjugated PAMAM dendrimer for biocompatible, high-efficiency DNA delivery. Macromolecules 35:3456–3462
18. Nitecki DE (1994) Preparation of an activated polymer ester for protein conjugation. United State Patent No. 5281698

19. Jatzkewitz H (1955) Peptamin (glycyl-L-leucyl-mescaline) bound to blood plasma expander (polyvinylpyrrolidone) as a new depot form of a biologically active primary amine (mescaline). Z Naturforsch 10:27–31
20. Muelhaupt R (2004) Hermann Staudinger and the origin of macromolecular chemistry. Angew Chem Int Ed 43:1054–1063
21. Shumikhina K, Panarin E, Ushakov S (1966) Experimental study of polymer salts of penicillins]. Antibiotiki 11:767
22. De Duve C, De Barsy T, Poole B, Trouet A, Tulkens P, Van Hoof F (1974) Commentary. Lysosomotropic agents. Biochem Pharmacol 23:2495
23. Regelson W, Parker G (1986) Clinical science review: the routinization of intraperitoneal (intracavitary) chemotherapy and immunotherapy. Cancer Investig 4:29–42
24. Pratten MK, Duncan R, Cable HC, Schnee R, Ringsdorf H, Lloyd JB (1981) Pinocytic uptake of divinyl ether-maleic anhydride (pyran copolymer) and its failure to stimulate pinocytosis. Chem Biol Interact 35:319–330
25. Qiu LY, Bae YH (2006) Polymer architecture and drug delivery. Pharmaceut Res 23:1–30
26. Kumar N, Ravikumar MN, Domb A (2001) Biodegradable block copolymers. Adv Drug Deliv Rev 53:23–44
27. Tomalia D, Uppuluri S, Swanson D, Brothers H, Piehler L, Li J et al (1998) Dendritic macromolecules: a fourth major class of polymer architecture–new properties driven by architecture. MRS Proceedings. Cambridge University Press
28. Tomalia DA (2005) The dendritic state. Mater Today 8:34–46
29. Buhleier E, Wehner W, Vogtle F (1978) Cascade and nonskid-chain-like synthesis of molecular cavity topologies. Synthesis 2:155–158
30. Tomalia D, Baker H, Dewald J, Hall M, Kallos G, Martin S et al (1985) A new class of polymers: starburst-dendritic macromolecules. Polym J 17:117–132
31. Tomalia DA, Naylor AM, Goddard WA III (1990) Starburst dendrimers: molecular-level control of size, shape, surface chemistry, topology, and flexibility from atoms to macroscopic matter. Angew Chem Int Ed Engl 29:138–175
32. de Brabander-van den Berg E, Meijer E (2003) Poly (propylene imine) dendrimers: large-scale synthesis by hetereogeneously catalyzed hydrogenations. Angew Chem Int Ed Engl 32:1308–1311
33. Sadler K, Tam JP (2002) Peptide dendrimers: applications and synthesis. Rev Mol Biotechnol 90:195–229
34. Hawker CJ, Frechet JMJ (1990) Preparation of polymers with controlled molecular architecture. A new convergent approach to dendritic macromolecules. J Am Chem Soc 112:7638–7647
35. Ihre H, Hult A, Söderlind E (1996) Synthesis, characterization, and 1H NMR self-diffusion studies of dendritic aliphatic polyesters based on 2, 2-bis (hydroxymethyl) propionic acid and 1, 1, 1-tris (hydroxyphenyl) ethane. J Am Chem Soc 118:6388–6395
36. Grinstaff MW (2002) Biodendrimers: new polymeric biomaterials for tissue engineering. Chemistry 8:2838–2846
37. Turnbull WB, Stoddart JF (2002) Design and synthesis of glycodendrimers. Rev Mol Biotechnol 90:231–255
38. D'Emanuele A, Attwood D (2005) Dendrimer–drug interactions. Adv Drug Deliv Rev 57:2147–2162
39. Quintana A, Raczka E, Piehler L, Lee I, Myc A, Majoros I et al (2002) Design and function of a dendrimer-based therapeutic nanodevice targeted to tumor cells through the folate receptor. Pharmaceut Res 19:1310–1316
40. Xie Z, Guan H, Chen X, Lu C, Chen L, Hu X et al (2007) A novel polymer–paclitaxel conjugate based on amphiphilic triblock copolymer. J Control Release 117:210–216
41. Yoo HS, Park TG (2004) Folate-receptor-targeted delivery of doxorubicin nano-aggregates stabilized by doxorubicin–PEG–folate conjugate. J Control Release 100:247–256
42. Greenwald R (2001) PEG drugs: an overview. J Control Release 74:159–171
43. Hooftman G, Herman S, Schacht E (1996) Review: poly (ethylene glycol)s with reactive end groups. II. Practical consideration for the preparation of protein-PEG conjugates. J Bioact Compat Polym 11:135–159

44. Khandare JJ, Jayant S, Singh A, Chandna P, Wang Y, Vorsa N et al (2006) Dendrimer versus linear conjugate: influence of polymeric architecture on the delivery and anticancer effect of paclitaxel. Bioconjug Chem 17:1464–1472
45. Rihova B (2009) Clinical experience with anthracycline antibiotics–HPMA copolymer–human immunoglobulin conjugates. Adv Drug Deliv Rev 61:1149–1158
46. Kopecek J, Kopeckova P, Minko T, Lu ZR (2000) HPMA copolymer–anticancer drug conjugates: design, activity, and mechanism of action. Eur J Pharm Biopharm 50:61–81
47. Sakuma S, Lu ZR, Kopeckova P, Kopecek J (2001) Biorecognizable HPMA copolymer–drug conjugates for colon-specific delivery of 9-aminocamptothecin. J Control Release 75:365–379
48. Etrych T, Kovar L, Strohalm J, Chytil P, Ríhova B, Ulbrich K (2011) Biodegradable star HPMA polymer–drug conjugates: biodegradability, distribution and anti-tumor efficacy. J Control Release 154:241–248
49. Dvorak M, Kopeckova P, Kopecek J (1999) High-molecular weight HPMA copolymer–adriamycin conjugates. J Control Release 60:321–332
50. Chytil P, Etrych T, Konak C, Sírova M, Mrkvan T, Ríhova B et al (2006) Properties of HPMA copolymer-doxorubicin conjugates with pH-controlled activation: effect of polymer chain modification. J Control Release 115:26
51. Etrych T, Subr V, Strohalm J, Sirova M, Rihova B, Ulbrich K (2012) HPMA copolymer-doxorubicin conjugates: the effects of molecular weight and architecture on biodistribution and in vivo activity. J Control Release 164(3):346–354
52. Stern R, Asari AA, Sugahara KN (2006) Hyaluronan fragments: an information-rich system. Eur J Cell Biol 85:699–715
53. Oh EJ, Park K, Kim KS, Kim J, Yang JA, Kong JH et al (2010) Target specific and long-acting delivery of protein, peptide, and nucleotide therapeutics using hyaluronic acid derivatives. J Control Release 141:2–12
54. Campo GM, Avenoso A, Campo S, D'Ascola A, Nastasi G, Calatroni A (2010) Molecular size hyaluronan differently modulates toll-like receptor-4 in LPS-induced inflammation in mouse chondrocytes. Biochimie 92:204
55. Harish Prashanth K, Tharanathan R (2007) Chitin/chitosan: modifications and their unlimited application potential—an overview. Trends Food Sci Technol 18:117–131
56. Synytsya A, Blafkova P, Copíkova J, Spevacek J, Uher M (2008) Conjugation of kojic acid with chitosan. Carbohydr Polym 72:21–31
57. Kovar M, Strohalm J, Ulbrich K, Rihova B (2002) In vitro and in vivo effect of HPMA copolymer-bound doxorubicin targeted to transferrin receptor of B-cell lymphoma 38C13. J Drug Target 10:23–30
58. Khandare J, Minko T (2006) Polymer–drug conjugates: progress in polymeric prodrugs. Prog Polym Sci 31:359–397
59. Duncan R (2011) Polymer therapeutics as nanomedicines: new perspectives. Curr Opin Biotechnol 22:492–501
60. Hermanson GT (1996) Bioconjugate techniques. Academic
61. Leriche G, Chisholm L, Wagner A (2012) Cleavable linkers in chemical biology. Bioorg Med Chem 20:571–582
62. Greco F, Vicent MJ (2009) Combination therapy: opportunities and challenges for polymer-drug conjugates as anticancer nanomedicines. Adv Drug Deliv Rev 61:1203–1213
63. Klee JE (2005) Mass spectrometry of step-growth polymers. Eur J Mass Spectrom 11:591–610
64. Paul A, Vicent MJ, Duncan R (2007) Using small-angle neutron scattering to study the solution conformation of N-(2-hydroxypropyl) methacrylamide copolymer-doxorubicin conjugates. Biomacromolecules 8:1573–1579
65. Ashok B, Arleth L, Hjelm RP, Rubinstein I, Önyüksel H (2004) In vitro characterization of PEGylated phospholipid micelles for improved drug solubilization: effects of PEG chain length and PC incorporation. J Pharmaceut Sci 93:2476–2487
66. Ferguson EL, De Luca E, Heenan RK, King SM, Griffiths PC (2010) Time-resolved small-angle neutron scattering as a tool for studying controlled release from liposomes using polymer-enzyme conjugates. Macromol Rapid Commun 31:1685–1690

67. Duncan R, Vicent MJ, Greco F, Nicholson RI (2005) Polymer-drug conjugates: towards a novel approach for the treatment of endrocine-related cancer. Endocr Relat Cancer 12:S189–S199
68. Li C, Wallace S (2008) Polymer-drug conjugates: recent development in clinical oncology. Adv Drug Deliv Rev 60:886–898
69. Danson S, Ferry D, Alakhov V, Margison J, Kerr D, Jowle D et al (2004) Phase I dose escalation and pharmacokinetic study of pluronic polymer-bound doxorubicin (SP1049C) in patients with advanced cancer. Br J Cancer 90:2085–2091
70. Schluep T, Cheng J, Khin KT, Davis ME (2006) Pharmacokinetics and biodistribution of the camptothecin-polymer conjugate IT-101 in rats and tumor-bearing mice. Cancer Chemother Pharmacol 57:654–662
71. Yang D, Yu L, Van S (2011) Clinically relevant anticancer polymer paclitaxel therapeutics. Cancers 3:17–42
72. Hayashi Y, Skwarczynski M, Hamada Y, Sohma Y, Kimura T, Kiso Y (2003) A novel approach of water-soluble paclitaxel prodrug with no auxiliary and no byproduct: design and synthesis of isotaxel. J Med Chem 46:3782–3784
73. Campone M, Rademaker-Lakhai JM, Bennouna J, Howell SB, Nowotnik DP, Beijnen JH et al (2007) Phase I and pharmacokinetic trial of AP5346, a DACH-platinum-polymer conjugate, administered weekly for three out of every 4 weeks to advanced solid tumor patients. Cancer Chemother Pharmacol 60:523–533
74. Greenwald RB, Choe YH, McGuire J, Conover CD (2003) Effective drug delivery by PEGylated drug conjugates. Adv Drug Deliv Rev 55:217–250
75. Schluep T, Hwang J, Cheng J, Heidel JD, Bartlett DW, Hollister B et al (2006) Preclinical efficacy of the camptothecin-polymer conjugate IT-101 in multiple cancer models. Clin Cancer Res 12:1606–1614
76. Soepenberg O, de Jonge MJA, Sparreboom A, de Bruin P, Eskens FALM, de Heus G et al (2005) Phase I and pharmacokinetic study of DE-310 in patients with advanced solid tumors. Clin Cancer Res 11:703–711
77. Ochi Y, Shiose Y, Kuga H, Kumazawa E (2005) A possible mechanism for the long-lasting antitumor effect of the macromolecular conjugate DE-310: mediation by cellular uptake and drug release of its active camptothecin analog DX-8951. Cancer Chemother Pharmacol 55:323–332
78. Springett GM, Takimoto C, McNamara M, Doroshow JH, Syed S, Eastham E et al (2004) Phase I study of CT-2106 (polyglutamate camptothecin) in patients with advanced malignancies. J Clin Oncol 22:3127
79. Kumazawa E, Ochi Y (2005) DE-310, a novel macromolecular carrier system for the camptothecin analog DX-8951f: potent antitumor activities in various murine tumor models. Cancer Sci 95:168–175
80. Homsi J, Simon GR, Garrett CR, Springett G, De Conti R, Chiappori AA et al (2007) Phase I trial of poly-L-glutamate camptothecin (CT-2106) administered weekly in patients with advanced solid malignancies. Clin Cancer Res 13:5855–5861
81. Hoes CJT, Grootoonk J, Duncan R, Hume IC, Bhakoo M, Bouma JMW et al (1993) Biological properties of adriamycin bound to biodegradable polymeric carriers. J Control Release 23:37–53
82. Sinha R, Kim GJ, Nie S, Shin DM (2006) Nanotechnology in cancer therapeutics: bioconjugated nanoparticles for drug delivery. Mol Cancer Ther 5:1909–1917
83. O'Hare K, Duncan R, Strohalm J, Ulbrich K, Kopeckova P (1993) Polymeric drug-carriers containing doxorubicin and melanocyte-stimulating hormone: in vitro and in vivo evaluation against murine melanoma. J Drug Target 1:217–229
84. Omelyanenko V, Kopeckova P, Prakash RK, Ebert CD, Kopecek J (1999) Biorecognition of HPMA copolymer-adriamycin conjugates by lymphocytes mediated by synthetic receptor binding epitopes. Pharmaceut Res 16:1010–1019
85. Line BR, Mitra A, Nan A, Ghandehari H (2005) Targeting tumor angiogenesis: comparison of peptide and polymer-peptide conjugates. J Nucl Med 46:1552–1560

86. Ríhova B, Strohalm J, Prausova J, Kubackova K, Jelinkova M, Rozprimova L et al (2003) Cytostatic and immunomobilizing activities of polymer-bound drugs: experimental and first clinical data. J Control Release 91:1–16
87. Kopecek J (2013) Polymer-drug conjugates: origins, progress to date and future directions. Adv Drug Deliv Rev 65(1):49–59
88. Nan A, Croft SL, Yardley V, Ghandehari H (2004) Targetable water-soluble polymer-drug conjugates for the treatment of visceral leishmaniasis. J Control Release 94:115–127
89. Ehrenfreund-Kleinman T, Golenser J, Domb AJ (2004) Conjugation of amino-containing drugs to polysaccharides by tosylation: amphotericin B–arabinogalactan conjugates. Biomaterials 25:3049–3057
90. Ehrenfreund-Kleinman T, Domb AJ, Jaffe CL, Nasereddin A, Leshem B, Golenser J (2005) The effect of amphotericin b derivatives on Leishmania and immune functions. J Parasitol 91:158–163
91. Sokolsky-Papkov M, Domb AJ, Golenser J (2006) Impact of aldehyde content on amphotericin B-dextranimine conjugate toxicity. Biomacromolecules 7:1529–1535
92. Bouhlal R, Haslin C, Chermann JC, Colliec-Jouault S, Sinquin C, Simon G et al (2011) Antiviral activities of sulfated polysaccharides isolated from Sphaerococcus coronopifolius (Rhodophytha, Gigartinales) and Boergeseniella thuyoides (Rhodophyta, Ceramiales). Mar Drugs 9:1187–1209
93. Wang D, Sima M, Mosley RL, Davda JP, Tietze N, Miller SC et al (2006) Pharmacokinetic and biodistribution studies of a bone-targeting drug delivery system based on N-(2-hydroxypropyl) methacrylamide copolymers. Mol Pharmaceut 3:717–725
94. Miller SC, Pan H, Wang D, Bowman BM, Kopeckova P, Kopecek J (2008) Feasibility of using a bone-targeted, macromolecular delivery system coupled with prostaglandin E 1 to promote bone formation in aged, estrogen-deficient rats. Pharmaceut Res 25:2889–2895
95. Gabriel D, Zuluaga MF, Van Den Bergh H, Gurny R, Lange N (2011) It is all about proteases: from drug delivery to in vivo imaging and photomedicine. Curr Med Chem 18:1785–1805
96. Liu XM, Miller SC, Wang D (2010) Beyond oncology – application of HPMA copolymers in non-cancerous diseases. Adv Drug Deliv Rev 62:258–271
97. Liu XM, Quan LD, Tian J, Alnouti Y, Fu K, Thiele GM et al (2008) Synthesis and evaluation of a well-defined HPMA copolymer-dexamethasone conjugate for effective treatment of rheumatoid arthritis. Pharmaceut Res 25:2910–2919
98. Folkman J (1995) Angiogenesis in cancer, vascular, rheumatoid and other disease. Nat Med 1:27–30
99. Maeda H (2001) The enhanced permeability and retention (EPR) effect in tumor vasculature: the key role of tumor-selective macromolecular drug targeting. Adv Enzyme Regul 41:189–207
100. Harris JM, Martin NE, Modi M (2001) Pegylation: a novel process for modifying pharmacokinetics. Clin Pharmacokinet 40:539–551
101. Rajender Reddy K, Modi MW, Pedder S (2002) Use of peginterferon alfa-2a (40 KD) (Pegasys) for the treatment of hepatitis C. Adv Drug Deliv Rev 54:571–586
102. Zeuzem S, Feinman SV, Rasenack J, Heathcote EJ, Lai MY, Gane E et al (2000) Peginterferon alfa-2a in patients with chronic hepatitis C. New Engl J Med 343:1666–1672
103. Parkinson C, Scarlett J, Trainer PJ (2003) Pegvisomant in the treatment of acromegaly. Adv Drug Deliv Rev 55:1303–1314
104. Molineux G (2004) The design and development of pegfilgrastim (PEG-rmetHuG-CSF, Neulasta). Curr Pharmaceut Des 10:1235–1244
105. Nori A, Kopecek J (2005) Intracellular targeting of polymer-bound drugs for cancer chemotherapy. Adv Drug Deliv Rev 57:609–636
106. Jiang YH, Emau P, Cairns JS, Flanary L, Morton WR, McCarthy TD et al (2005) SPL7013 gel as a topical microbicide for prevention of vaginal transmission of SHIV89.6P in macaques. AIDS Res Human Retroviruses 21:207–213
107. Liu Y, Li J, Shao K, Huang R, Ye L, Lou J et al (2010) A leptin derived 30-amino-acid peptide modified pegylated poly-L-lysine dendrigraft for brain targeted gene delivery. Biomaterials 31:5246–5257

108. Santos JL, Oliveira H, Pandita D, Rodrigues J, Pego AP, Granja PL et al (2010) Functionalization of poly (amidoamine) dendrimers with hydrophobic chains for improved gene delivery in mesenchymal stem cells. J Control Release 144:55–64
109. He H, Li Y, Jia XR, Du J, Ying X, Lu WL et al (2011) PEGylated poly (amidoamine) dendrimer-based dual-targeting carrier for treating brain tumors. Biomaterials 32:478–487
110. Jevprasesphant R, Penny J, Attwood D, McKeown NB, D'Emanuele A (2003) Engineering of dendrimer surfaces to enhance transepithelial transport and reduce cytotoxicity. Pharmaceut Res 20:1543–1550
111. Jevprasesphant R, Penny J, Jalal R, Attwood D, McKeown N, D'Emanuele A (2003) The influence of surface modification on the cytotoxicity of PAMAM dendrimers. Int J Pharm 252:263–266
112. Najlah M, Freeman S, Attwood D, D'Emanuele A (2007) Synthesis and assessment of first-generation polyamidoamine dendrimer prodrugs to enhance the cellular permeability of P-gp substrates. Bioconjug Chem 18:937–946
113. Yuan Q, Lee E, Yeudall WA, Yang H (2010) Dendrimer-triglycine-EGF nanoparticles for tumor imaging and targeted nucleic acid and drug delivery. Oral Oncol 46:698–704
114. Thomas TP, Shukla R, Kotlyar A, Kukowska-Latallo J, Baker JR (2010) Dendrimer-based tumor cell targeting of fibroblast growth factor-1. Bioorg Med Chem Lett 20:700–703
115. Yang W, Cheng Y, Xu T, Wang X, L-p W (2009) Targeting cancer cells with biotin–dendrimer conjugates. Eur J Med Chem 44:862–868
116. Zhang L, Zhu S, Qian L, Pei Y, Qiu Y, Jiang Y (2011) RGD-modified PEG–PAMAM–DOX conjugates: in vitro and in vivo studies for glioma. Eur J Pharm Biopharm 79:232–240
117. Bhadra D, Bhadra S, Jain S, Jain N (2003) A PEGylated dendritic nanoparticulate carrier of fluorouracil. Int J Pharm 257:111–124
118. Kihara F, Arima H, Tsutsumi T, Hirayama F, Uekama K (2003) In vitro and in vivo gene transfer by an optimized alpha-cyclodextrin conjugate with polyamidoamine dendrimer. Bioconjug Chem 14:342–350
119. Venuganti VVK, Perumal OP (2008) Effect of poly (amidoamine)(PAMAM) dendrimer on skin permeation of 5-fluorouracil. Int J Pharm 361:230–238
120. Chauhan AS, Sridevi S, Chalasani KB, Jain AK, Jain SK, Jain N et al (2003) Dendrimer-mediated transdermal delivery: enhanced bioavailability of indomethacin. J Control Release 90:335–343
121. Borowska K, Laskowska B, Magon A, Mysliwiec B, Pyda M, Wołowiec S (2010) PAMAM dendrimers as solubilizers and hosts for 8-methoxypsoralene enabling transdermal diffusion of the guest. Int J Pharm 398:185–189
122. Yiyun C, Na M, Tongwen X, Rongqiang F, Xueyuan W, Xiaomin W et al (2006) Transdermal delivery of nonsteroidal anti-inflammatory drugs mediated by polyamidoamine (PAMAM) dendrimers. J Pharmaceut Sci 96:595–602
123. Wiwattanapatapee R, Lomlim L, Saramunee K (2003) Dendrimers conjugates for colonic delivery of 5-aminosalicylic acid. J Control Release 88:1–9
124. Spataro G, Malecaze F, Turrin C-O, Soler V, Duhayon C, Elena P-P et al (2010) Designing dendrimers for ocular drug delivery. Eur J Med Chem 45:326–334
125. Vandamme TF, Brobeck L (2005) Poly (amidoamine) dendrimers as ophthalmic vehicles for ocular delivery of pilocarpine nitrate and tropicamide. J Control Release 102:23–38
126. Dong Z, Katsumi H, Sakane T, Yamamoto A (2010) Effects of polyamidoamine (PAMAM) dendrimers on the nasal absorption of poorly absorbable drugs in rats. Int J Pharm 393:245–253
127. Jain K, Kesharwani P, Gupta U, Jain N (2010) Dendrimer toxicity: let's meet the challenge. Int J Pharm 394:122–142
128. Duncan R, Izzo L (2005) Dendrimer biocompatibility and toxicity. Adv Drug Deliv Rev 57:2215–2237
129. Gee JMW, Howell A, Gullick WJ, Benz CC, Sutherland RL, Santen RJ et al (2005) Consensus statement. Endocr Relat Cancer 12:S1–S7

130. Santucci L, Mencarelli A, Renga B, Pasut G, Veronese F, Zacheo A et al (2006) Nitric oxide modulates proapoptotic and antiapoptotic properties of chemotherapy agents: the case of NO-pegylated epirubicin. FASEB J 20:765–767
131. Saad M, Garbuzenko OB, Ber E, Chandna P, Khandare JJ, Pozharov VP et al (2008) Receptor targeted polymers, dendrimers, liposomes: which nanocarrier is the most efficient for tumor-specific treatment and imaging? J Control Release 130:107–114
132. Nemunaitis J, Cunningham C, Senzer N, Gray M, Oldham F, Pippen J et al (2005) Phase I study of CT-2103, a polymer-conjugated paclitaxel, and carboplatin in patients with advanced solid tumors. Cancer Investig 23:671–676
133. Greco F, Vicent MJ (2008) Polymer-drug conjugates: current status and future trends. Front Biosci 13:2744–2756
134. Shiah JG, Sun Y, Kopeckova P, Peterson CM, Straight RC, Kopecek J (2001) Combination chemotherapy and photodynamic therapy of targetable N-(2-hydroxypropyl)methacrylamide copolymer-doxorubicin/mesochlorin e(6)-OV-TL 16 antibody immunoconjugates. J Control Release 74:249–253
135. Satchi R, Connors TA, Duncan R (2001) PDEPT: polymer-directed enzyme prodrug therapy. Br J Cancer 85:1070
136. Duncan R, Vicent MJ (2013) Polymer therapeutics-prospects for 21st century: the end of the beginning. Adv Drug Deliv Rev 65(1):60–70

Chapter 6
Pharmacokinetic and Pharmacodynamic Aspects of Focal and Targeted Delivery of Drugs

David Stepensky

Abbreviations

DDS	Drug delivery system
EPR	Enhanced permeability and retention effect
PBPK model	Physiologically based pharmacokinetic model
PD	Pharmacodynamics
PEG	Polyethylene glycol
PK	Pharmacokinetics

6.1 Quantitative Aspects of Drug Distribution and Pharmacological Effects

6.1.1 Pharmacokinetics

Pharmacokinetics (PK) refers to the time course of drug concentrations that are attained in different organs and tissues following drug administration. For the free (non-encapsulated) drug, the pharmacokinetics is governed by its physicochemical properties (size, shape, charge, lipid/water solubility, etc.) and the (patho)physiological factors (volumes of body fluids, size of body organs, their perfusion, presence of metabolic enzymes and transporters, etc.). However, encapsulation of the drug within the DDS masks its physicochemical properties and affects the pathways of its disposition (i.e., distribution and elimination) in the body (see Sect. 6.2 of this

D. Stepensky (✉)
Department of Clinical Biochemistry and Pharmacology, Faculty of Health Sciences, Ben-Gurion University of the Negev, POB 653, Beer-Sheva 84105, Israel
e-mail: davidst@bgu.ac.il

A.J. Domb and W. Khan (eds.), *Focal Controlled Drug Delivery*, Advances in Delivery Science and Technology, DOI 10.1007/978-1-4614-9434-8_6,

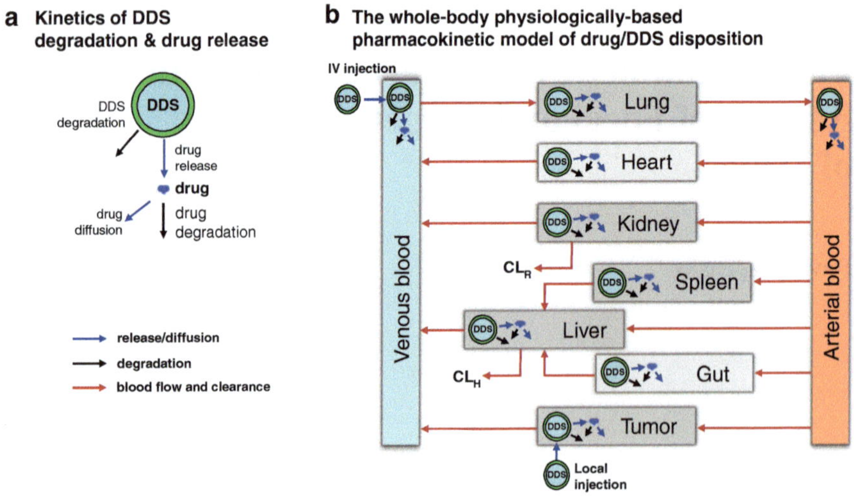

Fig. 6.1 Pharmacokinetics and pharmacodynamics of local vs. systemic drug delivery. (**a**) Kinetics of DDS degradation and drug release. Both the drug and the DDS can distribute and degrade in the systemic circulation or in specific organs and tissues. (**b**) The whole-body physiologically based pharmacokinetic model of drug/DDS disposition. Following systemic (IV) administration, drug/DDS can undergo extravasation, which can be reversible. Major organs of DDS accumulation are usually the lungs, liver, spleen, kidney, and the tumor. DDS degradation and drug release in the systemic circulation and in specific organs and tissues takes place leading to buildup of a gradient of drug concentrations within the body. The drug, DDS, and the products of their metabolism usually leave the body via the liver (CL_H—hepatic clearance) and the kidneys (CL_R—renal clearance). Following local (intratumoral) injection, part of the drug/DDS can leave the injection site, reach the other organs and tissues, and undergo the same processes that were described for the systemic administration. Cells (in the systemic circulation or within the specific organs/tissues) that are exposed to therapeutic levels of the drug for substantial period of time will undergo apoptosis and/or necrosis, giving rise to the desired or adverse drug effects. For the sake of clarity, only part of the organs/tissues and processes (release and degradation *arrows*) are shown

chapter). Specifically, drug encapsulated within the DDS can be a subject to the enhanced permeability and retention (EPR) effect [1] that increases the deposition of drug/DDSs in the organs/tissues with damaged/fenestrated endothelial lining of the blood vessels (e.g., tumors and inflamed tissues) [2]. Due to release of the drug from the DDS in the systemic circulation and at the deposition/extravasation sites (see Fig. 6.1), the resulting pharmacokinetics of the drug/DDS system follows a hybrid trend with both the drug's and DDS's properties affecting the resulting pharmacokinetic behavior of the drug.

It should be noted that the drug is present in the body in multiple forms following its administration within the specific DDS. Therefore, analysis of its pharmacokinetics should be based on analytical methods that can quantify separately the DDS-encapsulated vs. unencapsulated (free) drug. However, analytical methods that are frequently used for assessment of drug/DDS pharmacokinetics not always can differentiate between the free and encapsulated drug and frequently are suitable for drug quantification only in the systemic circulation (and not in the specific organs and body tissues). Use of these analytical methods can lead to biased or erroneous conclusions regarding the pharmacokinetic behavior of the drug/DDS.

6.1.2 Pharmacodynamics

Pharmacodynamics (PD) refers to the mechanisms of pharmacological effects of the drug/DDS following their administration. These effects derive from the drug concentrations at the individual locations within the body (i.e., the pharmacokinetics) and the concentration–response (PK–PD) relationship that reflects the sensitivity of the cells and body structures/components to the drug.

Accumulation of the drug in the target organ/tissue (e.g., the tumor) is expected to lead to the desired pharmacological responses, while accumulation of the drug in other organs and tissues can lead to adverse effects (see Fig. 6.1b). It is generally assumed that the pharmacological effects are derived from the drug concentrations alone. However, the DDS can also contribute to the pharmacodynamics of the drug-encapsulating DDS. Specifically, some of the DDSs are known to be toxic (e.g., cationic liposomes, carbon nanotubes, etc.), and their PK and PD properties should be taken into account for design of efficient and safe DDSs.

6.1.3 Targeted Drug Delivery for Enhancing the Administration Efficiency and Clinical Effectiveness

Controlling the pathways of drug/DDS disposition can be an efficient way to increase the drug concentrations and prolong the retention time of the drug at the target site and to decrease the drug concentrations at the sites where toxicity may occur. This will lead to enhanced magnitude of the desired effects (i.e., higher effectiveness of drug treatment), reduced magnitude of adverse effects (i.e., less toxicity), and higher efficiency of drug administration [3]. To attain this goal, different approaches for enhanced drug delivery to the site of action have been developed. The drug/DDS can be administered systemically and can be designed to accumulate preferentially at the target site (e.g., in the solid tumor, see Fig. 6.1b). Alternatively, drug/DDS can be focally injected into a specific target site (i.e., focal drug delivery) to form a depot of locally released drug. Both these approaches have specific advantages and drawbacks that can affect their clinical applicability. Specifically, ability of the systemically administered DDSs to reach the target site can be limited, but these formulations can be more suitable for management of disseminated, multifocal disease. On the other hand, focally injected DDSs can deliver high loads of the drug to the specific location, but they are not suitable for application to the deep sites that are not readily accessible for injection and are less suited for management of multifocal disease.

6.1.4 Drug Targeting Terminology

The major aim of both focal and targeted drug delivery approaches is to increase drug accumulation in the target tissue. However, it is difficult to attain this goal due to the processes of drug release and DDSs degradation in the body that lead to accumulation of the drug in the organs and tissues that are outside the target site of action (see Fig. 6.1). As a result, many of the DDSs that are intended for targeted drug delivery fail to attain this goal.

In general, there is certain ambiguity and lack of clarity in the use of term "targeting" in the context of drug delivery [4]. In this chapter, the term "*targeted drug delivery*" will be used to describe *preferential accumulation of the drug at the target site*. This implies quantification and comparison of the drug concentrations in the target tissue vs. other organs and tissues using suitable analytical methods. To attain targeted drug delivery, *active and passive targeting approaches* can be used. *Passive targeting* can be based on the EPR effect for enhanced drug/DDS accumulation in the tumor or inflamed tissue, while *active targeting* can be based on the formulation properties of the DDS (e.g., decoration of the drug or DDS with specific targeting residues) or application of external targeting signals (see Sect. 6.1.6). Use of active and passive targeting approaches not necessarily would lead to targeted drug delivery, and many of the currently available "targeted" DDSs apparently fail to attain preferential drug accumulation at the target site [4].

6.1.5 Mathematical Modeling for Analysis of Drug Targeting and Optimization of Drug Treatment

Assessment of drug targeting should be based on quantitative assessment of drug/DDS disposition following focal or systemic administration. The classical pharmacokinetic assessment of drug disposition is based on quantification of drug concentrations in the central circulation only and calculation and analysis of the major pharmacokinetic parameters, such as volume of drug distribution (V) and the drug clearance (CL). Unfortunately, these parameters do not provide direct information on the targeting efficiency and on the extent of drug accumulation at the target site vs. other tissues.

Extent of drug targeting can be established based on the sampling and measurement of drug/DDS concentrations in different organs and tissues, including the target site (see detailed description of the targeting indexes in [3]). Due to the ethical and practical limitations, this sampling strategy is limited to the preclinical studies and cannot be applied in the clinical settings. Some studies, indeed, report quantification of the drug in samples of different organs and tissues collected in experimental animals at one or more time points after the drug/DDS administration. Based on these measurements, the contribution of the individual formulation and (patho) physiological factors that shape the kinetics of drug disposition (see Fig. 6.1 and Sect. 6.1.1) can generally be revealed using pharmacokinetic modeling techniques.

Pharmacokinetic modeling is based on set of equations that describe the major processes that drug/DDS undergo in the body. The pharmacokinetic models for targeted DDSs can be simple, consisting of only two compartments (i.e., the central compartment and the target site) [3, 5]. However, the more detailed, whole-body physiologically based pharmacokinetic (PBPK) models are being increasingly used to describe the disposition of the drug/DDSs [6]. An example of such PBPK model is graphically shown on Fig. 6.1b. It can be seen that this model consists of several compartments that represent the major organs and reflects the physiology of the blood flow in the body. The parameters that are integrated into the equations that underlie this model derive from (a) the drug/DDS formulation properties (e.g., drugs' physicochemical properties, DDS composition, drug release, and degradation rate constants; see Fig. 6.1a), (b) physiologic factors (e.g., weight and perfusion of the individual organs), and (c) parameters that derive from both these groups of factors (e.g., drug/DDS affinity to the individual organs and tissues, elimination rate constants, etc.).

If the pharmacokinetic model reflects properly the disposition behavior of the drug/DDS, it can be used to analyze the formulation and (patho)physiological parameters that affect the drug/DDS targeting to the site of action [7]. Simulations based on this model can be used to predict the effect of formulation changes (e.g., choice of the drug or change of the drug release rate) or of specific pathological conditions (e.g., patients with liver or cardiac diseases) on the extent of drug targeting. Currently, pharmacokinetic models are seldom used for rational design of targeted DDSs, and selection of the formulation properties is usually based on the empirical trial and error approach. It is expected that PBPK models will be increasingly used for development of new targeted DDSs in the future and that their complexity will increase to reflect the accumulation of the detailed data on drug/DDSs disposition. Moreover, such PBPK models will be useful for prediction of effect of the combined (i.e., multifactorial) targeting approaches of drug/DDSs to the site of action (see Sect. 6.1.6).

6.1.6 Major Approaches for Drug/DDSs Targeting

For the purpose of targeted delivery to the site of action, the drug can be encapsulated, adsorbed, or chemically conjugated to different types of DDSs (e.g., nanoparticles, liposomes, etc.; see Table 6.1). These DDSs can be administered either systemically (e.g., intravenously), or locally, into or in vicinity to the site of action (e.g., focal injection into the tumor, injection into the artery that supplies blood to the tumor). To enhance the targeting efficiency of the systemically or locally administered DDSs, different approaches have been proposed (see Table 6.1). These approaches can be based on the design of the DDS (self-triggered targeting) or mediated by external signals (such as magnetic field, ultrasound, local heating, and local activation/uncaging). Some of these approaches can be used in combination with additional techniques for enhancing the local permeability of the drug/DDS (such as use of promoter drugs that affect the local drug diffusion/convection).

Table 6.1 Major approaches for drug/DDS targeting to the site of action (on example of anticancer drug delivery to solid tumors)

Formulation types	Administration routes	Approaches for self-triggered targeting	Approaches for externally activated targeting	Approaches to enhance the intratumoral drug/DDS permeability
Nanoparticles Liposomes Micelles Bolavesicles Polymer conjugates Dendrimers Carbon nanotubes Pasty polymers Hydrogels	Intravenous Local (focal) injection Intra-arterial Intraperitoneal	DDS size Decoration with targeting residues that can interact with molecules overexpressed in the tumor (vascular receptors and factors, apoptosis markers, etc.) pH-dependent drug release Enzyme-dependent drug release	Magnetic field Ultrasound Local heating Light-induced chemical activation	Sheddable PEGylation Multistage drug release (e.g., DDS disintegration into smaller particles) Promotor drugs (vascular damage, apoptosis induction, TGF-beta blockade, etc.)

It should be noted that there is overlap between the different types/classes of approaches that are presented in Table 6.1. For instance, application of external forces, such as focused ultrasound, will increase the local drug/DDS permeability, in addition to increase of drug decapsulation. Unfortunately, enhanced local decapsulation of DDSs that can be attained using the above-described approaches will lead to increased elimination of the released drug and will not necessarily enhance the drug targeting to the site of action. Thus, detailed analysis of this "drug escape" and of the resulting targeting efficiency of the individual targeting approaches and their combinations is required for rational development and clinical application of targeted DDSs. The subsequent sections of this chapter analyze in detailed and quantitative fashion the major pharmacokinetic and pharmacodynamic factors that govern the efficiency of systemically and focally administered DDSs. This analysis is focused on the DDSs that are loaded with anticancer drugs and are intended for management of patients with solid tumors.

6.2 Systemically Administered Anticancer Drug Delivery Systems

Systemically administered anticancer DDSs is a rapidly growing and developing group of approaches for cancer treatment. The major types of formulations that are used to encapsulate the anticancer agents are based on liposomes, nanoparticles,

Table 6.2 Major pathways of drug/DDSs disposition following systemic administration

Pathway
Release of the encapsulated drug
Degradation in the systemic circulation
Aggregation in the bloodstream
Uptake by mononuclear phagocyte system (MPS) cells
Uptake by the endothelial cells
Uptake by the red blood cells
Permeation to the target organ (the tumor)
Permeation to other organs and tissues
Interaction with the endogenous compounds (formation of "corona")

polymer conjugates, dendrimers, and other carriers (see Table 6.1). For the purpose of targeted drug delivery, these formulations should encapsulate substantial amounts of anticancer drug, should be stable in the systemic circulation for sufficient amount of time that will allow their accumulation in the target tissue, should efficiently reach the target tissue, and eventually release the encapsulated drug. Many strategies have been proposed for the purpose of anticancer drug/DDS targeting to the solid tumors (see Table 6.1). As will be described below, the choice of the formulation type and the targeting approaches may significantly affect the systemic and local pharmacokinetics and pharmacodynamics of the drug/DDS following its systemic administration.

6.2.1 Pathways of DDS Disposition Following Systemic Administration

Intravenously injected DDSs are exposed to a plethora of processes in the bloodstream which can interfere with their targeting efficiency (see Table 6.2). Specifically, the DDSs can undergo degradation and decapsulation in the systemic circulation, prior to reaching the target tissue. They may aggregate in the bloodstream with subsequent precipitation in the lungs (the first organ that the DDSs encounter following IV administration) or in the other organs. The DDSs can be endocytosed and digested by the cells that are circulating in the blood or the cells that are lining the blood vessels. The remaining fraction of the DDSs that managed to avoid these processes can extravasate to reach the target site or other organs/tissues. For detailed analysis of the factors that affect the pathways of DDSs' disposition, the readers are referred to several excellent recent reviews [4, 8–10].

It can be seen, thus, that the chances of the individual DDS to reach the target tissue are low and drug targeting to the site of action requires blocking or avoiding the above mentioned interfering pathways. Efficiency of the individual pathways listed in Table 6.2 can be substantially affected by the dimensions (size and shape) and the surface properties (charge, surface residues) of the DDS. For instance, the cells of the mononuclear phagocyte system (MPS) cells have their "preferences" for the size, shape, and charge of the DDSs and will preferably endocytose the charged DDSs with diameter bigger than 200 nm [9, 11]. Similarly, the EPR efficiency (see Sect. 6.1.1) in the tumor can be substantial for the DDSs with the diameter of 20–200 nm and much less for formulations of other size [12]. Therefore, for the purpose of preferential drug/DDS delivery to the solid tumors, systemically administered formulations with the diameter of less than 200 nm should be designed.

It should be noted that intravenously administered DDSs can interact with and adsorb endogenous compounds on their surface, such as albumin, transferrin, and lipoproteins [9]. As a result of this interaction, protein coat or "corona" is formed on the surface of DDSs which affects its surface properties and efficiency of the disposition pathways that were listed in Table 6.2 and can further reduce the chances for efficient drug targeting. A popular strategy to block these pathways and to avoid "corona" formation is surface PEGylation of the DDSs. PEGylated DDSs are often referred to as "stealth" formulations, and PEGylation technology that forms inert steric barrier on the DDSs' surface can indeed be used to prolong their residence in the bloodstream and increase the tumor targeting efficiency [13].

6.2.2 Strategies to Enhance the Distribution of the Drug/DDSs to the Solid Tumors

In addition to the choice of the DDSs' size and PEGylation, several additional strategies have been proposed to increase the accumulation of the drug/DDS in the tumors following systemic administration. These strategies can exploit the differences that exist between the tumor and other tissues, for instance, overexpression of specific surface molecules or enzymes, enhanced apoptosis, and reduced intracellular pH of the tumor cells (see Table 6.1). Alternatively, formulations can be designed to accumulate and/or decapsulate locally under the influence of the specific external signal, such as magnetic field, ultrasound, local heating, or light-induced chemical reaction. All the above-mentioned strategies can potentially enhance the extent of drug/DDS accumulation in the solid tumor. However, it is estimated that currently available DDSs accumulate only to a low extent in the tumor tissue, and more than 95 % of the IV-administered DDSs accumulate in other organs, in particular the liver, spleen, and lungs, and account for toxicity [4]. Moreover, even if the drug/DDS is able to reach the tumor, this will not necessarily lead to sufficient exposure of the tumor cells to the drug and effective anticancer effects (see below).

6.2.3 Intratumoral Disposition of the Drug Following Release from the DDS

In the case that the DDS successfully reaches the target tissue (the tumor) and extravasates there, it can exert the desired pharmacological effect. To this end, the drug should be released from the DDS and should reach its target: either permeate

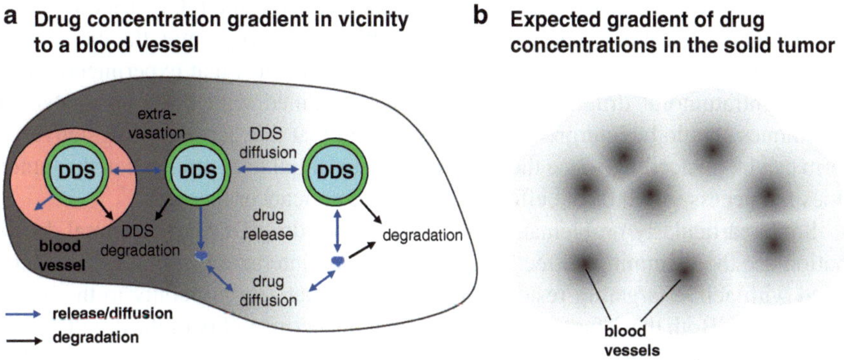

Fig. 6.2 Intratumoral disposition of systemically administered drug/DDS. Both the drug and the DDS can extravasate, distribute, and degrade in the tumor. Subsequently, a gradient of drug concentrations is generated within the tumor with high drug concentration in vicinity to the DDSs deposition near the blood vessels and low concentrations in the "deep" parts of the tumor. Cells that are exposed to therapeutic levels of the drug for substantial period of time will undergo apoptosis and/or necrosis. For the sake of clarity, only part of the morphological components (e.g., blood vessels) and processes (release and degradation *arrows*) are shown

to (or accumulate inside) the target cells (i.e., the cancer, stroma, or infiltrating immune cells) or interact with soluble factors (e.g., anti-VEGF or anti-TNF-β antibodies). It has been believed previously that efficient delivery of drug/DDS into the tumor is sufficient for mounting efficient anticancer effects. However, recently it become evident that intratumoral drug disposition (i.e., ability to reach and interact with the target cells or soluble factors) is the critical factor for the effectiveness of the drug/DDS that has reached the tumor.

The term "intratumoral disposition" refers to the processes of drug/DDS transport and elimination in the tumor tissue. Drug/DDS can be transported in/from the tumor by diffusion (according to the gradient of the drug concentrations) and convection (movement with bulk of interstitial fluid). In addition, drug can be metabolized by the cells in the tumor or can be captured by the soluble factors outside the cells. In addition to these processes, expression of efflux transporters by the tumor cells (e.g., Pgp, BCRP, etc.) can reduce the transport of the drug into the cells, enhance the efficiency of its intracellular metabolism, and reduce effectiveness of the anticancer treatment.

Vascularization features of the solid tumors can promote DDSs accumulation in vicinity to the blood vessels due to the EPR effect, with subsequent degradation of the DDSs and release of the active drug (see Fig. 6.2). However, these features do not lead to homogeneous drug distribution through the tumor, but instead to generation of a gradient of the drug intratumoral concentrations which is higher in vicinity to the blood vessels (at the micrometer–millimeter ranges), and lower/negligible in the cells that are located far away from the blood vessels (see Fig. 6.2). Visual demonstration of this intratumoral gradient based on the fluorescent labeling of the drug with subsequent imaging can be found in several publications [8, 14, 15].

The resulting heterogeneity in the drug/DDS intratumoral distribution and inefficient exposure of "deep" parts of the tumor to the drug limits the therapeutic efficiency of the anticancer treatment. It also demonstrates that experimental data on the intratumoral drug/DDS concentrations obtained with different analytical techniques should be interpreted with caution and that the "targeted drug delivery" term can be misleading, since the presence of substantial drug amounts in the tumor does not necessarily indicate effectiveness of the anticancer therapy.

Some particularly problematic aspects of limited drug/DDS intratumoral distribution are the "jamming" effect [16] and increased interaction of the DDSs decorated with active targeting residues with the cells that are in vicinity to the blood vessels [10]. Both these factors can further limit the permeability of the drug/DDS to the "deep" parts of the tumor and decrease the effectiveness of anticancer therapy.

6.2.4 Ways to Enhance the Exposure of the "Deep" Tumor Parts to the Anticancer Drugs

To overcome the barriers for efficient drug distribution within the tumor, several strategies are being investigated (see Table 6.1). Firstly, local application of the external targeting signals (see Sect. 6.2.2) can affect the blood flow and the temperature within the tumor and enhance permeability of the drug/DDS to the "deep" parts of the tumor. More sophisticated approaches include deprotection of the DDSs in the tumor via sheddable PEGylation [17], intratumoral degradation of the DDSs to release more permeable drug-encapsulating particles (i.e., multistage drug release approaches [18]). A promising approach is pre-treatment or co-administration of the anticancer drug/DDS with promoter drugs that affect the intratumoral blood flow and vascular permeability, induce apoptosis of cancer cells, or affect the function of the stroma cells [19].

It should be noted that enhanced permeability of the tumor that is induced by many of the above mentioned approaches leads not only to enhanced drug/DDS permeability to the "deep" parts of the tumor but also to increased elimination of the DDSs and of the released drug from the tumor and may lead to reduction of the drug targeting efficiency.

6.2.5 Modeling for Rational Design of Tumor-Targeted DDSs

Design of the DDSs with high targeting efficiency to the tumor tissue and effective anticancer activity is a challenging task that should take into account the complex interplay of parameters that govern the systemic and intratumoral drug/DDS disposition. Change of the individual formulation parameter can have profound effects on several stages/pathways of drug/DDS disposition. For instance, surface PEGylation of liposomes or nanoparticles reduces the magnitude of undesired pathways of

Table 6.3 Major parameters that govern the intratumoral drug/DDS disposition and should be taken into account during design of efficient anticancer therapies

Group	Parameter
Drug properties	MW
	Lipophilicity (LogP)
DDS properties	Type and composition
	Drug encapsulation efficiency
	Rate of drug release from the DDS
	Surface properties of the DDS (zeta potential, decoration with targeting residues, PEGylation, etc.)
Tumor properties	Type and grade
	Location and size
	Vascularization and its heterogeneity, presence of necrotic regions
	Fenestration of the capillaries
	Overexpression of surface receptors/molecules
	Presence of drug and DDS-metabolizing enzymes
	Sensitivity to the drug

systemic disposition (e.g., aggregation, uptake by the MPS and endothelial cells; see Table 6.2) but also negatively affects the drug permeability to the "deep" parts of the tumor. In similar fashion, change of dimensions (diameter) of the nano-DDSs can systemically affect the biopharmaceutical, pharmacokinetic, and pharmacodynamic parameters (e.g., the drug loading efficiency and release kinetics, the efficiency of the individual systemic and local disposition pathways, and anticancer treatment effectiveness).

These complex multifaceted effects of the formulation changes call for systemic and quantitative analysis of factors that govern the systemic and local PK and PD of anticancer drug/DDSs. To this end, mathematical modeling analysis that incorporates the major drug, DDS, and tumor parameters (see Table 6.3) is warranted. In addition to these parameters, the analysis should take into account the spatial heterogeneity of the tumor tissue (see Fig. 6.2) and tumor and patient variability factors [20]. For instance, it is well known that the extent of fenestration in the tumor vasculature is highly dependent on the tumor growth vs. vascular growth rate and the morphology of the specific tumor. Therefore, magnitude of the EPR effect in the specific tumor can range from high (Kaposi sacroma) to negligible (prostate or pancreatic cancer) [10, 12] and will affect the clinical effectiveness of anticancer DDSs.

There are several examples of studies that attempted quantitative modeling analysis of drug/DDS disposition and reached conclusions regarding the design of anticancer drug/DDSs. Wittrup et al. analyzed the effects on drug/DDS size and decoration with targeting residues on their tumor uptake and provided quantitative predictions on the combined effects of these parameters [21]. Specifically, intermediate-sized targeting agents (MW ~ 25 kDa) were predicted to have the lowest tumor uptake, as compared to agents with smaller and bigger sizes. Based on the applied analysis, smaller agents can accumulate rapidly in the tumor but require high affinity to the tumor antigens in order to be retained, whereas larger agents can have high retention even if their affinity to the tumor is low [22]. Thurber and Weissleder applied systems approach to classify drug/ DDSs into four categories depending on whether their tumor uptake is limited by blood flow, extravasation, interstitial diffusion, or local binding and metabolism [23].

These and other modeling-based studies provide important insights into the interplay of parameters that govern the disposition of antitumor drug/DDSs. It is expected that more and more complex and detailed models will be applied to describe the complex interplay of these factors and to guide the design of antitumor DDSs. Specifically, it is expected that whole-body physiologically based models (see Fig. 6.1b) will be increasingly used to analyze the drug/DDSs disposition (such as applied by Qin et al. [24]) and that these models will incorporate quantitative two- and three-dimensional analysis of drug/DDSs disposition based on noninvasive imaging techniques (e.g., PET, MRI, SPECT).

Based on the currently available data, it seems that quantitative modeling analysis will be helpful for analysis and interpretation of the experimental data and for prediction of effect of the formulation changes on the drug/DDS disposition and will aid to the rational design of the anticancer treatments. However, it appears that the currently existing approaches for drug/DDS targeting to the tumor tissue have limited efficiency and that new more potent targeting approaches are required to increase the effectiveness of anticancer treatment.

6.3 Focal Anticancer Drug Delivery

To increase the drug concentrations in the tumor and limit the extra-tumoral effects, anticancer drugs can be delivered directly to the site of tumor. In order to form depot of the drug at the tumor site, focal DDSs are usually based on application of either pre-formed or injectable implants, alone or in combination with tumor resection. For the general description of the types of implants that were proposed for focal drug delivery, the reader is referred to several recent reviews [25–28]. Majority of the focal DDSs that are presented in these reviews were evaluated in in vitro and preclinical settings, and only several types of focal DDSs have been approved for clinical use. The implants for focal drug delivery differ in their chemical composition, size, shape, number of layers, type of the encapsulated drug/s, and the mechanisms and rate of drug release. For example, the Gliadel® wafers consist of carmustine-loaded microparticles (with the size of 1–20 μm) that are compression-molded into wafers (14 mm diameter, 1 mm thick). Up to eight such wafers can be locally implanted into a brain cavity following resection of a brain tumor [29]. On the other hand, OncoGel® system is an aqueous solution of paclitaxel in biocompatible polymers that undergoes a reversible thermal gelation upon local injection to generate a thermosensitive, water-soluble implant [30].

6.3.1 *The Drug/DDSs Disposition Following Focal Administration*

All the above mentioned formulation-related factors affect the intratumoral (and systemic) disposition of the anticancer drug following administration of the specific focal DDSs. The focal DDSs can be designed to release the encapsulated drug over a shorter or longer period of time (usually several hours up to several weeks or months). Moreover, the drug release can be constant (zero-order kinetics) or may be exhibiting a "burst effect" and/or multistage release kinetics. For example, release of carmustine from the Gliadel® wafers exhibits two phases: the induction period ("burst" release by diffusion of ~60 % of the drug over the first ~10 h) followed by the erosion period (~5 days) [31]. Different types of mathematical models have been developed for analysis and prediction of the drug release from the focal implants (reviewed in [25]).

The intratumoral disposition behavior of the released drug is governed by the same processes, as previously described for the systemically administered DDSs (see Sect. 6.2.3). However, the resulting intratumoral gradient of anticancer drug concentrations is highly dependent on the DDS type. For systemically administered DDSs, therapeutic drug concentrations (i.e., drug concentrations that are sufficient to exert cytotoxic effect) are obtained in vicinity to the DDSs deposition close to the blood vessels. On the other hand, for focally administered DDSs, therapeutic drug concentrations are attained in close vicinity to the DDSs (see Fig. 6.3 and images of intratumoral drug distribution in preclinical settings [32–34]).

The major limitation of the currently existing focal DDSs is the very narrow (up to several millimeters) layer of the tumor tissue in vicinity to the DDSs that is exposed to the therapeutic drug concentrations [28]. The thickness of this layer is dependent on the balance between the release and disposition kinetics of the drug and can be reliably analyzed using noninvasive imaging-based or destructive analytical techniques (e.g., fluorescence, MRI, radiolabeling) [34]. The inefficient intratumoral disposition of the drug poses a major limitation for the clinical effectiveness of the focally administered DDSs. On the other hand, the drug that is escaping from the tumor to other organs and tissues becomes efficiently diluted. Therefore, the overall ratio of tumor vs. non-tumor drug concentrations for the focal DDSs is usually high (see examples in [35, 36]), and these systems are usually characterized by high safety (high therapeutic index).

6.3.2 *Modeling for Rational Design of Focal DDSs*

Modeling analysis of the major factors that affect the intratumoral disposition of focally administered DDSs has been applied by several research groups, and several mathematical models have been developed for this purpose (reviewed in [28, 34]). For example, Fleming and Saltzman analyzed the intratumoral distribution of

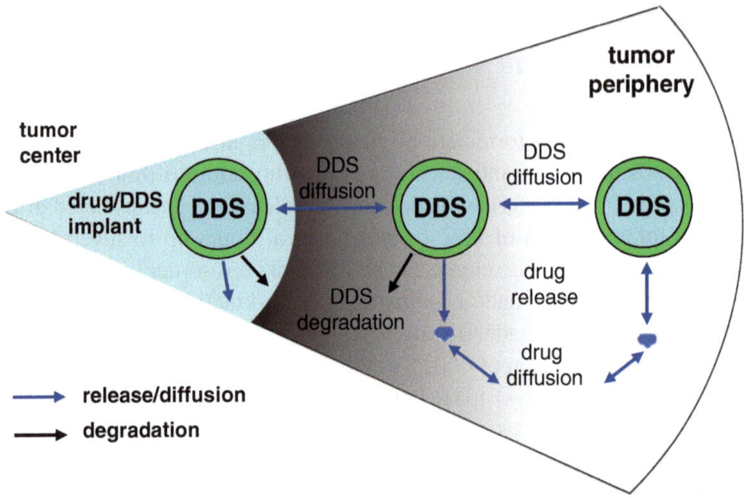

Fig. 6.3 Intratumoral disposition of the locally injected drug/DDS implant. This figure demonstrates the major processes of intratumoral drug disposition following focal administration of an implant that degrades to form drug-loaded particles (DDSs), such as Gliadel®. The input of the drug from the implant and from the particles and its subsequent distribution and degradation lead to buildup of a gradient of drug concentrations within the tumor with high drug concentration in vicinity to the implant (usually, up to several millimeters from the implant) and low concentrations in the parts that are distant from the implant. Cells that are exposed to therapeutic levels of the drug for substantial period of time will undergo apoptosis and/or necrosis. For the sake of clarity, only part of the morphological components (e.g., blood vessels) and processes (release and degradation *arrows*) are shown

carmustine eluted from Gliadel® implants [31]. Three pathways of drug transport were identified: diffusion, convection derived from postoperative edema after the Gliadel® implantation (that resolved by 3 days after implantation), and an additional pathway (that was present in nonhuman primates, but not in rodents) that produced diffuse distribution of the drug through the entire brain. Simulations based on the two- and three-dimensional finite element models were consistent with enhanced drug penetration into the tumor after the implantation (5 mm at the end of the first day) that reduced on subsequent days.

Tan et al. used a finite element 3D model of brain based on MRI reconstruction of brain geometry to compare the delivery of etanidazole to brain tumor from PLGA wafers with different drug release profiles [37]. Based on this model, zero-order release from the implant results in higher drug penetration depth and therapeutic index as compared to the double-burst profile. A similar 3D model of brain has been used by Arifin et al. to analyze the role of convective transport in carmustine intratumoral distribution following Gliadel® implantation [38]. The authors concluded that convection is crucial for intratumoral distribution of carmustine and contributes to attaining therapeutic concentrations in the tissues adjacent to the implant during the first postimplantation days. In additional publication, the authors used the same model to predict the effect of the drug physicochemical properties on the efficiency of intratumoral disposition [39]. Under the same experimental conditions, the drug penetration into the tumor and buildup of efficient concentrations was in the following order: paclitaxel > 5-FU > carmustine > MTX. Simulations of intratumoral drug distribution using 2D model indicated that paclitaxel released from hydrogel (OncoGel®) and carmustine released from Gliadel® wafers are characterized by similar therapeutic penetration depth(1–2 mm) but lead to different duration of therapeutically effective concentrations (30 days vs. 4 days, respectively) [40].

Thus, the above mentioned mathematical models appear to be useful for analysis of the intratumoral drug disposition following focal drug/DDS administration, for identification of the rate-limiting factors for drug disposition, and for choice and design drugs and implants with enhanced penetration depths and therapeutic indexes.

6.3.3 Ways to Enhance Intratumoral Penetration of Focally Administered Drug/DDSs

The focal DDSs, which have been approved for clinical use (e.g., Gliadel® and OncoGel®), can substantially prolong the patient survival and are a valuable addition to the arsenal of anticancer treatments. Nevertheless, the currently existing focal DDSs have suboptimal performance which stems from the limited penetration depth of anticancer drugs from these implants into the surrounding cancer tissue.

To overcome this limitation, focal DDSs can be combined with additional approaches for enhanced intratumoral drug distribution (see Table 6.1). For example, ultrasound may provide a useful tool to enhance delivery of therapeutics in the brain [41]. Similarly, radiofrequency ablation can be used to increase the penetration depth of anticancer agents to the brain and to prolong the duration of effective therapeutic concentrations in vicinity of the implant [34]. It is expected that mathematical models will be useful for the rational design of efficient combination strategies for focal anticancer drug delivery based on detailed analysis of intratumoral drug distribution. For instance, it has been predicted that antiangiogenic treatment has minimal effect on the convection by interstitial fluid in the treatment domain and is not expected to enhance the penetration depth and duration of effective therapeutic concentrations in vicinity of the implant [39].

6.4 Summary

Different approaches for targeted drug delivery for treatment of solid tumors have been developed. The types of formulations that are used for this purpose are based either on focal drug/DDS injection or on systemic administration of tumor-targeted drug/DDS. These two approaches can be used in combination with a plethora of targeting techniques and with treatments promoting the drug/DDS intratumoral disposition. Currently available focally or systemically administered DDSs are characterized by low clinical effectiveness due to inefficient exposure of the target cells to the drug.

New strategies to enhance drug/DDS accumulation in the tumor (for the systemically administered DDSs) and the intratumoral drug distribution (for both focally or systemically administered DDSs) are required to attain therapeutic drug concentrations in the target cells for sufficient period of time and to increase the effectiveness of anticancer therapy. For rational design of these strategies, more detailed analysis of drug disposition following administration of focal or targeted DDS is required.

To this end, selective and sensitive analytical methods should be developed for analysis of the encapsulated and the free drug at the target site vs. other organs/tissues/body fluids and the resulting pharmacological effects. These methods should have high spatial resolution (i.e., be suitable for analysis of drug concentrations and effects in the different parts of the tumor, including the cells that are distant from the blood vessels). Based on these analytical methods, systemic analysis of the formulation properties and their effect on the local and systemic pharmacokinetics and pharmacodynamics of the drug/DDS should be performed. Whole-body physiologically based mathematical models can be used to reveal the quantitative relationships between the studied parameters, to design new DDSs, to plan the preclinical and clinical studies, and to predict the anticancer effect for new types of DDSs and for different subsets of patients. It is expected that these analytical methods and mathematical models will be useful in design and development of new combined methods for enhanced permeation of the drug/DDS to the "deep" parts of the solid tumor that will increase the effectiveness of anticancer treatment.

References

1. Maeda H (2012) Macromolecular therapeutics in cancer treatment: the EPR effect and beyond. J Control Release 164(2):138–144
2. Azzopardi EA, Ferguson EL, Thomas DW (2013) The enhanced permeability retention effect: a new paradigm for drug targeting in infection. J Antimicrob Chemother 68(2):257–274
3. Proost JH (2001) Pharmacokinetic/pharmacodynamic modelling in drug targeting. In: Molema G, Meijer DKF (eds) Drug targeting organ-specific strategies. Wiley-VCH Verlag GmbH
4. Bae YH, Park K (2011) Targeted drug delivery to tumors: myths, reality and possibility. J Control Release 153:198–205
5. Siegel RA, MacGregor RD, Hunt CA (1991) Comparison and critique of two models for regional drug delivery. J Pharmacokinet Biopharm 19:363–373
6. Rowland M, Peck C, Tucker G (2011) Physiologically-based pharmacokinetics in drug development and regulatory science. Annu Rev Pharmacol Toxicol 51:45–73
7. Liu C, Krishnan J, Stebbing J, Xu XY (2011) Use of mathematical models to understand anticancer drug delivery and its effect on solid tumors. Pharmacogenomics 12:1337–1348
8. Ruenraroengsak P, Cook JM, Florence AT (2010) Nanosystem drug targeting: facing up to complex realities. J Control Release 141:265–276
9. Moghimi SM, Hunter AC, Andresen TL (2011) Factors controlling nanoparticle pharmacokinetics: an integrated analysis and perspective. Annu Rev Pharmacol Toxicol 52:481–503

10. Lammers T, Kiessling F, Hennink WE, Storm G (2012) Drug targeting to tumors: principles, pitfalls and (pre-) clinical progress. J Control Release 161:175–187
11. Decuzzi P, Godin B, Tanaka T, Lee SY, Chiappini C, Liu X, Ferrari M (2010) Size and shape effects in the biodistribution of intravascularly injected particles. J Control Release 141:320–327
12. Danhier F, Feron O, Preat V (2010) To exploit the tumor microenvironment: passive and active tumor targeting of nanocarriers for anti-cancer drug delivery. J Control Release 148:135–146
13. Barenholz Y (2012) Doxil(R)–the first FDA-approved nano-drug: lessons learned. J Control Release 160:117–134
14. Primeau AJ, Rendon A, Hedley D, Lilge L, Tannock IF (2005) The distribution of the anticancer drug Doxorubicin in relation to blood vessels in solid tumors. Clin Cancer Res 11: 8782–8788
15. Minchinton AI, Tannock IF (2006) Drug penetration in solid tumours. Nat Rev Cancer 6:583–592
16. Florence AT (2012) "Targeting" nanoparticles: the constraints of physical laws and physical barriers. J Control Release 164(2):115–124
17. Li SD, Huang L (2010) Stealth nanoparticles: high density but sheddable PEG is a key for tumor targeting. J Control Release 145:178–181
18. Wong C, Stylianopoulos T, Cui J, Martin J, Chauhan VP, Jiang W, Popovic Z, Jain RK, Bawendi MG, Fukumura D (2011) Multistage nanoparticle delivery system for deep penetration into tumor tissue. PNAS 108:2426–2431
19. Marcucci F, Corti A (2011) How to improve exposure of tumor cells to drugs – promoter drugs increase tumor uptake and penetration of effector drugs. Adv Drug Delivery Rev 64:53–68
20. Caron WP, Song G, Kumar P, Rawal S, Zamboni WC (2012) Interpatient pharmacokinetic and pharmacodynamic variability of carrier-mediated anticancer agents. Clin Pharmacol Ther 91:802–812
21. Wittrup KD, Thurber GM, Schmidt MM, Rhoden JJ (2012) Practical theoretic guidance for the design of tumor-targeting agents. Methods Enzymol 503:255–268
22. Schmidt MM, Wittrup KD (2009) A modeling analysis of the effects of molecular size and binding affinity on tumor targeting. Mol Cancer Ther 8:2861–2871
23. Thurber GM, Weissleder R (2011) A systems approach for tumor pharmacokinetics. PLoS One 6:e24696
24. Qin S, Seo JW, Zhang H, Qi J, Curry FR, Ferrara KW (2010) An imaging-driven model for liposomal stability and circulation. Mol Pharm 7:12–21
25. Exner AA, Saidel GM (2008) Drug-eluting polymer implants in cancer therapy. Expert Opin Drug Deliv 5:775–788
26. De Souza R, Zahedi P, Allen CJ, Piquette-Miller M (2010) Polymeric drug delivery systems for localized cancer chemotherapy. Drug Deliv 17:365–375
27. Wadee A, Pillay V, Choonara YE, du Toit LC, Penny C, Ndesendo VM, Kumar P, Murphy CS (2011) Recent advances in the design of drug-loaded polymeric implants for the treatment of solid tumors. Expert Opin Drug Deliv 8:1323–1340
28. Wolinsky JB, Colson YL, Grinstaff MW (2012) Local drug delivery strategies for cancer treatment: gels, nanoparticles, polymeric films, rods, and wafers. J Control Release 159:14–26
29. Lin SH, Kleinberg LR (2008) Carmustine wafers: localized delivery of chemotherapeutic agents in CNS malignancies. Expert Rev Anticancer Ther 8:343–359
30. Elstad NL, Fowers KD (2009) OncoGel (ReGel/paclitaxel)–clinical applications for a novel paclitaxel delivery system. Adv Drug Deliv Rev 61:785–794
31. Fleming AB, Saltzman WM (2002) Pharmacokinetics of the carmustine implant. Clin Pharmacokinet 41:403–419
32. Fung LK, Shin M, Tyler B, Brem H, Saltzman WM (1996) Chemotherapeutic drugs released from polymers: distribution of 1,3-bis(2-chloroethyl)-1-nitrosourea in the rat brain. Pharm Res 13:671–682
33. Weinberg BD, Ai H, Blanco E, Anderson JM, Gao J (2007) Antitumor efficacy and local distribution of doxorubicin via intratumoral delivery from polymer millirods. J Biomed Mater Res A 81:161–170

34. Weinberg BD, Blanco E, Gao J (2008) Polymer implants for intratumoral drug delivery and cancer therapy. J Pharm Sci 97:1681–1702
35. Kang YM, Kim GH, Kim JI, da Kim Y, Lee BN, Yoon SM, Kim JH, Kim MS (2011) In vivo efficacy of an intratumorally injected in situ-forming doxorubicin/poly(ethylene glycol)-b-polycaprolactone diblock copolymer. Biomaterials 32:4556–4564
36. Shikanov A, Shikanov S, Vaisman B, Golenser J, Domb AJ (2008) Paclitaxel tumor biodistribution and efficacy after intratumoral injection of a biodegradable extended release implant. Int J Pharm 358:114–120
37. Tan WH, Wang F, Lee T, Wang CH (2003) Computer simulation of the delivery of etanidazole to brain tumor from PLGA wafers: comparison between linear and double burst release systems. Biotechnol Bioeng 82:278–288
38. Arifin DY, Lee KY, Wang CH, Smith KA (2009) Role of convective flow in carmustine delivery to a brain tumor. Pharm Res 26:2289–2302
39. Arifin DY, Lee KY, Wang CH (2009) Chemotherapeutic drug transport to brain tumor. J Control Release 137:203–210
40. Torres AJ, Zhu C, Shuler ML, Pannullo S (2011) Paclitaxel delivery to brain tumors from hydrogels: a computational study. Biotechnol Prog 27:1478–1487
41. Liu Y, Paliwal S, Bankiewicz KS, Bringas JR, Heart G, Mitragotri S, Prausnitz MR (2010) Ultrasound-enhanced drug transport and distribution in the brain. AAPS PharmSciTech 11:1005–1017

Part II
Delivery to Body Sites/Systems

Chapter 7
Treatment of Brain Tumors

Betty M. Tyler, Gustavo Pradilla, Uri Hadelsberg, Hansen Bow, Ian Suk, and Henry Brem

7.1 Introduction

Gliomas are the most common primary tumor in the central nervous system. They arise from astrocytes, oligodendrocytes, or their corresponding precursors. They can be classified histologically as astrocytomas, oligodendrogliomas, or tumors with both features known as oligoastrocytomas. Gliomas are divided into subcategories or clinical grades with grades I and II being benign tumors and grades III and IV (known as GBM) as the more malignant. GBMs not only develop from lower grade gliomas but also are thought to arise de novo [1]. The Cancer Genome Atlas Research Network determined that GBMs are heterogeneous intraparenchymal masses with a few outstanding and defining trademarks. These include a dysregulation of growth factor signaling by amplification and mutational activation of receptor tyrosine kinase genes, the activation of the phosphatidylinositol-3-OH kinase (PI(3)K)

B.M. Tyler (✉) • G. Pradilla • H. Bow
Department of Neurosurgery, School of Medicine, The Johns Hopkins University, 1550 Orleans St/CRB2-2M45, Baltimore, MD 21231, USA
e-mail: btyler@jhmi.edu

U. Hadelsberg
Technion-Israel Institute of Technology, Haifa, Israel

I. Suk
Department of Neurosurgery, School of Medicine, The Johns Hopkins University, Baltimore, MD 21231, USA

H. Brem
Department of Neurosurgery, School of Medicine, The Johns Hopkins University, 1550 Orleans St/CRB2-2M45, Baltimore, MD 21231, USA

Department of Oncology, School of Medicine, The Johns Hopkins University, 1550 Orleans St/CRB2-2M45, Baltimore, MD 21231, USA

Department of BioMedical Engineering, School of Medicine, The Johns Hopkins University, 1550 Orleans St/CRB2-2M45, Baltimore, MD 21231, USA

A.J. Domb and W. Khan (eds.), *Focal Controlled Drug Delivery*, Advances in Delivery Science and Technology, DOI 10.1007/978-1-4614-9434-8_7,

pathway, and the inactivation of the p53 and retinoblastoma tumor suppressor pathways [2]. These data may lead to systematic and stratified research and subsequent treatment options. The physiological barriers of the brain as well as the heterogenic properties and invasiveness of this tumor type have made brain tumor therapy challenging. Recurrence of malignant gliomas occurs most often within 2 cm of the original site of resection [3], leading to reoperations and additional treatments.

7.2 Epidemiology of Primary Malignant Brain Tumors

Glioblastoma multiforme (GBM), a type of malignant brain tumor, is the most common brain tumor in adults [4], accounting for 60 % of all primary brain tumors [5]. The annual incidence of GBM in North America is 4–5 cases per 100,000 persons with the number of newly diagnosed patients being 10,000 people in North America and three million worldwide and with a peak incidence occurring in people over the age of 65 [6]. Recent advances in diagnostic neuroimaging, surgical treatment, radiation therapy, and chemotherapy have improved median survival rates for people with GBM, prolonging survival from 9 months in the 1990s to 15–21 months currently [7, 8].

7.3 Current Treatment Regimens for Malignant Primary Brain Tumors

The standard treatment for patients with newly diagnosed glioblastoma consists of maximal safe surgical removal of tumor mass, followed by 6 weeks of postoperative radiotherapy with concomitant systemic chemotherapy with the alkylating agent temozolomide (75 mg/m^2 daily), and complemented by 6 months of adjuvant temozolomide(150–200 mg/m^2/day for 5 days of every 28 days). As discussed below, intraoperative implantation of BCNU-loaded chemotherapy (Gliadel) wafers further complements treatment in selected patients. A synergistic effect can be achieved when these strategies are combined effectively [8].

7.3.1 Surgical Resection

Despite multiple advances in our understanding of the pathophysiology, genetic pathways, and multifactorial nature of the tumor microenvironment of malignant gliomas, noninvasive therapies remain ineffective to date, and maximal safe microsurgical resection of accessible lesions still constitutes a vital component in the treatment of malignant gliomas. Besides providing definitive diagnosis through pathological analysis of the resected tumor and alleviating symptoms related to mass effect, the benefits of gross total resection when feasible have been shown to correlate with prolonged overall survival, increased progression-free survival, and decreased dependence on supportive medications such as corticosteroids, thereby avoiding significant side effects, among others.

Recently, several studies have provided further evidence of the advantageous effect of maximal safe gross total resection of gliomas. A retrospective large series of malignant gliomas including 416 patients surgically treated for initial or recurrent malignant glioma showed that removal of greater than 98 % of the visible tumor on MR imaging significantly increased overall patient survival (median survival 13 months vs. 8.8 months median survival for resections of less than 98 %, $p<0.0001$). These results remained significant when controlling for accessibility of the lesion for resection and patient performance status. Furthermore, the authors suggested that subtotal resections that achieve at least 89 % removal of the lesion could also result in symptomatic benefits [9]. In 2008, Sanai and Berger [10] reported the results of a retrospective review of studies assessing the role of extent of resection of high- and low-grade gliomas. This analysis revealed 18 positive studies supporting the benefits of maximal safe gross total resection in high-grade gliomas. These results were further supported by the latest evaluation of the European Organization for Research and Treatment of Cancer-National Cancer Institute of Canada (EORTC-NCIC) study 26981–22981 (radiotherapy versus chemotherapy with temozolomide (TMZ)) that evaluated current "standard" treatments of glioblastoma [7]. In this study, patients who underwent radical resection and combined chemoradiotherapy presented significantly longer overall median survival (18.8 months) than patients submitted to partial removal (13.5 months) or biopsy without resection and chemoradiotherapy (9.4 months). Similarly, Sanai et al. [11] recently presented a single institution, retrospective study of 500 patients with newly diagnosed glioblastoma treated with surgical resection, followed by standard chemotherapy and radiation. Multivariate analysis showed that the extent of resection of at least 78 % was an independent prognostic factor for improved outcome and resulted in median survival of 12.5 months. Progressive volumetric extent of resection determined by early postoperative MR imaging continued to be an independent prognostic factor for longer median survival, with 100 % extent of resection resulting in 16-month median survival.

Novel technologies such as intraoperative MRI (iMRI) [12] and fluorescence-guided microsurgical resection [13] continue to increase our ability to achieve higher rates of maximal safe resection. Advantages of iMRI include identification of residual tumor during resection, clarification of anatomical landmarks distorted by manipulation during surgery, and accurate neuronavigational systems to guide further microsurgical resection. Fluorophores linked to tumor-specific antigens or proteins such as 5-aminolevulinic acid (ALA) have also been used with increasing frequency during glioma surgery, maximizing resection.

7.3.2 *Chemotherapy Regimens*

In addition to surgery and radiation therapy, several chemotherapeutic agents have been used previously in the treatment of malignant gliomas, including lomustine, procarbazine, etoposide, carboplatin, tamoxifen, and irinotecan [14, 15], among many others, with only modest effects in survival. Clinical use of temozolomide, based on the 2005 multicenter, randomized trial by Stupp and colleagues, provided concrete evidence on the safety and efficacy of an orally administered chemotherapeutic agent in patients with histologically confirmed glioblastoma [16]. Temozolomide acts by methylating DNA bases, which results in the formation of O6-methylguanine that activates and subsequently inhibits the cellular DNA mismatch repair mechanism. These changes produce DNA double-strand breaks that result in apoptosis. The cellular repair enzyme O6-methylguanine-DNA methyltransferase (MGMT) removes these adducts and can counteract the effect of temozolomide [17, 18]. Overexpression of MGMT therefore results in increased resistance to the drug. Resistance decreases with lower MGMT levels and is virtually absent in tumors in which the enzyme has been silenced by methylation of its promoter. This correlation has also been shown to carry prognostic significance [19]. Temozolomide is generally well tolerated, with patients in one study receiving as many as 44 cycles of second-line treatment without cumulative toxicity [20]. In addition to temozolomide, BCNU-impregnated wafers can be placed intracranially at the site of tumor resection (Fig. 7.1).

Antiangiogenic agents, i.e., bevacizumab, have been evaluated as an adjunct to chemotherapy. Bevacizumab is a humanized monoclonal antibody that inhibits vascular endothelial growth factor A (VEGF-A). This agent received FDA approval as monotherapy for the treatment of recurrent GBM [21], following the results of a randomized phase II study in patients with recurrent GBM at first or second relapse (BRAIN) [22] and of a phase II single-arm study using bevacizumab monotherapy in heavily pretreated patients with recurrent GBM (NCI 06-C-0064E) [23]. There is an objective response rate of 26 %, PFS at 6 months of 36 %, and a median overall survival of 9.2 months for the monotherapy arm, as well as a 20 % response rate and a median duration of response of 3.9 months for the non-comparative study [24]. Treatment also resulted in a 50 % decrease in the use of corticosteroids due to reduced peritumoral edema [23]. A retrospective review of the efficacy of bevacizumab monotherapy as salvage treatment for alkylating agent refractory recurrent GBM reported PFS rates at 6 and 12 months of 42 and 22 %, with a partial radiographic response in 42 % of patients. Toxicities associated with bevacizumab and other antiangiogenic agents include hypertension, proteinuria, fatigue, thromboembolic events, and wound healing complications, among others [24]. Although these adverse events are rarely severe, rare life-threatening intracranial hemorrhages associated with bevacizumab have been reported in 4.9 % of patients [24]. In addition, the antiangiogenic effect of bevacizumab on vascular permeability results in diminished contrast enhancement on CT or MRI scans, which may lead to overestimation of true antitumor activity and difficulties in evaluating treatment responses.

7.3.3 Radiation Therapy

A modest effect in survival has also been shown when chemotherapy is administered concomitantly with radiation therapy [25]. When treating gliomas with radiation

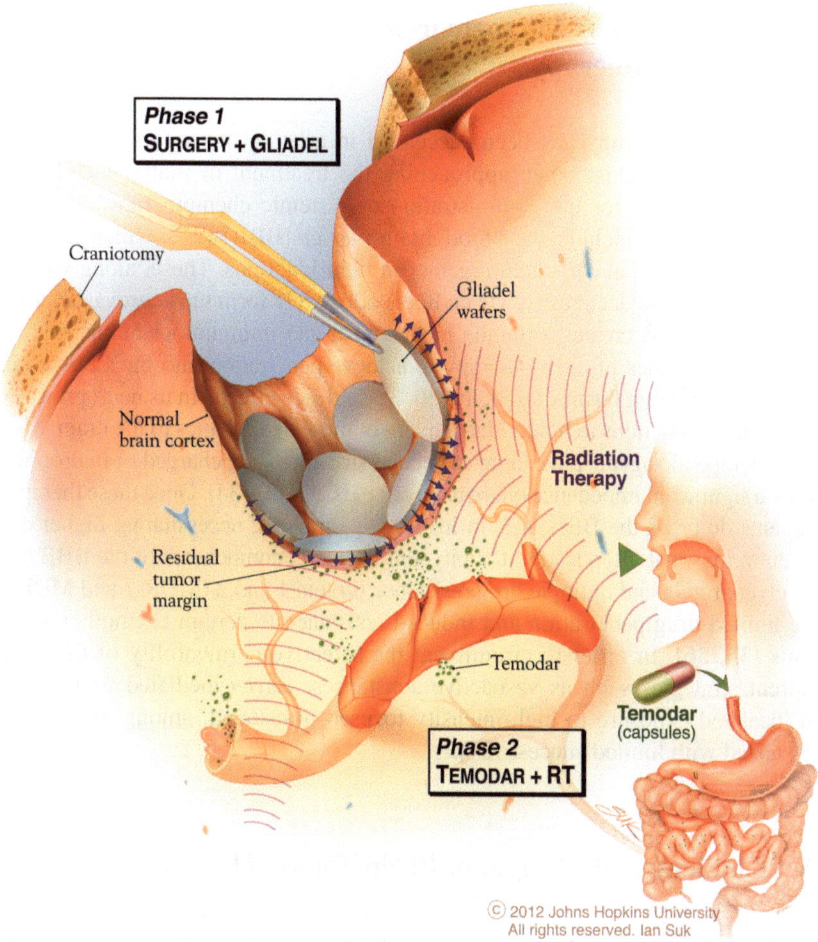

Fig. 7.1 An illustration of combinatorial treatment for glioblastoma multiforme. Patients treated with temozolomide and Gliadel in addition to radiation therapy and surgical resection showed a median survival of 20.7 months [8]. For Domb's Focal Controlled Drug Delivery (86143900)

therapy, however, local recurrence results from biological intracellular and microenvironmental resistance of the tumor, as well as from inadequate targeting of the tumor site [26]. In addition, radiation toxicity is not infrequent and may result in necrosis, with associated injury to cortical and white matter structures leading to severe cognitive deficits and in some cases focal neurological dysfunction from mass effect. Fractionated radiotherapy remains a vital component in the clinical treatment of malignant gliomas when used in combination with surgical resection and chemotherapy. Improvements in tumor-targeting technologies have increased its accuracy while minimizing the exposure of normal tissue resulting in fewer side effects. Experimental data of locally delivered radiation therapy has been encouraging but has not resulted in clinically tested efficacious treatments.

7.4 Role of Blood–Brain Barrier (Neurovascular Unit) and the Tumor Microenvironment in Brain Tumor Pharmacotherapy

Although chemotherapy has been successful in treating many forms of cancers, several challenges exist when approaching the treatment of malignant gliomas. The main obstacles for the administration of systemic chemotherapy in treating malignant gliomas include the blood–brain barrier (BBB) [27], which normally consists of endothelial cells held together in tight junctions. These, along with the basal lamina and adjacent astrocyte endfeet collectively form the neurovascular unit [28, 29]. The BBB creates a unique biochemical and immunological environment requiring high doses of chemotherapy to achieve penetration into the CNS, which can result in different forms of toxicity [30–32] and side effects. In its neuroprotective role, the BBB functions to hinder the delivery of these therapeutics to the brain. Many chemotherapeutic agents have large molecular structures, are charged or hydrophilic, and face significant impedance on their way to the CNS [33, 34]. Once these therapeutic agents do cross the BBB, it is in inadequate amounts, necessitating higher drug concentrations that often lead to toxicity. Furthermore, components of the BBB such as P-glycoprotein and multiple drug resistance protein 1 and 2 (MDR-1 and MDR-2) have limited drug penetration and induced resistance to certain chemotherapeutic agents [35, 36]. In order to circumvent the selective permeability of the BBB, different strategies such as vasoactive substances, carrier-mediated transporters, and localized exposure to high-intensity focused ultrasound, among others, have been tested with limited success [37].

7.5 Challenges of Malignant Brain Tumor Therapy

Significant and unique challenges are faced when treating malignant gliomas. Of particular note is this tumor type's cellular heterogeneity which leads to rapid and aggressive relapse [38] after surgical resection. Radiotherapy is somewhat effective but limited by normal tissue toxicity. In addition, due to the blood–brain barrier's protective and selective characteristics, standard chemotherapy has provided only modest improvements for glioma patients [38]. It is also thought that chemoresistance and radioresistance may be attributable to a glioma stem cell population which may be responsible for this tumor type's characteristically aggressive recurrence [38, 39]. Gliomas typically recur at the site of the initial tumor, usually within 2 cm.

7.5.1 Treatment of Recurrent Tumors

Recurrent gliomas predominantly occur at the site of the original tumor [40]. Repeat surgery is conducted when feasible, in addition to chemotherapy and radiotherapy.

Radiotherapy can be in the form of traditional radiotherapy or stereotactic radiosurgery. Fractionated stereotactic radiosurgery can offer an accurate focused way to deliver radiation and sometimes spares normal tissue that would otherwise be at risk with conventional radiation. The combination of radiotherapy and chemotherapy is also utilized for the treatment of recurrent tumor. There have been several clinical trials using this combination with varying degrees of efficacy. A retrospective analysis by Cuneo et al. which included [41] 63 patients with recurrent glioma who received stereotactic radiosurgery in addition to bevacizumab treatment showed that the combination was well tolerated and seemed associated with improved outcomes. A modified temozolomide dosing regimen in combination with stereotactic radiosurgery has also been explored for patients with recurrent glioma [42]. This combination can be associated with modest survival benefits as well as a low risk of complications. Many other agents that have shown promise in preclinical studies have been utilized for the treatment of recurrent glioma, such as gefitinib, vandetanib, and sunitinib, with varying degrees of success [43]. However, due to the complexity of this tumor type, clinical trials have not led to one overwhelming agent that has shown uniform or impressive efficacy.

7.6 Focal Drug Delivery Systems for Malignant Primary Brain Tumors

7.6.1 Implantable Catheter Systems

Multiple alternatives to systemic delivery routes have been developed. Implantable intracranial catheter systems were among the first to be tested clinically. These catheter/pump systems are surgically implanted into the brain at the resection cavity and permit the constant infusion of drugs over extended periods of time to the tumor bed but are prone to mechanical failure, infection, and obstruction from tissue debris and clots, with these complications often requiring a second intervention. Furthermore, when placed in solution, several drugs may change their chemical properties before reaching the desired site. Intracavitary balloon catheters have also been tested clinically. This system consists of an inflatable balloon catheter that is placed within the surgical cavity with a radiation source in solution that is delivered through the catheter into the surrounding brain tissue. After completion of therapy, the balloon is then removed and the catheter surgically explanted. Two studies found this type of catheter to be useful in the treatment of new and recurring GBM [44, 45], while another found the catheter to achieve the same median patient survival time and duration of functional independence similar to those achieved with resection plus whole-brain radiation therapy [46]. Another study explored the possibility of leakage from the balloon and raised significant safety concerns [47]. Another catheter system used for brain tumor therapy is the Ommaya reservoir. This system introduced by Dr. Ayub Ommaya in 1963 consists of a catheter placed in one of the anterior horns of the lateral ventricles attached to a reservoir implanted under the scalp. Chemotherapeutic agents are injected percutaneously into the reservoir and upon compression of the reservoir are delivered to the target. The Ommaya reservoir is commonly used in the context of meningeal dissemination, when tumor cells migrate to the brain surface and subarachnoid space via the cerebrospinal fluid (CSF) and start to proliferate there [48]. This device allows for injection of therapeutic agents over an extended period of time but requires surgical intervention for removal in cases of infection or at the end of the therapeutic cycle and is not commonly used in the treatment of malignant gliomas.

7.6.2 Convection-Enhanced Drug Delivery

Convection-Enhanced Drug Delivery (CEDD) is also a catheter-based delivery system that requires implantation into the resection cavity and removal at the end of the treatment. Drugs are delivered through one of the several catheters placed stereotactically within a tumor mass or around the resected cavity [49, 50]. CEDD utilizes continuous injection of a fluid containing a therapeutic agent under positive pressure. Several preclinical studies in animal models of GBM have shown prolonged survival using a variety of chemotherapeutics. For instance, Lopez and colleagues showed that topotecan, a topoisomerase I inhibitor, administered by CEDD, significantly improved survival [51]. In this study, a rat model of GBM was made by selectively targeting adult glial progenitors to overexpress platelet-derived growth factor (PDGF). These PDGF-driven induced tumors recapitulated the histological features of the human GBM. Topotecan was then directly infused into the tumor and surrounding brain through the interstitial space via CEDD. Median survival of the control animals was 20 days, and animals given topotecan on days 1,4, and 7 by CEDD showed a median survival of 23, 31, and 54 days, respectively. In another study by Dickinson and colleagues [52], CEDD of liposomal nanoparticles containing gadolinium and the chemotherapeutic agent irinotecan (CPT-11) was examined in four laboratory dogs. Monitoring of CEDD was done by real-time sequential MRI, and volumes of distribution were calculated from these MR images and histology. Infusion variability was observed in the gray matter, and leakage into the ventricular and subarachnoid spaces was observed. Other complications included mild transient proprioceptive deficits, focal hemorrhage in one dog, and focal and mild perivascular, nonsuppurative encephalitis in another dog. Although consistent delivery and minimal adverse effects were shown in this study, the authors also commented on problems associated with scale of delivery, chemotherapeutic efficacy, and the potential for local and systemic toxicity. The pressure gradient and heterogeneity of drug distribution within the tumor itself might act as limiting factors in drugs delivered by this method [53]. These and other preclinical studies were the basis for completed and ongoing clinical trials.

A randomized phase III trial evaluated the survival of patients treated with cintredekin besudotox (CB), a fusion protein comprised of human IL-13 and a mutated form of Pseudomonas aeruginosa exotoxin A (IL13-PE), delivered by CEDD and compared it with the survival of patients treated with Gliadel wafers [54]. This study enrolled patients at first recurrence of GBM in 52 clinical centers worldwide. In the 184 patients given CB, a median overall survival of 36.4 weeks was observed in comparison to a median overall survival of 35.3 weeks for the 92 patients in the control group given Gliadel wafers. No survival difference between CB administered by CEDD and treatment with Gliadel wafers was observed, but future studies should include drug distribution assessment. Experience with the CEDD trials indicates that a rigorously standardized and executed protocol must be applied in different centers to ensure exact and reproducible drug delivery [55] and to ascertain that a biologically significant effect can be accomplished by this method. An ongoing study is evaluating a novel antibody-based treatment, Cotara, delivered via CEDD that utilizes a radiolabeled antibody specific for a DNA/histone (H1) complex; the results of the phase I study showed feasibility and safety of this therapy, and further testing is currently ongoing [56, 57]. A phase I/II trial with 15 patients led by Lidar and colleagues analyzed CEDD of paclitaxel in patients with recurrent malignant glioma. Of the 15 patients, 5 had a complete response, and 6 had partial responses, resulting in a 73 % response rate. The treatment, however, was associated with a significant incidence of treatment-associated complications [58].

Despite encouraging successes in preclinical studies and early clinical testing, penetration of drugs delivered via CEDD remains limited as delivery occurs primarily into the inner necrotic or nonmigratory tumor core [49], with limited penetration to the outer edge of the tumor where gliomas tend to recur most frequently. Furthermore, infused agents may leave targeted areas and leak into either the ventricles or healthy sulci [59]. Improved penetration has been attempted with implantation of multiple catheter, but which also increases the potential for surgical morbidity [49].

7.7 Controlled-Release Polymers and the Development of Gliadel

Another alternative method for local drug delivery involves the use of controlled-release polymers. The first generation of these polymers involved nonbiodegradable diffusing matrices capable of incorporating and releasing macromolecules in a sustained fashion. The second generation involved biodegradable polymers with similar release properties. In 1976, Langer and Folkman [60] reported the first successful release of macromolecules from a polymer system in vivo. Nonbiodegradable implantable polymer systems are currently used in the treatment of GBM [61–63], glaucoma [64], dental disease prevention [65], contraception [66], and delivery of leuprolide, a GnRH analog used in the treatment of advanced hormone-responsive prostate cancers [67]. A biodegradable polymer permits drug delivery through a combination of polymer degradation and drug diffusion. One of the early biodegradable polymers to be introduced was the poly-lactide-co-glycolide (PLGA), which was rapidly introduced for clinical use in the development of absorbable surgical suture [68, 69] The PLGA polymers can also be shaped into injectable microspheres, thus allowing direct intra-tumoral administration [70]. A biodegradable polymer with time-dependent erosion in addition to drug diffusion had to be developed in order to implement it in clinical trials for brain tumor therapy. This would eliminate the need for surgical removal and control the drug concentration throughout a given time period. In 1985, Leong and colleagues [71] reported the development of a poly(bis(p-carboxyphenoxy)-propane (PCPP): sebacic acid (SA) or PCPP:SA matrix. This copolymer could release incorporated drugs in a sustained fashion through the formation of dicarboxylic acids via a spontaneous reaction with water. The prevention of hydrolysis and enzymatic degradation is due to the polymer's extremely hydrophobic properties. Furthermore, zero kinetic degradation of the polymer matrix is conducive to the release of a biologically active drug for prolonged periods at steady concentrations. Without this quality, drugs would have half-lives of a few minutes, similar to systemically administered agents. The degradation rate of the PCPP:SA polymer can be altered by changing the ratio between the monomers of carboxyphenoxypropane (CPP) and sebacic acid (SA). Another important quality is that by varying temperature and pressure, this polymer can take on different forms such as wafers, rods, sheets, and microspheres [72–75]. These polymers caused only minimal inflammatory reactions which are similar to the inflammation observed with common surgical hemostatic implants, and a study done on cynomolgus monkeys showed no signs of behavioral, neurological, or hematological changes [76].

Carmustine (BCNU) has been used as a systemic agent in the treatment of brain tumors with limited survival benefit [77, 78]. BCNU can alkylate the nitrogen bases of DNA and has dose-limiting side effects such as bone marrow suppression and pulmonary fibrosis with a relative short half-life of less than 15 min. These properties made BCNU an ideal candidate for incorporation into a polymer matrix for intracranial delivery. In preclinical studies, BCNU was delivered via PCPP:SA polymers in order to determine the biodistribution of the drug [79]. Further studies in cynomolgus monkeys documented tumoricidal concentrations of intracranially released BCNU delivered by 20 % PCPP:SA polymers at 4 cm from the implantation site at 24 h, 2 cm on day 7, and 1.3 cm on day 30 post-administration [80]. A study done by Tamargo and colleagues [81] demonstrated that Fischer 344 rats implanted with subcutaneous and intracranial 9L gliosarcoma tumors experienced a statistically significant improvement in survival after local delivery of BCNU when compared with control animals treated with empty polymers or intraperitoneal injections of BCNU. All of these findings paved the way for clinical testing of BCNU-loaded PCPP:SA polymers.

7.7.1 *Clinical Testing of Gliadel*

Starting in 1987, a multicenter phase I/II clinical trial was initiated to assess the safety of 20:80 PCPP:SA BCNU-loaded polymers intracranially implanted in humans [82]. Only patients with recurrent malignant gliomas and previous surgical debulking were included. Other inclusion criteria included failure of standard therapy, a Karnofsky performance score ≥60, and no exposure to nitrosoureas during the 6 weeks prior to polymer implantation. Twenty-one patients were enrolled and each given a maximum of eight wafers, each weighing 200 mg. The BCNU polymer was given in three different formulations: 1.93, 3.85, and 6.35 %. There was no evidence of systemic toxicity or signs of neurological deterioration following polymer implantation. Further blood chemistry and urinalysis tests did not reveal any signs of bone marrow, hepatic, or renal injury. Treated patients experienced a median survival of 46 weeks after polymer implantation and 87 weeks after initial diagnosis. Of 21 patients with recurrent malignant gliomas, 8 (38 %) survived >1 year. These positive results led to a phase III clinical trial [61]. This study investigated 3.8 % BCNU PCPP:SA polymers (Gliadel) for treatment of recurrent malignant gliomas in 222 patients at 27 medical centers throughout North America. The selection criteria were identical to those used for the phase I/II study, with the additional provision that no chemotherapy was permitted 4 weeks preoperatively and no nitrosoureas were allowed for 6 weeks prior to polymer implantation. One hundred and ten patients received BCNU wafers, while 112 patients received the placebo. Overall postoperative median survival was 31 weeks for the BCNU-treated group and 23 weeks for the placebo group. The 6-month survival rate was 60 % in the treatment group and 47 % in the placebo group. There was a 50 % increase in the survival for glioblastoma patients treated with Gliadel in comparison with those treated with placebo. As shown in prior studies, the BCNU polymer treatment was confirmed as being safe and effective, with no evidence of systemic toxicity. This was the pivotal study leading to the FDA approval of 3.85 % BCNU-loaded PCPP:SA wafer (Gliadel) for the treatment of recurrent glioblastoma in 1996.

In order to determine the safety and efficacy of 3.85 % BCNU-loaded PCPP:SA polymers as an initial treatment of malignant brain tumors, a phase I/II study was carried out with 22 patients who underwent surgical debulking and implantation of up to eight wafers [83]. All patients received standardized external beam radiation therapy postoperatively. The study showed a median survival of 42 weeks for the treatment group, with four patients surviving >18 months. This clinical study established that Gliadel was safe and well tolerated when combined with radiation therapy for patients with newly diagnosed malignant gliomas. These results prompted a phase III clinical trial [62]. The study, originally planned for 100 patients, was interrupted due to temporary drug unavailability; therefore, 32 patients were entered in the study. The applied admission criteria were similar to the phase I/II study except that a histopathological diagnosis of a malignant glioma was required by intraoperative frozen sections. Five anaplastic astrocytoma and 27 glioblastoma patients were randomized to receive either placebo or BCNU wafers (61.6 mg of BCNU) after maximal surgical resections; patients also underwent radiation therapy. The median survival was 58.1 weeks for the Gliadel-treated patients, whereas it was 39.9 weeks for the placebo arm ($p=0.012$). The subpopulation of glioblastoma patients had a median survival of 53.3 weeks when treated with Gliadel versus a median survival of 39.9 weeks when treated with the placebo. The results further indicated the 1-, 2- and 3-year survival rates to be 63, 31, and 25 %, respectively, for the Gliadel group in comparison with 19, 6, and 6 % for the placebo group. As with prior studies, no signs of systemic or local toxicity attributable to the polymer were noted. A third randomized, prospective, placebo-controlled clinical study was then carried out to further define the role of Gliadel as an initial therapeutic modality [63]. A total of 240 adult patients who underwent initial surgical resection for a high-grade malignant glioma were entered into the study. BCNU wafers (Gliadel) or placebo wafers were surgically implanted at the site of initial resection and followed by radiation therapy 2–3 weeks later. The primary end point for this study was survival; the median survival was 13.9 months for the intent-to-treat group compared with 11.6 months for the placebo group. The study indicated that the overall risk of death during the 3–4 years posttreatment was reduced in the Gliadel wafer treatment group. Treatment with Gliadel resulted in a 27 % reduction in the risk of death over the course of the study. In the Gliadel group, 16 % of patients were alive at 2 years, and 9 % were alive at 3 years, compared with 9 and 2 %, respectively, in the placebo group. A subsequent preclinical study was conducted to improve the therapeutic efficacy of Gliadel wafers [84]. This study showed that efficacy can be enhanced by increasing the BCNU loading dose to 20 % without generating local or systemic toxicity. Based on these findings, on February 26, 2003, the FDA approved Gliadel for use in newly diagnosed patients with high-grade malignant gliomas as an adjunct to surgery and radiation therapy. A National Institute of Health-funded dose-escalation trial at 11 medical centers in the USA was designed to evaluate the safety of Gliadel wafers between 6.5 and 20 % BCNU in patients with recurrent malignant brain tumors. This phase I/II escalation study established that the maximal nontoxic loading dose was 20 % [85].

7.7.2 Further Clinical Experience with Gliadel

The relatively wide use of Gliadel has given rise to several recent studies further discussing its efficacy in the management of malignant gliomas. Attenello and colleagues reported the 10-year institutional experience at Johns Hopkins on 288 patients undergoing resection for malignant glioma and found that the Gliadel wafer was not associated with an increase in perioperative morbidity after surgical treatment for malignant astrocytoma [86] with Gliadel patients reaching a median survival of >13 months [62, 63]. Moreover, Gliadel increased median survival after revision resection for recurrent glioma to a median survival of 11.3 months, higher than was reported previously [61] and higher than the control in this study. Furthermore, rates of morbidity between Gliadel and non-Gliadel groups were similar, despite patients being slightly older in the Gliadel group. The report also mentions that the incidence of meningitis and CSF leaks with Gliadel are similar to those of control patients. There was no statistically significant difference in the pulmonary embolism rate between Gliadel-treated patients and controls nor any difference in perioperative complications, CSF leaks, wound healing complications, postoperative seizures, symptomatic malignant edema, tumor bed cysts, or meningitis.

Chaichana and colleagues retrospectively examined the efficacy of Gliadel in prolonging survival for patients aged 65 years or older [87]. With an average age of

73 ± 5 years, 45 patients treated with Gliadel were compared to 45 matched historical control patients. No differences were seen in pre- and perioperative variables, although patients treated with Gliadel demonstrated an 8.7-month survival rate, while median patient survival was 5.5 months in the matched control group. This prolonged survival of patients treated with Gliadel was also demonstrated in a subgroup of the study with patients older than 70 years and 75 years who were given carmustine wafers than same-aged control patients. McGirt et al. reported that in those patients treated with temozolomide and Gliadel in addition to radiation therapy and surgical resection, a median survival of 20.7 months was reached with a 2-year survival rate of 36 % [8].

Gliadel has also been tested against metastatic cell lines in animal models with good results [88], and a study in which Gliadel and radiation therapy were given together to patients with a single brain metastasis following surgical resection has also been reported [89]. This study was comprised of 25 patients with a single supratentorial metastatic lesion, who underwent surgical resection, Gliadel treatment, and whole brain radiation therapy. There were no local recurrences, four patients relapsed in a different anatomical area in the brain, and two patients relapsed in the spinal cord. Median survival was 33 weeks; 33 % of patients survived 1 year, and 25 % survived 2 years.

7.8 Other Polymer-Based Therapies

7.8.1 Microspheres, Nanospheres, and Other Nanoparticles

Micro- and nanospheres have been tested as potential delivery systems for brain tumor therapy [90–92]. Ranganath and colleagues [93] demonstrated that paclitaxel microspheres achieved significant tumor inhibition in glioblastoma xenograft models in mice. A 30-fold tumor inhibition and low proliferation index after 41 days of treatment was seen in nano-fiber paclitaxel discs in mice when compared to controls. In a different study, Gil-Alegre and colleagues manufactured biodegradable microspheres of BCNU at a mean size of 35 μm [94]. In vitro studies treating human glioblastoma cells showed a sustained release of BCNU from microspheres over a period of 21 days. The authors noted that 320 mg of microspheres contain 61.6 mg of BCNU, which is the same amount of BCNU contained in 1,600 mg or eight wafers of Gliadel. The authors proposed a stereotactic intracranial administration by needle injection, which could benefit patients with lesions not amenable for resection or debulking.

An in vivo study in mice carried out by Benny and colleagues tested injections of 225 μg of imatinib mesylate, a small-molecule tyrosine kinase inhibitor, loaded into PLGA microspheres, in mice orthotopically challenged with U87 and the GL261 gliomas [90]. A single local injection of PLGA microspheres loaded with imatinib mesylate loaded at a low concentration led to 88 and 79 % reduction in subcutaneous (s.c.) U87 and GL261 tumors, respectively. Intracranial injections of the same formulation led to a 79 % reduction in tumor volume in U87-implanted animals. Immunohistochemistry showed marked decrease in proliferation indices and tumor vessel density in the s.c. model and induction of apoptosis in an intracranial model [90].

While the therapeutic use of nanospheres remains in developmental stages, the use of nanospheres as biomarkers in gliomas is rapidly evolving. Kantelhardt and colleagues were able to use target probes and demonstrate highly specific staining of tumor tissue compared with normal brain both in biopsies of high-grade and low-grade gliomas. They were also able to clearly distinguish tumor cells in low-grade biopsies where no enhanced MRI image was obtained [95]. By using gadolinium-containing endohedral fullerenes, which are highly efficient nanosphere-particle contrast agents for MRI, and conjugating them to a tumor-specific peptide, Fillmore and colleagues [96] successfully targeted brain tumor cells overexpressing the IL-13 receptor. A successful demonstration of the potential specific therapeutic targeting of tumor cells was achieved. This enhanced uptake reflects increased specificity of these targeted nanospheres to human glioma cells overexpressing the IL-13 receptor. Another biomarker example which might help in the future treatment of gliomas is quantum dots, nanoparticles resistant to chemical and metabolic degradation, demonstrating long-term photostability that are detectable on MRI. Nance et al. describe how modifying the poly(ethylene glycol) coating of nanoparticles can increase penetration into brain tissue allowing for more rapid and efficient diffusion [97]. Ardnt-Jovin and colleagues have shown a clear demarcation between brain and tumor tissue [98] by using glioma mouse models and human brain tumor biopsies. A further demonstration of effective delivery with significant concentrations into brain tumor tissue was seen in orthotopically implanted mice.

Nanoparticles can be injected systemically and are able to cross the BBB to deliver a host of efficacious substances, such as DNA plasmids, protein, vectors, RNA, and siRNA [99]. Synthetic siRNAs have been shown to silence genes in vivo that are important for the pathogenesis of GBM [100]. There are currently several early-stage clinical trials testing nanoparticles and liposomes in patients with malignant gliomas.

7.9 The Use of Microchips for Drug Delivery

Microchips are implantable reservoirs capable of delivering multiple agents at predetermined times. There are two types of microchips, active and passive. Both allow for programmed release of individual compounds, and neither manipulates the compound to be delivered chemically, thereby allowing the compound to remain in a bioactive and bioavailable state. Both chips are biocompatible and small enough in size for intracranial implantation in the rodent.

Active microchips, microelectromechanical systems (MEMS), utilize implantable solid-state silicon constructs that provide controlled release of multiple microreservoirs [101–103]. This system allows programming of delivery of different contents from separate microreservoirs at specific periods of times, providing multiple combinations of drugs working together or being released at different time intervals [101]. In a study by Li et al., the MEMS device was used to locally deliver the chemotherapeutic agent, BCNU, to experimental tumors in rats [104]. This delivery resulted in dose-dependent inhibitory effects on tumor growth, as effective as equipotent subcutaneous injections of the same drug. An intracranial MEMS-based device has been developed to deliver the clinically utilized chemotherapeutic temozolomide (TMZ) in a rodent glioma model [103]. The safety of implanting the device intracranially was confirmed, and TMZ delivered from the device was effective at prolonging animal survival in a 9L rodent glioma model. Further optimization using this MEMS device to deliver chemotherapeutic drugs in combination could be examined because of the unique capability of the device to precisely control the temporal release profiles of multiple substances.

Passive microchips are based on slow degradation of a thin polymeric membrane covering each reservoir of drug. This microchip could deliver multiple drugs on demand based on the specific brain tumor pathology of the patient. Kim et al. used a syngeneic Fischer 344 9L gliosarcoma rat model to study the tumoricidal effects of a microchip incorporating BCNU [105]. Tumors treated with 1.24 mg of BCNU showed significant tumor reduction compared to empty microchip controls. More recently, Scott et al. used the passive microchip to locally release temozolomide to treat the rodent 9L intracranial gliosarcoma model [106]. These in vivo efficacy results showed that the intracranial delivery of temozolomide from the microcapsule device prolonged animal survival. This drug delivery method may offer a unique form of treatment for brain tumors.

7.10 Other Non-focal Therapies for Malignant Primary Brain Tumors

7.10.1 Immunotherapy and Tumor Vaccines

Another proposed antiglioma strategy involves activation of the patient's immune system against their cancer cells. First proposed by Coley in the late 1800s for the treatment of systemic sarcoma, heat-neutralized bacteria were injected to generate a potent inflammatory response that would recruit sufficient immune cells to favor recognition of tumor antigens by the native immune system resulting in acquired immunological memory and vigorous antitumor responses [107]. Following multiple promising preclinical experimental results, well-controlled clinical trials involving immunotherapy against cancer are currently underway. In 2010, the therapeutic cancer vaccine sipuleucel-T was shown to increase median survival from 21.7 to 25.8 months in men with metastatic castration-resistant prostate cancer [108]. This prostate cancer vaccine involves harvesting the patient's peripheral blood mononuclear cells, activating them with a recombinant fusion protein, and reintroducing the activated cells back into the patient. Another recent immunotherapy trial involved a monoclonal antibody (ipilimumab) against cytotoxic T-lymphocyte-associated antigen 4 (CTLA-4), which downregulates pathways of T-cell activation. Administration of ipilimumab in unresectable stage III or IV melanoma resulted in an increase in median survival from 6.4 to 10 months [109].

There are numerous ongoing and completed clinical trials investigating the role of immunotherapy in the treatment of primary malignant brain tumors [110, 111]. These approaches can be grouped into three main categories. The first involves harvesting the patient's dendritic cells, activating them with tumor-associated antigens, and reinfusing them back into the patient. One example of this approach involves exposing harvested dendritic cells to peptides from the surface of autologous glioma cells [112]. Reintroduction of these dendritic cells resulted in T-cell infiltration of the glioma in two of four patients who underwent reoperation after vaccination. The second involves harvesting the patient's lymphocytes, nonspecifically activating them, and reinfusing them back into the patient. In a 40-patient trial using this approach, harvested peripheral blood mononuclear cells were incubated with interleukin-2 and reinfused back into the patient [113]. The last category involves directly injecting into the patient immune-stimulatory materials, such as inactivated tumor cells [114] and/or cytokines [115]. Although immunotherapy for primary malignant brain tumors seems promising based on preliminary results, randomized controlled trials are required to determine whether this approach can indeed increase median survival.

7.10.2 Angiogenesis Inhibitors

In the early 1970s, Folkman and colleagues postulated that formation of new blood vessels or "angiogenesis" was a vital process for tumor growth and proliferation [116]. In 1989, the primary signaling factor responsible for angiogenesis was identified and named vascular endothelial growth factor (VEGF) [117, 118]. Eight years later, a humanized monoclonal antibody against VEGF was developed [119]. In clinical trials, addition of the anti-VEGF antibody bevacizumab to standard therapies increased median survival in metastatic colon cancer from 15.6 to 20.3 months [120] and in non-small-cell lung cancer from 10.3 to 12.3 months [121].

Glioblastoma is recognized as one of the most angiogenic tumors [122], and many angiogenesis inhibitors [123], including DC101 (a monoclonal antibody against VEGF receptor 2) [124], the tetracycline antibiotic minocycline [125], angiostatin [126], and endostatin, have been shown to be effective against brain tumors in animal models [127]. One of the first clinical trials of an angiogenesis inhibitor in the treatment of primary brain cancer was performed in the early 2000s and involved thalidomide [128, 129]. However, it was unclear whether the addition of thalidomide increased survival, and enthusiasm for this approach diminished. More recently, phase II clinical trials of bevacizumab and irinotecan in recurrent glioblastoma were published [22, 130, 131], and the United States Food and Drug Administration (FDA) approved bevacizumab for the treatment of recurrent glioblastoma [23]. However, it is currently controversial whether bevacizumab is effective at increasing median survival in glioblastoma [132]. Bevacizumab is known to alter the BBB, resulting in changes of the enhancing tumor area on MR imaging [133], which is used to determine progression-free survival in phase II trials; therefore, potential masking of residual or recurrent tumor cells could occur. Although promising, further randomized controlled studies involving bevacizumab with median survival as the primary end point are required to clarify its beneficial effects on glioblastoma.

7.10.3 Gene Therapy

Gene therapy strategies have been investigated as possible candidates for brain tumor therapy. Gene therapy is defined as the "targeted transfer of genetic material into tumor cells for therapeutic purposes" [134]. There have been many promising gene therapy candidates over the last decade. Of these, retroviral and adenoviral vectors have been the most commonly used for delivery of antiglioma therapeutic genes [99]. One of the most commonly used strategies is the enzyme–prodrug suicide gene therapy system, which is designed to spare normal CNS cell types while maintaining high specificity for tumor targeting with potential tumoricidal effects. Suicide gene therapy inhibits cell division by blocking DNA replication. Tumor cells are transfected with a gene that encodes for an enzyme that converts a systemically delivered prodrug into an active drug that is toxic to tumor cells [99]. This therapy may also induce the "bystander effect" which allows the toxic metabolites to kill tumor cells distal from the original tumor site and which were not originally transduced with the therapeutic gene [99]. While this has shown promise in preliminary animal models, that success has not yet translated as effectively in human clinical trials. A large phase III clinical evaluation of the herpes simplex virus type 1 thymidine kinase in combination with systemically administered ganciclovir included 248 newly diagnosed GBM patients [135]. The therapy was shown to be well tolerated, but there was no significant difference in survival between the group that received the gene therapy and the control group. The authors attributed the limited efficacy to a poor rate of delivery of the HSV-tk gene therapy to tumor cells and commented on the need for new noninvasive methods for in vivo assessment of transduction rates as well as for improved delivery of the prodrug across the BBB to the transduced tumor cells.

To increase transduction efficiency, conditionally replicating tumor-selective oncolytic viruses have been developed. These are tumor-targeting and replication-competent viruses that have also been extensively studied in both preclinical and clinical settings. The first-generation trials have shown viral replication but with limited evidence of antitumor efficacy [136]. These have now been modified and are currently in phase I safety studies with optimism for future efficacy.

7.10.4 *Small-Molecule Inhibitors and Monoclonal Antibodies*

Epidermal growth factor receptor (EGFR) is the most amplified gene product seen in GBMs [137]. EGFR inhibition in GBM has included the use of small-molecule tyrosine kinase inhibitors, monoclonal antibodies, and immunotoxin conjugates. Although erlotinib, a reversible single-agent small-molecule tyrosine kinase inhibitor, showed promising preclinical results, it has not been proven efficacious in clinical trials [138]. Despite the crucial role EGFR plays in GBMs, it remains a puzzle as to why these targeted therapies have not proven successful. Poor penetration across the BBB has been proposed as the main factor limiting the efficacy of this approach in clinical trials.

Bevacizumab is a humanized IgG1 monoclonal antibody that selectively inhibits vascular endothelial growth factor (VEGF) and thereby inhibits angiogenesis. Bevacizumab was granted accelerated approval by the Food and Drug Administration (FDA) for the treatment of recurrent GBM in May 2009. Combining this antiangiogenic agent with active chemotherapies such as TMZ or irinotecan may be an effective strategy to improve clinical outcomes for patients with GBM [139–141].

7.11 Future Drug Delivery Strategies in Brain Tumor Therapy

In the past several years, a plethora of new devices, techniques, treatment strategies, and drug delivery methods have been investigated. Gliadel demonstrated the feasibility of using local controlled drug delivery to treat malignant brain tumors. Newer technologies and better therapeutic agents are currently being investigated. The developments of targeted micro- and nanospheres and the optimization of microchips loaded with different chemotherapeutic agents, antiangiogenic molecules, and immune system modulators will enable multimodal targeted therapies with potential higher efficacy and minimal toxicity and side effects [142]. A more selective treatment regimen may be envisioned in which a patient has a tumor biopsy that is then analyzed in the laboratory, which results in personalized adjuvant therapy.

References

1. Hulleman E, Helin K (2005) Molecular mechanisms in gliomagenesis. Adv Cancer Res 94:1–27
2. Cancer Genome Atlas Research Network (2008) Comprehensive genomic characterization defines human glioblastoma genes and core pathways. Nature 455(7216):1061–1068, Epub 2008 Sep 4
3. Gaspar LE, Fisher BJ, Macdonald DR, LeBer DV, Halperin EC, Schold SC Jr, Cairncross JG (1992) Supratentorial malignant glioma: patterns of recurrence and implications for external beam local treatment. Int J Radiat Oncol Biol Phys 24:55–57

4. Palanichamy K, Erkkinen M, Chakravarti A (2006) Predictive and prognostic markers in human glioblastomas. Curr Treat Options Oncol 7:490–504
5. Kornblith PL, Walker M (1988) Chemotherapy for malignant gliomas. J Neurosurg 68:1–17
6. Wrensch M, Minn Y, Chew T, Bondy M, Berger MS (2002) Epidemiology of primary brain tumors: current concepts and review of the literature. Neuro Oncol 4:278–299
7. Stupp HME, Mason WP et al (2009) Effects of radiotherapy with concomitant and adjuvant temozolomide versus radiotherapy alone on survival in glioblastoma in a randomised phase III study: 5-year analysis of the EORTC-NCIC trial. Lancet Oncol 10:459–466
8. McGirt MJ, Than KD, Weingart JD, Chaichana KL, Attenello FJ, Olivin A, Laterra J, Kleinberg LR, Grossman SA, Brem H et al (2009) Gliadel (BCNU) wafers plus concomitant temozolomide therapy after primary resection of glioblastoma multiforme. J Neurosurg 110:583–588
9. Lacroix M, Abi-Said D, Fourney DR et al (2001) A multivariate analysis of 416 patients with glioblastoma multiforme: prognosis, extent of resection, and survival. J Neurosurg 95:190–198
10. Sanai N, Berger MS (2008) Glioma extent of resection and its impact on patient outcome. Neurosurgery 62:753–766
11. Sanai N, Polley M, McDermott MW et al (2011) An extent of resection threshold for newly diagnosed glioblastomas. J Neurosurg 115:3–8
12. Schneider JP, Trantakis C, Rubach M, Schulz T, Dietrich J, Winkler D, Renner C, Schober R, Geiger K, Brosteanu O et al (2005) Intraoperative MRI to guide the resection of primary supratentorial glioblastoma multiforme—a quantitative radiological analysis. Neuroradiology 47(7):489–500. doi:10.1007/s00234-005-1397-1
13. Stummer W, Pichlmeier U, Meinel T, Wiestler OD, Zanella F, Reulen HJ, ALA-Glioma Study Group (2006) Fluorescence-guided surgery with 5-aminolevulinic acid for resection of malignant glioma: a randomised controlled multicentre phase III trial. Lancet Oncol 7(5):392–401
14. Butowski NA, Sneed PK, Chang SM (2006) Diagnosis and treatment of recurrent high-grade astrocytoma. J Clin Oncol 24:1273–1280
15. Newton HB, Junck L, Bromberg J et al (1990) Procarbazine chemotherapy in the treatment of recurrent malignant astrocytomas after radiation and nitrosourea failure. Neurology 40:1743–1746
16. Stupp R, Mason WP, van den Bent MJ, Weller M, Fisher B, Taphoorn MJ, Belanger K, Brandes AA, Marosi C, Bogdahn U, Curschmann J, Janzer RC, Ludwin SK, Gorlia T, Allgeier A, Lacombe D, Cairncross JG, Eisenhauer E, Mirimanoff RO, European Organisation for Research and Treatment of Cancer Brain Tumor and Radiotherapy Groups, National Cancer Institute of Canada Clinical Trials Group (2005) Radiotherapy plus concomitant and adjuvant temozolomide for glioblastoma. N Engl J Med 352(10):987–996
17. Pegg AE (1990) Mammalian O6-alkylguanine-DNA alkyltransferase: regulation and importance in response to alkylating carcinogenic and therapeutic agents. Cancer Res 50:6119–6129
18. Weller M, Stupp R, Reifenberger G et al (2010) MGMT promoter methylation in malignant gliomas: ready for personalized medicine? Nat Rev Neurol 6:39–51
19. Hansen RJ, Nagasubramanian R, Delaney SM, Samson LD, Dolan ME (2007) Role of O6-methylguanine-DNA methyltransferase in protecting from alkylating agent-induced toxicity and mutations in mice. Carcinogenesis 28:1111–1116
20. Hau P, Koch D, Hundsberger T et al (2007) Safety and feasibility of long-term temozolomide treatment in patients with high-grade glioma. Neurology 68:688–690
21. National Comprehensive Cancer Network. NCCN guidelines version 2.2011: central nervous system cancers. Available at: http://www.nccn.org. Accessed 16 Aug 2011
22. Friedman HS, Prados MD, Wen PY et al (2009) Bevacizumab alone and in combination with irinotecan in recurrent glioblastoma. J Clin Oncol 27:4733–4740
23. Kreisl TN, Kim L, Moore K et al (2009) Phase II trial of single-agent bevacizumab followed by bevacizumab plus irinotecan in recurrent glioblastoma. J Clin Oncol 27:740–745

24. FDA Briefing Document Oncology Drug Advisory Committee Meeting, March 31, 2009. http://www.fda.gov/ohrms/dockets/ac/09/briefing/2009-4427b1-01-FDA.pdf
25. Guckenberger M, Mayer M, Buttmann M, Vince GH, Sweeney RA, Flentie M (2011) Prolonged survival when temozolomide is added to accelerated radiotherapy for glioblastoma multiforme. Strahlenther Onkol 187(9):548–554
26. Cohen-Jonathan ME (2011) Optimization of the radiotherapy for the gliomas: hopes and research axis for the next future. Rev Neurol (Paris) 167(10):656–660
27. Guerin C, Laterra J, Hruban RH, Brem H, Drewes LR, Goldstein GW (1990) The glucose transporter and the blood–brain barrier of human brain tumors. Ann Neurol 28:758–765
28. Lee J, Lund-Smith C, Borboa A, Gonzalez AM, Baird A, Eliceiri BP (2009) Glioma-induced remodeling of the neurovascular unit. Brain Res 1288:125–134, Epub 2009 Jul 9
29. Simard M, Nedergaard M (2004) The neurobiology of glia in the context of water and ion homeostasis. Neuroscience 129(4):877–896, Review
30. Goodman LS, Hardman JG, Limbird LE, Gilman AG (eds) (2001) Goodman and Gilman's the pharmacological basis of therapeutics. McGraw-Hill, New York, NY. pp xxvii, 2148. [2141] fold leaf of plates
31. Hannigan E, Green S, Alberts DS, O'toole R, Surwit E (1993) Results of a southwest oncology group phase III trial of carboplatin plus cyclophosphamide versus cisplatin plus cyclophosphamide in advanced ovarian cancer. Oncology 50(Suppl 2):2–9
32. Slichenmyer WJ, Rowinsky EK, Dohenower RC, Kaufmann SH (1993) The current status of camptothecin analogues as antitumor agents. J Natl Cancer Inst 85(4):271–291
33. Abbott NJ, Romero IA (1996) Transporting therapeutics across the blood–brain barrier. Mol Med Today 2(3):106–113
34. Grieg NH (1987) Optimizing drug delivery to brain tumors. Cancer Treat Rev 14(1):1–28
35. Rautioa J, Chikhale PJ (2004) Drug delivery systems for brain tumor therapy. Curr Pharm Des 10(12):1341–1353
36. Kushara H, Sugiyama Y (2001) Efflux transport systems for drugs at the blood–brain barrier and blood-cerebrospinal fluid barrier (Part 1). Drug Discov Today 6(3):150–156
37. McDannold N, Vykhodtseva N, Hynynen K (2008) Blood–brain barrier disruption induced by focused ultrasound and circulating performed microbubbles appears to be characterized by the mechanical index. Ultrasound Med Biol 34(5):834–840
38. Auffinger B, Thaci B, Nigam P, Rincon E, Cheng Y, Lesniak MS (2012) New therapeutic approaches for malignant glioma: in search of the Rosetta stone. F1000 Med Rep 4:18, Epub 2012 Sep 5
39. Bao S, Wu Q, McLendon RE, Hao Y, Shi Q, Hjelmeland AB, Dewhirst MW, Bigner DD, Rich JN (2006) Glioma stem cells promote radioresistance by preferential activation of the DNA damage response. Nature 444(7120):756–760, Epub 2006 Oct 18
40. Biswas T, Okunieff P, Schell MC, Smudzin T, Pilcher WH, Bakos RS, Vates GE, Walter KA, Wensel A, Korones DN, Milano MT (2009) Stereotactic radiosurgery for glioblastoma: retrospective analysis. Radiat Oncol 4:11
41. Cuneo KC, Vredenburgh JJ, Sampson JH, Reardon DA, Desjardins A, Peters KB, Friedman HS, Willett CG, Kirkpatrick JP (2012) Safety and efficacy of stereotactic radiosurgery and adjuvant bevacizumab in patients with recurrent malignant gliomas. Int J Radiat Oncol Biol Phys 82(5):2018–2024, Epub 2011 Apr 12
42. Minniti G, Scaringi C, De Sanctis V, Lanzetta G, Falco T, Di Stefano D, Esposito V, Enrici RM (2013) Hypofractionated stereotactic radiotherapy and continuous low-dose temozolomide in patients with recurrent or progressive malignant gliomas. J Neurooncol 111(2): 187–194
43. Schwer AL, Damek DM, Kavanagh BD, Gaspar LE, Lillehei K, Stuhr K, Chen C (2008) A phase I dose-escalation study of fractionated stereotactic radiosurgery in combination with gefitinib in patients with recurrent malignant gliomas. Int J Radiat Oncol Biol Phys 70(4):993–1001
44. Welsh J, Sanan A, Gabavan AJ, Green SB, Lustig R, Burri S, Kwong E, Stea B (2007) GliaSite brachytherapy boost as part of initial treatment of glioblastoma multiforme: a retrospective multi-institutional pilot study. Int J Radiat Oncol Biol Phys 68(1):159–165

45. Wernicke AG, Sherr DL, Schwartz TH, Pannullo SC, Stieg PE, Boockvar JA, Molterno JA, Invandize J, Trichter S, Sabbas AM, Parashar B, Nori D (2010) The role of dose escalation with intracavitary brachytherapy in the treatment of localized CNS malignancies: outcomes and toxicities of a prospective study. Brachytherapy 9(1):91–99
46. Rogers LR, Rock JP, Sills AK, Vogelbaum MA, Suh JH, Ellis TL, Stieber VW, Asher AL, Fraser RW, Billingsley JS, Lewis P, Schellingerhout D, Shaw EG, Brain Metastasis Study Group (2006) Results of a phase II trial of the GliaSite radiation therapy system for the treatment of newly diagnosed, resected single brain metastases. J Neurosurg 105(3):375–384
47. Adkison JB, Thomadsen B, Howard SP (2008) Systemic iodine 125 activity after GliaSite brachytherapy: safety considerations. Brachytherapy 7(1):43–46
48. Hodozuka A, Hayashi Y, Annei R, Hiroshima S, Saito M, Orimoto R, Sato M, Tanaka T (2008) Intrathecal infusion of the antineoplastic agents for meningeal dissemination. Gan To Kagaku Ryoho 35(6):900–905
49. Stukel JM, Caplan MR (2009) Targeted drug delivery for treatment and imaging of glioblastoma multiforme. Expert Opin Drug Deliv 6(7):705–718
50. Debinski W, Tatter SB (2009) Convection-enhanced delivery for the treatment of brain tumors. Expert Rev Neurother 9(10):1519–1527
51. Lopez KA, Tannenbaum AM, Assanah MC, Linskey K, Yun J, Kangarlu A, Gill OD, Canoll P, Bruce JN (2011) Convection-enhanced delivery of topotecan into a PDGF-driven model of glioblastoma prolongs survival and ablates both tumor-initiating cells and recruited glial progenitors. Cancer Res 71(11):3963–3971
52. Dickinson PJ, Lecouteur RA, Higgins RB, Bringas JR, Roberts B, Larson RF, Yamashita Y, Krauze M, Noble CO, Drummond D, Kirpotin DB, Bankiewicz KS (2008) Canine model of convection-enhanced delivery of liposomes containing CPT-11 monitored with real-time magnetic resonance imaging. J Neurosurg 108:989–998
53. Vogelbaum MA (2005) Convection enhanced delivery for the treatment of malignant gliomas: symposium review. J Neurooncol 73(1):57–69
54. Kunwar S, Chang S, Westphal M, ZVogelbaum M, Sampson J, Barnett G, Shaffrey M, Ram Z, Piepmeier J, Prados M et al (2010) Phase III randomized trial of CED of Il13-PE38QQR vs Gliadel wafers for recurrent glioblastoma. Neuro Oncol 12(8):871–881
55. Gallia GL, Brem S, Brem H (2005) Local treatment of malignant brain tumors using implantable chemotherapeutic polymers. J Natl Compr Canc Netw 3:721–728
56. Patel SJ, Shapira WR, Laske DW, Jensen RL, Asher AL, Wessels BW, Carpenter SP, Shan JS (2005) Safety and feasibility of convection-enhanced delivery of Cotara for the treatment of malignant glioma: initial experience in 51 patients. Neurosurgery 56(6):1243–1252
57. Shapiro WR, Carpenter SP, Roberts K, Shan JS (2006) 1311-chTNT-1/B mAb: tumour necrosis therapy for malignant astrocytic glioma. Expert Opin Biol Ther 6(5):539–545
58. Lidar Z, Mardor Y, Jonas T, Pfeffer R, Faibel M, Nass D, Hadani M, Ram Z (2004) Convection enhanced delivery of paclitaxel for the treatment of recurrent malignant glioma: a phase I/II clinical study. J Neurosurg 47(6):472–479
59. Van Vlerken LE, Amiji MM (2006) Multi-functional polymeric nanoparticles from tumour-targeted drug delivery. Expert Opin Drug Deliv 3(2):205–216
60. Langer R, Folkman J (1976) Polymers for the sustained release of proteins and other macromolecules. Nature 263:797–800
61. Brem H, Piantadosi S, Burger PC, Walker M, Selker R, Vick NA, Black K, Sisti M, Brem S, Mohr G et al (1995) Placebo-controlled trial of safety and efficacy of intraoperative controlled delivery by biodegradable polymers of chemotherapy for recurrent gliomas. The Polymer-brain Tumor Treatment Group. Lancet 345(8964):1008–1012
62. Valtonen S, Timonen U, Toivanen P, Kalimo H, Kivipelto L, Heiskanen O, Unsgaard G, Kuurne T (1997) Interstitial chemotherapy with carmustine-loaded polymers for high-grade gliomas: a randomized double-blind study. Neurosurgery 41(1):44–48
63. Westphal M, Hilt DC, Bortey E, Delavault P, Olivares R, Warnke PC, Whittle IR, Jääskeläinen J, Ram Z (2003) A phase III trial of local chemotherapy with biodegradable carmustine (BCNU) wafers (Gliadel wafers) in patients with primary malignant glioma. Neuro oncol 5(2):79–877

64. Smith SG, Galanis JC (1995) One-year results of the intrascleral glaucoma implant. J Cataract Refract Surg 21(4):453–456
65. Yue IC, Poff J, Cortés ME, Sinisterra RD, Faris CB, Hildgen P, Langer R, Shastri VP (2004) A novel polymeric chlorhexidine delivery device for the treatment of periodontal disease. Biomaterials 25(17):3743–3750
66. Friend DR (1990) Transdermal delivery of contraceptives. Crit Rev Ther Drug Carrier Syst 7(2):149–186
67. Spitz A, Young JM, Larsen L, Mattia-Goldberg C, Donnelly J, Chwalisz K (2012) Efficacy and safety of leuprolide acetate 6-month depot for suppression of testosterone in patients with prostate cancer. Prostate Cancer Prostatic Dis 15(1):93–99. doi:10.1038/pcan.2011.50. Epub 2011 Oct 25
68. Brady JM, Cutright DE, Miller RA, Barristone GC (1973) Reabsorption rate, route, route of elimination and ultrastructure of the implant site of polylactic acid in the abdominal wall of the rat. J Biomed Mater Res 7(2):155–156
69. Frazza EJ, Schmitt EE (1971) A new absorbable suture. J Biomed Mater Res 5(2):43–58
70. Spenlehauer G, Vert M, Benoit JP, Boddaert A (1989) In vitro and in vivo degradation of Poly (D, L lactide/glycolide) type microspheres made by solvent evaporation method. Biomaterials 10(8):557–563
71. Leong KW, Brott BC, Langer R (1985) Bioerodible polyanhydrides as drug-carrier matrices. I: characterization, degradation, and release characteristics. J Biomed Mater Res 19(8):941–955
72. Chasin M, Domb A, Rone E (1990) Polyanhydrides as drug delivery systems. In: Chasin M, Langer R (eds) Biodegradable polymers as drug delivery systems. Macel Dekker, New York, NY, pp 43–70
73. Bindschadeler C, Leong K, Mathiowitz E, Langer R (1988) Polyanhydride microsphere formulation by solvent extraction. J Pharm Sci 77(8):696–698
74. Levy-Nissenbaum E, Khan W, Pawar RP, Tabakman R, Naftali E, Winkler I, Kaufman O, Klapper L, Domb AJ (2012) Pharmacokinetic and efficacy study of cisplatin and paclitaxel formulated in a new injectable poly(sebacic-co-ricinoleic acid) polymer. Eur J Pharm Biopharm 82(1):85–93. doi:10.1016/j.ejpb.2012.06.004, Epub 2012 Jun 23
75. Leong KW, Kost J, Mathiowitz E, Langer R (1989) Polyanhydrides for controlled release of bioactive agents. Biomaterials 7(5):364–371
76. Brem H, Tamargo RJ, Olivi Pinn M, Weingart JD, Wharam M, Epstein JI (1994) Biodegradable polymers for controlled delivery of chemotherapy with and without radiation therapy in the monkey brain. J Neurosurg 80(2):283–290
77. Green SB, Byar DP, Walker MD, Pistenmaa DA, Alexander E Jr, Batzdorf U, Brooks WH, Hunt WE, Mealey J Jr, Odom GL et al (1983) Comparisons of carmustine, procarbazine and, high-dose methylprednisolone as additions to surgery and radiotherapy for the treatment of malignant glioma. Cancer Treat Rep 67(2):121–132
78. Walker MD, Green SB, Byar DP, Alexander E Jr, Batzdorf U, Brooks WH, Hunt WE, MacCarty CS, Mahaley MS Jr, Mealey J Jr et al (1980) Randomized comparisons of radiotherapy and nitrosoureas for the treatment of malignant glioma after surgery. N Engl J Med 303(23):1323–1329
79. Grossman SA, Reinhard C, Colvin OM, Chasin M, Brundrett R, Tamargo RJ, Brem H (1992) The intracerebral distribution of BCNU delivered by surgically implanted biodegradable polymers. J Neurosurg 76(4):640–647
80. Fung LK, Ewend MG, Sills A, Sipos EP, Thompson R, Watts M, Colvin OM, Brem H, Saltzman WM (1998) Pharmacokinetics of interstitial delivery of carmustine 4-hydroperoxycyclophosphamide, and paclitaxel from a biodegradable polymer implant in the monkey brain. Cancer Res 58(4):672–684
81. Tamargo RJ, Myseros JS, Epstein JI, Yang MB, Chasin M, Brem H (1993) Interstitial chemotherapy of the 9L gliosarcoma: controlled release polymers for drug delivery in the brain. Cancer Res 53(2):329–333
82. Brem H, Mahaley MS Jr, Vick NA, Black KL, Schold SC Jr, Burger PC, Friedman AH, Ciric IS, Eller TW, Cozzens JW et al (1991) Interstitial chemotherapy with drug polymer implants for the treatment of recurrent gliomas. J Neurosurg 74(3):441–446

83. Brem H, Ewend MG, Piantadosi S, Greenhoot J, Burger PC, Sisti M (1995) The safety of interstitial chemotherapy with BCNU-loaded polymer followed by radiation therapy in the treatment of newly diagnosed malignant gliomas: phase I trial. J Neurooncol 26(2):111–123
84. Sipos EP, Tyler B, Piantadosi S, Burger PC, Brem H (1997) Optimizing interstitial delivery of BCNU from controlled release polymers for the treatment of brain tumors. Cancer Chemother Pharmacol 39(5):383–389
85. Olivi A, Grossman SA, Tatter S, Barker F, Judy K, Olsen J, Bruce J, Hilt D, Fisher J, Piantadosi S (2003) New approaches to brain tumor therapy CNS consortium: dose escalation of carmustine in surgically implanted polymers in patients with recurrent malignant glioma: a new approaches to brain tumor therapy CNS consortium trial. J Clin Oncol 21(9):1845–1849
86. Attenello FJ, Mukherjee D, Datoo G, McGirt MJ, Bohan E, Weingart JD, Olivi A, Quinones-Hinojosa A, Brem H (2008) Use of gliadel (BCNU) wafer in the surgical treatment of malignant glioma: a 10-year institutional experience. Ann Surg Oncol 15(10):2887–2893
87. Chaichana KL, Zaidi H, Pendelton C, McGrit MJ, Grossman R, Weingart JD, Olivi A, Quiniones-Hinojosa A, Brem H (2011) The efficacy of carmustine wafers for older patients with glioblastoma multiforme: prolonging survival. Neurol Res 33(7):759–764
88. Ewend MG, Sampath P, Williams JA, Tyler BM, Brem H (1998) Local delivery of chemotherapy prolongs survival in experimental brain metastases from breast carcinoma. Neurosurgery 43(5):1185–1193
89. Ewend MG, Brem S, Gilbert M, Goodkin R, Penar PL, Varia M, Cush S, Carey LA (2007) Treatment of single brain metastasis with resection, intracavity carmustine polymer wafers, and radiation therapy is safe and provides excellent local control. Clin Cancer Res 13(12):3637–3641
90. Benny O, Menon LG, Ariel G, Goren E, Kim SK, Stewman C, Black PM, Carroll RS, Machluf M (2009) Local delivery of poly lactic-co-glycolic acid microspheres containing imatinib mesylate inhibits intracranial xenograft glioma growth. Clin Cancer Res 15(4):1222–1231
91. You J, Shao R, Wei X, Gupta S, Li C (2010) Near-infrared light triggers release of paclitaxel from biodegradable microspheres: photothermal effect and enhanced antitumor activity. Small 6(9):1022–1031
92. Tahara K, Kato Y, Yamamoto H, Kreuter J, Kawashima Y (2011) Intracellular drug delivery using polysorbate 80-modified poly(D, L-lactide-co-glycolide) nanospheres to glioblastoma cells. J Microencapsul 28(1):29–36
93. Ranganath SH, Fu Y, Arifin DY, Kee I, Zheng L, Lee HS, Chow PK, Wang CH (2010) The use of submicron/nanoscale PLGA implants to deliver paclitaxel with enhanced pharmacokinetics and therapeutic efficacy in intracranial glioblastoma in mice. Biomaterials 31(19):5199–5207
94. Esther Gil-Alegre M, González-Alvarez I, Gutiérrez-Paúls L, Torres-Suárez AI (2008) Three weeks release BCNU loaded hydrophilic-PLGA microspheres for interstitial chemotherapy: development and activity against human glioblastoma cells. J Microencapsul 25(8):561–568
95. Kantelhardt SR, Caarls W, de Vries AH, Hagen GM, Jovin TM, Schulz-Schaeffer W, Rohde V, Giese A, Arndt-Jovin DJ (2010) Specific visualization of glioma cells in living low-grade tumor tissue. PLoS One 5(6):e11323
96. Fillmore HL, Shultz MD, Henderson SC, Cooper P, Broaddus WC, Chen ZJ, Shu CY, Zhang J, Ge J, Dorn HC, Corwin F, Hirsch JI, Wilson J, Fatouros PP (2011) Conjugation of functionalized gadolinium metallofullerenes with IL-13 peptides for targeting and imaging glial tumors. Nanomedicine (Lond) 6(3):449–458
97. Nance EA, Woodworth GF, Sailor KA, Shih TY, Xu Q, Swaminathan G, Xiang D, Eberhart C, Hanes J (2012) A dense poly(ethylene glycol) coating improves penetration of large polymeric nanoparticles within brain tissue. Sci Transl Med 4(149):149ra119. doi:10.1126/scitranslmed.3003594
98. Arndt-Jovin DJ, Kantelhardt SR, Caarls W, de Vries AH, Giese A, Jovin Ast TM (2009) Tumor-targeted quantum dots can help surgeons find tumor boundaries. IEEE Trans Nanobioscience 8(1):65–71

99. Tobias A, Ahmed A, Moon KS, Lesniak MS (2013) The art of gene therapy for glioma: a review of the challenging road to the bedside. J Neurol Neurosurg Psychiatry 84(2): 213–222
100. Germano IM, Binello E (2009) Gene therapy as an adjuvant treatment for malignant gliomas: from bench to bedside. J Neurooncol 93(1):79–87, Epub 2009 May 9
101. Richards Grayson AC, Choi IS, Tyler BM, Wang PP, Brem H, Cima MJ, Langer R (2003) Multi-pulse drug delivery from a resorbable polymeric microchip device. Nat Mater 2(11):767–772
102. Santini JTJR, Cima MJ, Langer R (1999) A controlled-release microchip. Nature 397(6717):335–338
103. Masi B, Tyler BM, Bow H, Wicks RT, Xue Y, Brem H, Langer R, Cima M (2012) Intracranial MEMS-based temozolomide delivery in a rat gliosarcoma model. Biomaterials 33(23): 5768–5775, Epub 2012 May 14
104. Li Y, Ho Duc HL, Tyler B, Williams T, Tupper M, Langer R, Brem H, Cima MJ (2005) In vivo delivery of BCNU from a MEMS device to a tumor model. J Control Release 106(1–2):138–145
105. Kim GY, Tyler BM, Tupper MM, Karp JM, Langer RS, Brem H, Cima MJ (2007) Resorbable polymer microchips releasing BCNU inhibit tumor growth in the rat 9L flank model. J Control Release 123(2):172–178
106. Scott A, Tyler B, Masi B, Upadhyay U, Patta Y, Grossman R, Basaldella L, Langer R, Brem H, Cima M (2011) Intracranial microcapsule drug delivery device for the treatment of an experimental glioma model. Biomaterials 32(10):2532–2539, Epub 2011 Jan 8
107. Hobohm U (2001) Fever and cancer in perspective. Cancer Immunol Immunother 50: 391–396
108. Kantoff P, Higano C, Shore N et al (2010) Sipuleucel-T immunotherapy for castration-resistant prostate cancer. N Engl J Med 363(5):411–422
109. Hodi F, O'Day S, McDermott D et al (2010) Improved survival with ipilimumab in patients with metastatic melanoma. N Engl J Med 363(8):711–723
110. Vauleon E, Avril T, Collet B et al (2010) Overview of cellular immunotherapy for patients with glioblastoma. Clin Dev Immunol 2010:1–18
111. Aguilar L, Arvizu M, Aguilar-Cordova E et al (2012) The spectrum of vaccine therapies for patients with glioblastoma multiforme. Curr Treat Options Oncol 13(4):437–450
112. Yu J, Wheeler C, Zeltzer P et al (2001) Vaccination of malignant glioma patients with peptide-pulsed dendritic cells elicits systemic cytotoxicity and intracranial T-cell infiltration. Cancer Res 62:842–847
113. Dillman R, Duma C, PSchiltz P et al (2004) Intracavitary placement of autologous lymphokine-activated killer (LAK) cells after resection of recurrent glioblastoma. J Immunother 27(5):398–404
114. Bloom H, Peckham M, Richardson A et al (1973) Glioblastoma multiforme: a controlled trial to assess the value of specific active immunotherapy in patients treated by radical surgery and radiotherapy. Br J Cancer 27(3):253–267
115. Sobol R, Fakhrai H, Shawler D et al (1995) Interleukin-2 gene therapy in a patient with glioblastoma. Gene Ther 2(2):164–167
116. Folkman J (1971) Tumor angiogenesis: therapeutic implications. N Engl J Med 285: 1182–1186
117. Levy AP, Tamargo R, Brem H, Nathans D (1989) An endothelial cell growth factor from the mouse neuroblastoma cell line NB41. Growth Factors 2(1):9–19
118. Ferrara N, Henzel W (1989) Pituitary follicular cells secrete a novel heparin-binding growth factor specific for vascular endothelial cells. Biochem Biophys Res Commun 161(2):851–858
119. Presta L, Chen H, O'Connor S et al (1997) Humanization of an anti-vascular endothelial growth factor monoclonal antibody for the therapy of solid tumors and other disorders. Cancer Res 57:4593–4599
120. Hurwitz H, Fehrenbacher L, Novotny W et al (2004) Bevacizumab plus irinotecan, fluorouracil, and leucovorin for metastatic colorectal cancer. N Engl J Med 305:2335–2342

121. Sandler A, Gray R, Perry M et al (2006) Paclitaxel–carboplatin alone or with bevacizumab for non–small-cell lung cancer. N Engl J Med 355:2542–2550
122. Brem S, Cotran R, Folkman J (1972) Tumor angiogenesis: a quantitative method for histologic grading. J Natl Cancer Inst 48(2):347–356
123. Purow B, Fine H (2004) Progress report on the potential of angiogenesis inhibitors for neuro-oncology. Cancer Invest 22(4):577–587
124. Kunkel P, Ulbricht U, Bohlen P et al (2001) Inhibition of glioma angiogenesis and growth in vivo by systemic treatment with a monoclonal antibody against vascular endothelial growth factor receptor-2. Cancer Res 62:6624–6628
125. Weingart J, Sipos E, Brem H (1995) The role of minocycline in the treatment of intracranial 9L glioma. J Neurosurg 82:635–640
126. Kirsch M, Strasser J, Allende R et al (1998) Angiostatin suppresses malignant glioma growth in vivo. Cancer Res 58:4654–4659
127. Pradilla G, Legnani FG, Petrangolini G, Francescato P, Chillemi F, Tyler BM, Gaini SM, Brem H, Olivi A, DiMeco F (2005) Local delivery of a synthetic endostatin fragment for the treatment of experimental gliomas. Neurosurgery 57(5):1032–1040, discussion 1032–40
128. Fine H, Figg W, Jaeckle K et al (2000) Phase II trial of the antiangiogenic agent thalidomide in patients with recurrent high-grade gliomas. J Clin Oncol 18(4):708–715
129. Fine H, Wen P, Maher E et al (2003) Phase II trial of thalidomide and carmustine for patients with high-grade gliomas. J Clin Oncol 21(12):2299–2304
130. Vredenburgh J, Desjardins A, Herndon J et al (2007) Bevacizumab plus irinotecan in recurrent glioblastoma multiforme. J Clin Oncol 25(30):4722–4729
131. Cohen M, Shen Y, Keegan P et al (2009) FDA drug approval summary: bevacizumab (avastin) as treatment of recurrent glioblastoma multiforme. Oncologist 14:1131–1138
132. Wick W, Weller M, van den Bent M et al (2010) Bevacizumab and recurrent malignant gliomas: a European perspective. J Clin Oncol 28(12):188–189
133. van den Bent M, Vogelbaum M, Wen P et al (2009) End point assessment in gliomas: novel treatments limit usefulness of classical Macdonald's criteria. J Clin Oncol 27(18):2905–2908
134. Bansal K, Engelhard HH (2000) Gene therapy for brain tumors. Curr Oncol Rep 2(5):463–472, Review
135. Rainov NG (2000) A phase III clinical evaluation of herpes simplex virus type 1 thymidine kinase and ganciclovir gene therapy as an adjuvant to surgical resection and radiation in adults with previously untreated glioblastoma multiforme. Hum Gene Ther 11(17): 2389–2401
136. Markert JM, Liechty PG, Wang W, Gaston S, Braz E, Karrasch M, Nabors LB, Markiewicz M, Lakeman AD, Palmer CA, Parker JN, Whitley RJ, Gillespie GY (2009) Phase Ib trial of mutant herpes simplex virus G207 inoculated pre-and post-tumor resection for recurrent GBM. Mol Ther 17(1):199–207, Epub 2008 Oct 28
137. Patel M, Vogelbaum MA, Barnett GH, Jalali R, Ahluwalia MS (2012) Molecular targeted therapy in recurrent glioblastoma: current challenges and future directions. Expert Opin Investig Drugs 21(9):1247–1266. doi:10.1517/13543784.2012.703177, Epub 2012 Jun 25
138. van den Bent MJ, Brandes AA, Rampling R, Kouwenhoven MC, Kros JM, Carpentier AF, Clement PM, Frenay M, Campone M, Baurain JF, Armand JP, Taphoorn MJ, Tosoni A, Kletzl H, Klughammer B, Lacombe D, Gorlia T (2009) Randomized phase II trial of erlotinib versus temozolomide or carmustine in recurrent glioblastoma: EORTC brain tumor group study 26034. J Clin Oncol 27(8):1268–1274, Epub 2009 Feb 9
139. Desjardins A, Reardon DA, Coan A, Marcello J, Herndon JE 2nd, Bailey L, Peters KB, Friedman HS, Vredenburgh JJ (2012) Bevacizumab and daily temozolomide for recurrent glioblastoma. Cancer 118(5):1302–1312. doi:10.1002/cncr.26381. Epub 2011 Jul 26
140. Møller S, Grunnet K, Hansen S, Schultz H, Holmberg M, Sorensen M, Poulsen HS, Lassen U (2012) A phase II trial with bevacizumab and irinotecan for patients with primary brain tumors and progression after standard therapy. Acta Oncol 51(6):797–804, Epub 2012 May 1

141. Burkhardt JK, Riina H, Shin BJ, Christos P, Kesavabhotla K, Hofstetter CP, Tsiouris AJ, Boockvar JA (2012) Intra-arterial delivery of bevacizumab after blood–brain barrier disruption for the treatment of recurrent glioblastoma: progression-free survival and overall survival. World Neurosurg 77(1):130–134, Epub 2011 Nov 21
142. Lawson HC, Sampath P, Bohan E, Park MC, Hussain N, Olivi A, Weingart J, Kleinberg L, Brem H (2007) Interstitial chemotherapy for malignant gliomas: the Johns Hopkins experience. J Neurooncol 83(1):61–70

Chapter 8
Intranasal Delivery of Neuropeptide-Loaded Nanoparticles and Their Application to Nervous System Therapeutics

Michael J. Kubek, Abraham J. Domb, Daniel J. Kubek, and Michael C. Veronesi

8.1 Introduction

The ongoing discovery, isolation, characterization, and synthesis of an ever-growing cohort of brain-derived neuropeptides have considerably transformed our knowledge base of the CNS. Neuropeptides represent a large class of potent neurotransmitters/neuromodulators that play a critical role in diverse neurological and/or behavioral disorders that can be utilized in mono- or combinational therapies. Unfortunately, the

M.J. Kubek (✉)
Department of Anatomy and Cell Biology, Indiana University School of Medicine, Indianapolis, IN 46202, USA

Department of Psychiatry, Indiana University School of Medicine, Indianapolis, IN 46202, USA

Program in Medical Neuroscience, Indiana University School of Medicine, Indianapolis, IN 46202, USA
e-mail: mjkubek@iupui.edu

A.J. Domb
Faculty of Medicine, Institute for Drug Research, School of Pharmacy, The Hebrew University, Jerusalem 91120, Israel

D.J. Kubek
Department of Anatomy and Cell Biology, Indiana University School of Medicine, Indianapolis, IN 46202, USA

School of Dentistry, University of Louisville, 501 South Preston, Louisville, KY 40202-1701, USA
e-mail: djkube01@louisville.edu

M.C. Veronesi
Department of Radiology, University of Chicago School of Medicine, 5841 S. Maryland Avenue- MC 2026, Chicago, IL 60637, USA
e-mail: mv6t@virginia.edu

A.J. Domb and W. Khan (eds.), *Focal Controlled Drug Delivery*, Advances in Delivery Science and Technology, DOI 10.1007/978-1-4614-9434-8_8,

delivery of neuropeptides directly to specific brain sites has been a major hindrance in their therapeutic advancement. Rapid metabolism in nearly all tissue compartments and a paucity of blood–brain barrier penetration appear to be the two major hurdles in providing sustained neuropeptide bioavailability. Interestingly, primary olfactory neurons within the neuroepithelium constitute the only nerves in direct contact with the external environment exploitable for direct central nervous system (CNS) access. However, the nasal epithelium provides adequate CNS and systemic protection against airborne pathogens and other unwanted substances, yet the nasal cavity and its mucosal lining can serve as a locus for drug delivery to the brain and systemic circulation. Historically, the nasal administration of tobacco snuff, cocaine, and various hallucinogens and psychotropic drugs are well-known examples [1]. Utilization of the nasal epithelium for the delivery of clinically relevant peptides to the systemic circulation is a more recent accomplishment [2–7].

8.2 Intranasal Nanoparticles: A Novel Way to Deliver Neuropeptides to the Brain

The delivery of neuropeptides directly to specific CNS targets using nanoparticles (NPs) represents the newest advance in drug delivery. Advantages of this means of circumventing the blood–brain barrier include ease of use, long-term compliance, uninterrupted delivery, ease of dosing, and treatment schedules [4, 8–16]. However, several significant transolfactory barriers must be traversed. Solutes entering the nasal cavity are destined for three regions: (1) vestibular, (2) respiratory, and (3) olfactory. The olfactory region is the most functionally crucial site for direct access to the brain. Three major barriers to neuropeptide bioavailability exist in this region: (1) the presence of tight junctions between sensory and supporting cells, preventing epithelial transport to the submucous space; (2) a mucous layer containing protective proteolytic/hydrolytic enzymes that impart an enzymatic barrier to nasally administered drugs and neuropeptides; and (3) a mucous layer clearance that influences time-dependent neuropeptide absorptive (uptake) availability. Following olfactory neuronal uptake, neuropeptides are susceptible to further degradation during anterograde transneuronal transport as they are presumably carried by axonal microtubules in the olfactory neuron to primary CNS targets, namely amygdala, piriform, and entorhinal cortices. Next, sufficient sustained neuropeptide release at these targets is necessary for the desired pharmacological effect [9, 10, 17–21].

8.3 Nasal Pathways to the Brain

The vestibular region in the nasal cavity functions primarily in filtering airborne particles from entering the nasal passages and plays a minor role in the eventual transport of odorants and drugs. The respiratory area is the largest of the three areas, and its ciliary epithelium serves as a removal system to particles that have been deposited on the mucous layer, and the filtered air will be delivered by the trachea to the lungs. This region facilitates transport of drugs and peptides to the systemic circulation via the lungs. For targeted delivery, the olfactory epithelium is the most functionally crucial site for direct access to the brain. This region is the key objective for intranasal drug delivery to specific sites within the CNS. The olfactory region is surprisingly free of significant respiratory airflow and is responsible for the sensory function of smell that relies mostly on diffusion of odorant to this region [9, 14, 16, 20]. The olfactory mucosa is composed of two components, olfactory epithelium and lamina propria. The olfactory epithelium consists of four principal cell types: receptor cells (primary olfactory neurons—cranial nerve I), supporting (sustentacular) cells, basal (stem) cells, and duct cells (Bowman's gland). The receptor cell is a unique bipolar neuron (5–7 μm) whose dendrites form a terminal knob-like swelling from which nonmotile cilia extend into an overlying layer of mucus to become the site of interaction between the olfactory stimulus and receptive cell. The unmyelinated axons of olfactory neurons are between 200 and 400 nm in diameter and are among the smallest in the nervous system. These neurons project directly into the CNS through the cribriform plate at the skull base to their primary synapse in the olfactory bulb. Consequently, they are the only neurons directly contacting both the external environment and the CNS. Remarkably, receptor neurons also undergo continuous turnover every 30–60 days by some estimations and as long as 1 year in others and are derived from a transformed basal (stem) cells that are capable of regeneration after damage. These receptor cells are responsible for primary olfactory transduction of chemosensory input to the CNS. This is accomplished through odorant-binding protein interactions with G-protein-coupled receptors and second-messenger signaling cascades that have been recently characterized [22–26]. Supporting cells are columnar and attach by foot-like processes at the basal lamina. Their apical pole is covered with microvilli that extend into the overlying mucus. Human supporting cells interact with mucus by releasing material into and/or absorbing material from the mucous layer. Additionally, sustentacular cells surround the receptor neurons making tight junctions with them forming a physical transport barrier. Supporting cells appear to play a variety of functions including receptor cell insulation and structural support, transport of substances, regulation of potassium concentration of the extracellular fluid, and maintaining a transmembrane permeability boundary. Basal (stem) cells are pyramidal shaped and generally found near the basal membrane. These are true stem cells capable of postnatal neurogenesis as evidenced by mitotic figures in the lower epithelial region. Two types of basal cells have been identified, and it is thought that the globose basal cells (GBC) replenish the primary olfactory receptor cell. Finally, duct cells of Bowman's glands secrete xenobiotic enzymes to the mucous layer. Indeed, the majority of xenobiotic-metabolizing enzymes in the olfactory epithelium have been localized to the duct and acinar cells of Bowman's glands and sustentacular and basal cells. This component of the olfactory mucosa provides an essential neuroprotective function much like the blood–brain barrier while at the same time providing for detoxification, metabolism, clearance, and chemosensation. The mucous layer is renewed every 10–15 min, and the pH of mucosal secretions ranges between 5.5 and 6.5 in adults and 5.0 and 6.7 in children. The mucous layer entraps substances that are cleared from the nasal cavity via cilia. Mucus moves through the nasal cavity at approximately 5–6 mm/min with a particle clearance nearly every 20 min [17–19, 26]. This role is vital for survival in most animal species. The second component of the olfactory mucosa is the lamina propria. This region contains axon fascicles, blood vessels, connective tissue, and Bowman's gland acini. Bowman's glands are the primary source of proteolytic/metabolic enzymes in mucous and serous secretions in the mucosa. Myoepithelial cells encompass the acini and contain actin filaments that aid in moving secretory product toward the duct. The unmyelinated receptor axons form larger unbranched bundles, pass through the basal lamina becoming ensheathed by Schwann cells as they terminate axodendritically in spherical neuropils called glomeruli in the olfactory bulb [22, 23].

Following nasal administration, anatomical distribution of neuropeptide and/or nanoparticle uptake can be classified as (1) olfactory nerve pathway, (2) olfactory epithelial pathway, and (3) systemic pathway. The first two pathways are direct specific (receptor cells, olfactory neurons) and nonspecific (nonneuronal olfactory epithelium, paracellular) CNS routes, respectively. The third (respiratory epithelium, support cells, etc.) is systemic and is not essential to the present discussion since unprotected neuropeptides do not cross the blood–brain barrier appreciably unless given in very high doses.

8.4 Cellular Pathway of Olfactory Transduction and Transneuronal Transport

There have been significant advances in the molecular biology of olfactory transduction. It begins when known or novel volatile odor molecules are inhaled and contact the mucous layer. Odorants bind to odorant-binding proteins that deliver the molecules to the receptor cell. Molecules that bind to metabotropic (G-protein-coupled) receptors can be internalized at the time of receptor activation, and others can be transported to the cytosol by endocytosis [8, 25, 27]. Data on the uptake mechanism(s) for non-odorant compounds is limited [8, 27]. Like other neurons, solutes that enter the primary olfactory dendrites are transported anterogradely by either slow (200 mm/day) or fast (410 mm/day) mechanisms via microtubules through the axon to the nerve terminal in the olfactory bulb [9, 28]. Within the layered bulb, an extensive network of axodendritic and dendo-dendritic convergence takes place especially in the glomerulus wherein continued anterograde transneuronal transport occurs [24, 29]. The axons of efferent mitral and tufted neurons and afferent neurons from locus coeruleus, raphe, and anterior olfactory nucleus form the fascicles of the olfactory tract that emerges from the glomerular layer. Solutes transported anterogradely (mitral and tufted cells) travel at the approximate rate of 410 mm/day [9, 14, 28, 29], whereas retrogradely transported solutes perhaps to locus coeruleus and raphe nuclei travel at the rate of approximately 200 mm/day

[28, 29]. During intraneuronal transport to the axon terminal, peptides are susceptible to further degradation by cytosolic and lysosomal peptidases through the ubiquitin-proteasome pathway [30–32]. The fate of transported solutes through the olfactory pathway is presently speculative. The question of sufficient basal solute release verses synaptically induced terminal release remains to be determined. Several studies have confirmed the process of transneuronal transport, but the mechanism has yet to be defined. Collectively, the data suggest that transneuronal transport requires receptor-mediated uptake or nonspecific endocytosis into olfactory sensory cells, followed by microtubular transport through the axon, with subsequent release and uptake in association with synaptic specializations [8, 9, 29, 33]. Primary limbic sites innervated by olfactory tract fibers include the amygdala, piriform cortex, entorhinal cortex, and secondarily the adjacent hippocampal formation through the perforant pathway [9, 29]. The relationship between these structures in human neurodegenerative and affective disorders is well documented. Many types of solutes are thought to follow this pathway, such as viruses, dyes, metals, and proteins [8, 9, 29, 35]. For the olfactory epithelial pathway, solute is thought to enter the olfactory epithelium somewhere other than the receptor neuron. Here the solute may enter supporting cells or Bowman's gland via pinocytosis or diffusion, or it may enter via paracellular transport through cell junctions into the intercellular fluid [9, 13, 20, 27, 33, 36]. If the solute crosses the basal membrane and enters the lamina propria, it can enter the perineural space around the olfactory nerve where it can travel to the subarachnoid space containing CSF or enter capillaries of the forebrain circulation. Solutes in the CSF can enter the brain parenchyma through the pial surface before absorption through the arachnoid villi of the sagittal sinus [9, 13, 20, 27, 33]. Neuropeptides entering the systemic circulation are actively metabolized. Thus, these two large compartments would in effect limit neuropeptide CNS bioavailability by metabolism and dilution.

8.5 Impediments to Intranasal Neuropeptide Uptake and Transport

Successful intranasal CNS delivery requires reasonable amounts of neuropeptide uptake by receptor cells through the neuroepithelial mucosal surface. Key barriers to neuropeptide bioavailability exist in this region of the nasal cavity. The first barrier is the mucous layer that is viscous and stationary. This seromucous layer, produced primarily from Bowman's glands, contains several proteolytic/hydrolytic enzymes that provide an enzymatic barrier to nasally administered drugs and especially peptides which are rapidly degraded by specific and nonspecific exo- and endopeptidases [14, 20, 22, 27, 37]. The second barrier is related to mucous layer clearance. The longer a drug or peptide can remain intact in the mucous layer, the greater the probability of being taken up by neural elements of the olfactory epithelium [9, 20, 22, 27]. Several approaches to improve nasal absorption have been examined. These include physicochemical (molecular weight, osmolarity, pH, drug distribution), pharmaceutical (lipophilicity, peptidase and protease inhibitors, absorption enhancers), and physical (mucociliary clearance reduction, droplet size, site of deposition) strategies. The delicate balance between neuroprotection and enhanced uptake is similar in concept to opening and closing the blood–brain barrier [14, 20, 38, 39]. Since it is well established that viruses can gain access to the CNS intranasally, it is imperative that the mucosal barrier be maintained. Ideally, neuropeptide delivery should mimic chemosensory initiation.

8.6 Problems Related to Neuronal and Transneuronal Neuropeptide Delivery

Two major issues relevant to neuropeptide delivery involve the metabolism or inactivation of the neuropeptide during transport. Intraneuronal transport mechanisms have been well defined [28, 40]. The subneuronal organelle most likely responsible for both fast and slow axonal transport is the microtubule. In anterograde transport, materials being moved include membrane-associated enzymes, vesicle-packaged neurotransmitters, and neuropeptides. The vesicles are attached to microtubules through kinesins. These molecular motors are specific for anterograde transport along the tubules [40]. During transport it is likely that unprotected (non-vesicular) neuropeptides would be continuously exposed to cytosolic degradative enzymes [32]. This would significantly diminish neuropeptide bioavailability and storage at the axon terminal. Thus, a delivery mechanism that could perhaps mimic vesicular protection would enhance bioavailability. Another major delivery question involves transneuronal transport. The mechanism(s) underlying transneuronal transport is at best speculative. One key issue is whether the neuropeptide crosses from one neuron to another at synapses between the two or if the second neuron takes up the peptide that diffuses from the presynaptic neuron at extra-synaptic sites. In the case of the primary olfactory cell afferent to mitral/tufted cell synapse, it has been suggested that the neuroanatomical relationship favored transfer primarily or most efficiently at the primary afferent to mitral/tufted cell dendritic synapse. However, the possibility of non-synaptic diffusion must also be considered [29, 35, 41, 42]. A more difficult problem involves the mechanism of retrograde transneuronal transport as seen from the receptor to such distant pathways as the raphe and diagonal band. It has been suggested that raphe and diagonal band fibers synapse upon primary afferent fibers or upon dendrites of mitral/tufted cells and thus are in a position to take up peptide from one of these two neuro-elements [29]. Perhaps even more significant to this process is the amount of neuropeptide inactivation during transneuronal transport. It is known that specific and nonspecific membrane-bound and extracellular enzymes present in the synaptic environment are responsible for signal termination through rapid neuropeptide metabolism [30, 32, 43]. This barrier would most likely impact the quantity of neuropeptide available for transneuronal postsynaptic uptake. One method to overcome these metabolic difficulties during transport is to develop metabolically stable peptide analogs. Another would be to protect the neuropeptides during transport.

8.7 Issues Related to Neuropeptide Release

Direct evidence for neuropeptide release in specific target tissues following intranasal application has not appeared. The major difficulty here is obviously related to tissue levels derived from direct olfactory pathways or indirectly through the CSF. Moreover, increased tissue levels of a given neuropeptide do not reflect increased release following intranasal administration. It has also been shown that the transneuronal pathway is variable and that agents reach the CNS as late as 24 h after administration in the nasal cavity [29, 41, 42, 44]. Of equal significance is the presence of sufficient densities of specific receptors at the primary and secondary olfactory tract target sites. The supposition being that neuropeptides mediate their physiological and pharmacological effects through G-protein-coupled receptors and that sufficient ligand receptor interaction must occur to elicit the desired pharmacological effect and to sustain it. Accordingly, pharmacokinetics of doses is also an important consideration. No information is currently available suggesting methods of enhancing transported neuropeptide release at olfactory pathway target sites. Evidence for defining and influencing these parameters is meager and only beginning to emerge.

8.8 Proof of Concept

8.8.1 Thyrotropin-Releasing Hormone (Protirelin)

Clinically, thyrotropin-releasing hormone (TRH) has shown efficacy in the treatment of affective disorders and suicidal ideation [45–49]. Moreover, what is noteworthy in these studies is that the TRH effects in both depression and suicidal ideation were rapid in onset showing clinical effects within hours of intrathecal TRH delivery [45–47, 49]. This rapid onset of effect is particularly significant since it is well known that no drugs are presently available to provide rapid intervention against suicidality [50]. Extensive preclinical and clinical studies strongly support the role of TRH as a novel antidepressant/mood stabilizer [48, 49] and antiepileptic in patients with infantile spasms, Lennox-Gastaut syndrome, and myoclonic seizures, suggestive of a novel antiepileptogenic drug. Moreover, this anticonvulsant effect has lasted up to 5 years posttreatment [51]. TRH has shown positive cognitive effects in Alzheimer's patients as well as other forms chemically induced neurodegeneration [48]. Such disorders arise in the brain and are potentially treatable with TRH therapy. Thus, the simple TRH tripeptide would serve as a prototype for the delivery of more potent TRH analogs and other larger more complex neuropeptides to the brain.

Methods developed to deliver neuropeptides via liposomes and capsular carriers have not been effective. We incorporated TRH into a surface-eroding biodegradable polyanhydride copolymer as a sustained-release microdisk carrier for stereotaxic implantation. We used the rat kindling model of focal epilepsy to assess sustained TRH release in situ. A single TRH microdisk (3.6 μg TRH) implanted stereotaxically into the seizure focus (amygdala) significantly suppressed kindling expression when evaluated by the number of stimulations required to reach each seizure stage and to become fully kindled (8.63 ± 0.92 vs. 16.17 ± 1.37; mean ± SEM). Two indices of seizure severity, after discharge duration (stimulated, 87.40 ± 5.47 vs. 51.80 ± 15.65; unstimulated amygdala, 89.60 ± 5.55 vs. 48.67 ± 15.8) and clonus duration (71.2 ± 5.94 vs. 29.40 ± 8.87), were also significantly reduced by a single microdisk implant. Fifty days after initiation of the study, a significant reduction in clonus duration (53.90 ± 3.27 vs. 40.09 ± 4.14) still remained in the TRH-implanted groups. These data support the use of sustained-release carriers for potential neuropeptide delivery to site-specific CNS loci [52]. We have recently demonstrated that intranasal delivery of a TRH analog was able to attenuate seizures in the amygdala-kindled rat. Kindled rats received a TRH analog (3Me-H TRH; 10^{-9}, 10^{-8}, 10^{-7} M) intranasally 60 and 30 min prior to amygdala stimulation. The afterdischarge duration (ADD) and seizure stage was compared to control kindled animals receiving physiological saline intranasally. We demonstrated that intranasal application of 3Me-H TRH at 30 and 60 min, but not at 90 or 120 min prior to stimulation, resulted in a concentration-dependent reduction in total seizure ADD. Additionally, the analog had significant concentration-dependent effects on behavioral stages I through IV (partial) and stage V (generalized) seizures when given at 30 and 60 min but not at 90 or 120 min. The data indicate that bioavailability of the TRH analog from the nose to the seizure focus (target) in the amygdale was sufficient to affect seizure activity when delivered 30 and 60 min prior to giving a seizure stimulus, but when the TRH analog was delivered 120 or 90 min before the seizure stimulus, bioavailability at the seizure focus (target) was insufficient to affect seizure activity (e.g., EEG). The results provide additional proof of principle that intranasal delivery of TRH analogs may be a viable means to suppress temporal lobe seizures and perhaps other seizure disorders. However, our results indicate that even using a metabolically stable TRH analog, unprotected TRH has unsustained target bioavailability beyond 60 min, most likely because of metabolism [53].

In order to determine if neuropeptide-loaded NPs delivered intranasally could have an impact on kindling development (epileptogenesis) as was seen with the microdisk implant, we pretreated two groups of rats with either TRH-NPs or control-nanoparticles (control-NPs) 7 days before kindling. On day 8, kindling stimulations were initiated along with intranasal NP treatment and continued once a day until the subjects became fully kindled (permanently epileptic) or up to 20 stimulations. As can be seen in Fig. 8.1, the afterdischarge duration (ADD) of the TRH-NP-treated group was significantly attenuated from 10 to 13 stimulations. Moreover, the number of stimulations required to reach the fully kindled state (four consecutive stage V seizures) was significantly prolonged in the TRH-NP group (Fig. 8.2). This trend was evident at stage IV ($P < 0.08$) and became significant by stage V with the transition to kindling permanence being most affected in the TRH-NP group (Fig. 8.2). These findings provide quantifiable electrophysiological data demonstrating that sufficient TRH was delivered by the nanoparticles from the nose to the seizure focus in the amygdala to significantly reduce the afterdischarge duration (EEG, ADD) recorded at that site. This EEG data appears in three reports in which we use the kindling paradigm. In each study we record EEG activity by

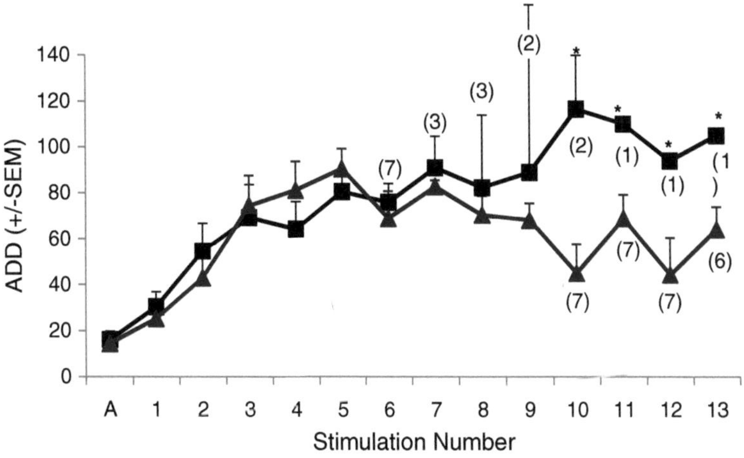

Fig. 8.1 Intranasal TRH-NP effect on ADD in the amygdala during kindling. Once an ADT (60–120 μA) was achieved, animals received daily treatments of either control-NPs ($n=8$) or TRH-NPs ($n=8$) for 7 days before initiation of kindling. Treatments were administered bilaterally into each posterior nasal cavity through implanted nasal ports using a sterile 50 μl Hamilton syringe. Animals received either control-NPs (100 nm; 200 μg PLA/50 μl) or TRH-NPs (100 nm; 20 μg TRH to 200 μg PLA/50 μl) over 4 min. On day 8 and for each day thereafter until either fully kindled or until day 20, the animals received daily intranasal treatments before receiving a kindling stimulus 3 h later. The ADD from each EEG was used to determine the duration of each seizure of both experimental and control animals. The ADD for the TRH-NP group (*red triangles*) was significantly lower than the control-NP group (*black squares*) as kindling progressed. Initially, both groups consisted of eight rats each, and subjects were removed from further stimulation once fully kindled (four consecutive stage V seizures), thus accounting for the decreasing subjects in () beyond the sixth stimulation. Statistical analysis was performed on data from stimulations 1–13 using repeated measures ANOVA. Vertical bars represent means ± SEM (*$P<0.05$) (from [34] with permission)

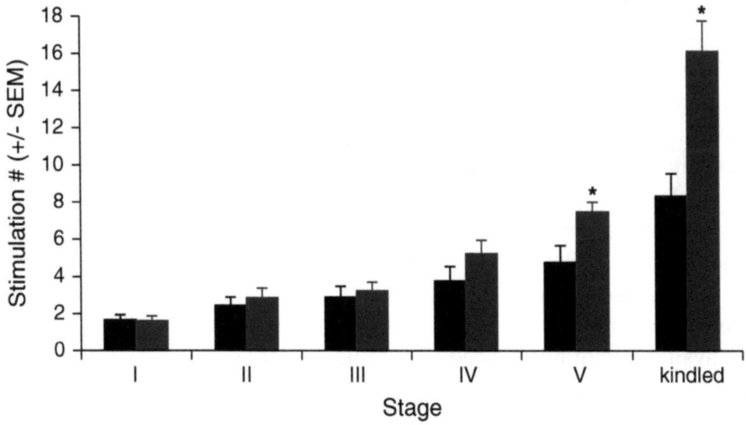

Fig. 8.2 Intranasal TRH-NP effect on kindling stage. The number of stimulations (#) required to become fully kindled (permanently epileptic) was significantly greater for the TRH-NP group (*red bars*) than for the control-NP group (*black bars*). The number (#) of stimulations needed to reach stage V kindling was significantly greater for the TRH-NP group (*red bars*) than the control-NP group (*black bars*). A trend toward suppression of kindling development was evident as early as stage IV ($P<0.08$) with this intranasal dose of TRH-NPs. Statistical analysis was performed on data using the Mann–Whitney test. Vertical bars represent means ± SEM kindled = number of stimulations to achieve four consecutive stage V seizures (*$P<0.05$) (from [34] with permission)

depth electrodes in both the ipsilateral (stimulated, kindling) amygdala and the contralateral amygdala (mirror focus). Our initial report demonstrated that direct implantation of sustained-release TRH from a polyanhydride microdisk in the kindling focus (amygdala) provided proof of concept that sustained delivery of sufficient TRH to the seizure focus (assessed by bilateral EEG recording in the amygdalae) could significantly attenuate kindling development by recording reduced ADD as well as kindling stage (behavior) [52]. Our more recent report demonstrated that intranasal administration of TRH-loaded biodegradable sustained-release nanoparticles given 3 h before a kindling stimulus provided sufficient TRH to the kindling seizure focus (assessed by bilateral EEG recording in the amygdala) to significantly attenuate kindling development via ADD and kindling stage [54]. By administering TRH nanoparticles intranasally and producing an attenuation of kindling development, similar to our direct delivery of TRH microdisks to the seizure (kindling) focus, provides additional proof of concept that the intranasal administered NPs were capable of delivering sufficient TRH to the seizure focus (amygdala), the site from which the seizure is stimulated. This electrophysiological and behavioral outcome provides convincing physiological data for TRH localization to the seizure focus after nasal nanoparticle delivery. Additionally, the data show that intranasal delivery of an unprotected TRH analog can attenuate seizures in the amygdala-kindled rat, also demonstrating more acute nose to amygdala delivery of enough TRH to affect seizure EEGs (ADD) from that focus [53]. Moreover, our nanoparticle results show that we have increased TRH bioavailability at the target since the anticonvulsant effect is evident 3 h after nasal deliver, while with the TRH analog, the effect lasted only 60 min following intranasal application. Another important outcome related to enhanced bioavailability in the nanoparticle study was seen when TRH-NPs were switched to control-NP, wherein it took an additional 3–10 stimulations for the TRH-treated animals to become fully kindled [54].

Collectively, our results reveal that the effective dose at the amygdale (target) is in the microgram range. Moreover, the microdisks contained ~4 μg TRH at the seizure focus, while with the intranasal nanoparticles, we delivered 20 μg TRH/day over a similar time frame; the ratio of delivery dose to target level should be in ratio of about 5:1.

Taken together, our results strongly suggest that intranasal administration of sustained-release anticonvulsant neuropeptide nanoparticles may be a viable new means to suppress temporal lobe seizures. Moreover, these results provide in vivo proof of concept for intranasal nanoparticle drug delivery [34, 54].

The availability of NPs loaded with a fluorescent dye has been a significant tool in determining uptake and distribution in nose-to-brain studies. Our goal was to determine the location of Nile red released from 560 and 100 nm-sized poly(lactic acid) (D-PLA) NPs within the brain parenchyma following intranasal delivery and the time course for its presence using fluorescence microscopic visualization. Intranasal delivery of 88–102 and 560 nm NPs loaded with Nile red fluorescent dye was given to a series of animals. Uptake and transport into the brain occurred with the 88 and 102 nm particles but not the 560 nm particles (control = 560 nm particles

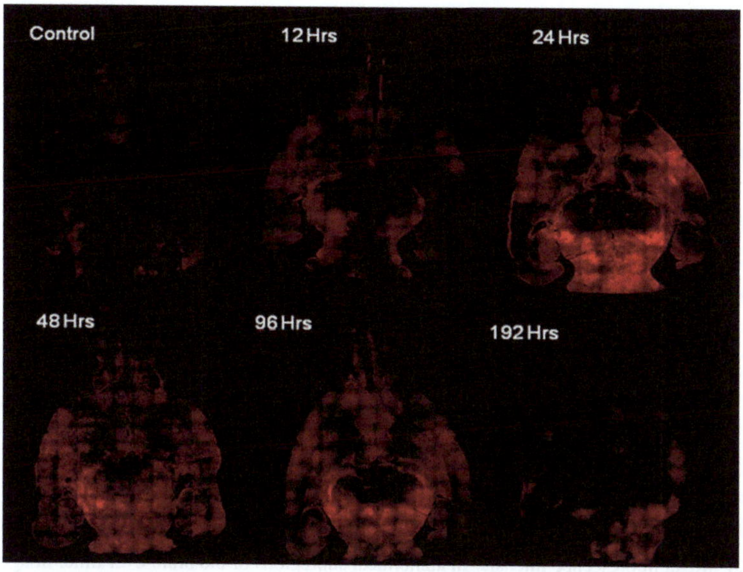

Fig. 8.3 Representative whole brain horizontal sections of Nile red fluorescence in the rat brain following intranasal nanoparticle delivery (control = fluorescence from 560 nm NPs 24 h post-intranasal delivery). For the intranasal uptake analysis, 130 digital images were captured for each sample at 4× magnification and montaged to reconstruct an entire horizontal brain section (Bioquant Imaging, Nashville, TN). The NIH ImageJ pixel-based analysis software program (NIH, Bethesda, MD) was used to analyze fluorescence intensity generated from a single image (from [34] with permission)

at 24 h postdelivery) (Fig. 8.3). Using the NIH ImageJ software, it was determined that the smaller particles were observed to transport and deliver the lipophilic Nile red dye to several seizure-sensitive areas of the rat brain at different rates (Fig. 8.4). These included the olfactory bulb (OB) and tract, septal nuclei (SN), insular cortex (GI), hippocampus (HC), and thalamus (TH) to cite a few. It was also found that the peak fluorescence occurred between 24 and 48 h after delivery (Figs. 8.3 and 8.4). The data also indicate that the peak delivery with these nanoparticles is variable, suggesting different transport pathways and intracellular trafficking. On the other hand, the apparent lack of uptake of the larger NPs is interesting. Given that Nile red is a lipophilic dye routinely used for specifically staining intracellular lipid-containing organelles, the possibility that Nile red was first released into the mucus and then transferred to the olfactory epithelium is reasonable. Moreover, since the surface area-to-volume ratio is much larger for the smaller NPs than the larger NPs, it's possible that dye release from the larger particles was much less and therefore undetected. To address this question, an in vitro study by Xu et al. [55] incubated Nile red-loaded PLGA NPs (NR/NP_{300}, 283.7 ± 2.2 nm) in phosphate-buffered saline (PBS, pH 7.4) containing hydrophobic or amphiphilic components that NR/NP300 would encounter in the cell-culture medium. NR/

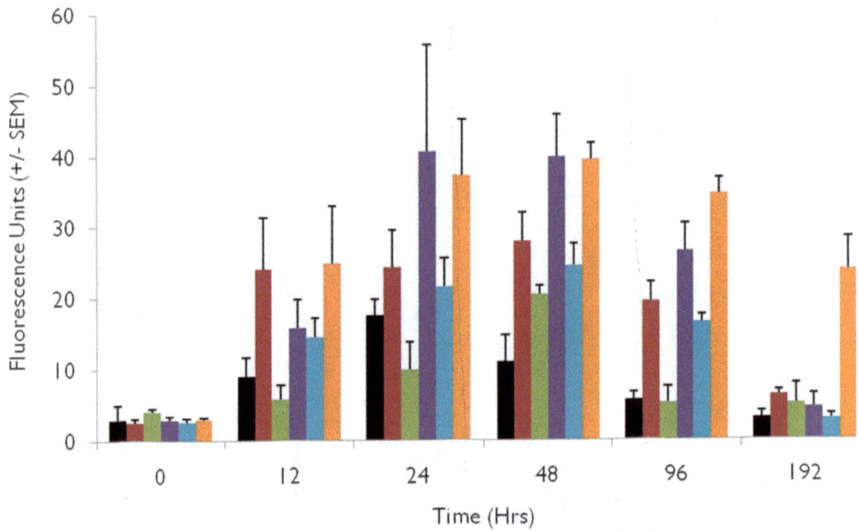

Fig. 8.4 Mean fluorescence intensity in anatomical subregions of the brain over time. Images were converted into 8-bit images and background was subtracted. Anatomical subregions of interest (ROI) were generated within ImageJ to allow for analysis of pixels contained within the olfactory bulb (OB: *black bars*), septal nuclei (MSLS: *red bars*), thalamus (TH: *green bars*), granular insular cortex (GI: *purple bars*), hippocampus (HP: *blue bars*), and deep mesencephalic nucleus (DpMe: *orange bars*) of all brains (n=3 for 0, 12, 24, 48, 72, and 96 h). A region of interest was defined using polarized light images of the sections corresponding to horizontal section from Paxinos and Watson. For consistency, we chose to analyze the horizontal sections corresponding to −5.32 mm ventral to bregma on the atlas. From each ROI, the mean and maximum fluorescence intensity was generated from a histogram depicting 256 possible grayscale Units (0=darkest, 256=brightest). The mean value of fluorescence indicates the overall fluorescence within an ROI [34] (from [34] with permission)

NP300 was incubated in PBS, PBS containing 10 % fetal bovine serum (FBS/PBS), or PBS containing 2.4 mg/ml of liposomes (liposome/PBS). FBS was included to represent the amphiphilic components in the culture medium. Liposomes were used to mimic the lipid cellular membrane. These release buffers were separated from NR/NP300 after timed incubation, and the Nile red fluorescence in each buffer solution was measured. A moderate level of Nile red fluorescence was detected in FBS/PBS. Much stronger Nile red fluorescence was detected in liposome/PBS as early as in 1 h and significantly increased after 3 h. On the other hand, Nile red fluorescence was negligible in PBS, which indicated lack of Nile red release in PBS in the absence of FBS or liposomes. To determine whether Nile red was directly transferred from NR/NP300 to liposomes or serum proteins via contact or released from NR/NP300 in PBS first and then picked up by them, NR/NP300 was incubated in PBS first, and the supernatant was separated from NR/NP300 and then incubated with liposomes. No fluorescence was detected in the liposome solution, which confirmed that Nile red release in PBS was indeed negligible. To evaluate the Nile red release relative to the total encapsulated dye, the release experiment was repeated with NR/NP100 and NR/NP300 in the same media for 3 h, and the released Nile red was extracted to chloroform and subjected to the HPLC analysis. Since it was noted in a previous experiment that a significant fraction of aggregated proteins and liposomes were excluded during filtration, which was performed to ensure complete removal of NR/NPs from the release medium, the extraction was performed without filtration. NR/NP100 showed a modest release of Nile red in PBS but 50 % (in the first hour) and additional 25 % (in 1–3 h) of Nile red in FBS/PBS or liposome/PBS. The % Nile red from NR/NP300 was relatively lower than NR/NP100 in all three media, which would be explained by the lower surface area-to-volume ratio. Importantly, Nile red release from NR/NP300 was negligible in PBS. In FBS/PBS and liposome/PBS, NR/NP300 released 16.4 and 25 % in 3 h, respectively. The two release experiments consistently show that Nile red is not released into aqueous solutions.

Since it is known that the mucus that coats the olfactory epithelium contains mostly H_2O, and mucopolysaccharides, antibodies, enzymes, salts, and odorant-binding proteins, but no lipids or albumin-like proteins, our results suggest that the lipophilic dye is not significantly released from the particles in the nasal cavity and absorbed by the olfactory epithelium but that the smaller particles are most likely taken up transcellularly or paracellularly and transported to various sites and deliver the dye while the larger NPs are shunted to lymphoid tissue.

The data also suggest that the peak delivery with these nanocapsules is variable.

Also, we have conducted uptake studies on cultured hippocampal neurons (Fig. 8.5). Both fluorescence microscopy and confocal microscopy of dye-loaded and dye-attached NPs provided additional data on the uptake and transport of NPs in neuronal processes. Since we used serum- and lipid-free culture medium in these studies, we would infer, based on the work by Xu et al. [55] cited above, that the fluorescent staining in our cultured neurons was primarily the result of NP endocytosis and transport, while some of the dye could be transferred to a small proportion of glia that remain in the culture.

By utilizing a highly specific antibody to TRH that has been used for both immunocytochemistry and radioimmunoassay [56], we have developed a novel immunocytochemical method to detect the transport and distribution of the TRH-NPs in the brain. With our specific TRH antibody, we have been able to visualize TRH-NPs in the brain, while control-NPs are undetectable (Fig. 8.6). This new method provides us with the technology to specifically determine the uptake, time course, distribution, and concentration of TRH-NPs as well as other neuropeptide nanoparticles throughout the CNS. This technology provides a very powerful semiquantitative tool that can help determine dose–response relationships necessary to reach pharmacological effects. It also provides a method to follow the distribution of intranasally delivered nanoparticles in tissue compartments outside the brain.

Additionally, we have developed in vitro assays to test the bioactivity of the TRH-NPs using cultured hippocampal neurons (Fig. 8.7). This system provides a means to continuously assess both the bioactivity and the potency of the nanoparticle preparations over time in storage.

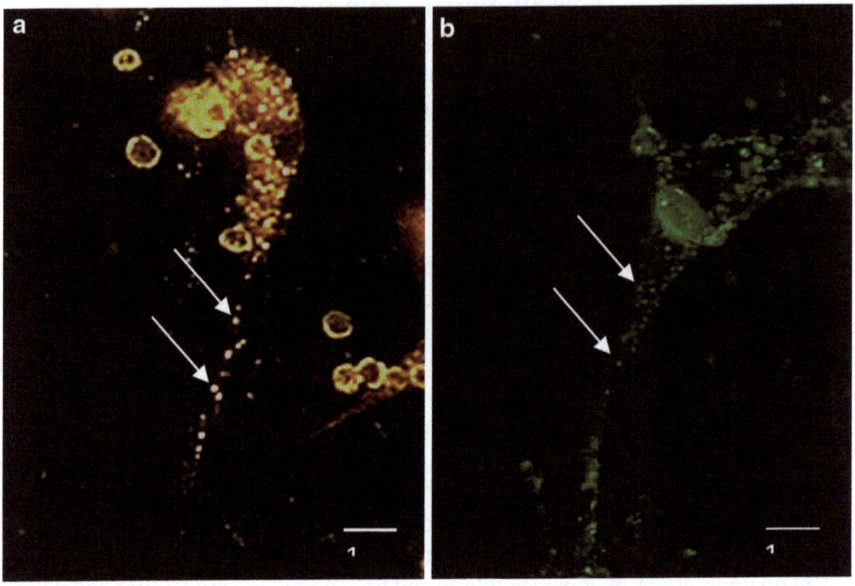

Fig. 8.5 Uptake of 100 nm Nile red NPs (**a**) and 20 nm polystyrene green fluorescent beads (**b**) into individual hippocampal neurons in vitro. Fetal hippocampal neurons were grown in serum-free medium containing Neurobasal supplemented with 2 % B-27 (Gibco, Grand Island, NY). This combination has been shown to reduce glia to less than 0.5 %. Media 1 consisted of Neurobasal medium, B-27 supplement, 200 mM L-glutamine (Sigma), basic fibroblast growth factor (BFGF) (50 μM/12 ml), and Normocin (2 μg/ml), an antibiotic, as additional components, which further optimize neuronal growth and select against microbial contamination. After 14 days in culture, the differentiated neurons were removed from the incubator, and the media was carefully aspirated from the culture dishes to remove cellular debris. A 1 mg/ml stock solution of Nile red NPs (10 mg/ml PLA) was diluted with Media 1 to provide a final amount of 50 ng/ml media (500 ng/ml PLA). A 10 mg/ml stock solution of control-NPs was also diluted to provide a final amount of 500 ng/ml PLA and served as a nonfluorescent nanoparticle control. Neurons were incubated with either Nile red NPs (100 nm) or 1×10^{14} polystyrene FluoSpheres (20 nm) for 5 h and then washed with PBS. Cells were fixed using 10 % formalin and viewed using a fluorescence microscope (Texas Red/FITC filter) at 60× magnification. Panel (**a**) A representative image of neuronal uptake of 100 nm Nile red NPs before or during release of Nile red NPs (*arrows* indicate individual NPs before or during release of Nile red within a process of a neuron). Panel (**b**) A representative image of neuronal uptake of 20 nm yellow–green fluorescent polystyrene FluoSpheres (*arrows* indicate individual FluoSpheres within a process of a neuron). Control-NPs were unable to be detected (data not shown) [34] (from [34] with permission)

In summary, the data provide proof of concept for intranasal nanoparticle delivery of clinically relevant neuropeptides, where the target receptors are concentrated, particularly in the forebrain, and the therapeutic effect is directly related to CNS function, such as in affective illness, PTSD, PTE, and neural recovery. Moreover, our results show that this approach increases transport through the BBB and that release of neuropeptides/drugs at specific sites can be achieved using intranasal delivery of biodegradable neuropeptide-loaded nanoparticles to induce a rapid and

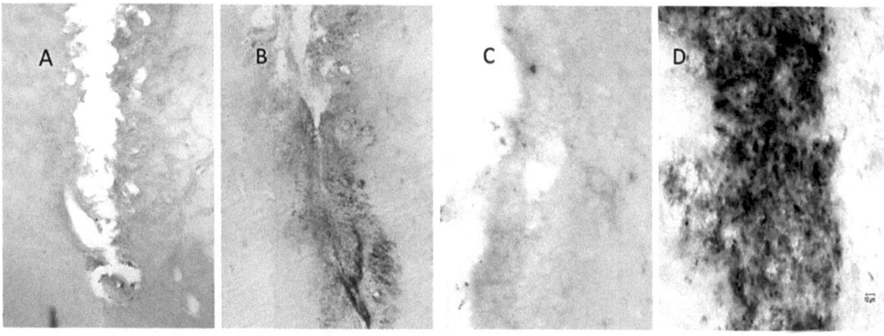

Fig. 8.6 Immunostaining of TRH nanoparticles 30 min following intra-amygdalar injection. A 10 μl Hamilton syringe (Hamilton Co. Inc., Whittier, CA) loaded with 10 μl (1 μg/1 μl) TRH-NPs was inserted into the *left basolateral amygdala (BLA)* using stereotactic coordinates from the atlas of Paxinos and Watson (incisor bar: −3.3 mm; anterior–posterior). (**A**) Image from a 30 μm section depicting injection site of control-NPs (100 nm) in the right basolateral amygdala (BLA) of a rat at 5× magnification. (**B**) Image from the same section depicting the injection site of TRH-NPs (100 nm) in the left BLA at 5× magnification. (**C**) Image of the injection site of control-NPs in the right BLA shown at 40× magnification. (**D**) Image of the injection site of TRH-NPs in the left BLA shown at 40× magnification. Note the intense staining exclusively at the injections sites of the TRH-NPs [34] (from [34] with permission)

sustained therapeutic effect. Levels of unprotected intranasal neuropeptides reported to access the CNS are very low, with concentrations in the brain and CSF compartments in the nM and pM range, or from 0.1 to 0.01 % bioavailability [4, 33, 57]. Although these levels can be biologically active (as we have shown) [53], it is important to develop ways to deliver and release sufficient sustained and controlled quantities preferentially to specific CNS targets while avoiding potential dosing that may yield CNS and/or systemic side effects in order for the nose-to-brain delivery route to be successful [10, 58]. We now have published data to support this concept using TRH-NPs [34, 54].

The mechanism whereby TRH exerts its rapid onset on affect in man is not clear. We and others have published data showing that TRH can modulate release of such "classical" neurotransmitters as glutamate, serotonin, dopamine, and acetylcholine. Additionally, since TRH binds to a G-protein-coupled receptor, the probability that it affects downstream events through second-messenger signaling is considerable.

Collectively, our preclinical studies provide proof of concept for intranasal delivery of biodegradable NPs as a novel platform to bring heretofore undeliverable neuropeptide candidate drugs to therapeutic use in the CNS. Our method is to combine a biodegradable surface-eroding polyanhydride and TRH into a TRH-loaded nanostructure (±100 nm) matrix formulation for intranasal delivery. The TRH is released from the NP surface in a controlled manner that is proportional to the rate of polymer hydrolysis. The NPs will be delivered to the olfactory epithelium using a refillable nasal atomizer that deposits vapor to the olfactory neuroepithelium.

There are two unique features in our treatment strategy that should enable therapeutic success.

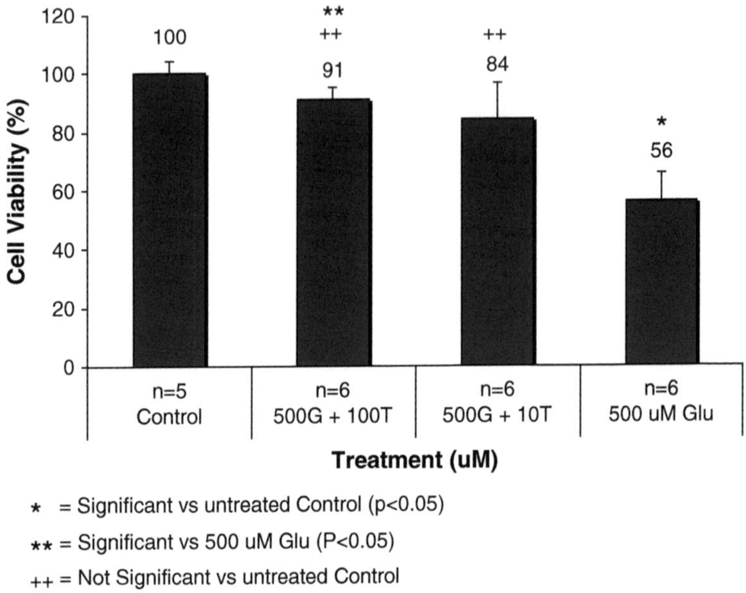

Fig. 8.7 Effect of 100 nm TRH-NPs on glutamate toxicity as measured by cell counts in cultured fetal hippocampal neurons. Cultured fetal (E17) hippocampal neurons were either untreated or exposed to varying concentrations of TRH-NPs (100 or 10 μM) and/or 500 μM Glu for 18 h. After 18 h, cell viability was assessed using the trypan blue exclusion method. Cells treated with 500 μM Glu alone experienced a significant reduction in cell viability relative to untreated control as indicated by an *asterisk*. Cell viability in samples co-treated with 500 μM Glu and either 100 μM or 10 μM TRH-NPs was not significantly ($P<0.05$) different compared to untreated control as indicated by a *double plus*. Samples treated with 100 μM TRH-NPs significantly (<0.05) protected neurons from cell death relative to the negative control (500 μM Glu alone) as indicated by a double *asterisk*, and although not significant, a tendency for protection was evident in the presence of 10 μM TRH-NPs. Data are expressed as mean ± SEM and represent a minimum of five replications for each treatment [54] (from [54] with permission)

First is the polymer. Although they are more complex to synthesize and load, polyanhydride NPs protect the peptide (TRH) from metabolism during uptake and neuronal transport. They also erode from the surface, releasing the peptide at a controlled rate specified by the polymer matrix undergoing hydrolysis. So the rate of peptide delivery is proportional to the erosion characteristics of the polymer surface. Thus, drug delivery and NP hydrolysis (degradation) are closely linked. On the other hand, with capsular-based NPs, such as polylactides (PLA) and poly(lactide-co-glycolides) (PLGA), most of the peptide is released at first as a burst and then at a slower rate, but the capsule biodegrades at a much slower rate. This results in residual polymer buildup that can become toxic in the brain over time after multiple applications. Also, the porosity of the capsular NPs makes them less capable of protecting the encapsulated peptide from metabolism. Presently, these polymers are not FDA approved for CNS use, but are used in proof of concept CNS delivery studies.

Second is the nasal applicator. The newer nasal atomizers are designed to selectively and reproducibly reach the human olfactory neuroepithelium while nasal breathing with little exposure to the respiratory nasal epithelium. This device-enabled property enhances olfactory epithelium NP residence time and improves the probability of NP uptake into the olfactory epithelium and hence into the CNS. Most other nasal sprays basically target the respiratory epithelium which is directed mainly toward the lungs.

References

1. Doty RL (1995) Introduction and historical perspective. Marcel Dekker, New York, NY, pp 1–32
2. Reger MA, Watson GS, Green PS, Baker LD, Cholerton B, Fishel MA, Plymate SR, Cherrier MM, Schellenberg GD, Frey WH, Craft S (2008) Intranasal insulin administration dose-dependently modulates verbal memory and plasma amyloid-beta in memory-impaired older adults. J Alzheimers Dis 13:323–331
3. Pirard C, Loumaye E (2005) GnRH agonist as novel luteal support: results of a randomized, parallel group, feasibility study using intranasal administration of buserelin. Hum Reprod 20:1798–1804
4. Born J, Lange T, Kern W, McGregor GP, Bickel U, Fehm HL (2002) Sniffing neuropeptides: a transnasal approach to the human brain. Nat Neurosci 5:514–516
5. Francis GJ, Martinez JA, Liu WQ, Xu K, Ayer A, Fine J, Tuor UI, Glazner G, Hanson LR, Frey WH, Toth C (2008) Intranasal insulin prevents cognitive decline, cerebral atrophy and white matter changes in murine type I diabetic encephalopathy. Brain 131:3311–3334
6. Thorne RG, Hanson LR, Ross TM, Tung D, Frey WH (2008) Delivery of interferon-beta to the monkey nervous system following intranasal administration. Neuroscience 152:785–797
7. Matsuoka Y, Gray AJ, Hirata-Fukae C, Minami SS, Waterhouse EG, Mattson MP, LaFerla FM, Gozes I, Aisen PS (2007) Intranasal NAP administration reduces accumulation of amyloid peptide and tau hyperphosphorylation in a transgenic mouse model of Alzheimer's disease at early pathological stage. J Mol Neurosci 31:165–170
8. Baker H (1995) Transport phenomena within the olfactory system. Marcel Dekker, New York, NY, pp 173–190
9. Mathison S, Nagilla R, Kompella UB (1998) Nasal route for direct delivery of solutes to the central nervous system: fact or fiction? J Drug Target 5:415–441
10. Kubek MJ, Ringel I, Domb AJ (2001) Issues related to intranasal delivery of neuropeptides to temporal lobe targets. Springer, New York, NY, pp 323–350
11. Lockman PR, Mumper RJ, Khan MA, Allen DD (2002) Nanoparticle technology for drug delivery across the blood–brain barrier. Drug Dev Ind Pharm 28:1–13
12. Gozes I, Brenneman DE, Geppetti P, Kastin AJ, Mains RE, Moody TW, Seroogy K, Spier AD, Zimmermann M (2001) Neuropeptides: brain messengers of many faces. Trends Neurosci 24:687–690
13. Illum L (2004) Is nose-to-brain transport of drugs in man a reality? J Pharm Pharmacol 56:3–17
14. Agarwal V, Mishra B (1999) Recent trends in drug delivery systems: intranasal drug delivery. Indian J Exp Biol 37:6–16
15. Egleton RD, Davis TP (2005) Development of neuropeptide drugs that cross the blood–brain barrier. NeuroRx 2:44–53
16. Mistry A, Stolnik S, Illum L (2009) Nanoparticles for direct nose-to-brain delivery of drugs. Int J Pharm 379:146–157

17. Mygind N, Anggard A (1984) Anatomy and physiology of the nose-pathophysiologic alterations in allergic rhinitis. Clin Rev Allergy 2:173–188
18. Chien YW, Chang S (1987) Intranasal drug delivery for systemic medication. Crit Rev Ther Drug Carrier Syst 4:67–194
19. Khanvilker K, Donovan MD, Flanagan DR (2001) Drug transfer through mucus. Adv Drug Deliv Rev 28:173–193
20. Wermeling DP, Miller JL, Rudy AC (2004) Systematic intranasal drug delivery: concept and applications. Drug Deliv Technol 2:22–30
21. Kosfeld M, Heinrichs M, Zak PJ, Fischbacher U, Fehr E (2005) Oxytocin increases trust in humans. Nature 435:673–676
22. Lewis J, Dahl AR (1995) Olfactory mucosa: composition, enzymatic localization, and metabolism. Marcel Dekker, New York, NY, pp 33–52
23. Morrison EE, Moran DT (1995) Anatomy and ultrastructure of the human olfactory neuroepithelium. Marcel Dekker, New York, NY, pp 75–101
24. Kratskin IL (1995) Functional anatomy, central connections, and neurochemistry of the mammalian olfactory bulb. Marcel Dekker, New York, NY, pp 103–126
25. Barnea G, O'Donnell S, Mancia F, Sun X, Nemes A, Mendelsohn M, Axel R (2004) Odorant receptors on axon termini in the brain. Science 304:1468
26. Watelet J-B, Katotomichelakis M, Eloy P, Danielidis V (2009) The physiological basics of the olfactory neuro-epithelium. B-ENT 5:11–19
27. Thorne RG, Emory CR, Ala TA, Frey WH (1995) Quantitative analysis of the olfactory pathway for drug delivery to the brain. Brain Res 692:278–282
28. Ochs S, Brimijoin WS (1993) Axonal transport, 3rd edn. WB Saunders, Philadelphia, PA, pp 331–360
29. Shipley MT (1985) Transport of molecules from nose to brain: transneuronal anterograde and retrograde labeling in the rat olfactory system by wheat germ agglutinin-horseradish peroxidase applied to the nasal epithelium. Brain Res Bull 15:129–142
30. Heuer H, Schafer MKH, Bauer K (1998) The thyrotropin-releasing hormone-degrading ectoenzyme: the third element of the thyrotropin-releasing hormone-signaling system. Thyroid 8:915–920
31. Griffiths EC, Kelly JA, Ashcroft A, Ward DJ, Robson B (1989) Part V. TRH metabolism. Comparative metabolism and conformation of TRH and its analogues. Ann NY Acad Sci 553:217–231
32. Rechsteiner M, Hill CP (2005) Mobilizing the proteolytic machine: cell biological roles of proteasome activators and inhibitors. Trends Cell Biol 15:27–33
33. Illum L (2000) Transport of drugs from the nasal cavity to the central nervous system. Eur J Pharm Sci 11:1–18
34. Kubek MJ, Domb AJ, Veronesi MC (2009) Attenuation of kindled seizures by intranasal delivery of neuropeptide-loaded nanoparticles. Neurotherapeutics 6:359–371
35. Kanayama Y, Enomoto S, Irie T, Amano R (2005) Axonal transport of rubidium and thallium in the olfactory nerve of mice. Nucl Med Biol 32:505–512
36. Wermeling DP (2009) Intranasal delivery of antiepileptic medications for treatment of seizures. Neurotherapeutics 6:352–358
37. Sarkar MA (1992) Drug metabolism in the nasal mucosa. Pharm Res 9:1–9
38. Tsuji A (2005) Small molecular drug transfer across the blood–brain barrier via carrier-mediated transport systems. NeuroRx 2:54–62
39. Pardridge WM (2005) The blood–brain barrier: bottleneck in brain drug development. NeuroRx 2:3–14
40. Cross R, Scholey J (1999) Kinesin: the tail unfolds [news; comment]. Nat Cell Biol 1:E119–E121
41. Cross DJ, Minoshima S, Anzai Y, Flexman JA, Keogh BP, Kim Y, Maravilla KR (2004) Statistical mapping of functional olfactory connections of the rat brain in vivo. Neuroimage 23:1326–1335

42. Takeda A, Kodama Y, Ishiwatari S, Okada S (1998) Manganese transport in the neural circuit of rat CNS. Brain Res Bull 45:149–152
43. O'Cuinn G, O'Connor B, Elmore M (1990) Degradation of thyrotropin-releasing hormone and luteinising hormone-releasing hormone by enzymes of brain tissue. J Neurochem 54:1–13
44. Kristensson K, Olsson Y (1971) Uptake of exogenous proteins in mouse olfactory cells. Acta Neuropathol 19:145–154
45. Marangell LB, George MS, Callahan AM, Ketter TA, Pazzaglia PJ, L'Herrou TA, Leverich GS, Post RM (1997) Effects of intrathecal thyrotropin-releasing hormone (protirelin) in refractory depressed patients. Arch Gen Psychiatry 54:214–222
46. Callahan AM, Frye MA, Marangell LB, George MS, Ketter TA, L'Herrou T, Post RM (1997) Comparative antidepressant effects of intravenous and intrathecal thyrotropin-releasing hormone: confounding effects of tolerance and implications for therapeutics. Biol Psychiatry 41:264–272
47. Szuba MP, Amsterdam JD, Fernando AT III, Gary KA, Whybrow PC, Winokur A (2005) Rapid antidepressant response after nocturnal TRH administration in patients with bipolar type I and bipolar type II major depression. J Clin Psychopharmacol 25:325–330
48. Nillni EA, Sevarino KA (1999) The biology of pro-thyrotropin-releasing hormone-derived peptides. Endocr Rev 20:599–648
49. Gary KA, Sevarino KA, Yarbrough GG, Prange AJ Jr, Winokur A (2003) The thyrotropin-releasing hormone (TRH) hypothesis of homeostatic regulation: implications for TRH-based therapeutics. J Pharmacol Exp Ther 305:410–416
50. Machado-Vieira R, Salvadore G, Luckenbaugh DA, Manji HK, Zarate CA Jr (2008) Rapid onset of antidepressant action: a new paradigm in the research and treatment of major depressive disorder. J Clin Psychiatry 69:946–958
51. Kubek MJ, Garg BP (2002) Thyrotropin-releasing hormone (TRH) in the treatment of intractable epilepsy. Pediatr Neurol 26:9–17
52. Kubek MJ, Liang D, Byrd KE, Domb AJ (1998) Prolonged seizure suppression by a single implantable polymeric-TRH microdisk preparation. Brain Res 809:189–197
53. Veronesi MC, Kubek DJ, Kubek MJ (2007) Intranasal delivery of a thyrotropin-releasing hormone analog attenuates seizures in the amygdala-kindled rat. Epilepsia 48:2280–2286
54. Veronesi MC, Aldouby Y, Domb AJ, Kubek MJ (2009) Thyrotropin-releasing hormone d, l polylactide nanoparticles (TRH-NPs) protect against glutamate toxicity in vitro and kindling development in vivo. Brain Res 1303:151–160
55. Xu P, Gullotti E, Tong L, Highley CB, Errabelli DR, Hasan T, Cheng JX, Kohane DS, Yeo Y (2009) Intracellular drug delivery by poly(lactic-co-glycolic acid) nanoparticles, revisited. Mol Pharm 6:190–201
56. Kubek MJ, Hill TG (1987) Methods of Thyrotropin-releasing hormone measurement. In: Hingtgen JN, Hellhammer DH, Huppmann G (eds) Advanced Methods In Psychology. CJ Hogrefe, Inc. Toronto, pp 261–279
57. Chen XQ, Fawcett JR, Rahman YE, Ala TA, Frey WH II (1998) Delivery of nerve growth factor to the brain via the olfactory pathway. J Alzheimer's Dis 1:35–45
58. Kubek MJ, Yard M, Lahiri DK, Domb AJ (2007) Characterization of novel intranasal sustained-release nanoparticles for delivery of neuropeptides to the brain. American Scientific Publishers, New York, NY, pp 73–84

Chapter 9
Focal Drug Delivery in Inner Ear Therapy

Jing Zou

9.1 Introduction

The inner ear is composed of the cochlea and vestibular system, which are isolated from the middle ear (tympanic cavity) by the round window and oval window (Fig. 9.1). The cochlea transduces sound as electric signals via the hair cells and sends electrical stimulation to the brain through the auditory nerve. The vestibular system senses linear acceleration through the hair cells in the utricle and saccule and senses head movement though the hair cells in the ampulla of the semicircular canals. Disorders of the inner ear can result in symptoms of the cochlea, including sensorineural hearing loss, tinnitus, and ear fullness, and a common symptom of the vestibular system caused by inner ear diseases is vertigo. The most common inner ear diseases, which can significantly reduce the life quality of human beings, are sudden sensorineural hearing loss, congenital hearing loss, and Meniere's disease, among others. Sudden sensorineural hearing loss has an estimated incidence between 5 and 20 per 100,000 persons per year [1]. Meniere's disease affects approximately 0.19 % of people in the USA, and it has a prevalence as high as 0.513 % in the Finnish population [2]. Large-scale newborn hearing screening studies have shown that the prevalence of congenital hearing loss among newborns in China varies from 0.5 to 3.8 % [3, 4]. The carrier rate of genetic mutations in the newborn population in China is 2.05–2.29 % [4, 5].

The two windows isolating the inner ear from the middle ear provide excellent approaches for inner ear drug delivery via the tympanic cavity, although the round

J. Zou, M.D., Ph.D. (✉)
Hearing and Balance Research Unit, Field of Otolaryngology,
School of Medicine, University of Tampere, Tampere, Finland

Department of Otolaryngology-Head & Neck Surgery, Center for Otolaryngology-Head & Neck Surgery of Chinese PLA, Changhai Hospital, Second Military Medical University, Shganghai, China
e-mail: Jing.Zou@uta.fi; zoujinghb@hotmail.com

A.J. Domb and W. Khan (eds.), *Focal Controlled Drug Delivery*, Advances in Delivery Science and Technology, DOI 10.1007/978-1-4614-9434-8_9,

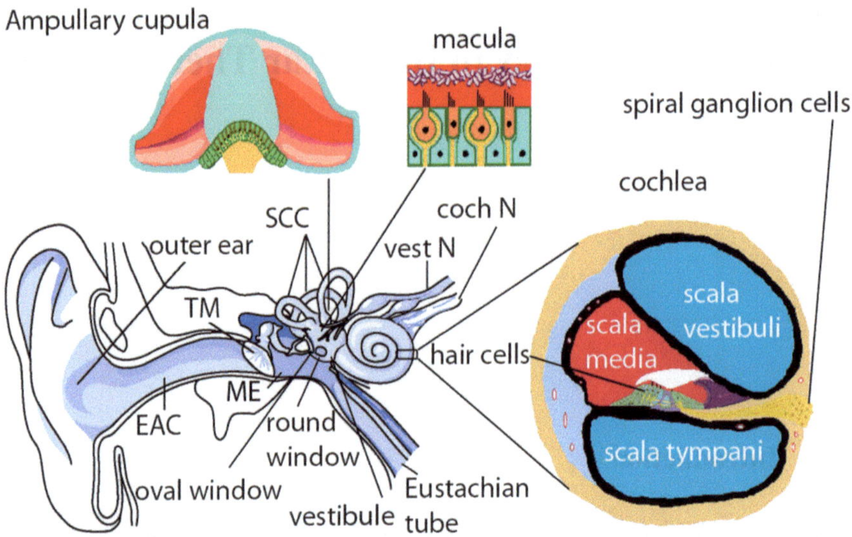

Fig. 9.1 Illustration of the ear. The ear is composed of the outer ear, middle ear, and inner ear. The middle ear (ME) is isolated from the outer ear by the tympanic membrane (TM), and it communicates with the pharynx via the Eustachian tube. The inner ear communicates with the middle ear through the oval and round windows. The hair cells in the cochlea transduce the sound into an electrical signal that is sent to the brain through the spiral ganglion cells and the cochlear nerve (coch N). The hair cells in the ampullary cupula of the semicircular canals (SCCs) and the macula of the utricle and saccule transduce the motion into an electric signal that is sent to the brain via the vestibular nerve (vest N). *EAC* external auditory canal

window was previously considered to be the only drug pathway from the middle ear to the inner ear because the oval window is occupied by the stapes (Fig. 9.1). In comparison to systemic administration, intratympanic delivery has a significantly lower risk of inducing adverse effects in the body because of the minimum exposed drug dosage. Focal drug delivery in inner ear therapy was first described by Schuknecht more than 50 years ago to induce chemical vestibulectomy using intratympanic administration of streptomycin for the control of intractable vertigo due to unilateral Meniere's disease [6]. Since then, intratympanic delivery of steroids was also first reported by Itoh to treat Meniere's disease [7]. Today, intratympanic delivery of steroids is a common therapy for idiopathic sudden sensorineural hearing loss. Recently, a multicentre, randomised clinical trial showed that intratympanic methylprednisolone is as effective as oral prednisone for the treatment of idiopathic sudden sensorineural hearing loss [8].

Direct cochlear drug delivery can be used in combination with cochlear implantation to bypass damaged cochlear hair cells, which is the only effective treatment for severe to profound hearing loss and is a critical intervention in deaf children to facilitate language acquisition. The efficacy of currently available cochlear implantation is far from perfect. Aimed to promote regeneration of the spiral ganglion cells and periphery processes, which constitute the first station to receive electrical signals from the electrodes of cochlear implants, modification of cochlear implant electrodes for direct drug delivery to the inner ear was reported [9]. Further, the biocompatibility of different biomaterials with potential application in the next generation of cochlear implantation has been reported in rats [10].

Details on different methods for focal drug delivery in inner ear therapy will be discussed in this chapter. The advantages and disadvantages of each technique will also be compared.

9.2 Intratympanic Perfusion

9.2.1 Transtympanic Injections Using Needles

Transtympanic injections using needles are simple and standard procedures in ear clinics, which were first reported by Weber-Liel in 1879, in which several compounds, including silver nitrate and common salt, among others, have been tried to treat otitis media [11]. As focal drug delivery in inner ear therapy, transtympanic injections using needles should be performed according to the following procedure. The patient sits in a chair, and the tympanic membrane is anaesthetised with a topical anaesthetic such as "EMLA" cream (lidocaine 2.5 % and prilocaine 2.5 %) or a drop of phenol. Penetration should be performed at the posterior-inferior quadrant of the tympanic membrane using a 25- or 26-gauge spinal needle, and 0.5–0.7 ml of drug should be injected (Fig. 9.2a). To prevent air bubble generation during injections, an additional perforation should be made at the anterior-inferior quadrant of the tympanic membrane (Fig. 9.2b). The patient should be asked to lie down for 15–30 min with the treated side up and to avoid swallowing or coughing.

Transtympanic injection of poly(e-caprolactone)-block-poly(ethylene glycol) polymersomes using a needle induced distribution of the polysomes into both the cochlea and vestibule of rats [12].

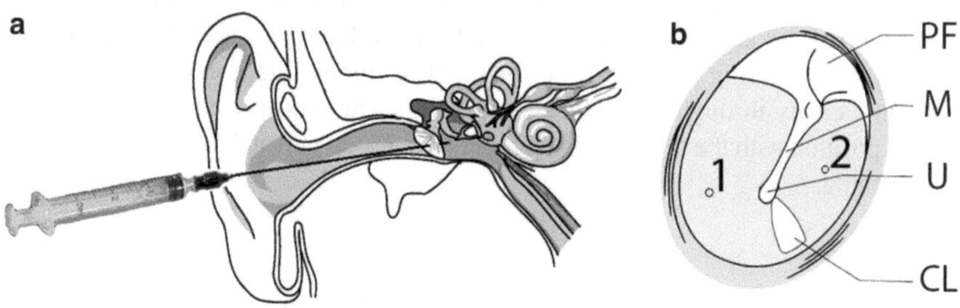

Fig. 9.2 Illustration of a transtympanic injection using a needle. A traditional transtympanic injection (**a**) is performed by penetrating the needle through the posterior-inferior quadrant of the tympanic membrane (1 in **b**). An additional perforation is created on the tympanic membrane (2 in **b**) to release any air bubbles generated during the injection

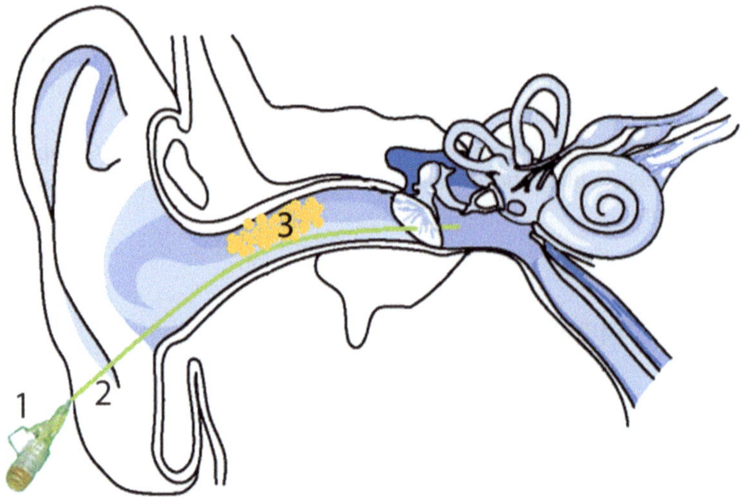

Fig. 9.3 Illustration of transtympanic catheter injection. The insertion point of the catheter (2) is at the posterior-inferior quadrant of the tympanic membrane, which is the same as that in transtympanic injections using needles. The external ear canal is packed with carbasus iodoformata (iodoform gauze) (3). The outer end of the catheter is connected to a vein-detained needle (1). *CL* cone of light, *M* malleus, *PF* pars flaccida, *U* umbo

9.2.2 Transtympanic Catheter Injections

In 2010, She et al. reported intratympanic methylprednisolone perfusion by transtympanic catheter injection to treat refractory sudden sensorineural hearing loss [13]. During the procedure, the patient should be placed in the supine position with the heads tilted approximately 45° away from the surgeon. After topical anaesthesia (1 % tetracaine solution), the tympanic membrane is penetrated at the posterior-inferior quadrant under a microscope or endoscope, and a microcatheter (1.4 mm in diameter) is inserted into the tympanic cavity through the perforation. The exterior end of the catheter is placed onto the mastoid process and fixed with plastic adhesive. The external ear canal is packed with carbasus iodoformata (iodoform gauze) to prevent infection (Fig. 9.3). Slow perfusion of 0.5 ml of drug is undertaken into the tympanic cavity through the catheter, and the patient is instructed to remain in the supine position with the treated side up for at least 15–30 min and to avoid swallowing or coughing.

9.2.3 Trans-Eustachian Tube Catheter Injections

Yamazaki et al. reported intratympanic gentamicin therapy for Meniere's disease in 1988 via trans-Eustachian tube catheter injection (the authors claimed that it began in 1974) [14]. Liu et al. applied an MRI contrast agent, and Zhang et al. delivered

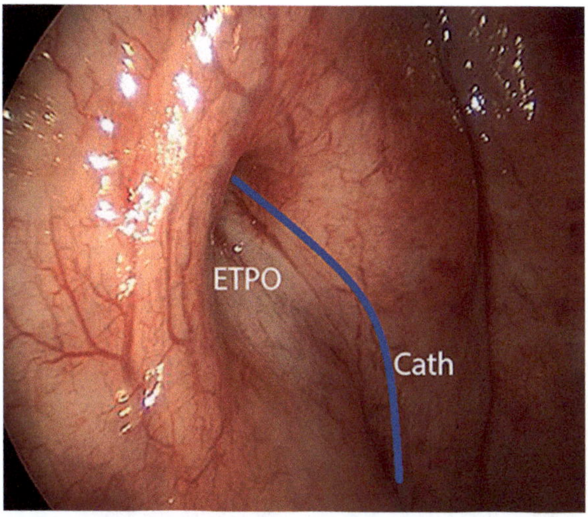

Fig. 9.4 Illustration of trans-Eustachian tube catheter injection. *ETPO* pharyngeal opening of the Eustachian tube, *Cath* catheter

dexamethasone into the tympanic cavities of patients with Meniere's disease using the same technique [15, 16]. In practice, the patient is placed in the supine position at 30° and then turns the head approximately 45° away from the sagittal line, towards the ipsilateral ear. After identifying the pharyngeal opening of the Eustachian tube, using either a 30 °C or 70°nasal endoscope, a catheter (made from 19-gauge SPIROL®Spring-reinforced epidural catheter, Becton Dickinson and Company, New Jersey, USA) is inserted, which is connected to a 1-ml syringe and is primed with the defined drug, into the pharyngeal opening of the Eustachian tube (Fig. 9.4). Then, 0.5–0.7 ml of drug solution is injected into the tympanic cavity. After the injection, the patient is instructed to remain in the supine position with the treated side up for at least 15–30 min and to avoid swallowing or coughing.

9.3 Organ-Targeted Delivery

9.3.1 Round Window Diffusion-MicroWick Device

In 1999, Silverstein reported a new inner ear therapeutic system for delivering medication directly to the round window membrane, which is called the MicroWick [17]. With this system, a small wick is inserted through a tympanic membrane vent tube into the round window niche (Fig. 9.5). Once the wick has been inserted, the patient can self-administer eardrops into the ear canal, where they are absorbed by the wick and are transported to the round window membrane and to the inner ear fluids.

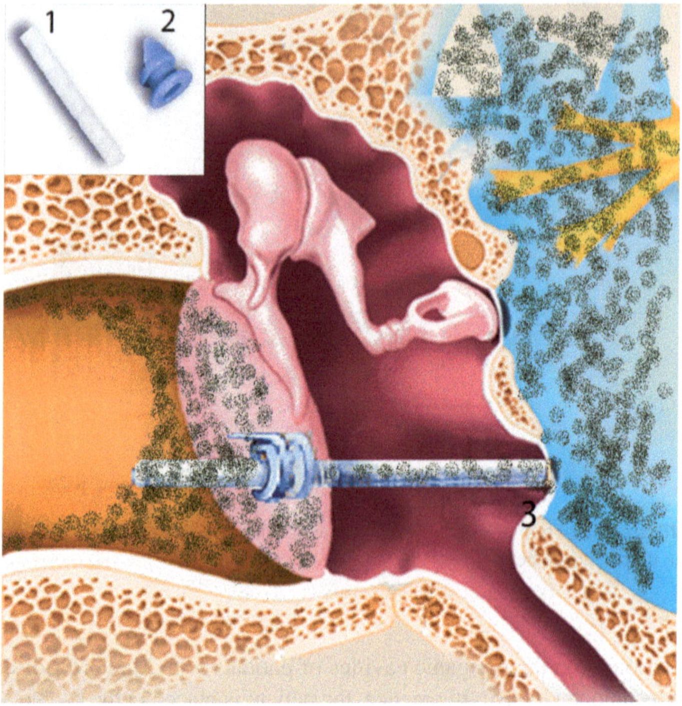

Fig. 9.5 Illustration of the round window diffusion-MicroWick device. The MicroWick device (1) is inserted through a ventilation tube (2) that is installed on the tympanic membrane and touches the round window niche (3). The drug (*green*) is delivered via the MicroWick device and is thought to enter the inner ear through the round window

9.3.2 Round Window Diffusion-Gelatine Sponge

In 1998, Arriaga and Goldman reported administration of dexamethasone onto the round window membrane with an absorbable gelatine sponge to treat Meniere's disease [18]. Additionally, Husmann et al. applied a gentamicin-soaked gelatine sponge to the round window membrane and established an animal model for the study of ototoxicity [19]. In the application, surgery is performed. Following tympanotomy and lysis of round window adhesions, a gelatine sponge pledget soaked in drug is placed in the round window niche, and the wound is sealed. The tympanic cavity can also be filled with drug solution. In 2008, Zou et al. first reported that round window delivery of lipid nanocapsules using a gelatine sponge pledget induced a sufficient distribution of nanoparticles into the inner ear in rats [20]. A biocompatibility study showed that administration of lipid nanocapsules did not cause hearing loss, nanoparticle application-related cell death, or morphological changes in the inner ear at up to 28 days of observation. The cochlear neural elements, such as synaptophysin, ribbon synapses, and S-100, were not affected by the administration of lipid nanocapsules [21].

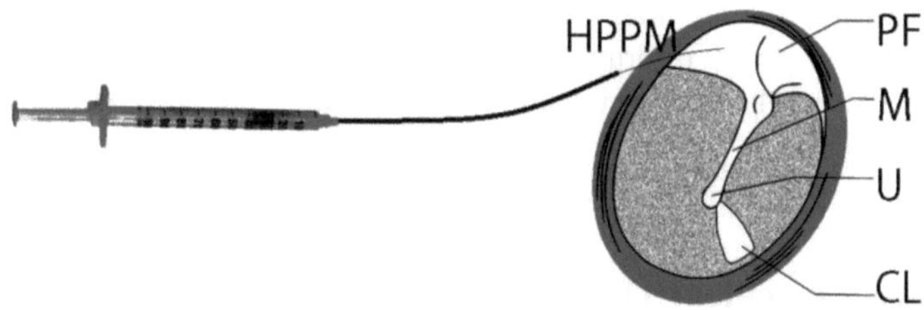

Fig. 9.6 Illustration of vestibular selective drug delivery via oval window diffusion. An epitympanic injection is performed by inserting high-performance polyimide microlumen (HPPM) through the pars flaccida (PF). *CL* cone of light, *M* malleus, *U* umbo

9.3.3 *Oval Window Diffusion-Polyimide Microlumen*

Zou et al. reported novel vestibular selective delivery of the MRI contrast agent gadolinium-tetra-azacyclo-dodecane-tetra-acetic acid (Gd-DOTA) via oval window diffusion by epitympanic injection using high-performance polyimide microlumen (HPPM) in 2012. The mechanism of the procedure is that the drug that is injected into the epitympanum diffuses along the thin epithelium, coating the ossicles to the oval window, and then enters the vestibule [22]. Rather than a watertight seal, the annular ligament across the stapediovestibular joint in the oval window is a porous structure, composed of fibrillin, 36-kDa microfibril-associated glycoprotein, and hyaluronic acid [23]. To prepare the injection device, the HPPM is cut (inner diameter 121.92 μm, outer diameter 180.34 μm, MicroLumen Inc., Florida, USA) into 30-mm pieces with a bevelled tip at one end, which is connected to polyethylene tubing (inner diameter 0.28 mm, outer diameter 0.61 mm, PEIO, Becton Dickinson and Company, New Jersey, USA), and sealed with Loctite Super Attak Plastic glue (Henkel, Germany). The other end of the catheter is connected to an insulin syringe (Becton Dickinson and Company, USA) through a 27-gauge needle. The tubing is primed with the drug to be administered. Under a microscope, the tympanic membrane is penetrated on the lateral wall of the lateral attic compartment using a 25-gauge needle, and the HPPM is inserted through the perforation (Fig. 9.6). Approximately 2.5 μl (half scale of the syringe) of drug is manually injected into the epitympanum over 1–2 s.

9.3.4 *Tympanic Medial Wall Diffusion*

To further reduce the volume of drug that is delivered into the tympanic cavity, Zou et al. reported a novel method of ultrasmall-volume administration of Gd-DOTA to the medial wall of the middle ear cavity of rats using HPPM in 2011 [24]. The injection device was created in the same manner as that used in oval window diffusion (see that in Sect. 9.3.3). The tympanic membrane was penetrated at the upper edge of the posterior upper quadrant with a 25-gauge needle, and the HPPM was inserted through the hole until it touches the medial wall of the middle ear cavity. A total of 2.5 μL of drug solution was injected onto the medial wall of the middle ear cavity. The animals remained in the lateral position with the injected ear upward for 15 min.

9.4 Direct Cochlear Drug Delivery: Next-Generation Cochlear Implant

In 1999, Carvalho and Lalwani first reported that cochleostomy followed by intracochlear infusion, as a model system for continuous direct cochlear drug delivery, was well tolerated by the auditory system of guinea pigs. Different therapeutic agents, including neurotrophins, growth factors, anti-inflammatory drugs, and viral gene therapy vectors, were infused into the cochlea over a prolonged period of time, resulting in auditory insult limited largely to the region of surgical trauma [25]. Trying to improve the effect of cochlear implantation, Paasche et al. modified the cochlear implant electrode for direct drug delivery to the inner ear [9]. A novel cochlear implant electrode is under development with the support of the European Union Framework Programme projects, NanoEar and NANOCI [26, 27]. The NANOCI project is a continuation of NanoEar, aimed to develop a neuroprosthesis with a gapless interface with auditory nerve fibres. The neurites will be attracted and guided by an innovative, nanostructured gel matrix, containing diffusible and surface-bound neurotrophic compounds towards the functionalised, neurotrophic electrode array surface. Within the NanoEar project, Tet1-functionalised PEG-b-PCL polymersomes showed cochlear nerve targeting after direct cochlear administration [28]. Superparamagnetic iron oxide nanoparticles, hierarchically coated with oleic acid and Pluronic F127 copolymers (POA@SPION), were demonstrated as a sensitive contrast agent in inner ear MRI after direct cochlear delivery [29]. SPION will be used as an MRI tag to visualise dynamic release of BDNF and BDNF mimetic molecules from the next-generation cochlear implant electrode into the cochlear perilymph as part of the NANOCI project.

9.5 Advantages and Disadvantages of Different Delivery Methods

Transtympanic injections using needles constitute a simple and efficient method for focal drug delivery in inner ear therapy, but the injection volume is the largest among focal drug delivery methods. The MicroWick device is capable of inducing sustained, selective cochlear drug delivery, but with a high risk of middle ear infection and obvious trauma. Trans-Eustachian tube catheter injections are noninvasive, but the patient can suffer from the insertion of tubing through the Eustachian tube, and the procedure is complicated. Transtympanic catheter injections are capable of mediating sustained delivery, but the volume is large. Round window diffusion using a gelatine sponge makes it possible to deliver a drug constantly and selectively to the cochlea, but an operation is required to enter the middle ear. Oval window diffusion by an epitympanic injection using HPPM is an excellent, minimally invasive, and selective vestibular drug delivery method, but it lacks clinical application. Tympanic medial wall diffusion using HPPM is minimally invasive with an ultrasmall volume of drug being administered, but it requires clinical data to support its efficacy. Direct cochlear drug delivery, in combination with next-generation cochlear implantable electrodes, is the future of deafness therapy, but no product has yet appeared on the market.

Acknowledgements I want to thank Professor Ilmari Pyykkö for his positive comments on the chapter and providing the image of Fig. 9.4. The writing was supported by the 7th Framework Programme project of the European Union, NANOCI (NMP4-SL-2012-281056).

References

1. Byl FM Jr (1984) Sudden hearing loss: eight years' experience and suggested prognostic table. Laryngoscope 94(5 Pt 1):647–661
2. Havia M, Kentala E, Pyykko I (2005) Prevalence of Meniere's disease in general population of Southern Finland. Otolaryngol Head Neck Surg 133(5):762–768
3. Chen G et al (2012) A large-scale newborn hearing screening in rural areas in China. Int J Pediatr Otorhinolaryngol 76(12):1771–1774
4. Zhang Z et al (2012) Auditory screening concurrent deafness predisposing genes screening in 10,043 neonates in Gansu province, China. Int J Pediatr Otorhinolaryngol 76(7):984–988
5. Wang QJ et al (2011) Newborn hearing concurrent gene screening can improve care for hearing loss: a study on 14,913 Chinese newborns. Int J Pediatr Otorhinolaryngol 75(4):535–542
6. Schuknecht HF (1956) Ablation therapy for the relief of Meniere's disease. Laryngoscope 66(7):859–870
7. Itoh A, Sakata E (1991) Treatment of vestibular disorders. Acta Otolaryngol Suppl 481: 617–623
8. Rauch SD et al (2011) Oral vs intratympanic corticosteroid therapy for idiopathic sudden sensorineural hearing loss: a randomized trial. JAMA 305(20):2071–2079
9. Paasche G et al (2003) Technical report: modification of a cochlear implant electrode for drug delivery to the inner ear. Otol Neurotol 24(2):222–227
10. Zou J et al (2008) Biocompatibility of different biopolymers after being implanted into the rat cochlea. Otol Neurotol 29(5):714–719
11. Weber L (1879) On intratympanic injections in catarrhal affections of the middle ear. Br Med J 2(975):364–365
12. Zhang Y et al (2011) Comparison of the distribution pattern of PEG-b-PCL polymersomes delivered into the rat inner ear via different methods. Acta Otolaryngol 131(12):1249–1256
13. She W et al (2010) Hearing evaluation of intratympanic methylprednisolone perfusion for refractory sudden sensorineural hearing loss. Otolaryngol Head Neck Surg 142(2):266–271
14. Yamazaki T, Hayashi M, Hayashi N, Kozaki H (1988) Intratympanic gentamicin therapy for Meniere's disease placed by tubal catheter with systemic isosorbide. Arch Otorhinolaryngol 245(3):170–174
15. Liu F, Huang WN, Song HT, Zhang QY (2008) Endolymphatic visualization in patients with Meniere's disease. Zhongguo Yi Xue Ke Xue Yuan Xue Bao 30(6):651–654

16. Zhang Q et al (2012) Noninvasive intratympanic dexamethasone treatment for sudden sensorineural hearing loss. Acta Otolaryngol 132(6):583–589
17. Silverstein H (1999) Use of a new device, the MicroWick, to deliver medication to the inner ear. Ear Nose Throat J 78(8):595–598, 600
18. Arriaga MA, Goldman S (1998) Hearing results of intratympanic steroid treatment of endolymphatic hydrops. Laryngoscope 108(11 Pt 1):1682–1685
19. Husmann KR, Morgan AS, Girod DA, Durham D (1998) Round window administration of gentamicin: a new method for the study of ototoxicity of cochlear hair cells. Hear Res 125(1–2):109–119
20. Zou J et al (2008) Distribution of lipid nanocapsules in different cochlear cell populations after round window membrane permeation. J Biomed Mater Res B Appl Biomater 87(1):10–18
21. Zhang Y et al (2011) Inner ear biocompatibility of lipid nanocapsules after round window membrane application. Int J Pharm 404(1–2):211–219
22. Zou J, Poe D, Ramadan UA, Pyykko I (2012) Oval window transport of Gd-dOTA from rat middle ear to vestibulum and scala vestibuli visualized by in vivo magnetic resonance imaging. Ann Otol Rhinol Laryngol 121(2):119–128
23. Ohashi M et al (2008) Histochemical localization of the extracellular matrix components in the annular ligament of rat stapediovestibular joint with special reference to fibrillin, 36-kDa microfibril-associated glycoprotein (MAGP-36), and hyaluronic acid. Med Mol Morphol 41(1):28–33
24. Zou J, Yoshida T, Ramadan UA, Pyykko I (2011) Dynamic enhancement of the rat inner ear after ultra-small-volume administration of Gd-DOTA to the medial wall of the middle ear cavity. ORL J Otorhinolaryngol Relat Spec 73(5):275–281
25. Carvalho GJ, Lalwani AK (1999) The effect of cochleostomy and intracochlear infusion on auditory brain stem response threshold in the guinea pig. Am J Otol 20(1):87–90
26. Pyykkö I (2010) Nanoear: 3g-Nanotechnology based targeted drug delivery using the inner ear as a model target organ. (Nanoear Project), pp An Integrated Project funded by the European Commission under the "nanotechnologies and nano-sciences, knowledge-based multifunctional materials and new production processes and devices" (NMP) thematic priority of the sixth framework programme
27. Senn P (2012) NANOCI: nanotechnology based cochlear implants. (NANOCI), p The NANOCI Project is funded under the 7th framework programme of the European Union
28. Zhang Y et al (2012) Targeted delivery of Tet1 peptide functionalized polymersomes to the rat cochlear nerve. Int J Nanomedicine 7:1015–1022
29. Zou J et al (2010) MRI manifestation of novel superparamagnetic iron oxide nanoparticles in the rat inner ear. Nanomedicine (Lond) 5(5):739–754

Chapter 10
Nanotechnology-Based Ophthalmic Drug Delivery System

Fahima Dilnawaz and Sanjeeb Kumar Sahoo

10.1 Anatomy and Function of the Eye

The eye is the organ of sight and the most precious part of a human body, which is spherical hollow globe in structure. Broadly we can categorize its structure into two subheadings: (1) anterior segment and (2) posterior segment. The anterior segment is the front site of the eye, which mainly consists of the pupil, cornea, iris, ciliary body, aqueous humor, and lens, whereas the posterior segment is the back of the eye that consists of the vitreous humor, retina, choroid, macula, and optic nerve (Fig. 10.1). The cornea is the clear front window of the eye, which transmits and focuses light into the eye. The iris is the colored part of the eye, which helps to regulate the amount of light that enters the eyes. The pupil is the dark center in the middle of the iris, which determines how much light will be permitted into eye, and also it changes its size to accommodate the amount of light that is available. The lens, which is the transparent structure inside the eye, focuses the light rays onto the retina. The aqueous humor is a thin watery fluid that fills the space between the cornea and the iris (anterior chamber). This fluid nourishes the cornea and lens and gives the eye its shape. In the posterior chamber, the vitreous humor is a transparent, colorless, gelatinous mass which fills the middle of the eye. The retina is the nerve layer that lines the back of the eye and senses light which creates impulses that are sent through the optic nerve to the brain. Macula is a small area in the retina that contains special light-sensitive cells. The macula allows us to see fine details clearly. The optic nerve connects the eye to the brain. The optic nerve carries the impulses formed by the retina to the brain, which interprets it to image. The choroid is the middle layer of the eye that contains blood vessels and connective tissues which supply nutrients to the inner portion of the eye.

F. Dilnawaz • S.K. Sahoo, Ph.D. (✉)
Nanomedicine Laboratory, Institute of Life Sciences,
Nalco Square, Chandrasekharpur, Bhubaneswar, Orissa 751023, India
e-mail: sanjeebsahoo2005@gmail.com

A.J. Domb and W. Khan (eds.), *Focal Controlled Drug Delivery*, Advances in Delivery Science and Technology, DOI 10.1007/978-1-4614-9434-8_10,

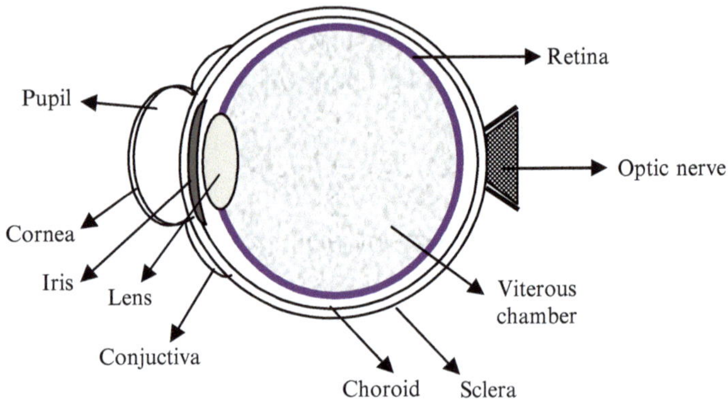

Fig. 10.1 Structure of the eye

10.2 Delivery Routes for Ocular Drug Delivery

Ocular drug delivery is one of the most challenging delivery system encountered by a pharmaceutical scientist due to its unique anatomy and physiology. The distinctive and complex structure of the eye restricts the entry of any foreign substances including drug. The anterior and posterior positions though juxtaposed to each other are very different in their anatomical and physiological features. Apart from that both of them function independently and in tandem upon application of an ocular preparation. The drug delivery to the eye is mainly classified into anterior and posterior segments. Conventional drug administration systems such as eye drops, suspensions, and ointments are not considered favorable for vision-threatening ocular diseases. Nearly about 90 % marketed ophthalmic formulations are in the form of eye drops which mainly target the anterior segment of the eye disease [1]. These topically applied drugs are generally washed off from the eye by various defensive mechanisms such as blinking, lid closure due to reflex blinking, lacrimation, and nasolacrimal drainage which results to low ocular bioavailability of drugs. In addition, drug binding to tear proteins and to conjunctival mucin also inactivates a portion of the administered dose. Further loss can arise through physical means. During administration, a part of an aqueous drop instilled is inevitably lost by over flow/drainage, as the conjunctival pouch can accommodate only few microliters of added fluid [2, 3]. Apart from the human cornea comprising of epithelium, substantia propria also restricts the ocular entry of drug molecules [2]. As a result of these factors, <5 % of administered drug enters the eye. Moreover, alternative approaches like incorporation of permeation enhancers/cyclodextrins and increasing the viscosity of solutions did not provide any significant improvement. Also many drug efflux pumps have been identified and significant enhancement in ocular drug absorption was achieved following their inhibition or evasion. But prolonged use of such inhibitors may result in undesirable effects [4].

10.3 Nanotechnology-Based Ocular Drug Delivery System

For ophthalmic drug delivery, controlled and sustained release of the drug delivery has become one of the major focuses of the pharmaceutical industry. Additionally, in clinical therapeutics, maintaining adequate concentration of drugs at the site of action is the main concerned factor. But the physiological constraints imposed by the protective mechanisms of the eye lead to poor absorption of drugs with a very small fraction of the instilled dose penetrating the cornea and reaching the intraocular tissues. The anatomy, physiology, and biochemistry of the eye render this precious organ to be highly impervious to foreign substances [5]. A significant challenge to the pharmaceutical scientist is to circumvent the protective barriers of the eye without causing permanent tissue damage; therefore, much attention has been given towards the development of newer, more accurate delivery system with high therapeutic efficacy. In this regard, attention has been shifted towards the recent emergence of nanotechnology-based drug delivery approach. Nanotechnology plays a crucial role in the development of drug delivery vehicle, which has the potentiality to revolutionize the prevention and treatment of various ocular diseases. Nanoparticles are solid, colloidal particles loaded with macromolecular substances that vary in size from 10 to 1,000 nm. Smaller particles are better tolerated by the patients than larger particles; therefore nanoparticles of smaller size (<200 nm) may represent very comfortable ophthalmic prolonged action delivery systems [6]. The major objective of clinical therapeutics is to provide and maintain adequate concentration of drugs at the site of action. In ocular drug delivery, the physiological constraints imposed by the protective mechanisms of the eye lead to poor absorption of drugs with very small fractions of the instilled dose penetrating the cornea and reaching the intraocular tissues [5]. Conventional ophthalmic solution, suspension, and ointment dosage forms no longer constitute optimal therapy for these indications. In this aspect, nanoparticulate formulation represents one of the promising drug carriers for ophthalmic applications. These nanoparticulate formulations are showing a better application as compared to conventional drug delivery systems [5, 6]. The benefits of having the drug in the form of nanoparticulate forms are reduction in the amount of dose, drug release for a prolonged period of time, higher drug concentrations in the infected tissue, longer residence time of nanoparticles on the cornea surface, and reduction systemic toxicity of drug [5, 6]. Various nanoparticulate formulations such as nanoparticles, nanosuspensions, cationic nanoemulsions, iontophoresis, dendrimers, nanocapsules, liposomes, solid lipid nanocarriers, niosomes, and cyclodextrins are developed for the significant improvement of ophthalmic drug delivery (Fig. 10.2).

10.3.1 Nanoparticles

Nanoparticle-based drug delivery system has demonstrated promising results in ophthalmic drug delivery over the last decade, consisting of various synthetic

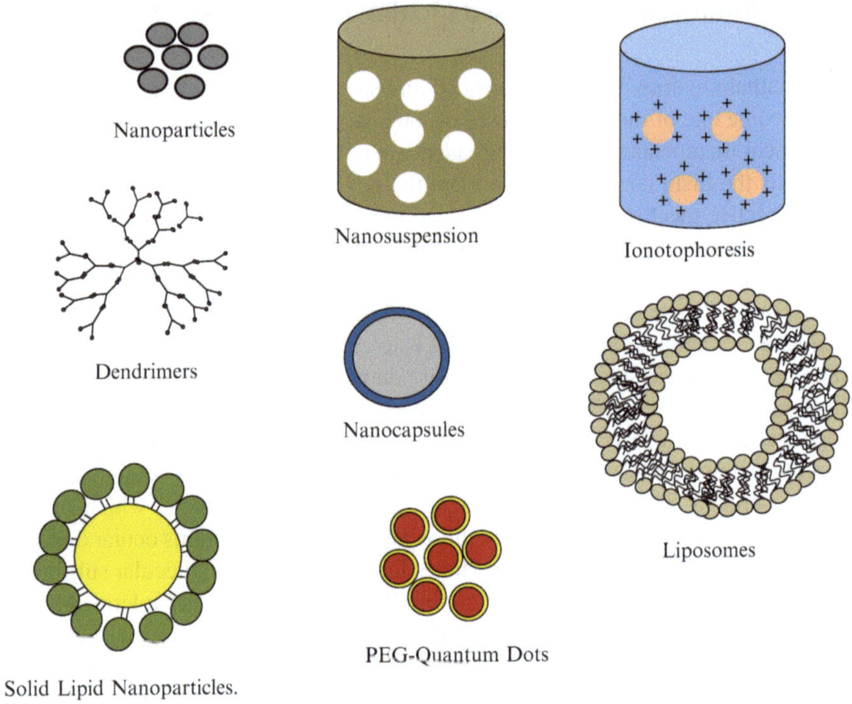

Fig. 10.2 Nanoparticulate formulations for ocular drug delivery

biodegradable materials, such as natural or synthetic polymers, lipids, phospholipids, and even metals. Mostly various biodegradable polymers, polylactides (PLAs), polycyanoacrylate, and poly (D, L-lactides), and natural polymers, like chitosan, gelatine, sodium alginate, and albumin, can be used effectively for efficient drug delivery to the ocular tissues [7, 8]. Drugs can be either integrated in the matrix or attached to the surface. At cellular level the nanoparticles are endocytosed/phagocytosed by cells, which results in the internalization of the encapsulated drug. Studies conducted by Bourges et al. in rabbits have shown that nanoparticles of different sizes and with different surface charges when injected into the vitreous migrate through the retinal layers and tend to accumulate in the retinal pigment epithelium (RPE) cells. After a single intravenous injection, they observed the presence of nanoparticles for a period of 4 months [9]. The movement of nanoparticles takes place because of rupture of the internal limiting membrane (ILM) along with the modification of the vitreous interface structure secondary to the presence of the PLA and poly (D, L-lactide-co-glycolide) (PLGA). The inflammatory reactions following the injection contribute the transretinal movement of the nanoparticles into vitreous cavity and settling on the ILM. With these findings novel drug delivery systems targeting to the posterior segment of the eye and RPE cells of the retina in particular were thought for the development. Moreover, the bioavailability of the drug increases when encapsulated in nanoparticulate formulations. Also there is report of retention of polystyrene nanospheres (made from nonbiodegradable polymers) in neuro-retina and RPE even after 2 months of single intravenous injection in rabbits [10]. Nanoparticles made up from natural polymers such as chitosan are quite effective in intraocular penetration of some specific drugs, because of its intimate contact between corneal and conjunctival surfaces [5]. Polyethylene glycol (PEG)-coated poly-epsilon-caprolactone (PECL) nanoparticles also demonstrated greater corneal penetration [11]. For improvement of adhesion to the corneal surface, the nanoparticles were coated with different polymers. PECL nanoparticles were coated with chitosan; the bioavailability of encapsulated indomethacin was doubled [12]. Furthermore, the biodegradable, nontoxic, and nonantigenic properties of albumin were taken into account for an efficient drug delivery system. For the treatment of chronic cytomegalovirus retinitis (CMV), albumin nanoparticles serve as a very efficient drug delivery system because they did not induce inflammatory reactions in the retinal tissue nor disturb the organization of the surrounding ocular tissues, as greater amount of charged amino acid allows the adsorption of drug ganciclovir (GCV) [13].

10.3.2 Nanosuspensions

Nanosuspension is a submicron colloidal dispersion used for poorly water-soluble drugs suspended in appropriate dispersion medium. The nanosuspension method is exploited for the drug compounds that form crystals with high energy, which makes them insoluble in either organic (lipophilic) or hydrophilic media. Nanoparticle suspensions prepared from inert polymeric resins can be utilized for drug delivery vehicle as they are capable of prolonging drug release and enhancing bioavailability. As these carriers do not irritate the cornea, iris, or conjunctiva, they act as an inert carrier for ophthalmic drugs [14]. Polymeric nanoparticle suspensions prepared from Eudragit RS 1001 and RL 1001 polymer resins loaded with flurbiprofen (FLU) prevent myosis during extracapsular cataract surgery. FLU is a nonsteroidal anti-inflammatory drug (NSAID) that inhibits cyclooxygenase and therefore antagonizes papillary constriction during intraocular surgery. The positive charge on the nanoparticle surface facilitates their adhesion to the corneal surface [15]. Therefore, a nanosuspension in ophthalmic pharmaceutical formulations is an attractive area, which offers a great possibility to overcome the inherent difficulties associated with ocular drug delivery [16]. In an experimental approach, Sakurai et al. (2001), studied the efficacy of the ibuprofen (IBU)-loaded nanoparticles in vivo in rabbit eye after inducing ocular trauma. After application of nanoparticle suspension, the drug level of the aqueous humor was higher as compared to control aqueous eye drop showing no signs of toxicity towards ocular tissues [10].

10.3.3 Cationic Nanoemulsions

The cationic nanoemulsion technology is more efficient at delivering the appropriate concentrations of bioactive molecules to the eye. The bioadhesiveness of nanosystems and the electrostatic interaction prolong the residence time on the ocular surface [17]. The electrostatic interaction was created between the negatively charged cells of the ocular surface and with positively charged delivery vehicles. Most widely used cationic lipids for the manufacturers of the cationic liposomes or cationic emulsions are stearylamine, oleylamine, poly(ethylenimine) (PEI), and poly-L-lysine (PLL) [18, 19]. Oleylamine, the cationic lipid, has been used to manufacture ophthalmic emulsions [20], but this lipid has stability concerns due to its primary amine function and an unsaturated site in the aliphatic chain. However, PEI and PLL organic polymer has a high density of amino groups that can be protonated at physiological pH; the polycation is very effective in binding DNA and can mediate the transfection of eukaryotic cells [21]. Cationic lipids, DOTAP (*N*-(1-(2,3-dioleoyloxy) propyl)-*N*,*N*,*N* trimethylammonium) chloride, and DOPE (dioleoyl phosphatidylethanolamine) represent another potential class of cationic agents that are well tolerated and biodegradable. Hagigit and colleagues showed that using DOTAP was better than the seminatural lipid oleylamine to make stable cationic emulsions. Moreover, DOTAP cationic emulsion enhanced the penetration of antisense oligonucleotides after either topical ocular instillations or intravitreous injection [22]. Cyclosporine A, a highly lipophilic immunomodulatory drug, is widely used by ophthalmologists for its recognized therapeutic potential for the treatment of ocular diseases (dry eye, allergy, and inflammation) [23]. Clinical trials of cationic emulsions loaded with cyclosporine A have demonstrated higher efficacy and safety in patients with dry eye disease. Using the cationic nanoemulsion technology, Novagali Pharma has developed few products, i.e., Cyclokat and Cationorm (treatment of dry eye symptoms), Vekacia (treatment of vernal keratoconjunctivitis), and Catioprost (treatment of elevated intraocular pressure (IOP) due to glaucoma) [19].

10.3.4 Iontophoresis

Ocular iontophoresis recently has received significant interest due to its noninvasive nature of delivery of drugs across cornea and sclera. For iontophoresis the ions of importance should be charged molecules of the drug, which requires a mild electric current to drive the ions into cells or tissues. Following application of iontophoresis technique, the positively and negatively charged nanoparticles were shown to distribute to the inner ocular tissues such as retina and vitreous humor by increasing the bioavailability and decreasing the adverse effects of the drugs. Studies conducted reveals that positively charged nanoparticles have shown higher penetration effect than negatively charged nanoparticles [24]. With this mode of delivery, the potential side effects associated with intraocular injections and implants can also be overcomed. Encouraging results are obtained with the use of transcorneal iontophoresis of ciprofloxacin hydrochloride (ocular infection), gentamicin (pseudomonas keratitis), and antisense oligonucleotides (treatment of angiogenesis in cornea) in ocular therapy [25, 26]. Using iontophoresis technique, dexamethasone phosphate, methylprednisolone (posterior segment inflammation), carboplatin (retinoblastoma), and methotrexate (inflammatory diseases and intraocular lymphoma) were also successfully delivered in affected ocular tissues [24, 27]. For transscleral iontophoresis, a system OcuPhor™ has been designed with an applicator, dispersive electrode, and a dose controller for release of drug [28]. This device can also release the active drug into the retina–choroid. A similar device called Visulex™ has also been designed to allow selective transport of ionized molecules through sclera [29]. Iontophoretic application of antibiotics to the eye also increases the bactericidal activity; as a result the severity of disease is reduced. Moreover, application of anti-inflammatory agents also reduced the vision-threatening side effects [30, 31].

10.3.5 Dendrimers

Dendrimers are a macromolecular system made up of a sequence of branches around an inner core. They are attractive macromolecular compounds for drug delivery because of their nanometer size range, ease of preparation and functionalization, and their ability to display multiple copies of surface groups for biological recognition processes [6, 32–34]. Because of these multifaceted properties, they are used as a valuable vehicle for ophthalmic drug delivery. Recently much attention was given to poly(amidoamine) (PAMAM) dendrimers, which are liquid or semisolid polymers having several amine, carboxylic, and hydroxyl surface groups, which increase with the generation number (G0, G1, G2, and so on). Its characteristic branched a unique structure; PAMAM dendrimers are able to solubilize efficiently poorly water-soluble drugs into their inner core tiers of branch cells with radial connectivity to the initiator core and an exterior or surface region of terminal moieties [35–37]. Research efforts of Vandamme and coworkers revealed that for improving residence time of pilocarpine in the eye was increased with PAMAM dendrimers having carboxylic or hydroxyl surface groups [38]. In another study, Shaunak et al. used carboxylated PAMAM dendrimer (G3.5) to synthesize dendrimer-glucosamine (DG) and dendrimer-glucosamine 6-sulfate (DGS) conjugates with immunomodulatory and anti-angiogenic properties, respectively. These two conjugates were used together for the clinical validation of wound healing after glaucoma filtration in rabbit model to increase the long-term success of the surgery and prevent scar tissue formation. Rabbits treated with DG and DGS showed minimal scar tissue formation compared to placebo-treated animals with no clinical, hematological, biochemical toxicity, or microbial infections during the whole experimental period. These results suggest that DG and DGS are effective for safe ocular administration [39]. For normal eye development ocular neovascularization is essential, but it leads to blindness when it is not well controlled. The effective angiogenic factor involved in ocular neovascularization is vascular endothelial growth factor (VEGF). A sense oligonucleotide named ODN-1 was reported to have potent anti-VEGF activity. Wimmer et al. designed and synthesized lipid–lysine dendrimers for the delivery of ODN-1 into the nuclei of retinal cells [40]. In another study Marano et al. reported the ability of lipid–lysine dendrimers to deliver ODN-1 into cells which causes a reduction in VEGF expression in a rat model. This dendrimer/ODN-1 complexes even remained active for up to the period of 2 months after injection, indicating the prolonged delivery of ODN-1 dendrimer complex [41]. Furthermore, ophthalmological examinations of injected rat eyes illustrated no significant toxicity and damage. These results altogether demonstrated the lipid–lysine dendrimers to be biocompatible ocular gene carriers for prevention of ocular neovascularization. Furthermore greater possibilities utilizing dendrimers as ophthalmic drug delivery vehicles can be explored.

10.3.6 Nanocapsules

Polymeric nanocapsules were developed for ocular delivery to increase the corneal penetration of drugs delivered and on the other hand to prevent any systemic side effects. Calvo et al. studied the efficacy of chitosan (CS)-coated and PLL-coated-PECL nanocapsules as ocular drug carrier. The high positive surface charge of CS and PLL along with mucoadhesiveness of CS significantly increased the concentration of indomethacin cornea and aqueous humor, compared to uncoated particles as well as commercial eye drops. The above formulated nanocapsules also displayed good ocular tolerance [42]. The poly(epsilon-caprolactone) (PCL) polymer was used for the formulation of nanocapsule for testing of IOP in intraocular hypertensive-induced rabbits. These formulated nanocapsules displayed better therapeutic results by decreasing of IOP as compared to commercial aqueous eye drops (carteol eye drops). Also, the incorporation of the carteol drug into nanocapsules decreased cardiovascular side effects in comparison with aqueous eye drops, thus showing reduction of undesired noncorneal absorption [43]. In another study, nanocapsule formulated with an oily core of *Migliol 840* (M840) coated with PECL was evaluated for topical ocular administration of cyclosporin A (CyA). The thick spongeous polymer coating around the oily nanodrops increased the levels of CyA up to five times than that of oily solution of CyA up to 3 days [12]. Using polyisobutylcyanoacrylate (PIBCA) and PECL polymers, nanocapsule of metipranolol base (a β blocker) was formulated for ocular delivery to prevent the systemic side effects of glaucoma [44].

10.3.7 Liposomes

Liposomes are membrane-like vesicles that consist of one or more concentric phospholipid or cholesterol bilayers designed to carry lipophilic drugs either into the core or into the bilayer [45]. Depending upon the chemical composition, liposomes can have positive, negative, or neutral surface charge. However, the positively charged liposomes have greater affinity towards precorneal drug retention and drug bioavailability, as the corneal epithelium is thinly coated with negatively charged mucin to which the positively charged surface of the liposome absorbs strongly. Also coating with bioadhesive polymers to liposomes prolongs the precorneal retention of liposomes [46]. Thermosensitive liposomes were also used for light-targeted diagnosis and therapy of chorioretinal neovascularization in age-related macular degeneration (AMD), showing the possibility of targeting neovascular areas of the eye [46]. The main problem associated with the anterior portion of the eye is the mucus barrier that incorporates the ocular tear film. This defensive mechanism represents the main limitation to the liquid formulations for ophthalmic therapy. Moreover, liposome- and chitosan-based nanoformulation increases the retention of drug on the ocular surface to facilitate penetration across the corneal epithelium. Diebold et al. (2007), formulated a complex of liposomes and chitosan (CS) nanoparticles to overcome the ocular mucosal barrier which also exhibited negligible toxicity in vitro and a good tolerance in vivo [47]. Similarly immunoliposomes are generated by conjugating antibodies to the lipid bilayer for site-specific and sustained release; immunoliposomes could act as improved vehicles for drug delivery in treatment of ocular drug delivery [48]. For ocular disease antisense oligonucleotides can be encapsulated in the liposomes which are efficiently targeted to the retina to treat CMV retinitis. Studies done by Bochot et al. demonstrated retention of larger amount of administered phosphodiester (16-mer oligothymidylate) (pdT16) oligonucleotides into the vitreous and retina–choroid even after 15 days as compared to release from solution [49]. In another study, fluconazole liposomal formulation was evaluated in the candidal keratitis model in rabbits. Rabbits treated with fluconazole-encapsulated liposomes were having higher healing rate than rabbits treated with fluconazole solution [50]. Kawakami et al. (2001), formulated liposome for ocular delivery of tilisolol for topical administration as well as intravitreally to the rabbit eye. After administration retention of liposomes was increased which resulted in higher drug concentration in the vitreous body compared to free tilisolol [51]. Similarly, Abrishami et al. (2009) prepared nanoliposomes of bevacizumab which illustrated greater efficacy than free drug, showing sustained release of bevacizumab for a period of 6 weeks in rabbit model [52]. Therefore liposomal formulation showed potentiality in ocular drug delivery system.

10.3.8 Solid Lipid Nanocarriers

Solid lipid nanoparticles (SLNs) are colloidal particles developed as an alternative to emulsions, liposomes, and polymeric nanoparticles for the delivery of hydrophilic, lipophilic, and/or poorly water-soluble drugs [53]. The formation of SLNs from a single type of lipid (homolipid) suffered low encapsulation efficiency due to drug bursting due to crystallization of the lipid into ordered β modification; however to overcome this condition, a mixture of lipids (heterolipids) is used for SLNs which do not form a highly ordered crystalline arrangement by modifying the polymorphic properties of the single lipid [54]. Youshia et al. (2012) formulated SLN loaded with methazolamide (MZA) using the Compritol® and cetostearyl alcohol (CSA) for treating glaucoma. Administration of MZA-loaded SLN in rabbit eye illustrated higher pharmacological response and more sustained release effect with decrease in ocular pressure as well as no visual irritation than the MZA solution [55]. Liu et al. (2011), synthesized SLN of baicalin (BA) (BA-SLN) for ocular drug delivery system. The BA-SLN illustrated corneal permeability with sustained release activity better than the ophthalmic solution of baicalin in isolated rabbit cornea [56], whereas the in vivo ocular irritation of BA-SLN was tested by pathological section observation using rabbits. Cavalli et al. (2002) used SLN of tobramycin (TOB) as a carrier for topical ocular delivery. The TOB-SLN showed significantly higher bioavailability in aqueous humor than the TOB commercial eye drops, illustrating the efficacy of the formulation [57]. Başaran et al. (2010), formulated cyclosporine-A-loaded cationic SLN for ocular delivery, which is mainly used for corneal transplantation and dry eye syndrome. The in vivo studies in sheep revealed availability of the formulation in aqueous and vitreous humor for 48 h depicting ocular penetration with good controlled prolonged release profile [58].

10.4 Ocular Photodynamic Therapy

Photodynamic therapy (PDT) is an evolving and innovative approach for treating ocular neovascular disorders of the eye. In treating neovascularization or abnormal endothelial proliferation in the back of the eye, a colloidal carrier containing a photosensitizer is to be injected intravenously to facilitate the localization of the carrier in the neovascular region due to enhanced permeability and slow clearance of colloidal carriers from such regions (Fig. 10.3). Thereafter, subsequent activation of the photosensitizing compound via a nonthermal laser light generates highly reactive singlet oxygen ($^1O_2$) in order to prompt light-induced cytotoxicity [59]. Effectiveness of PDT is dependent upon an abundant supply of molecular O_2, which ensures its usefulness in living tissue [60]. The photosensitizer molecule after activation can be administered either intravenously or as a topical drop. The AMD is the

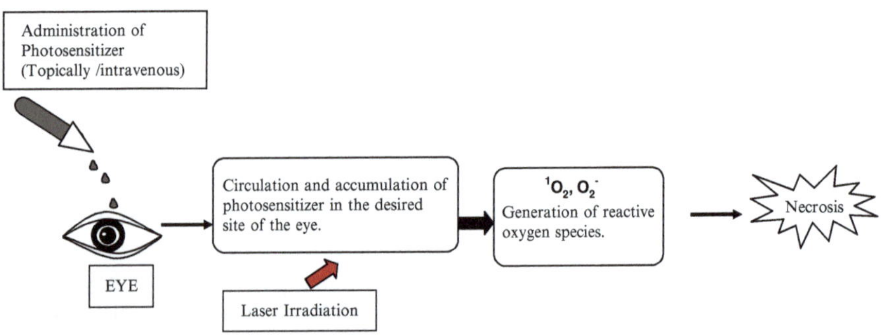

Fig. 10.3 Photodynamic therapy of eye

leading cause of loss of vision, for which a long-term therapeutic strategy is required because they are chronic and progressive [61]. Liposomes possess characteristics that are favorable for ocular PDT. Liposomal formulation are also established nanocarriers for drug and gene delivery. However, the liposomes can also entrap photosensitizers with different physicochemical properties [62]. Clinically PDT is now facilitated through a liposomal formulation Visudyne®, an FDA-approved light-sensitive delivery system. The photosensitizer verteporfin (a benzoporphyrin-derivative monoacid ring A) is the principal ingredient of Visudyne®. Liposomal formulation of Visudyne® is selectively retained in neovascular spots of the eye, which allows targeted therapy. Likewise, Photofrin® is currently approved by the FDA for treatment of other cancers including ocular with safety profile. The active ingredient of Photofrin® is the photosensitizer, porfimer sodium, often referred to as dihematoporphryin ether (DHE). In preclinical trials, DHE demonstrated a distinct ability of localization in ocular malignancies, particularly corneal neovascularization (CNV) [63]. Integration of DHE into liposomal formulation will enhance the therapeutic approach. Ionic dendritic porphyrin loaded in PEG-block-poly(L-lysine) micelles has been shown to be a viable delivery system for ophthalmic applications [59]. In a rat model PDT with dendritic porphyrin-loaded PEG-block-poly(L-lysine) micelles was administered for the treatment of CNV. These dendritic porphyrin-loaded micelles specifically accumulated in the CNV sites as early as 15 min after the injection and continued till 24 h, whereas free dendritic porphyrin concentrated in CNV sites up to 4 h following injection but disappeared within 24 h due to its negatively charged periphery indicating the success of using dendritic porphyrin-loaded micelles for PDT [64]. In another approach light-induced gene transfer may be delivered through the use of dendritic porphyrin-coated non-viral vectors. Nishiyama et al. (2005), formulated pDNA/polycation polyplex which was enveloped with the photosensitizer, anionic dendritic phthalocyanine, and functionalized with a nuclear localization signal. The nuclear localization signal enabled the transport of the complexes to the nuclei [65]. The formulated complex was injected into the eyes of rat and laser irradiated, as a result of which there was appreciable gene expression in the conjunctival tissue with reduced level of light-induced cytotoxicity [59].

10.5 Nano-Vehicles Ocular Imaging

Nanotechnology-based therapeutic applications have great potential for improving clinical management of ocular disease. Engineering particles to the nanoscale can influence the manner and efficiency in which organic or inorganic materials interact with biological systems. For this reason, significant efforts in the field of nanomedicine have also been directed towards the development of nanoscale imaging agents. The clinical approaches take the advantage of the unique properties such as controlled release, site specificity, and stability. These features reduce the therapeutic dose as well as off-target effects which usually contribute to toxicity. For this purpose imaging agents are manipulated on the nanoscale level to home specifically to the dysfunctional tissues to aid in the early detection of disease [66]. In ocular therapeutics, the specific arena in which the nanoscale contrast agents are poised to improve disease detection and management involves vascular diseases of the retina. Retinal vascular diseases such as diabetic retinopathy and neovascular AMD are the leading causes of blindness, and efforts to improve diagnostic and prognostic ocular imaging techniques are needed to improve the efficacy of clinical management [67]. In diabetic retinopathy the endothelial dysfunction and subsequent retinal capillary dropout is the critical event in disease progression. Colloidal semiconductor quantum dots (QDs) have received much attention due to their unique optical properties which can be efficiently used for highly sensitive cellular imaging. Using nanotechnological approach the molecular level can be imaged by QD. Tantrum et al. (2012) used PEG-coated QD to prolong the circulation time and reduce immunorecognition and clearance by the reticuloendothelial tissues such as liver and spleen [67]. Further the pegylated QD were surface functionalized by specific biomarker of endothelial inflammatory responses, such as intercellular adhesion molecule (ICAM-1) and vascular cell adhesion molecule (VCAM-1), which are known to be upregulated in streptozotocin (STZ) rat model. Multispectral molecular fluorescence imaging of retinal vessels illustrated increased QD binding in STZ-induced diabetic rat model due to enhanced expression of ICAM-1 and VCAM-1 [68]. This combination of QD surface design elements offers a promising new in vivo approach to specifically label vascular cell and biomolecules of interest in retinal imaging. Another consequence outcome of endothelial dysfunction in diabetic retinopathy is leukostasis, in which endothelial cell surface inflammatory molecules provoke the cascade on rolling, arrest, and extravasation of leukocytes on and across retinal vessel walls [69]. To study this particular event, Jaygopal et al. formulated antibody-functionalized QD probes. The QD was decorated with RP-1 (marker of rat neutrophils) which was used for targeting circulating leukocyte subpopulations to study their role in retinal vascular diseases such as diabetic retinopathy and uveitis using STZ animal models. Apart from that, another major vision-threatening complication of diabetes is proliferative diabetic retinopathy, which is characterized by aberrant growth of blood vessels (angiogenesis) above the inner limiting membrane into the vitreous. To preserve the vision of the patient, early detection of this disease component therefore becomes crucial. To study this process Boron PLA nanoparticles were formulated which exhibited oxygen-sensitive properties; specifically these nanoparticles were capable of emitting long-lived, green phosphorescence (i.e., "glow in the dark" emission) which is responsive to oxygen concentrations, and also these probes can also be useful for imaging hypoxia in retinal tissue noninvasively [70, 71]. To ensure optimum delivery of therapeutic agents in the eye and detailed information of the transport mechanism and elimination pathways, Kim et al. (2005) studied Gd-DTPA (Magnevist) release from a polymer-based implant in rabbit vitreous using T1-weighted magnetic resonance imaging (MRI). The concentration of Gd-DTPA varies from high values near implant to low values away from implant. This regional concentration throughout the vitreous will have clinical significance to treat eye diseases using a single positional implant [36]. Studies were also carried out for improving the imaging properties of poly (*N*-isopropylacrylamide) (PNIPAM) nanoparticles by physically embedding quantum dots in hydrogel nanoparticles. These quantum dots encapsulated PNIPAM nanoparticles that have superior intratumoral accumulation ability in subcutaneous melanoma tumor implantation model. These bifunctional nanodevices with a quantum dot core and hydrogel shell would aid tumor imaging and drug delivery, respectively, through which intraocular tumor imaging and treatment could be explored. Spherical gold nanoparticles are used as potential photoacoustic imaging (PAI) contrast agents [72]. These nanoparticles show distinctive absorption bands in the visible range; when the gold nanoparticles are extended to a rod shape, the absorption wavelength shifts to the NIR region. The intense optical absorption, minute dimension, nontoxicity, and inertness of gold make the formulated nanoparticles ideal candidates for imaging due to surface plasmon oscillation of free electrons. These gold nanorods can be targeted to specific biomolecules or can be coated with PEG to avoid agglomeration or attachment to the tissues.

10.6 Conclusions and Future Perspectives

Nanotechnology based approach towards the development of injectable, topical, implantable, transscleral etc. holds significant promise for opthalmic therapy. The nanocarrier vehicles have successfully improved the drug bioavailability in a controlled and sustained release manner. The advantages enforced by nanotechnological approach in improving imaging agent for ophthalmic disease might expand the clinician's tool kit for therapeutic diagnosis which may be helpful in the assessment of therapeutic response.

References

1. Bharath S, Hiremath SR (1999) Ocular delivery systems of pefloxacin mesylate. Pharmazie 54(1):55–58
2. Bourlais CL, Acar L, Zia H, Sado PA, Needham T, Leverge R (1998) Ophthalmic drug delivery systems-recent advances. Prog Retin Eye Res 17(1):33–58
3. Kaur IP, Garg A, Singla AK, Aggarwal D (2004) Vesicular systems in ocular drug delivery: an overview. Int J Pharm 269:1–14
4. Jain R, Majumdar S, Nashed Y, Pal D, Mitra AK (2004) Circumventing P-glycoprotein-mediated cellular efflux of quinidine by prodrug derivatization. Mol Pharm 1(4):290–299
5. Das S, Suresh PK (2010) Drug delivery to eye: special reference to nanoparticles. Int J Drug Deliv 2:12–21
6. Sahoo SK, Dilnawaz F, Krishnakumar S (2008) Nanotechnology in ocular drug delivery. Drug Discov Today 13(3–4):144–151
7. Mitra M, Dilnawaz F, Misra R, Harilal A, Verma R, Sahoo SK et al (2011) Toxicogenomics of nanoparticulate delivery of etoposide: potential impact on nanotechnology in retinoblastoma therapy. Cancer Nanotechnol 2(1–6):21–36
8. Parveen S, Sahoo SK (2011) Long circulating chitosan/PEG blended PLGA nanoparticle for tumor drug delivery. Eur J Pharmacol 670(2–3):372–383

9. Bourges J-L, Gautier S-E, Delie F, Bejjani RA, Jeanny J-C, Gurny R, BenEzra D, Behar-Cohen F-F (2003) Ocular drug delivery targeting the retina and retinal pigment epithelium using polylactide nanoparticles. Invest Ophthalmol Vis Sci 44:3562–3569
10. Sakurai E, Ozeki H, Kunou N, Ogura Y (2001) Effect of particle size of polymeric nanospheres on intravitreal kinetics. Ophthalmic Res 33:31–36
11. De Campos A, Sanchez A, Gref R, Calvo P, Alonso MJ (2003) The effect of a PEG versus a chitosan coating on the interaction of drug colloidal carriers with the ocular mucosa. Eur J Pharm Sci 20:73–81
12. Calvo P, Sanchez A, Martinez J, Lopez MI, Calonge M, Pastor JC, Alonso MJ (1996) Polyester nanocapsules as new topical ocular delivery systems for cyclosporin A. Pharm Res 13(2): 311–315
13. Irache JM, Merodio M, Arnedo A, Camapanero MA, Mirshahi M, Espuelas S (2005) Albumin nanoparticles for the intravitreal delivery of anticytomegaloviral drugs. Mini Rev Med Chem 5:293–305
14. Van Erdenbrugh B, Froyen L, Vanden HG (2008) Drying of crystalline drug nano-suspensions the importance of hydrophobicity on dissolution behaviour upon redispersion. Eur J Pharm Sci 35:127–135
15. Pignatello R, Bucolo C, Ferrara P, Maltese A, Puleo A, Puglisi G (2002) Eudragit RS100 nanosuspension for the ophthalmic controlled delivery of ibuprofen. Eur J Pharm Sci 16(1–2):53–61
16. Patel VR, Agrawal YK (2011) Nanosuspension: an approach to enhance solubility of drugs. J Adv Pharm Technol Res 2(2):81–87
17. Lovelyn C, Attama A (2011) Current state of nanoemulsions in drug delivery. J Biomater Nanobiotechnol 2:626–639
18. Abdulrazik M, Tamilvanan S, Khoury K, Benita S (2001) Effect of submicron emulsion's surface charge on ocular distribution of topical cyclosporin A. STP Pharma Sci 11(6):427–432
19. Lallemand F, Daull P, Benita S, Buggage R, Garrigue JS (2012) Successfully improving ocular drug delivery using the cationic nanoemulsion, novasorb. J Drug Deliv 2012:604204
20. Guilatt LR, Couvreur P, Lambert G, Goldstein D, Bentia S, Dubernet C (2004) Extensive surface studies help to analyse zeta potential data: the case of cationic emulsions.Chemistry and Physics of Lipids 131:1–13
21. Boussif O, Lezoualc'h F, Zanta M, Mergny M, Scherman D, Demeneix B et al (1995) A versatile vector for gene and oligonucleotide transfer into cells in culture and in vivo: polyethylenimine. Proc Natl Acad Sci U S A 92(16):7297–7301
22. Hagigit T, Abdulrazik M, Orucov F, Valamanesh F, Hagedorn M, Lambert G et al (2010) Topical and intravitreous administration of cationic nanoemulsions to deliver antisense oligonucleotides directed towards VEGF KDR receptors to the eye. J Control Release 145(3): 297–305
23. Lallemand F, Felt-Baeyens O, Besseghir K, Behar-Cohen F, Gurny R (2003) Cyclosporine A delivery to the eye: a pharmaceutical challenge. Eur J Pharm Biopharm 56(3):307–318
24. Eljarrat-Binstock E, Domb AJ, Orucov F, Dagan A, Frucht-Pery J, Peer J (2008) In vitro and in vivo evaluation of carboplatin delivery to the eye using hydrogel-iontophoresis. Curr Eye Res 33(3):269–275
25. Frucht-Pery J, Raiskup F, Mechoulam H, Shapiro M, Eljarrat-Binstock E, Domb A (2008) Iontophoretic treatment of experimental pseudomonas keratitis in rabbit eyes using gentamicin-loaded hydrogels. Cornea 25(10):1182–1186
26. Vaka SR, Sammeta SM, Day LB, Murthy SN (2008) Transcorneal iontophoresis for delivery of ciprofloxacin hydrochloride. Curr Eye Res 33(8):661–667
27. Raiskup-Wolf F, Eljarat-Binstock E, Rehak M, Domb A, Fruct-Perry J (2007) Transcorneal and transscleral iontophoresis of the dexamethasone phosphate into the rabbit eye. Cesk Slov Oftalmol 63(5):360–368
28. Parkinson TM, Ferugoson E, Febraro S, Bakhtiary A, King M, Mundasad M (2003) Tolerance of ocular iontophoresis in healthy volunteers. J Ocul Pharmacol Ther 19:145–151

29. Shastri DH, Patel PB, Shelat PK, Shukla AK (2010) Ophthalmic drug delivery system: challenges and approaches. Syst Rev Pharm 1:113–120
30. Engel LS, Callegan MC, Hobden JM, Hill JM, O'Callaghan RJ (1995) Ocular drug delivery: a comparison of transcorneal iontophoresis to corneal collagen shields. Int J Pharm 123:173–179
31. Rootman DS, Jantzen JA, Gonzalez JR, Fischer MJ, Beuerman R, Hill JM (1998) Pharmacokinetics and safety of transcorneal iontophoresis of tobramycin in the rabbit. Invest Ophthalmol Vis Sci 29:1397–1401
32. Ihre HR, Padilla De Jesús OL, Szoka FC Jr, Frechet JM (2002) Polyester dendritic systems for drug delivery applications: design, synthesis, and characterization. Bioconjug Chem 13: 443–452
33. Padilla De Jesús OL, Ihre HR, Gagne L, Fréchet JM, Szoka FC Jr (2002) Polyester dendritic systems for drug delivery applications: in vitro and in vivo evaluation. Bioconjug Chem 13:453–461
34. Quintana A, Raczka E, Piehler L, Lee I, Myc A, Majoros I, Patri AK, Thomas T, Mule J, Baker JR Jr (2002) Design and function of a dendrimer-based therapeutic nanodevice targeted to tumor cells through the folate receptor. Pharm Res 19:1310–1316
35. Bhadra D, Bhadra S, Jain S, Jain NK (2003) A PEGylated dendritic nanoparticulate carrier of fluorouracil. Int J Pharm 257:111–124
36. Kim H, Lizak MJ, Tansey G, Csaky KG, Robinson MR, Yuan P, Wang NS, Lutz RJ (2005) Study of ocular transport of drugs released from an intravitreal implant using magnetic resonance imaging. Ann Biomed Eng 33(2):150–164
37. Ooya T, Lee J, Park K (2003) Effects of ethylene glycol-based graft, star-shaped, and dendritic polymers on solubilization and controlled release of paclitaxel. J Control Release 93: 121–127
38. Vandamme TF, Brobeck L (2005) Poly(amidoamine) dendrimers as ophthalmic vehicles for ocular delivery of pilocarpine nitrate and tropicamide. J Control Release 102:23–38
39. Shaunak S, Thomas S, Gianasi E, Godwin A, Jones E, Teo I, Mireskandari K, Luthert P, Duncan R, Patterson S, Khaw P, Brocchini S (2004) Polyvalent dendrimer glucosamine conjugates prevent scar tissue formation. Nat Biotechnol 22:977–984
40. Wimmer N, Marano RJ, Kearns PS, Rakoczy EP, Toth I (2002) Syntheses of polycationic dendrimers on lipophilic peptide core for complexation and transport of oligonucleotides. Bioorg Med Chem Lett 12:2635–2637
41. Marano RJ, Wimmer N, Kearns PS, Thomas BG, Toth I, Brankov M, Rakoczy PE (2004) Inhibition of in vitro VEGF expression and choroidal neovascularization by synthetic dendrimer peptide mediated delivery of a sense oligonucleotide. Exp Eye Res 79:525–535
42. Calvo P, Vita-Jato JL, Alonso MJ (1997) Evaluation of cationic polymer-coated nanocapsules as ocular drug carriers. Int J Pharm 153:41–50
43. Marchal-Heussler L, Sirbat D, Hoffman M, Maincent P (1993) Poly(epsilon-caprolactone) nanocapsules in carteolol ophthalmic delivery. Pharm Res 10(3):386–390
44. Losa C, Marchal-Huessler L, Orallo F, Vila Jato JL, Alonso MJ (1993) Design of new formulations for topical ocular administration: polymeric nanocapsules containing metipranolol. Pharm Res 10(1):80–87
45. Mainardes RM, Silva LP (2004) Drug delivery systems: past, present, and future. Curr Drug Targets 5:449–455
46. Monem AS, Ali FM, Ismail MW (2000) Prolonged effect of liposomes encapsulating pilocarpine HCl in normal and glaucomatous rabbits. Int J Pharm 198:29–38
47. Diebold Y, Jarrın M, Saez V, Carvalho EL, Orea M, Calonge M, Seijo B, Alonso MJ (2007) Ocular drug delivery by liposome–chitosan nanoparticle complexes (LCS-NP). Biomaterials 28:1553–1564
48. Norley SG, Huang L, Rouse BT (1986) Targeting of drug loaded immunoliposomes to herpes simplex virus infected corneal cells: an effective means of inhibiting virus replication in vitro. J Immunol 136:681–685
49. Bochot A, Fattal E, Boutet V, Deverre J, Jeanny J, Chacun H et al (2002) Intravitreal delivery of oligonucleotides by sterically stabilized liposomes. Invest Ophthalmol Vis Sci 43:253–259

50. Habib FS, Fouad EA, Abdel-Rhaman MS, Fathalla D (2010) Liposomes as an ocular delivery system of fluconazole: in vitro studies. Acta Ophthalmol 88:901–904
51. Kawakami S, Yamamura K, Mukai T, Nishida K, Nakamura J, Sakaeda T, Nakashima M, Sasaki H (2001) Sustained ocular delivery of tilisolol to rabbits after topical administration or intravitreal injection of lipophilic prodrug incorporated in liposomes. J Pharm Pharmacol 53(8):1157–1161
52. Abrishami M, Zaeri-Ghanavati S, Soroush D, Rouhbakhsh M, Jaafari M, Malaekeh-Nikouei B (2009) Preparation, characterization, and in vivo evaluation of nanoliposomes-encapsulated bevacizumab (avastin) for intravitreal administration. Retina 29(5):699–703
53. Müller RH, Maeder K, Gohla S (2000) Solid lipid nanoparticles (SLN) for controlled drug delivery – a review of the state of the art. Eur J Pharm Biopharm 50(1):161–177
54. Attama AA, Schicke BC, Müller-Goymann CC (2006) Further characterization of theobroma oil-bees wax admixtures as lipid matrices for improved drug delivery systems. Eur J Pharm Biopharm 64(3):294–306
55. Youshia J, Kamel AO, El Shamy A, Mansour S (2012) Design of cationic nanostructured heterolipid matrices for ocular delivery of methazolamide. Int J Nanomedicine 7:2483–2496
56. Liu Z, Zhang X, Wu H, Li J, Shu L, Liu R, Li L, Li N (2011) Preparation and evaluation of solid lipid nanoparticles of baicalin for ocular drug delivery system in vitro and in vivo. Drug Dev Ind Pharm 37(4):475–481
57. Cavalli R, Gasco MR, Chetoni P, Burgalassi S, Saettone MF (2002) Solid lipid nanoparticles (SLN) as ocular delivery system for tobramycin. Int J Pharm 238(1–2):241–245
58. Başaran E, Demirel M, Sirmagül B, Yazan Y (2010) Cyclosporine-A incorporated cationic solid lipid nanoparticles for ocular delivery. J Microencapsul 27(1):37–47
59. Nishiyama N, Kataoka K (2006) Current state, achievements, and future prospects of polymeric micelles as nanocarriers for drug and gene delivery. Pharmacol Ther 112(3):630–648
60. Pervaiz S, Olivo M (2006) Art and science of photodynamic therapy. Clin Exp Pharmacol Physiol 33(5–6):551–556
61. Bloquel C, Bourge JL, Touchard E, Berdugo M, BenEzra D, Behar-Cohen F (2006) Non-viral ocular gene therapy: potential ocular therapeutic avenues. Adv Drug Deliv Rev 58(11): 1224–1242
62. Derycke AS, de Witte PA (2004) Liposomes for photodynamic therapy. Adv Drug Deliv Rev 56(1):17–30
63. Sheppard JD Jr, Epstein RJ, Lattanzio FA Jr, Marcantonio D, Williams PB (2006) Argon laser photodynamic therapy of human corneal neovascularization after intravenous administration of dihematoporphyrin ether. Am J Ophthalmol 141:524–529
64. Prasmickaite LHA, Berg K (2001) Evaluation of different photosensitizers for use in photochemical gene transfection. Photochem Photobiol 73(4):388–395
65. Nishiyama N, Iriyama A, Jang WD, Miyata K, Itaka K, Inoue Y, Takahashi H, Yanagi Y, Tamaki Y, Koyama H, Kataoka K (2005) Light-induced gene transfer from packaged DNA enveloped in a dendrimeric photosensitizer. Nat Mater 4(12):934–941
66. Cormode DP, Skajaa GO, Delshad A, Parker N, Jarzyna PA, Calcagno C, Galper MW, Skajaa T, Briley-Saebo KC, Bell HM, Gordonn RE, Fayad ZA, Woo SL, Mulder WJ (2011) A versatile and tunable coating strategy allows control of nanocrystal delivery to cell types in the liver. Bioconjug Chem 22:3
67. Tantrum JR, Penn JS, Jayagopal A (2012) Nanotechnology-guided imaging of retinal vascular disease. CRC, Boca Raton, FL, pp 365–381, Chapter 19
68. Jayagopal A, Russ PK, Haselton FR (2007) Surface engineering of quantum dots for in vivo vascular imaging. Bioconjug Chem 18:1424–1433
69. Adamis AP, Berman AJ (2008) Immunological mechanisms in the pathogenesis of diabetic retinopathy. Semin Immunopathol 30(2):65–84
70. Palmer GM, Fontanella AN, Zhang G, Hanna G, Fraser CL, Dewhirst MW (2010) Optical imaging of tumor hypoxia dynamics. J Biomed Opt 15(6):066021

71. Pfister A, Zhang G, Zareno J, Horwitz A, Fraser C (2008) Boron polylactide nanoparticles exhibiting fluorescence and phosphorescence in aqueous medium. ACS Nano 2(6):1252–1258
72. Chen PC, Mwakwari S, Oyelere AK (2008) Gold nanoparticles: from nanomedicine to nanosensing. Nanotechnol Sci Appl 1:45–66

Chapter 11
Site-Specific Ocular Nucleic Acid Delivery

Ravi S. Shukla and Kun Cheng

11.1 Introduction

Nucleic acid-based therapy has drawn much attention as a novel approach for the management of various ocular diseases. Studies carried out over the last 2 decades have clearly demonstrated the promise of using nucleic acids, such as DNA, siRNA, AS-ODN, and aptamer, in treating acquired as well as inherited ocular diseases. Rapid progress in nucleic acid therapy for ocular disease treatment has brought two products in the market: Vitravene and Macugen [1, 2]. Vitravene is an AS-ODN that was approved by the US Food and Drug Administration (FDA) in 1998 for the treatment of cytomegalovirus (CMV) retinitis in AIDS patients [1]. Macugen is an aptamer-based therapeutics, which was approved by the FDA in 2000 for treating wet age-related macular degeneration (AMD). The reasons that the eye is advantageous over other organs for nucleic acid therapy is due to the well-defined structure of the ocular cavity, the presence of the blood-retinal barrier, ease to anesthetize ocular tissues, and less immune reaction which enables the use of immunogenic or proinflammatory vectors [3].

During the last 20 years, wide varieties of viral and nonviral vectors have been employed for nucleic acid delivery to the ocular cavity. Nonviral vectors are relatively safe and include cationic lipids, polymers, or peptides that can spontaneously condense negatively charged nucleic acids to nanoscale complexes. In this regard, cationic lipids along with neutral lipids have been frequently utilized to form "lipoplex" on condensation of nucleic acids. Due to positive charges on its surface, lipoplex can facilitate nucleic acid transfer through the negatively charged cell membrane. Although nonviral vectors do not elicit immune reactions, their application

R.S. Shukla • K. Cheng, Ph.D. (✉)
Division of Pharmaceutical Sciences, School of Pharmacy, University of Missouri-Kansas City, 2464 Charlotte Street, Kansas City, MO 64108, USA
e-mail: chengkun@umkc.edu

A.J. Domb and W. Khan (eds.), *Focal Controlled Drug Delivery*, Advances in Delivery Science and Technology, DOI 10.1007/978-1-4614-9434-8_11,

is compromised due to the presence of their inherent toxicity and low transfection efficiency. In contrast, viral vectors are very efficient in delivering nucleic acids and are able to prolong the production of therapeutic proteins in ocular tissues. Although effective, their use has always been limited due to immune reactions [3]. In addition, both nonviral and viral vectors can only deliver nucleic acids in a nonspecific manner to the whole ocular cavity, leading to undesired toxicity and reduced therapeutic effect.

Another approach for ocular nucleic acid therapy is physical delivery, which also provides site-specific delivery into a specific compartment in the eye. Over the past few years, physical delivery methods have made significant strides in their efforts to increase the delivery of nucleic acids to affected sites in the eye. These delivery methods apply electric field or high-speed bombardment to facilitate nucleic acid entry into target ocular tissues. The most commonly used physical delivery methods include electroporation, sonoporation, iontophoresis, and gene gun [3, 4].

The aim of this chapter is to introduce the structure of the eye, the major types of nucleic acids for the treatment of ocular diseases, and various strategies that have been used to achieve site-specific delivery of nucleic acids to the eye.

11.2 Structure of the Eye

The spherical shape of the eye can be divided into the anterior and posterior segments. The anterior segment consists of the conjunctiva, iris, cornea, ciliary body, and lens. The posterior segment includes the sclera, vitreous humor, retina, and choroid. Each part of the two segments has a unique functionality for vision, and each of them affects drug delivery to specific site of the ocular cavity in their own unique way (Fig. 11.1).

11.2.1 Anterior Segment

Drug applied topically on the eye is absorbed either via corneal or non-corneal (conjunctival-sclera) route and then translocated towards the intraocular tissues. The cornea plays a vital role in the absorption of lipophilic drugs, whereas the conjunctiva exhibits propensity for hydrophilic drugs.

11.2.1.1 Cornea

The cornea is the exterior, avascular, and transparent part of the eye which forms an immune and mechanical barrier, thus shielding the eye from pathogens and physical injuries. It also assists in transmitting light to the back side of the eye, particularly to the retina by transmitting visible light through the lens [3]. The cornea consists of

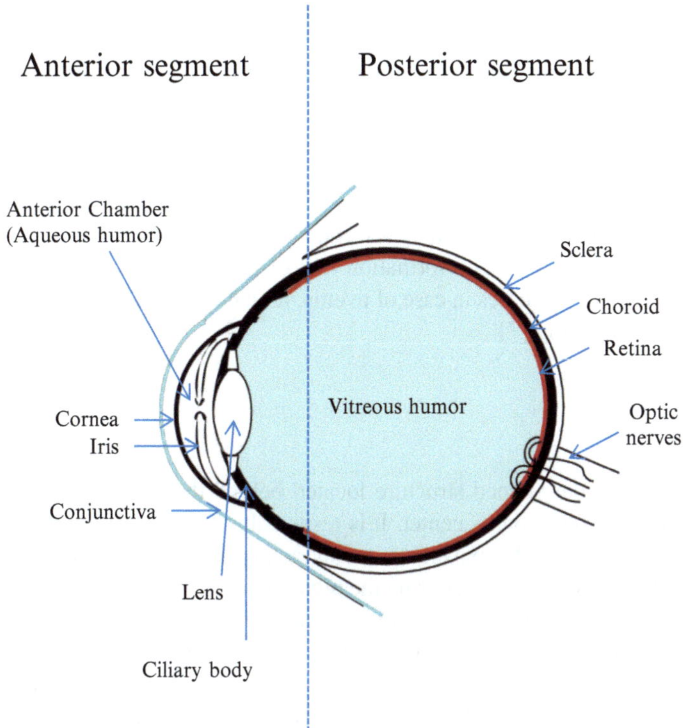

Fig. 11.1 Schematic structure of the eye

five layers: epithelium, stroma, Descemet's membrane, Bowman's membrane, and endothelium. Each layer possesses a different polarity that ultimately affects drug permeation rate to the anterior chamber [3, 5]. Among them, the epithelium, stroma, and endothelium layers play essential roles in drug permeation. The outermost epithelium layer (5 μm thickness) has tight junctions and is lipophilic in nature, which restricts the entry of drugs, particularly hydrophilic macromolecules such as nucleic acids. The stroma is responsible for the maintenance of transparency of the cornea by synthesis of collagens and proteoglycans. Its hydrophilic nature is a hurdle for the absorption of hydrophobic drugs. Another important layer is the endothelium, which is composed of hexagonal cells and presents underneath the stroma. It allows transport of macromolecules between the aqueous humor and the stroma [5].

11.2.1.2 Conjunctiva

The conjunctiva is a thin transparent membrane that covers the outer surface of the eyeball and is a key player for the absorption of hydrophilic drugs. It prevents the absorption of hydrophobic drugs. For example, it has been observed that the absorption of hydrocortisone through the conjunctiva was ~70-fold lower than its corneal absorption [6]. In addition, compared to the cornea, the larger surface area and the presence of leaker tight junctions in the conjunctiva make it a preferable absorption route for hydrophilic drugs, especially nucleic acids.

11.2.1.3 Aqueous Humor

The anterior chamber is filled with aqueous humor produced by the ciliary epithelium in the posterior chamber. The formation rate of aqueous humor is approximately 2–3 μL/min which decreases in case of uveitis, retinal detachment and ciliochoroidal detachment, and aging [7].

11.2.1.4 Iris

The iris is a circular-shaped structure located behind the cornea but ahead of the lens, holding the pupil on its center. It is responsible for controlling the amount of light reaching the retina by expanding or contracting the pupil. The color of an eye is the color of the iris which depends on the pigment of the iris stroma, the cell density of the iris stroma, and the pigment of iris pigment epithelium (IPE). The iris stroma contains numerous blood vessels, connective tissues containing collagen fibers, and myelinated and nonmyelinated nerve fibers. The irides of different colors contain the same amount of melanin pigment. Therefore, certain therapeutic molecules that have higher binding affinity to melanin requires careful consideration in ocular delivery because melanin binding may affect drug absorption in the anterior ocular tissues [8].

11.2.1.5 Lens

The lens is a biconvex structure in the anterior chamber that helps focus images on the retina. It is located directly behind the pupil and is enclosed in an elastic and fibrous capsule suspended by suspensory ligaments from the ciliary body. The lens capsule is a transparent basement membrane that completely surrounds the lens and is rich in type IV collagen and other matrix proteins [9, 10].

11.2.2 Posterior Segment

11.2.2.1 Sclera

The sclera is a 0.5–1 mm thick layer above the choroid and composed of collagens and proteoglycans. It protects inner ocular organelles and plays an essential role in drug absorption. The major part of the sclera is present in the back of the eye and is mostly composed of loose collagen fibers. The particle radius of a drug plays a critical role in its absorption in the sclera. It is postulated that the bigger the radius, the less the scleral absorption [5]. The absorption of therapeutic molecules through the sclera is also inversely proportional to its thickness. Although the surface area of the sclera is approximately 16–17 cm^2, only the thin part near the equator can be utilized for transscleral nucleic acid delivery [11].

11.2.2.2 Bruch's Membrane: Choroid

The choroid is a layer consisting of blood vessels and is located between the sclera and the retina. It is made of different types of collagens, particularly collagen type I–V, IX, XI, and XII. The Bruch's membrane and the choroid layers are difficult to separate and therefore they are studied together in in-vivo drug delivery. Presence of melanin and lipoidal plasma membrane in the choroid, Bruch's membrane offers more resistance than the sclera. In addition, the continuous blood flow in the choroid also limits the entry of nucleic acids into the retina [12–14].

11.2.2.3 Retina

The retina is a neurosensory component present in the posterior segment of the eye. The retina consists of five major cell types including photoreceptor cells, glial cells, retinal pigment epithelium (RPE) cells, interneurons, and ganglion cells [15]. These cells are arranged in a very systematic way in separate layers that are involved in visual functions. Histologically, the retina consists of the following layers: the RPE layer, photoreceptor layer, external limiting membrane, outer nuclear layer, outer plexiform layer, inner nuclear layer, inner plexiform layer, ganglion cell layer, nerve fiber layer, and inner limiting membrane [15]. Among them, the RPE, photoreceptor, and ganglion layers play important roles in drug delivery. The RPE cells make a single layer of cuboidal epithelium that consists of tight junctions and work as a blood-retinal barrier between the outer retina and choroid blood supply. Photoreceptors help in phototransduction, a process of transforming visual photo-waves into nerve impulses. These photoreceptors consists of rods (~77–107 million) and cones (~5 million). On the other hand, the ganglion layer contains approximately 1.2 million ganglion cells along with other cell types, such as astrocytes, amacrine, and pericytes [15, 16].

The macula is an oval-shaped, yellow spot that is highly pigmented and present in the middle of the retina. It helps in maintaining detailed and sharp vision. The fovea, located in the center part of the macula, contains the highest density of cones which are responsible for the most visual acuity.

To deliver macromolecules (>100 kDa) to the retina, intravitreal injection can be used to directly deliver the therapeutic cargos to the vitreous fluid. For example, the anti-VEGF monoclonal antibody Avastin (mw ~150 kDa) exhibits good absorption in the RPE when injected in the vitreous fluid. In addition, it has been postulated that nanoparticles with a diameter of 200 nm can easily penetrate into the RPE after intravitreal injection [5]. Drugs are absorbed through the RPE by intracellular and paracellular pathways.

11.2.2.4 Vitreous Humor

The vitreous humor is a clear, gel-like fluid that makes up for more than 70 % of the volume of the eye. Hyaluronic acid and some inorganic and sugar components are the major components of the fluid [17]. The presence of collagen fibers and hyaluronic acid makes the vitreous humor more viscous compared to water.

11.3 Nucleic Acids for Ocular Delivery

11.3.1 Naked DNA

Circular plasmids or naked DNAs are negatively charged macromolecules with a molecular weight larger than 2,000 kDa [3]. Although these macromolecules can be directly injected into the ocular tissues, the negatively charged surface and susceptibility to nuclease degradation limit their therapeutic efficacy. As a result, these molecules are either condensed into nanosized complex using cationic materials or transfected by the use of gene gun, a physical method of delivering nucleic acids to desired cells. Cationic lipids and polymers not only neutralize the negative charges of DNA to facilitate cell entry but also improve the in vivo stability of DNA.

Although cationic biomaterials and physical methods are less immunogenic and safe, their transfection efficiency is usually low and vanishes in a short period of time. By contrast, several viral vectors, such as adenovirus (AdV), adeno-associated virus (AAV), and lentivirus, have been successfully utilized because of their high capability to transduce ocular tissues [18].

11.3.2 Antisense Oligonucleotide

AS-ODNs are 15–20 nucleotide DNA sequences that can bind to their complementary mRNAs encoding undesired proteins [19]. Upon hybridization by reverse complimentary, AS-ODNs either block the translation or induce the degradation of the target mRNA.

11.3.3 siRNA

RNA interference (RNAi) is a process of silencing a target gene by small interfering RNA (siRNA) of 21–23 nucleotides, which can bind to their complementary mRNA and trigger the degradation. Briefly, in the cytoplasm of cells, siRNA molecules are unwound by the Dicer enzyme, and the antisense strand matches with the complementary mRNA sequence with the help of the RNA-induced silencing complex

(RISC), leading to degradation of the mRNA [1]. Due to its specific and potent knockdown of target genes, siRNA holds considerable promise as a novel therapeutic approach. However, the clinical application of siRNA is hampered due to its poor serum stability, poor cellular uptake, and rapid renal clearance.

11.3.4 Aptamer

Aptamers are nucleic acids, either RNA or DNA, screened in vitro for high binding to a vast variety of targets, such as proteins, cells, and nucleic acids, by a process called Systematic Evolution of Ligands by Exponential Enrichment (SELEX). The binding affinity of aptamers to their targets is comparable to antibodies. Aptamers can also exert inhibitory effects by binding to the functional domain of a protein [1]. The stability and bioavailability of aptamers in the ocular tissues can be improved by PEGylation and backbone modifications, such as 2′-*O*-methylation [1, 20].

11.4 Nucleic Acid Delivery to the Anterior Segment of the Eye

The efficacy and selectivity of ocular nucleic acid delivery depend on two important factors: the type of vector and the mode of administration. The most commonly utilized vectors include viral vectors, such as AAV, AdV, retrovirus and lentivirus, and nonviral vectors, such as polyethylenimine, cationic lipoplexes, and polymeric micelles. The selection of a suitable vector is determined by the inherent traits of the vector and its ability to transfect the desired ocular cells. In general, viral vectors are more effective in delivering nucleic acids as compared to nonviral vectors [21]. In addition, the capability of viral vectors to accommodate tissue-specific and inducible promoters makes them more selective to target-specific tissues or disease conditions. For instance, adenoviruses containing the E-selectin promoter (a tumor necrosis factor-inducible promoter) specifically deliver nucleic acids to the corneal epithelium [22]. Despite the effectiveness and selectivity of viral vectors, their inherent toxicity promotes the use of nonviral vectors for nucleic acid delivery.

Administration mode also has a considerable impact on the delivery and selectivity of nucleic acids. In principle, the administration mode is solely dependent on the site of action in the eye. Two administration modes have been applied so far for ocular delivery of nucleic acids: systemic and direct ocular administrations. Clinical investigations have demonstrated that systemic delivery is mainly compromised by the blood-retinal barrier which impedes entry of nucleic acids to the eye. By contrast, direct ocular administration is advantageous due to its local delivery, and it can be further categorized into invasive and noninvasive methods. The noninvasive administration methods include topical instillation and physical methods. Invasive administrations include various ocular injections into different sites of the eye.

11.4.1 Topical Administration Using Penetration Enhancers

Topical instillation is the most convenient and patient-friendly method for ocular nucleic acid delivery. This method allows direct instillation of nucleic acids on the ocular surface for absorption. Owing to their negative charge and high molecular weight, nucleic acids do not effectively enter into the corneal epithelium layer through passive diffusion, leading to limited therapeutic efficacy of topical instillation [23]. To overcome this hurdle, several penetration enhancers, such as ethylenediaminetetraacetic acid (EDTA) and cytochalasin B, have been utilized to open the tight junctions to a certain extent to allow better drug absorption [24, 25]. These penetration enhancers are applied before topical instillation of nucleic acids. For instance, polymeric micelles containing the pKera 3.2-LacZ gene were successfully delivered into the stroma cells by topical instillation after applying 5 mM EDTA to the eye [26]. This study suggested the role of EDTA in opening the tight junctions of the corneal epithelium layer, which facilitates micelles to enter the stroma.

11.4.2 Ocular Injection

Another frequently used approach for ocular nucleic acid delivery is ocular injection, which is capable of delivering both naked and formulated nucleic acids. Subconjunctival, intrastromal, and intracameral injections are the most commonly employed injections for nucleic acid delivery into the anterior segment of the eye.

11.4.2.1 Subconjunctival Injection

Subconjunctival injection is less invasive and delivers drugs under the conjunctival membrane that lines the inner surface of the eye lid. This injection has been used to deliver nucleic acids to the conjunctiva and cornea. For instance, Wen et al. have shown high expression levels of rAd-p21 DNA in conjunctival fibroblast (the target cells in glaucoma filtration surgery) after administering the DNA through subconjunctival injection [27]. Similarly, subconjunctival delivery of siRNA against vascular endothelial cell growth factor (VEGF) gene exhibited significant inhibition of herpes simplex virus (HSV)-induced angiogenesis and keratitis [28].

11.4.2.2 Intrastromal Injection

Intrastromal and intracameral injections are other frequently used injections to deliver genetic materials to the cornea in the anterior eye [3]. Intrastromal injection delivers drugs into the stromal layer of the cornea, which is difficult for nucleic acids to reach by topical instillation. Intrastromal injection can also be used to deliver vectors containing a tissue-specific promoter, which allows long-term and robust gene expression in the corneal stroma. As the largest layer of the cornea, the stroma mainly consists of keratocyte cells, which express a stroma-specific keratin sulfate proteoglycan, keratocan. The 3.2 kb keratocan promoter can specifically drive the transgene expression in the corneal stroma cells. Carlson et al. constructed an adenoviral vector containing the 3.2-kb keratocan promoter to transfect the EGFP gene into the stromal keratocytes using intrastromal injection [29]. EGFP expression was observed in the corneal stroma and lasted for 3 weeks. By contrast, the naked pCMV-EGFP plasmid DNA that does not contain the keratocan promoter did not exhibit EGFP expression in the corneal stroma [29].

11.4.2.3 Intracameral Injection

The corneal endothelium layer is generally transfected with nucleic acids either ex vivo or by the use of intracameral injection. The endothelium layer can be maintained for several weeks in ex vivo culture, thus allowing easy transfection [30]. Recent progresses in intracameral injection have allowed researchers to selectively deliver nucleic acids to the endothelium layer without disrupting the stromal and epithelium layers of the cornea [31]. Bainbridge et al. have shown a 12-week GFP expression in mouse corneal endothelium after an intracameral injection of a lentivirus encoding the GFP reporter gene [32]. By contrast, the rat corneal endothelium layer was only transfected up to 25 % when an HSV vector encoding the LacZ reporter gene was delivered by intracameral injection. This observation indicates the importance of vectors in transfection efficiency while using the same injection method. Although HSV-mediated LacZ reporter gene transduction was low by intracameral injection, it was still higher compared to the ex vivo transfection in the cornea cells [31, 33, 34]. This might be attributed to the site-specific delivery of nucleic acids by the intracameral injection, which selectively delivers the drugs to the endothelium layer.

11.4.3 Physical Method

The potential immune risk of viral delivery systems and the low transfection efficiency of nonviral delivery systems have created an urgent need for alternative delivery strategies for nucleic acids. Physical methods including electroporation, iontophoresis, gene gun, and ultrasound-mediated nucleic acid delivery methods have emerged as promising alternatives for ocular nucleic acid delivery. Gene gun technology is used to transfect naked nucleic acids to the epithelium layer of the cornea, while electroporation, iontophoresis, and sonophoresis are used to deliver nucleic acids to the inner parts of the anterior portion of the eye, such as the corneal endothelium and the stromal layer.

11.4.3.1 Gene Gun

Gene gun technology is utilized for selective delivery of nucleic acids to target cells using high-velocity gold microparticles coated with the nucleic acids [35]. The ability of the gene gun to transfect a wide variety of cells, including dividing and nondividing cells, makes it a promising and versatile strategy in ocular nucleic acid delivery.

Regardless of the ease of using gene gun, its ability to penetrate only the upper surface of the tissue, uncontrolled delivery, and relatively high cost limit its clinical applicability. In addition, the use of gene gun is influenced by several other factors, such as the cell density, the amount of force, and the coating amount of nucleic acids on the particle surface [36]. There are several successful reports demonstrating the use of gene gun for ocular nucleic acid delivery. For example, IL-4 and IL-10 plasmid DNAs were successfully transfected into mouse corneal epithelium using gene gun to treat HSV-1 infection [37].

11.4.3.2 Electroporation

Electroporation utilizes a high voltage (>100 V/cm) for a very short period of time (100 μs to 20 ms) to induce reversible destabilization of cell membrane and consequently generates hydrophilic pores [4]. These pores open for a very short period of time (40–200 ms) and allow the entry of nucleic acids into the cells. The electroporation technique allows targeted delivery of nucleic acids to the anterior ocular cavity. For example, plasmid DNA encoding the lacZ gene was specifically delivered to the corneal endothelium layer by using electroporation. To achieve endothelium layer-specific delivery, a circular electrode was placed on the corneal surface, enabling current flow with varying intensities ranging from 5 to 40 V/cm. The galactosidase expression was restricted within the ring-shaped area of the corneal endothelial layer. The expression was visible from day 1 and lasted until day 21. In addition, the expression was voltage dependent (6.45 % expression at 20 V as compared to 0.09 % at 5 V) [38]. Similarly, a plasmid encoding the tissue plasminogen activator (tPA) gene was delivered to the corneal endothelium for the treatment of intracameral fibrin formation in YAG laser-generated animal model. High tPA expression was observed after 4 days in target area of the cornea without any inflammation, which significantly reduced the fibrin formation. In addition, the opacity in the cornea of the treated eye was significantly lower than the untreated eye [39].

In addition to endothelium, electroporation has also been utilized for targeted delivery of nucleic acids to the corneal stroma. For example, Oshima et al. injected a plasmid DNA encoding the GFP gene under a CMV promoter to the corneal stroma of adult Brown Norway rats, followed by applying electric pulses with various intensities using a bipolar linear stainless steel electrode. The transgene expression was only observed in the stromal keratocytes without apparent cell damage in the treated cells. The direct stereomicroscopy of the fluorescence using real-time imaging confirmed the gene expression in the corneal stroma as early as day 1 and lasted for 15 days [40].

11.4.3.3 Iontophoresis

Iontophoresis is a noninvasive method for the delivery of nucleic acids to the eye. In this method, a low electric voltage (~10 V or less) is applied to enhance the ionized drug penetration into ocular tissues. Ionized nucleic acids are applied to the eye with a cathode, while an anode is kept elsewhere in the body to complete the circuit [4]. The electrode arrangement makes an electric circuit in the body and facilitates the ionized drug penetration in ocular tissues. There are three mechanisms involved in drug penetration by using iontophoresis [4, 41]. The first mechanism is the Nernst–Planck effect, where drugs are internalized by electro-repulsion. For example, the internalization of negatively charged nucleic acids at cathode is caused by electro-repulsion which creates a flux and facilitates penetration of ionized molecules. The second mechanism is called the electroosmotic flow, in which a voltage difference is imposed across the negatively charged ocular membrane at physiological pH, inducing a bulk fluid flow. This mechanism is mostly employed for the penetration of non-charged or high-molecular-weight molecules. The third mechanism is the damage effect caused by electric currents (0.5–1.5 mA/cm^2 or less), leading to increased tissue permeability [41].

Ever since the advent of this technique, several modifications have been carried out in iontophoretic devices to minimize the risk of ocular damage caused by these devices. The three most commonly utilized devices for iontophoresis are coulomb-controlled iontophoresis (CCI), mini ion iontophoretic unit, and eyegate. CCI allows constant drug absorption regardless of the change in resistance caused by damaged ocular tissues. It has been observed that applying continuous current to a damaged tissue may change the hydration level and thus the resistance of the damaged tissues, which consequently alter the drug absorption. CCI is designed in such a way that can adjust changes in resistance and allow constant drug flow. Mini ion iontophoretic units are commercially available devices that use disposable hydrogels to deliver charged nucleic acid drugs following transscleral iontophoresis. Eyegate is an annular-shaped silicon iontophoretic device that is designed to fit closely to the eye contour and palpebral opening. The annular well of the device simultaneously accommodates electrode and drugs supplied through the silicon tubing. One of the two silicon tubes generates slight suction pressure to the device, which allows its attachment to the conjunctiva or sclera and consequently ensures constant drug flow [11, 42].

Generally, nucleic acids are inefficient to penetrate into the corneal deeper layer regardless of their formulation [43]. Therefore, iontophoresis technique has been explored as an alternative to deliver nucleic acids to the deep anterior segment of the eye. For instance, transcorneal iontophoresis of fluorescence (CY5)-labeled antisense oligonucleotides (AS-ODN) against the vascular endothelial growth factor (VEGF)-R2 receptor (KDR/Flk) to the rat cornea showed significant high fluorescence in all layers of the cornea [44]. The absorption of nucleic acids to the corneal layer can further be augmented by the use of positively charged nanoparticles. These nanoparticles are capable of condensing negatively charged nucleic acids and can further enhance the transfection efficiency through negatively charged membrane by the use of transcorneal iontophoresis. The positively charged nanoparticles can rapidly penetrate to the deep anterior tissues and retain up to 12 h [45]. However, highly negative glycosaminoglycan in the cornea stromal may impede the corneal iontophoresis of positively charged molecules [4, 46].

11.5 Nucleic Acid Delivery to the Posterior Segment of the Eye

There is an imperative need to develop effective delivery systems of nucleic acids to manage rapidly progressive ocular diseases, such as diabetic retinopathy, AMD, and retinitis pigmentosa. However, the complex physiological and anatomical structures of the eye make the delivery of nucleic acids to the posterior segment very challenging. The acellular nature of the vitreous fluid and long diffusion distance are few reasons that limit the use of delivery strategies designed for the anterior segment of the eye. Widely used delivery methods, such as intravitreal injection, are unable to deliver naked nucleic acids to retinal tissues to a sufficient extent. Intravitreal injection delivers naked nucleic acids into the vitreous humor, where the majority of the nucleic acids are degraded before reaching the retina. In addition, posterior-segment ocular diseases, such as AMD, require multiple injections which increase the chances of retinal detachment and endophthalmitis. Therefore, posterior-segment drug delivery is usually formulated by various novel controlled delivery systems, such as nanoparticles and liposomes, with the aid of ocular injection or physical delivery [11]. For efficient and targeted delivery of nucleic acids to the posterior segment of the eye, a number of factors, such as the type of injection, viral serotype, specific gene promoter, delivery methods, and specific formulation, need to be carefully investigated.

11.5.1 By the Aid of Injection

Subretinal, intravitreal, and periocular injections are mostly used for nucleic acid delivery to the posterior segment of the eye (Fig. 11.2).

11.5.1.1 Subretinal Injection

Subretinal injection is the most commonly used strategy for the efficient delivery of genetic materials into the photoreceptors and RPE cells. High degree of immune privilege and low inflammation at the injection site makes the subretinal space an ideal site for ocular nucleic acid delivery [47]. The fluid delivered using subretinal injection is easily absorbed by RPE cells without causing the deformation of anatomical structure. Both viral and nonviral nucleic acid delivery systems have been investigated for the treatment of posterior ocular malfunctions. Viral vectors have

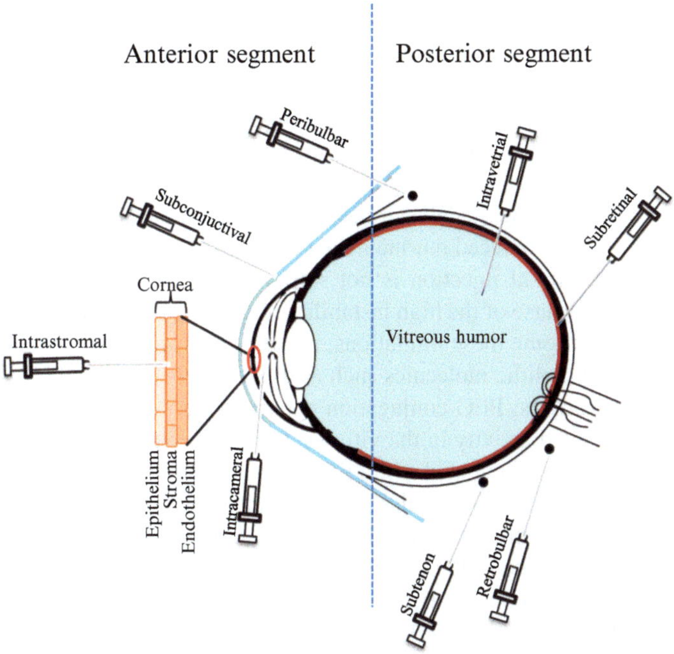

Fig. 11.2 Schematic representation of different injections used to deliver nucleic acids to specific ocular segments

shown relatively prominent effect and therefore been utilized to transfect different retinal cell types. For example, a rAAV vector encoding the RPE-specific protein 65-kD (RPE65) gene induced 3-year expression of the gene in the RPE65-deficient eyes of Briard dogs after a single subretinal injection. The RPE65 gene encodes for the RPE cell membrane-associated proteins, which are mainly involved in retinoid metabolism. The mutation in the RPE65 causes severe retinal dystrophy which is related with meager vision at the time of birth and complete loss of vision in early childhood. A similar study using rAAV expressing RPE65 complementary DNA (cDNA) was conducted in human patients. A significant improvement in visual functions was observed without causing any adverse effects [48–50].

11.5.1.2 Intravitreal Injection

Intravitreal injection is directly applicable in human patients and routinely applied for the delivery of Macugen, Avastin, and Lucentis for the management of AMD. Transfection of retinal ganglion cells (RGC) and Muller cells is efficiently achieved by use of intravitreal injections. However, delivery of AAV using intravitreal injections causes more host immune response as compared to subretinal injections. For instance, a significant high antibody response was observed in dogs when AAV vectors encoding the GFP gene was administered intravitreally [51].

Degeneration of the RGC layer results in Leber hereditary optic neuropathy (LHON), which is a maternally inherited visual dysfunction caused by mutations in several genes of the mitochondrial respiratory NADH-ubiquinone oxidoreductase complex (complex I). Obstruction in the complex I system results in decreased energy production with enhanced reactive oxygen species formation which ultimately leads to cell degeneration [52]. Intravitreal injection specifically delivered an AAV vector encoding the NDI1 gene to RGCs and protected them from degeneration in a rotenone-induced murine model of LHON [52].

A single intravitreal injection is not sufficient for the treatment of posterior ocular diseases because of the high instability of nucleic acids in the vitreous fluid. Therefore, to overcome these limitations, nucleic acids are either chemically conjugated with amphiphilic molecules such as polyethylene glycol (PEG) or formulated into nanocarriers. PEG conjugation enhances the stability of nucleic acids, leading to prolonged activity in the vitreous fluid. For example, an RNA aptamer against VEGF, when conjugated with PEG and administered intravitreally, showed 80 % decrease in retinal neovascularization. In addition, the half-life of this conjugate was 94 h in the vitreous fluid, and its active form was maintained for 28 days in the vitreous fluid. The PEG-modified RNA aptamer has shown significant therapeutic outcomes in human patients and was approved for marketing under the brand name Pegaptanib (EYE001®) [23, 53]. In another approach, the stability of nucleic acids was improved by formulating them into nanocarriers, such as liposomes, polyplexes, and nanoparticles. Liposomes have been used to enhance the efficacy of intravitreally administered nucleic acids by protecting them from degradation. It has been observed that liposomes can deliver oligonucleotides up to 50 % to the RGC and retinal cells for 3 days without any toxicity [54]. The only problem associated with liposomes is transient clouding which may interfere with the normal visual function [23]. Polyplexes and nanoparticles are the other nanocarrier systems that can condense negatively charged nucleic acids by electrostatic interactions.

11.5.1.3 Periocular Injections

Intravitreal injection, although effective, is limited due to endophthalmitis and retinal detachment. Therefore, the periocular route has emerged as a safe and least painful alternative to deliver nucleic acids to the posterior segment of the eye. The periocular injection delivers nucleic acids to the posterior segment of the eye by use of subtenon, peribulbar, and retrobulbar routes [55]. To evaluate the feasibility of this route, periocular injection of adenoviral vector (AdPEDF.11) encoding pigment epithelium-derived factor (PEDF) was utilized for the treatment of choroidal neovascularization (CNV) in a pig CNV model (thickness of the sclera is akin to human). High expression of PEDF protein was observed in choroid, which significantly reduced the extent of CNV [56].

11.5.2 By the Aid of Virus

It has been observed that viral vectors are more effective in nucleic acid delivery for a prolonged period of time. Among various viral vectors, AAV vectors show the most promising potential in this regard. AAV vectors are present in six different serotypes containing their unique virion shell proteins, which differ in their binding affinity to different cell types in the retina [57]. For instance, the transfection efficiency of AAV4 and AAV5 in the retina is relatively high as compared to other serotypes, such as AAV1, AAV2, and AAV3. This could be attributed to the different viral capsids which mediate the binding of AAV to target cells [57, 58].

11.5.3 By the Aid of Tissue-Specific Promoter

The promoter sequence is an important determinant of cell-specific nucleic acid delivery and has been utilized extensively in the case of most widely used viral vectors AAVs. In recent studies, three promoters have been investigated in terms of their specificity to transduce nucleic acids in specific cell types. These promoters are (1) CMV promoters which transduce high gene expression in retinal neurons, Muller cells, and vascular endothelial cells with less extent to RGCs [57], (2) opsin promoters which drive nucleic acid expressions in photoreceptors [59], and (3) chicken β-actin (CBA) promoters which are utilized for transduction in retinal neurons. Cell-specific transduction capability of these promoters allows controlled nucleic acid delivery to the affected retinal cells to achieve maximum outcome with minimal toxicity.

11.5.4 By the Aid of Physical Method

11.5.4.1 Ultrasound-Targeted Microbubble

Ultrasound contrast agents (microbubbles) are novel ultrasound-mediated nucleic acid delivery systems with reduced density as compared to water. These compressible and elastic microbubbles contain gases with low diffusivity and low blood solubility. Shell of the microbubbles is generally prepared by lipids, albumin, and polysaccharides. Among them, two of the albumin-based microbubbles, Albunex and Optison™, have already been approved as contrast agents [60–62].

These bubbles are filled with gases that allow the reflection of ultrasound waves and therefore can be utilized as contrast agents. The compressed gas in the microbubbles is expanded by certain predefined ultrasound waves, leading to the collapse of the bubbles and the creation of a cavitation effect. This process enables high permeability of cell membrane and can also be used to release entrapped genetic drugs inside the microbubbles [60, 63]. Cationic lipid-coated microbubbles have been mostly utilized for the delivery of negatively charged nucleic acids. These cationic lipids condense the nucleic acids either inside or on the surface of the microbubbles through electrostatic interaction [63].

CNV is a vision-threatening disease in the posterior segment that has been traditionally treated with genetic materials encoding the PEDF gene using either liposome or AAV vectors. However, these traditional delivery methods have shown toxicity and low transfection efficiency in targeted tissues [64, 65]. In a recent approach, ultrasound-mediated microbubbles destruction (UTMD) has been utilized to deliver the PEDF gene into the retina of rats. In this study, rats were treated by infusing the PEDF gene entrapped either in microbubbles followed by ultrasound exposure or in liposomes (PEDF+Lipofectamine). Western blot, quantitative real-time PCR (qPCR), and immunofluorescence staining revealed high PEDF expression mediated by ultrasound/microbubbles in the retina up to 28 days post-transfection. The PEDF gene transfer of the microbubbles was significantly higher, safer, and more effective as compared to that of the Lipofectamine. In addition the UTMD-mediated gene transfer showed effective inhibition of CNV in the rats [66].

11.5.4.2 Electroporation

Electroporation is a relatively safe and effective delivery method that has shown a great potential in the delivery of nucleic acids to the retina. The retina has the neurosensory function which can be defined as a part of the CNS. Therefore, development of an efficient nucleic acid delivery to the glial and neuronal cells could be a prolific approach to study gene function. The ventricular cells of the eye during its developmental phase differentiate into seven retinal cell types, which include one type of glial cell and six types of neurons. The glial cells are the only retinal cells whose axons leave the retina and form the optic nerve to transmit the visual information to the brain [67]. Therefore, electroporation could be a useful tool to deliver nucleic acids into the retinal cells. For instance, intravitreal injection of a plasmid encoding the GFP gene followed by 5-min electroporation was able to enhance the transfection efficiency in the RGC cells. The gene transfer was carried out using a round concave anode (kept on the sclera) and a cathode electrode (attached to corneal side) with five consecutive electric pulses of 12 V/cm and pulse span of 99 ms. Gene transfection using this technique was retained for 21 days with the maximum intensity on day 7 [68].

11.5.5 Iontophoresis

Iontophoresis is another physical technique that has been extensively investigated as an efficient, noninvasive, and safe approach for conveying nucleic acids to the retina.

A better understanding of the ocular structures, their response to electric current, and the availability of several best-fit devices that can penetrate drug into the deep tissues have made iontophoresis the most efficient technique for posterior ocular nucleic acid delivery. Several proof-of-concept experiments have been carried out to reveal the potential of this method. For example, Souied et al. have administered a plasmid encoding the GFP gene into the retina by the aid of iontophoresis. In this study, researchers first optimized various parameters of iontophoresis to sustain the GFP expression and consequently used the optimized parameters for the transfection of the cGMP-phosphodiesterase β-subunit (β-PDE) cDNA in the rd1 mouse, which is an animal model for autosomal recessive retinitis pigmentosa caused by mutation in the β-PDE gene. High expression of GFP and β-PDE was observed in the retina and photoreceptors of the mouse. The high expression of β-PDE in the photoreceptors has rescued the morphology of damaged photoreceptor cells without any adverse effect [69].

11.6 Conclusions

In recent years, the progress in the delivery of nucleic acids has taken a significant stride towards the treatment of various ocular diseases. The advent of several biotechnology techniques along with novel delivery systems has made tissue-specific delivery more convenient. Several viral and nonviral delivery systems have been used to deliver nucleic acid therapeutics. Nonviral systems include various formulations made of lipids and polymers. Although some of the formulations have shown enhanced and prolonged absorption in the anterior segment of the eye followed by topical instillation, the targetability of these formulations to the posterior segment is still dependent on invasive injections. Some of these injections, such as intravitreal injections, are patient incompatible and prone to cause endophthalmitis and retinal detachment. To minimize these side effects, physical delivery methods have emerged as an alternative. These methods are noninvasive and can be used alone or in combination to target nucleic acids to desired ocular tissues. However, the use of wave and current energy-generating devices makes these methods inconvenient. Therefore, developing small and patient-friendly devices becomes critical to the future success of the physical delivery strategy.

References

1. Fattal E, Bochot A (2006) Ocular delivery of nucleic acids: antisense oligonucleotides, aptamers and siRNA. Adv Drug Deliv Rev 58(11):1203–1223
2. Ali RR (2012) Ocular gene therapy: introduction to the special issue. Gene Ther 19(2):119–120
3. Klausner EA et al (2007) Corneal gene therapy. J Control Release 124(3):107–133
4. Bejjani RA et al (2007) Electrically assisted ocular gene therapy. Surv Ophthalmol 52(2):196–208

5. Kuno N, Fujii S (2011) Recent advances in ocular drug delivery systems. Polymers 3(1): 193–221
6. Hosoya K, Lee VH, Kim KJ (2005) Roles of the conjunctiva in ocular drug delivery: a review of conjunctival transport mechanisms and their regulation. Eur J Pharm Biopharm 60(2):227–240
7. Doolittle RF (1988) Lens proteins. More molecular opportunism. Nature 336(6194):18
8. Gaudana R et al (2010) Ocular drug delivery. AAPS J 12(3):348–360
9. (2006) Human Eye Physiology. Available from: http://cmp.felk.cvut.cz/~hlavac/TeachPresEn/15ImageAnalysis/61HumanEyePhysiology.ppt
10. Levin LA, Nilsson SFE, Hoeve JV, Wu S, Kaufman PL, Alm A (2012) Adler's physiology of the eye. 11th edn. US Elsevier Health Bookshop. Available from: http://www.us.elsevierhealth.com/Medicine/Ophthalmology/book/9780323057141/Adlers-Physiology-of-the-Eye/
11. Myles ME, Neumann DM, Hill JM (2005) Recent progress in ocular drug delivery for posterior segment disease: emphasis on transscleral iontophoresis. Adv Drug Deliv Rev 57(14):2063–2079
12. Cheruvu NP, Kompella UB (2006) Bovine and porcine transscleral solute transport: influence of lipophilicity and the Choroid-Bruch's layer. Invest Ophthalmol Vis Sci 47(10):4513–4522
13. Kim SH et al (2007) Transport barriers in transscleral drug delivery for retinal diseases. Ophthalmic Res 39(5):244–254
14. Robinson MR et al (2006) A rabbit model for assessing the ocular barriers to the transscleral delivery of triamcinolone acetonide. Exp Eye Res 82(3):479–487
15. Hildebrand G, Fielder A (2011) Anatomy and physiology of the retina. In: Reynolds J, Olitsky S (eds) Pediatric retina. Springer, Berlin, pp 39–65
16. Provis JM (2001) Development of the primate retinal vasculature. Prog Retin Eye Res 20(6):799–821
17. Trotter RR (1968) Cornea and sclera. Arch Ophthalmol 79(3):338–348
18. Colella P, Cotugno G, Auricchio A (2009) Ocular gene therapy: current progress and future prospects. Trends Mol Med 15(1):23–31
19. Crooke ST (2004) Antisense strategies. Curr Mol Med 4(5):465–487
20. Burmeister PE et al (2005) Direct in vitro selection of a 2'-O-methyl aptamer to VEGF. Chem Biol 12(1):25–33
21. Klebe S et al (2001) Gene transfer to ovine corneal endothelium. Clin Experiment Ophthalmol 29(5):316–322
22. Arancibia-Carcamo CV et al (1998) Lipoadenofection-mediated gene delivery to the corneal endothelium: prospects for modulating graft rejection. Transplantation 65(1):62–67
23. Andrieu-Soler C et al (2006) Ocular gene therapy: a review of nonviral strategies. Mol Vis 12:1334–1347
24. Liaw J, Chang SF, Hsiao FC (2001) In vivo gene delivery into ocular tissues by eye drops of poly(ethylene oxide)-poly(propylene oxide)-poly(ethylene oxide) (PEO-PPO-PEO) polymeric micelles. Gene Ther 8(13):999–1004
25. Hao J et al (2010) Gene delivery to cornea. Brain Res Bull 81(2–3):256–261
26. Tong YC et al (2007) Eye drop delivery of nano-polymeric micelle formulated genes with cornea-specific promoters. J Gene Med 9(11):956–966
27. Wen SF et al (2003) Characterization of adenovirus p21 gene transfer, biodistribution, and immune response after local ocular delivery in New Zealand white rabbits. Exp Eye Res 77(3):355–365
28. Kim B et al (2004) Inhibition of ocular angiogenesis by siRNA targeting vascular endothelial growth factor pathway genes: therapeutic strategy for herpetic stromal keratitis. Am J Pathol 165(6):2177–2185
29. Carlson EC et al (2004) In vivo gene delivery and visualization of corneal stromal cells using an adenoviral vector and keratocyte-specific promoter. Invest Ophthalmol Vis Sci 45(7): 2194–2200
30. Jun AS, Larkin DF (2003) Prospects for gene therapy in corneal disease. Eye (Lond) 17(8):906–911

31. Spencer B et al (2000) Herpes simplex virus-mediated gene delivery to the rodent visual system. Invest Ophthalmol Vis Sci 41(6):1392–1401
32. Bainbridge JW et al (2001) In vivo gene transfer to the mouse eye using an HIV-based lentiviral vector; efficient long-term transduction of corneal endothelium and retinal pigment epithelium. Gene Ther 8(21):1665–1668
33. George AJ et al (2000) Gene delivery to the corneal endothelium. Am J Respir Crit Care Med 162(4 Pt 2):S194–S200
34. Borras T (2003) Recent developments in ocular gene therapy. Exp Eye Res 76(6):643–652
35. Tanelian DL et al (1997) Controlled gene gun delivery and expression of DNA within the cornea. Biotechniques 23(3):484–488
36. Mohan RR et al (2012) Gene therapy in the cornea: 2005–present. Prog Retin Eye Res 31(1): 43–64
37. Bauer D et al (2006) Immunomodulation by topical particle-mediated administration of cytokine plasmid DNA suppresses herpetic stromal keratitis without impairment of antiviral defense. Graefes Arch Clin Exp Ophthalmol 244(2):216–225
38. Oshima Y et al (1998) Targeted gene transfer to corneal endothelium in vivo by electric pulse. Gene Ther 5(10):1347–1354
39. Sakamoto T et al (1999) Target gene transfer of tissue plasminogen activator to cornea by electric pulse inhibits intracameral fibrin formation and corneal cloudiness. Hum Gene Ther 10(15):2551–2557
40. Oshima Y et al (2002) Targeted gene transfer to corneal stroma in vivo by electric pulses. Exp Eye Res 74(2):191–198
41. Eljarrat-Binstock E, Domb AJ (2006) Iontophoresis: a non-invasive ocular drug delivery. J Control Release 110(3):479–489
42. Halhal M et al (2004) Iontophoresis: from the lab to the bed side. Exp Eye Res 78(3):751–757
43. Bochot A et al (1998) Comparison of the ocular distribution of a model oligonucleotide after topical instillation in rabbits of conventional and new dosage forms. J Drug Target 6(4):309–313
44. Berdugo M et al (2003) Delivery of antisense oligonucleotide to the cornea by iontophoresis. Antisense Nucleic Acid Drug Dev 13(2):107–114
45. Eljarrat-Binstock E et al (2008) Charged nanoparticles delivery to the eye using hydrogel iontophoresis. J Control Release 126(2):156–161
46. Yasukawa T et al (2004) Drug delivery systems for vitreoretinal diseases. Prog Retin Eye Res 23(3):253–281
47. Anand V et al (2000) Additional transduction events after subretinal readministration of recombinant adeno-associated virus. Hum Gene Ther 11(3):449–457
48. Acland GM et al (2001) Gene therapy restores vision in a canine model of childhood blindness. Nat Genet 28(1):92–95
49. Acland GM et al (2005) Long-term restoration of rod and cone vision by single dose rAAV-mediated gene transfer to the retina in a canine model of childhood blindness. Mol Ther 12(6):1072–1082
50. Bainbridge JW et al (2008) Effect of gene therapy on visual function in Leber's congenital amaurosis. N Engl J Med 358(21):2231–2239
51. Dudus L et al (1999) Persistent transgene product in retina, optic nerve and brain after intraocular injection of rAAV. Vision Res 39(15):2545–2553
52. Chadderton N et al (2013) Intravitreal delivery of AAV-NDI1 provides functional benefit in a murine model of Leber hereditary optic neuropathy. Eur J Hum Genet 21(1):62–68
53. Drolet DW et al (2000) Pharmacokinetics and safety of an anti-vascular endothelial growth factor aptamer (NX1838) following injection into the vitreous humor of rhesus monkeys. Pharm Res 17(12):1503–1510
54. Hangai M et al (1998) In vivo delivery of phosphorothioate oligonucleotides into murine retina. Arch Ophthalmol 116(3):342–348

55. Thrimawithana TR et al (2011) Drug delivery to the posterior segment of the eye. Drug Discov Today 16(5–6):270–277
56. Saishin Y et al (2005) Periocular gene transfer of pigment epithelium-derived factor inhibits choroidal neovascularization in a human-sized eye. Hum Gene Ther 16(4):473–478
57. Martin KR, Klein RL, Quigley HA (2002) Gene delivery to the eye using adeno-associated viral vectors. Methods 28(2):267–275
58. Rabinowitz JE et al (2002) Cross-packaging of a single adeno-associated virus (AAV) type 2 vector genome into multiple AAV serotypes enables transduction with broad specificity. J Virol 76(2):791–801
59. Flannery JG et al (1997) Efficient photoreceptor-targeted gene expression in vivo by recombinant adeno-associated virus. Proc Natl Acad Sci U S A 94(13):6916–6921
60. Zhou S et al (2010) Ultrasound-targeted microbubble destruction mediated herpes simplex virus-thymidine kinase gene treats hepatoma in mice. J Exp Clin Cancer Res 29:170
61. Hu YZ et al (2009) Ultrasound microbubble contrast agents: application to therapy for peripheral vascular disease. Adv Ther 26(4):425–434
62. Wood SC et al (2012) Effects of ultrasound and ultrasound contrast agent on vascular tissue. Cardiovasc Ultrasound 10(1):29
63. Unger EC et al (2001) Local drug and gene delivery through microbubbles. Prog Cardiovasc Dis 44(1):45–54
64. Lai CC et al (2001) Suppression of choroidal neovascularization by adeno-associated virus vector expressing angiostatin. Invest Ophthalmol Vis Sci 42(10):2401–2407
65. Takahashi T et al (2000) Inhibition of experimental choroidal neovascularization by overexpression of tissue inhibitor of metalloproteinases-3 in retinal pigment epithelium cells. Am J Ophthalmol 130(6):774–781
66. Zhou XY et al (2009) Ultrasound-mediated microbubble delivery of pigment epithelium-derived factor gene into retina inhibits choroidal neovascularization. Chin Med J (Engl) 122(22):2711–2717
67. Garcia-Frigola C et al (2007) Gene delivery into mouse retinal ganglion cells by in utero electroporation. BMC Dev Biol 7:103
68. Dezawa M et al (2002) Gene transfer into retinal ganglion cells by in vivo electroporation: a new approach. Micron 33(1):1–6
69. Souied EH et al (2008) Non-invasive gene transfer by iontophoresis for therapy of an inherited retinal degeneration. Exp Eye Res 87(3):168–175

Chapter 12
Topical Iontophoresis for Targeted Local Drug Delivery to the Eye and Skin

Taís Gratieri and Yogeshvar N. Kalia

12.1 Introduction

Iontophoresis is a noninvasive technique that involves the application of a mild electric current to enhance the penetration of hydrosoluble molecules into and through tissues [1]. It has been extensively investigated over many years for the transdermal delivery of several therapeutic agents including, among others, opioids [2, 3], anti-emetics [4], antihypertensives [5], and steroids [6]. The patient-friendly transdermal route avoids degradation in the gastrointestinal tract and potential first-pass metabolism. Although the skin does contain metabolizing enzymes, drug molecules encounter a significantly less challenging enzymatic barrier [7], which represents a major advantage for the delivery of molecules prone to metabolism including peptides and proteins [8].

A distinctive feature of this technique is the control over drug delivery achieved by regulating parameters such as the intensity of the applied current, the duration of current application, the drug concentration, and the surface area in contact with the active electrode compartment. In addition to fast onset and offset times, iontophoresis can be used to enable targeted delivery to specific tissues while minimizing systemic distribution of drugs with a narrow therapeutic index. The risk of irritation at the site of application is decreased by the use of low current densities. These properties make iontophoresis very attractive not only for systemic therapy through

T. Gratieri
Faculdade de Ciências da Saúde, Universidade de Brasília,
Campus Universitário Darcy Ribeiro, s/n, 70910-900 Brasília, DF, Brazil

School of Pharmaceutical Sciences, University of Geneva and University of Lausanne,
30 Quai Ernest Ansermet, 1211 Geneva, Switzerland

Y.N. Kalia (✉)
School of Pharmaceutical Sciences, University of Geneva and University of Lausanne,
30 Quai Ernest Ansermet, 1211 Geneva, Switzerland
e-mail: yogi.kalia@unige.ch

A.J. Domb and W. Khan (eds.), *Focal Controlled Drug Delivery*, Advances in Delivery Science and Technology, DOI 10.1007/978-1-4614-9434-8_12,

transdermal delivery but also for the targeted treatment of local conditions by topical delivery to the skin or to the eye. In addition to providing significant advantages for the treatment of dermatological conditions, e.g., those requiring very toxic therapeutic agents, such as skin cancers, targeted cutaneous delivery may also be useful for the delivery of cosmeceuticals. Iontophoresis allows the active ingredients to transit the outermost skin layer, the stratum corneum (SC)—the main barrier against molecular transport—and reach the viable epidermis and/or the dermis more rapidly and in greater amounts.

Local therapies are also preferred in the case of ophthalmic diseases due to possible toxicity following systemic administration and the low bioavailability due to the blood–retina barrier. Although local administration of eye drops is the most convenient way to deliver drugs to the human eye, topical treatment frequently displays poor efficacy due to the presence of protective mechanisms—lacrimal secretion and the blinking reflex cause rapid drainage of the formulation—resulting in a short residence time and necessitating multiple dosing. An even greater challenge is encountered when the drug must exert its action in the posterior structures of the eye such as the vitreous and retina. In addition to the side effects often associated with systemic administration, intravitreal injections present serious potential complications including retinal detachment, vitreous hemorrhage, endophthalmitis, and cataract. Given this scenario, ocular iontophoresis represents a promising and less invasive alternative to overcome these penetration barriers and to target drug delivery to the anterior or posterior segments of the eye using, respectively, the transcorneal or transscleral routes. However, in contrast to transdermal iontophoresis, which has been studied for several decades, it is only in the last 20 years that this technique has been investigated for drug delivery to ocular tissues. Nevertheless, several application devices have been developed, and in January 2012 an iontophoresis-based ocular therapy for the treatment of anterior uveitis entered Phase III clinical trials.

In this chapter the basic concepts and transport mechanisms behind iontophoresis are presented, and recent studies investigating topical iontophoretic delivery to the skin and transcorneal and transscleral iontophoresis are reviewed and discussed. Advantages, disadvantages, and limitations of this technique for cutaneous and ocular drug delivery are summarized in Table 12.1.

12.2 Basic Concepts and Transport Mechanisms

Molecular transport during iontophoresis can be attributed to three component mechanisms: (enhanced) passive diffusion, electromigration (EM), and electroosmosis (EO), an electrically induced convective solvent flow (Fig. 12.1). Assuming that each phenomenon is independent, the total iontophoretic flux of a molecule can be described by the sum of the fluxes due to these three processes (Nernst–Planck theory) [9]:

$$J_{TOT} = J_P + J_{EM} + J_{EO} \tag{12.1}$$

Table 12.1 Advantages and disadvantages/limitations of iontophoresis for targeted drug delivery

	Advantages	Disadvantages/limitations
Cutaneous	Noninvasive	Only for hydrophilic drugs
	Skin remains intact	Current limited to 0.5 mA/cm^2
	Controlled delivery	High costs
	Higher drug penetration across the stratum corneum	May need long application periods for some molecules
	Rapid onset and offset	
	Targeted delivery: reduction of possible toxic effects	
Ocular	Noninvasive, avoids risk of infections or ulcerations	Administration should be done in a clinic, which may increase cost
	Good drug penetration to anterior and posterior segment of the eye	No sustained half-life: frequent administrations may be needed
	Possibility of using higher current density due to low tissue resistance	Duration of current application should be short

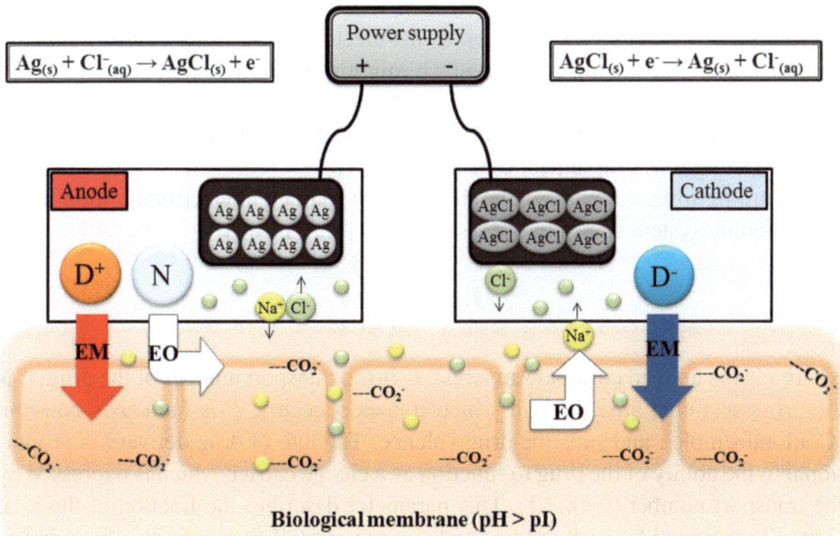

Fig. 12.1 Iontophoresis using an Ag/AgCl electrode system. The anodal compartment is used to deliver cationic (D$^+$) or neutral (N) drugs, while the cathodal compartment is host to negatively charged drug molecules (D$^-$). Application of an electric potential causes a current to flow through the circuit. At the anode–solution interface, the Ag and Cl$^-$ react to form insoluble AgCl (releasing an electron), which is deposited on the electrode surface; electroneutrality requires loss of a cation or entry of an anion into the electrode compartment. In the cathodal chamber, Cl$^-$ ions are released from the electrode and here electroneutrality requires either that an anion is lost from the chamber or that a cation enters the chamber from the skin. Electromigration (EM) transports the cations and anions from the anodal and cathodal compartments, respectively, and into the biological membrane. Under physiological conditions, electroosmosis (EO) transports neutral molecules along with cations in the direction of counterion motion to neutralize membrane charge

where J_{TOT} is the total flux, J_P is the passive flux, and J_{EM} and J_{EO} are the fluxes resulting from EM and EO, respectively. The role of passive diffusion in iontophoretic delivery is usually minor compared to the two other mechanisms [10].

12.2.1 Electromigration

Electromigration refers to the ordered movement of the ions in the presence of an electric field applied by a power source. The electric current is distributed with the aid of two electrodes. Most iontophoretic studies are performed employing reversible electrodes Ag/AgCl (Fig. 12.1). Inert electrodes, such as carbon or platinum, are also used but to a lesser extent and mainly in ocular iontophoresis. In this case, the surface electrochemistry involves the electrolysis of water:

$$\text{At the anode}: 2H_2O \rightarrow 4H^+ + O_2 + 4e^-$$

$$\text{At the cathode}: 2H_2O + 2e^- \rightarrow H_2 + 2OH^-$$

The problem with inert electrodes is that electrolysis of water causes the pH to drop/increase at the anode/cathode which might result in irritation or a significant decrease in delivery due to the generation of competing ions. Since ocular iontophoresis is performed for a very short time (usually <20 min), the impact of pH changes and competing ions is assumed to be less.

According to Faraday's law, the flux of each ion due to electromigration in the iontophoretic system at steady state is given by [1, 9, 11–13]

$$J_{EM} = \frac{I t_D}{A F z_D} \tag{12.2}$$

where I is the applied current (Amperes), t_D is the transport number of the drug, A is the cross-sectional area through which transport occurs, F is Faraday's constant (Coulombs/mole), and z_D is the drug valence. The rate of drug delivery is proportional to the ability of the drug to function as a charge carrier, which is expressed by the transport number ($0 < t_D < 1$). This parameter describes the fraction of the total charge transported by each species and is a measure of their relative efficiency as charge carriers. The transport number of a drug depends on its mobility and concentration and how these compare with the corresponding properties of the other charge carriers present in the system [14–16].

12.2.2 Electroosmosis

The skin has an isoelectric point of ~4–4.5 [17], and at physiological pH it is negatively charged and acts as a cation-selective ion-exchange membrane. The cornea and sclera are also cation selective under physiological conditions and have pIs of 3.2 [18] and between 3.0 and 4.0 [19, 20], respectively. As a consequence, under the influence of an electric field, a convective solvent flow is generated in the direction of counter-ion flow to neutralize membrane charge, i.e., in the anode to cathode direction [21–24] (Fig. 12.1). The electroosmotic flux (J_{EO}) of a drug may be expressed as

$$J_{EO} = V_w . c_D \qquad (12.3)$$

where V_W is the linear velocity of the solvent flow and c_D is the concentration of the drug in the vehicle [25]. The practical consequence of EO is that it contributes to the permeation of cations but opposes the movement of anions. Furthermore, under physiological conditions, neutral molecules can also be transported from the anode into the body [26].

12.3 Iontophoresis for Topical Skin Treatments

In both dermatological and cosmeceutical applications, the goal of cutaneous iontophoresis is to increase skin retention without increasing permeation so that the possible side effects of the active agents are reduced. Thus, drugs should ideally cross the stratum corneum and reach the viable epidermis in the desired amounts while avoiding entry into the circulatory system in the dermis (Fig. 12.2).

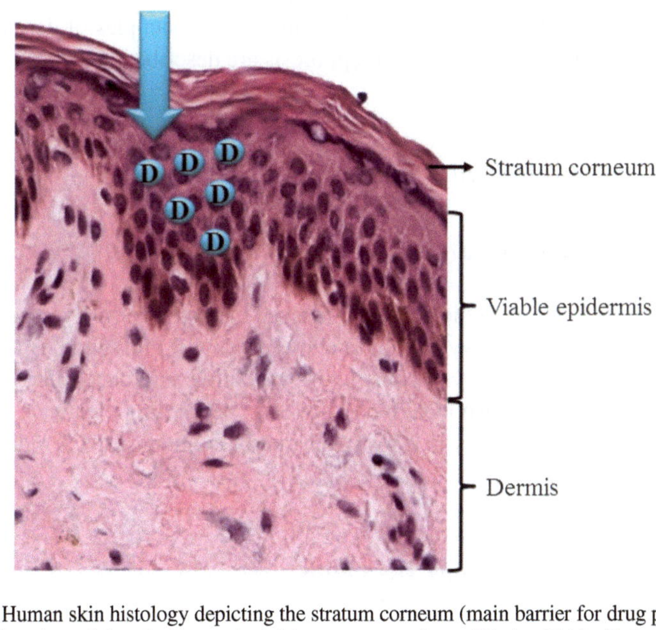

Fig. 12.2 Human skin histology depicting the stratum corneum (main barrier for drug penetration), viable epidermis, and dermis. For topical skin treatments, drugs (D) should ideally be localized in the epidermis

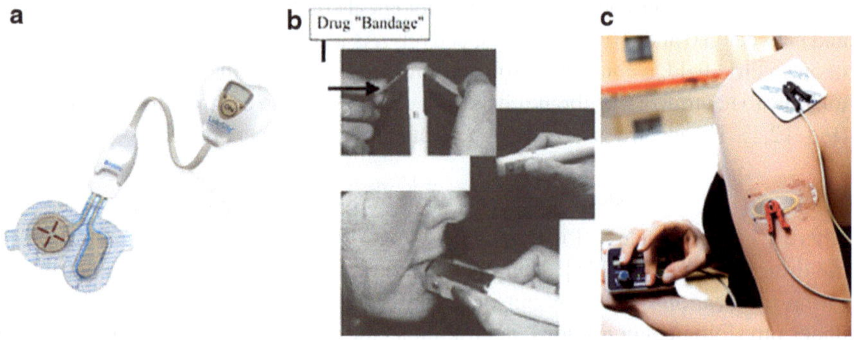

Fig. 12.3 Iontophoretic devices applied on the skin: (**a**) LidoSite® (lidocaine HCl/epinephrine topical iontophoretic patch) from Vyteris Inc., (**b**) handheld iontophoretic device used for the treatment of herpes labialis (reproduced with permission from [52]), and (**c**) active electrode filled prior to use and a separate return electrode is placed at a different body site

Fully integrated prefilled iontophoretic devices, containing the drug reservoir, electrode systems, and microelectronics, can be miniaturized and placed on the skin in a similar way to traditional passive transdermal patches. A second type of prefilled system consists of a single use disposable iontophoretic patch which is attached to a reusable controller (containing the microelectronics and power source). Alternatively, the compartment containing the active electrode can be filled with the drug formulation just before use. The return electrode can be held by the patient or placed anywhere in the body (Fig. 12.3). Some recent examples of dermatological and cosmeceutical applications of iontophoresis are described below.

12.3.1 Dermatological Applications

12.3.1.1 Local Anesthesia

The first prefilled iontophoretic patch that reached the market employed lidocaine for local anesthesia. LidoSite (Vyteris, Inc.) allowed fast (10 min) and deep anesthesia (6–10 mm). Epinephrine, a vasoconstrictor, was delivered concomitantly with the anesthetic in order to reduce local blood flow and hence drug clearance. The patch included Ag/AgCl electrodes in a hydrogel matrix containing 10 % lidocaine and 0.1 % epinephrine [27].

12.3.1.2 Psoriasis and Atopic Dermatitis

Psoriasis is an inflammatory condition of the skin characterized by (1) vascular changes where the papillary blood vessels become dilated, resulting in redness or erythema; (2) inflammation, with leukocyte influx to the epidermis and release of proinflammatory cytokines; (3) hyperproliferation of the keratinocytic layer (acanthosis); and (4) altered epidermal differentiation (parakeratosis) [28]. Atopic dermatitis is another inflammatory skin condition marked by intensely pruritic, eczematous changes that occur chronically with periods of remissions and flares [29]. Several recent studies have proposed iontophoresis for the topical delivery of drugs with anti-inflammatory properties that could act on at least one of the events that characterizes these diseases in order to minimize skin lesions and associated symptoms.

Topical corticosteroids are often prescribed for the treatment of psoriasis and atopic dermatitis due to their vasoconstrictive, anti-inflammatory, immunosuppressive, and antiproliferative effects [30]. Skin deposition of mometasone furoate, a potent corticosteroid, loaded in a sodium deoxycholate gel formulation was improved following anodal iontophoresis at 0.5 mA/cm^2 applied for 6 h in porcine skin in vitro (~1.7-fold enhancement over passive delivery; although the absolute values are not described in the report, it is possible to infer from the graph that the passive amounts deposited in the epidermis are approximately 1 μg/mg). Since drug permeation was not observed, iontophoresis was suggested as a suitable method for increasing skin accumulation of the molecule [30].

Methotrexate is a hydrosoluble drug existing mostly in the ionized form at physiological pH, which possesses antineoplastic activity and is also effective for treating psoriasis, since it inhibits mitotic activity and hence hyperproliferation of the epidermal cells and its iontophoretic delivery has been extensively studied [31–33]. The use of aqueous solutions (from 2.0 to 3.0 mg/ml) and different concentrations of NaCl in the donor compartment was reported to result in enhancement of the amounts permeated across healthy porcine skin in comparison to the amounts accumulated within the membrane after cathodal iontophoresis. Skin deposition values ranged from 2.3 to 6.25 μg/cm^2 after 24 h of passive diffusion and from 3.7 to 10.9 μg/cm^2 after 10 h iontophoresis at 0.5 mA/cm^2, while permeated amounts ranged from 0.6 to 2.9 μg/cm^2 and from 11.6 to 207.0 μg/cm^2 after, respectively, passive and iontophoretic delivery. The highest value (enhancement ratio of ~100) was obtained for cathodal iontophoresis at 0.5 mA/cm^2 in the absence of NaCl in the donor compartment [31]. It is obvious that the absence of NaCl in the formulation removes the competition from Cl^- ions which will now only be present in solution due to cathodal electrochemistry as AgCl is reduced and chloride ions are released (Fig. 12.1). Iontophoresis at 0.5 mA/cm^2 for 10 h from acrylic acid and 1:1 acrylic acid/acrylamide hydrogel resulted in approximately equivalent deposition and permeation, 1.5 μg/cm^2. The amount of methotrexate retained in the skin was not significantly affected by either hydrogel composition or drug concentration in the loading solution (120 or 200 μg/ml) [33]. A case of palmer psoriasis treated with iontophoresis of methotrexate has also been reported. The lesion on the right palm which received iontophoretic treatment showed very good (>75 %) improvement at the end of 4 weeks, whereas the left palm lesion which did not receive any treatment remained the same [34].

The cutaneous iontophoretic delivery of the E-selectin antagonist CGP69669A, a sialyl Lewis^x-glycomimetic, that lowers/blocks leukocyte trafficking, and hence has a potential therapeutic role for inflammatory disorders, has recently been demonstrated in porcine and human skin in vitro [35]. The experiments showed that while passive delivery was negligible, cumulative permeation of CGP69669A could be controlled by current density (69.73 ± 9.51, 113.97 ± 26.80, and 160.44 ± 13.79 μg/cm² at 0.1, 0.3, and 0.5 mA/cm², respectively) and drug concentration (37.42 ± 13.13, 78.96 ± 23.13, and 160.44 ± 13.79 μg/cm² at 1, 3, and 5 mg/ml, respectively). Moreover, skin deposition was enhanced (threefold) when using a 2 % hydroxyethyl cellulose gel [35].

Transcription factors are critical components in the machinery that regulates gene expression. By blocking transcription factors responsible for promoting the inflammatory pathway, topical decoy oligodeoxynucleotides (ODNs) may offer a targeted treatment for atopic dermatitis. In vitro transdermal delivery and epidermal accumulation of fluorescein isothiocyanate (FITC)-nuclear factor-kappa B decoy was demonstrated in murine skin using pulsed current iontophoresis. The iontophoretic delivery of NF-kappa B decoy ODN significantly reduced the increase in ear thickness in a murine model of skin inflammation, caused by phorbol ester as well as protein and mRNA expression levels of tumor necrosis factor-alpha (TNF-alpha) [36].

Cutaneous iontophoresis was also able to deliver naked anti-IL-10 siRNA effectively to the epidermis in a rat model of atopic dermatitis in which the skin was sensitized with ovalbumin to stimulate IL-10 mRNA expression; there was a significant reduction (73 %) in the level of IL-10 mRNA [37].

12.3.1.3 Skin Cancer

Topical Chemotherapy

Topical iontophoresis of antineoplastic agents is an interesting alternative for the local treatment of skin cancer with reduced systemic toxicity. It has been studied for the delivery of 5-fluorouracil [38], cisplatin [39], and methotrexate [31–33] (Sect. 12.3.1.2) and, more recently, the percutaneous absorption and retention of doxorubicin (DOX) [40]. Although passive permeation of DOX in porcine skin in vitro was negligible (values were under the limit of detection of the methodology—6.14 ng/ml), anodal iontophoresis at 0.5 mA/cm² for 6 h resulted in cumulative permeation of 150 ng/cm² across porcine skin in vitro. The amount of drug retained in the skin was also significantly improved from less than 5 μg/cm² of DOX in the stratum corneum after passive delivery to ~20 and 80 μg/cm² when a drug solution and a hydroxyethyl cellulose gel, respectively, were used with iontophoresis. The presence of a strong interaction between the gel and the negatively charged stratum corneum probably hindered permeation to the deep skin layers, and lower DOX levels were found in the viable epidermis (~1.5 μg/cm²). The incorporation of a cationic polymer (chitosan) in the drug formulation reduced retention in the stratum corneum (~5 μg/cm²), probably due to a decrease of the skin's negative charges when iontophoresis was applied. It was also demonstrated that the direct application of electrical current to melanoma cells in culture increased DOX cytotoxicity by nearly threefold [40].

Photodynamic Therapy

Topical Photodynamic Therapy (PDT) is a convenient and less invasive therapy for skin diseases such as actinic keratosis, psoriasis, and nonmelanoma skin cancers including basal cell carcinoma and squamous cell carcinoma. In topical PDT a photosensitizer (PS) is administered through the skin and accumulates in the tissue. The subsequent exposure of the tissue to light at the maximum absorption wavelength of the PS starts a photochemical reaction producing reactive oxygen species and ultimately killing the cells. The most studied drug for topical PDT is 5-aminolevulinic acid (5-ALA), which itself is not a PS agent but is a precursor of the photosensitive protoporphyrin IX. Its cutaneous permeation has been studied for more than 15 years [41]. 5-ALA is a hydrophilic molecule and a zwitterion at physiological pH. Using iontophoresis, therapeutic levels of PpIX can be achieved in the epidermis in 10 min or less (0.065 μmol/cm^2 of PpIX from a 2 % solution after 10 min iontophoresis at 0.255 mA/cm^2) [42]—a much shorter time than in the absence of current application [43].

The iontophoretic delivery of 5-ALA, through porcine skin in vitro, as a function of pH, and the principal mechanisms responsible for its electrotransport have been determined [44]. It was found that 5-ALA iontophoresis at pH 7.4 is a linear function of concentration over the range 1–100 mM. However, since 5-ALA exists principally as a zwitterion (pK_as of 4.05 and 8.9), at pH 7.4 the mechanism of electrotransport is primarily electroosmosis, with little or no contribution from electromigration. When the pH of the topically applied vehicle was reduced to 4.0, the net alA flux was similar to the flux at pH 7.4 (~ 50 nmol/cm^2 h). At pH 4.0, 50 % of the acidic groups are ionized and 5-ALA has greater cationic character and electromigration is enhanced. However, the net negative charge of the skin is neutralized and the electroosmotic flow is reduced. In this way the increased electromigration was balanced by a decrease in electroosmosis. In an attempt to optimize 5-ALA iontophoretic delivery, reduction of the NaCl concentration in the anode led to a three- to fourfold increase in 5-ALA flux [45]. The amount of 5-ALA delivered into the skin (epidermis + dermis) was fourfold greater with iontophoresis than after passive application of a DMSO formulation (~0.4 μmol and ~0.1 μmol, respectively) [45]. Another strategy to improve iontophoretic delivery was to use 5-ALA esters, which have a net positive charge under physiological conditions and may be transported via electromigration. The in vitro iontophoresis of a homologous series of 5-ALA esters (methyl, ethyl, butyl, hexyl, and octyl) has been evaluated. A more than 50-fold enhancement of delivery of the methyl ester compared to the zwitterionic parent 5-ALA was observed (~4.5 and 0.065 μmol/cm^2 h, respectively), an effect that gradually decreased with increasing chain length. It was also found that the delivery of methyl-ALA into the skin was also considerably enhanced over that of 5-ALA itself: a sevenfold increase in the amount found in the SC (42 nmol compared with 6 nmol) and an 18-fold enhancement in the viable tissue (180 nmol compared with 10 nmol) [46].

An alternative approach is to iontophorese the photosensitizing agent directly; this has the advantage that cutaneous levels of the photosensitizer are not dependent on the efficiency of skin metabolism for conversion of 5-ALA into the active molecule. This has been demonstrated for the meso-tetra-[4-sulfonatophenyl]-porphyrin (TPPS4) and meso-tetra-(*N*-methylpiridinium-4-yl)-porphyrin (TMPyP), negatively and positively charged porphyrin derivatives, respectively [47, 48]. In vivo experiments in male Wistar rats showed that 10 min of iontophoresis delivered TPPS4 more quickly and homogeneously to deeper skin layers than passive administration [47]. The positively charged TMPyP showed higher passive and iontophoretic delivery than the negatively charged TPPS4 from a hydroxyethyl cellulose gel containing 5.0 mg/g of porphyrin at pH 5.5 across porcine skin in vitro (23.4 and 276.9 nmol/cm^2, for TMPyP for passive and iontophoretic delivery at 0.5 mA/cm^2, respectively, in comparison with 1.2 and 4.1 nmol/cm^2 for TPPS4). Moreover, in vivo experiments indicated homogeneous distribution of TMPyP around and in the nuclei of the skin cells, suggesting its potential use in topical PDT [48].

Phthalocyanines are second-generation PS porphyrin derivatives that have a higher molar absorptivity than porphyrins at wavelengths that allow greater penetration of light in normal tissues (typically at 670–690 nm). Cathodal iontophoresis was used to transport significant amounts of zinc phthalocyanine tetrasulfonic acid (ZnPcS(4)) to the viable epidermis (55 μg/cm^2), and in vivo experiments showed homogeneous accumulation in rat skin even with only 15 min of iontophoresis [49].

All these studies show that iontophoresis offers an enormous potential for the topical treatment of skin cancers. However, more in vivo studies are necessary to evaluate both the efficiency of treatment and the potential risks. It is extremely important to ensure that antineoplastic drugs accumulate in the tumor without entering the systemic circulation in significant amounts [50].

12.3.1.4 Herpes Labialis

Recurrent herpes labialis is a relatively common disease affecting approximately 20–40 % of the adult population and is caused by the herpes simplex 1 virus. Treatments currently available include aciclovir (ACV) and related drugs (e.g., valaciclovir (VCV) and penciclovir) administered either orally or topically as ointment or creams [51]. As oral therapy normally results in low bioavailability and topical creams and ointments in inadequate penetration of the drug into the target site of infection, the basal epidermis, iontophoresis has been studied as a means to increase local drug concentration and provide faster healing. Indeed, a pilot study with 200 patients with an incipient cold sore outbreak at the erythema or papule/edema lesion stage showed that 10 min iontophoresis of a cream formulation of 5 % ACV resulted in a superior therapeutic outcome and 1.5 days reduction in healing time. Furthermore, among a subset of subjects treated in the erythema stage, there was a 3-day decrease in healing time compared to individuals who received a placebo formulation [52]. Previous studies demonstrated that ACV 5 % cream applied 5 times daily for 4 days (20 administrations), initiated within 1 h of the onset of signs or symptoms of a recurrent herpes labialis episode, resulted in only 0.5–0.6-day decrease in healing time [53]. Iontophoresis of a 5 % ACV gel formulation for

10 min in hairless rats was approximately fourfold greater than passive delivery (30.70±4.83 and 7.83±1.59 μg/cm^2, respectively). The drug level achieved in the underlying skin layers (epidermis and dermis) following 10 min of iontophoresis was 29.27±3.53 μg/cm^2, while no drug was detected in this skin compartment following 10 min of passive delivery. At 24 h post-iontophoresis, ACV levels in the SC decreased with a corresponding increase in the underlying skin due to drug migration. In this way, iontophoretic delivery of ACV resulted in a drug depot [54]. It has also been demonstrated that only the ACV in the water phase of the formulation is available for transport. In this way, when pH 11 gels (glycerin and water based) were applied to rabbit skin in vivo, they exhibited a statistically significant increase of ACV in the dermis in comparison to neutral cream formulations ($AUC_{0–60}$ of 143.7±33.3 and 8.83±1.47 mg/l min from the gel without preservatives and stabilizers and the neutral cream, respectively). This is because at pH 11 most of the drug (~98 %) in the formulation was negatively charged (pK_a 9.25) and therefore cathodal iontophoresis benefited from electromigration [51]. However, it is important to highlight that a pH of 11 is irritant to the skin and would not be tolerated, especially if the application site contains any lesions, which are very characteristic of herpes labialis.

Valaciclovir (VCV) is the L-valyl ester prodrug of ACV. It possesses three ionizable groups with pK_a values of 1.90, 7.47, and 9.43. Consequently, VCV is 50 % protonated at physiological pH [55]. A preliminary in vitro study across porcine skin indicated that iontophoresis of VCV was significantly more efficient than that of ACV. While the charged nature of the prodrug, VCV, enabled it to be more efficiently iontophoresed into the skin than the parent molecule, ACV, only the latter was detectable in the receptor chamber, suggesting that VCV was enzymatically cleaved into the active metabolite during skin transit. In this way, the cumulative permeation of ACV after 1, 2, and 3 h of VCV iontophoresis at 0.5 mA/cm^2 and using an aqueous 2 mM (similar to 0.06 %) formulation was 20±10, 104±47, and 194±82 μg/cm^2, respectively (cf. nonquantifiable levels, 0.1 and 1.0±0.7 μg/cm^2 after ACV iontophoresis) [55].

12.3.1.5 Hyperhidrosis

Hyperhidrosis is a disorder of excessive sweating beyond that required for thermoregulation and with respect to ambient conditions. Current therapeutic strategies include topical medications (most commonly aluminum chloride), tap water iontophoresis, injections of botulinum toxin, local surgical approaches, and systemic medications (including glycopyrrolate and clonidine) [56]. More recently, studies have also been performed that involve iontophoresis of therapeutic agents. The concept of using iontophoresis to treat hyperhidrosis was first introduced to the medical literature in 1936, and a report in 1952 demonstrated the efficacy of tap water iontophoresis for treating 113 patients with palmar and plantar hyperhidrosis [56]. The mechanism is currently believed to involve changes in the resorption of sodium ions by the sweat ducts as both the sodium concentration and sweat volume have been shown to decrease following iontophoresis [57]. In contrast, the mechanism of action of topical aluminum salts is related to obstruction of distal sweat gland ducts. While aluminum salts are used in regular antiperspirants, hyperhidrosis sufferers need a much higher concentration to be able to treat the symptoms of the condition effectively [58]. The need for greater amounts of aluminum salts and the different mechanism related to the effects of tap water iontophoresis justify the combining of these two treatment strategies. In a clinical study, iontophoresis of 1 % aluminum chloride applied for 30 min to one hand of 12 patients with palmar hyperhidrosis for four successive days showed significant hypohidrosis from the 3rd day until the 4th week posttreatment ($p<0.04$) which was lower than the other control hand treated topically with the same solution (but without iontophoresis) [58].

In addition, a case report on the successful iontophoresis of botulinum toxin A to treat two patients appeared in 2004 [59]; this was followed by a pilot study performed in 8 patients who were refractory to conventional therapy. In this study one hand was treated by iontophoresis of botulinum toxin type A, whereas the other received saline iontophoresis; thus, each patient served as its own control. Nine sites were treated on each hand with a total dose of 15 mA min charge passed per site (maximum current 2.5 mA). Palmar sweating was assessed by a starch iodine test and by gravimetry and was reduced in the botulinum-treated hand [60]. It is believed that the toxin works by inhibiting the release of acetylcholine at the neuromuscular junction and affecting the postganglionic sympathetic innervation of sweat glands [61]. Botulinum toxin A is a two-chain protein with a 100 kDa heavy chain joined by a disulphide bond to a 50 kDa light chain and has a pI of 6.06. An in vitro study found the toxin in hair roots, sebaceous glands, and arrector pili muscle fibers of Wistar rats after only 10 min of iontophoresis [62].

12.3.2 Cosmeceutical Applications

Iontophoresis can also be used to target the delivery of cosmeceutical active ingredients into the skin, increasing concentrations in the viable epidermis. Short-duration iontophoresis (20 min) was shown to enhance deposition of sodium ascorbyl phosphate (NaAP) porcine skin in vitro from a gel formulation (0.10 ± 0.04 and 0.74 ± 0.27 µg/mg tissue, for iontophoretic and passive treatments, respectively) [63]. The study was performed using an electrotreatment device (**tmt** system™, Mesoestetic Laboratories, Barcelona, Spain), which used a pulse duration of 320 µs and frequency of 1,200 Hz. The system combines the use of a low voltage to generate a constant direct current across the skin with the application of a high-frequency voltage to create microchannels in the skin that would also enhance passive diffusion [63]. In another study, iontophoresis using a commercially available apparatus (Aqua Puff, Toshiba Medical Supply Co., Tokyo Japan) using a pulsed waveform (1 kHz, 60 % duty cycle) at 10 V for 20 s enhanced percutaneous absorption of [14C] ascorbic acid (1 mg/ml) in rat skin, as demonstrated by the increase in radioactivity [64].

Despite the interest from the skin care industry in using iontophoresis for the topical cutaneous delivery of active cosmeceuticals and the positive results demonstrated either in vitro or in animals [63, 64], the criteria used to evaluate efficacy in many clinical studies are subjective, e.g., overall patient satisfaction. Some clinical studies describe the application of iontophoresis for the delivery of an active ingredient to one side of the face, the so-called split-face studies, but the control is solely the formulation without the active ingredient passively applied to the other side. That was the case for studies performed with ascorbic acid in a moisturizing formulation for the treatment of photoaged skin [65] and in solution for the treatment of melasma [66]. Even though in the latter study, the treated site showed a significant decrease in pigmentation, the effect of iontophoresis to improve the delivery of ascorbic acid and its contribution to the efficacy of the treatment could not be determined. The results obtained so far on the use of iontophoresis for cosmetic purposes merit further investigation into the potential of this technique to increase the penetration of other actives used for rejuvenation. Further research on vehicles that could facilitate administration as well as guarantee drug stability is also required.

12.4 Ocular Iontophoresis

Ocular iontophoresis was first reported in 1908 by the German investigator Wirtz, who passed an electric current through electrolyte-saturated cotton sponges placed over the eye globe for the treatment of corneal ulcers, keratitis, and episcleritis [67]. In the last 20 years, there have been an increasing number of studies using this technique for the delivery of antibacterials, antifungals, antimetabolites, corticosteroids, and even macromolecules, including siRNAs, to ocular tissues [20, 68–73].

There are two different approaches for ocular iontophoresis—transcorneal and transscleral—which are used to deliver drugs to the anterior or posterior segments of the eye, respectively (Fig. 12.4). The anterior segment consists of the front one-third of the eye that includes the pupil, cornea, iris, ciliary body, aqueous humor, and lens, while the posterior segment consists of the back two-thirds of the eye—the vitreous humor, retina, choroid, macula, and optic nerve [74].

As with cutaneous iontophoresis, the drug is placed in the electrode compartment bearing the same charge and the return electrode can be placed anywhere on the body to complete the electrical circuit; usually it is positioned on the forehead. Several applicators have been developed in the last few years. For transcorneal iontophoresis, typically either a modified eye cup is filled with the drug solution [75] or a drug-loaded hydrogel pad is used [76]. These systems are also used for transscleral iontophoresis, but the eye cups contain an opening to avoid contact with the cornea. Such annular-shaped probes deliver the drug through the sclera, surrounding the cornea, which is the case of the EyeGate® II system (EyeGate Pharmaceuticals) (Fig. 12.5) [77]. Alternatively, the OcuPhor™ system (Iomed) and Visulex™ system (Aciont Inc.) are lens-shaped devices that are inserted into the inferior cul-de-sac for transscleral iontophoresis [78].

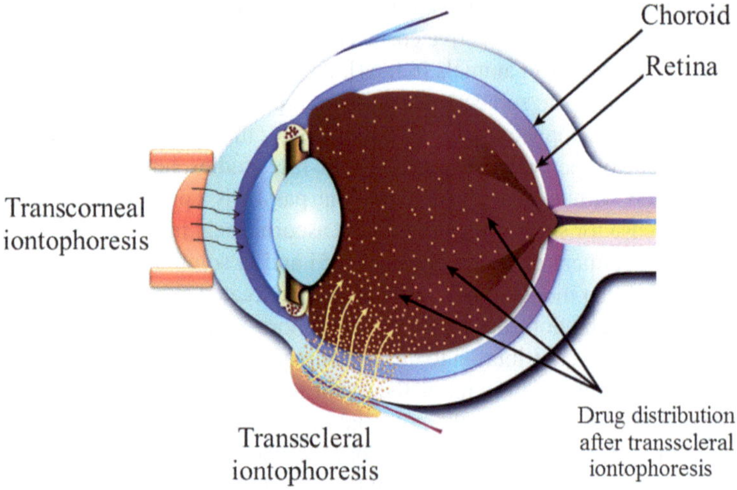

Fig. 12.4 Schematic illustration of transcorneal and transscleral iontophoresis. Adapted from Eljarrat-Binstock and Domb [67]

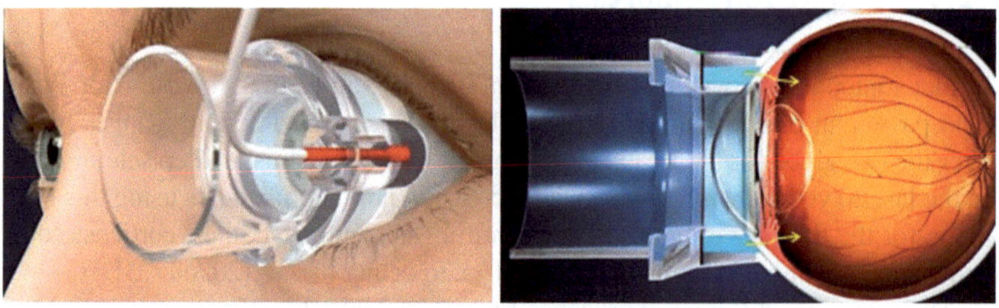

Fig. 12.5 Illustration of EyeGate® II delivery system used for transscleral ocular iontophoresis (EyeGate Pharmaceuticals, Inc., Waltham, MA)

Since the cornea and sclera have a lower resistance than the skin, higher current densities can be tolerated. Transscleral iontophoresis at 5.0 mA/cm^2 for 10 min in rabbits showed no retinal detachment, abnormal histological findings, or other intra-ocular complications [79]. It has been reported in mice that although application of an iontophoretic current density of 5.7 mA/cm^2 for 10 min was acceptable, an increase to 8.5 mA/cm^2 caused some lesions, scleral burns, and corneal vasculariza-tion [80]. Tolerance studies in humans gave similar outcomes to the studies in ani-mals and showed the maximum tolerated iontophoretic current to be 5.5 mA/cm^2 (applied for 20 min). It was shown in the same study that half this current density could be applied for double the time (i.e., 40 min) without clinically significant side effects [78]. Recent clinical studies into transscleral iontophoresis applied a current of up to 3.5 mA for 3 min. Over the course of the study, 87 % of the patients experienced at least one adverse event including ocular hyperemia, keratitis, ocular discomfort, and conjunctival edema. Nevertheless, the majority of adverse effects were reported as being resolved without sequelae within 24 h, and no severe adverse events or effects on visual acuity were observed [77]. In a similar clinical study using the same applicator device and iontophoresis for 4 min for the treatment of anterior uveitis, it was observed that a lower current (1.6 mA) provided greater efficacy, faster outcome, and the fewest number of patients who required rescue therapy. Therefore, although it might be acceptable to use much higher current densities, a clinical evaluation should be performed to optimize the iontophoretic parameters [69]. Although the current density is not provided in the two publications mentioned above, another paper describes the device as a hollow cylinder with inner diameter of 14 mm and active contact surface area between the eye and the applicator of approximately 1 cm^2 [20]; hence, it can be estimated that current densities of ~3.5 mA/cm^2 for 3 min or 4 min were used in the clinical trials.

12.4.1 Transcorneal Iontophoresis

Topical ocular treatments must overcome the challenge of the short residence time of a formulation due to the eye's protective mechanisms, such as lacrimal secretion and blinking reflex. In addition, they must also contend with the impermeability of the cornea, in which the tight junctions present in the epithelium form the major barrier to delivery. Transcorneal iontophoresis has therefore been investigated as a means to provide higher drug permeation through the cornea to target diseases affecting the cornea itself (keratitis, corneal ulcers) as well as the whole anterior segment of the eye (aqueous humor, ciliary body, iris, and lens), with the potential of treating conditions such as dry eye, glaucoma, and ocular inflammations [67].

Macromolecules such as oligonucleotides have been delivered by iontophoresis in proof-of-concept studies. Fluorescence microscopy was used to confirm delivery of labeled siRNA and dextrans of up to 70 kDa into the cornea of mice after iontophoresis for 1 min at 25 mA/cm^2. Even though the current density was extremely high, it was reported that visual examination did not show any corneal damage [75]. Another study reported the delivery of single-stranded cyanine 5 (Cy5)-labeled antisense oligonucleotides targeting vascular endothelial growth factor (VEGF-R2) receptor to all corneal layers in rat eyes using iontophoresis at 0.3 mA for 5 min [81]; the dimensions of the applicator were not reported.

Transcorneal iontophoresis has also been studied for the delivery of small molecules such as ciprofloxacin, commonly used to treat infections in the anterior chamber of the eye [82]. It has been shown that after a short time (5 min) of transcorneal iontophoresis, a drug reservoir is formed, which slowly releases the drug into the aqueous humor, eliciting a sustained therapeutic effect. The study also showed that after iontophoresis for 5 min, the drug concentration in the receiver compartment fluid in in vitro studies (using excised porcine corneas) or in aqueous humor in ex vivo studies (using whole porcine eyes) was not significantly higher than the control (in which electric current was not applied). However, 6–12 h after iontophoresis for 5 min, the concentrations of drug in aqueous humor in ex vivo studies were ~six- and ~fivefold higher than the control (130.12 ± 78.99 ng/ml), respectively, which may be assumed as evidence of a reservoir that slowly releases the drug when iontophoresis is applied. Moreover, it was also observed that increasing the current density from 0.75 mA/cm^2 to 6.25 mA/cm^2 increased the drug loaded in the cornea ~sixfold (~20 and 120 ng/mg, respectively). The application of the higher current did not show any sign of loss of vision and abnormal discharge, redness of eye, or edema [82]. Transcorneal iontophoresis of gentamicin sulfate in rabbits using a current intensity of 1 mA for 60 s showed that peak gentamicin concentrations in the cornea (363.1 ± 127.3 μg/g) and in the aqueous humor (29.4 ± 17.4 μg/ml) were reached at 0 and 2 h after the iontophoretic treatment, respectively. The peak gentamicin concentrations after a single iontophoresis treatment were 12–15 times higher than those obtained after subconjunctival gentamicin injection or after topical eye drop instillation every 5 min for 1 h and much higher than in no-current controls [70]. In a subsequent study using an experimental animal model of Pseudomonas keratitis, it was demonstrated that iontophoresis applied for just 60 s was able to lower the Pseudomonas count in the cornea (the logarithmic values of Pseudomonas colony-forming units (CFUs) were 2.96 ± 0.45 and 6.29 ± 0.45, respectively, for the iontophoretic and the control groups, the latter receiving eye drops of 1.4 % gentamicin every hour for 8 h) [83].

Despite the promising results, transcorneal iontophoresis has not yet been studied in randomized, clinically controlled studies.

12.4.2 *Transscleral Iontophoresis*

Transscleral iontophoresis overcomes the lens–iris barrier and delivers drugs directly into the vitreous and retina through the choroid or indirectly via the systemic circulation or the anterior chamber. The iontophoretic device is placed on the conjunctiva, over the pars-plana area to avoid current damage to the retina [84]. It represents a potential alternative to multiple intravitreal injections or systemic therapy used for posterior ocular disorders, such as endophthalmitis, uveitis, retinitis, optic nerve atrophy, pediatric retinoblastoma, and age-related macular degeneration (AMD) [67]. It has been studied for the delivery of a number of low molecular weight antibiotics—including gentamicin, cephazolin, ticarcillin, amikacin, and vancomycin—and steroids, including dexamethasone and methylprednisolone.

Transscleral delivery of dexamethasone phosphate and dexamethasone has shown promising results in healthy [71] and endotoxin-induced uveitis animal models [85], with short application times, on the order of a few minutes (4 min). Since dexamethasone lacks a charged group and has limited aqueous solubility (0.1 mg/ml), negatively charged micellar carrier systems prepared with taurocholate and egg lecithin have also been developed for its transscleral iontophoretic delivery [86]. Iontophoresis at 10 mA/cm^2 for 20 min using excised human sclera in vitro resulted in higher cumulative amounts in the sclera (from ~80 to 200 μg/cm^2) in comparison with the passive control using the same carrier system (~50 μg/cm^2) [86].

Dexamethasone phosphate, however, seems to be a more suitable candidate for iontophoresis. It is a prodrug of dexamethasone that is readily converted to dexamethasone in vivo. Dexamethasone phosphate possesses 2 acidic protons that allow the production of highly water-soluble formulations of a charged species within the pH range of 5.5–7.4. It has successfully been used in clinical trials for the treatment of dry eye [77] and noninfectious anterior uveitis [69]. Both studies were conducted using the EyeGate® II device (EyeGate Pharmaceuticals) and dexamethasone phosphate solution at 40 mg/ml (EGP-437). Previous studies in rabbits using this device demonstrated that iontophoresis of EGP-437 for 5 min and current densities of 2, 4, and 6 mA/cm^2 achieved significantly higher drug concentrations in the aqueous and vitreous humor (~20 nmol of drug in each) than those obtained with either topical or intravenous administration, but there was no increase in delivery upon further increasing the current density [20].

The penetration of methylprednisolone hemisuccinate into the ocular tissue in rabbits after cathodal iontophoresis at 2.6 mA/cm^2 for a maximum of 10 min was investigated using drug-loaded hydrogels mounted in a portable iontophoretic device. Significantly higher methylprednisolone levels were found 2 h after iontophoresis as compared to the passive control groups. It was reported that 178.59 ± 21.63 μg/g, 6.74 ± 2.38 μg/ml, and 2.71 ± 0.57 μg/ml were found in the retina, aqueous humor, and vitreous, respectively, while nondetectable concentrations were found 2 h after an i.v. infusion of 10 mg/kg of methylprednisolone in all evaluated ocular tissues and fluids [72].

Transscleral iontophoresis has also been investigated for the delivery of high molecular weight model molecules and therapeutics for which the only feasible administration route so far would be intravitreal injections or intraocular implants [19, 68, 87]. Bevacizumab (Avastin) (MW 149 kDa) is an essentially non-charged recombinant humanized IgG1 monoclonal antibody that inhibits angiogenesis by binding with high affinity to human vascular endothelial growth factor. It is used in ophthalmology (off-label) for the treatment of neovascularization in diseases such as diabetic retinopathy and age-related macular degeneration (wet form). Anodal iontophoresis of fluorescein isothiocyanate (FITC)-bevacizumab (2.5 mg/ml) using a current density of 3.8 mA/cm^2 for 2 h across excised human sclera in vitro resulted in a 7.5-fold enhancement of flux over passive permeation (37.01 ± 9.37 and 4.92 ± 6.73 μg/cm^2 h) [87]. Although it was suggested that, based on recent data from human volunteers, bevacizumab might be active at doses as low as 12.5 μg, the main limitation of the technique would be the long application time for iontophoresis to achieve a significant enhancement in drug transport (1 h). Although fluorescence microscopy showed that superior penetration was achieved after 30 min, it is still necessary to consider the dynamic barriers that would be present in vivo; therefore, more experiments are necessary [87].

Indeed, this is true for the majority of the studies performed so far. Despite the promising results obtained with transscleral iontophoresis, a possible limitation of the technique for the treatment of chronic eye diseases is the fast clearance of the drug from the eye to the systemic circulation through the episcleral vessels and the conjunctival lymphatic system, resulting in the need for repeated iontophoretic administration. Also, short-duration transscleral iontophoresis is not expected to deliver drugs directly to the posterior segment of eye due to the long path length from the application site to the back of the eye. As mentioned above, the applicator is normally placed on the conjunctiva, over the pars-plana area to avoid current damage to the retina, and therefore, the ocular dynamic barrier may be all the more relevant for drug delivery. Evidence of this are the huge differences in the amounts of drug delivered in vitro or ex vivo through excised membranes and the amount delivered in vivo [20, 88]. Nevertheless, it is clear that both the physicochemical properties of the drug molecules and iontophoretic parameters such as current density, area available for diffusion, and application time have a major impact on drug penetration and distribution in the vitreous. Each drug and protocol has to be evaluated separately in order to determine its potential in treating different ophthalmic disorders.

12.5 Conclusions

Targeted iontophoresis appears to offer significant benefits for local therapy. It can quickly and effectively deliver significant amounts of drug and active ingredients in a controlled manner to the upper layers of the skin through an intact stratum corneum, without compromising skin physiology. The ability to administer drugs noninvasively is of particular interest in the treatment of ocular diseases since the risk of serious complications and infections may be significantly reduced. The results from animal and clinical studies demonstrate that iontophoresis may enable effective treatment of patients with dermatological and ophthalmic diseases. Therefore, it is hoped that more systems will enter into clinical trials in the coming years.

References

1. Kalia YN, Naik A, Garrison J, Guy RH (2004) Iontophoretic drug delivery. Adv Drug Deliv Rev 56:619–658
2. Mattia C, Coluzzi F, Sonnino D, Anker-Moller E (2010) Efficacy and safety of fentanyl HCl iontophoretic transdermal system compared with morphine intravenous patient-controlled analgesia for postoperative pain management for patient subgroups. Eur J Anaesthesiol 27: 433–440
3. Takasuga S, Yamamoto R, Mafune S, Sutoh C, Kominami K, Yoshida Y, Ito M, Kinoshita M (2011) In-vitro and in-vivo transdermal iontophoretic delivery of tramadol, a centrally acting analgesic. J Pharm Pharmacol 63:1437–1445
4. Cazares-Delgadillo J, Ganem-Rondero A, Quintanar-Guerrero D, Lopez-Castellano AC, Merino V, Kalia YN (2010) Using transdermal iontophoresis to increase granisetron delivery across skin in vitro and in vivo: effect of experimental conditions and a comparison with other enhancement strategies. Eur J Pharm Sci 39:387–393
5. Nair A, Reddy C, Jacob S (2009) Delivery of a classical antihypertensive agent through the skin by chemical enhancers and iontophoresis. Skin Res Technol 15:187–194
6. Cazares-Delgadillo J, Balaguer-Fernandez C, Calatayud-Pascual A, Ganem-Rondero A, Quintanar-Guerrero D, Lopez-Castellano AC, Merino V, Kalia YN (2010) Transdermal

iontophoresis of dexamethasone sodium phosphate in vitro and in vivo: effect of experimental parameters and skin type on drug stability and transport kinetics. Eur J Pharm Biopharm 75:173–178
7. Zhang Q, Grice JE, Wang GJ, Roberts MS (2009) Cutaneous metabolism in transdermal drug delivery. Curr Drug Metab 10:227–235
8. Gratieri T, Kalaria D, Kalia YN (2011) Non-invasive iontophoretic delivery of peptides and proteins across the skin. Expert Opin Drug Deliv 8:645–663
9. Phipps JB, Padmanabhan RV, Lattin GA (1989) Iontophoretic delivery of model inorganic and drug ions. J Pharm Sci 78:365–369
10. Abla N, Naik A, Guy RH, Kalia YN (2005) Contributions of electromigration and electroosmosis to peptide iontophoresis across intact and impaired skin. J Control Release 108:319–330
11. Delgado-Charro MB, Guy RH (2001) Transdermal iontophoresis for controlled drug delivery and non-invasive monitoring. STP Pharma Sci 11:403–414
12. Phipps JB, Gyory JR (1992) Transdermal ion migration. Adv Drug Deliv Rev 9:137–176
13. Sage BH, Riviere JE (1992) Model systems in iontophoresis transport efficacy. Adv Drug Deliv Rev 9:265–287
14. Mudry B, Guy RH, Begona Delgado-Charro M (2006) Prediction of iontophoretic transport across the skin. J Control Release 111:362–367
15. Mudry B, Guy RH, Delgado-Charro MB (2006) Transport numbers in transdermal iontophoresis. Biophys J 90:2822–2830
16. Mudry B, Guy RH, Delgado-Charro MB (2006) Electromigration of ions across the skin: determination and prediction of transport numbers. J Pharm Sci 95:561–569
17. Marro D, Guy RH, Delgado-Charro MB (2001) Characterization of the iontophoretic permselectivity properties of human and pig skin. J Control Release 70:213–217
18. Rojanasakul YY, Robinson JR (1989) Transport mechanisms of the cornea—characterization of barrier permselectivity. Int J Pharm 55:237–246
19. Nicoli S, Ferrari G, Quarta M, Macaluso C, Santi P (2009) In vitro transscleral iontophoresis of high molecular weight neutral compounds. Eur J Pharm Sci 36:486–492
20. Gungor S, Delgado-Charro MB, Ruiz-Perez B, Schubert W, Isom P, Moslemy P, Patane MA, Guy RH (2010) Trans-scleral iontophoretic delivery of low molecular weight therapeutics. J Control Release 147:225–231
21. Pikal MJ, Shah S (1990) Transport mechanisms in iontophoresis.3. An experimental-study of the contributions of electroosmotic flow and permeability change in transport of low and high-molecular-weight solutes. Pharm Res 7:222–229
22. Pikal MJ (1990) Transport mechanisms in iontophoresis.1. A theoretical-model for the effect of electroosmotic flow on flux enhancement in transdermal iontophoresis. Pharm Res 7:118–126
23. Pikal MJ, Shah S (1990) Transport mechanisms in iontophoresis.2. Electroosmotic flow and transference number measurements for hairless mouse skin. Pharm Res 7:213–221
24. Pikal MJ (2001) The role of electroosmotic flow in transdermal iontophoresis. Adv Drug Deliv Rev 46:281–305
25. Pikal MJ (1992) The role of electroosmotic flow in transdermal iontophoresis. Adv Drug Deliv Rev 9:201–237
26. Green PG, Hinz RS, Kim A, Szoka FC Jr, Guy RH (1991) Iontophoretic delivery of a series of tripeptides across the skin in vitro. Pharm Res 8:1121–1127
27. Zempsky WT, Sullivan J, Paulson DM, Hoath SB (2004) Evaluation of a low-dose lidocaine iontophoresis system for topical anesthesia in adults and children: a randomized, controlled trial. Clin Ther 26:1110–1119
28. Su YH, Fang JY (2008) Drug delivery and formulations for the topical treatment of psoriasis. Expert Opin Drug Deliv 5:235–249
29. Chase EP, Armstrong AW (2012) Advances in management of atopic dermatitis: new therapies and novel uses of existing treatments. Semin Cutan Med Surg 31(1):17–24
30. Senyigit T, Padula C, Ozer O, Santi P (2009) Different approaches for improving skin accumulation of topical corticosteroids. Int J Pharm 380:155–160

31. Alvarez-Figueroa MJ, Delgado-Charro MB, Blanco-Mendez J (2001) Passive and iontophoretic transdermal penetration of methotrexate. Int J Pharm 212:101–107
32. Prasad R, Anand S, Khar RK, Dinda AK, Koul V (2009) Studies on in vitro and in vivo transdermal flux enhancement of methotrexate by a combinational approach in comparison to oral delivery. Drug Dev Ind Pharm 35:1281–1292
33. Alvarez-Figueroa MJ, Blanco-Mendez J (2001) Transdermal delivery of methotrexate: iontophoretic delivery from hydrogels and passive delivery from microemulsions. Int J Pharm 215:57–65
34. Tiwari SB, Kumar BCR, Udupa N, Balachandran C (2003) Topical methotrexate delivered by iontophoresis in the treatment of recalcitrant psoriasis—a case report. Int J Dermatol 42:157–159
35. Gratieri T, Wagner B, Kalaria D, Ernst B, Kalia YN (2012) Cutaneous iontophoretic delivery of CGP69669A, a sialyl Lewisx mimetic, in vitro. Exp Dermatol 21:126–128
36. Abu Hashim II, Motoyama K, Abd-ElGawad AEH, El Shabouri MH, Borg TM, Arima H (2010) Potential use of iontophoresis for transdermal delivery of NF-kappa B decoy oligonucleotides. Int J Pharm 393:127–134
37. Kigasawa K, Kajimoto K, Hama S, Saito A, Kanamura K, Kogure K (2010) Noninvasive delivery of siRNA into the epidermis by iontophoresis using an atopic dermatitis-like model rat. Int J Pharm 383:157–160
38. Merino V, Lopez A, Kalia YN, Guy RH (1999) Electrorepulsion versus electroosmosis: effect of pH on the iontophoretic flux of 5-fluorouracil. Pharm Res 16:758–761
39. Bacro TRH, Holladay EB, Stith MJ, Maize JC, Smith CM (2000) Iontophoresis treatment of basal cell carcinoma with cisplatin: a case report. Cancer Detect Prev 24:610–619
40. Taveira SF, Nomizo A, Lopez RFV (2009) Effect of the iontophoresis of a chitosan gel on doxorubicin skin penetration and cytotoxicity. J Control Release 134:35–40
41. Rhodes LE, Tsoukas MM, Anderson RR, Kollias N (1997) Iontophoretic delivery of ALA provides a quantitative model for ALA pharmacokinetics and PpIX phototoxicity in human skin. J Invest Dermatol 108:87–91
42. Bodde HE, Roemele PE, Star WM (2002) Quantification of topically delivered 5-aminolevulinic acid by lontophoresis across ex vivo human stratum corneum. Photochem Photobiol 75:418–423
43. Lesar A, Ferguson J, Moseley H (2009) A time course investigation of the fluorescence induced by topical application of 5-aminolevulinic acid and methyl aminolevulinate on normal human skin. Photodermatol Photoimmunol Photomed 25:191–195
44. Lopez RF, Bentley MV, Delgado-Charro MB, Guy RH (2001) Iontophoretic delivery of 5-aminolevulinic acid (ALA): effect of pH. Pharm Res 18:311–315
45. Lopez RF, Bentley MV, Begona Delgado-Charro M, Guy RH (2003) Optimization of aminolevulinic acid delivery by iontophoresis. J Control Release 88:65–70
46. Lopez RF, Bentley MV, Delgado-Charro MB, Salomon D, van den Bergh H, Lange N, Guy RH (2003) Enhanced delivery of 5-aminolevulinic acid esters by iontophoresis in vitro. Photochem Photobiol 77:304–308
47. Gelfuso GM, Figueiredo FV, Gratieri T, Lopez RF (2008) The effects of pH and ionic strength on topical delivery of a negatively charged porphyrin (TPPS(4)). J Pharm Sci 97:4249–4257
48. Gelfuso GM, Gratieri T, Souza JG, Thomazine JA, Lopez RFV (2011) The influence of positive or negative charges in the passive and iontophoretic skin penetration of porphyrins used in photodynamic therapy. Eur J Pharm Biopharm 77:249–256
49. Souza JG, Gelfuso GM, Simao PS, Borges AC, Lopez RFV (2011) Iontophoretic transport of zinc phthalocyanine tetrasulfonic acid as a tool to improve drug topical delivery. Anticancer Drugs 22:783–793
50. Taveira SF, Lopez RFV (2011) Topical administration of anticancer drugs for skin cancer treatment. In: La Porta CAM (ed) Skin cancers—risk factors, prevention and therapy, 1st edn. InTech, Croatia, pp 247–272
51. Shukla C, Friden P, Juluru R, Stagni G (2009) In vivo quantification of acyclovir exposure in the dermis following iontophoresis of semisolid formulations. J Pharm Sci 98:917–925

52. Morrel EM, Spruance SL, Goldberg DI (2006) Topical iontophoretic administration of acyclovir for the episodic treatment of herpes labialis: a randomized, double-blind, placebo-controlled, clinic-initiated trial. Clin Infect Dis 43:460–467
53. Spruance SL, Nett R, Marbury T, Wolff R, Johnson J, Spaulding T (2002) Acyclovir cream for treatment of herpes simplex labialis: results of two randomized, double-blind, vehicle-controlled, multicenter clinical trials. Antimicrob Agents Chemother 46:2238–2243
54. Siddoju S, Sachdeva V, Friden PM, Yu YY, Banga AK (2011) Acyclovir skin depot characterization following in vivo iontophoretic delivery. Skin Res Technol 17:234–244
55. Abla N, Naik A, Guy RH, Kalia YN (2006) Topical iontophoresis of valaciclovir hydrochloride improves cutaneous aciclovir delivery. Pharm Res 23:1842–1849
56. Walling HW, Swick BL (2011) Treatment options for hyperhidrosis. Am J Clin Dermatol 12:285–295
57. Ohshima Y, Shimizu H, Yanagishita T, Watanabe D, Tamada Y, Sugenoya J, Tsuda T, Matsumoto Y (2008) Changes in Na(+), K(+) concentrations in perspiration and perspiration volume with alternating current iontophoresis in palmoplantar hyperhidrosis patients. Arch Dermatol Res 300:595–600
58. Kalantari KK, Zeinalzade A, Kobarfard F, Moghadam SN (2011) The effect and persistency of 1% aluminum chloride hexahydrate iontophoresis in the treatment of primary palmar hyperhidrosis. Iran J Pharm Res 10:641–645
59. Kavanagh GM, Oh C, Shams K (2004) BOTOX (R) delivery by iontophoresis. Br J Dermatol 151:1093–1095
60. Kavanagh GM, Shams K (2005) Delivery of botox (R) by iontophoresis: reply from authors. Br J Dermatol 153:1076
61. Woolery-Lloyd H, Elsaie ML, Avashia N (2008) Inguinal hyperhidrosis misdiagnosed as urinary incontinence: treatment with botulinum toxin A. J Drugs Dermatol 7:293–295
62. Pacini S, Gulisano M, Punzi T, Ruggiero M (2007) Transdermal delivery of clostridium botulinum toxin type A by pulsed current iontophoresis. J Am Acad Dermatol 57:1097–1099
63. Marra F, Levy JL, Santi P, Kalia YN (2008) In vitro evaluation of the effect of electrotreatment on skin permeability. J Cosmet Dermatol 7:105–111
64. Ebihara M, Akiyama M, Ohnishi Y, Tajima S, Komata K, Mitsui Y (2003) Iontophoresis promotes percutaneous absorption of L-ascorbic acid in rat skin. J Dermatol Sci 32:217–222
65. Xu TH, Chen JZS, Li YH, Wu Y, Luo YJ, Gao XH, Chen HD (2012) Split-face study of topical 23.8% L-ascorbic acid serum in treating photo-aged skin. J Drugs Dermatol 11:51–56
66. Huh CH, Seo KI, Park JY, Lim JG, Eun HC, Park KC (2003) A randomized, double-blind, placebo-controlled trial of vitamin C lontophoresis in melasma. Dermatology 206:316–320
67. Eljarrat-Binstock E, Domb AJ (2006) Iontophoresis: a non-invasive ocular drug delivery. J Control Release 110:479–489
68. Chopra P, Hao JS, Li SK (2010) Iontophoretic transport of charged macromolecules across human sclera. Int J Pharm 388:107–113
69. Cohen AE, Assang C, Patane MA, From S, Korenfel M (2012) Evaluation of dexamethasone phosphate delivered by ocular iontophoresis for treating noninfectious anterior uveitis. Ophthalmology 119:66–73
70. Eljarrat-Binstock E, Raiskup F, Stepensky D, Domb AJ, Frucht-Pery J (2004) Delivery of gentamicin to the rabbit eye by drug-loaded hydrogel iontophoresis. Invest Ophthalmol Vis Sci 45:2543–2548
71. Eljarrat-Binstock E, Raiskup F, Frucht-Pery J, Domb AJ (2005) Transcorneal and transscleral iontophoresis of dexamethasone phosphate using drug loaded hydrogel. J Control Release 106:386–390
72. Eljarrat-Binstock E, Orucov F, Frucht-Pery J, Pe'er J, Domb AJ (2008) Methylprednisolone delivery to the back of the eye using hydrogel iontophoresis. J Ocul Pharmacol Ther 24:344–350
73. Gratieri T, Gelfuso GM, Thomazini JA, Lopez RFV (2010) Excised porcine cornea integrity evaluation in an in vitro model of iontophoretic ocular research. Ophthalmic Res 43: 208–216

74. Gaudana R, Jwala J, Boddu SHS, Mitra AK (2009) Recent perspectives in ocular drug delivery. Pharm Res 26:1197–1216
75. Hao JS, Li SK, Liu CY, Kao WWY (2009) Electrically assisted delivery of macromolecules into the corneal epithelium. Exp Eye Res 89:934–941
76. Eljarrat-Binstock E, Raiskup F, Frucht-Pery J, Domb AJ (2004) Hydrogel probe for iontophoresis drug delivery to the eye. J Biomater Sci Polym Ed 15:397–413
77. Patane MA, Cohen A, From S, Torkildsen G, Welch D, Ousler GW (2012) Ocular iontophoresis of EGP-437 (dexamethasone phosphate) in dry eye patients: results of a randomized clinical trial. Clin Ophthalmol 5:633–643
78. Parkinson TM, Ferguson E, Febbraro S, Bakhtyari A, King M, Mundasad M (2003) Tolerance of ocular iontophoresis in healthy volunteers. J Ocul Pharmacol Ther 19:145–151
79. Kralinger MT, Voigt M, Kieselbach GF, Hamasaki D, Hayden BC, Parel JM (2003) Ocular delivery of acetylsalicylic acid by repetitive Coulomb-controlled iontophoresis. Ophthalmic Res 35:102–110
80. Souied EH, Reid SNM, Piri NI, Lerner LE, Nusinowitz S, Farber DB (2008) Non-invasive gene transfer by iontophoresis for therapy of an inherited retinal degeneration. Exp Eye Res 87:168–175
81. Berdugo M, Valamanesh F, Andrieu C, Klein C, BenEzra D, Courtois Y, Behar-Cohen F (2003) Delivery of antisense oligonucleotide to the cornea by iontophoresis. Antisense Nucleic Acid Drug Dev 13:107–114
82. Vaka SRK, Sammeta SM, Day LB, Murthy SN (2008) Transcorneal iontophoresis for delivery of ciprofloxacin hydrochloride. Curr Eye Res 33:661–667
83. Frucht-Pery J, Raiskup F, Mechoulam H, Shapiro M, Eljarrat-Binstock E, Domb A (2006) Iontophoretic treatment of experimental pseudomonas keratitis in rabbit eyes using gentamicin-loaded hydrogels. Cornea 25:1182–1186
84. Eljarrat-Binstock E, Pe'er J, Domb AJ (2010) New techniques for drug delivery to the posterior eye segment. Pharm Res 27:530–543
85. BeharCohen FF, Parel JM, Pouliquen Y, ThillayeGoldenberg B, Goureau O, Heydolph S, Courtois Y, DeKozak Y (1997) Iontophoresis of dexamethasone in the treatment of endotoxin-induced-uveitis in rats. Exp Eye Res 65:533–545
86. Chopra P, Hao J, Li SK (2012) Sustained release micellar carrier systems for iontophoretic transport of Dexamethasone across human sclera. J Control Release 160:96–104
87. Pescina S, Ferrari G, Govoni P, Macaluso C, Padula C, Santi P, Nicoli S (2010) In-vitro permeation of bevacizumab through human sclera: effect of iontophoresis application. J Pharm Pharmacol 62:1189–1194
88. Molokhia S, Jeong EK, Higuchi W, Li K (2008) Examination of barriers and barrier alteration in transscleral iontophoresis. J Pharm Sci 97:831–844

Chapter 13
Local Drug Delivery to the Oral Cavity

Sheeba Qureshi, Wahid Khan, Michael Perez-Davidi, Ervin I Weiss, Nurit Beyth, and Abraham J. Domb

13.1 Introduction

The human mouth is a unique organ that hosts various functions for daily life. It is the portal to the gastrointestinal tract, responsible for chewing and swallowing food materials. It also coordinates with pharyngeal and laryngeal structures to generate various sounds for communication through language. For such diverse functions, oral cavity is equipped with heterogeneous tissue types such as teeth, tongue, lips, palate, gingiva, and vestibule. Regarding the microbial flora, oral cavity is one of the most densely populated sites of the human body and more than

S. Qureshi
Braun School of Public Health, The Hebrew University of Jerusalem, Jerusalem 91120, Israel

W. Khan
Department of Pharmaceutics, National Institute of Pharmaceutical Education and Research, Hyderabad 500037, India

M. Perez-Davidi • E.I. Weiss
Department of Prosthodontics, Hadassah School of Dental Medicine, Hebrew University, Jerusalem 91120, Israel

N. Beyth
Department of Prosthodontics, Hadassah School of Dental Medicine, The Hebrew University—Hadassah, Jerusalem 91120, Israel

Faculty of Dentistry, The Hebrew University—Hadassah, Jerusalem 91120, Israel

A.J. Domb (✉)
Faculty of Medicine, Institute of Drug Research, School of Pharmacy, The Hebrew University of Jerusalem, Jerusalem 91120, Israel
e-mail: avid@ekmd.huji.ac.il

A.J. Domb and W. Khan (eds.), *Focal Controlled Drug Delivery*, Advances in Delivery Science and Technology, DOI 10.1007/978-1-4614-9434-8_13,

500 isolated microorganism species. These microorganisms colonize oral surfaces where they form a microbial consortium referred as dental plaque or oral biofilm [1]. Due to its unique anatomic location and structure, the mouth is the home for diseases such as dental caries, gingivitis, periodontitis, oral ulcers, aphthous stomatitis (canker sores), oral mucositis, oral herpes (cold sores), oropharyngeal candidiasis, and vesiculobullous mucosal disease. There has been tremendous progress during the past few decades in improving treatment of these diseases/conditions.

Oral route is a favored site for the administration of drugs for patients and clinicians. The drug delivery in the oral cavity can be broadly divided into three categories: (a) sublingual, (b) buccal, and (c) local/topical drug delivery. In this chapter, we have discussed local drug delivery methods available to treat localized oral lesions and oral manifestations of systemic diseases. The main advantage of local/topical drug delivery is the ability to deliver the bioactive agent directly to the site; on the other hand, combating harsh physiological conditions and poor retention and maintaining required concentration of drug at the site are the major concerns for local/topical drug delivery. It is primarily due to the limited interaction of the drug with the host epithelium and mechanical removal due to the action of body fluids, mucosal turnover, and dislodging forces during processes such as chewing, drinking, and swallowing.

13.2 Requirements for Local Drug Delivery to the Oral Cavity

Retention of drug at the site/lesions with proper doses and duration to achieve therapeutic efficacy and with limited side effects are the major concerns for local drug delivery to the oral cavity. Oral cavity has been an interesting potential site for the drug delivery since past, but with the evolution of permeability enhancers, polymers, bioadhesive/mucoadhesive systems, etc., it is being explored more than ever and presents a possible method by which the retention of dosage forms at the site of application may be enhanced [2]. Formulations that are designed for the topical/local use in the oral cavity:

- Should be easy to administer and retention at the site of application for desired period of time
- Should offer controlled release of drug at the site
- Should release drug in an unidirectional way toward the mucosa or multidirectionally in saliva
- Should not cause any irritation or inconvenience to the patient
- Should not interfere with the normal physiological functions

13.3 Formulations for Local Drug Delivery to the Oral Cavity

The formulations for topical/local drug delivery are available in the form of (a) solids (tablets, troches, wafers, lozenges, and microparticles), (b) semisolids (gels, pastes, films, and patches), and (c) liquids (sprays, solutions, mouthwashes). The formulations for topical/local drug delivery are available in conventional, mucoadhesive, and other forms. Buccal mucoadhesive formulations are alternative to the conventional oral medications as they can be readily attached to the buccal cavity retained for a longer period of time and removed at any time. Buccal adhesive drug delivery systems includes matrix tablets, films, layered systems, discs, microspheres, ointments, and hydrogel systems [2, 3].

13.3.1 Solid Formulations

Solid formulations such as tablets and lozenges dissolve into the saliva utilizing the whole surface area of the oral cavity for local application and absorption. In general, they are used for oropharyngeal candidiasis, mucositis, xerostomia, etc. Drawbacks of tablets and lozenges include variation due to differences in saliva production and sucking intensity, accidental swallowing, and short exposure time. Mucoadhesive tablet formulations are better in this respect as they can be readily attached to the buccal cavity retained for a longer period of time and removed at any time. With this, they adhere to the mucosa and hence increase the exposure time [3].

13.3.1.1 Tablets

Several conventional and bioadhesive tablet formulations are developed for local or systemic drug delivery. While conventional tablets are sucked upon, the bioadhesive tablets are placed directly onto the mucosal surface and adhere to the buccal mucosa in the presence of saliva. Bioadhesive tablets are designed to release the drug either unidirectionally targeting buccal mucosa or multidirectionally into the saliva. Although tablets have been demonstrated excellent bioadhesiveness, their size is a major limitation for tablets due to the requirement for the dosage form to have intimate contact with the mucosal surface, e.g., OraMoist® (Fig. 13.1), Salix®, and Corlan® pellets. OraMoist® is a timed-release oral disc containing xylitol and enzymes that adheres to roof of the mouth and has a moisturizing effect for about 4 h [4], whereas Salix® is a saliva-stimulating lozenge containing sorbitol, malic acid, and sodium citrate. Corlan is a mucoadhesive buccal tablet containing hydrocortisone 2.5 mg and acacia gum as a bioadhesive polymer [5].

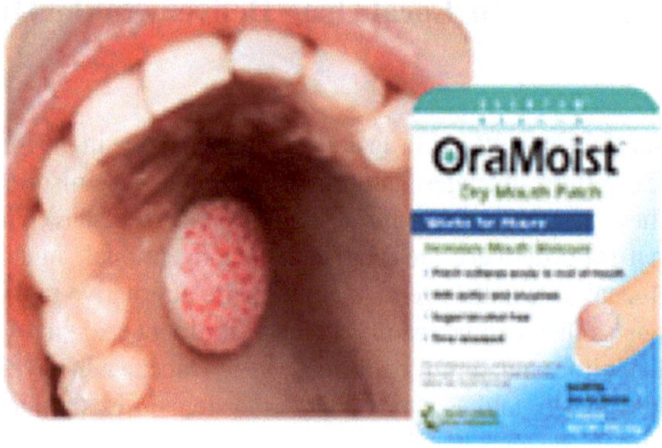

Fig. 13.1 OraMoist® mucoadhesive tablet for dry mouth

13.3.1.2 Microparticles/Microspheres

Over the last 25 years, research on microparticles has surfaced to attain the unmet needs in drug delivery. Microspheres are defined as homogeneous, monolithic particles in the size range of about 1–1,000 μm and are widely used as drug carrier for controlled-release formulations. They have greater encapsulation efficiency and greater stability. Microparticles can be made from biocompatible and biodegradable materials. Bioadhesive microparticles offer the same advantages as tablets in terms of drug delivery and adhesiveness, but their physical properties enable them to make intimate contact with a lager mucosal surface area. The small size of microparticles compared with tablets makes them less likely to cause local irritation at the site of adhesion and the uncomfortable sensation of a foreign object within the oral cavity. ARESTIN® is one such formulation for the treatment for periodontitis and contains microspheres that are filled with the minocycline hydrochloride.

13.3.1.3 Wafers

Wafers are also available as a periodontal drug delivery system for the treatment of microbial infections associated with periodontitis. The delivery system is a composite wafer with surface layers possessing adhesive properties, while the bulk layer consists of antimicrobial agents, biodegradable polymers, and matrix polymers, e.g., PerioChip® [6–8].

13.3.1.4 Lozenges

Lozenges are also used for the delivery of drugs that act topically within the mouth including antimicrobials, corticosteroids, local anesthetics, antibiotics, and antifungals.

Conventional lozenges produce a high initial release of drug in the oral cavity, which rapidly declines to subtherapeutic levels; thus, multiple daily dosing is required. Slow-release bioadhesive lozenge is available that offers the potential for prolonged drug release with improved patient compliance. Researchers have investigated bioadhesive lozenges as a means to deliver antifungal agents to the oral cavity, e.g., Mycelex® Troche, containing 10 mg clotrimazole, a synthetic antifungal agent for topical use in the mouth [9, 10].

13.3.2 Semisolid Dosage Forms

Gels, films, and patches are common forms of semisolid dosage forms for drug delivery in the oral cavity.

13.3.2.1 Gels

Gels have been widely used in the delivery of drugs to the oral cavity. These are mainly used as protecting barriers and drug vehicles for the damaged mucosa from applied stresses. Advantages of gel formulations include their ability to form intimate contact with the mucosal membrane and their rapid release of drug at the absorption site. A limitation of gel formulations lies on their inability to deliver a measured dose of drug to the site. They are therefore of limited use for drugs with narrow therapeutic window. Gel-forming bioadhesive polymers include cross-linked polyacrylic acid that has been used to adhere to mucosal surfaces for extended periods of time and provide controlled release of drugs [2].

13.3.2.2 Patches and Films

Flexible films may be used to deliver drugs directly to a mucosal membrane. They also offer advantages over creams and ointments in that they provide a measured dose of drug to the site. Buccal adhesive films are already in use commercially, for example, Zilactin used for the therapy of canker sores, cold sores, and lip sores [11, 12].

13.3.3 Liquid Dosage Forms

Various solutions in the form of solutions and mouthwashes are available for the use in the oral cavity for the treatment of halitosis, dentinal hypersensitivity, plaque inhibitors, etc. Viscous liquids are also used to coat buccal surface either as protectants or as drug vehicles for delivery to the mucosal surface [13–15]. Gel formulations have also been widely used in drug delivery to oral cavity. Advantages of gel formulations include their ability to form intimate contact with the mucosal membrane and their rapid release of drug at the absorption site. Corsodyl dental gel containing 1 % chlorhexidine digluconate in hydroxypropyl methylcellulose is used in the management of gingival and periodontal disease. Traditionally, pharmaceutically acceptable polymers were used to enhance the viscosity of products to aid their retention in the oral cavity. Dry mouth is treated with artificial saliva solutions that are retained on mucosal surfaces to provide lubrication [2]. Mouthwashes and oral rinses predominantly focus on their use in the local delivery of antimicrobial agents.

13.4 Disease-Specific Local Drug Delivery to the Oral Cavity

13.4.1 Periodontitis

The probabilities of some oral conditions are very high even in an otherwise healthy patient, e.g., gingivitis and periodontitis [16–19]. Periodontal disease is more severe condition in which the bone and the periodontal ligament are also affected. Periodontal diseases are a general term which encompasses several pathological conditions affecting the tooth supporting structures, characterized by destruction of the periodontal ligament, resorption of the alveolar bone, and the migration of the junctional epithelium along the tooth surface. The clinical signs of periodontitis are changes in the morphology of gingival tissues, bleeding upon probing, as well as periodontal pocket formation, which provides an ideal environment for the growth and proliferation of anaerobic pathogenic bacteria [20]. Periodontal diseases include conditions such as chronic periodontitis, aggressive periodontitis, systemic disease-associated periodontitis, and necrotizing periodontitis [1]. The treatment of this condition is complex and needs a multidirectional treatment approach, and it is usually treated with both drugs and surgical means. Systemic antimicrobial therapy has been advocated for the treatment of severe forms of periodontitis. However, side effects including hypersensitivity, gastrointestinal intolerance, and the development of bacterial resistance are also a matter of concern. There are several systemic and local formulations available for the treatment of periodontitis, but controlled drug delivery fits well into the treatment philosophy as it seeks to deliver sustained high concentrations of antimicrobial agents into the periodontal pocket at a fraction of the dose required for the systemic drug administration. Table 13.1 presents some of the commercially available local drug delivery products for the treatment of periodontitis.

Tetracycline fibers were the first commercial product as controlled-release formulation for the treatment of periodontitis (in 1994). Actisite® treatment is a 10-day site-specific course of antibiotic periodontal therapy, indicated as an adjunctive therapy to scaling and root planing for reduction of pocket depth and bleeding on probing in patients with adult periodontitis. Actisite® periodontal fiber is made up of

Table 13.1 Products in clinical use for periodontitis

Active agent	Product	Description	References
Tetracycline fiber	Actisite® (25 % tetracycline HCl)	Nonresorbable fiber	[21, 22]
	Periodontal Plus AB™	Resorbable fiber	[23]
	PerioCol-TC	Resorbable fiber	
Chlorhexidine chip	PerioChip® (2.5 mg)	Biodegradable chip	[6–8, 24, 25]
	PerioCol-CG (2.5 mg)	Biodegradable chip	[26, 27]
	Chlo-Site (1.5 % CHX)	Biodegradable gel	[28]
Doxycycline polymer	ATRIDOX® (10 %)	Biodegradable powder in syringe system	[29]
Minocycline	Dentomycin (2 %)	Biodegradable gel	[30–32]
	Periocline (2.1 %)	Biodegradable gel	[33–36]
	ARESTIN®	Biodegradable microspheres in syringe system	[37–40]
Metronidazole gel	Elyzol (25 %)	Biodegradable gel	[32, 41, 42]

biologically inert, nonresorbable flexible plastic copolymer of ethylene and vinyl acetate impregnated with 12.7 mg tetracycline hydrochloride, which is released continuously over a period of 10 days. In Actisite® treatment, the fiber is placed into the affected gingival pocket and secured by an adhesive which needs removal after 10 days. Actisite® delivers a continuous local concentration of tetracycline to the individual pockets for a specific therapeutic length of time, with little systemic exposure [20, 43].

As an alternative to Actisite®; Periodontal Plus AB™ and PerioCol-TC have also been introduced for the treatment of gingival and periodontal diseases as new biodegradable local drug delivery systems. These are tetracycline-impregnated fibrillar collagen that contains tetracycline hydrochloride in collagen fibrils. It is biodegradable and hence does not require removal and has shown to release tetracycline for a period of 8–12 days in vitro [44, 45].

Doxycycline is a broad-spectrum antibiotic synthetically derived from oxytetracycline. It is marketed as ATRIDOX® (doxycycline hyclate). This is a subgingival controlled-release product composed of a two-syringe mixing system. Syringe A contains 450 mg of the ATRIGEL® Delivery System, which is a bioabsorbable, flowable polymeric formulation composed of 36.7 % poly(D,L-lactide) (PLGA) dissolved in 63.3 % *N*-methyl-2-pyrrolidone. Syringe B contains 50 mg of doxycycline hyclate, which is equivalent to 42.5 mg doxycycline. The reconstituted product is a pale yellow to yellow viscous liquid with a concentration of 10 % doxycycline hyclate. Upon contact with the crevicular fluid, the liquid product solidifies and then allows for controlled release of the drug for a period of 7 days. However, during medication, the patient may be more sensitive to the sun; hence prolonged sun exposure, tanning booths, and sunlamps should be avoided [20, 45].

Another means of treating mild periodontal lesions is the use of PerioChip® (chlorhexidine gluconate). This is a small, orange–brown, rectangular chip (rounded at one end) for insertion into periodontal pockets. Each PerioChip® weighs about 6.9 mg and contains 2.5 mg of chlorhexidine gluconate in a biodegradable matrix of hydrolyzed gelatin (cross-linked with glutaraldehyde). PerioChip® also contains glycerin and purified water [46]. PerioChip® has been shown to be an effective and safe adjunctive treatment for reduction of pocket depth in patients with adult periodontitis with scaling and root planing. PerioChip® is recommended for use in periodontal pockets where the pocket depth is ≥5 mm. PerioChip® may be used as a part of a periodontal maintenance program, which includes good oral hygiene and scaling and root planing. Once in place, PerioChip® dissolves naturally in 7–10 days and does not need removal. PerioChip® releases chlorhexidine in vitro in a biphasic manner, releasing about 40 % of the molecule within the first 24 h and the remainder, in an almost linear fashion, for 7–10 days. The concentration of chlorhexidine released from the PerioChip® can be determined in the gingival crevicular fluid. Clinical results vary among patients, but clinical evidence has shown significant reduction of probing pocket depth, compared with those treated with scaling and root planing alone, at 9 months after initial treatment [20]. Other subgingival chlorhexidine products available in the market are PerioCol-CG (chip) and Chlo-Site (1.5 % chlorhexidine gel) [45].

ARESTIN® microspheres is a subgingival sustained-release product containing the antibiotic minocycline hydrochloride incorporated into a bioresorbable polymer, poly(glycolide-co-dl-lactide) or PGLA, for professional subgingival administration into periodontal pockets for the treatments for periodontitis. Each unit dose cartridge delivers minocycline hydrochloride equivalent to 1 mg of minocycline free base. These microspheres release the antibiotic over time. ARESTIN® is indicated as an adjunct to scaling and root planing procedures for reduction of pocket depth in patients with adult periodontitis. It may be used as part of a periodontal maintenance program, which includes good oral hygiene and scaling and root planing. Its application is painless and it is completely resorbed over time. Since it is bioadhesive preparation, it does not require any retention dressings [45, 47].

Some sustained-release gel formulations containing 2 % minocycline have also been commercialized such as Dentomycin® and Periocline®. These require multiple applications (3–4 applications every 2 weeks) [20]. Dentomycin is a bioabsorbable system which contains 2 % minocycline HCl in a matrix of hydroxyethyl cellulose, aminoalkyl methacrylate, triacetin, and glycerin in the form of a gel. Magnesium chloride is added to modify the drug release properties. It is categorized as a sustained-release system and provides high drug concentration, subgingivally, for 24 h [48]. A topical medication Elyzol® contains an oil-based metronidazole 25 % dental gel (glyceryl mono-oleate and sesame oil). It is applied in viscous consistency to the pocket, where it is liquidized by the body heat and then hardens again by forming crystals in contact with water. The vehicle biodegrades after the active drug has been released over a period of short time [45]. There is also a strip which is pre-impregnated with metronidazole for placement in pockets [20, 45].

13.4.2 Oral Ulcers

Mouth ulcers are small painful sores that form in the mouth [49, 50]. Most people have at least one attack of mouth ulcers in their lifetime, but mouth ulcers are more common in women and those under the age of 40. Up to one in five people has repeated attacks of mouth ulcers. More than a third of people with recurrent mouth ulcers have a family history of the sores. This figure rises to over 80 % if both parents suffer from frequent mouth ulcers.

There are three main types of aphthous ulcers: (a) minor ulcers are small and non-scarring, usually heal without any treatment within 2 weeks. Minor ulcers are the most common type of ulcers and constitute about 75 % of all mouth ulcers. (b) Major ulcers are 1 cm in diameter or larger; usually only one or two appear at a time. They can be very painful, cause difficulty in eating, heal slowly, and may leave scars. They have raised borders and may last for 2 weeks to several months, about 10 % of the ulcers are of this type, and (c) herpetiform ulcers are multiple tiny ulcers that can be very painful particularly if they fuse to form a large ulcer that can last for 1 week to 2 months. Despite their name, they have nothing to do with the herpes virus. There are also other similar conditions that cause problems in the mouth such as stomatitis (sore mouth) and yeast, bacterial, or viral infections.

Minor mouth ulcers are usually caused by damage to the mouth from accidental biting of the cheek, vigorous tooth brushing, sharp teeth, or fillings. Although there may be no evident reason for recurring ulcers, factors that can increase the risk include trauma due to excessive brushing or chewing of hard food, anxiety and stress, and certain foods such as chocolate, coffee, nuts, strawberries, and cheese. Many women also experience ulcers during their menstrual period. Quitting smoking can initially trigger mouth ulcers in the first few weeks after withdrawal, but this then tends to settle down. Certain medical conditions can also make mouth ulcers more likely such as vitamin B_{12} deficiency, viral infections, iron deficiency, celiac disease (intolerance to a protein called gluten, found in wheat, rye, and barley), Crohn's disease, Reiter's syndrome (urethral discharge followed by conjunctivitis and arthritis), and HIV infection. Occasionally, ulcers can be due to medication such as painkillers, such as ibuprofen and aspirin, nicorandil, and beta-blockers that are used in heart conditions [51].

Usually protective gels or oral pastes containing local anesthetics are used to form a protective covering over ulcers to promote healing and reduce the pain sensation for minor ulcers. ORABASE paste contains adrenocorticoid, hydrocortisone acetate in base composed of gelatin, pectin, and sodium carboxymethylcellulose in Plastibase (plasticized hydrocarbon gel) [52]. By virtue of its superior adhesive properties, ORABASE® paste adheres tenaciously and remains in intimate contact with mucous membranes of the mouth and gums, protecting the afflicted area in the mouth against further irritation from chewing, swallowing, and other normal mouth activity. It acts like an invisible bandage [53].

Triamcinolone belongs to the class of corticosteroids and used as adjunctive treatment for temporary relief of symptoms associated with oral inflammatory lesions and ulcerative lesions resulting from trauma. Marketed products are Oracort and Oralone® (triamcinolone acetonide dental paste, USP 0.1 %); when applied directly to the affected areas, it decreases inflammation and protect the damaged mucosa from further injury. Oracort E additionally contains 3 % lidocaine to further diminish the pain sensation due to ulcers. The ointment is adhesive and attaches itself to the mucosa, prolonging the action.

One of the products used to treat major ulcers or aphthous ulcers is Canker Cover™. The tablet-like patch contains herbal ingredients to soothe the sore. This is an adhesive patch that acts as a protective cover over the ulcer and promotes healing. It attaches to the ulcer and remains in place for 8–10 h, and then it liquefies to a gel. A published clinical study has showed that the Canker Cover oral patch healed 78 % of the participant's sores within 8–12 h (one patch), and the average overall healing time was 1.5 days, compared to a typical canker sore healing time of up to 14 days [54]. Some other marketed products for the treatment of canker sores are Cankermelts Discs, Canker-Rid®, Zilactin-B Canker Sore Gel, Canker-X, etc.

13.4.3 Cold Sores/Herpes Labialis

Approximately 20 and 40 % of the population is estimated to suffer from episodes of recurrent herpes labialis [55]. It is infection of the lip by herpes simplex virus (HSV-1). Herpes labialis infection occurs when the HSV-1 comes into contact with oral mucosal tissue or abraded skin of the mouth. Infection by the type 1 strain of HSV-1 is most common; however, cases of oral infection by the type 2 strain are increasing. An outbreak typically causes small blisters or sores on or around the mouth commonly known as cold sores or fever blisters. The sores typically heal within 2–3 weeks, but the herpes virus remains dormant in the facial nerves. Cold sores are the result of the virus's reactivating in the body; it reactivates periodically (in symptomatic people) to create sores in the same area of the mouth or face at the site of the original infection. During reactivation, the virus travels down the nerves to the skin where it may cause blisters (cold sores) around the lips, in the mouth, or, in about 10 % of cases, on the nose, chin, or cheeks. Reactivation may be influenced by stress, menstruation, sunlight, sunburn, fever, dehydration, or local skin trauma [56]. Surgical procedures such as dental or neural surgery, lip tattooing, or dermabrasion are also common triggers. Besides systemic antiviral therapy, once-daily valacyclovir (Valtrex®) therapy and few topical therapies are available; these include Zovirax® (acyclovir 5 %) [56], Abreva® (docosanol 10 %) [57], and Denavir® (penciclovir 1 %) to shorten the healing time and duration of tingling, pain, burning, and/or itching symptoms and are available as cream or pump [58–61]. Abreva conceal is a nonmedicated clear invisible patch that is placed on the sore, and it covers the cold sore and prevents spread of infection and helps people to use makeup to hide the sore [62]. It changes the cell membranes of healthy, uninfected cells. These changes help prevent the cold sore virus from getting into healthy cells so the viral particles cannot spread to new cells. This is the only FDA-approved drug to shorten the duration of cold sores and speed healing time [63]. Zovirax® Ointment 5 % is a formulation for topical administration. Each gram of Zovirax® Ointment 5 % contains 50 mg acyclovir in a polyethylene glycol base. Denavir (penciclovir 1 % cream) is a topical antiviral cream for the treatment of recurrent cold sores in adults. This prescription product is applied every 2 h while awake for 4 days, starting as soon as possible after the first sign or symptom of herpes labialis [64]. Zilactin cold sore gel is another pain-relieving bioadhesive gel, containing 10 % benzyl alcohol that relieves discomfort for up to 6 h. Orajel™ Single Dose is a new cold sore treatment that provides instant pain relief, and healing begins with just one dose. The Touch-free applicator helps clean the site of the cold sore and the affected area. Orajel™ Single Dose's one-time vial is mess-free, convenient, and easy to use. Topical pain reliever is used to numb the affected area. This numbing allows you to effectively treat the skin around the cold sore without pain or irritation. It contains benzalkonium chloride 0.13 % and benzocaine 5 % [59, 65].

13.4.4 Oral Malodor

Oral malodor is a common complaint worldwide among dental patients [66, 67]. Oral malodor reduces an individual's confidence during social interactions. The oral cavity has a suitable environment for the numerous bacteria which colonize the mouth to induce odor. This condition derives in most cases from the proteolytic activity of anaerobic Gram-negative oral bacteria such as *P. gingivalis*, *F. nucleatum*, and *P. intermedia*. These bacteria reside in various locations within the oral cavity (e.g., tongue dorsum, interdental space, periodontal pockets, faulty and leaky restorations, and tonsils) and break down salivary and oral proteins into their amino acid building blocks. Some of these amino acids (e.g., methionine and cysteine) are further metabolized, yielding malodorous volatile sulfide compounds (VSC) such as methyl mercaptan and hydrogen sulfide.

The tongue dorsum, especially its posterior portion, is considered the key location for this process. Therefore, the treatment regimen includes in most cases the daily use of tongue scrapers and mouthwashes. The treatment of halitosis can include a combination of mechanical and chemical strategies to neutralize or suppress odor, and rinsing in addition to gargling with an efficacious mouthwash is advised. Mechanical treatments such as tongue scraping or teeth brushing with oral preparations are recommended. A recent Cochrane systematic review found tongue scrapers to have short-term efficacy in controlling halitosis. Indeed, the use of tongue scrapers apparently has little effect on the bacterial load of the tongue, as well as being unpleasant and inducing in many cases a strong gag reflex [68–70]. Several over-the-counter mouthwashes and sprays are available that include chlorhexidine, triclosan, etc. to combat oral malodor. However, some mouthwashes containing chlorhexidine have adverse side effects such as tooth staining.

Active ingredients in oral care preparations play an important role in neutralizing or suppressing vomit odor and mainly rely on their antimicrobial efficacy toward

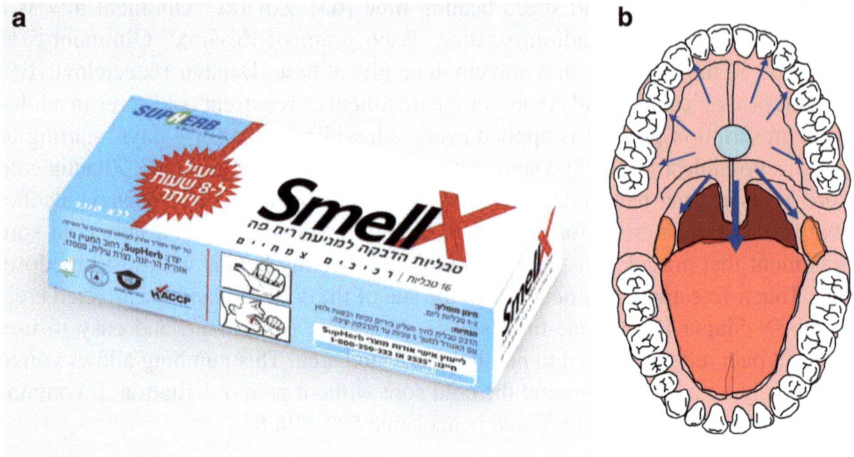

Fig. 13.2 (**a**) Clinically used SmellX and (**b**) location for placement of palatal patch

oral cavity microbes. However, some of these compounds, such as essential oils and botanical extracts, contribute to flavoring the preparations and are more beneficial than other ingredients as they do not have staining effect on teeth and are believed to be safer than synthetic agents. Furthermore, a combination of active ingredients enhances oral deodorant activity and stability as well as having an anticaries effect [71].

In addition to the incorporation of multifunctional ingredients into mouth rinses and dentifrices, other preparations have also been introduced. Breath refreshing lipsticks [72] and biofilms have been developed for more convenient use in order to regain the individual's confidence and to stop oral malodor [71]. SmellX (Fig. 13.2a) is a palatal patch designed to treat halitosis and to provide daylong relief from malodor. The patch is applied to the palate (Fig. 13.2b). This places it right above and in direct contact with the tongue dorsum, thus allowing sustained release of the active ingredients directly to the target site.

The adhesive polymers hydroxypropyl cellulose and carbopol are mixed in a ratio of 4:1, each patch weighing 250 mg. The active ingredients are echinacea (*Echinacea angustifolia*), mastic gum (*Pistacia lentiscus*), lavender (*Lavandula angustifolia*), and sage (*Salvia officinalis*). The adhesive tablet containing the herbal formulation is effective in reducing oral malodor and VSC levels. This new delivery system has two distinct advantages: adhesiveness and sustained release. The adhesiveness of the tablet enables it to stay in place, thus maintaining constant contact with the target site. Furthermore, it allows patients to carry on with their regular activities (e.g. eating, drinking, and talking) without interference, in contrast to breath mints and lozenges. The slow dissolution of the tablet enables sustained release of the active ingredients, facilitating prolonged exposure of bacteria to the active ingredients and improving their effectiveness [68].

13.4.5 Oral Infections

Oropharyngeal candidiasis is a commonly encountered problem in daily clinical practice. Oropharyngeal candidiasis is a very common localized infection of the mucus membranes of the oropharynx that is most commonly caused by the patient's own commensal *Candida albicans*. Candida is a fungus present in the mouths of up to 60 % of healthy people. In most people, untreated candidiasis persists for months or years unless associated risk factors are treated or eliminated. It is the most common opportunistic infection affecting patients with human immunodeficiency virus (HIV), hematological malignancies, diabetes, corticosteroid use, and long-term use of broad-spectrum antibiotics and patients with immunocompromised status [73]. Various systemic and topical treatment options are available for patients with oropharyngeal candidiasis. In general, systemic therapy is convenient and very effective treatment modality; it includes amphotericin B and azole antifungal agents. Topical therapies for oropharyngeal candidiasis are considered preferable to systemic therapies in most patient populations, available in form of troche, oral suspension, and buccal tablets. However, traditional topical therapies have limitations including short contact time with the oral mucosa and the need for multiple doses each day [73, 74].

Oral suspensions are marketed as Bio-Statin, Mycostatin, Mycostatin Pastilles, Nilstat (Nystatin), NOXAFIL® (Posaconazole), and Sporanox (Itraconazole) to swish and swallow. Patient is instructed to place half of the dose in one side of the mouth, swish it around the mouth, gargle, and swallow and then repeat with the remaining half of the dose in the other side of the mouth. Patient is advised to keep the liquid in mouth for as long as possible before swallowing and avoid eating for 5–10 min after using the medication [73, 75].

Mycelex Troche is a large, slowly dissolving tablet (lozenge) containing 10 mg of clotrimazole dispersed in dextrose, microcrystalline cellulose, povidone, and magnesium stearate. It is sucked five times per day until it is completely dissolved in the mouth. The troches should be allowed to dissolve slowly in the mouth; usually it takes 30 min.

Miconazole mucoadhesive tablet are being marketed as Loramyc®, Lauriad®, and Oravig™ for the treatment of oropharyngeal candidiasis. Each buccal tablet contains 50 mg miconazole and hypromellose as bioadhesive polymer; it is advised to use it once daily for 14 consecutive days. Once applied, it stays in position and gradually provides sustained local release of miconazole over a period of several hours with just one daily application [76–78].

Miconazole offered in 50 mg dosage strength and provides patients with a flavorless, odorless, and convenient treatment option that does not interfere with daily activities such as eating and drinking. These are found as effective as clotrimazole troches in the treatment of oropharyngeal candidiasis in patients with HIV and with head and neck cancer [76, 77]. A low-dose mucoadhesive miconazole nitrate 10 mg is available under the brand name Tibozole [79]. Study has shown that therapy with topical mucoadhesive tablet is efficient and well tolerated than the systemic antifungal therapy and oral gels [78, 80, 81].

13.4.6 Extraction Wounds

Following extraction of a tooth, the blood clot is formed within an hour in the socket. The raw open wound overlying the dental socket takes about 1 week to heal. Thereafter, the socket will gradually fill in with soft gum tissue over a period of about 1–2 months. Final closure of the socket with bony remodeling can take 6 months or more. Following extraction, systemic therapy is recommended for patient that includes antibiotics and anti-inflammatory drugs. Bleeding is common in first hour, but its likelihood decreases as time passes, and usually stops after 24 h. Besides systemic therapy, post-extraction local antibiotic therapy is also available in the form of cones that are fitted in the extraction socket and provide antibiotic coverage, local homeostasis, and socket preservation. These are resorbable cones that contain collagen. Examples of some available dental cones are as follows.

PARASORB® Cone Genta contains 22.4 mg equine, native collagen fibrils and 16 mg gentamicin sulfate. PARASORB® Fleece Genta HD is a sponge of 2.5×2.5×0.5 cm that contains 35 mg native equine collagen fibrils and 25 mg gentamicin sulfate. GENTA-COLL® resorb is a hemostyptic collagen sponge that contains the aminoglycoside antibiotic gentamicin for local protection from infections. These dental cones produced without chemical additives support and promote body's own bone regeneration in an effective way. The naturally structured collagen fibrils activate clotting like endogenous collagen. Within short period of time, blood vessels grow into the collagen of the cone, thus connecting it to the surrounding tissue. This leads to unhindered development of new bone due to sufficient vascularization [82].

13.4.7 Xerostomia

Saliva possesses many important functions including lubrication of the oral cavity, control of pH, mechanical cleansing action, antimicrobial activity, removal of food debris from the oral cavity, remineralization, and maintaining the integrity of the oral mucosa. Xerostomia is defined as dry mouth resulting from reduction or absence of salivary flow that may or may not be associated with decreased salivary gland function. It is not a disease but a symptom of various medical conditions, e.g., a side effect of a radiation to the head and neck, Sjögren's syndrome, or a side effect of a wide variety of medications. According to a systematic review, the prevalence of self-reported xerostomia in population-based samples ranged from 0.9 to 64.8 % in 2006 [83]. Xerostomia is often a contributing factor for both minor and serious health problems. It can affect dental, nutritional, as well as psychological health. Xerostomia decreases oral pH that can lead to dental caries, parotid gland enlargement, cheilitis, inflammation or ulcers of the tongue and buccal mucosa, oral candidiasis, salivary gland infection (sialadenitis), halitosis, and cracking and fissuring of the oral mucosa. Oral candidiasis is one of the most common oral infections seen in association with xerostomia. Other common problems associated with xerostomia include constant sore throat, burning sensation, difficulty speaking and swallowing, hoarseness, and/or dry nasal passages [84].

Table 13.2 Salivary stimulants

			References
Mechanical stimulants	Sugarless gums	Biotene	[89]
		Extra	[90]
		Trident	[90]
Chemical stimulants	Solutions	Optimoist	[91]
Electrical stimulants	Electronic stimulator of saliva	Salitron	[92]
		E-stim	[93]
Pharmacological stimulants	Drugs	Salagen (Pilocarpine HCl)	[94]
		Evoxac (Cevimeline HCl)	[94, 95]
Oral moisturizers/ salivary substitutes	Sprays	Xerolube	[96]
		Optimoist	[91]
		BioXtra	[97]
	Solutions	Salivart	[98]
		Oralube	[99–101]
		Xerolube	[96]
		Plax	[102]
		Biotene	[103]
	Gel	Biotene oral balance gel	[97]
		BioXtra	[97]
	Tablet	OraMoist® mucoadhesive tablet	[104]
		Salix®	[105]

Individuals with xerostomia often complain of problems with eating, speaking, taste disorders, painful tongue, swallowing, and wearing dentures. Denture wearers may have problems with denture retention, denture sores, and the tongue sticking to the palate. There are several products available to provide assistance in the management of xerostomia. These products range from saliva substitutes and stimulant products designed to minimize dental problems [85]. A variety of formulations can be seen in the form of chewing gums [86–88], solutions, sprays, gels, and lozenges as mentioned in Table 13.2 http://www.drymouth.info/consumer/TreatmentForDM.asp.

OraMoist® is a mucoadhesive patch that adheres to the roof of the mouth or inside the cheek. It slowly dissolves, releasing ingredients that moisten the mouth. In most cases, the patch fully dissolves in 2–4 h but can also last overnight [104].

13.5 Conclusions

The current market offers a wide variety of therapeutic products for local drug delivery in oral cavity to combat challenges in day-to-day practice for dental practitioners. The dynamic conditions of the oral cavity and the use of advanced research and available scientific solutions are presented in this chapter. Incorporation of bioadhesive polymers, gels, microspheres, and nanoparticles has improved the means of local drug delivery to the oral cavity, and it has enormous potential to counter the dynamic environment of the oral cavity.

References

1. Takahashi N (2005) Microbial ecosystem in the oral cavity: metabolic diversity in an ecological niche and its relationship with oral diseases. Int Congr Ser 1284:103–112
2. Sudhakar Y, Kuotsu K, Bandyopadhyay AK (2006) Buccal bioadhesive drug delivery–a promising option for orally less efficient drugs. J Control Release 114(1):15–40
3. Hearnden V, Sankar V, Hull K, Juras DV, Greenberg M, Kerr AR, Lockhart PB, Patton LL, Porter S, Thornhill MH (2012) New developments and opportunities in oral mucosal drug delivery for local and systemic disease. Adv Drug Deliv Rev 64(1):16–28
4. Dixit RP, Puthli SP (2009) Oral strip technology: overview and future potential. J Control Release 139(2):94–107
5. Roy S, Prabhakar B (2010) Bioadhesive polymeric platforms for transmucosal drug delivery systems—a review. Trop J Pharm Res 9(1):91–104
6. Heasman PA, Heasman L, Stacey F, McCracken GI (2001) Local delivery of chlorhexidine gluconate (PerioChip) in periodontal maintenance patients. J Clin Periodontol 28(1):90–95
7. Soskolne WA, Chajek T, Flashner M, Landau I, Stabholtz A, Kolatch B, Lerner EI (1998) An in vivo study of the chlorhexidine release profile of the PerioChip in the gingival crevicular fluid, plasma and urine. J Clin Periodontol 25(12):1017–1021
8. Stabholz A, Shapira L, Mahler D, Gellman Y, Ramon T, Dolev E, Schwartz M, Berger L, Proskin HM, Finkelman RD, Flashner M, Kolatch B, Soskolne A (2000) Using the PerioChip in treating adult periodontitis: an interim report. Compend Contin Educ Dent 21(4):325–328, 330, 332 passim; quiz 338
9. Dreizen S, Keating MJ, Beran M (1992) Orofacial fungal infections. Nine pathogens that may invade during chemotherapy. Postgrad Med 91(5):349–350, 353–344, 357–360 passim
10. Kovacs SO, Hruza LL (1995) Superficial fungal infections. Getting rid of lesions that don't want to go away. Postgrad Med 98(6):61–62, 68–69, 73–75
11. Carpenter W, Schiff T, Fat D (1997) Adherence of two film-forming medications to the oral mucosa. Gen Dent 45(5):478–480
12. Redding SW, Haveman CW (1999) Treating the discomfort of oral ulceration resulting from cancer chemotherapy. Compend Contin Educ Dent 20(4):389–392, 394, 396
13. Lindquist B, Gisselsson H, Wennerholm K (2010) Effect of chlorhexidine gel on approximal caries increment in adolescents with high caries risk using professional flossing compared to individual trays. Swed Dent J 34(1):17–25
14. Unsal E, Akkaya M, Walsh TF (1994) Influence of a single application of subgingival chlorhexidine gel or tetracycline paste on the clinical parameters of adult periodontitis patients. J Clin Periodontol 21(5):351–355
15. Wallman C, Birkhed D (2002) Effect of chlorhexidine varnish and gel on mutans streptococci in margins of restorations in adults. Caries Res 36(5):360–365
16. Bonito AJ, Lux L, Lohr KN (2005) Impact of local adjuncts to scaling and root planing in periodontal disease therapy: a systematic review. J Periodontol 76(8):1227–1236
17. Eberhard J, Jepsen S, Jervoe-Storm PM, Needleman I, Worthington HV (2008) Full-mouth disinfection for the treatment of adult chronic periodontitis. Cochrane Database Syst Rev (1):CD004622
18. Herrera D, Sanz M, Jepsen S, Needleman I, Roldan S (2002) A systematic review on the effect of systemic antimicrobials as an adjunct to scaling and root planing in periodontitis patients. J Clin Periodontol 29(Suppl 3):136–159, discussion 160-132
19. Jeffcoat MK, Bray KS, Ciancio SG, Dentino AR, Fine DH, Gordon JM, Gunsolley JC, Killoy WJ, Lowenguth RA, Magnusson NI, Offenbacher S, Palcanis KG, Proskin HM, Finkelman RD, Flashner M (1998) Adjunctive use of a subgingival controlled-release chlorhexidine chip reduces probing depth and improves attachment level compared with scaling and root planing alone. J Periodontol 69(9):989–997
20. Schwach-Abdellaoui K, Vivien-Castioni N, Gurny R (2000) Local delivery of antimicrobial agents for the treatment of periodontal diseases. Eur J Pharm Biopharm 50(1):83–99

21. Baer PN (1995) Actisite (tetracycline hydrochloride periodontal fiber) not a product for the pediatric dentist. J Clin Pediatr Dent 19(2):147
22. Litch JM, Encarnacion M, Chen S, Leonard J, Burkoth TL (1996) Use of the polymeric matrix as internal standard for quantitation of in vivo delivery of tetracycline HCl from Actisite tetracycline fiber during periodontal treatment. J Periodontal Res 31(8):540–544
23. Dodwad V, Ahuja S, Kukreja BJ (2012) Effect of locally delivered tetracycline hydrochloride as an adjunct to scaling and root planing on Hba1c, C-reactive protein, and lipid profile in type 2 diabetes: a clinico-biochemical study. Contemp Clin Dent 3(2):150–154
24. Ciancio SG (1999) Local delivery of chlorhexidine. Compend Contin Educ Dent 20(5):427–432, quiz 434
25. Killoy WJ (1998) The use of locally delivered chlorhexidine in the treatment of periodontitis. Clinical results. J Clin Periodontol 25(11 Pt 2):953–958, discussion 978-959
26. Grover V, Kapoor A, Malhotra R, Battu VS, Bhatia A, Sachdeva S (2011) To assess the effectiveness of a chlorhexidine chip in the treatment of chronic periodontitis: a clinical and radiographic study. J Indian Soc Periodontol 15(2):139–146
27. Kondreddy K, Ambalavanan N, Ramakrishna T, Kumar RS (2013) Effectiveness of a controlled release chlorhexidine chip (PerioCol-CG) as an adjunctive to scaling and root planing when compared to scaling and root planing alone in the treatment of chronic periodontitis: a comparative study. J Indian Soc Periodontol 16(4):553–557
28. Chandra C, Valavalkar N, Vandana KL (2011) The comparative evaluation of xanthan gel with chlorhexidine (Chlosite) in smokers and non-smokers: a clinical and microbiological assessment. J Indian Soc Periodontol 15(3):221–227
29. Zeidner NS, Massung RF, Dolan MC, Dadey E, Gabitzsch E, Dietrich G, Levin ML (2008) A sustained-release formulation of doxycycline hyclate (Atridox) prevents simultaneous infection of Anaplasma phagocytophilum and Borrelia burgdorferi transmitted by tick bite. J Med Microbiol 57(Pt 4):463–468
30. (1994) Antibiotic gels for periodontal disease. Drug Ther Bull 32(6):43–44
31. Jarrold CR, Allaker RP, Young KA, Heath MR, Hardie JM, Lynch E (1997) Clinical and microbiological effects of topical minocycline in the treatment of elderly patients with periodontitis. Br Dent J 183(2):51–56
32. Kinane DF (2000) Local antimicrobial therapies in periodontal disease. Ann R Australas Coll Dent Surg 15:57–60
33. Geng S, Cao C, Chen Z (2000) The effect of non-surgical periodontal and adjunctive minocycline-HCL treatments on collagenase activity. Zhonghua Kou Qiang Yi Xue Za Zhi 35(5):336–339
34. Hagiwara S, Iida M, Ishikawa I (1993) Clinical and microbiological study concerning effects of scaling and root planing. 2. The effects of mechanical debridement until one year and of minocycline application topically. Kokubyo Gakkai Zasshi 60(2):301–312
35. Hagiwara S, Umeda M, Tominaga Y, Takamatsu N, Yoshida N (1995) Subgingival distribution of periodontopathic bacteria in periodontic patients and susceptibility of these bacteria to minocycline-HCl. Kokubyo Gakkai Zasshi 62(4):495–505
36. Nakagawa T, Yamada S, Oosuka Y, Saito A, Hosaka Y, Ishikawa T, Okuda K (1991) Clinical and microbiological study of local minocycline delivery (Periocline) following scaling and root planing in recurrent periodontal pockets. Bull Tokyo Dent Coll 32(2):63–70
37. Lessem J, Hanlon A (2004) A post-marketing study of 2805 patients treated for periodontal disease with Arestin. J Int Acad Periodontol 6(4 Suppl):150–153
38. Oringer RJ, Van Dyke TE, Lessem J (2002) The challenge of treating periodontal patients who smoke–the efficacy of Arestin. J Int Acad Periodontol 4(3):89–94
39. Skaleric U, Schara R, Medvescek M, Hanlon A, Doherty F, Lessem J (2004) Periodontal treatment by Arestin and its effects on glycemic control in type 1 diabetes patients. J Int Acad Periodontol 6(4 Suppl):160–165
40. Van Dyke TE, Offenbacher S, Braswell L, Lessem J (2002) Enhancing the value of scaling and root-planing: Arestin clinical trial results. J Int Acad Periodontol 4(3):72–76
41. Al-Mubarak SA, Karring T, Ho A (2000) Clinical evaluation of subgingival application of metronidazole 25%, and adjunctive therapy. J Int Acad Periodontol 2(3):64–70

42. Montebugnoli L, Servidio D, Prati C (2002) Effectiveness of metronidazole gel on cyclosporine-induced gingival overgrowth in heart transplant patients. Clin Oral Investig 6(1):24–27
43. Candio S Actisite® (tetracycline hydrochloride) and the recare patient. Dentistry IQ, Tulsa, OK
44. PerioCol-TC Sterile medicated periodontal collagen fibers. Eucare Pharmaceuticals, India
45. Dodwad V, Vaish S, Mahajan A, Chhokra M (2012) Local drug delivery in periodontics: a strategic intervention. Int J Pharm Pharm Sci 4(4):30–34
46. PerioChip Dexcel Pharma Technologies, Fair Lawn, NJ
47. Arestin Minocycline microspheres. OraPharma, Inc., Horsham, PA
48. Jain R, Mohamed F, Hemalatha M (2012) Minocycline containing local drug delivery system in the management of chronic periodontitis: a randomized controlled trial. J Indian Soc Periodontol 16(2):179–183
49. Salamat-Miller N, Chittchang M, Johnston TP (2005) The use of mucoadhesive polymers in buccal drug delivery. Adv Drug Deliv Rev 57(11):1666–1691
50. Shinkar DM, Dhake AS, Setty CM (2012) Drug delivery from the oral cavity: a focus on mucoadhesive buccal drug delivery systems. PDA J Pharm Sci Technol 66(5):466–500
51. Segal AL, Katcher AH, Brightman VJ, Miller MF (1974) Recurrent herpes labialis, recurrent aphthous ulcers, and the menstrual cycle. J Dent Res 53(4):797–803
52. Gaffar A (2001) Recurrent oral ulcers—an overview. Compend Contin Educ Dent Suppl (32):4–6. quiz 22
53. Reznik D, O'Daniels CM (2001) Clinical treatment evaluations of a new topical oral medication. Compend Contin Educ Dent Suppl (32):17–21, quiz 23
54. Mizrahi B, Golenser J, Wolnerman JS, Domb AJ (2004) Adhesive tablet effective for treating canker sores in humans. J Pharm Sci 93(12):2927–2935
55. Pica F, Volpi A (2011) Public awareness and knowledge of herpes labialis. J Med Virol 84(1):132–137
56. Raborn GW, McGaw WT, Grace M, Percy J (1988) Treatment of herpes labialis with acyclovir. Review of three clinical trials. Am J Med 85(2A):39–42
57. Spruance SL (2002) N-docosanol (Abreva) for herpes labialis: problems and questions. J Am Acad Dermatol 47(3):457–458
58. Quong TL (1989) Oral acyclovir (Zovirax) in the treatment of recurrent herpetic gingivostomatitis and recurrent herpes labialis. Hawaii Dent J 20(10):10, 12, 14 passim
59. Poland JM (1994) Current therapeutic management of recurrent herpes labialis. Gen Dent 42(1):46–50, quiz 51-42
60. Raborn GW, Chan KS, Grace M (2004) Treatment modalities and medication recommended by health care professionals for treating recurrent herpes labialis. J Am Dent Assoc 135(1):48–54
61. Raborn GW, Grace MG (2003) Recurrent herpes simplex labialis: selected therapeutic options. J Can Dent Assoc 69(8):498–503
62. McCarthy JP, Browning WD, Teerlink C, Veit G (2012) Treatment of herpes labialis: comparison of two OTC drugs and untreated controls. J Esthet Restor Dent 24(2):103–109
63. Berry TM, Murphy JA, Montooth AK, Hing E, Cherry DK, Woodwell DA National Ambulatory Medical Care Survey: 2003 summary. Accessed at http://0-www.cdc.gov.mill1.sjlibrary.org/nchs/data/ad/ad365.pdf 9 Feb 2007
64. Pearson NL (2006) Product update. Can Pharm J/Revue des Pharmaciens du Canada 139(5):1–4
65. ORAJEL™ Single Dose Cold Sore Treatment http://www.orajel.com/adult-oral-care/cold-sore/Products/orajel-single-dose-cold-sore-treatment.aspx
66. Sterer N, Nuas S, Mizrahi B, Goldenberg C, Weiss EI, Domb A, Davidi MP (2008) Oral malodor reduction by a palatal mucoadhesive tablet containing herbal formulation. J Dent 36(7):535–539
67. Sterer N, Rubinstein Y (2006) Effect of various natural medicinals on salivary protein putrefaction and malodor production. Quintessence Int 37(8):653–658
68. Danser MM, Gomez SM, Van der Weijden GA (2003) Tongue coating and tongue brushing: a literature review. Int J Dent Hyg 1(3):151–158

69. Pedrazzi V, Sato S, de Mattos Mda G, Lara EH, Panzeri H (2004) Tongue-cleaning methods: a comparative clinical trial employing a toothbrush and a tongue scraper. J Periodontol 75(7):1009–1012
70. Quirynen M, Avontroodt P, Soers C, Zhao H, Pauwels M, van Steenberghe D (2004) Impact of tongue cleansers on microbial load and taste. J Clin Periodontol 31(7):506–510
71. Lourith N, Kanlayavattanakul M (2010) Oral malodour and active ingredients for treatment. Int J Cosmet Sci 32(5):321–329
72. Meyers AJ, Lutrario CA, Elliott M, Gallagher LA Breath freshening lipstick. US Patent 63834752002
73. Pankhurst CL (2012) Candidiasis (oropharyngeal). Clin Evid (Online) 2012:pii: 1304
74. Pienaar ED, Young T, Holmes H (2006) Interventions for the prevention and management of oropharyngeal candidiasis associated with HIV infection in adults and children. Cochrane Database Syst Rev (3):CD003940
75. Klotz SA (2006) Oropharyngeal candidiasis: a new treatment option. Clin Infect Dis 42(8):1187–1188
76. (2010) Miconazole (Oravig) for oropharyngeal candidiasis. Med Lett Drugs Ther 52(1352): 95–96
77. Lalla RV, Bensadoun RJ (2010) Miconazole mucoadhesive tablet for oropharyngeal candidiasis. Expert Rev Anti Infect Ther 9(1):13–17
78. Collins CD, Cookinham S, Smith J (2011) Management of oropharyngeal candidiasis with localized oral miconazole therapy: efficacy, safety, and patient acceptability. Patient Prefer Adherence 5:369–374
79. Tibozole. http://preprod.www.tibotec.com/bgdisplay.jhtml?itemname=products_characteristics
80. Van Roey J, Haxaire M, Kamya M, Lwanga I, Katabira E (2004) Comparative efficacy of topical therapy with a slow-release mucoadhesive buccal tablet containing miconazole nitrate versus systemic therapy with ketoconazole in HIV-positive patients with oropharyngeal candidiasis. J Acquir Immune Defic Syndr 35(2):144–150
81. Bensadoun RJ, Daoud J, El Gueddari B, Bastit L, Gourmet R, Rosikon A, Allavena C, Ceruse P, Calais G, Attali P (2008) Comparison of the efficacy and safety of miconazole 50-mg mucoadhesive buccal tablets with miconazole 500-mg gel in the treatment of oropharyngeal candidiasis: a prospective, randomized, single-blind, multicenter, comparative, phase III trial in patients treated with radiotherapy for head and neck cancer. Cancer 112(1):204–211
82. Resorba repair and regenerate Vol. 2013
83. Orellana MF, Lagravere MO, Boychuk DG, Major PW, Flores-Mir C (2006) Prevalence of xerostomia in population-based samples: a systematic review. J Public Health Dent 66(2):152–158
84. Napenas JJ, Brennan MT, Fox PC (2009) Diagnosis and treatment of xerostomia (dry mouth). Odontology 97(2):76–83
85. Rayman S, Dincer E, Almas K (2010) Xerostomia. Diagnosis and management in dental practice. N Y State Dent J 76(2):24–27
86. Dodds MW, Hsieh SC, Johnson DA (1991) The effect of increased mastication by daily gum-chewing on salivary gland output and dental plaque acidogenicity. J Dent Res 70(12):1474–1478
87. Jenkins GN, Edgar WM (1989) The effect of daily gum-chewing on salivary flow rates in man. J Dent Res 68(5):786–790
88. Odusola F (1991) Chewing gum as aid in treatment of hyposalivation. N Y State Dent J 57(4):28–31
89. Epstein JB, Emerton S, Le ND, Stevenson-Moore P (1999) A double-blind crossover trial of Oral Balance gel and Biotene toothpaste versus placebo in patients with xerostomia following radiation therapy. Oral Oncol 35(2):132–137
90. Yankell SL, Emling RC (1989) Clinical study to evaluate the effects of three marketed sugarless chewing gum products on plaque pH, pCa, and swallowing rates. J Clin Dent 1(3):70–74

91. Rhodus NL, Bereuter J (2000) Clinical evaluation of a commercially available oral moisturizer in relieving signs and symptoms of xerostomia in postirradiation head and neck cancer patients and patients with Sjogren's syndrome. J Otolaryngol 29(1):28–34
92. Lafaurie G, Fedele S, Lopez RM, Wolff A, Strietzel F, Porter SR, Konttinen YT (2009) Biotechnological advances in neuro-electro-stimulation for the treatment of hyposalivation and xerostomia. Med Oral Patol Oral Cir Bucal 14(2):E76–E80
93. Pattani KM, McDuffie CM, Morgan M, Armstrong C, Nathan CO (2010) Electrical stimulation of post-irradiated head and neck squamous cell carcinoma to improve xerostomia. J La State Med Soc 162(1):21–25
94. Braga MA, Tarzia O, Bergamaschi CC, Santos FA, Andrade ED, Groppo FC (2009) Comparison of the effects of pilocarpine and cevimeline on salivary flow. Int J Dent Hyg 7(2):126–130
95. al-Hashimi I, Taylor SE (2001) A new medication for treatment of dry mouth in Sjogren's syndrome. Tex Dent J 118(3):262–266
96. Walizer EM, Ephraim PM (1996) Double-blind cross-over controlled clinical trial of vegetable oil versus xerolube for xerostomia: an expanded study abstract. ORL Head Neck Nurs 14(1): 11–12
97. Shahdad SA, Taylor C, Barclay SC, Steen IN, Preshaw PM (2005) A double-blind, crossover study of Biotene Oralbalance and BioXtra systems as salivary substitutes in patients with post-radiotherapy xerostomia. Eur J Cancer Care (Engl) 14(4):319–326
98. Devlin H, Bassiouny MA, Boston D (2006) Hardness of enamel exposed to Coca-Cola and artificial saliva. J Oral Rehabil 33(1):26–30
99. Nieuw Amerongen AV, Veerman EC (2003) Current therapies for xerostomia and salivary gland hypofunction associated with cancer therapies. Support Care Cancer 11(4):226–231
100. Shannon IL, Trodahl JN, Starcke EN (1978) Remineralization of enamel by a saliva substitute designed for use by irradiated patients. Cancer 41(5):1746–1750
101. Meyer-Lueckel H, Schulte-Monting J, Kielbassa AM (2002) The effect of commercially available saliva substitutes on predemineralized bovine dentin in vitro. Oral Dis 8(4): 192–198
102. Schaeken MJ, van der Hoeven JS (1990) Plax more effective than water? Ned Tijdschr Tandheelkd 97(7):315–317
103. Guneri P, Alpoz E, Epstein JB, Cankaya H, Ates M (2011) In vitro antimicrobial effects of commercially available mouth-wetting agents. Spec Care Dentist 31(4):123–128
104. Kerr AR, Corby PM, Shah SS, Epler M, Fisch GS, Norman RG (2010) Use of a mucoadhesive disk for relief of dry mouth: a randomized, double-masked, controlled crossover study. J Am Dent Assoc 141(10):1250–1256
105. Senahayake F, Piggott K, Hamilton-Miller JM (1998) A pilot study of Salix SST (saliva-stimulating lozenges) in post-irradiation xerostomia. Curr Med Res Opin 14(3):155–159

Chapter 14
Focal Drug Delivery for Management of Oral Infections

Nurit Beyth, Orith Haramaty, and David Polak

14.1 Introduction

The term "biofilm" is used to describe communities of microorganisms attached to a surface, biofilm. The organisms are usually spatially arranged in a three-dimensional structure and enclosed in a matrix of extracellular material derived both from the cells themselves and from the environment [1]. Research in recent years has revealed that cells growing as biofilms have unique properties, some of which are of clinical significance. Biofilm bacteria are distinguished from planktonic bacteria by a number of features, one of which is hyper-resistance to antimicrobial agents [2]. In part, this can be attributed to alterations in the expression of a large number of genes in response to the proximity of a specific surface [3]. Furthermore, biofilm bacteria, as opposed to bacteria in the planktonic phase, often survive under extreme conditions, including an anoxic microenvironment, extreme pH, and ionic strength.

A wide range of microorganisms, including bacteria, yeast, and viruses, constitute the microflora found in the oral cavity. All of these groups, the principal being bacteria, may be related to oral infections. It is estimated that more than 1,300

N. Beyth (✉)
Faculty of Dentistry, The Hebrew University—Hadassah, Jerusalem, Israel

Department of Prosthodontics, Hebrew University—Hadassah School of Dental Medicine, P.O. Box 12272, Jerusalem 91120, Israel
e-mail: nuritb@ekmd.huji.ac.il

O. Haramaty
Department of Prosthodontics, Hebrew University - Hadassah School of Dental Medicine, P.O. Box 12272, Jerusalem 91120, Israel
e-mail: orith@hadassah.org.il

D. Polak
Department of Peridontics, Hebrew University-Hadassah School of Dental Medicine, P.O Box 12272, Jerusalem 91120, Israel
e-mail: polak@cc.huji.ac.il

A.J. Domb and W. Khan (eds.), *Focal Controlled Drug Delivery*, Advances in Delivery Science and Technology, DOI 10.1007/978-1-4614-9434-8_14,

different bacterial species may be present in the oral cavity, including *Spirochaetes*, *Fusobacteria*, *Actinobacteria*, *Firmicutes*, *Bacteroidetes,* and *Proteobacteria*. The extreme diversity can be attributed to the variety of nutrients; the different forms of colonization environments, including surfaces on the teeth, mucosa, and tongue; and the possibility to survive as a biofilm. The latter offers mainly three advantages: (1) a physical barrier to external insults, (2) protection from the host immune system, and (3) nutrient availability.

Biofilm formation may have crucial implications in medicine and dentistry. One example is the serious complications encountered in implant surgery caused by implant-associated infections.

14.2 Biofilms and Oral Infections

Biofilm is the main cause of localized diseases in the oral cavity, including caries, gingivitis, periodontitis, candidiasis, endodontic infections, orthodontic infections, and peri-implantitis [3]. In the oral cavity, as in most natural environments, microorganisms exist predominantly in multispecies biofilms, making it difficult to specify the role of each species in oral infections. Such infections stem mostly from microorganisms normally found in the mouth, seldom is the origin an exogenous source. Regardless whether bacterial accumulation and biofilm formation occur on natural surfaces or on artificial materials, dental caries or periodontal disease may evolve.

14.2.1 Formation and Characteristics of Oral Biofilms

Formation of the biofilm depends on different variables, such as the species of bacteria, the physical and chemical properties of the underlying surface, environmental factors, and essential gene products, and may be regulated by the quorum sensing system [4]. In the oral cavity organisms are not distributed randomly, but selectively attach and grow on certain surfaces, forming dental plaque. The latter is an example of a microbial biofilm, sharing most of the features of other currently known biofilms, with antimicrobial resistance being of special relevance [5]. Potential bacterial attachment sites include the tooth and epithelial surfaces. Conditions on these surfaces change with respect to oxygen levels, nutrient availability, saliva and gingival crevicular fluid secretion, masticatory forces, and other variables such as oral hygiene. Generally, adhesion is facilitated in the initial stage by nonspecific interactions, followed by the production of specific molecular interactions. Primary bacterial adhesion entails a random encounter between an acquired pellicle or a conditioning film composed of proteins drawn from the saliva and planktonic bacteria. The initial colonizers of the oral biofilm are Gram-positive species, primarily from the streptococcal group. The bacteria include facultative anaerobes, which are part of the resident microbial flora of the oral cavity and upper respiratory tract but are also opportunistic pathogens in human diseases. These initial colonizers can influence the subsequent sequence of microbial

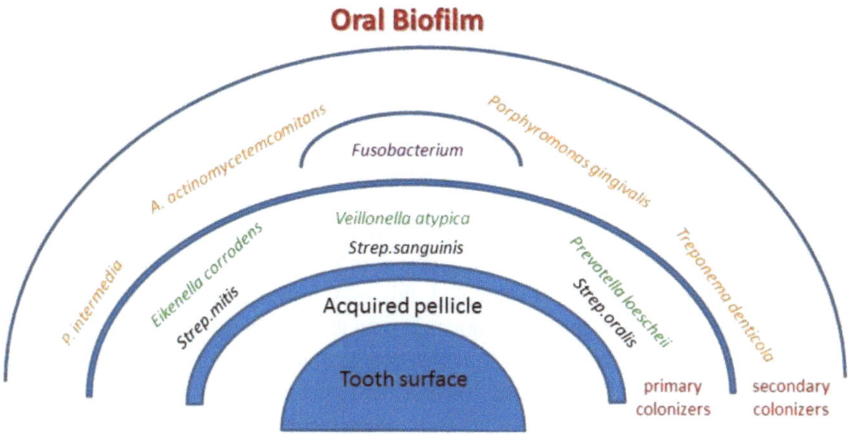

Fig. 14.1 Oral biofilm

colonization. Members of this group can stimulate host cells, and the biofilm itself generates a habitat for additional species, some of which are closely associated with the initiation and progression of dental caries and periodontal disease. The formation of oral biofilm has a pattern of colonization called autogenic succession; i.e., the microorganisms themselves generate or induce local physicochemical changes that modify the microbial composition of the biofilm. The initial stage of biofilm formation depends on the strength of the bacterial attachment, as this step is reversible. Once attachment is established, major mass growth occurs mainly by bacterial cell division within the biofilm, rather than by co-aggregation at the surface of the developing biofilm [6]. The first colonizers in the supragingival biofilm are of a relatively low diversity compared with those present in the mature biofilm. The initial colonizers include *Streptococcus sanguinis*, *Streptococcus mitis,* and *Streptococcus oralis*. Other bacteria, predominantly Gram-negative species, such as *Eikenella corrodens, Veillonella atypica,* and *Prevotella loescheii*, then emerge and co-aggregate with them. After multiplication of the primary colonizers, additional bacteria, known as secondary or late colonizers, are incorporated. *Fusobacterium,* a late colonizer also recognized as a bridging microorganism, binds by co-aggregating with the pioneer species. Late colonizers include *Aggregatibacter actinomycetemcomitans*, *Prevotella intermedia, Treponema denticola,* and *Porphyromonas gingivalis* [7] (Fig. 14.1). Depending largely on the effectiveness of dental hygiene, other organisms join the biofilm. The late arrivers have little ability to co-aggregate with the pioneers, but do bind with intermediates such as *fusobacterium*. Biofilm development and maturation include interactions between bacteria such as physical contact, metabolic exchange, molecular communication, and exchange of genetic material. Biofilm may accumulate on both the hard and soft oral tissues, growth and maturation continuing until the biofilm reaches a critical mass.

14.2.2 Biofilms and Oral Disease

14.2.2.1 Dental Caries

Dental caries, also known as tooth decay, is a destructive condition of the dental hard tissues (enamel, dentin, and cementum). It originates in bacterial infection causing demineralization of the tissues due to fermentation of food debris on the tooth surface. As the hard tissues progressively break down, inflammation and death of vital pulp tissue may occur. Moreover, if untreated, this may lead to periapical infection. The disease process involves mostly acidogenic plaque bacteria, including *S. mutans*, *S. sobrinus,* and *Lactobacillus* spp. [8], which produce acid in the presence of fermentable carbohydrates such as sucrose, fructose, and glucose. The acidity of the mineral content of the tooth surface tends to increase when lactic acid is produced, resulting in demineralization. Today, caries remains one of the most common diseases in the world.

14.2.2.2 Periodontal and Peri-Implant Diseases

Periodontal diseases can involve one or more of the periodontal tissues, including alveolar bone, periodontal ligament, cementum, and gingiva. The vast majority of the periodontal diseases are initiated by plaque biofilm that develops on the hard root surface neighboring upon the soft tissues of the supporting periodontium. Periodontal diseases range from simple gingival inflammation to severe disease that results in major damage to the soft tissue and bone supporting the teeth, leading in some cases to tooth loss. When the deeper supporting structures, including the periodontal ligament and the alveolar bone, are harmed, pocket formation, bone loss, and tooth loosening occur. In periodontal disease bacterial infection is considered the initiating and maintaining factor for the destructive inflammatory response. Host response, nutrient availability, and environmental changes in the gingival sulcus and periodontal pocket can affect disease progression. Moreover, compromised tissue metabolism and repair may enhance microbial virulence factors. *P. gingivalis, Tannerella forsythia,* and *Treponema denticola* are regarded as the major pathogens.

Implant systems are routinely used in the reconstruction of fully or partially edentulous individuals. Like natural teeth, the implants may be susceptible to the contaminated oral cavity. Biofilms formed on implant surfaces may result in a peri-implant tissue response very similar to that of periodontal tissues in a prone host, i.e., an inflammatory reaction, with the loss of supporting bone in the tissues surrounding a functioning implant. Anaerobic Gram-negative organisms play a major role in peri-implantitis [9], inflammatory alterations in tissues adjacent to the implant leading to bone destruction, and progressive loss of support and terminating in implant failure.

14.2.2.3 Endodontic Infection

Endodontic infections are polymicrobial, predominantly involving anaerobic bacteria and some facultative bacteria. A tooth with an infected nonvital pulp is a reservoir of infection that is isolated from the patient's immune response and will eventually produce a periradicular inflammatory response. When the microbes invade the periradicular tissues, abscesses and cellulitis may develop. The severity of the infection depends on the pathogenicity of the microbes and the resistance of the host. These not only may give rise to an immunopathogenic and protective response but also can be destructive to the surrounding tissues and contribute to the adverse signs and symptoms.

14.2.3 The Challenge of Controlling Oral Biofilm

Biofilm bacteria exhibit significantly reduced susceptibility to antimicrobials compared with cultures grown in suspension [10]. Various factors such as biofilm impermeability [11, 12], low uptake of antimicrobial molecules [13, 14], formation of an antagonizing microenvironment against antimicrobial activity, and horizontal gene transfer of antibiotic resistance genes [15] have been related to the increased resistance of the biofilm bacteria. Frequently, prevention of dental caries and periodontal diseases is accomplished by mechanical or nonspecific control of the plaque biofilm. Nevertheless, the use of antimicrobial agents may serve as a beneficial complement to mechanical plaque control [16]. New approaches aim at rapid bacterial destruction, minimizing the risk of evolving resistance, enhancing the safety of the adjacent host tissue, and maintaining the biological equilibrium of the microflora. Moreover, real-time exposure to antimicrobial agents during toothbrushing and mouth rinsing can be extremely brief and can amount to about 30 s, rather than the recommended 2 min [17]. Thus, adjunctive antimicrobial treatment may be beneficial.

Innovative technologies and drug delivery systems for the prevention of colonization and biofilm formation have been proposed. Antiplaque agents, generally classified as preventive, removing, or disrupting agents, do not necessarily eradicate the biofilm microorganisms. Antimicrobial agents can be bacteriostatic when acting as bacterial growth inhibitors or bactericidal when causing bacterial death. Such therapeutic strategies, particularly in the oral cavity, are expected to reach less accessible sites where plaque mostly accumulates. The uptake and penetration of antimicrobial agents into biofilm are key considerations in the administration of therapeutics [18].

As conventional plaque control basically relies on the patient, it is strongly affected by noncompliance. Thus, many new antiplaque strategies requiring minimal patient compliance and minimal professional healthcare intervention are advantageous. Proposed strategies include surface modification of tissues and materials to reduce adhesion and biofilm formation, antimicrobial-impregnated matrices and enhancement of drug penetration into biofilm, and using carrier systems to target antimicrobials to the biofilm surface [19]. The antiplaque potential of various antimicrobials, including chlorhexidine (CHX), hexetidine, delmopinol, amine fluoride/stannous fluoride, triclosan, and phenolic compounds, has been well characterized, and they are frequently found in mouthwashes and toothpastes [16]. Additionally, the use of nanoparticles incorporated into topical agents or into various dental materials has been gaining interest in the past decade [20–22]. As the nanoparticle surface-area-to-volume ratio is extremely high, small quantities of the therapeutic agent are required. Also, it is possible to modify and adjust the physical and chemical characteristics of the nanoparticles by altering their surface charge and hydrophobicity [23].

14.3 Focal Drug Agents Against Caries Lesions

14.3.1 Remineralization Agents

14.3.1.1 Toothpastes and Mouthwashes

Plaque formation is conventionally controlled through regular toothbrushing and mouth rinse. Various ingredients are incorporated into toothpastes and mouthwashes to maintain appropriate oral hygiene and thus attain oral health. Most toothpaste formulations are a combination of abrasives, fluorides, and detergents, mainly sodium dodecyl sulfate (SDS), to enhance the efficacy of brushing. Antibacterial fluoride mouthwash, whitening mouthwash, and bad breath mouthwash, among others, are available on the market. Both mouthwashes and toothpastes contain active and inactive ingredients and are recommended according to the prevailing oral conditions. Toothpaste is considered the most popular local drug delivery system for preventing and treating caries-related diseases. Fluoridated toothpastes are the most commonly used types. Like toothpastes, mouthwashes may contain remineralizing ingredients such as fluoride and casein phosphopeptides (CPPs). Common fluoride derivatives incorporated in toothpastes include sodium fluoride (NaF), amine fluoride, stannous fluoride, and monofluorophosphate.

14.3.1.2 Fluorides

Fluorides are highly effective in preventing caries. The controlled release of fluoride devices is based on the counter-equilibrium existing between intraoral fluoride levels and the prevalence of dental caries [24]. The mechanisms of action by which fluoride prevents dental caries development include a reduction in enamel solubility under acidic conditions, promotion of enamel remineralization, inhibition of glucose uptake and utilization by acidogenic bacteria, and possibly bacteriostatic or bactericidal effects [25]. It is widely accepted that constant application of appropriate levels of fluoride in the oral cavity, specifically at the biofilm–tooth interface, is an extremely important factor in preventing dental caries [26]. Moreover, it is well recognized that the clearance time of most drugs from the oral cavity is rapid because of salivation and food. Thus, treatment or prevention of caries, which is considered a chronic disease, requires optimum levels of the drug for prolonged periods of time. Studies in animals and humans demonstrated that the use of fluoride slow-release devices of the copolymer membrane and glass bead type efficiently protect masticatory surfaces, which are normally not protected by conventional fluoride regimens. However, retention rates are the main problem related to these devices and require further improvement [27]. Slow-release fluoride devices include the copolymer membrane, glass beads, and a mixture of NaF and hydroxyapatite. The copolymer membrane device consists of a small pellet with an inner core of hydroxyethyl methacrylate (HEMA)/methyl methacrylate (MMA) copolymer mixture, containing NaF [27]. This device is a membrane-controlled reservoir of fluoride owing to the HEMA/MMA copolymer membrane surrounding the core. The fluoride glass bead system, once exposed to saliva, gradually releases fluoride. The latest device has the form of a disk that can be placed within a plastic bracket, facilitating replacement and installation. Both the copolymer membrane and the glass bead systems are usually attached to the buccal surface of the first permanent molar. An additional diffusion control device is the Hydroxyapatite-Eudragit RS100 F system, which is a mixture of hydroxyapatite, NaF, and Eudragit RS100.

Focal fluoride treatment, in conjunction with antiplaque treatments, may overcome compliance difficulties and be beneficial in patients with a high caries risk.

Fluoride has been utilized in restorative materials for caries prevention and treatment. Examples include dental cements such as polycarboxylate, zinc oxide, and zinc phosphate in which SnF_2 is incorporated. These materials were shown to increase fluoride levels and decrease plaque scores. Another example is glass ionomer cements in which aluminum-fluorosilicate glass is incorporated. Not only do these cements release fluoride for extended periods of time, but also they are capable of taking up fluoride from toothpastes and releasing it later, thereby serving as a fluoride reservoir. Despite the numerous advantages of fluoride-impregnated dental materials, disadvantages include the difficulty in controlling the rate of fluoride release. However, the fluoride slow-release devices have proved to be beneficial and cost-effective [26, 28]. Moreover, specifically in patients with a high risk for caries, fluoride slow-release systems can lead to a decrease in susceptibility.

14.3.1.3 Calcium Phosphate-Based Systems

CPP, a cariostatic derivative of dairy products such as milk [29], was shown to increase substantially the calcium and phosphate content in proximity to the tooth [30]. CPPs are able to form an amorphous calcium phosphate (ACP) complex with a phosphoseryl residue that stabilizes the calcium phosphate ions in solution. It has been suggested that upon binding to hydroxyapatite, CPP–ACP diffuses into the dental plaque, thereby raising the calcium phosphate levels, inhibiting enamel demineralization, and enhancing remineralization [30].

Enamel subsurface remineralization was reported to take place in the presence of CPP–ACP added to tooth paste [31], dental cream [32], and sugar-free chewing gum [33]. It has been speculated that CPP–ACP has both a short-term remineralization effect and a long-term caries-preventing effect [34]. Additionally, ACP–CPP can induce surface remineralization of enamel when calcium fluoride phosphate (CPP–ACFP) is produced [35] (Fig. 14.2).

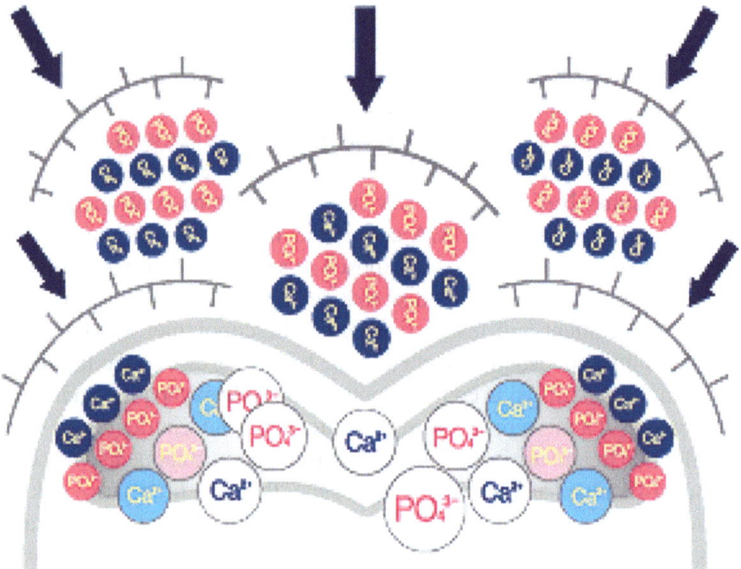

Fig. 14.2 CPP–ACP (Recaldent™)

Fig. 14.3 Chlorhexidine

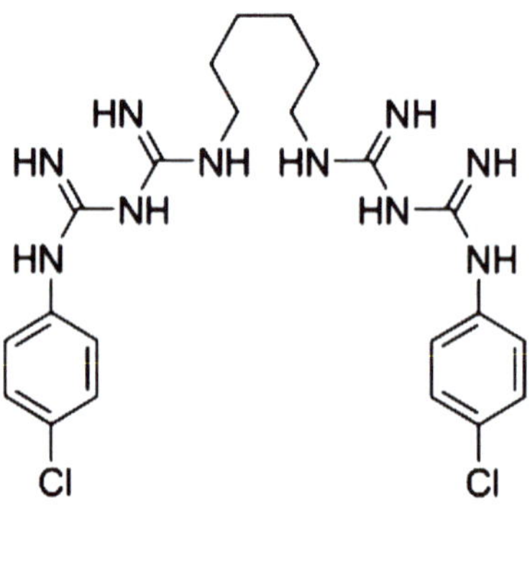

14.3.2 *Antiplaque Agents*

14.3.2.1 Chemoprophylactic Agents

Chemoprophylactic agents can be delivered in toothpastes, mouthwashes, gels, and varnishes. These agents are commonly used as a slow-release modality, where a prolonged inhibitory effect can be achieved. Chemoprophylactic agents include CHX and cetylpyridinium chloride [36], SDS, triclosan, antibiotics such as penicillin and vancomycin [37, 38], and sanguinaria extract [39].

Of the cationic agents, CHX (Fig. 14.3) is the most tested anticariogenic one, its antiplaque properties including inhibition of bacterial biofilm formation [40].

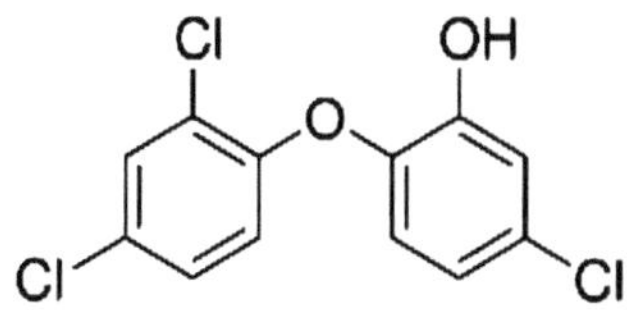

Fig. 14.4 Triclosan

The most effective reduction in mutans streptococci was achieved using CHX varnishes being most effective, followed by gels, and, lastly, mouth rinses [41]. CHX is considered anticariogenic as it was shown to suppress *mutans streptococci* growth [42]. CHX has been referred to as a nonsurgical approach for caries management and represents the modern medical model of caries treatment. Nevertheless, there is a lack of consensus on evidence-based treatment protocols and controversy among dental educators regarding the role of CHX in caries prevention [41].

Triclosan (Fig. 14.4), a nonionic antibacterial compound, is a potent chlorinated diphenyl ether. Clinically, toothpastes containing triclosan and zinc citrate were shown to reduce significantly plaque and gingival scores [43]. Although effective in periodontal disease, triclosan is restricted in its ability to prevent dental caries, owing to the compound's poor water solubility and its short-term half-life clearance. To overcome these drawbacks, several approaches have been suggested including triclosan emulsion, triclosan rinse and dentifrice compositions containing cyclodextrin, and a triclosan oral composition containing sodium lauryl sulfate as surfactant [44, 45]. Indeed, a polyvinylmethyl ether/maleic acid copolymer delivery system was reported to prolong triclosan retention in the oral cavity [46, 47].

Recently, plant-derived natural antibacterial substances, mainly polyphenols (Fig. 14.5), have been gaining interest. These substances, mostly incorporated in mouthwashes, include tea tree oil, miswak extracts, green tea, and manuka honey [48]. All of them have an antibacterial effect against *S. mutans*. These substances inhibit the activity of glucosyltransferases (GTFs), the synthesis of insoluble glucan, and the production of acid from sucrose and glucose [49]. The polyphenol mode of action may reduce some of the metabolic activity of *S. mutans,* although it is not bactericidal.

14.3.2.2 Antimicrobial Peptides

It is well recognized that biofilms respond poorly to conventional antibiotics and may develop resistance to antibiotics. Interestingly, despite constant challenges by commensal and pathogenic bacteria, normal mucosal surfaces are resistant to biofilm infections. Oral mucosal tissue expresses antimicrobial peptides (AMPs), the innate immune system's own "antibiotic," with a broad-spectrum capacity to kill invading microbes [50]. The majority of these molecules exhibit similar physical hallmarks, including amphipathic mixtures of α-helical and β-sheet structures and an overall cationic charge. It is generally accepted that AMP-mediated killing typically occurs through microbial membrane disruption. Whereas conventional antibiotics are becoming less effective in the treatment of biofilm infections, bacteria do not appear to develop resistance to AMPs.

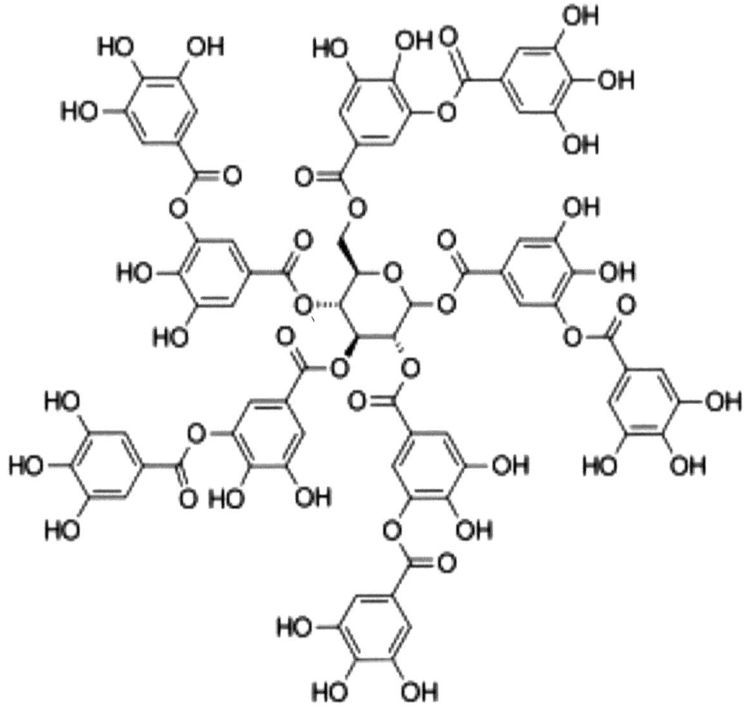

Fig. 14.5 Polyphenol

Although the potential importance of AMPs as the first line of mucosal defense against biofilm-associated infections has been long appreciated, to date no approach has been effective in translating their activity therapeutically. Application has been limited due to the difficulty and expense of manufacture and to their short half-lives owing to proteolytic degradation. A possible solution would be to use peptide mimetics [51] and exogenous modifiers to regulate AMP expression at the transcriptional level [52].

A peptide mimetic based on the structure of magainin, meta-phenylene ethynylene, was found to inhibit *S. mutans* biofilm formation at nanomolar concentrations [53]. Another approach is to target bacteria based on acid production. Histidine and phenylalanine AMP-based peptides showed a significant increase in antimicrobial activity at low pH values compared with that at neutral conditions [54]. The aim should be to produce more stable, easier to synthesize peptides at a lower cost that could be used in various oral care products.

14.3.2.3 Vaccines

Prevention and control of dental caries is a major public health concern. Consequently, much effort is being put into the development of vaccines aimed at

Fig. 14.6 Xylitol

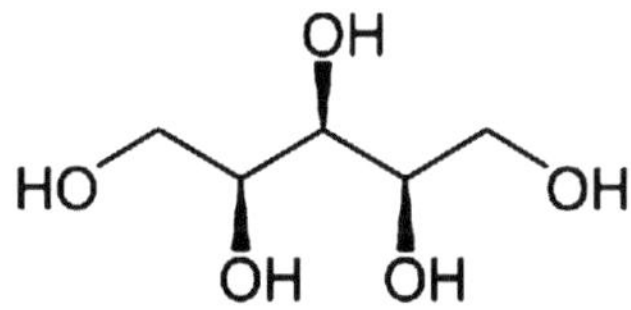

prevention. The feasibility of immunizing experimental rodents or primates with protein antigens derived from *S. mutans* or *S. sobrinus* against oral colonization has been demonstrated [55]. Salivary IgA antibodies inhibit sucrose-independent and sucrose-dependent mechanisms of streptococcal accumulation on tooth surfaces, thus preventing caries development. Maintenance of high levels of salivary antibodies can induce immunization [56]. Similarly, in humans both salivary antibodies to mutans streptococci and passively applied antibodies suppressed oral recolonization by mutans streptococci [57, 58]. Progress towards practical vaccine development is still in its early stages, and further clinical studies are required.

14.3.2.4 Probiotics Therapy

Probiotic therapy aims at focal treatment by disrupting dental plaque formation. By selectively inhibiting oral pathogens, the equilibrium of the oral microbial flora can be preserved. Probiotics are defined by the World Health Organization as live microorganisms which, when administered in adequate amounts, promote health in the host. Probiotic therapy utilizes several approaches for this purpose, including competition for binding sites and nutrients with cariogenic bacteria and inhibition of oral bacteria by production of antimicrobial compounds such as organic acids, hydrogen peroxide, low-molecular-weight antimicrobial compounds, bacteriocins, and adhesion inhibitors produced by lactic acid bacteria [59].

14.3.2.5 Sugar Substitutes

Sugar is a key component in the induction and progression of dental caries. It has been shown that several sugar substitutes are potentially effective against caries. For example, intake of xylitol, a 5-carbon sugar alcohol (polyol) (Fig. 14.6), that cannot be utilized and fermented by mutans streptococci or other microorganisms, does not cause a drop in plaque pH, thus preventing demineralization. Moreover, xylitol consumption can significantly reduce saliva and dental plaque *S. mutans* levels [60]. Xylitol is frequently used in chewing gum, which has also been proven to protect against caries by stimulating salivary flow. Conventionally, delivery systems such as chewing gum, hard candy, and mints are used to provide the sugar substitutes.

Regrettably, as the time of exposure to the tooth surface is limited to a few minute or even seconds, innovative delivery systems are needed for the effective delivery of sugar substitutes. Xylitol and other sugar substitutes such as sorbitol and palatinit can inhibit glucan synthesis from sucrose, decrease mutans streptococcus number, and increase buffering capacity and pH. Consequently, sugar substitutes can delay enamel demineralization and promote remineralization. Nonetheless, as their antibacterial properties are weak, only prolonged exposure to sugar substitutes can be effective in caries prevention.

14.4 Focal Delivery Systems Against Periodontal and Peri-Implant Infection

Periodontal diseases are a group of infectious diseases manifesting as a continuous destruction of tooth-supporting tissues (Fig. 14.7). The estimated prevalence of periodontal disease in the USA is 47 % of the general population, of which 8.5 % suffer from the severe form of the disease [61]. The etiology of the disease is not fully understood, although its association with biofilm and bacterial infection is well established [62]. Most treatment modalities aim to eradicate the infection by means of mechanical or chemical treatment.

Peri-implant diseases are relatively new pathologies that are emerging along with the growing popularity of dental implants. The clinical characteristics of the disease are similar to those seen in periodontal diseases [63], but its etiology is mostly unclear. Therapeutic modalities for peri-implantitis are drawn mainly from the field of periodontics, but often these modes are of limited success.

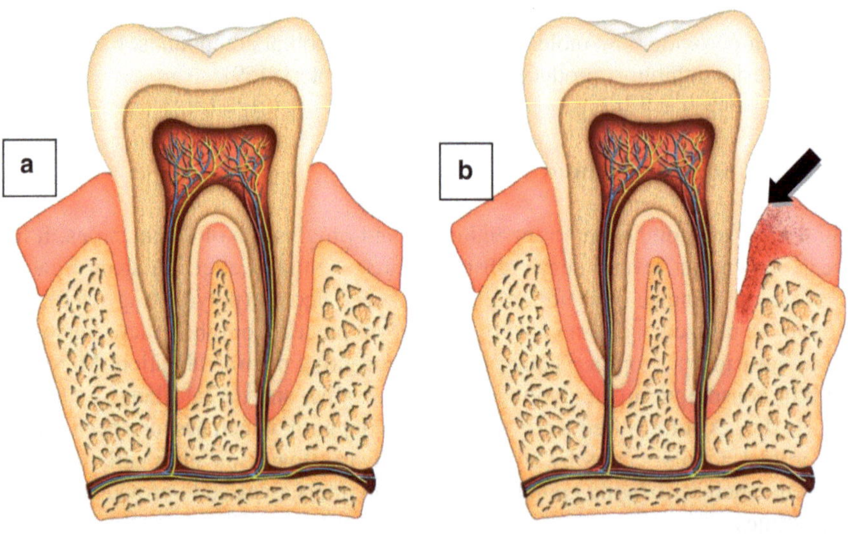

Fig. 14.7 (**a**) Healthy tooth and periodontium morphology. (**b**) Periodontally affected tooth exhibiting alveolar bone loss and pocket formation (*arrow*)

14.4.1 *Traditional Periodontal and Peri-Implant Therapy*

Periodontal treatment consists of an anti-infective phase followed by corrective and maintenance phases. The anti-infective phase aims to eliminate the periodontal infection and to reestablish non-inflamed periodontal tissues. The classic treatment modality consists of mechanical plaque debridement, along with strict patient oral hygiene practice, allowing physical clearance of biofilm located beneath the gingival margin. The use of systemic antibiotics in periodontitis has been, and still is, a controversial issue. In the chronic form of the disease, the use of systemic antibiotics is not commonly recommended, although recent insights show such treatment has a beneficial effect [64]. In the aggressive form of the disease, the use of amoxicillin with metronidazole treatment is recommended and commonly used, mainly because of the resistance of *Aggregatibacter actinomycetemcomitans* (the main pathogen associated with the disease) to treatment with either antibiotic alone [65].

In periodontal treatment, all chemical agents used in the oral cavity without systemic exposure fall under the definition of focal delivery systems. Classic and commonly used focal delivery systems are dentifrices and mouthwashes. Dentifrices serve as a detergent during tooth brushing, providing ease of use during biofilm removal by mechanical action. Additional active materials in the dentifrice have a role in maintaining and preventing pathological processes around the teeth (see Sect. 14.3).

14.4.1.1 Dentifrices Used in Periodontics

The use of dentifrices containing triclosan/copolymer has been shown to reduce gingival inflammation compared with dentifrices containing fluoride [66]. Long-term clinical studies that included over 1,200 subjects examined the effect of the daily use of a triclosan/copolymer/fluoride dentifrice on periodontitis [67–69]. The results showed that dentifrice effectively retarded disease progression, reducing the frequency of deep periodontal pockets and the number of sites that exhibited additional probing attachment and bone loss. Dentifrices also provided a significant periodontal health benefit in subjects with recurrent periodontal disease who are at increased risk of developing further breakdown of periodontal tissues. It was noteworthy that even in the 5-year study, the findings were consistent with the lack of development of resistance by any of the microbial organisms studied over this period and the absence of significant adverse effects. Dentifrices containing essential oils have also shown clinical and microbiological benefits in patients with periodontitis [70].

14.4.1.2 Mouth Rinses Used in Periodontics

The use of chemical mouth rinse agents with antiplaque or anti-gingivitis action as adjuncts to oral hygiene shows specific benefits in controlling gingival inflammation, especially in acute situations, postsurgically, and during periods of interrupted hygiene. On the downside, it is also recognized that some oral hygiene products may have the potential to cause harm in the mouth ranging from the production of a cosmetic nuisance (such as tooth staining) to more permanent damage to the dental hard tissues (erosion and abrasion). Of serious concern is the controversial ability to produce carcinogenic changes in the oral mucosa through the use of alcoholic mouth rinses.

Among the first-generation antimicrobials, the phenolic compounds are known to prevent and reduce supragingival plaque accumulation and gingivitis. Short-term studies have shown a reduction in plaque and gingivitis averaging 35 % [71], and long-term studies have found a 13.8–56.3 % reduction in plaque and a 14–35.9 % reduction in gingivitis [72, 73]. Possible adverse effects reported in the literature include a burning sensation, bitter taste, and tooth staining.

CHX gluconate (0.12 %) was the first antimicrobial found to inhibit plaque formation and the development of chronic gingivitis [74]. CHX is more effective against Gram-positive than against Gram-negative bacteria. It is of very low toxicity, as it is poorly absorbed from the gastrointestinal tract and 90 % is excreted in the feces. CHX (0.12 %) is indicated for short-term (<2 months) use, intermittent short-term (alternating on and off every 1–2 months) use, and long-term (>3 months to indefinite) use, depending on clinical indications. CHX appears to be the most effective agent for reducing both plaque and gingivitis, with short-term reduction averaging 60 % [75]. Long-term plaque reduction averaged 40.4 % and gingivitis 28.7 % [76]. Reported adverse effects include tooth staining, mucositis and reversible epithelial desquamation, alteration of taste, and increased supragingival calculus [75, 77].

Several other agents have been evaluated for their effect on bacterial plaque and gingivitis, but they are inferior to CHX and phenolic compounds [76]. Mouthwash containing a combination of triclosan/Gantrez and sodium bicarbonate was shown to have in vitro antimicrobial activity superior to that of placebo, although it was less effective than CHX [78].

Mouthwash containing oxygenating agents has anti-inflammatory properties, and its use results in less bleeding on probing (major sign of periodontal inflammation), although the bacteria causing the disease are not necessarily reduced in number [79]. Safety issues such as tissue injury and cocarcinogenicity have been raised with the chronic use of hydrogen peroxide [80].

Several studies compared different mouth rinses that were used to reduce plaque and gingivitis [77, 79, 81]. CHX is reportedly the gold standard, with superior effectiveness when compared with other mouth rinses and when the possible adverse effects are taken into consideration. It is effective in two thirds of the cases; the phenolic compounds are next in order, reducing by about 35 % plaque formation and gingivitis. Sanguinarine and the quaternary ammonium compounds follow, with 18 % and 15 %, respectively. The oxygenating agents are ineffective against both plaque formation and gingivitis.

14.4.1.3 Controlled Focal Drug Delivery

As discussed above, antibiotics are occasionally added to the mechanical treatment modality against periodontal disease. In light of the known shortcomings of systemic antibiotics, local delivery of antibacterial agents into periodontal pockets has been developed and extensively studied. This mode of drug delivery avoids most of the problems associated with systemic therapy, limiting the drug to its target site and thereby achieving a much higher concentration. The fact that periodontal diseases are localized to the immediate environment of the pocket renders it a natural site for treatment with local sustained delivery systems.

Historically, the first such local antibiotic therapy for periodontal disease was nonabsorbable fibers filled with tetracycline. The fiber was inserted into the pocket; wrapped repeatedly around the tooth, keeping the fiber in the pocket; and removed 10 days later by the operator. Following removal, the soft tissues generally showed shrinkage, reduced pocket depth, and a reduction in the clinical signs of inflammation. However, use of this system was tedious and fueled the development of absorbable systems for antibiotic delivery.

The first resorptive local antibiotic system was doxycycline carried in a matrix that was absorbed after 7 days and did not require a second visit for removal of the material. However, application of this system was somewhat problematic, as the material tends to pull out of the pocket when the syringe is removed.

Further development of absorbable local antibiotic systems led to minocycline in a microspheric configuration. The sphere is a bio-absorbable polymer of polyglycolide-co-DL lactide, which is hydrolyzed into CO_2 and H_2O. The antibiotic maintains therapeutic drug levels and remains in the pocket for 14 days. This configuration of the material allows placement at the depths of most pockets and is easier to use than the solid fiber systems. A similar system is based on 25 % metronidazole in a glyceryl mono-oleate and sesame oil gel. Another sustained local delivery system in periodontitis is based on a chip containing CHX gluconate (Fig. 14.8). The chip is a gelatin matrix containing a controlled-release CHX cross-linked with glutaraldehyde. It is reported to be self-retentive and to biodegrade over the subsequent 7- to 10-day period.

Meta-analysis compared the effects of the different sustained focal delivery systems on periodontal pocket reduction following periodontal treatment adjunct to local sustained-release agents versus mechanical treatment alone ([82], Table 14.1). The weighted mean difference for probing depth reduction with adjunctive therapy was 3.06 ± 1.06 mm (mean ± SD), the mean reduction score for bleeding on probing was 0.92 ± 0.3 (mean ± SD), and the plaque score was 1.16 ± 0.57 (mean ± SD). A significant reduction in probing depth was found upon adjunct CHX chip use, and the effect size (the magnitude of intervention impact on the outcome) was 10.4. This is very high, indicating the very large additive effect of the chip in pocket reduction of chronic periodontitis cases. Similarly, the effect size for the tetracycline fiber was 5.4, which was again very large. Minocycline microspheres and

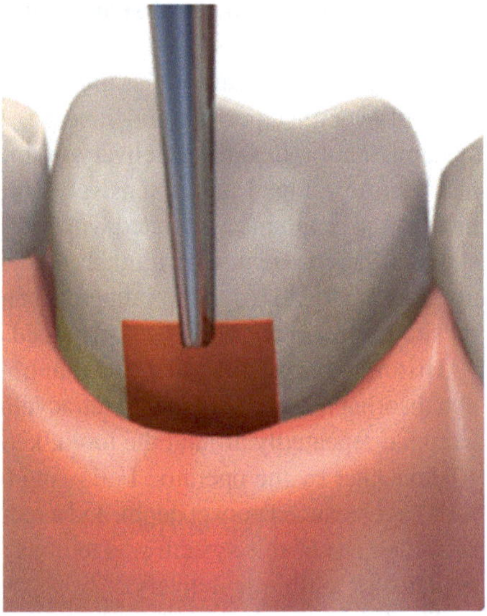

Fig. 14.8 Chlorhexidine gluconate chip system (PerioChip®)

Table 14.1 Meta-analysis of long-term clinical effectiveness of local controlled-release systems treatments [82]

	Probing pocket depth (mean ± SD)			
Treatment modality	Control group	Test group	*P* value	Effect size
Doxycycline hyclate	0.39 ± 0.42	0.57 ± 1.22	NS	0.2
Minocycline hydrochloride	1.50 ± 1.24	2.10 ± 1.25	NS	0.5
Conventional treatment	2.0 ± 1.3	9.3 ± 1.4	NS	5.4
Chlorhexidine gluconate	2.7 ± 0.13	4.15 ± 0.15	$P<0.01$	10.4
Pooled data	1.42 ± 0.71	3.06 ± 1.06	$P<0.001$	1.9

Effect size = magnitude of intervention impact on the outcome
NS nonsignificant

doxycycline gel were reported to have moderate effect sizes of 0.5 and 0.2, respectively. Overall, the effect size after meta-analysis of pooled data revealed a value of 1.9, pointing to the large impact of adjunctive therapy in local drug delivery with mechanical treatment periodontal pocket reduction in chronic periodontitis subjects.

14.4.1.4 Challenges in Peri-Implant Infection

As discussed above, peri-implantitis is a relative new pathology and its etiology is still vague. Treatment modalities used to treat peri-implantitis are those applied in the periodontal field and include nonsurgical (debridement using ultrasonic or hand instruments) and surgical strategies.

A clinical trial comparing the effectiveness of nonsurgical mechanical therapy methods in the treatment of peri-implantitis lesions showed poor success [83, 84].

Adjunct antibiotic therapy, local or systemic, has been suggested in the nonsurgical management of peri-implantitis. Some improvement in terms of probing depth and bleeding index was noted in a case series in which tetracycline fibers were added to mechanical debridement [85]. The adjunct use of slow-release doxycycline revealed a clinical benefit over mechanical therapy alone [86], as did the addition of minocycline microspheres [83, 87]. However, the effectiveness of focal sustained drug delivery systems in peri-implantitis is mostly obscure, and further studies are needed to establish a successful treatment modality.

14.5 Focal Delivery Systems Against Endodontic Infection

Endodontic treatment is performed once the dental pulp is infected or undergoes necrosis. The treatment consists of tissue debridement and disinfection, followed by sealing the tooth's root canal system (Fig. 14.9). Most often the endodontic treatment is performed locally, without any need for systemic treatment. Only when the infection results in abscess formation or in systemic illness is administration of antibiotics indicated. Therefore, most endodontic treatment modalities are considered focal chemical systems.

Elimination of microorganisms from infected root canal systems is complicated. Numerous measures have been described to reduce the number of microorganisms, including the use of various mechanical instrumentation techniques, irrigation

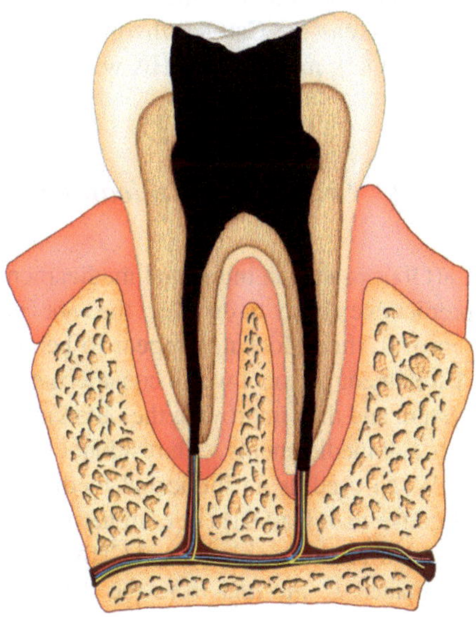

Fig. 14.9 Root canal system cavity (marked in *black*)

regimes, and intracanal medicaments. There is no definitive evidence in the literature indicating that mechanical instrumentation alone will result in a bacteria-free root canal system. This is because of the complexity of the root canal system and the fact that mechanical instrumentation leaves significant portions of the root canal walls untouched [88]. Therefore, some form of chemical irrigation and disinfection is necessary to remove tissue and other debris from the root canal system and to eliminate any remaining microorganisms.

14.5.1 Irrigation Regimes Against Endodontic Infection

CHX possesses adequate antimicrobial properties for use as an antimicrobial endodontic irrigant. In an in vitro model, CHX at different concentrations (0.2 %, 1 %, and 2 %) and in various forms (gel and liquid) effectively killed the endodontic-associated pathogen *Enterococcus faecalis* [89]. In the liquid form, CHX at these concentrations and sodium hypochlorite (NaOCI) (5.25 %) were the most effective. Although all the tested irrigants exhibited antibacterial activity, the time required to eliminate *E. faecalis* depended on the concentration and type of irrigant used. The application of 2 % CHX gel as an endodontic irrigant resulted in a clean root canal surface and had an antimicrobial ability comparable with that of other solutions [90]. In an in vivo antimicrobial study, the addition of 2 % CHX rinse to the conventional treatment protocol increased the effectiveness of disinfection compared with that of the control [91]. Four intracanal medications ($Ca(OH)_2$/glycerin, $Ca(OH)_2$/0.12 % CHX, $Ca(OH)_2$/hydroxide/camphorated paramonochlorophenol/glycerin, and 0.12 % CHX/zinc oxide) were used to disinfect the root dentine in bovine teeth that had been infected with Candida albicans [92]. The specimens exposed to pastes containing the latter two medications were completely disinfected after 1 h of exposure, whereas the $Ca(OH)_2$/hydroxide/glycerin paste required 7 days of exposure and $Ca(OH)_2$ mixed with CHX was ineffective even after 1 week.

14.5.2 Intracanal Medicaments Against Endodontic Infections

$Ca(OH)_2$ is also an intracanal medicament which is commonly used because of its predictable ability to disinfect the root canal system. However, after its mechanical removal, there does not appear to be any residual antimicrobial effect [93, 94]. A treatment protocol worth considering for preventing reinfection of the root canal system and thus improving the outcome of endodontic treatment is the use of irrigants and medicaments with residual antimicrobial activity.

When used as an intracanal medicament, CHX has been reported to be more effective than $Ca(OH)_2$ in eliminating *E. faecalis* from inside the dentinal tubules [95]. Different formulations of CHX, including a CHX/$Ca(OH)_2$ 1:1 mix, efficiently eliminated *E. faecalis* from the dentinal tubules with a 1 % CHX gel working slightly better than the other preparations [96]. These findings were corroborated in bovine [97] and human dentine [98]. The 2 % CHX gel was also significantly more effective than the $Ca(OH)_2$/2 % CHX mix against *C. albicans* after 7 days, whereas $Ca(OH)_2$ alone was completely ineffective.

CHX is unique in that the dentine medicated with it acquires antimicrobial substantivity. The positively charged CHX ions adsorb to dentine and prevent microbial colonization on its surface for some time beyond the actual period of medication [95]. In an in vivo periodontal study, Stabholz et al. [99] evaluated the substantivity of the human root surface after in situ subgingival irrigation with tetracycline HCL and CHX. They found that the substantivity of 50 mg mL^{-1} tetracycline was significantly greater than that of CHX over a period of 12 days and greater than that of saline over 16 days. A 5 min application of a 2 % CHX solution induced substantivity for up to 4 weeks [100], and a 10 min application of a 2 % CHX solution resulted in the drug's retention in the root canal dentine in antimicrobially effective amounts for up to 12 weeks [101].

MTAD, a relatively new root canal irrigant, introduced by Torabinejad et al. in 2003 [102]. It is a solution containing 3 % doxycycline, 4.25 % citric acid, and a detergent (0.5 % polysorbate 80). Several studies evaluated its effectiveness in disinfection of root canals and found that it is able to remove the smear layer [102] and is effective against *E. faecalis* [103].

MTAD was found to be more effective than 5.25 % NaOCl alone in disinfecting root canals [104]. As stated above, tetracyclines (including doxycycline) readily attach to dentine and are subsequently released without losing their antibacterial activity. The presence of doxycycline in MTAD suggests that the latter may have some enduring antimicrobial action. An in vitro study that evaluated the substantivity of NaOCl, CHX, and MTAD, using a bovine dentine tube model, showed that the substantivity of MTAD was significantly greater than that of CHX and NaOCl [100].

14.6 Conclusions

In the oral cavity, focal drug delivery may offer many advantages as a therapeutic means for preventing or treating dental diseases. Moreover, slow release and focal treatment may overcome disadvantages such as low dosage and limited penetration into the surrounding tissues while minimizing the risk of undesired side effects.

References

1. Watnick P, Kolter R (2000) Biofilm, city of microbes. J Bacteriol 182(10):2675–2679
2. Amano A, Nakagawa I, Hamada S (1999) Studying initial phase of biofilm formation: molecular interaction of host proteins and bacterial surface components. Methods Enzymol 310: 501–513
3. Marsh PD, Bradshaw DJ (1995) Dental plaque as a biofilm. J Ind Microbiol 15(3):169–175

4. Hannig C, Hannig M (2009) The oral cavity–a key system to understand substratum-dependent bioadhesion on solid surfaces in man. Clin Oral Investig 13(2):123–139
5. Herrera D et al (2008) Antimicrobial therapy in periodontitis: the use of systemic antimicrobials against the subgingival biofilm. J Clin Periodontol 35(8 Suppl):45–66
6. Kolenbrander PE (2000) Oral microbial communities: biofilms, interactions, and genetic systems. Annu Rev Microbiol 54:413–437
7. Kolenbrander PE et al (2006) Bacterial interactions and successions during plaque development. Periodontol 2000 42:47–79
8. Hardie JM (1992) Oral microbiology: current concepts in the microbiology of dental caries and periodontal disease. Br Dent J 172(7):271–278
9. Zitzmann NU, Berglundh T (2008) Definition and prevalence of peri-implant diseases. J Clin Periodontol 35(8 Suppl):286–291
10. Donlan RM (2000) Role of biofilms in antimicrobial resistance. ASAIO J 46(6):S47–S52
11. Lewis K (2001) Riddle of biofilm resistance. Antimicrob Agents Chemother 45(4):999–1007
12. Mah TF, O'Toole GA (2001) Mechanisms of biofilm resistance to antimicrobial agents. Trends Microbiol 9(1):34–39
13. Xu KD, McFeters GA, Stewart PS (2000) Biofilm resistance to antimicrobial agents. Microbiology 146(Pt 3):547–549
14. Walters MC 3rd et al (2003) Contributions of antibiotic penetration, oxygen limitation, and low metabolic activity to tolerance of Pseudomonas aeruginosa biofilms to ciprofloxacin and tobramycin. Antimicrob Agents Chemother 47(1):317–323
15. Roberts AP, Mullany P (2006) Genetic basis of horizontal gene transfer among oral bacteria. Periodontol 2000 42:36–46
16. Baehni PC, Takeuchi Y (2003) Anti-plaque agents in the prevention of biofilm-associated oral diseases. Oral Dis 9(Suppl 1):23–29
17. van der Ouderaa FJ (1991) Anti-plaque agents. Rationale and prospects for prevention of gingivitis and periodontal disease. J Clin Periodontol 18(6):447–454
18. Stewart PS (2003) Diffusion in biofilms. J Bacteriol 185(5):1485–1491
19. Smith AW (2005) Biofilms and antibiotic therapy: is there a role for combating bacterial resistance by the use of novel drug delivery systems? Adv Drug Deliv Rev 57(10): 1539–1550
20. Beyth N et al (2008) Surface antimicrobial activity and biocompatibility of incorporated polyethylenimine nanoparticles. Biomaterials 29(31):4157–4163
21. Beyth N et al (2006) Antibacterial activity of dental composites containing quaternary ammonium polyethylenimine nanoparticles against Streptococcus mutans. Biomaterials 27(21):3995–4002
22. Beyth N et al (2010) Polyethyleneimine nanoparticles incorporated into resin composite cause cell death and trigger biofilm stress in vivo. Proc Natl Acad Sci U S A 107(51):22038–22043
23. Nel AE et al (2009) Understanding biophysicochemical interactions at the nano-bio interface. Nat Mater 8(7):543–557
24. Marinho VC et al (2003) Systematic review of controlled trials on the effectiveness of fluoride gels for the prevention of dental caries in children. J Dent Educ 67(4):448–458
25. Featherstone JD (2000) The science and practice of caries prevention. J Am Dent Assoc 131(7):887–899
26. Featherstone JD (1999) Prevention and reversal of dental caries: role of low level fluoride. Community Dent Oral Epidemiol 27(1):31–40
27. Pessan JP et al (2008) Slow-release fluoride devices: a literature review. J Appl Oral Sci 16(4):238–246
28. Featherstone JD (2006) Delivery challenges for fluoride, chlorhexidine and xylitol. BMC Oral Health 6(Suppl 1):S8
29. Reynolds EC (1998) Anticariogenic complexes of amorphous calcium phosphate stabilized by casein phosphopeptides: a review. Spec Care Dentist 18(1):8–16
30. Reynolds EC (1987) The prevention of sub-surface demineralization of bovine enamel and change in plaque composition by casein in an intra-oral model. J Dent Res 66(6):1120–1127

31. Rao SK et al (2009) Study of the efficacy of toothpaste containing casein phosphopeptide in the prevention of dental caries: a randomized controlled trial in 12- to 15-year-old high caries risk children in Bangalore, India. Caries Res 43(6):430–435
32. Bailey DL et al (2009) Regression of post-orthodontic lesions by a remineralizing cream. J Dent Res 88(12):1148–1153
33. Morgan MV et al (2008) The anticariogenic effect of sugar-free gum containing CPP-ACP nanocomplexes on approximal caries determined using digital bitewing radiography. Caries Res 42(3):171–184
34. Yengopal V, Mickenautsch S (2009) Caries preventive effect of casein phosphopeptide-amorphous calcium phosphate (CPP-ACP): a meta-analysis. Acta Odontol Scand 67(6): 321–332
35. Cross KJ, Huq NL, Reynolds EC (2007) Casein phosphopeptides in oral health–chemistry and clinical applications. Curr Pharm Des 13(8):793–800
36. Anderson MH (2003) A review of the efficacy of chlorhexidine on dental caries and the caries infection. J Calif Dent Assoc 31(3):211–214
37. Jordan HV, De Paola PF (1974) Effect of a topically applied 3 percent vancomycin gel on Streptococcus mutans on different tooth surfaces. J Dent Res 53(1):115–120
38. Zander HA, Bibby BG (1947) Laboratory and animal studies on the effect of penicillin on caries activity. J Dent Res 26(6):454
39. Walker C (1990) Effects of sanguinarine and Sanguinaria extract on the microbiota associated with the oral cavity. J Can Dent Assoc 56(7 Suppl):13–30
40. Shapiro S, Giertsen E, Guggenheim B (2002) An in vitro oral biofilm model for comparing the efficacy of antimicrobial mouthrinses. Caries Res 36(2):93–100
41. Autio-Gold J (2008) The role of chlorhexidine in caries prevention. Oper Dent 33(6): 710–716
42. van Rijkom HM, Truin GJ, Van't Hof MA (1998) A meta-analysis of clinical studies on the caries-inhibiting effect of fluoride gel treatment. Caries Res 32(2):83–92
43. Svatun B et al (1993) The effects of three silica dentifrices containing Triclosan on supragingival plaque and calculus formation and on gingivitis. Int Dent J 43(4 Suppl 1):441–452
44. Panagakos FS et al (2005) Advanced oral antibacterial/anti-inflammatory technology: a comprehensive review of the clinical benefits of a triclosan/copolymer/fluoride dentifrice. J Clin Dent 16(Suppl):S1–S19
45. Gaffar A, et al. (1990) Antiplaque effects of dentifrices containing triclosan/copolymer/NaF system versus triclosan dentifrices without the copolymer. Am J Dent 3 Spec No:S7–14
46. Nabi N, et al. (1989) In vitro and in vivo studies on triclosan/PVM/MA copolymer/NaF combination as an anti-plaque agent. Am J Dent 2 Spec No:197–206
47. Chen F et al (2009) Tooth-binding micelles for dental caries prevention. Antimicrob Agents Chemother 53(11):4898–4902
48. Petti S, Scully C (2009) Polyphenols, oral health and disease: a review. J Dent 37(6): 413–423
49. Brighenti FL et al (2012) Effect of Psidium cattleianum leaf extract on enamel demineralisation and dental biofilm composition in situ. Arch Oral Biol 57(8):1034–1040
50. Diamond G et al (2009) The roles of antimicrobial peptides in innate host defense. Curr Pharm Des 15(21):2377–2392
51. Rennie J et al (2005) Simple oligomers as antimicrobial peptide mimics. J Ind Microbiol Biotechnol 32(7):296–300
52. Wang TT et al (2004) Cutting edge: 1,25-dihydroxyvitamin D3 is a direct inducer of antimicrobial peptide gene expression. J Immunol 173(5):2909–2912
53. Beckloff N et al (2007) Activity of an antimicrobial peptide mimetic against planktonic and biofilm cultures of oral pathogens. Antimicrob Agents Chemother 51(11):4125–4132
54. Li L et al (2010) Design and characterization of an acid-activated antimicrobial peptide. Chem Biol Drug Des 75(1):127–132
55. Russell MW et al (2004) A caries vaccine? the state of the science of immunization against dental caries. Caries Res 38(3):230–235

56. Sun Y et al (2012) Flagellin-PAc fusion protein is a high-efficacy anti-caries mucosal vaccine. J Dent Res 91(10):941–947
57. Smith DJ, Mattos-Graner RO (2008) Secretory immunity following mutans streptococcal infection or immunization. Curr Top Microbiol Immunol 319:131–156
58. Ma JK et al (1998) Characterization of a recombinant plant monoclonal secretory antibody and preventive immunotherapy in humans. Nat Med 4(5):601–606
59. Meurman JH (2005) Probiotics: do they have a role in oral medicine and dentistry? Eur J Oral Sci 113(3):188–196
60. Milgrom P et al (2006) Mutans streptococci dose response to xylitol chewing gum. J Dent Res 85(2):177–181
61. Eke PI et al (2012) Prevalence of periodontitis in adults in the United States: 2009 and 2010. J Dent Res 91(10):914–920
62. Paper I (1999) The pathogenesis of periodontal diseases. J Periodontol 70(4):457–470
63. Lang NP, Berglundh T (2011) Periimplant diseases: where are we now?–Consensus of the Seventh European Workshop on Periodontology. J Clin Periodontol 38(Suppl 11):178–181
64. Cionca N et al (2010) Microbiologic testing and outcomes of full-mouth scaling and root planing with or without amoxicillin/metronidazole in chronic periodontitis. J Periodontol 81(1):15–23
65. Pavicic MJ et al (1994) Microbiological and clinical effects of metronidazole and amoxicillin in Actinobacillus actinomycetemcomitans-associated periodontitis. A 2-year evaluation. J Clin Periodontol 21(2):107–112
66. Davies RM, Ellwood RP, Davies GM (2004) The effectiveness of a toothpaste containing triclosan and polyvinyl-methyl ether maleic acid copolymer in improving plaque control and gingival health: a systematic review. J Clin Periodontol 31(12):1029–1033
67. Rosling B et al (1997) The use of a triclosan/copolymer dentifrice may retard the progression of periodontitis. J Clin Periodontol 24(12):873–880
68. Ellwood RP et al (1998) Effect of a triclosan/copolymer dentifrice on the incidence of periodontal attachment loss in adolescents. J Clin Periodontol 25(5):363–367
69. Cullinan MP et al (2003) The effect of a triclosan-containing dentifrice on the progression of periodontal disease in an adult population. J Clin Periodontol 30(5):414–419
70. Javed F, Al-Hezaimi K, Romanos GE (2012) Role of dentifrices with essential oil formulations in periodontal healing. Am J Med Sci 343(5):411–417
71. Fornell J, Sundin Y, Lindhe J (1975) Effect of listerine on dental plaque and gingivitis. Scand J Dent Res 83(1):18–25
72. DePaola LG et al (1989) Chemotherapeutic inhibition of supragingival dental plaque and gingivitis development. J Clin Periodontol 16(5):311–315
73. Gordon JM, Lamster IB, Seiger MC (1985) Efficacy of Listerine antiseptic in inhibiting the development of plaque and gingivitis. J Clin Periodontol 12(8):697–704
74. Loe H, Schiott CR (1970) The effect of mouthrinses and topical application of chlorhexidine on the development of dental plaque and gingivitis in man. J Periodontal Res 5(2):79–83
75. Flotra L et al (1972) A 4-month study on the effect of chlorhexidine mouth washes on 50 soldiers. Scand J Dent Res 80(1):10–17
76. Gunsolley JC (2010) Clinical efficacy of antimicrobial mouthrinses. J Dent 38(Suppl 1): S6–S10
77. Overholser CD et al (1990) Comparative effects of 2 chemotherapeutic mouthrinses on the development of supragingival dental plaque and gingivitis. J Clin Periodontol 17(8):575–579
78. Pires JR, Rossa Junior C, Pizzolitto AC (2007) In vitro antimicrobial efficiency of a mouthwash containing triclosan/gantrez and sodium bicarbonate. Braz Oral Res 21(4):342–347
79. Gusberti FA et al (1988) Microbiological and clinical effects of chlorhexidine digluconate and hydrogen peroxide mouthrinses on developing plaque and gingivitis. J Clin Periodontol 15(1):60–67
80. Weitzman SA et al (1984) Chronic treatment with hydrogen peroxide. Is it safe? J Periodontol 55(9):510–511

81. Brecx M et al (1992) Efficacy of listerine, meridol and chlorhexidine mouthrinses as supplements to regular tooth cleaning measures. J Clin Periodontol 19(3):202–207
82. Kalsi R, Vandana KL, Prakash S (2011) Effect of local drug delivery in chronic periodontitis patients: a meta-analysis. J Indian Soc Periodontol 15(4):304–309
83. Renvert S et al (2009) Mechanical non-surgical treatment of peri-implantitis: a double-blind randomized longitudinal clinical study. I: clinical results. J Clin Periodontol 36(7):604–609
84. Karring ES et al (2005) Treatment of peri-implantitis by the Vector® system: a pilot study. Clin Oral Implants Res 16(3):288–293
85. Mombelli A et al (2001) Treatment of peri-implantitis by local delivery of tetracycline: clinical, microbiological and radiological results. Clin Oral Implants Res 12(4):287–294
86. Büchter A et al (2004) Sustained release of doxycycline for the treatment of peri-implantitis: randomised controlled trial. Br J Oral Maxillofac Surg 42(5):439–444
87. Renvert S et al (2004) Treatment of incipient peri-implant infections using topical minocycline microspheres versus topical chlorhexidine gel as an adjunct to mechanical debridement. J Int Acad Periodontol 6(4 Suppl):154–159
88. Peters OA et al (2001) Changes in root canal geometry after preparation assessed by high-resolution computed tomography. J Endod 27(1):1–6
89. Gomes BPFA et al (2001) In vitro antimicrobial activity of several concentrations of sodium hypochlorite and chlorhexidine gluconate in the elimination of Enterococcus faecalis. Int Endod J 34(6):424–428
90. Ferraz CCR et al (2001) In vitro assessment of the antimicrobial action and the mechanical ability of chlorhexidine gel as an endodontic irrigant. J Endod 27(7):452–455
91. Zamany A, Safavi K, Spångberg LSW (2003) The effect of chlorhexidine as an endodontic disinfectant. Oral Surg Oral Med Oral Pathol Oral Radiol Endod 96(5):578–581
92. Siqueira JF Jr et al (2003) Elimination of Candida albicans infection of the radicular dentin by intracanal medications. J Endod 29(8):501–504
93. Messer HH, Chen R-S (1984) The duration of effectiveness of root canal medicaments. J Endod 10(6):240–245
94. Byström A, Claesson R, Sundqvist G (1985) The antibacterial effect of camphorated paramonochlorophenol, camphorated phenol and calcium hydroxide in the treatment of infected root canals. Dent Traumatol 1(5):170–175
95. Athanassiadis B, Abbott PV, Walsh LJ (2007) The use of calcium hydroxide, antibiotics and biocides as antimicrobial medicaments in endodontics. Aust Dent J 52:S64–S82
96. Almyroudi A et al (2002) The effectiveness of various disinfectants used as endodontic intracanal medications: an in vitro study. J Endod 28(3):163–167
97. Gomes BPFA et al (2003) Effectiveness of 2% chlorhexidine gel and calcium hydroxide against Enterococcus faecalis in bovine root dentine in vitro. Int Endod J 36(4):267–275
98. Schäfer E, Bössmann K (2005) Antimicrobial efficacy of chlorhexidine and two calcium hydroxide formulations against enterococcus faecalis. J Endod 31(1):53–56
99. Stabholz A et al (1993) Retention of antimicrobial activity by human root surfaces after in situ subgingival irrigation with tetracycline HCl or chlorhexidine. J Periodontol 64(2):137–141
100. Khademi AA, Mohammadi Z, Havaee A (2006) Evaluation of the antibacterial substantivity of several intra-canal agents. Aust Endod J 32(3):112–115
101. Rosenthal S, Spångberg L, Safavi K (2004) Chlorhexidine substantivity in root canal dentin. Oral Surg Oral Med Oral Pathol Oral Radiol Endod 98(4):488–492
102. Torabinejad M et al (2003) A new solution for the removal of the smear layer. J Endod 29(3):170–175
103. Shabahang S, Torabinejad M (2003) Effect of MTAD on Enterococcus faecalis–contaminated root canals of extracted human teeth. J Endod 29(9):576–579
104. Shabahang S, Pouresmail M, Torabinejad M (2003) In vitro antimicrobial efficacy of MTAD and sodium hypochlorite. J Endod 29(7):450–452

Chapter 15
Segmental-Dependent Drug Absorption and Delivery: The Stomach

Omri Wolk and Arik Dahan

15.1 Introduction

Oral administration is by far the most popular and convenient way of drug delivery. Oral dosage forms are easy to use and usually do not require hospitalization in a medical facility or the assistance of medical professionals. In addition, oral uptake of drugs prevents both the local side effects and the risk of systemic infections that are associated with injections.

However, the majority of drugs that are taken orally must enter the enterocytes and cross through the gastrointestinal tract (GIT) membrane in order to be absorbed to the systemic bloodstream and exert their pharmacological effect. The GIT proves to be a formidable barrier, to the point where many drug substances, such as antibodies, hormones, and even small molecules, fail to penetrate through it and have to be administered parenterally. Drugs that are taken through the oral route may be heavily influenced by the physiological conditions in the GIT, resulting in significant diversions from the predicted/desired pharmacokinetic and pharmacodynamic patterns.

The GIT can be broadly divided to several regions: the stomach, the small intestine (subdivided to duodenum, jejunum, and ileum), and the colon. The environment in each of these regions is dependent on a multitude of factors; among them are pH values, membrane transporters expression, metabolic enzymes activity, and resident microflora. Each of these variables, alone or in combination with others, can fundamentally alter the solubility and/or permeability of drugs [1–3].

In light of this great variability in conditions across the intestine, many attempts were made to deliver drugs to specific regions of the GIT. The goal of many of these

O. Wolk • A. Dahan, Ph.D. (✉)
Department of Clinical Pharmacology, School of Pharmacy, Faculty of Health Sciences, Ben-Gurion University of the Negev, P.O. Box 653, Beer-Sheva 84105, Israel
e-mail: arikd@bgu.ac.il

A.J. Domb and W. Khan (eds.), *Focal Controlled Drug Delivery*, Advances in Delivery Science and Technology, DOI 10.1007/978-1-4614-9434-8_15,

attempts was to optimize systemic absorption of drugs, while others were intended to retain the drugs within the lumen or epithelial cells and treat local ailments [4].

The focus of this chapter is gastroretentive dosage forms (GRDF), which are designed to resist the normal tendency of the stomach to clear its content and to allow targeted and controlled release of the drug to the upper GI tract. A special emphasis will be given to the rationale for GRDF, different GRDF technologies, and their unique applications. A specific case of GRDFs for the treatment of local *Helicobacter pylori*-related pathologies will be reviewed.

15.2 GRDF

15.2.1 Definition and Purpose of GRDF

Controlled release oral drug delivery systems offer several advantages over immediate release dosage forms. First, they reduce fluctuations in drug concentrations in the plasma and possibly at the site of action; thus, they may increase the therapeutic effect of the drug while reducing side effects. Moreover, controlled release formulations decrease the administration frequency of the drug, leading to improved patient compliance [5].

Once an oral dosage form is taken, it passes through the esophagus to the stomach, where it lingers before continuing down the intestinal tract.

While transit time through the small intestine is fairly constant (3 ± 1 h) [6], the gastric residence time of dosage forms is highly variable and can range from several minutes to several hours [7]. This variability is easily explained considering the numerous factors affecting gastric emptying. These would include physiological factors such as body posture, age, gender, osmolarity, and pH [7]; pathological factors, for example, *H. pylori* infection [8]; and pharmaceutical agents, e.g., antacids containing Mg+ [9] and erythromycin [10].

Once emptied from the stomach, the drug enters the upper small intestine. This region is the main absorption site for a number of drugs: small hydrophilic drugs exploit the relatively large pores in the epithelial junctions at this region to passively penetrate the blood stream [11]. In addition, the upper intestine is characterized by high expression of several transporters (e.g., PEPT1 and CNT) [12], making it an ideal absorption site for their substrates. Finally, the P-gp efflux pump is relatively absent from the proximal GI tract [1, 13, 14]. This allows P-gp substrates that are absorbed in this region to avoid efflux.

Transit time through the upper intestinal region, i.e., the duodenum, tends to be constant and rapid. As a result, "standard" controlled release formulation would allow only limited amount of drug to be released at this site and permeate into the bloodstream. It is therefore advantageous in some cases to prolong the residence time of the controlled release dosage forms in the stomach, thus maintaining the benefits of both the controlled release formulation and the upper small intestine [5].

Table 15.1 Drugs that were studied in gastroretentive dosage forms

Drug	References
Acyclovir	[16]
Atenolol	[16]
Bisphosphonates	[17]
Calcium	[18]
Captopril	[19]
Ciprofloxacin	[20]
Cisapride	[21]
Dextromethorphan HBr	[22]
Diltiazem	[23]
Dipyridamole	[24]
Glipizide	[25]
Ketoprofen	[26]
Metformin	[27]
Metoprolol	[16]
Propranolol HCl	[28]
Ranitidine HCl	[29]
Tetracycline	[30]
Valacyclovir	[16]
Verapamil	[31]

Table 15.2 Examples for GRDFs currently marketed

Brand name	Drug	GRDF type	References
MadoparHBS (PropalHBS)	Levodopa and benserazide	Floating CR capsules	[32]
Valrelease	Diazepam	Floating capsules	[33]
Amalgate float coat	Antacid	Floating dosage form	[32]
Topalkan	Aluminum and magnesium mixture	Effervescent floating liquid alginate preparation	[32]
Conviron	Ferrous sulfate	Colloidal gel-forming FDDS	[34]
Cifran OD	Ciprofloxacin	Gas-generating floating form	[34]
Cytotech	Misoprostol	Bilayer floating capsule	[34]
Liquid gaviscone	Mixture of alginate	Suppress gastroesophageal reflux and alleviate the heart burn	[34]

The term GRDF encompasses several technologies with one common feature: they are designed to resist the normal tendency of the stomach to clear its content, resulting in a targeted and controlled release of the active moiety to the upper GI tract [11]. The improved release profile from GRDFs may augment absorption to the systemic bloodstream for drugs with a narrow absorption window in the upper part of the GIT or drugs with a poor stability in the colon. Furthermore, once the drug is released from the GRDF, it can act within the stomach and treat local disease conditions with increased efficacy [15]. A few examples for drugs that were tested for GRDFs are listed in Table 15.1. Examples for GRDFs currently marketed can be found in Table 15.2.

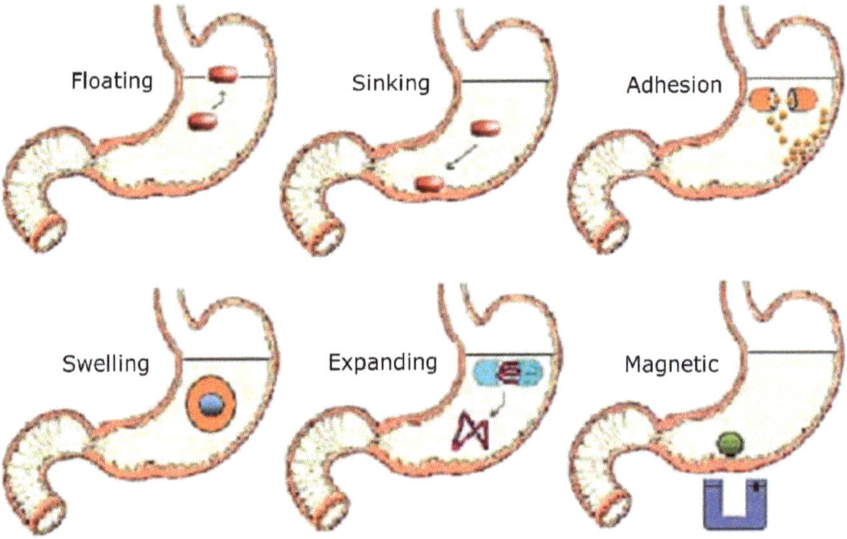

Fig. 15.1 Schematic description of the various GRDF types. Modified from [35] with permission

15.2.2 General Considerations in the Development of GRDF

The two most important parameters to be considered when developing gastric retentive dosage forms are their size and density. Size is especially important in designing indigestible solid dosage forms (single-unit systems). When open, the average human pyloric sphincter diameter is 12 ± 7 mm, and so a practical diameter >15 mm is necessary for useful prolonged retention. A single-unit system can be removed from the stomach at its entirety at any particular gastric emptying cycle. Multiple-unit systems, such as those based on microparticles, avoid this "all-or-nothing" behavior [15].

Density determines the location of the system in the stomach. Systems with density lower than gastric contents can float to the surface, while high-density systems sink to the bottom of the stomach. Both positions may isolate the dosage system from the pylorus [15].

15.2.3 Types of GRDF

Gastroretentive formulation may be broadly classified into high-density (sinking) systems, floating systems, expandable systems, superporous hydrogel systems, mucoadhesive systems, and magnetic systems [15]. The different approaches are graphically portrayed in Fig. 15.1.

15.2.3.1 High-Density Systems

Gastric contents have a density close to water (1.004 g/cm^3). Barium sulfate, zinc oxide, iron powder titanium, and titanium dioxide are used to create small and high-density (d>2.5 g/cm^3) pellets that sink to the bottom of the stomach, where they become entrapped in the folds of the antrum and withstand the peristaltic waves of the stomach wall [15]. This method has yet to produce evidence of efficiency in human subjects [15].

15.2.3.2 Floating Systems

These have a lower density than the gastric content and float in the stomach for a prolonged period of time while the drug is released from the dosage form at a constant rate. Eventually, the residual system is emptied from the stomach. These systems rely heavily on the presence of food to retard emptying [36–38] and increase liquid volume for effective buoyancy [38]. Floating systems are subdivided to several groups, as detailed below.

Hydrodynamically Balanced Systems

Hydrodynamically balanced systems (HBS) systems are single-unit dosage forms, containing one or more gel-forming hydrophilic polymers (e.g., HPMC, HEC, HPC). The polymer is mixed with drug and usually encapsulated in gelatin. The capsule rapidly dissolves in the gastric fluid, and hydration and swelling of the surface polymers produces a floating mass. Drug release is controlled by the formation of a hydrated boundary at the surface. Continuous erosion of the surface allows water penetration to the inner layers, maintaining surface hydration and buoyancy [39]. The main drawback in the original method was that both buoyancy and drug release rate depended on the intrinsic features of the polymer, and controlling both variants at the same time proved to be difficult [40]. In order to overcome this problem, several formulations were developed in which one phase provided floatability and the other controlled drug release, so each could be manipulated independently [15]. Ali et al. designed a formulation containing 500 mg of metformin granulated with 5 % of ethyl cellulose and coated with 150 mg of HPMC K4M. This dosage form gave an in vitro release of 97 % in 12 h in simulated gastric fluid at pH 3, as well as a prolonged period of stomach retention and increased in vivo AUC when compared to immediate release capsules of metformin [41].

Gas-Forming Systems

Floatability can also be achieved by generation of entrapped gas bubbles. CO_2 can be generated in situ by incorporation of carbonates or bicarbonates, which react with acid, either the natural gastric acid or co-formulated as citric or tartaric acid [15, 42]. Alternatively, it is possible to entrap liquid within the formula, which forms a gas at body temperature [43]. It is important to note, however, that this type of formulation has a lag time while the gas-forming reaction is taking place, in which the dosage form may be emptied from the stomach.

Raft-Forming Systems

These formulations are based on a gel-forming solution (e.g., sodium alginate solution containing carbonates or bicarbonates). Once they encounter the gastric acid, they swell and form a viscous cohesive gel containing entrapped CO_2 bubbles. Raft-forming systems produce a layer on the top of gastric fluids and therefore are useful in the treatment of gastroesophageal reflux [44, 45] as was recently demonstrated for the raft-forming formulation Gaviscon Advance [46].

Low-Density Systems

This system solves the lag time problem of the gel- and gas-based formulas by producing light, instantly floating hollow microspheres (microballoons) [47] foam particles [48] or matrix [49].

This is done by simple solvent evaporation or solvent diffusion/evaporation methods. Polycarbonate, Eudragit S, cellulose acetate, calcium alginate, agar, and low-methoxylated pectin are commonly used as polymers [47, 50]. Buoyancy and drug release are dependent on the quantity of polymer, the plasticizer–polymer ratio, and the solvent used [39]. Talukder and Fassihi developed a multiple-unit system based on cross-linked beads. They were made by using Ca^{+2} and low-methoxylated pectin (anionic polysaccharide), with or without sodium alginate. Riboflavin, tetracycline, and methotrexate were used as model drugs, and drying was done by two methods: air convection oven at 40 °C for 6 h and freeze drying. The freeze-dried beads remained buoyant over 12 h in USP hydrochloride buffer (pH 1.5), while the air-dried beads sank. Calcium–pectinate–alginate beads released their contents at much faster rates than did calcium–pectinate beads (100 % vs. 50 % in 10 h) [51]. Hollow microspheres are considered to be one of the most promising buoyant systems because they combine the advantages of multiple-unit systems and good floating properties [15].

15.2.3.3 Expendable Systems

This dosage form consists of a capsule, incorporating a compressed biodegradable polymer which expands in the stomach to a large rigid structure, followed by erosion that releases the drug and removes the residual formulation. Several such formulations were investigated [15, 52, 53]. Klausner et al. described a levodopa GRDF that combines extended dimensions (5 cm ± 2.5 cm) with high rigidity and is folded into a large gelatin capsule. In vitro and in vivo studies showed that the drug delivery system reached its unfolded form in 15 min and was maintained for at least 2 h [53]. In humans, 67 % of drug delivery systems containing levodopa were retained in the stomach during 5 h. The plasma concentration–time curve was very similar to that of the reference drug (Sinemet CR®) but showed an extended absorption phase. Rigidity of the system was a crucial parameter; a system with an extended size but with a lack of high rigidity was not retained in the stomach [53].

While expendable systems are an interesting and innovative approach for stomach retention, this kind of dosage form is probably the most difficult to industrialize and may not be cost-effective [34]. Furthermore, it is extremely important for expandable systems not to have sharp edges or cause local damage on prolonged retention and be completely biodegradable, since permanent retention of rigid, large single-unit forms may cause bowel obstruction, intestinal adhesion, and gastropathy [15].

15.2.3.4 Superporous Hydrogels

Superporous hydrogels have a network of large (>100 Am) interconnected open pores. Water rapidly enters the gel through the pores, inflaming the gel to 100 times or more its original volume within a minute. The adding of croscarmellose sodium to the gel gives it the mechanical strength to withstand pressure by gastric contraction. After several hours the gel is degraded and cleared from the stomach [54].

15.2.3.5 Mucoadhesive/Bioadhesive Systems

Mucoadhesion refers to the ability of a dosage form to stick to the mucosal surface by different mechanisms, such as attractive electrostatic forces, Van der Waals forces, hydrogen bonding, and physical penetration of mucin strands into the porous polymer.

Poly(acrylic acid) (Carbopol, polycarbophil), chitosan, Gantrez (Polymethyl vinyl ether/maleic anhydride copolymers), cholestyramine, tragacanth, sodium alginate, HPMC, sephadex, sucralfate, polyethylene glycol, dextran, poly(alkyl cyanoacrylate), and polylactic acid are commonly used for bioadhesion. Even though some of these polymers are effective at producing bioadhesion, it is difficult to maintain its effectivity because of the rapid turnover of mucus in the GIT and the hydration of the stomach content [55, 56]. In spite of that, Singh et al. were recently able to develop an effervescent floating-bioadhesive formulation that was able to retain the antiviral drug lamivudine for at least 5 h and possessed adequate drug release control and pharmacokinetic extension of plasma levels [57].

15.2.3.6 Magnetic Systems

This system is based on a simple idea: the dosage form contains a small internal magnet, and a magnet is placed on the abdomen over the position of the stomach. A magnetic tablet containing acyclovir produced significantly higher plasma concentrations when compared to immediate release tablets [58]. The external magnet must be positioned with great precision which might prove to be difficult for the patients [15].

15.2.4 *Specific Case: GRDFs for Treatment of Local* H. pylori-*Related Pathologies*

H. pylori is one of the most prevalent bacterial pathogens, affecting as many as 50 % and 80 % of the population in industrialized and developing countries, respectively [59, 60]. The bacteria colonize the surface mucus and mucosa of the lower part of the stomach, adjacent to the pylorus [11]. It is associated with many gut-related disorders, including peptic ulcer, gastric cancer, and mucosa-associated lymphoma [11, 59, 61–63]. The treatment of *H. pylori* infections usually consists of at least two different types of antibiotics and proton pump inhibitors or H_2 blockers [64].

In order to successfully eradicate the bacteria, high concentrations of antibiotics need to be maintained in this area for a substantial period of time. However, since the residence time of conventional dosage forms is relatively short, achieving adequate local levels of antibiotics is proving to be a major challenge. Indeed, the short retention time of conventional dosage forms in the stomach may account for the recent reports of *H. pylori's* incomplete eradication and subsequent recurrence [65] and its increasing resistance to antibiotics [65–67]. GRDF-based formulations have the potential to both enhance the effectiveness of antibiotics and simplify their administration regimens, thus improving compliance and increasing the probability for total bacteria eradication.

Several approaches were attempted in the development of stomach-targeting GRDFs with variable success. Yang et al. proposed a gas-generating system consisting of a triple-layer tablet. One layer was a swellable gas-generating layer of polyethylene oxide, HPMC, and sodium bicarbonate/calcium carbonate (1:2 w/w). The second one was the expandable/sustainable drug-containing layer of polyethylene oxide, tetracycline hydrochloride, and metronidazole. The third one was a rapidly dissolving drug layer (bismuth salts). The group obtained in vitro (in a 37 °C, 0.1 M HCl solution) duration of buoyancy and sustained release of metronidazole and tetracycline over 6–8 h, with a buoyancy lag time in the range of 17–28 min. However, no in vivo data are available concerning the floating characteristics of the drug delivery system or its effect against *H. pylori* [68]. Liu et al. studied mucoadhesive microspheres containing amoxicillin. They prepared them by an emulsification/evaporation method, using ethyl cellulose as matrix and Carbopol 934P as a mucoadhesive polymer and demonstrated that free amoxicillin was rapidly degraded in acidic medium; however, amoxicillin entrapped in the microspheres kept stable. The in vitro release test showed that about 90 % of the amoxicillin was released in the pH 1.0 HCl solution within 4 h, while in vivo evaluation of mucoadhesiveness showed that, during the same time, 63.6 % (±21.9) of the microspheres still remained in the rat stomach. Moreover, higher amoxicillin concentrations in gastric tissue of rats were reached after oral administration of mucoadhesive microspheres vs. amoxicillin powder at the same dose (43 mg/kg). The mucoadhesive microspheres where three folds more effective in eradicating *H. pylori* (expressed by the ratio of colony counts between amoxicillin powder and microspheres) compared to amoxicillin powder. The authors concluded that larger groups of animal are required to confirm these results [69]. Finally, Umamaheshwari's research group developed a receptor-mediated drug delivery system, based on phosphatidylethanolamine (PE) containing lipid. PE seems to be a major receptor promoting *H. pylori* adhesion to gastric cells. This system consisted of a lipid bilayer shell PE that is anchored on the surface of a hydrogel polymer core that contained the urease inhibitor acetohydroxamic acid (AHA). Specific and strong binding between lipobeads and PE surface receptor of two different *H. pylori* strains was demonstrated. In vitro studies achieved complete growth inhibition of the bacteria within 6 h, while the control formulations required 24 h to meet this goal. This system might be helpful in the treatment of *H. pylori*, but further in vivo studies are required in order to prove its effectiveness [70].

There are several points that should be taken into consideration when developing a gastroretentive system targeted for the treatment of *H. pylori* conditions:

- It seems that expandable large dimension systems may be inadequate and pose some safety problems in cases of mucosal inflammation or ulcer. A multiparticulate system, however, could be advantageous.
- Mucoadhesion-based delivery systems seem promising for the treatment of *H. pylori* as close contact with the mucus layer may significantly increase the local antibiotic concentrations. This in spite of some obstacles such as the high turnover rate of gastric mucus, unspecific binding of the dosage form to the intestinal mucus [11] and destruction of gastric mucus by the bacteria [15].
- A combination of approaches (like low density and mucoadhesion) can lead to longer retention and improved reproducibility [11].
- Specific binding of gastroretentive formulations to *H. pylori* might play a role in successful eradication of the bacteria. However, characterization of this pathogen is still incomplete. Further research is required; in particular, identification of a bacterium surface receptor expressed by all *H. pylori* strains with a strong affinity to a ligand is crucial for the implementation of this approach [15].
- *H. pylori* gastritis can produce marked alterations in gastric acid secretion [59, 71]. In addition, some of the drugs that are implemented in *H. pylori* treatment significantly elevate gastric pH [11]. As a result, GRDFs formulations that rely on acidic environment may fail to work.

15.3 Conclusions

This chapter set out to map the various technologies currently used to target drugs to the upper GIT by GRDF. While there are numerous promising research directions in this field, not all systems proved to be effective in vivo. Two key points should be accounted for in this regard: the dosage form might remain in the stomach for too long, possibly causing local side effects. Alternatively, the GRDF might be evacuated from the stomach too rapidly, resulting in reduced release of the drug in the desired site. Therefore, a more complete understanding of each variable, as well as their interactions, is necessary for the successful targeting of this specific gastrointestinal region.

References

1. Dahan A, Amidon GL (2009) Segmental dependent transport of low permeability compounds along the small intestine due to P-glycoprotein: the role of efflux transport in the oral absorption of BCS class III drugs. Mol Pharm 6(1):19–28
2. Dahan A, Lennernäs H, Amidon GL (2012) The fraction dose absorbed, in humans, and high jejunal human permeability relationship. Mol Pharm 9(6):1847–1851
3. Dahan A, West BT, Amidon GL (2009) Segmental dependent membrane permeability along the intestine following oral drug administration: evaluation of a triple single-pass intestinal perfusion (TSPIP) approach in the rat. Eur J Pharm Sci 36(2–3):320–329
4. Dahan A, Amidon GL, Zimmermann EM (2010) Drug targeting strategies for the treatment of inflammatory bowel disease: a mechanistic update. Expert Rev Clin Immunol 6(4):543–550
5. Streubel A, Siepmann J, Bodmeier R (2006) Drug delivery to the upper small intestine window using gastroretentive technologies. Curr Opin Pharmacol 6(5):501–508
6. Kararli TT (1995) Comparison of the gastrointestinal anatomy, physiology, and biochemistry of humans and commonly used laboratory animals. Biopharm Drug Dispos 16(5):351–380
7. Washington N, Washington C, Wilson CG (eds) (2003) Physiological pharmaceutics barriers to drug absorption. Taylor & Francis e-Library, London
8. Lahner E, Annibale B, Delle Fave G (2009) Systematic review: Helicobacter pylori infection and impaired drug absorption. Aliment Pharmacol Ther 29(4):379–386
9. Ogawa R, Echizen H (2011) Clinically significant drug interactions with antacids: an update. Drugs 71(14):1839–1864
10. Berthet S, Charpiat B, Mabrut JY (2010) Erythromycin as a prokinetic agent: risk factors. J Visc Surg 147(2):e13–e18
11. Kagan L, Hoffman A (2008) Systems for region selective drug delivery in the gastrointestinal tract: biopharmaceutical considerations. Expert Opin Drug Deliv 5(6):681–692
12. Englund G et al (2006) Regional levels of drug transporters along the human intestinal tract: co-expression of ABC and SLC transporters and comparison with Caco-2 cells. Eur J Pharm Sci 29(3–4):269–277
13. Chan LM, Lowes S, Hirst BH (2004) The ABCs of drug transport in intestine and liver: efflux proteins limiting drug absorption and bioavailability. Eur J Pharm Sci 21(1):25–51
14. Dahan A, Sabit H, Amidon GL (2009) Multiple efflux pumps are involved in the transepithelial transport of colchicine: combined effect of P-gp and MRP2 leads to decreased intestinal absorption throughout the entire small intestine. Drug Metab Dispos 37(10):2028–2036
15. Bardonnet PL et al (2006) Gastroretentive dosage forms: overview and special case of Helicobacter pylori. J Control Release 111(1–2):1–18

16. Kagan L, Hoffman A (2008) Selection of drug candidates for gastroretentive dosage forms: pharmacokinetics following continuous intragastric mode of administration in a rat model. Eur J Pharm Biopharm 69(1):238–246
17. Chauhan B et al (2005) Preparation and evaluation of floating risedronate sodium-Gelucire 43/01 formulations. Drug Dev Ind Pharm 31(9):851–860
18. Li S et al (2001) Statistical optimization of gastric floating system for oral controlled delivery of calcium. AAPS PharmSciTech 2(1):E1
19. Nur AO, Zhang JS (2000) Captopril floating and/or bioadhesive tablets: design and release kinetics. Drug Dev Ind Pharm 26(9):965–969
20. Chavanpatil M et al (2005) Development of sustained release gastroretentive drug delivery system for ofloxacin: in vitro and in vivo evaluation. Int J Pharm 304(1–2):178–184
21. Wei Z, Yu Z, Bi D (2001) Design and evaluation of a two-layer floating tablet for gastric retention using cisapride as a model drug. Drug Dev Ind Pharm 27(5):469–474
22. Hu L et al (2011) Floating matrix dosage form for dextromethorphan hydrobromide based on gas forming technique: in vitro and in vivo evaluation in healthy volunteers. Eur J Pharm Sci 42(1–2):99–105
23. Gambhire MN et al (2007) Development and in vitro evaluation of an oral floating matrix tablet formulation of diltiazem hydrochloride. AAPS PharmSciTech 8(3):E73
24. Patel VF, Patel NM (2007) Statistical evaluation of influence of viscosity and content of polymer on dipyridamole release from floating matrix tablets: a technical note. AAPS PharmSciTech 8(3):E69
25. Chowdary KP, Rao YS (2003) Design and in vitro and in vivo evaluation of mucoadhesive microcapsules of glipizide for oral controlled release: a technical note. AAPS PharmSciTech 4(3):E39
26. El-Kamel AH et al (2001) Preparation and evaluation of ketoprofen floating oral delivery system. Int J Pharm 220(1–2):13–21
27. Stepensky D et al (2001) Preclinical evaluation of pharmacokinetic-pharmacodynamic rationale for oral CR metformin formulation. J Control Release 71(1):107–115
28. Strubing S, Metz H, Mader K (2008) Characterization of poly(vinyl acetate) based floating matrix tablets. J Control Release 126(2):149–155
29. Dave BS, Amin AF, Patel MM (2004) Gastroretentive drug delivery system of ranitidine hydrochloride: formulation and in vitro evaluation. AAPS PharmSciTech 5(2):e34
30. Hejazi R, Amiji M (2003) Stomach-specific anti-H. pylori therapy. II. Gastric residence studies of tetracycline-loaded chitosan microspheres in gerbils. Pharm Dev Technol 8(3): 253–262
31. Sawicki W (2002) Pharmacokinetics of verapamil and norverapamil from controlled release floating pellets in humans. Eur J Pharm Biopharm 53(1):29–35
32. Bagul US et al (2011) Stomach specific drug delivery systems: a review. Int J Pharm Res Dev 4(4):147–150
33. Deshpande AA et al (1997) Development of a novel controlled-release system for gastric retention. Pharm Res 14(6):815–819
34. Makwana A et al (2012) Advancements in controlled release gastroretentive drug delivery systems: a review. J Drug Deliv Ther 2(3):12–21
35. Jaimini M, Rana AC, Tanwar YS (2007) Formulation and evaluation of famotidine floating tablets. Curr Drug Deliv 4(1):51–55
36. Mazer N et al (1988) Intragastric behavior and absorption kinetics of a normal and "floating" modified-release capsule of isradipine under fasted and fed conditions. J Pharm Sci 77(8):647–657
37. Saito N et al (2003) Plural transformation-processes from spiral to coccoid Helicobacter pylori and its viability. J Infect 46(1):49–55
38. Singh BN, Kim KH (2000) Floating drug delivery systems: an approach to oral controlled drug delivery via gastric retention. J Control Release 63(3):235–259
39. Reddy LH, Murthy RS (2002) Floating dosage systems in drug delivery. Crit Rev Ther Drug Carrier Syst 19(6):553–585

40. Hwang SJ, Park H, Park K (1998) Gastric retentive drug-delivery systems. Crit Rev Ther Drug Carrier Syst 15(3):243–284
41. Ali J et al (2007) Formulation and development of hydrodynamically balanced system for metformin: in vitro and in vivo evaluation. Eur J Pharm Biopharm 67(1):196–201
42. Hamdani J et al (2006) In vitro and in vivo evaluation of floating riboflavin pellets developed using the melt pelletization process. Int J Pharm 323(1–2):86–92
43. Ritschel WA (1991) Targeting in the gastrointestinal tract: new approaches. Methods Find Exp Clin Pharmacol 13(5):313–336
44. Erdeve O et al (2011) Efficacy and safety of sodium alginate for GERD in preterm infants. Aliment Pharmacol Ther 33(8):981–982, author reply 982–3
45. Kwiatek MA et al (2011) An alginate-antacid formulation (Gaviscon Double Action Liquid) can eliminate or displace the postprandial 'acid pocket' in symptomatic GERD patients. Aliment Pharmacol Ther 34(1):59–66
46. Strugala V et al (2009) The role of an alginate suspension on pepsin and bile acids - key aggressors in the gastric refluxate. Does this have implications for the treatment of gastro-oesophageal reflux disease? J Pharm Pharmacol 61(8):1021–1028
47. Sato Y et al (2004) Pharmacoscintigraphic evaluation of riboflavin-containing microballoons for a floating controlled drug delivery system in healthy humans. J Control Release 98(1):75–85
48. Streubel A, Siepmann J, Bodmeier R (2002) Floating microparticles based on low density foam powder. Int J Pharm 241(2):279–292
49. Streubel A, Siepmann J, Bodmeier R (2003) Floating matrix tablets based on low density foam powder: effects of formulation and processing parameters on drug release. Eur J Pharm Sci 18(1):37–45
50. Streubel A, Siepmann J, Bodmeier R (2003) Multiple unit gastroretentive drug delivery systems: a new preparation method for low density microparticles. J Microencapsul 20(3): 329–347
51. Talukder R, Fassihi R (2004) Gastroretentive delivery systems: hollow beads. Drug Dev Ind Pharm 30(4):405–412
52. Kedzierewicz F et al (1999) Evaluation of peroral silicone dosage forms in humans by gamma-scintigraphy. J Control Release 58(2):195–205
53. Klausner EA et al (2003) Novel levodopa gastroretentive dosage form: in-vivo evaluation in dogs. J Control Release 88(1):117–126
54. Chen J et al (2000) Gastric retention properties of superporous hydrogel composites. J Control Release 64(1–3):39–51
55. Huang Y et al (2000) Molecular aspects of muco- and bioadhesion: tethered structures and site-specific surfaces. J Control Release 65(1–2):63–71
56. Vasir JK, Tambwekar K, Garg S (2003) Bioadhesive microspheres as a controlled drug delivery system. Int J Pharm 255(1–2):13–32
57. Singh B et al (2012) Formulation development of gastroretentive tablets of lamivudine using the floating-bioadhesive potential of optimized polymer blends. J Pharm Pharmacol 64(5):654–669
58. Groning R, Berntgen M, Georgarakis M (1998) Acyclovir serum concentrations following peroral administration of magnetic depot tablets and the influence of extracorporal magnets to control gastrointestinal transit. Eur J Pharm Biopharm 46(3):285–291
59. Bhandari A, Crowe SE (2012) Helicobacter pylori in gastric malignancies. Curr Gastroenterol Rep 14(6):489–496
60. Suerbaum S, Michetti P (2002) Helicobacter pylori infection. N Engl J Med 347(15):1175–1186
61. Conway BR (2005) Drug delivery strategies for the treatment of Helicobacter pylori infections. Curr Pharm Des 11(6):775–790
62. Noonan B, Alm RA (2002) Novel intervention strategies for Helicobacter pylori treatment. Curr Drug Targets Infect Disord 2(4):331–338
63. Touati E (2010) When bacteria become mutagenic and carcinogenic: lessons from H. pylori. Mutat Res 703(1):66–70

64. Malfertheiner P et al (2007) Current concepts in the management of Helicobacter pylori infection: the Maastricht III Consensus Report. Gut 56(6):772–781
65. Niv Y (2008) H pylori recurrence after successful eradication. World J Gastroenterol 14(10):1477–1478
66. Megraud F (2004) H pylori antibiotic resistance: prevalence, importance, and advances in testing. Gut 53(9):1374–1384
67. De Francesco V et al (2010) Worldwide H. pylori antibiotic resistance: a systematic review. J Gastrointestin Liver Dis 19(4):409–414
68. Yang L, Eshraghi J, Fassihi R (1999) A new intragastric delivery system for the treatment of Helicobacter pylori associated gastric ulcer: in vitro evaluation. J Control Release 57(3):215–222
69. Liu Z et al (2005) In vitro and in vivo studies on mucoadhesive microspheres of amoxicillin. J Control Release 102(1):135–144
70. Umamaheshwari RB, Jain NK (2004) Receptor-mediated targeting of lipobeads bearing acetohydroxamic acid for eradication of Helicobacter pylori. J Control Release 99(1):27–40
71. Iijima K et al (2004) Long-term effect of Helicobacter pylori eradication on the reversibility of acid secretion in profound hypochlorhydria. Aliment Pharmacol Ther 19(11):1181–1188

Chapter 16
Segmental-Dependent Drug Absorption and Delivery: The Intestinal Tract

Omri Wolk and Arik Dahan

16.1 Introduction

Oral administration is by far the most popular and convenient way of drug delivery. Oral dosage forms are easy to use and usually do not require hospitalization in a medical facility or the assistance of medical professionals. In addition, oral uptake of drugs prevents both the local side effects and the risk of systemic infections that are associated with injections.

However, the majority of drugs that are taken orally must enter the enterocytes and cross through the intestinal tract membrane in order to be absorbed to the systemic bloodstream and exert their pharmacological effect. The intestine proves to be a formidable barrier, to the point where many drug substances, such as antibodies, hormones, and even small molecules, fail to penetrate through it and have to be administered parenterally. Drugs that are taken through the oral route may be heavily influenced by the physiological conditions in the intestine, resulting in significant diversions from the predicted/desired pharmacokinetic and pharmacodynamic patterns.

The intestinal tract can be broadly divided to several regions: the small intestine, subdivided to duodenum, jejunum, and ileum, and the colon. The environment in each of these regions is dependent on a multitude of factors, among them are pH values, membrane transporters expression, metabolic enzymes activity, and resident micro-flora. Each of these variables, alone or in combination with others, can fundamentally alter the solubility and/or permeability of drugs [1–3].

In light of this great variability in conditions across the intestine, many attempts were made to deliver drugs to specific regions of the intestine. The goal of many of

O. Wolk • A. Dahan, Ph.D. (✉)
Department of Clinical Pharmacology, School of Pharmacy, Faculty of Health Sciences, Ben-Gurion University of the Negev, P.O. Box 653, Beer-Sheva 84105, Israel
e-mail: arikd@bgu.ac.il

A.J. Domb and W. Khan (eds.), *Focal Controlled Drug Delivery*, Advances in Delivery Science and Technology, DOI 10.1007/978-1-4614-9434-8_16,

these attempts was to optimize systemic absorption of drugs, while others were intendant to retain the drugs within the lumen or epithelial cells and treat local ailments [4].

In this chapter, several of these factors are reviewed, and their influences on segmental differences in drug absorption are discussed. In addition, technologies and methods of drug targeting to specific location within the intestinal tract are presented.

16.2 pH Considerations

Luminal pH values change extensively across the GIT (Fig. 16.1). The pH of the stomach can be as low as 1–2 due to acidic secretions. In the duodenum, luminal content is quickly neutralized by the secretion of the pancreatic bicarbonate and bile. As luminal content progresses from the proximal to the distal intestine, it becomes

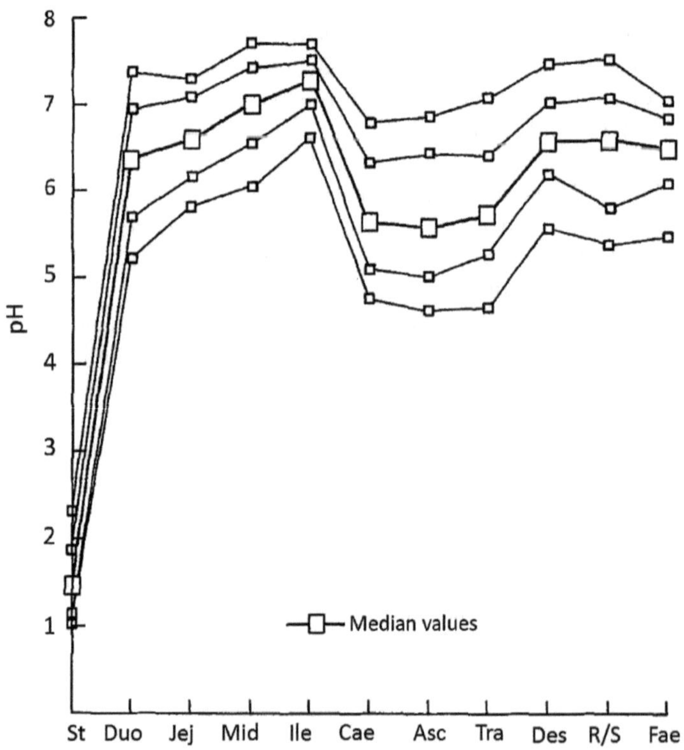

Fig. 16.1 Percentiles (90, 75, 50, 25, and 10) of pH determinations from 39 patients in relation to the intestinal location, using a pH-sensitive radio-transmitting capsule. *St* stomach, *Duo* duodenum, *Jej* proximal small intestine, *Mid* mid small intestine, *Ile* distal small intestine, *Cae* caecum, *Asc* ascending colon, *Tra* transverse colon, *Des* descending colon, *R/S* sigmoid colon or rectum, *Fae* faeces. Modified from [6] with permission

progressively more alkaline [5]. However, the pH values of the large intestine are generally more acidic than the pH observed in the small intestine in humans, possibly due to the fermentation mediated by the microbial flora.

The environmental pH directly affects the degree of drug ionization. Since the extent of ionization is a major factor in determining the drug dissolution rate and passive permeability, it is clear that pH at the absorption site may facilitate or inhibit both the dissolution and the absorption of various ionizable drug molecules [5].

The drug intestinal permeability measure is one of the main factors governing the rate and extent of drug absorption (F_{abs}) in humans following oral administration [7]. Starting in the year 2000, the FDA has adopted the "biowaiver" concept, and for BCS class I drugs, i.e., when a drug is proven to be both highly soluble and highly permeable, the FDA may grant an exemption to pharmaceutical manufacturers of the need to conduct in vivo bioequivalence studies and be satisfied with in vitro dissolution tests for the determination of bioequivalence [8]. FDA guidelines determine that a drug must be absorbed (F_{abs}) to >90 % in order to be classified as high permeability [8]. F_{abs} is usually determined by pharmacokinetic or mass-balance studies in humans [7, 9, 10]. The complexity and cost of carrying out these studies are very high, as they require radiolabeled API for the validation of high drug and metabolite recovery. Extensive scientific research has established that a good correlation exists between the human jejunal permeability (P_{eff}) and F_{abs} [9], and so an alternative high-permeability classification channel of measuring P_{eff} in validated cell-culture or animal intestinal perfusion techniques is suggested by the FDA BCS guidance [8]. The β-blocker metoprolol has been widely used as a marker for the low-/high-permeability class boundary; that is, if a compound shows higher P_{eff} than metoprolol, then it is considered to be high-permeability, and vice versa [8]. Examination of the human intestinal absorption of metoprolol reveals that it is in fact completely (100 %) absorbed [11], thus making metoprolol an overly conservative reference drug for the low-/high-permeability class boundary when compared to the corresponding FDA's F_{abs}-based high-permeability benchmark of 90 %. Moreover, we have recently shown that the use of metoprolol as the permeability marker carries an additional complexity. Metoprolol's intestinal permeability in the rat was shown to be segmental dependent; as a basic drug (pKa = 9.5) with Log P value ~2, the pH changes along the small intestine were significant enough to lead to fivefold higher permeability in the ileum (pH 7.5) than in the jejunum (pH 6.5). This raises the question of which of metoprolol's P_{eff} values represents the true low-/high-permeability boundary reference, the P_{eff} in the upper small intestine or the higher P_{eff} obtained in the lower small intestine? Several studies have shown that an IR oral dose of metoprolol is completely absorbed already from the upper small intestine, in both human and animal models [12, 13]. It follows, hence, that metoprolol's P_{eff} value at pH 7.5 (ileum) is not likely to be physiologically relevant to the absorption of the drug from an IR dosage form; rather, the permeability at pH 6.5, the average pH of the human jejunum, would govern metoprolol's in vivo intestinal absorption from an IR formulation. This analysis points out a possible extension to the regulatory high-permeability criterion: taking metoprolol's jejunal P_{eff} at pH 6.5 as the benchmark for high permeability, it is suggested that if a compound matches/exceeds this

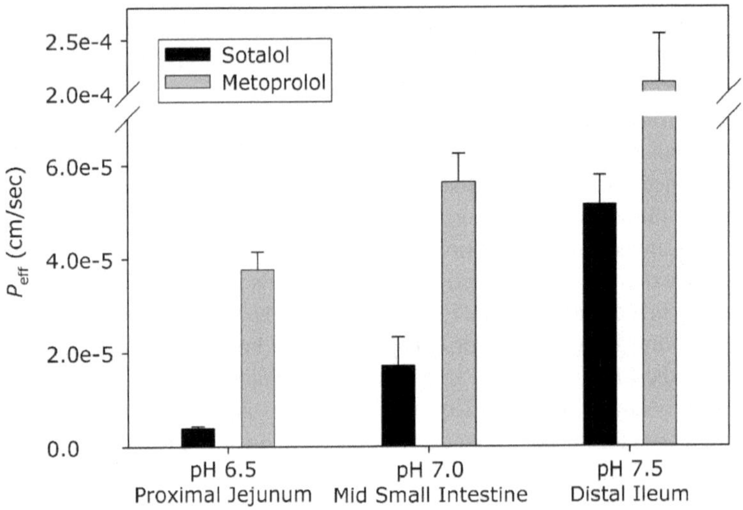

Fig. 16.2 Effective permeability values (Peff; cm/s) obtained for sotalol and metoprolol after in situ single-pass perfusion to the rat proximal jejunum at pH 6.5, mid small intestine at pH 7.0, and distal ileum at pH 7.5. Modified from [14] with permission

threshold anywhere in the intestine, and not necessarily in the jejunum, it is a high-permeability compound with close to complete (F_{abs}>90 %) absorption [2, 14].

This realization was recently implemented to address the distinctive absorption characteristics of the β-blocker sotalol [14], for which low apparent permeability across Caco-2 cell monolayers has been observed, but also a high fraction dose absorbed (F_{abs}>90 %) in humans has been reported [15–17]. Similarly to metoprolol, sotalol's intestinal permeability in the rat was shown to be segmental dependent, with higher P_{eff} at distal small intestinal regions (with higher average pH) than in proximal segments. At any given small intestinal segment (and pH), sotalol's permeability was lower than that of metoprolol. However, most significantly, sotalol's permeability in the ileum at pH 7.5 exceeded that of metoprolol in the jejunum at pH 6.5 and matched metoprolol's P_{eff} in the middle small intestine at pH 7.0 (Fig. 16.2). Since the ileum accounts for more than half of the human small intestinal length, an IR oral dose of sotalol would have an apparent high permeability (i.e., greater than that of metoprolol in the jejunum) throughout a significant portion of the small intestinal residence time (~2 h), resulting in its high fraction dose absorbed [14]. Significantly, the human t_{max} values of sotalol (~4 h) and metoprolol (~1 h) from IR products support this analysis, indicating an absorption in the distal and proximal intestine, respectively. This study demonstrates the importance of the intestinal pH gradient to segmental and overall absorption of drugs. Sotalol may have been falsely classified as low-permeability drug based merely on jejunal P_{eff}, but when more thoroughly evaluated, the permeability basis for sotalol's nearly complete oral absorption is apparent [2, 14].

The pH changes along the gastrointestinal tract have been exploited for the purpose of drug delivery. One application is the enteric coating of drugs which are unstable in acidic conditions or may cause gastric side effects. These coatings are generally made from pH-responsive polymers which remain unionized and intact at the low pH of the stomach but dissolve at the higher pH of the small intestine [18]. Insulin is a hormone required by many diabetic patients in order to regulate their blood sugar. Insulin is degraded in the stomach and so it must be administered by injections, which are often unpleasant and reduce compliance. Recently, an enteric coating of pH-sensitive hydroxypropyl methylcellulose phthalate (HP55) was applied to a capsule containing cationic nanoparticles of insulin. The new formulation was able to deliver insulin to the bloodstream and reduce blood glucose levels [19]. Of course, insulin also suffers from enzymatic degradation by various peptidases within the intestinal tract, a problem that further complicates its oral delivery. Enteric-coated formulations have also been developed for erythromycin [20] and omeprazole [21] to overcome their gastric instability and for mycophenolate in order to avoid gastric side effects [22] and more. This pH-triggered approach has been extended to ileo-colonic delivery and formed the basis for the development of Eudragit S-coated 5-ASA tablets marketed as Asacol® for ulcerative colitis. This and other preparations based on the same concept (Mesren, Lialda, and Mezavant) are in clinical use [18]. All of these formulations were based on the dissolution of Eudragit S at pH >7. However, it was demonstrated that Eudragit S-based dosage forms might fail to dissolve [23]. This may be due to the target pH not being reached in some subjects or not being high enough for a sufficiently long time for the pH-responsive film coating to dissolve [23]. In addition, it was shown that disease states, such as ulcerative colitis, may dramatically reduce colonic pH [24, 25]. A thorough investigation to the many factors which might influence the dissolution of pH-sensitive polymers may be critical to the optimization of pH-dependent formulations.

A more sophisticated colonic drug delivery system, CODES, was developed, which is both time and pH dependent. A typical configuration of CODES consists of a core tablet coated with three polymer layers. The first coating layer, surrounding the core tablet, is an acid-soluble polymer (e.g., Eudragit E®), and the outer coating is enteric, with an HPMC barrier layer in between to prevent any possible interactions between the oppositely charged polymers. The core tablet is comprised of the active moiety, one or more polysaccharides, and other desirable excipients. During its transit through the stomach, CODES remains intact due to the enteric protection. The enteric and barrier coating dissolves in the small intestine where the pH is above 6, while the inner layer only slightly swells. Upon entry into the colon, the polysaccharide inside the core tablet dissolves and diffuses through the acid-soluble coating. The bacteria enzymatically degrade the polysaccharide and produce organic acids. This lowers the local pH and results in the dissolution of the acid-soluble coating and subsequent drug release. It was clearly shown that CODES provides rapid colonic release regardless of the ingestion of food [26]. A schematic depiction of CODES is presented in Fig. 16.3.

Overall, the pH gradient throughout the gastrointestinal tract may represent both a complication and an opportunity for successful drug delivery and absorption.

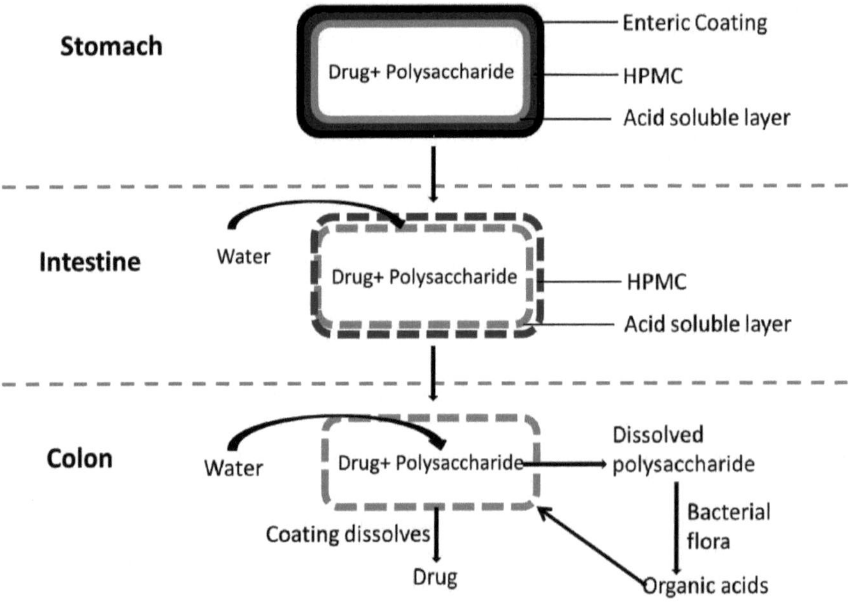

Fig. 16.3 Schematic description of drug release of CODES™ in the gastrointestinal tract

An in-depth understanding of the intestinal pH environment and its determinants, considering the whole of the human intestine in both physiological and pathophysiological conditions, is crucial for successful exploitation of the variable intestinal conditions for better drug delivery and therapeutic effect.

16.3 Membrane Transporters

Membrane transporters act as "checkpoints" in the cells, actively inserting important nutrients into the cell while removing potentially toxic substances [27]. In addition to their role in cellular uptake of endogenous molecules, membrane transporters also have a major impact on the efficacy and toxicity of drugs [27, 28]. Transporters can be classified as influx and efflux transporters and are located either at the basolateral or apical membrane of polarized cells [29]. In this chapter, two major transporter families will be discussed: the ATP-binding cassette (ABC) superfamily and the solute carrier family (SLC). Transporters belonging to the ABC family are primarily active efflux transporters and use the energy derived from ATP hydrolysis to drive the active cellular efflux of drugs [30], while SLC family members usually facilitate the cellular influx of substrates by allowing diffusion or by cotransporting of ions to provide the driving force [31]. Certain SLC transporters may exhibit efflux or bidirectional transport, depending on the concentration gradients of substrate and coupled ions across the membrane [32].

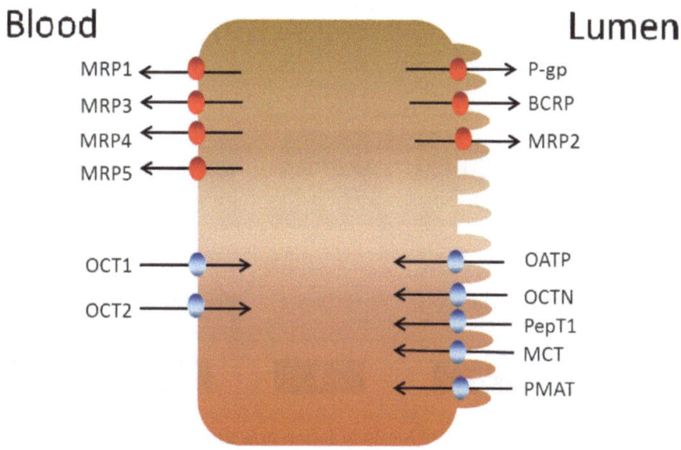

Fig. 16.4 Diagram of major drug transporter proteins expressed at the intestinal epithelia including intestinal uptake (*blue*) and efflux (*red*) transporters

Like most of the physiological barriers in the human body, the intestinal wall also expresses various transports from both families [27] (Fig. 16.4).

Intestinal transporter activity often affects the absorption of drugs, though BCS class 1 compounds, i.e., high-solubility high-permeability drugs, are not likely to be affected [1, 27, 33]. In this section we will review a few of the main transporters expressed in the intestinal epithelium, discuss their segmental-dependent expression pattern, and present their potential use to increase drug absorption.

16.3.1 ABC Superfamily

16.3.1.1 P-Glycoprotein (P-gp)

Initially discovered as a result of its interaction with multiple anticancer drugs, P-gp is able to interact with many kinds of chemical groups, including hydrophobic, amphiphilic, or cationic molecules containing a planar ring system ranging in size from 200 to 1,900 Da, negatively charged carboxyl groups, and hydrophilic molecules [34]. As a result, P-gp possesses an enormous spectrum of substrate drugs, ranging from anthracyclines (doxorubicin, daunorubicin) to alkaloids (reserpine, vincristine, vinblastine), peptides (valinomycin, cyclosporine), steroid hormones (aldosterone, hydrocortisone), local anesthetics (dibucaine), immunosuppressive agents (cyclosporine, tacrolimus), talinolol, digoxin, tandutinib and more [27]. This diverse substrate spectrum, along with its localization in the apical membrane of enterocytes (Fig. 16.4), promotes the potential role of P-gp as an obstacle to drug absorption. Indeed, numerous drug–drug interactions were found to be mediated by P-gp [29, 35].

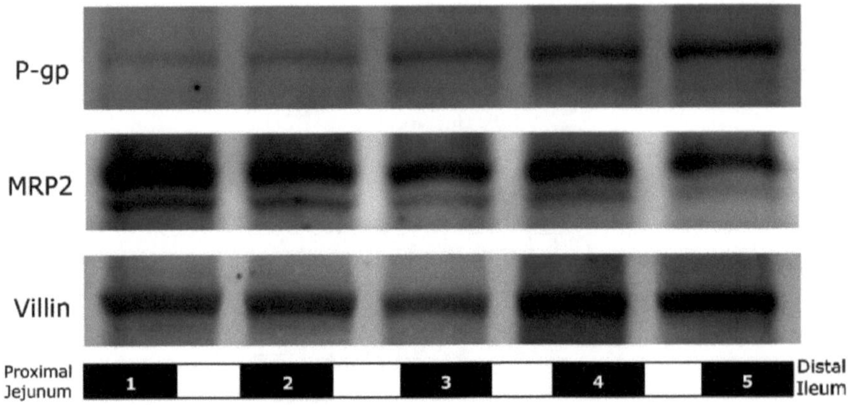

Fig. 16.5 Analysis of P-gp and MRP2 levels in different segments along the rat small intestine. Segments 1–5 correspond to equal, 10-cm-long, symmetrically spaced segments, starting from the ligament of Treitz (segment 1) and ending at the terminal ileum (segment 5). Modified from [36] with permission

The expression of P-gp was shown to increase from proximal to distal regions of the small intestine in rats [1, 36, 37], and also in humans although with significantly higher variability [38–41] (Fig. 16.5), and absorption from different segments of the rat intestine was demonstrated to be in inverse correlation to P-gp levels [1, 36]. A study of the absorption pattern of the BCS class III drugs, P-gp substrates cimetidine and famotidine, revealed that both drugs exhibit segmental-dependent absorption, with permeability decreasing from proximal to distal intestinal segments in rats. However, in the presence of the P-gp inhibitor verapamil, permeability of both drugs in the ileum equalized the permeability in the jejunum. Therefore, it has been suggested that targeting of P-gp substrates to the upper intestine might enhance their absorption [42–44]. This expression pattern may indicate a less effective, saturable P-gp-mediated efflux in the proximal intestine when compared to the distal intestine [45]. Therefore, gastroretentive dosage forms (GRDF) may be adequate to increase the absorption and bioavailability of P-gp substrates, provided CYP450 3A4, which is more abundant at the upper small intestine, is not involved, as will be further discussed hereinafter.

16.3.1.2 MRP2

MRP2 was originally cloned from rat liver as cMRP and from human liver as the canalicular multispecific organic anion transporter (cMOAT) [38]. MRP2 is expressed mainly in liver, intestine, and kidney tubules. The substrate specificity of MRP2 includes endogenous molecules such as LTC4 and 2,4-dinitrophenyl-*S*-glutathione and bilirubin glucuronides. In addition, MRP2 also secretes various conjugated drug metabolites, as well as unmodified drugs [27]. MRP2 shares some

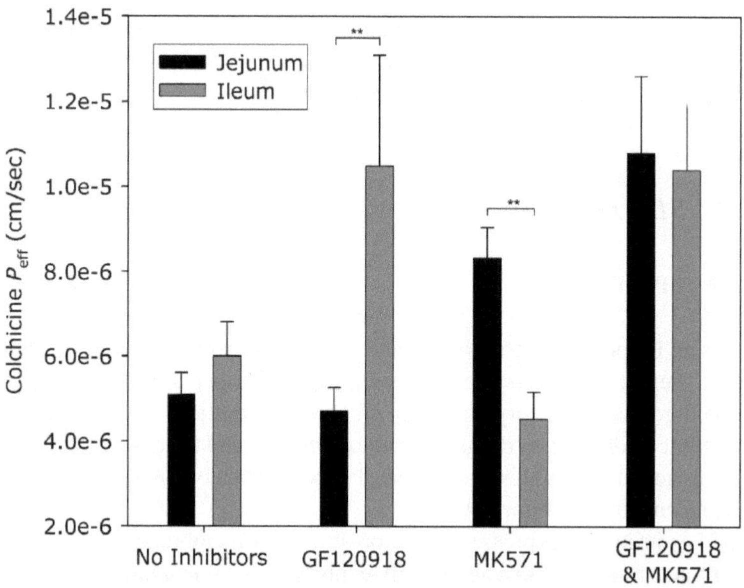

Fig. 16.6 P_{eff} (cm/s) values obtained for colchicine after perfusion to the rat proximal jejunum and to the distal ileum, without inhibitors, in the presence of GF120918 or MK571 and in the presence of both inhibitors simultaneously. Modified from [36] with permission

degree of overlapping substrate specificity with P-gp and is localized to the apical membrane, with expression levels increasing from crypt to villus, in a pattern similar to that of P-gp [38]. However, in contrast to P-gp, MRP2 expression is highest at the proximal segments of the intestine and decreases towards the distal ileum (Fig. 16.5) [36, 38].

The microtubule polymerization inhibitor colchicine is a known substrate of P-gp [46]. However, inhibition of P-gp consistently failed to completely abolish the efflux of colchicine in Caco-2 experiments, suggesting the involvement of at least one additional transporter. This issue was recently addressed, where it was shown that in addition to P-gp, MRP2 also plays a role in colchicine's efflux, and when both transporters were inhibited simultaneously, efflux ratio (P_{app} BL→AP/P_{app} AP→BL) was 1, indicating complete abolishment of efflux [36]. These results indicate that P-gp and MRP2 "collaborate" to decrease the intestinal absorption of colchicine. These findings were further validated by intestinal perfusion rat model; baseline permeability of colchicine was low and constant along the entire rat small intestine. When P-gp inhibitor was added (GF120918), ileal permeability increased dramatically, with no significant effect on jejunal permeability. The addition MRP2 inhibitor (MK571) mirrored this pattern, displaying a marked increase in permeability from the jejunum and no change in the ileum. A combination of both inhibitors increased permeability throughout the entire rat small intestine (Fig. 16.6) [36].

Taken together, these results suggest that while targeting MRP2-exclusive substrates to the distal small intestine can, in theory, be advantageous, common substrates of P-gp and MRP2 are vulnerable to efflux by one transporter or the other in any segment of the intestine. Targeting such drugs to specific intestinal regions might prove to be ineffective in overcoming the efflux.

16.3.2 SLC Family

16.3.2.1 hPepT1

Pept-1 is an H^+-peptide cotransporter and depends on proton gradient for its uphill transport. PepT1 transports mainly dipeptides and tripeptides that are produced from the digestion of dietary and body proteins. In addition, its substrates include peptidomimetic drugs, i.e., drugs that have dipeptide- and tripeptide-like structures, such as β-lactam antibiotics, the antiviral prodrug valacyclovir, and angiotensin-converting enzyme inhibitors [47].

PepT1 is primarily expressed in the brush-border membranes of enterocytes of the small intestine. It is most abundantly expressed in the jejunum, followed by the ileum, and little or no expression is evident in the normal colon, stomach, or esophagus [47]. Indeed, the PepT1 marker peptide GlySar displayed a regional-dependent intestinal absorption in wild-type mice, with high permeability in the duodenum and jejunum, which decreased twofold in the ileum, and negligible absorption from the colon. In contrast, knockout mice displayed a low, invariable absorption of GlySar throughout the intestinal tract [48]. However, PepT1 distribution may be dramatically altered as conditions change. PepT1 intestinal expression was shown to be susceptible to many natural peptides, hormones, and pharmacological agents [47]. Furthermore, colonic expression of PepT1 was found to be significantly increased in Crohn's disease and ulcerative colitis [49]. The elevated PepT1 level in the inflamed colon was exploited for the delivery of the anti-inflammatory tripeptide KPV in a mice model of inflammatory bowel disease [50].

PepT1 has been considered as a prime target for oral drug delivery. This can be done by attaching the drug to a peptide or peptide-like carrier, creating a prodrug that utilizes the transporter to facilitate its absorption. PepT1 is an attractive target for prodrug design for several reasons. First, peptide transporters have broad substrate specificity and high capacity; theoretically, hundreds of dipeptides and thousands of different tripeptides can be generated from 20 amino acids that are chemically and structurally diverse, thus allowing the engineering of numerous different prodrugs [51, 52]. Second, peptide transporters have been more extensively studied than other transporters, and considerable information is available about them. Therefore, structural modifications targeting peptide transporters can be optimized to a much greater degree compared to other transporters [51]. Finally, cloning and controlled expression in mammalian cell systems allow us to attempt rational drug design to target the peptide transporters [51]. A good example for this approach is valaciclovir, the L-valine ester of acyclovir, which was developed as a prodrug of acyclovir and is transported by PepT1. The mean bioavailability of valaciclovir is 54 %, which is 3–5 times greater than that of oral acyclovir [53]. Other examples are L-a-methyldopa-L-Phe, L-Val-AZT, and L-Pro-L-Phe and conjugated alendronate and pamidronate [52]. In general, carrier moieties which depend on PepT1 influx share several features: They are predominantly L-enantiomers (since PepT1 is highly stereoselective). Many of them possess a free N-terminal and C-terminal groups, as both groups are important for binding and transport of the moiety by PepT1. However, this is not an absolute demand for PepT1 substrates, and some modifications are possible [52]. The carrier is usually connected to the parent compound by a peptidic bond [52]. Certain side chains can be inserted to the carrier, while others were shown to inhibit transportation of the moiety.

In addition to the development of chemically modified compounds, there is another way to further enhance the transportation of PepT1 substrates. This strategy relies on the dependency of the PepT1 transporter on proton gradient for its activity; by including an agent in the formulation that would acidify the local microenvironment, H^+ concentration rises, thereby increasing the driving force of PepT1 [54]. A combination of the two strategies, i.e., chemical modification of parent compounds and acidification of the local microenvironment, might further enhance intestinal absorption of drugs via PepT1-mediated transportation [55].

16.3.2.2 OATP

Organic anion-transporting polypeptides (OATPs) are involved in the cellular uptake of endogenous and xenobiotic organic anions in various tissues and affect absorption and disposition of their substrate drugs. Their substrate spectrum includes statins, angiotensin II receptor blockers, beta-lactam antibiotics, peptides, fluoroquinolones, angiotensin-converting enzyme inhibitors, H_1-antagonists, troglitazone sulfate, digoxin, bosentan, rifampicin, methotrexate, and glibenclamide, and so their pharmacological relevance is very high [56].

Several family members of the OATP family are distributed in various tissues. In the intestine, OATP2B1 is the predominant transporter [56]. OATP2B1 protein is localized at the apical membrane of intestinal epithelial cells, as shown in Fig. 16.4. Therefore, it is probable that OATP2B1 contributes to the absorption of its substrates from intestinal lumen. Similarly to some other transporters, OATP2B1 is not evenly distributed along the intestine: the expression level of the transporter increases from the duodenum to the ileum and drops again in the colon [57]. The relevance of this expression pattern to in vivo oral drug absorption should be further investigated.

16.4 CYP3A4 Gradient

Drug metabolizing enzymes (DME) play a central role in the bioavailability, metabolism, elimination, and detoxification of endogenous and exogenous compounds. DME include phase I and II metabolizing enzymes. The hydroxylation activity of phase I DME increases the hydrophilicity of the molecules and produces a chemically active site for phase II conjugation [58].

The prominent phase I DME group is the Cytochrome P450 (CYP) family. The most abundant CYP isoenzyme in the intestine is 3A4. Indeed, it was shown that CYP3A4 expression in mature enterocytes is comparable or may even exceed the expression in hepatocytes [59, 60].

While present throughout the GI tract, CYP3A4 enzymes are unevenly distributed: The enzyme expression in the stomach is relatively low. The highest levels of CYP3A4 are found in the duodenum and jejunum before decreasing dramatically in the ileum and colon [41, 61]. These intra-intestinal differences in CYP3A4 expression may contribute to changes in drug exposure: CYP3A4 substrates might be metabolized when passing through the proximal intestine, resulting in reduced bioavailability of the drug. In other cases, CYP3A4 metabolism can result in the production of unpredictable amounts of active [62] or toxic [63] metabolites.

In order to evade the complexities of presystemic metabolism, it might be beneficial for some CYP3A4 substrates to be targeted to the distal small intestine and colon, where expression of the enzyme is minimal [64]. So far, only a few attempts were made to use delivery to the distal GI tract as a way to avoid intestinal metabolism [65]. Tubic-Grozdanis et al. designed a delayed release tablet of the CYP3A4 substrate simvastatin in order to target intestinal regions with lower CYP3A4 expression level. Threefold higher simvastatin AUC was obtained following the delayed release tablet compared to administration of an immediate release capsule. Overall, the interplay between gastrointestinal physiology (lower CYP3A4 expression in the distal ileum and the colon) and formulation design (zero-order controlled release after a predetermined lagtime) resulted in successful absorption and bioavailability improvement and represent a viable strategy to reduce the dose of CYP3A4 drugs [65].

Many compounds are substrates to both CYP3A4 enzymes and P-gp efflux pumps. Since P-gp and CYP3A4 expression along the intestine is complimentary, common substrates may exhibit a reduced absorption, due to either metabolism or active efflux, all along the intestine. Such is the case of tacrolimus, a substrate to both P-gp and CYP3A4; tacrolimus uptake in the jejunum was twofold greater than in the ileum, as was measured by rat intestinal perfusions. However, a double perfusion test, in which outflow from the intestinal vasculature was also measured, revealed that absorption to the bloodstream remained unchanged along the intestinal tract [66, 67]. These results indicate that both transport and metabolism considerations should be accounted for when trying to target P-gp/CYP3A4 cosubstrates to specific intestinal region.

16.5 Conclusions

This chapter aimed to map the key determinants of segmental-dependent intestinal absorption of drugs and to present the various technologies currently used to target drugs to a specific site within the intestine. While there are several promising research directions in this field, not all systems proved to be effective in vivo. Any one of the factors within the intestinal tract (e.g., pH, membrane transporters, and metabolic enzymes expression) may vary extensively due to physiological and pathological causes, and thus altering the intestinal environment. Also, cross talk between the various factors may influence overall intestinal conditions. Therefore, a more complete understanding of each variable, as well as their interactions, is necessary for the successful targeting of specific intestinal regions.

References

1. Dahan A, Amidon GL (2009) Segmental dependent transport of low permeability compounds along the small intestine due to P-glycoprotein: the role of efflux transport in the oral absorption of BCS class III drugs. Mol Pharm 6(1):19–28
2. Dahan A, Lennernäs H, Amidon GL (2012) The fraction dose absorbed, in humans, and high jejunal human permeability relationship. Mol Pharm 9(6):1847–1851
3. Dahan A, West BT, Amidon GL (2009) Segmental-dependent membrane permeability along the intestine following oral drug administration: evaluation of a triple single-pass intestinal perfusion (TSPIP) approach in the rat. Eur J Pharm Sci 36(2–3):320–329
4. Dahan A, Amidon GL, Zimmermann EM (2010) Drug targeting strategies for the treatment of inflammatory bowel disease: a mechanistic update. Expert Rev Clin Immunol 6(4):543–550
5. Washington N, Washington C, Wilson CG (eds) (2003) Physiological pharmaceutics barriers to drug absorption. Taylor & Francis e-Library, London
6. Fallingborg J et al (1989) pH-profile and regional transit times of the normal gut measured by a radiotelemetry device. Aliment Pharmacol Ther 3(6):605–613
7. Dahan A, Miller JM, Amidon GL (2009) Prediction of solubility and permeability class membership: provisional BCS classification of the world's top oral drugs. AAPS J 11(4):740–746
8. CDER/FDA (2000) Guidance for Industry: waiver of in vivo bioavailability and bioequivalence studies for immediate-release solid oral dosage forms based on a biopharmaceutics classification system. Center for Drug Evaluation and Research (CDER)
9. Lennernäs H (2007) Intestinal permeability and its relevance for absorption and elimination. Xenobiotica 37(10–11):1015–1051
10. Lennernas H (2007) Modeling gastrointestinal drug absorption requires more in vivo biopharmaceutical data: experience from in vivo dissolution and permeability studies in humans. Curr Drug Metab 8(7):645–657
11. Regardh CG et al (1974) Pharmacokinetic studies on the selective beta1-receptor antagonist metoprolol in man. J Pharmacokinet Biopharm 2(4):347–364
12. Jobin G et al (1985) Investigation of drug absorption from the gastrointestinal tract of man. I. Metoprolol in the stomach, duodenum and jejunum. Br J Clin Pharmacol 19(Suppl 2):97S–105S
13. Masaoka Y et al (2006) Site of drug absorption after oral administration: assessment of membrane permeability and luminal concentration of drugs in each segment of gastrointestinal tract. Eur J Pharm Sci 29(3–4):240–250
14. Dahan A et al (2010) High-permeability criterion for BCS classification: segmental/pH dependent permeability considerations. Mol Pharm 7(5):1827–1834
15. Alt A et al (2004) Biopharmaceutical characterization of sotalol-containing oral immediate release drug products. Eur J Pharm Biopharm 58(1):145–150
16. Bachmakov I et al (2006) Characterization of beta-adrenoceptor antagonists as substrates and inhibitors of the drug transporter P-glycoprotein. Fundam Clin Pharmacol 20(3):273–282
17. Yang Y et al (2007) Biopharmaceutics classification of selected beta-blockers: solubility and permeability class membership. Mol Pharm 4(4):608–614
18. McConnell EL, Fadda HM, Basit AW (2008) Gut instincts: explorations in intestinal physiology and drug delivery. Int J Pharm 364(2):213–226

19. Wu ZM et al (2012) HP55-coated capsule containing PLGA/RS nanoparticles for oral delivery of insulin. Int J Pharm 425(1–2):1–8
20. Lai X et al (2008) Evaluation of poly(styrene-alt-maleic anhydride)-ethanol as enteric coating material. Int J Pharm 352(1–2):66–73
21. He W et al (2009) Design and in vitro/in vivo evaluation of multi-layer film coated pellets for omeprazole. Chem Pharm Bull (Tokyo) 57(2):122–128
22. Budde K et al (2010) Enteric-coated mycophenolate sodium. Expert Opin Drug Saf 9(6):981–994
23. Ibekwe VC et al (2006) An investigation into the in vivo performance variability of pH responsive polymers for ileo-colonic drug delivery using gamma scintigraphy in humans. J Pharm Sci 95(12):2760–2766
24. Nugent SG et al (2001) Intestinal luminal pH in inflammatory bowel disease: possible determinants and implications for therapy with aminosalicylates and other drugs. Gut 48(4): 571–577
25. Fallingborg J et al (1993) Very low intraluminal colonic pH in patients with active ulcerative colitis. Dig Dis Sci 38(11):1989–1993
26. Katsuma M et al (2004) Scintigraphic evaluation of a novel colon-targeted delivery system (CODES) in healthy volunteers. J Pharm Sci 93(5):1287–1299
27. Estudante M et al (2013) Intestinal drug transporters: an overview. Adv Drug Deliv Rev 15;65(10):1340–1356
28. Keogh JP (2012) Membrane transporters in drug development. Adv Pharmacol 63:1–42
29. Han HK (2011) Role of transporters in drug interactions. Arch Pharm Res 34(11):1865–1877
30. Szakacs G et al (2008) The role of ABC transporters in drug absorption, distribution, metabolism, excretion and toxicity (ADME-Tox). Drug Discov Today 13(9–10):379–393
31. Fredriksson R et al (2008) The solute carrier (SLC) complement of the human genome: phylogenetic classification reveals four major families. FEBS Lett 582(27):3811–3816
32. Hediger MA et al (2004) The ABCs of solute carriers: physiological, pathological and therapeutic implications of human membrane transport proteins Introduction. Pflugers Arch 447(5):465–468
33. Cao X et al (2005) Permeability dominates in vivo intestinal absorption of P-gp substrate with high solubility and high permeability. Mol Pharm 2(4):329–340
34. Giacomini KM et al (2010) Membrane transporters in drug development. Nat Rev Drug Discov 9(3):215–236
35. Muller F, Fromm MF (2011) Transporter-mediated drug-drug interactions. Pharmacogenomics 12(7):1017–1037
36. Dahan A, Sabit H, Amidon GL (2009) Multiple efflux pumps are involved in the transepithelial transport of colchicine: combined effect of P-gp and MRP2 leads to decreased intestinal absorption throughout the entire small intestine. Drug Metab Dispos 37(10):2028–2036
37. MacLean C et al (2008) Closing the gaps: a full scan of the intestinal expression of P-glycoprotein, breast cancer resistance protein, and multidrug resistance-associated protein 2 in male and female rats. Drug Metab Dispos 36(7):1249–1254
38. Chan LM, Lowes S, Hirst BH (2004) The ABCs of drug transport in intestine and liver: efflux proteins limiting drug absorption and bioavailability. Eur J Pharm Sci 21(1):25–51
39. Kunta JR, Sinko PJ (2004) Intestinal drug transporters: in vivo function and clinical importance. Curr Drug Metab 5(1):109–124
40. Mouly S, Paine M (2003) P-glycoprotein increases from proximal to distal regions of human small intestine. Pharm Res 20(10):1595–1599
41. Thorn M et al (2005) Cytochromes P450 and MDR1 mRNA expression along the human gastrointestinal tract. Br J Clin Pharmacol 60(1):54–60
42. Friend DR (2004) Drug delivery to the small intestine. Curr Gastroenterol Rep 6(5):371–376
43. Hoffman A et al (2004) Pharmacokinetic and pharmacodynamic aspects of gastroretentive dosage forms. Int J Pharm 277(1–2):141–153
44. Kagan L, Hoffman A (2008) Systems for region selective drug delivery in the gastrointestinal tract: biopharmaceutical considerations. Expert Opin Drug Deliv 5(6):681–692

45. Kagan L et al (2010) Role of p-glycoprotein in region-specific gastrointestinal absorption of talinolol in rats. Drug Metab Dispos 38(9):1560–1566
46. Bittner B et al (2002) Improvement of the bioavailability of colchicine in rats by co-administration of D-alpha-tocopherol polyethylene glycol 1000 succinate and a polyethoxylated derivative of 12-hydroxy-stearic acid. Arzneimittelforschung 52(9):684–688
47. Adibi SA (2003) Regulation of expression of the intestinal oligopeptide transporter (Pept-1) in health and disease. Am J Physiol Gastrointest Liver Physiol 285(5):G779–G788
48. Jappar D et al (2010) Significance and regional dependency of peptide transporter (PEPT) 1 in the intestinal permeability of glycylsarcosine: in situ single-pass perfusion studies in wild-type and Pept1 knockout mice. Drug Metab Dispos 38(10):1740–1746
49. Ingersoll SA et al (2012) The role and pathophysiological relevance of membrane transporter PepT1 in intestinal inflammation and inflammatory bowel disease. Am J Physiol Gastrointest Liver Physiol 302(5):G484–G492
50. Dalmasso G et al (2008) PepT1-mediated tripeptide KPV uptake reduces intestinal inflammation. Gastroenterology 134(1):166–178
51. Han HK, Amidon GL (2000) Targeted prodrug design to optimize drug delivery. AAPS PharmSci 2(1):E6
52. Brodin B et al (2002) Transport of peptidomimetic drugs by the intestinal Di/tri-peptide transporter, PepT1. Pharmacol Toxicol 90(6):285–296
53. Sugiura T, Kato Y, Tsuji A (2006) Role of SLC xenobiotic transporters and their regulatory mechanisms PDZ proteins in drug delivery and disposition. J Control Release 116(2):238–246
54. Nozawa T et al (2003) Enhanced intestinal absorption of drugs by activation of peptide transporter PEPT1 using proton-releasing polymer. J Pharm Sci 92(11):2208–2216
55. Dahan A et al (2012) Targeted prodrugs in oral drug delivery: the modern molecular biopharmaceutical approach. Expert Opin Drug Deliv 9(8):1001–1013
56. Tamai I (2012) Oral drug delivery utilizing intestinal OATP transporters. Adv Drug Deliv Rev 64(6):508–514
57. Meier Y et al (2007) Regional distribution of solute carrier mRNA expression along the human intestinal tract. Drug Metab Dispos 35(4):590–594
58. Muntane J (2009) Regulation of drug metabolism and transporters. Curr Drug Metab 10(8):932–945
59. Watkins PB (1992) Drug metabolism by cytochromes P450 in the liver and small bowel. Gastroenterol Clin North Am 21(3):511–526
60. von Richter O et al (2004) Cytochrome P450 3A4 and P-glycoprotein expression in human small intestinal enterocytes and hepatocytes: a comparative analysis in paired tissue specimens. Clin Pharmacol Ther 75(3):172–183
61. Zhang QY et al (1999) Characterization of human small intestinal cytochromes P-450. Drug Metab Dispos 27(7):804–809
62. Boocock DJ et al (2002) Identification of human CYP forms involved in the activation of tamoxifen and irreversible binding to DNA. Carcinogenesis 23(11):1897–1901
63. Maseneni S et al (2012) Toxicity of clopidogrel and ticlopidine on human myeloid progenitor cells: importance of metabolites. Toxicology 299(2–3):139–145
64. Basit AW (2005) Advances in colonic drug delivery. Drugs 65(14):1991–2007
65. Tubic-Grozdanis M et al (2008) Pharmacokinetics of the CYP 3A substrate simvastatin following administration of delayed versus immediate release oral dosage forms. Pharm Res 25(7):1591–1600
66. Tamura S et al (2002) Tacrolimus is a class II low-solubility high-permeability drug: the effect of P-glycoprotein efflux on regional permeability of tacrolimus in rats. J Pharm Sci 91(3):719–729
67. Tamura S et al (2003) The site-specific transport and metabolism of tacrolimus in rat small intestine. J Pharmacol Exp Ther 306(1):310–316

Chapter 17
Nanotechnology Applications in Local Arterial Drug Delivery

Dipti Deshpande, Aziza Jamal-Allial, Kinjal Sankhe, and Mansoor Amiji

17.1 Introduction

Cardiovascular disorders are the leading cause of mortality in the USA as well as other nations, despite the introduction of several blockbuster drugs and other breakthrough technologies [1]. The term cardiovascular disorders comprise diseases of the heart as well as the blood vasculature [2]. However, the major cause of the morbidity and mortality associated with cardiovascular diseases are the disorders that affect the circulatory system [1]. These include atherosclerosis, arteriosclerosis, arterial (coronary, carotid, renal, and peripheral arteries) thrombosis and ischemia, venous thrombosis and occlusion, varicose veins, aneurysms, dysregulation of vascular homeostasis, and so on, and most of these disorders have overlapping molecular and cellular mechanisms [1]. The pharmacotherapy of vascular disorders can be broadly classified into two categories (a) systemic drug delivery and (b) local drug delivery. Systemic approach involves several limitations like suboptimal delivery at the site of damage, rapid clearance and hence frequent dosing, low therapeutic index, and lastly side effects like hemorrhage and coagulopathy [1]. Local drug delivery can overcome many of these limitations; however, this approach has its own challenges such as the cost and the amount and duration of drug delivery that can be achieved through this method [3].

The diagnosis and treatment of cardiovascular diseases has been severely impacted by the field of nanomedicine. Nanomedicine has witnessed exponential progress over the past few years and extends over a broad spectrum of applications

D. Deshpande • A. Jamal-Allial • K. Sankhe
Department of Pharmaceutical Sciences, School of Pharmacy, Northeastern University, Boston, MA 02115, USA

M. Amiji (✉)
Department of Pharmaceutical Sciences, School of Pharmacy, Northeastern University, 140 The Fenway Building, Room 156, 360 Huntington Avenue, Boston, MA 02115, USA
e-mail: m.amiji@neu.edu

A.J. Domb and W. Khan (eds.), *Focal Controlled Drug Delivery*, Advances in Delivery Science and Technology, DOI 10.1007/978-1-4614-9434-8_17,

including in vitro diagnostics, in vivo imaging, and synthesis of biomaterials, therapeutic implants, and nontargeted as well as targeted nano-sized therapeutics [4, 5]. Using nanocarriers for the treatment of cardiovascular disorders, local or site-specific delivery, prolonged duration of action, targeted drug delivery, and a significant lowering of the shear effects of blood flow can be achieved [2].

The current review will focus on the development and pathophysiology of arterial diseases, a comparative analysis of local versus systemic approach for treatment of arterial diseases and the advances in nanotechnology that can impact the field of local drug delivery in arterial diseases.

17.2 Arterial Structure and Physiology

17.2.1 Tissue Structure and Function

Blood and all its physiological components are transported throughout the body via the circulatory system that comprises of the heart and the blood vessels—arteries and veins (Fig. 17.1). The diameter of the arteries varies depending on its anatomical location, and overall the arteries have a thicker wall and a narrower lumen as compared to the veins. This is mainly because the arteries not only control the blood flow but also bear the powerful force generated from the cardiac contractions [6].

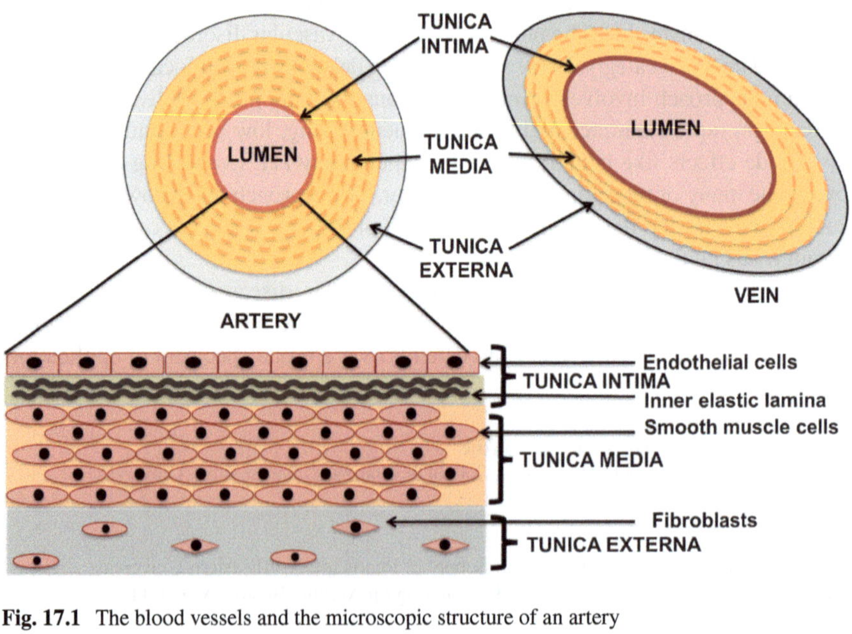

Fig. 17.1 The blood vessels and the microscopic structure of an artery

Based on the anatomical location and arterial function, the arteries can be broadly classified into different types—systemic artery, pulmonary artery, aorta, arteriole, and capillary. Microscopically, the artery is made up of three layers—*tunica externa, tunica media,* and *tunica intima* (Fig. 17.1). Tunica externa, formerly known as tunica adventitia, is composed of the connective tissue, which contains elastin, collagen, fibroblasts, and occasionally smooth muscle cells [7]. The second layer, the tunica media, is the middle layer and is composed of the smooth muscle cells and the elastic tissue on the external side [7]. The number of layers of the smooth muscle cells varies depending on the size of the artery. The innermost layer is the tunica intima that comprises of the endothelial cell layer at the luminal surface of the arteries. The endothelium plays a major role in the maintenance of the vascular homeostasis in the ever-changing pathophysiological environment of the circulatory system [7].

17.2.2 Role of the Endothelium in Vascular Homeostasis

The intact functioning endothelial layer is essential in the maintenance of the vascular homeostasis by releasing various autocrine and paracrine substances that control the blood vessel diameter and the fluidity of blood [8]. Its functions include maintenance of the vascular tone, maintaining anticoagulant nature, promoting angiogenesis in response to inflammation or injury of the blood vessel, and permitting metabolic activity and lipid transport in the cardiovascular system [8]. All the above mentioned functions are performed by sensing the change in the microenvironment and releasing biologically active factors such as endothelium-derived growth factors (EDGF), platelet-derived growth factor (PDGF), nitric oxide (NO), and prostaglandins (PGI) [8]. The modulation of the vascular tone is monitored through the interaction between the relaxing and the contracting factors. The relaxing factors are mainly lipids—prostaglandins, leukotrienes, and proteins such as endothelium-derived growth factors and platelet-derived growth factors [8]. The endothelium surface has receptors for these factors and they are released in response to oxidative stress, sheer stress, thrombus formation, inflammatory markers, toxins, and hormones [8]. The most important endothelium-derived relaxing factor is the nitric oxide (NO) which is mainly synthesized by the activation of endothelial nitric oxide synthase enzyme [8]. Nitric oxide mainly brings about the relaxation of the vascular smooth muscle by elevating the cAMP levels. Recently, a novel theory has been proposed to bring about the relaxation of the vascular smooth muscles—the activation of the K^+ channels by hyperpolarization of the membrane [8] which is induced by the activation of the endothelium-derived hyperpolarizing factors (EDHF). Similarly, the contraction of the vascular smooth muscles is brought about by endothelium-derived contracting factors (EDCF). The release of the EDCFs is mainly initiated by the released vasoconstrictor metabolites of arachidonic acid (e.g., thromboxanes) and peptidergic compound (e.g., endothelin) [8]. The rennin-angiotensin system (RAS) plays a key role in bringing about the contraction of the blood vessels. It is localized on the endothelium, and angiotensin II (Ang II), which is the final product of the RAS, acts as a trigger for the release of endothelium-derived contracting factor [8]. Similar to the EDRFs and EDHFs, the release of the EDCFs is also triggered by sheer stress, hypoxia, thrombin, and several pharmacological agents [8]. The release of the endothelium-derived relaxing and contracting factors is tightly regulated by the feedback regulatory mechanism, and their regulation and release is ATP dependent. The endothelium also has certain mechanoreceptors, which detect the changes in the hemodynamic pressure, the sheer stress, and flow rate. Increase in flow rate brings about the synthesis or release of the EDRF from the endothelium. Similarly, increase in the intraluminal pressure induces release of the EDCF, and the homeostatic balance is maintained by the negative feedback mechanism [8]. Similarly, the endothelium maintains the fluidity of the blood primarily by keeping the luminal surface nonadhesive, anticoagulant, antithrombotic, and fibrinolytic. The endothelium secrets heparin-like glycosaminoglycans that bind to the thrombin III and factor X, thus preventing coagulation [8]. The endothelium also synthesizes thrombomodulin that binds thrombin and converts it to protein C. This in turn inactivates factors VIII and acts as an anticoagulant [8]. The endothelium also modulates the plasminogen activation by activating the endothelial cell tissue plasminogen activator, which converts it to plasmin thereby preventing clot formation [8]. These factors are released in response to the pathological stimuli such as the presence of histamine, venous occlusion, and proinflammatory cytokine [8]. Thus, the endothelium keeps a regulated balance in the vascular tone and maintains blood fluidity necessary for the homeostasis in the dynamic cardiovascular environment.

17.3 Pathophysiology of Arterial Diseases

The onset of most arterial diseases is mainly triggered by endothelial dysfunction. Endothelial dysfunction occurs at the site of injury mainly as an outcome of the imbalance between the contracting and relaxing factors, anti- and pro-coagulating factors, and growth-inhibiting or growth-promoting factors [9]. The imbalance results to endothelial activation which leads to inflammation, immune stimulation, oxidative stress, platelet activation, and aggregation which bring about pathological changes in the healthy cardiovascular system leading to arterial diseases (Fig. 17.2). The primary causes for the arterial diseases are higher levels of low-density lipids (LDL) and very-low-density lipids (VLDL), hypoxia, ischemia, sheer stress, smoking, diabetes, hypertension, immune disorders, and infections which are related to the current lifestyles [10, 11]. The disruption of the endothelial homeostasis leads to certain phenotypic changes and thereby makes it unresponsive to the regulatory mediators. Overall, the progression of arterial diseases is a multifactorial process. Along with endothelial activation and dysfunction, oxidative stress, inflammation, and vessel thrombosis are also implicated in the pathophysiology of these disorders.

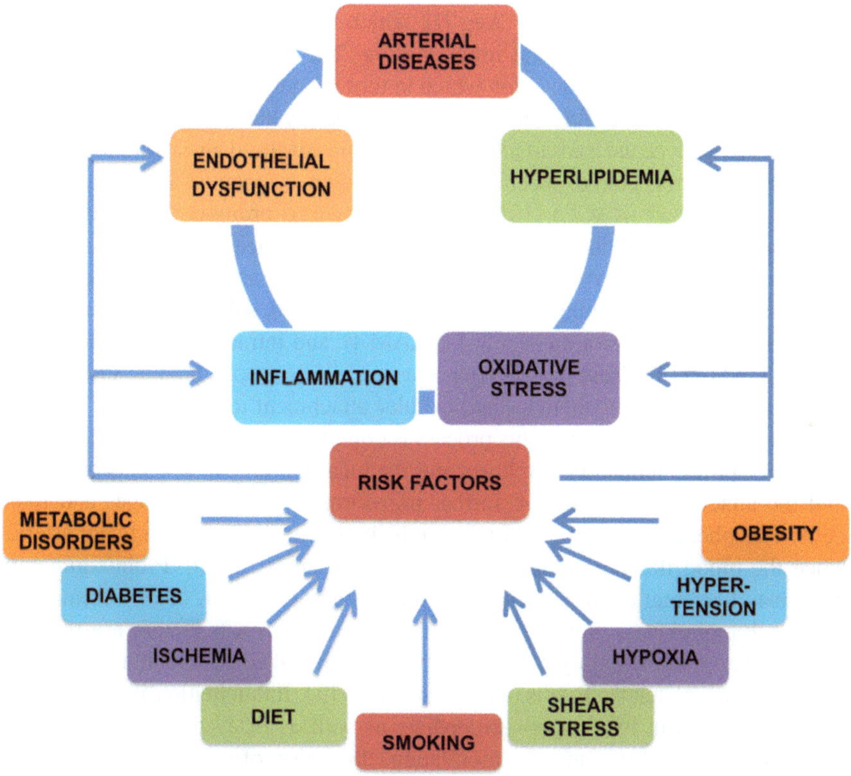

Fig. 17.2 Schematic illustration of risk factors and pathological changes involved in arterial diseases

17.3.1 Inflammation and Thrombosis in Arterial Diseases

As mentioned previously, inflammation is triggered due to lipids such as LDL and VLDL. These lipids undergo oxidation and further cause oxidative stress along the arterial microenvironment which acts as a potential atherogenic trigger [10]. In case of elevated LDL and VLDL levels, high-density lipids (HDL) are known to counteract their effects through upregulation of the reverse cholesterol transport mechanism and transportation of excess lipids across the arterial wall to the liver, bile, and feces for excretion [12]. In addition, HDL can prevent expression of adhesion molecules, inhibit migration of monocytes into subendothelial space, inhibit the procoagulant pathways, and also upregulate nitric oxide synthase activity of the endothelial cells, thus promoting cardiovascular homeostasis [12]. However, in case of an inflammatory and oxidative microenvironment, the protective HDL is rendered dysfunctional and proinflammatory through increased catabolism of these lipids [12]. Similarly, hypertension acts as a sheer stress component that triggers the EDCFs and brings about release of angiotensin II (Ang II), which in addition to its vasoconstrictor properties can instigate intimal inflammation. The release of Ang II elicits the production of the superoxide anion which in turn triggers the inflammatory cascade [10]. Another stimulus for inflammation is hyperglycemia observed in diabetic patients. Hyperglycemia leads to formation of certain advanced glycation end products. These end products bind to specific receptors known as receptors for advanced glycation end product (RAGE) [10]. They are known to promote the production of proinflammatory cytokines. Most inflammatory stimuli lead to endothelial activation which lead to the expression of adhesive cell membrane glycoproteins like the vascular cell adhesion molecule-I (VCAM I), GMP 140, selectins, integrins, endothelial leukocyte adhesion molecule-I (ELAM-I), and intracellular adhesion molecule-l (ICAM-I) [8]. These protein molecules act as a neutrophil binding site and thus attract the circulating monocytes and cellular attachment to the normally resistant endothelial cell surface. Once this process is complete, the lymphocyte attachment, release of proinflammatory cytokines and the local immune response is promoted, and the attached monocytes are further transformed into macrophages and foam cells. This phenomenon is predominant at the branching points of the major arteries since the blood flow variation is highest in these areas [8, 10]. Proinflammatory cytokines like interleukins, interferons, tumor necrosis factor, colony-stimulating factors, lymphokines, monokines, and transforming growth factors are released during inflammation mainly by the activated endothelial cells, smooth muscle cells, macrophages, platelets, T cells, and monocytes [13]. Inflammation in the vascular arteries happens in two stages—immediate as well as delayed. As an immediate response, inflammation induces vasodilation, disruption of cellular tight junctions, and thus an increase in the permeability of the damaged artery [13]. In cases of delayed inflammation, hallmark events like protease-activated receptor (PAR) signaling, plasminogen activator inhibitor-1 (PAI-1) release, CD40/CD40 ligand interactions and expression cytokines such as interleukins, tumor necrosis factor-α (TNF-α), C-reactive protein (CRP), and interferon-γ (INF-γ) dominate the damage site [13]. As an outcome of these events, increase in extracellular matrix composition, granular tissue formation, and proliferation of the quiescent connective tissue is observed at the sight of injury [13].

Along with the proinflammatory cytokines, the activated endothelial cell surface also releases the procoagulant factors leading to activation of tissue factors and release of thrombin, which is inhibited under normal conditions [8]. This in turn promotes release of factor VIIa which mediated the release of factors IX and X [8]. When these factors bind to the membrane, they stimulate the platelet-activating factor (PAF) and platelet-derived growth factor (PDGF) [8]. All these cumulatively lead to thrombus formation as a response to the initial trigger. The anticoagulant activated protein kinase C synthesis system is downregulated, and there is a suppression of thrombomodulin responsible for anticoagulation under the inflammatory and oxidative microenvironments [8, 10]. Together, inflammation and thrombus formation contribute significantly towards the progression of arterial diseases and finally to heart disorders as elaborated in the following section.

17.3.2 Atherosclerosis, Restenosis, and Arteriogenesis

Atherosclerosis is an outcome of all the thrombotic and inflammatory changes occurring in the injured blood vessel leading to a lipid laden intimal mass. At molecular level, atherosclerosis is triggered by hyperlipidemia and inflammation, which leads to the expression of the adhesion proteins such as vascular cell adhesion molecule-1 (VCAM-1), monocyte chemoattractant protein-1 (MCP-1), macrophage colony-stimulating factor (M-CSF), and angiotensin II (Ang II). These along with selectins and integrins mediate the leukocyte binding to the endothelial surface and promote release of the proinflammatory cytokines. The proinflammatory cytokines release chemotactic substances which in turn attract the inflammatory mediators and increase the expression of macrophages loaded with lipids and T and B lymphocytes at the site of injury [10]. The whole assembly acts as a signal for proliferation of the smooth muscle cells. Once the proliferation of the smooth muscle cells reaches the medial layer of the blood vessel, they release certain enzymes which degrade elastin and collagen present [10]. The degraded arterial cell wall permits the penetration of the SMCs through the elastic laminae and collagenous matrix of the growing plaque. The result of the entire process is the formation of an atherosclerotic plaque [10]. In case of an atherosclerotic plaque, the alterations in extracellular matrix metabolism cause thinning of the fibrous cap, rendering it weak and susceptible to rupture. The cross talk between T lymphocytes and macrophages heightens the expression of the potent procoagulant tissue factor. When the plaque ruptures the tissue factor induced by the inflammatory signaling triggers the thrombus that causes most acute complications of atherosclerosis [10].

Atherosclerotic plaque is generally removed by minimally invasive techniques like percutaneous transluminal coronary angioplasty (PTCA or balloon angioplasty), stent therapy, or atherectomy, which are together termed as percutaneous transluminal intervention (PTI) [14]. However, patients undergoing either of these procedures experience relapse of vessel occlusion and restricted blood flow, and this phenomenon is known as restenosis [15]. In case of coronary arteries, vessels that show greater than 50 % narrowed lumen in an angiogram are identified as restenotic arteries [16, 17]. About 70 % of the patients undergoing PTCA or atherectomy eventually experience restenosis, while the rate of restenosis post-stent therapy is lower—about 15–50 % [15]. Restenosis occurs in two stages—neointimal hyperplasia which is characterized by smooth muscle cell migration and proliferation and extracellular matrix deposition, and the second stage is vascular remodeling [18]. The primary response at the cellular and molecular level is deposition of a layer of platelets and fibrin [18]. Especially after stent placement, the strut of the stents invades the tissue and exposes the tissue collagen, which further stimulates the release of several inflammatory signals [15]. Similar to atherosclerosis, there is expression of the adhesion molecules and eventually binding of the leukocytes. The main difference from atherosclerotic plaque in this case is the absence of the lipid laden macrophages and foam cells which give the plaque a hardened unresponsive nature [8]. Once the leukocyte migration and diapedesis is complete due to the chemokines released by the SMCs, the granulation or the cell proliferation phase begins. The growth factors released aid in the migration of the SMCs from the media to the neointima. The final composition of this layer is made of SMCs, macrophages, and the extracellular matrix. This process continues for several weeks. Cellular division continues throughout this process. The next step in the process of restenosis is remodeling of the neo-blood vessel. The process involves continuous extracellular matrix protein degradation and resynthesis. In this phase there are fewer cellular elements produced and an increase in the production of extracellular matrix (ECM). Due to this there is regulated growth of the neointimal tissue, which prevents excessive thickening of the new layer formed. ECM is composed of various collagen subtypes and proteoglycans, and constitutes the major component of the mature restenotic plaque [18]. Thus, in contrast to atherosclerosis, restenosis is a response to injury and a process similar to wound healing and scar tissue formation.

Arteriogenesis is the process of the growth of collateral arteries as a response to cardiovascular compromise in case of diseased conditions to compensate for the lack of blood supply [11]. The source for the neo-arteries are the preexisting arterioles—mainly arterioles from the upper leg [11]. The trigger for the arteriogenesis is sheer stress and the inflammatory factors. The process of arteriogenesis is seen when the mechanoreceptors in the endothelium detect the changes in the hemodynamic pressure [11]. They upregulate the expression of the adhesion molecules (ICAM family) that assist in docking of the monocytes and leukocytes at the site [11]. The presence of monocytes marks the difference from atherosclerotic plaque formation. There is also a release of the vasoactive NO at the site which leads to activation of the FGF, EDGF, and PDGF and increases the permeability of the membrane [11]. Mast cells also aggregate and bring about the release of the VEGF and bFGF [11]. The whole process of neointimal plaque formation as described previously occurs with simultaneous release of colony-stimulating factors (CSF) and transforming growth factor β [11]. The CSF reduces the serum cholesterol levels in this phase. After the acute phase of arteriogenesis that is dominated by the inflammatory events, remodeling begins, i.e., the much slower consolidation of the arterial structure after the final diameter is almost reached. A new elastic lamina is synthesized by the SMCs. The rebuilding of the media and the formation of an intima begins with the downregulation of the tissue inhibitor of matrix-metalloproteinases (TIMP and MMP) [11]. This is followed by an upregulation of the expression and activity of the MMPs that digest the matrix and provide the space for new cells and enable SMCs to migrate towards the intima [11]. Many SMCs of the old medial layer die an apoptotic death and are replaced by new ones. Thus, new arteries are formed in order to compensate the compromised cardiovascular output. Surprisingly, arteriogenesis and atherosclerosis have many features in common except one: collateral vessels increase in diameter in arteriogenesis, but decrease in atherosclerotic vessel [11]. They also share the inflammatory component, in particular the T lymphocyte involvement; the upregulation of MCP-1, adhesion molecules, and matrix proteases; the change in phenotype and migration of SMCs; and the formation of an intima [11]. The main difference is that in arteriogenesis, small preexistent arterioles become wider under the concerted actions of physical forces, growth factors, and matrix proteases, whereas in case of the larger arteries, consisting of numerous layers of smooth muscle and larger amounts of matrix proteins, react to endothelial activation with a locally restricted intimal growth, because the thick intima and media inhibit the progression of the growth stimuli and are a barrier to the invasion of monocytes [11]. Thus, while atherosclerosis and restenosis both involve negative remodeling of the diseased blood vessel, arteriogenesis involves positive remodeling of blood vessels [19]. Several researchers have hypothesized that therapeutic strategies promoting arteriogenesis may prove beneficial in the treatment of occlusive arterial diseases, and such therapies have been investigated under clinical trials [20]. Approaches like intra-arterial injections of MCP-1, colony-stimulating factors (GM-CSF and G-CSF), fibroblast growth factors, transformation growth factors, and statin therapy have been studied extensively to promote arteriogenesis in animal models/patients with occlusive vascular diseases [21]. However, due to the overlap of molecular mechanisms underlying arteriogenesis and inflammatory disorders like atherosclerosis, the success of these approaches has been limited by unwanted side effects [21]. In addition, drug dosage, delivery method, and selection of suitable end-points also need to be optimized and validated for the success of these therapies [21]. The benefits and limitations of arteriogenesis as a therapeutic strategy have been reviewed extensively elsewhere [20, 21]. For the clinical implementation of therapeutic arteriogenesis, certain detailed aspects of the molecular mechanisms underlying arteriogenesis need to be elucidated, and this knowledge needs to be extended towards the study design of the future clinical trials [20].

17.3.3 Carotid, Coronary, and Peripheral Arterial Disease

The carotid artery is located at the base of the heart closer to the aorta and is subjected to severe blood pressure and sheer stress. Injury to the intimal layer of the blood vessel is the primary site of activation of the atherosclerosis plaque formation and eventually the plaque occludes the blood vessel leading to the carotid artery disease. These occlusions mainly reduce the blood flow to the brain and multiply the chances of stroke and ischemia [22, 23]. In case of stable plaques within the carotid arteries, the larger the plaque area the greater is the risk of stroke [23]. In case of unstable plaques, the plaque fragments can travel to the smaller vessels within the brain thus blocking blood flow in those regions [23]. Usually, carotid artery disease has no symptoms and is detected only after the blood flow to some region of the brain has been blocked [23].

In case of the peripheral artery disease, the arteries of the peripheral circulation are affected. The disease is mainly known as the atherosclerosis of the extremities [24]. Similar to the development of atherosclerosis, the disease progression occurs in three stages—initiation of the lesion, progression of the lesion, and formation of the unstable plaque [25]. This results in obstruction of blood flow mainly to the coronary and intracranial vessels, and to the arteries at the extremities, mesenteric, and renal arteries [25]. The symptoms include pain in muscles, compromised cardiac output, and tingling sensation of the affected area [25].

Vascular injury and thrombus formation are once again the etiology for the coronary artery disease. The severity of the injury dominates the progression and extent of the disease. There are mainly three types of lesion—Type I, Type II, and Type III. In case of Type I, there is functional alteration of the endothelium without significant morphological changes [22]. In type II, the denudation progresses to the intimal layer and in type three the denudation progresses to the tunic media [22]. The clinical effects of the disease are seen prominently in case of types II and III injury, and they eventually lead to thrombus formation and occlusion of the artery leading to cardiac angina and heart failure [22]. CAD is mainly associated with two phases—chronic phase with occlusive plaque and the acute phase with nonobstructive lesions [26]. While the localized lesions are associated with symptoms like pain, shortness of breath, weakness and dizziness, the silent plaques show no symptoms and can directly lead to angina or myocardial infarction on disruption [26]. Disruption of the unstable plaques can lead to many new solid-state stimuli for lesion formation and result in acute coronary syndrome [26]. Thus, for effective treatment of CAD, it is important to first target the occlusive plaque followed by treating the other vulnerable, unstable nonocclusive lesions to prevent recurrent plaques elsewhere in the blood vessel [26].

Several approaches have been employed in the treatment of arterial diseases and they can broadly be classified into three groups (a) lifestyle changes, (b) use of pharmacological agents, and (c) use of minimally invasive techniques. When the arterial damage is beyond the scope of these treatment options, bypass surgery is performed as the final alternative. Lifestyle changes include options like lowering body weight, aerobic exercise, and diet control. The use of systemic pharmacological agents in the treatment of arterial diseases has been in practice for decades; however, the treatment's effect might not be in parallel with improvement of the existing arterial disease especially when treating advanced disease cases. In case of severely occluded vessels, where systemic therapy alone cannot improve the lesions, invasive techniques like balloon angioplasty and stent therapy are used for the treatment of the disease. The final goal of all treatment options is to restore the arterial structure and regular blood circulation. Since the vascular endothelium plays a vital role in maintaining the health of the cardiovascular systems and is primarily implicated in the development of the disease, the endothelium has become an attractive therapeutic target for the treatment of arterial diseases.

17.4 Systemic Versus Local Arterial Therapeutic Strategies

17.4.1 Pharmacological Drug Classes Used for Systemic Arterial Therapy

The use of systemic pharmacological agents in the treatment of arterial diseases is not new; however, systemic approaches may not result in the improvement of the existing arterial disease, especially when treating advanced disease cases.

The angiotensin-converting enzyme inhibitors (ACE-1) and angiotensin receptor blockers (ARBs) that are primarily used for the treatment of hypertension and congestive heart failure are also prescribed for arterial disease treatment. These drugs are mainly associated with the increase in NO bioavailability and superoxide dismutase as well as reduction in oxidative stress. Several clinical trials have shown that ACE-1 inhibitors can improve endothelial function as measured by coronary blood flow (CBF) as well as flow-mediated dilation (FMD) [27, 28]. Another popular class of drugs includes the antilipidemics that are used to lower abnormally high blood levels of lipids, such as cholesterol, triglycerides, and phospholipids [29, 30]. Statins or HMG-CoA reductase inhibitors are a subclass of antilipemics drugs that lower cholesterol levels by inhibiting the enzyme HMG-CoA reductase as well as improve endothelial functions as measured by FMD [31, 32]. In addition, statins can inhibit the onset of ischemic heart disease. However, their use has been associated with muscle weakness and, in rare situations, myositis and myopathy [30, 33]. Similarly, the beta-adrenergic blockers (β-blockers) used in managing angina can also prove beneficial for the treatment of arterial diseases. Unlike the classical β-blockers, the newer and improved third-generation β-blockers, such as carvedilol, exert endothelium-dependent vasodilation properties that can help restore vascular balance [27, 28, 34]. Other therapies include nitrates, such as nitroglycerin, which can be given orally, transdermally, or sublingually. Nitrates improve the blood supply to the heart through artery dilation [35]. Another therapeutic option is calcium channel blockers that are used in treating stable angina and hypertension. They produce arterial vasodilation that can help in the prevention of arterial damage [27, 28, 34]. Many of the systemic arterial therapies are fairly effective in the improvement of endothelial dysfunction mainly when used to treat discrete lesions with low restenosis rate. However, not all of the systemic drugs are effective in treating coronary artery disease or preventing restenosis. Of the various probable explanations, efficacy has been the most commonly cited reason for the failure of systemic drug therapy in treating advanced arterial diseases. In addition, the fact that several of these drugs have low potency and narrow therapeutic window limits their application as systemic therapeutics for arterial diseases [3, 29].

17.4.2 Rationale for Local Drug Delivery

Local drug delivery (LDD) is a direct and targeted application of pharmacotherapy to one specific area or the site of damage. The development of LDD is mainly driven by the inefficiency of the systemic drug delivery as well as the need to find more sophisticated therapeutics [3, 36]. Several applications have been developed including the drug-eluting stents (DES) for local drug delivery and treatment of arterial diseases. A recent review suggests that a successful LDD approach should (1) deliver an effective therapeutic agent that has a wide therapeutic window and can be delivered consistently, (2) be capable of loading an adequate concentration of the drug, and (3) produce little or no systemic toxicity [37]. In theory, local drug delivery allows a greater concentration of the drug to accumulate at the local site using a lower drug dose [3]. However, the drug retention time is a critical aspect that can limit the use of the local drug delivery especially in the case of rapidly metabolized molecules. Several approaches can be used to prolong the drug retention at the site of damage; one example is the use of lipophilic molecules, which have rapid permeation and longer residence in cells due to the lipid bilayer structure of cellular membranes [38]. For instance, paclitaxel is a hydrophobic drug that is characterized with long retention time. Paclitaxel was selected as a therapeutic component for TAXUS™ drug-eluting stent, and several researchers have reported the benefits of local paclitaxel delivery in case of arterial disorders [39–42]. Another way to improve local drug retention is using nanotechnology-based endothelium targeted drug delivery systems, which will be discussed further in this chapter. The most commonly used devices for local therapeutic administration are the stents and perfusion balloons, however, use of the later shows significantly lower drug uptake at the site of angioplasty as compared to the drug-eluting stents [29]. Examples of the newer balloon-based technologies include the (a) occlusive balloon technology, GENIE™ catheter system (Acrostak, Winterthur, Switzerland); (b) microporous balloon technology, Clearway™ RX system (Atrium Medical Corporation, New Hampshire, USA); and (c) drug-eluting balloons, SeQuent® Please balloon catheter (Braun, Germany) [3]. However, the balloon-based systems are associated with a higher rate of ISR and much lower drug retention and variable drug release at the tissue site as compared to the stent therapy [3].

17.4.3 *Stent Therapy for Arterial Diseases*

Stent or bare-metal stent (BMS) is generally a hollow cylinder-shaped tube that can be placed within the artery to provide support and allows uninterrupted blood flow [36, 43]. Prior to the emergence of drug-eluting stents, BMS were implanted within the arterial wall to prevent acute elastic recoil, which in turn reduces the risk of vessel occlusion and prevents negative remodeling [3, 44]. Even though the STRESS and the BENESTENT trials demonstrated that the use of stents lowers the risk of angiographic restenosis compared to angioplasty, the use of BMS initiated other mechanical injures within the arterial wall [3, 45]. BMS implantation initiates localized inflammation within the surrounding area, which in turn results in subacute stent thrombosis, neointimal hyperplasia, and ultimately leads to in-stent restenosis (ISR) [3, 46]. Even though, the concurrent use of systemic antiplatelet drugs can lower the risk of the subacute stent thrombosis, ISR still remains as a major limitation of the use of BMS. Considerable attempts have been made to further modify BMS to lower the incidence of in-stent neointimal hyperplasia; however, noticeable reduction of this new problem was only attained with the advancement of BMS into a drug-eluting stents (DES) [3].

DES is basically a bare-metal stent, but modified through coating the surface with a polymer. It has an intricate mesh-like design to allow flexibility. DES is composed of three main components: the stent platform, which can be metallic or polymeric; the delivery matrix that contains the carrier material and delivery construct; and the therapeutic agents that will be eluted into the arterial wall [3, 36]. Several researchers have discussed some key advantages of DES including efficient local site-specific drug delivery, minimal systemic toxicity, and bypass of the first-pass metabolism [3, 29, 36]. The composite metals of this particular type of stents differs with regard of strength and flexibility; they are usually made of variety of material such as 316L stainless steel, tantalum (Ta), cobalt-chromium alloy (Co–Cr), nickel–titanium alloy (Ni–Ti), magnesium alloy (Mg), and platinum–chromium alloy (Pt–Cr) [36, 47]. The cobalt-chromium alloy is more radiopaque and durable material compared to the 316L stainless steel stents, which allows these stents to be thinner, thus lowers the risk of restenosis [36, 48].

On the other hand, the polymeric stents are the most used type of stents for drug delivery within the arterial system [3, 36]. These stents possess a backbone of biocompatible or bioinert polymers that is designed to biodegrade after the drug elution. Table 17.1 illustrates the assortment of stent applications with various carrier materials. Durable material ranges from poly(ethylene-co-vinyl acetate) (PEVA),

Table 17.1 Commercial stents used for the treatment of arterial diseases

Stent	Components	Coating characteristics
Cyper™	Poly(ethylene-co-vinyl acetate) (PEVA) and poly(butyl methacrylate) (PBMA)	Base layer: sirolimus (PEVA and PBMA) Topcoat layer: drug-free PEVA [49]
TAXUS™	Styrene-isobutylene-styrene copolymer (SIBS) and polyurethane (PU)	SIBS coating contains paclitaxel [50]
Echinomycin-eluting stent		Pre-coated BMS (bare-metal stent) with PU and topcoated with heparin polymer [51]
Endeavor® stent Zomaxx™ Genous™	PC (phosphorylcholine)	Compatible with the blood and tissues [52]
XIENCE™	PVDF-HFP (poly(vinylidene fluoride-cohexafluoropropene)	Two-layer coating system (acrylate primer and fluorinated copolymer) [53]
Elixir DESyne Novolimus™	(Poly[lactide]) PLA and poly-*N*-butyl methacrylate	Polymers sprayed not coated [54]
Endeavor® Resolute	Biolinx polymer system (blend of hydrophilic C19 polymer, polyvinylpyrrolidone (PVP), and hydrophobic C10 polymer)	C19 polymer rapid drug elution; PVP initial drug burst and sustain overall elution rate C10 polymer provides the blend with hydrophobicity and rigidity [55]
BioMatrix™ stent Nobori™ stent JACTAX HD stent BVS stent Axxess™ Plus Stent Excel™	Poly-L-lactide (PLLA) and poly-D,L-lactide (PDLLA) [56]	–

(continued)

Table 17.1 (continued)

Stent	Components	Coating characteristics
Nevo™ stent MAHOROBA® stent Conor CoStar™ stent Stellium™ stent eucatTAX stent	PLGA (poly[glycolic-co-lactic acid]) [54]	–
BioMime™ stent	Copolymer-poly(L-lactic acid) (PLLA) and poly(D,L-lactide-co-glycolide) (PLGA) [57]	–
Sparrow® stent	SynBiosys™ polymer [36]	
Supralimus® stent	Mixture of PLLA, PVP, and PLGA	Base: PLLA and PLGA (50/50), PVP, and 35 % sirolimus; topcoat PVP [58]
Infinnium™ stent	Mixture of PLA, PVP, and PLGA	Three different layers of coatings. Each coat has different release profile [59]
Ideal™ stent	Salicylate-based poly-anhydride ester	Drug coated on bioresorbable backbone [56]
Metal bioresorbable stent	Blend of poly(L-lactic acid) and poly(4-hydroxybutyrate)	Drug incorporated within the whole degradable stent [56]
ReZolve™ bioresorbable stent	ReZorb™ polymer (tyrosine-derived polycarbonate polymer)	Balloon-expandable stent with mechanical locking technique [56]
Polymer-free sirolimus-eluting stent	Hydroxyapatite (HA)	HA microporous coat [60]
Tacrolimus-eluting-coated stent	Aluminum oxide (Al_2O_3)	Nanoporous-coating structure incorporated with tacrolimus [61]

Adapted from Lei et al. [36]

poly(*n*-butyl methacrylate) (PBMA), poly(styrene-block-isobutylene-block-styrene) (SIBS), polyurethane (PU), phosphorylcholine (PC), poly(vinylidene fluoride-co-hexafluoropropylene) (PVDF-HFP), and poly-*N*-butyl methacrylate [36, 62, 63]. Meanwhile, many of the newly available commercial stents have a biodegradable polymer coating such as polyanhydrides and polysalicylates; polymers like polylactide (PLA), poly(lactide-*co*-glycolide) (PLGA), and poly(ε-caprolactone) (PCL) have shown to exert greater advantages [36, 43, 56]. With regard to the therapeutic agents, they range from antiproliferative, antimetabolic, immunosuppressive, antibiotics, and cardiovascular known drugs, which are mainly to inhibit cellular proliferation, reduce the neointimal growth that is due to smooth muscle cell proliferation, and so profoundly reduce restenosis development (Table 17.2). The four different types of the metallic stents most widely used include drug-containing membrane-covered stents, drug-polymer layer-coated stents, drug reservoir-based metallic, and polymer-free stents [36].

Table 17.2 Different drug types and their stent applications in arterial disease treatments

Drug type	Drug	Mechanism of action	Application	Commercial stents
Antiproliferative	Paclitaxel	Inhibits cell processes including mitosis, cell proliferation, and cell migration	Eluting stent [50, 59]	Taxus™ stent Infinnium™ stent
	7-Hexanoyl-ataxol	Reduces neointimal growth (taxane analogue)	Eluting stent [64]	QP2-eluting Quanam QueST stent
	Actinomycin	Inhibits DNA synthesis	Pre-mounted L-605cobalt-chromium alloy (Co–Cr) [65]	Multi-Link Tetra™ -D Guidant
	Angiopeptin	Reduces the effect of growth hormone, insulin-like growth factor, and interleukin-1 on cell adhesions	Angiopeptin-eluting coated with phosphorylcholine stents [66]	BiodivYsio DDS
Antimetabolite	Methotrexate	Folate antagonist	Methotrexate-coated stent [67]	
Immunosuppressive	Sirolimus (rapamycin)	Blocks cell cycle at G_1 phase through the inhibition of mTOR	Polymer-free [41, 68]	Cyper™ stent Nevo™ stent Yokun™ stent
	Biolimus A9	Inhibits T cell and smooth muscle proliferation	Derivative of sirolimus [69, 70]	Biometrix™ stent BioFreedom™ stent
	Novolimus		Metabolite of sirolimus (11)	Elixir DESyne
	Myolimus		Eluting stent	Elixir myolimus-eluting stent
	Tacrolimus (FK506)	Inhibits neointimal proliferation	Tacrolimus-eluting stent (12)	Janus® stent
	Zotarolimus (ABT-578)		Lipophilic analogue of sirolimus [52, 71]	Endeavor™ stent Zomaxx™ stent
	Dexamethasone	Anti-inflammatory drug	Stent [72]	Dexamet™ stent
	Everolimus	Analogue of sirolimus	Biodegradable stent [72, 73]	XIENCE™ stent BVS stent
	Prednisolone acetate	Anti-inflammatory and antiproliferative	Helical film-based biodegradable PLGA stent [73]	
	Cyclosporine A	Immunosuppressive	Metal stent coated with PLLLA/cyclosporine A [74]	
	Leflunomide	A pyrimidine synthesis inhibitor with anti-inflammatory and antiproliferative property	Custom-made eluting stent [75]	Yokun™ stent

(continued)

Table 17.2 (continued)

Drug type	Drug	Mechanism of action	Application	Commercial stents
Antibiotic	Actinomycin D	Inhibits DNA-primed RNA synthesis	Custom-made eluting stent [65]	Multi-link Tetra™ stent
	Echinomycin	Binds to DNA and acts as a stapler	Custom-made eluting stent topcoated with heparin-containing polymer [51]	
Protein	Anti-CD34 antibody	Improves endothelial healing and protect against thrombosis	Bioengineered stent [76]	Combo™ stent
Cardiovascular/blood system agent	Cerivastatin	Inhibits neointimal hyperplasia	Stainless steel coated with servastatin and polymers [47]	
		Reduce stent-induced inflammation	Polymer-free cerivastatin-eluting stent [77]	
	Atorvastatin	Lowers blood cholesterol	Atrovastatin-coated stent [78]	BiodivYsio™ stent
		Stabilizes plaque and preclude strokes		
	Abciximab	Blocks platelet aggregation	Abciximab-coated stent [79]	MAC™ stent
		Inhibits vascular smooth muscle proliferation and inflammation		
Hormone/steroid	17β-estradiol	Inhibits intimal proliferation	Drug-eluting stent [80]	BiodivYsio™ stent

Adapted from Lei et al. [36]

17.4.3.1 Metallic Stents

Drug-Containing Membrane-Covered Stents

Metallic stents can be developed into drug-containing membrane-covered stents by wrapping a drug-containing membrane around it. Drugs loaded within the membrane include anti-inflammatory drugs or antitumor agents, which makes them ideal even for diseases other than coronary artery disease such as nonvascular benign/malignant tumor; vessel narrowing and repair of vessel ruptures [36].

Drug-Eluting/Polymer-Layer-Coated Stent

These consist of a drug-polymer coating a metal struts, which make them the most sought after stent type for drug delivery. Taxus® stent (Boston Scientific, MA, USA) and Cypher® stent (Johnston & Johnston, NJ, USA) are considered as the successful first generation DES. Taxus® is coated with copolymer SIBS incorporated with the drug paclitaxel [56, 81]. Cypher® is coated with two polymers PEVA and PBMA incorporated with the drug sirolimus (rapamycin) and one topcoat of PBMA [56, 82]. Despite the advancement of the first generation DES, they are associated with several challenges including late thrombosis and delayed healing. Thus, improved inert polymers with enhanced compatibility have been a part of the second-generation DES. One example is the Endeavor stent (Medtronic, USA), where the thin strut made from cobalt-chromium, is coated with biometric polymer based on phosphorylcholine, and incorporated with the drug zotarolimus (ABT-578). One distinct feature is that the permanent coating of phosphorylcholine is hydrophilic while most polymers are hydrophobic [56, 83]. Another example is the use of fluorinated copolymer on XIENCE™ (Abbott, USA). It has the permanent PVDF-HFP incorporated with the drug everolimus [36, 84]. These stents coated with durable material also encounter issues, such as endothelial dysfunction and chronic inflammation. However, the use of biodegradable polymers as coating show better results in lowering inflammation and improving endothelial dysfunction. Many commercially available stents (Table 17.1) are coated with biodegradable polymer coating that primarily consists of polyesters such as PLA, PLGA, and PCL. One main advantage of the biodegradable drug-loaded coatings is while releasing the drug, these polymers degrade into absorbable small molecules such as carbon dioxide and water [36, 42, 56]. The action of drug release can be modified through various formulation as well as coating patterns. For example, the drug-eluting stent DREAMS™ (AMS-3) (Biotronik, Berlin, Germany) has a rapid degradable polymer matrix on a magnesium alloy stent which slows the release of pimecrolimus. Another example is the BioMatrix stent with albumin coating that allows the releases of Biolimus A9 into the vessel wall [48]. One of the major differences between the metallic and the polymer-coated stents is much smaller drug surface area because of the meshes between the metal struts, which might translate into inconsistent drug distribution [36].

Drug Reservoir-Based Metallic

Different from the prior types of DES, the reservoir-based metallic stents contain holes or grooves that are filled with drug alone or a drug-polymer mixture [36]. One advantage of this particular stent type is the reduced polymer surface area, which might reduce tissue exposure to these polymers. Moreover, this design allows the use of different drugs, each in a groove(s), which allows numerous ways for drug delivery. In the case of multiple drugs, these reservoirs will allow each drug to be delivered independently [3, 36].

Polymer-Free Stents

This type of stent utilizes rough microporous surfaces to avoid the use of polymer through dip-dry coating method, which is simply can be achieved by dipping the stents into a drug solution followed by solvent evaporation. For example, paclitaxel is coated on the polymer-free ACHIEVE™ (Cook, IN, USA) stent by proprietary coatings process on the 316L stainless steel surface. Another example is the Yokun® Choice stent (Translumina, Hechingen, Germany), which has sirolimus within its PEARL surface-a microporous surface allowing for better control of the dose of the drug [41].

17.4.3.2 Bioabsorbable Stents

Different from the biodegradable metal stents, the bioabsorbable polymeric stents are entirely made of biodegradable polymer without any metal in its backbone. After completing the drug release, these stents have the ability of complete degradation into smaller molecules that can be further metabolized and hence, removed from the body. Some of the biodegradable polymers being used within this type of stents are the following: poly(L-lactide) (PLLA), poly(DL-lactide) (PDLLA), PLGA, and PCL [85]. Compared to metallic stent, one of the advantages of these stents is that they can carry larger dose of the drug since its backbone is fully polymeric; however, its mechanical properties are not inferior to the prior. Another advantage is that this type can be further developed through the use of biodegradable coating [36, 85]. Currently, bioresorbable vascular (BVS) everolimus-eluting stent by Abbott is the only approved bioabsorbable stent that is fully absorbed within 2 years. BVS backbone contains PLLA and PDLLA as a thin coating material, which controls the release of antiproliferative drug, everolimus [43, 86, 87]. Others under clinical trials include REVA stent (REVA Medical, CA, USA), Combo™ Bioengineered (OrbusNeich, Hong Kong, China), and Ideal™ stent (Bioabsorbable Therapeutics, CA, USA) [43]. The bioabsorbable stent type has three main subtypes including drug-polymer layer coated, monolithic drug-containing stent, and multi-layered film-based stent platforms. Analogous to the drug-polymer layer-coated metallic stent, the drug-polymer layer-coated bioabsorbable stent is enclosed with biodegradable drug-loaded coating material. One example of this type of stent is the

4-week biodegradable paclitaxel-eluting polylactide stent, which is coated with tyrosine kinase antagonist and PDLLA stent [88]. Secondly, the monolithic drug-containing stent and this type in particular is characterized with the integration of the drug within the stent's backbone, which allows greater drug loading [36]. Finally, the multilayered film-based stent platform is one promising type of stent, where it provides a matrix of multilayers of various formulations that can be modified to achieve a desired release profile of the loaded drug. One example of that is the PLLA/PLGA bi-/trilayer-film-based stents that is being used to study the loading efficiency and the release of sirolimus [89].

17.4.3.3 Gene-Eluting Stents

Gene therapy is one new approach that looks very promising in the area of local drug delivery. This sophisticated approach is based on the principle of coupling adenoviral vectors to promoter sequences to alter the expression of some endothelial genes like the human vascular endothelial growth factor (VEGF2), thus exerting specific yet local vascular response [27]. One of the most successful gene delivery systems is the BiodivYsio stent that contains phosphorylcholine, which is positively charged, as a coating material to interact with the negatively charged DNA (plasmid encoding the gene) through electrostatic reaction [90]. Because of the specificity of DNA and its unique physical properties, this particular stent can be adapted to adsorb certain amount of DNA through stalking of the DNA and the positively coating material in layers [91]. Another point is that in contrast to the metallic stents, the use of ordinary polymer coating might not be appropriate because of the size of the DNA; hence, the use of the highly permeable nano/microporous polymer is preferable. While DNA is susceptible to degradation within a physiological environment, gene delivery stents are modified to release the DNA in the form of DNA/vector complex, which increases DNA stability within these environments. Some examples of the DNA/vector are viruses and polycations, such as poly-L-lysine (PLL), poly(ethylenimine) (PEI), and cationic lipids [92].

In conclusion, stent development has been one of the most efficient therapies for arterial diseases. Stents serve not only as a drug delivery apparatus that offers a wide range of therapeutics to treat the disease locally but also act as a mechanical device that provides support to occluded structures.

17.5 Nanotechnology Approaches for Arterial Drug Delivery

The use of nanotechnology for designing newer platforms for prevention, diagnosis, and treatment of diseases has proved as a technological breakthrough in the field of medical research. Nanomedicine can be considered as the foundation for patient-specific personalized medicine, which is envisioned as the next revolution in the field of medical research [1]. In particular, cancer and vascular disorders have benefitted the most with the advent of nanomedicine in medical research. In case of arterial diseases, the use of nanotechnology-based approaches have the following advantages (a) prolonged circulation and increased half life of the therapeutic agent, (b) ability to target specific disease sites based on cell specific interactions, (c) increased drug retention at the site of injury and minimal systemic toxicity, (d) ability to alter release profile of therapeutic molecules, and (e) use of biodegradable and safe components in the formulation of these systems [1]. As described previously, arterial diseases involve a complex interplay of several pathological events, and at every stage of the disease different cell types dominate the site of injury. Hence, based on the treatment approach, nanoparticles can be targeted towards a specific cell type leading to increased site-specific retention of the nanoparticle payload. For example, nanomedicine approaches targeting integrins, selectins, and cellular adhesion molecules on endothelial cell surface; integrins and selectins on platelet cell surface, scavenger, and oxidized LDL receptors on macrophages; and other protein targets, like tissue factors, MMPs fibrin, collagen, lipoproteins, and annexins which are overexpressed in regions of arterial disease can enhance site-specific delivery of several therapeutic molecules [1].

17.5.1 Nanotechnology in Small Molecule Drug Delivery

As described in the previous sections, several antiproliferative, antianginal, antilipemic, and antihypertensive agents have been used in the treatment of arterial diseases. Based on the cellular target, several targeted nanoparticle/liposome approaches have been investigated for the delivery of these pharmacological agents. Liposomes with surface modification using hydrophilic polyethylene glycol are a well established drug delivery platform that promote prolonged circulation and residence of the payload and prevent rapid clearance through reticuloendothelial system (RES) [1]. In vivo delivery of RGD-conjugated liposomes specific to integrin GPIIb–IIIa expressed on the surface of activated platelets in a catheter injured carotid artery rat model of restenosis showed successful platelet targeting and payload delivery using these modified liposomes [93]. This approach can, thus, prove beneficial for the delivery of antiplatelet, anticoagulant, fibrinolytic, and thrombolytic agents for the treatment of arterial diseases. Similarly, in vivo delivery of the anticoagulant drug hirulog using modular multimodal fibrin-targeting micelles to high fat diet induced atherosclerotic ApoE$^{-/-}$ mice showed increased concentration of the drug in the plaques as compared to the nontargeted micelles [94]. In addition a greater accumulation of the micelles was observed in the shoulder region of the plaque, which is prone to plaque disruption [94].

17.5.2 Nanotechnology-Based Nucleic Acid Delivery

As mentioned earlier, gene delivery is a powerful tool for the treatment of arterial disorders. A recent study reported significant inhibition of neointimal hyperplasia and restenosis post-angioplasty in an atherosclerotic rat model using an amino acid-based nanoparticle system for local delivery of siRNA targeting NOX2 gene [95]. NOX2 is a component of the nicotinamide adenine dinucleotide phosphate (NADPH) oxidase enzyme system, which is upregulated in cases of arterial damage [96]. It is well known that oxidative stress occurs in the arterial microenvironment during atherosclerosis as well as angioplasty and stenting and is partly mediated through the NADPH oxidases [97]. Thus, intra-arterial delivery of the nanoparticles containing siRNA targeting NOX2, in atherosclerotic rats post-angioplasty, showed a significantly lower neointima to media ratio and an increased lumen to whole artery ratio as compared to the rats who received angioplasty alone [95]. While viral delivery of the siRNA can lead to systemic toxicity, the nanoparticle-based delivery showed no signs of toxicity in the animals [95]. Another interesting study has reported the use of non-viral lipopolyplexes coated onto stents for delivering the endothelial nitric oxide synthase gene encoding plasmid DNA for the treatment of restenosis [98]. While most gene delivery stents use either polymer coats or adeno-viral constructs for encapsulation of plasmids, this system used pre-complexed DNA loaded cationic liposomes for coating the stents and delivering the gene of interest at the site of injury [98]. Deployment of these stents in the iliac artery of the rabbit model of restenosis showed a significant reduction of neointimal hyperplasia and increase in the re-endothelialization at the lumen of the iliac artery after 14 days of stent placement [98]. Both examples suggest that in case of cardiovascular gene therapy, the nanotechnology platform can help overcome challenges like toxicity associated with viral vectors and promote increased stability, release, and residence of the therapeutic payload.

17.5.3 Nanotechnology-Based Imaging and Image-Guided Therapy

Along with the therapeutic approach of nanotechnology in the management of arterial diseases, nanomedicine-based targeted imaging of arterial damage and lesions has become a rapidly emerging area of cardiovascular research [99]. The potential benefit of targeted imaging involves early diagnosis and better understanding of the disease progression [99]. Thus, the current research is focused on combining both the therapeutic and the diagnostic therapy into a single platform using nanomedicine to design a "theranostic" approach [1]. An interesting approach involving image-guided therapy is the conjugation of gadolinium (Gd)-loaded PEG-lipid micelles with specific antibodies that recognize epitopes specific to oxidized lipoproteins in the areas of vascular injury [100]. As mentioned earlier, oxidized lipoproteins are one of the triggers for the onset and progression of atherosclerosis. The study demonstrates increased residence, enhanced MRI signal, and longer circulation of these targeted micelles in vivo in atherosclerotic mice as compared to the nontargeted micelles [100]. The use of ultrasound-based imaging modality has always been the primary choice for vascular imaging. It has been reported that combining ultrasound with gas-filled nanobubbles and microbubbles leads to a significant contrast enhancement and improved ultrasound-based imaging of vascular regions [1]. Definity®—the lipid microsphere containing perfluropropane/air—is one such technology that has been clinically approved for cardiac imaging [101]. Another study has reported the use of polyvinyl-based nanobubbles that can be loaded with nitric oxide and used for imaging and the delivery of nitric oxide to vascular cells and tissues [102]. Thus, combining different imaging and therapeutic modalities using nanotechnology can help in the development of several theranostic agents that are highly efficient in focal arterial delivery.

17.6 Conclusions

This chapter explains the molecular and cellular pathways that govern the development of arterial disorders and the current therapeutic and diagnostic therapies used for local arterial delivery. It also provides a brief description of the different nanotechnology-based approaches that are used in arterial drug delivery. As discussed in the chapter, the development of arterial diseases is multifactorial and complex which involves several pathological conditions including inflammation, oxidative stress, thrombosis, endothelial activation, hyperlipidemia, and so on. Even the use of drug-eluting stents is associated with the risk of in-stent restenosis, which limits the success of several stent-based therapies. Thus, novel drug delivery techniques that address all the key factors affecting arterial disorders need to be designed and tested for better management of the arterial diseases. Further development in the applications of nanomedicine can aid in better diagnosis and treatment through improved site-specific delivery and can thus address the current challenges in the field of arterial drug delivery.

References

1. Gupta AS (2011) Nanomedicine approaches in vascular disease: a review. Nanomed Nanotechnol Biol Med 7:763–779
2. Godin B, Sakamoto JH, Serda RE, Grattoni A, Bouamrani A, Ferrari M (2010) Emerging applications of nanomedicine for the diagnosis and treatment of cardiovascular diseases. Trends Pharmacol Sci 31:199–205
3. Sharma S, Christopoulos C, Kukreja N, Gorog DA (2011) Local drug delivery for percutaneous coronary intervention. Pharmacol Ther 129:260–266
4. Wagner V, Dullaart A, Bock AK, Zweck A (2006) The emerging nanomedicine landscape. Nat Biotechnol 24:1211–1217
5. Lobatto ME, Fuster V, Fayad ZA, Mulder WJ (2011) Perspectives and opportunities for nanomedicine in the management of atherosclerosis. Nat Rev Drug Discov 10:835–852
6. Lawson ND, Weinstein BM (2002) Arteries and veins: making a difference with zebrafish. Nat Rev Genet 3:674–682
7. Lusis AJ (2000) Atherosclerosis. Nature 407:233–241
8. Rubanyi GM (1993) The role of endothelium in cardiovascular homeostasis and diseases. J Cardiovasc Pharmacol 22(Suppl 4):S1–S14

9. Meurice T, Vallet B, Bauters C, Dupuis B, Lablanche JM, Bertrand ME (1996) Role of endothelial cells in restenosis after coronary angioplasty. Fundam Clin Pharmacol 10:234–242
10. Libby P, Ridker PM, Maseri A (2002) Inflammation and atherosclerosis. Circulation 105:1135–1143
11. Schaper W, Scholz D (2003) Factors regulating arteriogenesis. Arterioscler Thromb Vasc Biol 23:1143–1151
12. Natarajan P, Ray KK, Cannon CP (2010) High-density lipoprotein and coronary heart disease: current and future therapies. J Am Coll Cardiol 55:1283–1299
13. Sprague AH, Khalil RA (2009) Inflammatory cytokines in vascular dysfunction and vascular disease. Biochem Pharmacol 78:539–552
14. Dangas G, Kuepper F (2002) Cardiology patient page. Restenosis: repeat narrowing of a coronary artery: prevention and treatment. Circulation 105:2586–2587
15. Deshpande D, Devalapally H, Amiji M (2008) Enhancement in anti-proliferative effects of paclitaxel in aortic smooth muscle cells upon co-administration with ceramide using biodegradable polymeric nanoparticles. Pharm Res 25:1936–1947
16. Karthikeyan G, Bhargava B (2004) Prevention of restenosis after coronary angioplasty. Curr Opin Cardiol 19:500–509
17. Goldberg SL, Loussararian A, De Gregorio J, Di Mario C, Albiero R, Colombo A (2001) Predictors of diffuse and aggressive intra-stent restenosis. J Am Coll Cardiol 37:1019–1025
18. Costa MA, Simon DI (2005) Molecular basis of restenosis and drug-eluting stents. Circulation 111:2257–2273
19. van Royen N, Piek JJ, Buschmann I, Hoefer I, Voskuil M, Schaper W (2001) Stimulation of arteriogenesis; a new concept for the treatment of arterial occlusive disease. Cardiovasc Res 49:543–553
20. van Royen N, Piek JJ, Schaper W, Fulton WF (2009) A critical review of clinical arteriogenesis research. J Am Coll Cardiol 55:17–25
21. Grundmann S, Piek JJ, Pasterkamp G, Hoefer IE (2007) Arteriogenesis: basic mechanisms and therapeutic stimulation. Eur J Clin Invest 37:755–766
22. Fuster V, Badimon L, Badimon JJ, Chesebro JH (1992) The pathogenesis of coronary artery disease and the acute coronary syndromes. New Engl J Med 326:242–250
23. Sobieszczyk P, Beckman J (2006) Carotid artery disease. Circulation 114:e244–e247
24. Gornik HL, Beckman JA (2005) Cardiology patient page. Peripheral arterial disease. Circulation 111:e169–e172
25. Ouriel K (2001) Peripheral arterial disease. Lancet 358:1257–1264
26. Libby P, Theroux P (2005) Pathophysiology of coronary artery disease. Circulation 111: 3481–3488
27. Deshpande DD, Janero DR, Amiji MM (2011) Therapeutic strategies for endothelial dysfunction. Expert Opin Biol Ther 11:1637–1654
28. Dobarro D, Gomez-Rubin MC, Sanchez-Recalde A, Moreno R, Galeote G, Jimenez-Valero S, Calvo L, López de Sa E, Lopez-Sendon JL (2009) Current pharmacological approach to restore endothelial dysfunction. Cardiovasc Hematol Agents Med Chem 7:212–222
29. Faxon DP (2002) Systemic drug therapy for restenosis. Circulation 106:2296–2298
30. Law M, Rudnicka AR (2006) Statin safety: a systematic review. Am J Cardiol 97:52C–60C
31. Fabian E, Varga A, Picano E, Vajo Z, Ronaszeki A, Csanady M (2004) Effect of simvastatin on endothelial function in cardiac syndrome X patients. Am J Cardiol 94:652–655
32. Kwak BR, Mulhaupt F, Mach F (2003) Atherosclerosis: anti-inflammatory and immunomodulatory activities of statins. Autoimmun Rev 2:332–338
33. Sathasivam S, Lecky B (2008) Statin induced myopathy. BMJ 337:a2286
34. Tomasoni L, Sitia S, Borghi C, Cicero AF, Ceconi C, Cecaro F, Morganti A, De Gennaro Colonna V, Guazzi M, Morricone L, Malavazos AE, Marino P, Cavallino C, Shoenfeld Y, Turiel M (2010) Effects of treatment strategy on endothelial function. Autoimmun Rev 9:840–844
35. Stewart DJ, Holtz J, Bassenge E (1987) Long-term nitroglycerin treatment: effect on direct and endothelium-mediated large coronary artery dilation in conscious dogs. Circulation 75:847–856

36. Lei L, Guo S-R, Chen W-L, Rong H-J, Lu F (2011) Stents as a platform for drug delivery. Expert Opin Drug Deliv 8:813–831
37. Herdeg C, Oberhoff M, Siegel-Axel DI, Baumbach A, Blattner A, Kuttner A, Schroder S, Karsch KR (2000) Paclitaxel: a chemotherapeutic agent for prevention of restenosis? Experimental studies in vitro and in vivo. Z Kardiol 89:390–397
38. Herdeg C, Oberhoff M, Baumbach A, Blattner A, Axel DI, Schroder S, Heinle H, Karsch KR (2000) Local paclitaxel delivery for the prevention of restenosis: biological effects and efficacy in vivo. J Am Coll Cardiol 35:1969–1976
39. Jabr-Milane L, van Vlerken L, Devalapally H, Shenoy D, Komareddy S, Bhavsar M, Amiji M (2008) Multi-functional nanocarriers for targeted delivery of drugs and genes. J Control Release 130:121–128
40. Waugh J, Wagstaff AJ (2004) The paclitaxel (TAXUS)-eluting stent: a review of its use in the management of de novo coronary artery lesions. Am J Cardiovasc Drugs 4:257–268
41. Ruef J, Storger H, Schwarz F, Haase J (2008) Comparison of a polymer-free rapamycin-eluting stent (YUKON) with a polymer-based paclitaxel-eluting stent (TAXUS) in real-world coronary artery lesions. Catheter Cardiovasc Interv 71:333–339
42. Grube E, Schofer J, Hauptmann KE, Nickenig G, Curzen N, Allocco DJ, Dawkins KD (2010) A novel paclitaxel-eluting stent with an ultrathin abluminal biodegradable polymer: 9-month outcomes with the JACTAX HD stent. JACC Cardiovasc Interv 3:431–438
43. Garg S, Serruys PW (2010) Coronary stents: looking forward. J Am Coll Cardiol 56:S43–S78
44. Virmani R, Farb A (1999) Pathology of in-stent restenosis. Curr Opin Lipidol 10:499–506
45. Serruys PW, de Jaegere P, Kiemeneij F, Macaya C, Rutsch W, Heyndrickx G, Emanuelsson H, Marco J, Legrand V, Materne P, Belardi J, Sigwart U, Colombo A, Goy JJ, van den Heuvel P, Delcan J, Morel M-a (1994) A comparison of balloon-expandable-stent implantation with balloon angioplasty in patients with coronary artery disease. New Engl J Med 331:489–495
46. Gertz ZM, Wilensky RL (2011) Local drug delivery for treatment of coronary and peripheral artery disease. Cardiovasc Ther 29:e54–e66
47. Miyauchi K, Kasai T, Yokayama T, Aihara K, Kurata T, Kajimoto K, Okazaki S, Ishiyama H, Daida H (2008) Effectiveness of statin-eluting stent on early inflammatory response and neointimal thickness in a porcine coronary model. Circ J 72:832–838
48. Waksman R (2009) Current state of the absorbable metallic (magnesium) stent. EuroIntervention 5(Suppl F):F94–F97
49. Acharya G, Park K (2006) Mechanisms of controlled drug release from drug-eluting stents. Adv Drug Deliv Rev 58:387–401
50. Kamath KR, Barry JJ, Miller KM (2006) The Taxus drug-eluting stent: a new paradigm in controlled drug delivery. Adv Drug Deliv Rev 58:412–436
51. Lee Y-K, Hyung Park J, Tae Moon H, Yun Lee D, Han Yun J, Byun Y (2007) The short-term effects on restenosis and thrombosis of echinomycin-eluting stents topcoated with a hydrophobic heparin-containing polymer. Biomaterials 28:1523–1530
52. Burke SE, Kuntz RE, Schwartz LB (2006) Zotarolimus (ABT-578) eluting stents. Adv Drug Deliv Rev 58:437–446
53. Ding NI, Pacetti SD, Tang F-W, Gada M, Roorda W (2009) XIENCE V™ stent design and rationale. J Interv Cardiol 22:S18–S27
54. de Ribamar Costa J, Jr AA (2010) Novolimus™-eluting coronary stent system. Interv Cardiol 2:645–649
55. Udipi K, Melder RJ, Chen M, Cheng P, Hezi-Yamit A, Sullivan C, Wong J, Wilcox J (2007) The next generation endeavor resolute stent: role of the BioLinx polymer system. EuroIntervention 3:137–139
56. Grabow N, Martin DP, Schmitz K-P, Sternberg K (2010) Absorbable polymer stent technologies for vascular regeneration. J Chem Technol Biotechnol 85:744–751
57. Seth A (2010) Moving towards biomimicry—the development of the Novel BioMime™ sirolimus-eluting coronary stent system. Eur Cardiol 6:78–82

58. Dani S, Kukreja N, Parikh P, Joshi H, Prajapati J, Jain S, Thanvi S, Shah B, Dutta JP (2008) Biodegradable-polymer-based, sirolimus-eluting Supralimus stent: 6-month angiographic and 30-month clinical follow-up results from the series I prospective study. EuroIntervention 4:59–63
59. Vranckx P, Serruys PW, Gambhir S, Sousa E, Abizaid A, Lemos P, Ribeiro E, Dani SI, Dalal JJ, Mehan V, Dhar A, Dutta AL, Reddy KN, Chand R, Ray A, Symons J (2006) Biodegradable-polymer-based, paclitaxel-eluting Infinnium stent: 9-month clinical and angiographic follow-up results from the SIMPLE II prospective multi-centre registry study. EuroIntervention 2:310–317
60. Costa JR Jr, Abizaid A, Costa R, Feres F, Tanajura LF, Abizaid A, Mattos LA, Staico R, Siqueira D, Sousa AG, Bonan R, Sousa JE (2008) Preliminary results of the hydroxyapatite nonpolymer-based sirolimus-eluting stent for the treatment of single de novo coronary lesions a first-in-human analysis of a third-generation drug-eluting stent system. JACC Cardiovasc Interv 1:545–551
61. Wieneke H, Dirsch O, Sawitowski T, Gu YL, Brauer H, Dahmen U, Fischer A, Wnendt S, Erbel R (2003) Synergistic effects of a novel nanoporous stent coating and tacrolimus on intima proliferation in rabbits. Catheter Cardiovasc Interv 60:399–407
62. Parker T, Dave V, Falotico R (2010) Polymers for drug eluting stents. Curr Pharm Des 16:3978–3988
63. Zhao J, Alquier L (2012) Drug-eluting stents. In: Wright JC, Burgess DJ (eds) Long acting injections and implants. Springer, New York, NY, pp 383–407
64. Kataoka T, Grube E, Honda Y, Morino Y, Hur SH, Bonneau HN, Colombo A, Di Mario C, Guagliumi G, Hauptmann KE, Pitney MR, Lansky AJ, Stertzer SH, Yock PG, Fitzgerald PJ (2002) 7-hexanoyltaxol-eluting stent for prevention of neointimal growth: an intravascular ultrasound analysis from the study to COmpare REstenosis rate between QueST and QuaDS-QP2 (SCORE). Circulation 106:1788–1793
65. Serruys PW, Ormiston JA, Sianos G, Sousa JE, Grube E, den Heijer P, de Feyter P, Buszman P, Schomig A, Marco J, Polonski L, Thuesen L, Zeiher AM, Bett JH, Suttorp MJ, Glogar HD, Pitney M, Wilkins GT, Whitbourn R, Veldhof S, Miquel K, Johnson R, Coleman L, Virmani R, ACTION investigators (2004) Actinomycin-eluting stent for coronary revascularization: a randomized feasibility and safety study: the ACTION trial. J Am Coll Cardiol 44: 1363–1367
66. Kwok OH, Chow WH, Law TC, Chiu A, Ng W, Lam WF, Hong MK, Popma JJ (2005) First human experience with angiopeptin-eluting stent: a quantitative coronary angiography and three-dimensional intravascular ultrasound study. Catheter Cardiovasc Interv 66:541–546
67. Huang Y, Salu K, Liu X, Li S, Wang L, Verbeken E, Bosmans J, De Scheerder I (2004) Methotrexate loaded SAE coated coronary stents reduce neointimal hyperplasia in a porcine coronary model. Heart 90:195–199
68. Shand JA, Menown IBA (2010) Drug-eluting stents: the next generation. Interv Cardiol 2:341–350
69. Grube E, Buellesfeld L (2006) BioMatrix Biolimus A9-eluting coronary stent: a next-generation drug-eluting stent for coronary artery disease. Expert Rev Med Devices 3:731–741
70. Tada N, Virmani R, Grant G, Bartlett L, Black A, Clavijo C, Christians U, Betts R, Savage D, Su SH, Shulze J, Kar S (2010) Polymer-free biolimus a9-coated stent demonstrates more sustained intimal inhibition, improved healing, and reduced inflammation compared with a polymer-coated sirolimus-eluting cypher stent in a porcine model. Circ Cardiovasc Interv 3:174–183
71. Nakazawa G, Finn AV, John MC, Kolodgie FD, Virmani R (2007) The significance of preclinical evaluation of sirolimus-, paclitaxel-, and zotarolimus-eluting stents. Am J Cardiol 100:36M–44M
72. Konig A, Leibig M, Rieber J, Schiele TM, Theisen K, Siebert U, Gothe RM, Klauss V (2007) Randomized comparison of dexamethasone-eluting stents with bare metal stent implantation

in patients with acute coronary syndrome: serial angiographic and sonographic analysis. Am Heart J 153:979.e1–979.e8
73. Sarisozen C, Arica B, Hincal AA, Calis S (2009) Development of biodegradable drug releasing polymeric cardiovascular stents and in vitro evaluation. J Microencapsul 26:501–512
74. Sternberg K, Kramer S, Nischan C, Grabow N, Langer T, Hennighausen G, Schmitz KP (2007) In vitro study of drug-eluting stent coatings based on poly(L-lactide) incorporating cyclosporine A—drug release, polymer degradation and mechanical integrity. J Mater Sci Mater Med 18:1423–1432
75. Deuse T, Erben RG, Ikeno F, Behnisch B, Boeger R, Connolly AJ, Reichenspurner H, Bergow C, Pelletier MP, Robbins RC, Schrepfer S (2008) Introducing the first polymer-free leflunomide eluting stent. Atherosclerosis 200:126–134
76. den Dekker WK, Houtgraaf JH, Onuma Y, Benit E, de Winter RJ, Wijns W, Grisold M, Verheye S, Silber S, Teiger E, Rowland SM, Ligtenberg E, Hill J, Wiemer M, den Heijer P, Rensing BJ, Channon KM, Serruys PWJC, Duckers HJ (2011) Final results of the HEALING IIB trial to evaluate a bio-engineered CD34 antibody coated stent (Genous™Stent) designed to promote vascular healing by capture of circulating endothelial progenitor cells in CAD patients. Atherosclerosis 219:245–252
77. Pendyala L, Yin X, Li J, Shinke T, Xu Y, Chen JP, King SB III, Colley K, Goodchild T, Chronos N, Hou D (2010) Polymer-free cerivastatin-eluting stent shows superior neointimal inhibition with preserved vasomotor function compared to polymer-based paclitaxel-eluting stent in rabbit iliac arteries. EuroIntervention 6:126–133
78. Scheller B, Schmitt A, Bohm M, Nickenig G (2003) Atorvastatin stent coating does not reduce neointimal proliferation after coronary stenting. Z Kardiol 92:1025–1028
79. Hong YJ, Jeong MH, Lee SR, Hong SN, Kim KH, Park HW, Kim JH, Kim W, Ahn Y, Cho JG, Park JC, Kang JC (2007) Anti-inflammatory effect of abciximab-coated stent in a porcine coronary restenosis model. J Korean Med Sci 22:802–809
80. New G, Moses JW, Roubin GS, Leon MB, Colombo A, Iyer SS, Tio FO, Mehran R, Kipshidze N (2002) Estrogen-eluting, phosphorylcholine-coated stent implantation is associated with reduced neointimal formation but no delay in vascular repair in a porcine coronary model. Catheter Cardiovasc Interv 57:266–271
81. Ranade SV, Miller KM, Richard RE, Chan AK, Allen MJ, Helmus MN (2004) Physical characterization of controlled release of paclitaxel from the TAXUS Express2 drug-eluting stent. J Biomed Mater Res A 71:625–634
82. Regar E, Sianos G, Serruys PW (2001) Stent development and local drug delivery. Br Med Bull 59:227–248
83. Kandzari DE, Leon MB, Popma JJ, Fitzgerald PJ, O'Shaughnessy C, Ball MW, Turco M, Applegate RJ, Gurbel PA, Midei MG, Badre SS, Mauri L, Thompson KP, LeNarz LA, Kuntz RE, Investigators EI (2006) Comparison of zotarolimus-eluting and sirolimus-eluting stents in patients with native coronary artery disease: a randomized controlled trial. J Am Coll Cardiol 48:2440–2447
84. Brieger D, Topol E (1997) Local drug delivery systems and prevention of restenosis. Cardiovasc Res 35:405–413
85. Zilberman M, Eberhart RC (2006) Drug-eluting bioresorbable stents for various applications. Ann Rev Biomed Eng 8:153–180
86. Ormiston JA, Webster MW, Armstrong G (2007) First-in-human implantation of a fully bioabsorbable drug-eluting stent: the BVS poly-L-lactic acid everolimus-eluting coronary stent. Catheter Cardiovasc Interv 69:128–131
87. Ormiston JA, Serruys PWS (2009) Bioabsorbable coronary stents. Circ Cardiovasc Interv 2:255–260
88. Vogt F, Stein A, Rettemeier G, Krott N, Hoffmann R, Vom Dahl J, Bosserhoff AK, Michaeli W, Hanrath P, Weber C, Blindt R (2004) Long-term assessment of a novel biodegradable paclitaxel-eluting coronary polylactide stent. Eur Heart J 25:1330–1340
89. Wang X, Venkatraman SS, Boey FY, Loo JS, Tan LP (2006) Controlled release of sirolimus from a multilayered PLGA stent matrix. Biomaterials 27:5588–5595

90. Walter DH, Cejna M, Diaz-Sandoval L, Willis S, Kirkwood L, Stratford PW, Tietz AB, Kirchmair R, Silver M, Curry C, Wecker A, Yoon YS, Heidenreich R, Hanley A, Kearney M, Tio FO, Kuenzler P, Isner JM, Losordo DW (2004) Local gene transfer of phVEGF-2 plasmid by gene-eluting stents: an alternative strategy for inhibition of restenosis. Circulation 110:36–45
91. Wang F, Wang J, Zhai Y, Li G, Li D, Dong S (2008) Layer-by-layer assembly of biologically inert inorganic ions/DNA multilayer films for tunable DNA release by chelation. J Control Release 132:65–73
92. Sun X, Zhang N (2010) Cationic polymer optimization for efficient gene delivery. Mini Rev Med Chem 10:108–125
93. Srinivasan R, Marchant RE, Gupta AS (2010) In vitro and in vivo platelet targeting by cyclic RGD-modified liposomes. J Biomed Mater Res A 93:1004–1015
94. Peters D, Kastantin M, Kotamraju VR, Karmali PP, Gujraty K, Tirrell M, Ruoslahti E (2009) Targeting atherosclerosis by using modular, multifunctional micelles. Proc Natl Acad Sci U S A 106:9815–9819
95. Li JM, Newburger PE, Gounis MJ, Dargon P, Zhang X, Messina LM (2010) Local arterial nanoparticle delivery of siRNA for NOX2 knockdown to prevent restenosis in an atherosclerotic rat model. Gene Ther 17:1279–1287
96. Li JM, Zhang X, Nelson PR, Odgren PR, Nelson JD, Vasiliu C, Park J, Morris M, Lian J, Cutler BS, Newburger PE (2007) Temporal evolution of gene expression in rat carotid artery following balloon angioplasty. J Cell Biochem 101:399–410
97. Vendrov AE, Hakim ZS, Madamanchi NR, Rojas M, Madamanchi C, Runge MS (2007) Atherosclerosis is attenuated by limiting superoxide generation in both macrophages and vessel wall cells. Arterioscler Thromb Vasc Biol 27:2714–2721
98. Brito LA, Chandrasekhar S, Little SR, Amiji MM (2010) Non-viral eNOS gene delivery and transfection with stents for the treatment of restenosis. Biomed Eng Online 9:56
99. Mulder WJ, Cormode DP, Hak S, Lobatto ME, Silvera S, Fayad ZA (2008) Multimodality nanotracers for cardiovascular applications. Nat Clin Pract Cardiovasc Med 5(Suppl 2): S103–S111
100. Briley-Saebo KC, Shaw PX, Mulder WJ, Choi SH, Vucic E, Aguinaldo JG, Witztum JL, Fuster V, Tsimikas S, Fayad ZA (2008) Targeted molecular probes for imaging atherosclerotic lesions with magnetic resonance using antibodies that recognize oxidation-specific epitopes. Circulation 117:3206–3215
101. Unger EM, Matsunaga TO, Schumann PA, Zutshi R (2003) Microbubbles in molecular imaging and therapy. Medicamundi 47:8
102. Cavalieri F, Finelli I, Tortora M, Mozetic P, Chiessi E, Polizio F, Brismar T, Paradossi T (2008) Polymer microbubbles as diagnostic and therapeutic gas delivery device. Chem Matter 20:8

Chapter 18
Drug-Eluting Stents

Wahid Khan, Rajesh Thipparaboina, Shady Farah, Judah Z. Weinberger, and Abraham J. Domb

18.1 Introduction

Cardiovascular diseases are the major cause of death in middle- and high-income countries. They account for 20% of deaths occurring globally [1]. Coronary arteries supply a constant flow of oxygen-rich blood to the heart. If plaque builds up in coronary arteries, blood flow may be reduced, leading to symptomatic coronary artery disease (CAD). Currently, over 16 million Americans have CAD. Consequently, almost eight million Americans have suffered a myocardial infarction (MI). One method of treating CAD and MI is the implantation of an expandable stent in a stenosed coronary artery (e.g. one partially blocked by atherosclerotic plaque). In 2007 alone approximately 560,000 Americans underwent a coronary stent implantation [2]. Cardiovascular procedures performed in the United States have more than tripled in the last decade. This increased trend is expected to continue with the ageing of the population, coupled with epidemics of obesity and diabetes mellitus [3].

W. Khan • R. Thipparaboina
Department of Pharmaceutics, National Institute of Pharmaceutical Education and Research (NIPER), Hyderabad 500037, India

S. Farah • A.J. Domb (✉)
Faculty of Medicine, Institute of Drug Research, School of Pharmacy, The Hebrew University of Jerusalem, Jerusalem 91120, Israel
e-mail: avid@ekmd.huji.ac.il

J.Z. Weinberger
Cardiovascular Division, Department of Medicine, Columbia University, New York, NY, USA

A.J. Domb and W. Khan (eds.), *Focal Controlled Drug Delivery*, Advances in Delivery Science and Technology, DOI 10.1007/978-1-4614-9434-8_18,

Bare-metal stent (BMS) is a small, tubular, metallic device which is preloaded in a collapsed form onto a balloon catheter, advanced to the narrowed section of the artery, and expanded within the vessel. Once expanded, the BMS acts as a mechanical scaffold, reducing elastic recoil and maintaining posttreatment vessel patency. BMS generally results in extremely favourable initial clinical results; however re-narrowing of the treated artery is commonly observed in 20–30 % of patients. This re-narrowing of the treated artery is due to in-stent restenosis (ISR). It is mainly due to excessive neointimal proliferation within the stented segment [4–6]. ISR was commonly treated using a repeat procedure angioplasty, until the advent of DES; no therapy consistently prevented this difficult problem. Direct local administration of pharmacological agents to the site of injured arterial tissue was prescribed to counter the failure of systemic drug delivery in preventing restenosis. Extensive research led to the discovery that the immobilisation of pharmacological agents on the stent surface and their sustained release from the surfaces could successfully reduce the frequency of ISR [7].

Clinical evaluation has proven the efficacy of drug-eluting stents (DES) in reduction of ISR rates compared to BMS, even in complicated situations [4]. In patients with acute myocardial infarction, treatment with DES is associated with decreased 2-year mortality rates and a reduction in the need for repeat revascularisation procedures, as compared to treatment with BMS [8]. Research in this area is currently centred on the development and evaluation of new improved DES which maintain the impressive clinical benefits while reducing long-term safety concerns [9].

18.2 Development of Drug-Eluting Stents

In an effort to deal with the ISR problem associated with BMS (introduced in 1994), DES was introduced in 2002. Since then DES has transformed the practice of interventional cardiology by drastically reducing rates of ISR and the need for repeat revascularisation.

18.2.1 First-Generation DES

First-generation DES was mainly composed of a stainless steel platform with a slotted-tube design. Drugs such as sirolimus (CYPHER) and paclitaxel (TAXUS) were used along with a durable polymer coating (Table 18.1). Compared to BMS, first-generation DES was superior in reducing neointimal proliferation and restenosis. Safety of first-generation DES regarding late stent thrombosis, especially after discontinuation of dual antiplatelet therapy, was a major concern.

Table 18.1 FDA-approved DES for coronary artery diseases[a]

Trade name	Manufacturer	Drug	FDA approval year
CYPHER™	Cordis Corporation	Sirolimus	2003
TAXUS™ Express2™	Boston Scientific Corporation	Paclitaxel	2004
TAXUS® Liberte™ Long	Boston Scientific Corporation	Paclitaxel	2009
TAXUS® Liberté® Atom™	Boston Scientific Corporation	Paclitaxel	2009
Medtronic and Endeavor	Abbott Vascular	Zotarolimus	2011
XIENCE nano™	Abbott Vascular	Everolimus	2011
PROMUS® Element™ Plus	Boston Scientific Corporation	Everolimus	2012
TAXUS® Liberte®	Boston Scientific Corporation	Paclitaxel	2012
TAXUS® Express2®	Boston Scientific Corporation	Paclitaxel	2012
ION™	Boston Scientific Corporation	Paclitaxel	2012
Resolute MicroTrac	Medtronic Vascular, Inc.	Zotarolimus	2012

[a]http://www.fda.gov/MedicalDevices/ProductsandMedicalProcedures/DeviceApprovalsandClearances/Recently-ApprovedDevices/ucm286493.htm

18.2.2 Second-Generation DES

Second-generation DES, such as zotarolimus-eluting (Endeavor) and everolimus-eluting (XIENCE V) stents, have been introduced with promising anti-restenotic efficacy as well as improved long-term safety. Thrombosis rates with these stents were lower when compared to first-generation DES [10]. They are composed of a cobalt–chromium platform, which exhibits superior radial strength and improved radio-opacity, allowing for thinner stent struts that act themselves in favour of a lower restenosis rate [10]. They differ from the first-generation stents with respect to the antiproliferative agent, the polymer layer, and the stent frame [11–15].

18.2.3 Third-Generation DES

As an extension of second-generation DES, a third-generation DES was introduced using biodegradable polymers. The ION stent (Boston Scientific) is a platinum–chromium platform, poly(styrene-*b*-isobutylene-*b*-styrene)-based paclitaxel-eluting stent. The drug/polymer coating adheres to the entire surface of the stent to allow consistent and controlled release of the drug [16]. Various late thrombosis rates with biodegradable polymeric DES were observed in a recent meta-analysis including eight comparative studies of DES with durable polymers [17].

18.2.4 Fourth-Generation DES

Fourth-generation DES is characterised by fully erodible stents, which are still under development. They are composed of bioabsorbable and polymer-free DES.

It is too early to know whether these stents will eradicate late thrombosis. The Bioabsorb pilot study of 30 patients with a 4-year follow-up showed no stent thrombosis. These studies using optical coherence tomography show promising findings [18]. BioFreedom (Biosensors), VESTAsync (MIV Therapeutics), and Yukon (Translumina) are few examples of polymer-free DES [19].

18.3 Design of Drug-Eluting Stents

Computational fluid dynamics (CFD) have been used as a valuable tool for analysing the haemodynamic effects of stent geometry, since associated indices (i.e. pressure, velocity, wall shear stress) are difficult to quantify. Computational studies of idealised stent geometries have shown that thinner struts and those more aligned with the primary flow direction decrease the amount of low wall shear stress at the arterial wall [20, 21]. Kolachalama et al. employed computational fluid dynamics modelling tools and explained that flow patterns within the milieu of the stent strut are significantly affected by net luminal flow and strut geometry [22]. Pant et al. described three variable geometry parameterisations of a CYPHER (Cordis Corporation, Johnson & Johnson Co.). The performance of each design is measured by six figures of merit (objectives/metrics) representing (1) acute recoil, (2) tissue stresses, (3) haemodynamic disturbance, (4) drug delivery, (5) uniformity of drug distribution, and (6) flexibility. They used a multi-objective surrogate modelling approach using a non-dominated sorting genetic algorithm (NSGA-II) for design improvement [23]. DES normally consists of three components: (a) stent platform, (b) stent coating, and (c) therapeutic agent.

18.3.1 The Stent Platform

Design of stent plays a critical role in the development of DES and affects both immediate and long-term clinical outcomes. An ideal stent should be radio-opaque, biocompatible, trackable, deliverable, generate significant radial force and corrosion resistant and possess good fatigue properties. The above requirements are achieved by proper selection of material, dimensions, geometry, and manufacturing process. Typically stents are manufactured using biologically inert materials such as 316L stainless steel [24]. In recent years, however, metallic alloys such as cobalt–chromium have proven superiority over steel as the material of choice for stent design [25]. These metallic alloys have been developed with increased levels of strength and X-ray attenuation, allowing stents to be more biologically inert and designed with significantly thinner struts. They are available in variety of geometries like coil, helical spiral, woven, individual rings, sequential rings, closed cell, and open cell [26]. Modular or slotted-tube configurations are most suitable and are employed for manufacturing stents [5, 27]. After implantation of the stent, it should experience minimum shortening during expansion. Upon deployment it should conform to the vessel geometry without any unnatural straightening of the vessel. Additionally, it should provide optimum vessel coverage and possess high radial strength to undergo minimal radial recoiling. The identification of long-term safety issues with the clinical experience of first-generation DES has increased clinical interest in the development of stents that are more biologically based, including fully biodegradable stents, stents using biomimetic and biodegradable polymers, and polymer-free stent systems [4, 27, 28].

18.3.2 The Stent Coating

As discussed above, to address the problem of restenosis, DES was introduced. These DES are made by coating a standard coronary stent with a thin polymer film containing a drug, which is released in controlled amounts locally, thereby preventing the formation of neointima at the site of coronary intervention. Adverse reactions may be caused by the polymers used in stent coating [29]. Thus, biocompatibility of the polymers used for coating stents is vital. For effective suppression of intimal growth, the ideal DES polymer should be non-thrombotic, noninflammatory, and nontoxic to cells and should encourage arterial healing by re-endothelialisation [30]. DES coating must also be capable of being stretched without flaking or delaminating. The inflation of the balloon during implantation may lead to cracking of the polymer coating partly with delamination [31]. Also, the polymer needs to be able to deliver the drug at a sustained, controlled, and predictable rate [32, 33]. Polymers used for coating stents can be broadly classified [29, 34] as:

- *Biostable/durable* (non-biodegradable) polymers, e.g. polyethylene-co-vinyl acetate, poly-*n*-butyl methacrylate, and poly(styrene-*b*-isobutylene-b-styrene)
- *Biodegradable polymers*, e.g. polyglycolic or polylactic acid or their copolymers
- *Biological polymers*, e.g. phosphorylcholine, hyaluronic acid, and fibrin

First-generation DES were coated with a biostable permanent polymer to provide controlled release of the anti-restenotic drug [35]. Subsequently, these non-biodegradable polymers were replaced by advanced biocompatible permanent polymers such as phosphorylcholine and copolymers [4]. Now, a new generation of DES is being developed using bioabsorbable polymers which degrade over time [19, 33].

18.3.3 Drugs

Delivering the drugs locally using DES is an established approach for preventing ISR. Drugs delivered using DES counter ISR. They fall under four classes (anti-inflammatory, antithrombogenic, antiproliferative, and immunosuppressive) as listed in Table 18.2. These drugs inhibit one or more biochemical pathways

Table 18.2 List of drugs used in DES

Drug	Comment	Mechanism	References
Paclitaxel	Common component of cancer chemotherapy; it reduces neointimal hyperplasia after balloon- and stent-mediated injury	Polymerisation of the α- and β-units of tubulin, thereby stabilising microtubules which are needed for G2 transition into M-phase	[36–38]
Sirolimus/ rapamycin	Macrocyclic antibiotic with potent immunosuppressive properties inhibits several phases of the restenosis cascade, such as inflammation, neointimal hyperplasia, and total protein synthesis	Prodrug that binds to specific cytosolic proteins (FK-506-binding protein-12), which blocks the cell proliferation	[28, 37–39]
Zotarolimus and everolimus	Analogues of rapamycin with higher n-octanol/water partition coefficients, which favour slow release from the stents, lipophilic character, favour crossing cell membranes to inhibit neointimal proliferation of target tissue	They bind to cytosolic FK-506-binding protein-12 and inhibit the proliferation of smooth muscle cells and T-cells	[28, 38, 40]
Tacrolimus	Tacrolimus is less potent than sirolimus for inhibiting vascular smooth muscle cell proliferation with less antiproliferative effects on endothelial cell. But, with its potent anti-inflammatory effects, tacrolimus may represent a promising compound for the use in DES	Immunosuppressive agent that also binds to cytosolic FK-506-binding protein-12. The resulting complex interacts and inhibits calcineurin and in this way inhibits the T-lymphocyte signal transduction and IL-2 transcription	[38]
Biolimus A9	Semi-synthetic sirolimus analogue. Biolimus possesses enhanced anti-inflammatory and antiproliferative activity with an improved pharmacokinetic profile	At a cellular level, biolimus forms a complex with intracellular FK-506-binding protein-12, with a similar potency to sirolimus	[41, 42]
Dexamethasone	Preclinical data and limited observations in humans using DES for local drug delivery have suggested beneficial effects of dexamethasone on neointimal proliferation	Dexamethasone and more general the corticosteroids are well-established anti-inflammatory drugs, used systemically for a broad range of inflammatory diseases and inhibiting proliferation of fibroblasts, smooth muscle cells, and macrophages	[43, 44]
Curcumin	Anticoagulation properties of curcumin-eluting stents were found to be superior to those of BMS and PLGA only coated stents	Curcumin can prevent platelet aggregation by platelet-activated factors and inhibits the formation of thromboxane A2 (TXA2) by platelets	[45]

leading to restenosis. Research is also being conducted using antibodies and genes as active compounds [46, 47]. For local drug delivery to be successful, challenges to be addressed include the following: (1) decision on the most appropriate agent to be used, (2) determination of the proportion of the systemic dose needed locally, and (3) identification of a biocompatible vehicle that can deliver drug for the required therapeutic window at a controlled rate [48, 49].

18.4 Safety and Efficacy of DES

There is interest in continuing to improve the safety and long-term efficacy of DES persists, and therefore, DES should be chosen selectively and correctly [50]. Stent thrombosis is a potentially fatal event that often leads to myocardial infarction and/or death. The exact cause of stent thrombosis is not yet fully understood; a number of patient, lesion, and procedural factors are associated with an increased risk of stent thrombosis [51]. The Academic Research Consortium recommended standardised definitions of stent thrombosis to provide uniform description of thrombotic events in various studies. These definitions were adopted in 2007 [4].

First-generation DES were widely adopted by interventional cardiologists with up to 90 % of stent procedures carried out in the USA using DES by late 2005 [52]. Durable polymers used in first-generation DES have been linked to local inflammatory reaction, positive vessel remodelling, late incomplete stent apposition, and, in some cases, stent thrombosis [53]. Therefore, use of polymers raises an important question about the safety of DES. It is not certain whether the polymers used are stable and inert over a longer period of time. Curcio et al. reported that methacrylate coating induces vascular smooth muscle cell apoptosis. This apoptotic property of methacrylate coating should be taken into account during the safety evaluation [37, 54]. Coronary artery aneurysm and stent infection are a few of the rarest problems encountered after coronary stent implantation. Schoenkerman and Lundstrom reported three cases of coronary stent infections; two were with mycotic aneurysms that ruptured into an adjacent cardiac chamber, and one had purulent pericarditis [55].

18.5 Guidelines for use of DES

Despite the success of DES over the years, optimal duration of dual antiplatelet therapy following DES implantation is still unclear. A number of national and international bodies have attempted to provide guidance to aid the interventional cardiologist with the decision whether to use DES or not, taking into account the risk–benefit ratio to the patient, and the cost-effectiveness of undergoing percutaneous coronary intervention therapy. International guidelines are available regarding the use of DES in those patients with CAD requiring revascularisation.

18.5.1 US Guidelines

- *Class I recommendations.* Drug-eluting stents should be used in patients requiring percutaneous coronary intervention, if clinical trial evidence shows effectiveness/safety. Prior to drug-eluting stents, the patient should be informed of the risks versus the benefits of dual antiplatelet therapy. If patients requiring percutaneous coronary intervention are awaiting surgical intervention, bare-metal stent or balloon angioplasty only should be considered.
- *Class IIa recommendations.* If there is any doubt regarding bleeding risk, aspirin can be used in a lower dose (75–162 mg).
- *Class IIb recommendations.* Drug-eluting stents may be used in situations where they are deemed to be effective and safe, even if clinical trial evidence has not yet fully proven this.

18.5.2 UK Guidelines

In the UK, the balance between clinical effectiveness and cost-effectiveness is determined by the National Institute for Health and Clinical Excellence (NICE). The most recent guidance regarding the use of DES was published in July 2008. Following thorough evaluation, therefore, DES were recommended for coronary artery lesions with a diameter less than 3 mm, a length greater than 15 mm, and cost-effectiveness dependent on the price differential between BMS and DES. It was additionally stated that patients who had suffered an acute myocardial infarction within 24 h and those with evidence of thrombosis angiographically are not eligible for reimbursement from the NHS, due to lack of evidence [56].

18.5.3 European Guidelines

In 2010 the Task Force of the European Society of Cardiology and the European Association for Cardio-Thoracic Surgery (ESC/EACTS) published their guidelines on myocardial revascularisation, including recommendations for the use of DES. They stated that "DES with proven efficacy should be considered by default in almost all clinical conditions and lesion subsets requiring revascularisation". In particular, DES is recommended for reduction of restenosis/reocclusion in the absence of contraindications to extended DAPT. The Task Force emphasises the only exception to this very general statement is when there are concerns or contraindications regarding the prolonged use of DAPT, including the careful consideration of the bleeding risk to the patient and the likely compliance of the individual. Contraindications stated by the committee on use of DES are listed in Table 18.3. It also suggested that, under some circumstances or with some DES, DAPT for 3 months could be sufficient, but the evidence is not robust. Diabetics may require a longer duration of DAPT [57].

Table 18.3 Relative clinical contraindications to the use of drug-eluting stents (ESC/EACTS)

S. No.	Clinical contraindications
1	Clinical history difficult to obtain, especially with acute severe clinical conditions (ST-segment elevation myocardial infarction or cardiogenic shock)
2	Expected poor compliance with DAPT, including patients with multiple comorbidities and polypharmacy
3	Nonelective surgery required in the short term that would require interruption of DAPT
4	Increased risk of bleeding
5	Known allergy to acetylsalicylic acid or clopidogrel/prasugrel/ticagrelor
6	Absolute indication for long-term anticoagulation

18.6 Future of Drug-Eluting Stents

The revolution brought about by DES in the treatment of symptomatic coronary artery disease led to widespread clinical use of DES, but safety issues have been identified in later follow-up. This prompted efforts to develop new DES, with the ultimate goal to provide safe and durable coronary patency [58]. The next-generation (newer) DES are being developed, which involves the optimisation of the three major components of DES: the stent platform, the polymer coating, and the drug. New technologies already being developed include the use of polymer-free DES and the use of biodegradable polymers and stents, use of nanoscale modifications on metal surfaces, dual delivery and gene delivery approaches, [50].

18.6.1 Polymer-Free DES

The continuous presence of a durable polymer in the coronary vasculature is believed to be associated with impaired vascular healing and a moderate increase of life-threatening late or very late stent thrombosis; thus, substantial efforts are currently under way to identify alternatives to biostable polymeric coating [59]. The use of polymer-free stents may have a potential long-term benefit over traditional polymeric coated DES. The polymer-free Biolimus A9-coated stent demonstrates equivalent early and superior late reduction of intimal proliferation compared with SES in a porcine model. After implantation of a BioFreedom stent, delayed arterial healing was minimal, and there was no increased inflammation at 180 days compared with SES implantation [60]. The safety and efficacy of the novel VESTAsync-eluting stent combining a stainless steel platform with a nanothin-microporous hydroxyapatite surface coating impregnated with a polymer-free low dose of sirolimus was assessed. The novel VESTAsync-eluting stent was effective in reducing late-lumen loss and neointimal hyperplasia at 4 and 9 months, with no evidence of late catch-up by quantitative coronary angiography or intravascular ultrasound [53]. Levi et al. report a new crystallisation methodology involving a temperature-induced crystallisation process for rapamycin to crystallise and adhere to metallic stents.

Table 18.4 Polymer-free drug-eluting stent

Stent (manufacturer)	Stent platform	Drug	Remark	References
Amazonia PAX (MINVASYS)	Cobalt–chromium	Paclitaxel	Abluminal coating was applied on crimped stent using the polymer-free PAX technology	[64]
BioFreedom (Biosensors Inc.)	Stainless steel	Biolimus	Composed of 316L stainless steel platform modified with a microstructured, abluminal surface. This modification allows Biolimus A9 adhesion to the abluminal surface of the stent without the use of a polymer or binder	[19, 60]
Optima (CID S.r.l.)	Stainless steel	Tacrolimus	Patented polymer-free drug reservoir and the proven antithrombotic and potentially prohealing integral Carbofilm coating	[64]
VESTAsync (MIV Therapeutics)	Stainless steel	Sirolimus	Microporous hydroxyapatite-based DES. Effective in reducing lumen loss and neointimal with no evidence of late catch-up	[19, 53, 65]
YUKON Choice (Translumina)	Stainless steel	Sirolimus and probucol	Shellac resin was used for loading of drugs. Inclusion of resin allows improved adherence of the drug mixture to the stent surface and enhances the structural integrity of the coating	[19, 66]

This carrier-free DES coating minimises the use of synthetic substances. These coatings displayed stability and biocompatibility. Coating obtained from this process allows drug release gradually over a period of a several weeks [61–63] (Table 18.4).

18.6.2 Bioabsorbable Drug-Eluting Stent

Despite all the benefits of using a metallic DES, limitations have generated interest towards biodegradable technology. The realisation about the probability of DES polymers causing the inflammatory reactions led to the development of a bioabsorbable polymer-based DES. Bioabsorbable DES is a device that can achieve excellent acute and long-term results, and it can disappear completely within months, thereby avoiding the need for prolonged dual antiplatelet therapy. These biodegradable stents, made of polymers or metal alloys with or without a drug coating, have the potential to scaffold the artery to allow natural healing and then biodegrade.

Table 18.5 Bioabsorbable DES

Stent (manufacturer)	Stent platform	Drug	Remark	References
Biotronik stent (Biotronik)	Absorbable metal stent 93 % magnesium and 7 % rare earth metals, zigzag helical coil design	Nil	Degrades into inorganic salts, not releasing an antiproliferative drug to counter the intimal hyperplastic response to stenting	[68, 69]
Igaki-Tamai stent	Poly-L-lactic acid	Nil	The deployment of the stent was rather complex, requiring thermal balloon expansion to actuate the device	[70–72]
BVS stent (Abbot Vascular)	Poly-L-lactic acid, cohort A, out-of-phase sinusoidal hoops with straight and direct links; cohort B, in-phase hoops with straight links	Everolimus	Coating of poly-D,L-lactic acid that contains and controls the release of the antiproliferative agent everolimus, 80 % release in 30 days	[72, 73]
The REVA (REVA Medical)	Polymer tyrosine-derived polycarbonate polymer, slide and lock design	Nil	The absorption time can be modified	[68, 72]
Bioabsorbable Therapeutics stent (Bioabsorbable Therapeutics)	Polymer salicylate + adipic acid linker molecules, Tube with laser cut voids	Sirolimus salicylate	Polymer backbone that gives the stent physical structure and a polymer coating that contains and controls the release of the antiproliferative agent	[72]

Such stents would obviate the need for long-term dual antiplatelet therapy. Since no foreign material would be left, future surgical options will not be limited, and follow-up with noninvasive imaging such as CT angiography would be possible [7, 67]. Table 18.5 shows several biodegradable stents that have entered into clinical trials. Many more are at the preclinical stage of development [74].

Two broad categories of materials are generally used: those made from organic biopolymers and those made from corrodible metals. To date, however, none of the materials/stents tested have been able to establish a perfect balance between biocompatibility, the kinetics of degradation needed to maintain mechanical strength to limit recoil, and inflammation [68]. Compared with biodegradable polymers, there are fewer metals used in the manufacture of biodegradable stents. The only metal biodegradable stent in trials is the Biotronik absorbable magnesium stent.

Unlike magnesium stents, there has been little progress with iron stents, which remain in the preclinical phase. This may be partly due to longer degradation times needed and potential issues related to iron clearance [68, 69].

18.6.3 Dual Drug-Eluting Stent

Dual DES was introduced to promote stent-based therapies that can collectively reduce neointimal hyperplasia and promote endothelial healing. Limited clinical data available for the delivery of multiple drugs from a DES, a key for successful treatment of restenosis and late thrombosis, restricts their use. The advent of new technologies and additional clinical trials are needed to ascertain that dual DES is clinically useful. The current dual DES delivers two drugs simultaneously, but the most efficient use of dual DES may come from controlling the drug release kinetics. Ideally, the first drug for preventing vascular smooth muscle cell proliferation is released for the first few weeks, and then the second drug for promoting re-endothelialisation is released after a month or so. Achieving sequential release from a thin layer in the range of 5–30 μm on a stent with predictable kinetics is very difficult. However, with the advent of new technologies in controlled drug delivery, researchers are cautiously optimistic to achieve such sequential release in the near future [75]. Song et al. grafted α-lipoic acid (ALA) with abciximab, an effective inhibitor of stent restenosis, on a bare-metal stent coated with a polymer layer by plasma polymerisation of 1,2-diaminocyclohexane (DACH) to prepare a dual-drug-eluting stent. They found that the dual DES has shown smooth and uniform morphology with an improved blood compatibility releasing ALA and abciximab simultaneously for 3 weeks [76]. Byrne et al. studied polymer-free dual DES of rapamycin and probucol; they reported an anti-restenotic efficacy comparable with that of the SES and superior to that of the zotarolimus stent [77]. Ma et al. studied the combination of sirolimus and paclitaxel and reported that the combination can be completely eluted within 21 days. The combination of two drugs in DES does not affect their individual release kinetics. They concluded that both drugs can be pharmacokinetically combined in DES for the treatment of CAD [78]. Kleinedler et al. studied the effect of nanocomposite stent releasing resveratrol and quercetin; they demonstrated that the synergistic combination can successfully inhibit neointimal hyperplasia and arterial inflammation while promoting vascular re-endothelialisation in a rat model [79].

18.6.4 Nanoscale Surface Modifications and Gene Delivery

Problems associated with current DES are potential cracking/detachment of the polymeric coating during balloon expansion, delayed healing, and hypersensitivity reaction to polymeric components [80]. Direct nanoscale modifications of surface without use of polymeric systems can be an effective strategy to overcome adverse events. Nanostructured surfaces possess the unique capacity of directly affecting the molecular and cellular events that ultimately determine the overall biological response to an implanted material. Variola et al. thoroughly reviewed mechanical, chemical, and physical methods that are being used to alter the physicochemical features of the surfaces of metallic biomaterials, thus modifying their impact on different cellular activities and functions. They conclude that most effective approaches to create a new generation of biometals are those that (1) confer enhanced biocompatibility directly onto material surfaces; (2) create synergistic effects; (3) selectively influence cells and guide stem cell differentiation; (4) result in surfaces that have more than one medical application, such as orthopaedic and cardiovascular; (5) simultaneously reach all surfaces in devices with complex geometries; and (6) can be manufactured by large-scale processes [81]. Youssefian et al. studied nanoscale adhesion in multilayered DES using atomistic simulation, and then they compared them with experiments to study the effects of aqueous environment with higher temperature inside the body on the adhesion between layers in the structure of DES. In this study they used poly(o-chloro-p-xylylene) (parylene C) for stainless steel 316L coating with γ-methacryloxypropyltrimethoxysilane as an adhesion promoter. They found that the effect of temperature on the adhesion is found to be regressive; as water molecules permeate, polymer adhesion decreases. They concluded that the effect of silane on the adhesion between parylene C and steel is modest [82]. Yang et al. attempted to combine gene therapy and DES drug delivery for effective inhibition of smooth muscle cell proliferation while simultaneously promoting re-endothelialisation and repair. They designed bilayered PLGA nanoparticles, which contain a VEGF165 DNA plasmid in the outer layer and paclitaxel in the inner core. They spray-coated these NPs onto bare-metal stents that had been laser treated to have a nonporous surface. In vitro experiments and in vivo animal studies both confirmed that the bilayer system can promote early endothelium healing and inhibit smooth muscle cell proliferation through sequential release of selected genes and drugs [47].

References

1. Hirsch AT, Gotto AM Jr (2002) Undertreatment of dyslipidemia in peripheral arterial disease and other high-risk populations: an opportunity for cardiovascular disease reduction. Vasc Med 7:323–331
2. Eggum JH, Howard SA, Goff RP, Iaizzo PA (2012) Imaging of a coronary artery stent implantation within an isolated human heart. J Cardiovasc Transl Res 5:73–74
3. Rao SV, Califf RM, Kramer JM, Peterson ED, Gross TP, Pepine CJ et al (2008) Postmarket evaluation of breakthrough technologies. Am Heart J 156:201–208
4. Martin DM, Boyle FJ (2011) Drug-eluting stents for coronary artery disease: a review. Med Eng Phys 33:148–163
5. Butt M, Connolly D, Lip GY (2009) Drug-eluting stents: a comprehensive appraisal. Future Cardiol 5:141–157
6. Alfonso F, Perez-Vizcayno MJ, Cruz A, Garcia J, Jimenez-Quevedo P, Escaned J et al (2009) Treatment of patients with in-stent restenosis. EuroIntervention 5(Suppl D):D70–D78

7. Wykrzykowska JJ, Onuma Y, Serruys PW (2009) Advances in stent drug delivery: the future is in bioabsorbable stents. Expert Opin Drug Deliv 6:113–126
8. Mauri L, Silbaugh TS, Garg P, Wolf RE, Zelevinsky K, Lovett A et al (2008) Drug-eluting or bare-metal stents for acute myocardial infarction. N Engl J Med 359:1330–1342
9. Stefanadis CI (2011) Are all drug-eluting stents the same? Hellenic J Cardiol 52:96
10. Akin I, Schneider H, Ince H, Kische S, Rehders TC, Chatterjee T et al (2011) Second- and third-generation drug-eluting coronary stents: progress and safety. Herz 36:190–196
11. Doostzadeh J, Clark LN, Bezenek S, Pierson W, Sood PR, Sudhir K (2010) Recent progress in percutaneous coronary intervention: evolution of the drug-eluting stents, focus on the XIENCE V drug-eluting stent. Coron Artery Dis 21:46–56
12. Sheiban I, Villata G, Bollati M, Sillano D, Lotrionte M, Biondi-Zoccai G (2008) Next-generation drug-eluting stents in coronary artery disease: focus on everolimus-eluting stent (Xience V). Vasc Health Risk Manag 4:31–38
13. Joner M, Nakazawa G, Finn AV, Quee SC, Coleman L, Acampado E et al (2008) Endothelial cell recovery between comparator polymer-based drug-eluting stents. J Am Coll Cardiol 52:333–342
14. Lange RA, Hillis LD (2010) Second-generation drug-eluting coronary stents. N Engl J Med 362:1728–1730
15. Lopez JJ, Keyes MJ, Nathan S, Piana R, Pencina M, Dhar G et al (2010) Rapid adoption of drug-eluting stents: clinical practices and outcomes from the early drug-eluting stent era. Am Heart J 160:767–774
16. Kereiakes DJ, Cannon LA, Ormiston JA, Turco MA, Mann T, Mishkel GJ et al (2011) Propensity-matched patient-level comparison of the TAXUS Liberté and TAXUS element (ION) paclitaxel-eluting stents. Am J Cardiol 108:828–837
17. Navarese EP, Kubica J, Castriota F, Gibson CM, De Luca G, Buffon A et al (2011) Safety and efficacy of biodegradable vs. durable polymer drug-eluting stents: evidence from a meta-analysis of randomised trials. EuroIntervention 7:985–994
18. de la Torre Hernandez JM, Windecker S (2012) Very late stent thrombosis with newer drug-eluting stents: no longer an issue? Rev Esp Cardiol (Engl Ed) 65:595–598
19. Garg S, Serruys PW (2010) Coronary stents: looking forward. J Am Coll Cardiol 56:S43–S78
20. LaDisa JF, Olson LE, Hettrick DA, Warltier DC, Kersten JR, Pagel PS (2005) Axial stent strut angle influences wall shear stress after stent implantation: analysis using 3D computational fluid dynamics models of stent foreshortening. Biomed Eng Online 4:59
21. LaDisa JF, Olson LE, Guler I, Hettrick DA, Audi SH, Kersten JR et al (2004) Stent design properties and deployment ratio influence indexes of wall shear stress: a three-dimensional computational fluid dynamics investigation within a normal artery. J Appl Physiol 97:424–430
22. Kolachalama VB, Tzafriri AR, Arifin DY, Edelman ER (2009) Luminal flow patterns dictate arterial drug deposition in stent-based delivery. J Control Release 133:24–30
23. Pant S, Limbert G, Curzen NP, Bressloff NW (2011) Multiobjective design optimisation of coronary stents. Biomaterials 32:7755–7773
24. Colombo A, Stankovic G, Moses JW (2002) Selection of coronary stents. J Am Coll Cardiol 40:1021–1033
25. Sketch MH, Ball M, Rutherford B, Popma JJ, Russell C, Kereiakes DJ (2005) Evaluation of the Medtronic (Driver) cobalt-chromium alloy coronary stent system. Am J Cardiol 95:8–12
26. Patel MJ, Patel SS, Patel NS, Patel NM (2012) Current status and future prospects of drug eluting stents for restenosis. Acta Pharm 62:473–496
27. Ako J, Bonneau HN, Honda Y, Fitzgerald PJ (2007) Design criteria for the ideal drug-eluting stent. Am J Cardiol 100:3M–9M
28. Pendyala LK, Yin X, Li J, Chen JP, Chronos N, Hou D (2009) The first-generation drug-eluting stents and coronary endothelial dysfunction. JACC Cardiovasc Interv 2:1169–1177
29. Puskas JE, Munoz-Robledo LG, Hoerr RA, Foley J, Schmidt SP, Evancho-Chapman M et al (2009) Drug-eluting stent coatings. Wiley Interdiscip Rev Nanomed Nanobiotechnol 1: 451–462

30. Mrowietz C, Franke RP, Seyfert UT, Park JW, Jung F (2005) Haemocompatibility of polymer-coated stainless steel stents as compared to uncoated stents. Clin Hemorheol Microcirc 32:89–103
31. Matschke K, Tugtekin SM, Kappert U, Jung F, Park JW, Knaut M (2004) Do drug-eluting stents influence the spectrum of coronary artery bypass surgery? Herz 29:201–207
32. Levy Y, Mandler D, Weinberger J, Domb AJ (2009) Evaluation of drug-eluting stents' coating durability–clinical and regulatory implications. J Biomed Mater Res B Appl Biomater 91:441–451
33. Parker T, Dave V, Falotico R (2011) Polymers for drug eluting stents. Curr Pharm Des 16:3978–3988
34. Mani G, Feldman MD, Patel D, Agrawal CM (2007) Coronary stents: a materials perspective. Biomaterials 28:1689–1710
35. Moore JE Jr (2009) Biomechanical issues in endovascular device design. J Endovasc Ther 16(Suppl 1):I1–I11
36. Stone GW, Ellis SG, Cox DA, Hermiller J, O'Shaughnessy C, Mann JT et al (2004) A polymer-based, paclitaxel-eluting stent in patients with coronary artery disease. N Engl J Med 350: 221–231
37. van der Hoeven BL, Pires NM, Warda HM, Oemrawsingh PV, van Vlijmen BJ, Quax PH et al (2005) Drug-eluting stents: results, promises and problems. Int J Cardiol 99:9–17
38. Deconinck E, Sohier J, De Scheerder I, Van den Mooter G (2008) Pharmaceutical aspects of drug eluting stents. J Pharm Sci 97:5047–5060
39. Tamburino C, Capodanno D (2009) Evolution of stents: past, present and future. Expert Rev Cardiovasc Ther 7:443–446
40. Burke SE, Kuntz RE, Schwartz LB (2006) Zotarolimus (ABT-578) eluting stents. Adv Drug Deliv Rev 58:437–446
41. Wykrzykowska JJ, Raber L, de Vries T, Bressers M, Buszman P, Linke A et al (2009) Biolimus-eluting biodegradable polymer versus sirolimus-eluting permanent polymer stent performance in long lesions: results from the LEADERS multicentre trial substudy. EuroIntervention 5: 310–317
42. Grube E, Buellesfeld L (2006) BioMatrix Biolimus A9-eluting coronary stent: a next-generation drug-eluting stent for coronary artery disease. Expert Rev Med Devices 3:731–741
43. Radke PW, Weber C, Kaiser A, Schober A, Hoffmann R (2004) Dexamethasone and restenosis after coronary stent implantation: new indication for an old drug? Curr Pharm Des 10:349–355
44. Pesarini G, Ferrero V, Tomai F, Paloscia L, De Cesare N, Tamburino C et al (2009) Steroid-eluting stents in patients with acute coronary syndromes. Angiographic results of DESIRE: Dexamethasone-Eluting Stent Italian REgistry. J Invasive Cardiol 21:86–91
45. Pan CJ, Tang JJ, Weng YJ, Wang J, Huang N (2006) Preparation, characterization and anticoagulation of curcumin-eluting controlled biodegradable coating stents. J Control Release 116:42–49
46. Grube E, Gerckens U, Muller R, Bullesfeld L (2002) Drug eluting stents: initial experiences. Z Kardiol 91(Suppl 3):44–48
47. Yang J, Zeng Y, Zhang C, Chen YX, Yang Z, Li Y et al (2013) The prevention of restenosis *in vivo* with a VEGF gene and paclitaxel co-eluting stent. Biomaterials 34:1635–1643
48. Teomim D, Fishbien I, Golomb G, Orloff L, Mayberg M, Domb AJ (1999) Perivascular delivery of heparin for the reduction of smooth muscle cell proliferation after endothelial injury. J Control Release 60:129–142
49. Kavanagh CA, Rochev YA, Gallagher WM, Dawson KA, Keenan AK (2004) Local drug delivery in restenosis injury: thermoresponsive co-polymers as potential drug delivery systems. Pharmacol Ther 102:1–15
50. Vorpahl M, Virmani R, Ladich E, Finn AV (2009) Vascular remodeling after coronary stent implantation. Minerva Cardioangiol 57:621–628
51. Park DW, Park SW, Park KH, Lee BK, Kim YH, Lee CW et al (2006) Frequency of and risk factors for stent thrombosis after drug-eluting stent implantation during long-term follow-up. Am J Cardiol 98:352–356

52. Jeremias A, Kirtane A (2008) Balancing efficacy and safety of drug-eluting stents in patients undergoing percutaneous coronary intervention. Ann Intern Med 148:234–238
53. Costa JR Jr, Abizaid A, Costa R, Feres F, Tanajura LF, Maldonado G et al (2009) 1-year results of the hydroxyapatite polymer-free sirolimus-eluting stent for the treatment of single de novo coronary lesions: the VESTASYNC I trial. JACC Cardiovasc Interv 2:422–427
54. Curcio A, Torella D, Cuda G, Coppola C, Faniello MC, Achille F et al (2004) Effect of stent coating alone on in vitro vascular smooth muscle cell proliferation and apoptosis. Am J Physiol Heart Circ Physiol 286:H902–H908
55. Schoenkerman AB, Lundstrom RJ (2009) Coronary stent infections: a case series. Catheter Cardiovasc Interv 73:74–76
56. de Belder MA (2008) NICE guidelines for the use of drug-eluting stents: how do we establish worth? Heart 94:1646–1648
57. The Task Force on Myocardial Revascularization of the European Society of Cardiology (ESC) and the European Association for Cardio-Thoracic Surgery (EACTS), Wijns W, Kolh P, Danchin N, Di Mario C, Falk V, Folliguet T et al (2010) Guidelines on myocardial revascularization. Eur Heart J 31:2501–2555
58. Kereiakes DJ (2010) Safety of drug-eluting stents. Rev Cardiovasc Med 11:187–199, quiz 200
59. Webster MW, Ormiston JA (2007) Drug-eluting stents and late stent thrombosis. Lancet 370:914–915
60. Tada N, Virmani R, Grant G, Bartlett L, Black A, Clavijo C et al (2010) Polymer-free biolimus a9-coated stent demonstrates more sustained intimal inhibition, improved healing, and reduced inflammation compared with a polymer-coated sirolimus-eluting cypher stent in a porcine model. Circ Cardiovasc Interv 3:174–183
61. Levi Y, Domb AJ, Amir N, Eliyahu N, Cohn U, Tal N. Crystalline drug-containing coatings. WO2010086863 A3, pp 1–55
62. Khan W, Farah S, Domb AJ (2012) Drug eluting stents: developments and current status. J Control Release 161:703–712
63. Levy Y, Khan W, Farah S, Domb AJ (2012) Surface crystallization of rapamycin on stents using a temperature induced process. Langmuir 28:6207–6210
64. Abizaid A, Costa JR Jr (2010) New drug-eluting stents: an overview on biodegradable and polymer-free next-generation stent systems. Circ Cardiovasc Interv 3:384–393
65. Costa JR Jr, Abizaid A, Costa R, Feres F, Tanajura LF, Mattos LA et al (2008) Preliminary results of the hydroxyapatite nonpolymer-based sirolimus-eluting stent for the treatment of single de novo coronary lesions a first-in-human analysis of a third-generation drug-eluting stent system. JACC Cardiovasc Interv 1:545–551
66. Mehilli J, Kastrati A, Wessely R, Dibra A, Hausleiter J, Jaschke B et al (2006) Randomized trial of a nonpolymer-based rapamycin-eluting stent versus a polymer-based paclitaxel-eluting stent for the reduction of late lumen loss. Circulation 113:273–279
67. Di Mario C, Ferrante G (2008) Biodegradable drug-eluting stents: promises and pitfalls. Lancet 371:873–874
68. Ramcharitar S, Serruys PW (2008) Fully biodegradable coronary stents: progress to date. Am J Cardiovasc Drugs 8:305–314
69. Erbel R, Di Mario C, Bartunek J, Bonnier J, de Bruyne B, Eberli FR et al (2007) Temporary scaffolding of coronary arteries with bioabsorbable magnesium stents: a prospective, non-randomised multicentre trial. Lancet 369:1869–1875
70. Erbel R, Bose D, Haude M, Kordish I, Churzidze S, Malyar N et al (2007) Absorbable coronary stents. New promising technology. Herz 32:308–319
71. Tamai H, Igaki K, Kyo E, Kosuga K, Kawashima A, Matsui S et al (2000) Initial and 6-month results of biodegradable poly-l-lactic acid coronary stents in humans. Circulation 102:399–404
72. Ormiston JA, Serruys PW (2009) Bioabsorbable coronary stents. Circ Cardiovasc Interv 2:255–260
73. Ormiston JA, Serruys PW, Regar E, Dudek D, Thuesen L, Webster MW et al (2008) A bioabsorbable everolimus-eluting coronary stent system for patients with single de-novo coronary artery lesions (ABSORB): a prospective open-label trial. Lancet 371:899–907

74. Garg S, Serruys P (2009) Biodegradable stents and non-biodegradable stents. Minerva Cardioangiol 57:537–565
75. Park K (2012) Dual drug-eluting stent. J Control Release 159:1
76. Song SJ, Kim KS, Park YJ, Jeong MH, Ko YM, Cho DL (2009) Preparation of a dual-drug-eluting stent by grafting of ALA with abciximab on a bare metal stent. J Mater Chem 19:8135–8141
77. Byrne RA, Mehilli J, Iijima R, Schulz S, Pache J, Seyfarth M et al (2009) A polymer-free dual drug-eluting stent in patients with coronary artery disease: a randomized trial vs. polymer-based drug-eluting stents. Eur Heart J 30:923–931
78. Ma X, Oyamada S, Gao F, Wu T, Robich MP, Wu H et al (2011) Paclitaxel/sirolimus combination coated drug-eluting stent: in vitro and in vivo drug release studies. J Pharm Biomed Anal 54:807–811
79. Kleinedler JJ, Foley JD, Orchard EA, Dugas TR (2012) Novel nanocomposite stent coating releasing resveratrol and quercetin reduces neointimal hyperplasia and promotes re-endothelialization. J Control Release 159:27–33
80. Kukreja N, Onuma Y, Daemen J, Serruys PW (2008) The future of drug-eluting stents. Pharmacol Res 57:171–180
81. Variola F, Vetrone F, Richert L, Jedrzejowski P, Yi JH, Zalzal S et al (2009) Improving biocompatibility of implantable metals by nanoscale modification of surfaces: an overview of strategies, fabrication methods, and challenges. Small 5:996–1006
82. Youssefian S, Rahbar N (2013) Nano-scale adhesion in multilayered drug eluting stents. J Mech Behav Biomed Mater 18:1–11

Chapter 19
Drug-Eluting Vascular Grafts

Jingjia Han and Peter I. Lelkes

19.1 Introduction

According to the American Heart Association, coronary vascular disease (CVD) accounts for one of every six deaths in the USA in 2009. On average, approximately every single minute, someone in the USA will die from a coronary-associated disease. For 2009, the total cost for treating CVD in the USA was estimated at \$312.6 billion, more than for any other disease, such as the cost for all cancer and benign neoplasms together (\$228 billion) [1]. Although the rate of death attributed to CVD has been declining in the past decade, the burden of disease in terms of mortality and healthcare cost remains unacceptably high [1].

A highly effective and well-established method of combating CVD, caused by blockage of small-diameter coronary arteries (<5 mm) on the surface of the heart muscle, is to surgically bypass the blocked arteries using a procedure called coronary artery bypass grafting (CABG) [2]. The traditional gold standard of CABG is to use autologous blood vessels, such as internal mammary artery (IMA) or saphenous vein. Of all autologous blood vessels, IMA has the longest patency. As a drawback, the use of IMA is often associated with donor-site morbidity, size mismatch, or high surgical risk. Therefore, saphenous vein grafts have become the most common choice [3]. However, due to its limited patency and relatively high rate of restenosis, the effectiveness of even the saphenous vein graft is restricted.

J. Han
Department of Pediatrics, School of Medicine, Emory University
and Parker H. Petit Institute of Biotechnology and Biosciences,
Georgia Institute of Technology, Atlanta, GA, USA
e-mail: jingjia.han@emory.edu

P.I. Lelkes (✉)
Department of Bioengineering, College of Engineering, Temple University,
Philadelphia, PA, USA
e-mail: pilelkes@temple.edu

A.J. Domb and W. Khan (eds.), *Focal Controlled Drug Delivery*, Advances in Delivery Science and Technology, DOI 10.1007/978-1-4614-9434-8_19,

When there is inadequate supply of suitable autologous blood vessels, engineered vascular grafts have to be considered as alternative approaches [4].

Progress in surgical approaches tailored to different subgroups of patients and in CABG techniques, such as unique techniques used for grafting saphenous veins or synthetic grafts, respectively, has reduced bypass-related immediate morbidity and late mortality. However, the incidence of late graft restenosis remains a pervasive problem both in the CABG procedures and in the use of percutaneous transluminal coronary angioplasty (PTCA), which significantly impedes the efficacy and safety of these treatment modalities. The cause of late restenosis is the inevitable damage of the vessel wall during the surgical procedure in CABG/PTCA, which induces endothelial denudation, platelet adherence/activation, and leukocyte infiltration [5, 6]. Subsequently, smooth muscle cells (SMCs) in the *tunica media* switch from a quiescent, contractile phenotype to a proliferative phenotype: the cells proliferate and migrate inwardly along with elaboration of abundant extracellular matrix (ECM). As a result of these events, also known as neointimal hyperplasia, the vessel wall thickens and the vascular lumen narrows, causing vascular restenosis, which severely limits the patency and durability of vascular grafts [7].

Despite intensive investigations and successful treatment of bypass graft stenosis in animal models, no optimal therapy for bypass graft diseases has been achieved as yet in the clinical setting [7]. To prevent bypass neointimal hyperplasia and restenosis, numerous antiproliferative and anti-migratory drugs were first administrated systemically, yet without effective clinical outcome; some even led to increased mortality. For example, heparin, an antiproliferative drug identified in the middle of the 1970s [8], was systemically administrated in animal models or in human trials. However, severe complications were observed in the heparin-treated rats, such as hemorrhage, while in the human trials, the systemic prolonged administration of heparin failed to prevent restenosis and led to severe vascular injury [9–11]. This is, at least partially, due to the difficulty of systemic treatments to achieve high local drug concentration while minimizing systemic side effects. Subsequently, local drug delivery systems were developed, including perfusion balloon catheters and stents coated with drug-releasing polymers. For example, Camenzind et al. [12] used a balloon catheter to infuse large volumes of heparin solution following coronary angioplasty. The unfavorable clinical long-term effect seen in this and similar studies has been blamed mainly on low tissue uptake and/or a rapid washout of the administered drugs [7].

A localized and sustained drug delivery system is thus required for effective and safe focal treatment of CVD. In the early 1990s, Edelman et al. [13] placed poly(ethylene-co-vinyl acetate) (EVAc) matrices next to denuded rat carotid arteries to deliver the matrix-contained heparin to the adjacent blood vessel wall. This local and sustained delivery system demonstrated effective inhibition of arterial occlusion by preventing SMC proliferation without systemic side effects. However, EVAc matrices are nonbiodegradable requiring postsurgical removal of the implant, thus preventing its usefulness in the clinical setting. More recently, local drug delivery systems made of biocompatible and biodegradable films or particles have been extensively studied and achieved great success [7, 14, 15]. One good example is a heterogeneous system designed also by the Edelman group, where heparin was encapsulated in hydrophobic Poly(D,L-lactic-co-glycolide (PLGA) microspheres that were sequestered in a hydrophilic calcium alginate matrix [14]. Controlled release of heparin from this system was obtained over a period of 25 days in vitro [14]. However, the long-term safety and efficacy of this and similar localized and sustained drug delivery system remains challenging and requires a careful choreography of the combination of drug, drug carrier/polymer, and the release kinetics [2, 15].

In this chapter, we summarize some of the drug delivery technologies applied in vascular grafts for treating CVDs. The technologies described in this chapter can be extended to a variety of interventions of PTCA, such as drug-eluting stents (DES), as well as interposition grafts, arteriovenous (AV) bridge grafts, and bypass grafting in renal implantation. Of note, degradable drug-eluting vascular grafts, in contrast to DES, can be engineered to match the rate of scaffold degradation to that of natural tissue regeneration, thereby avoiding a "late catch-up phenomenon," i.e., in-stent restenosis, as observed in DES [16, 17]. Following a very brief overview of the historical developments related to drug-eluting vascular grafts, the technological aspects of fabrication (choices of drug and drug carrier), system design, and pharmacokinetics of various drug-releasing graft systems are discussed. In focusing on electrospinning as a platform technology for generating drug-eluting fibers for vascular grafts, we opinion that electrospun fibrous drug delivery systems may be a promising approach for creating drug-eluting vascular grafts with improved clinical safety and patency.

19.2 Anatomy, Histology, and Pathology

Before we discuss some technological aspects of the various drug-eluting vascular grafts, we will provide a brief overview of the anatomy and histology of coronary arteries as well as the main etiologies associated with bypass grafting. The understanding of these aspects will greatly help in the design of the next generation of drug-eluting vascular grafts with improved efficacy and safety.

19.2.1 Anatomy of Coronary Arteries

Originating from the base of the aorta, coronary arteries consist of two main branches: the left and the right arteries. Dividing into the left anterior descending artery and the circumflex branch, the left coronary artery supplies blood to the left ventricle and left atrium, while the right coronary artery divides into the right posterior descending and acute marginal arteries and is responsible for blood supply to the right ventricle, right atrium, sinoatrial node, and atrioventricular node (Fig. 19.1a) [18, 19]. Because of their critical role in delivering oxygen and nutrients to the myocardium, occlusion of the coronary arteries can lead to severe clinical complications, such as heart attack and possibly death [19].

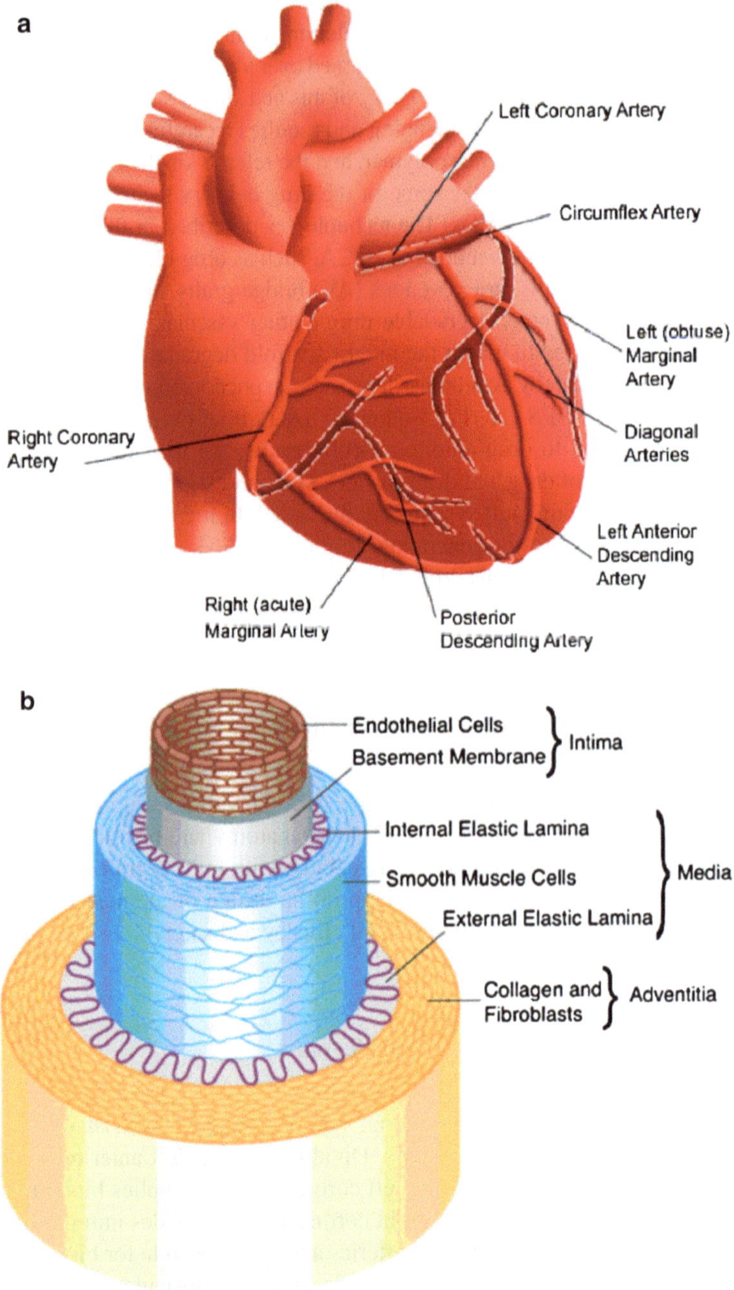

Fig. 19.1 Anatomy of coronary arteries of the heart (**a**) [19] and schematics rendering of tissue architecture in arterial wall (**b**) (reproduced with permission from Sarkar et al. [20]

19.2.2 Histology of Coronary Arteries

The wall of an artery is composed of three concentric layers, the intima, media, and adventitia (Fig. 19.1b) [20]. In close contact with the circulating blood, the innermost layer (*tunica intima*) consists a single-cell layer of confluent endothelial cells (ECs) resting on a basement membrane (BM), mainly comprised of type IV collagen and laminin [4, 21]. The quiescent endothelium is also critical for preventing infection and inflammation of the underlying tissue as well as for signaling to the middle layer. The middle layer, namely, *tunica media*, is composed of multiple layers of SMCs. Bidirectional signaling between ECs and SMCs is essential for maintaining the quiescent phenotype of both cell types. The SMCs in the *tunica media* are intercalated with elastic fibers along the vessel axis in either a circumferentially aligned fashion or a spiral pattern, thus conferring essential recoil elasticity to the vessel walls [22]. The outmost layer, also called *tunica adventitia*, is primarily composed of fibroblasts and collagen. The *tunica adventitia* primarily anchors the blood vessels to the neighboring tissue and harbors the microscopic capillary blood supply (*vasa vasorum*) as well as the innervations of the artery. Of note, between each concentric layer, there are two elastin layers, namely, the internal elastic lamina (between *tunica intima* and *tunica media*) and the external elastic lamina (between *tunica media* and *tunica adventitia*). These two elastin layers are deposited by ECs and SMCs with continual interaction during development [20].

19.2.3 Pathology of Vascular Grafts

Broadly speaking, there are three main clinical complications associated with bypass grafting: graft restenosis, acute thrombosis, and postoperative graft infection. We describe them here in a sequence depending on the pervasiveness and severity of the related diseases, not following the timeline of the associated etiology.

19.2.3.1 Restenosis

(Re)stenosis, caused by (neo)intimal hyperplasia and the thickening of the vessel wall and the narrowing of the vascular lumen, is a direct consequence of the hyperproliferation of SMCs in the *tunic media.* It is the main etiology in the subacute postoperative period (1–12 months) and a prevalent problem following graft/stent implantation for CAD: approximately 30 % of all arterial bypass grafts and 50 % of venous grafts fail within the first few years postoperatively, owing to neointimal hyperplasia [5, 23]. Restenosis originates from the denudation/loss of the endothelium and, specifically for vein grafts, from the graft's adaptation to arterial pressures. Upon vascular injury, the endothelial lining is damaged, which interrupts the EC–SMC signaling required for maintaining a quiescent SMC phenotype [24, 25]. Moreover, endothelial denudation exposes the underlying matrix that in turn stimulates the adhesion/activation of platelets and leads to the release of various mitogenic and chemotactic growth factors, such as platelet-derived growth factor (PDGF) and basic fibroblast growth factor (bFGF). These growth factors stimulate SMCs in the *tunica media* to proliferate and migrate into the *tunica intima*, causing intimal hyperplasia at the site of injury [5].

19.2.3.2 Thrombosis

Thrombosis, i.e., blood clot formation in the vascular lumen at the site of injury or distally, is the dominant etiology in the early (<1 month) postsurgical period [3]. This clinical complication is related to factors such as disruption/activation of the endothelial layer caused by mechanical trauma and manual distention, size/compliance mismatch between the graft and the target vessel creating turbulent flow, and graft ischemia. The loss of endothelial layer integrity can promote blood coagulation, platelet adhesion, thrombosis, as well as vasospasm resulting from lowered nitric oxide (NO) levels [3, 26, 27].

19.2.3.3 Infection

Infection of vascular grafts remains a serious complication following vascular surgery. Graft infection can induce graft adhesion and obstruction and, in the worst-case scenario, death due to rupture of anastomotic aneurysm or sepsis [28]. The incidence of this problem remains low (1–6 %) depending on the recipient site and the necessary graft length, but the prognosis following graft infection is poor: mortality has been reported as high as 36 % [29]. In more than 50 % of the cases, *Staphylococcus epidermidis* is responsible for the early (with 3 months) graft infection that occurs postoperatively [30]. Sound surgical technique and the use of antibiotics are regarded as the most important factors for the prevention of graft infection [31].

19.3 Current Technological Development and Challenges

The choice of appropriate pharmaceutical compounds (drugs), the selection of a biodegradable and biocompatible polymer/drug carrier, and the determination of suitable drug dosage at the precise time and site are some of the critical parameters determining the efficacy and safety of a local drug delivery system. Current focus is on the creation of novel, safer, and more selective drugs, biodegradable and biocompatible drug carriers/polymers, smart and efficacious drug delivery system design, etc. [15, 32]. In this section, we focus on describing a few of the major trends in the development revolving around these aspects, with our apologies to the many other contributions and researchers, whose outstanding work we could not cite or discuss due to the page limits of this chapter.

19.3.1 Drugs

In general, three classes of drugs targeting each of the main etiologies associated with vascular diseases (antiproliferative/anti-migratory, antithrombotic, and antibiotic) are suitable candidates that are widely used in the clinics to prevent or ameliorate graft restenosis and occlusion. A number of them are rather nonspecific multiple-target drugs, inhibiting one or more biochemical pathways, and thus of use in addressing one or more manifestations of CVDs. In fact, controlled release of appropriate drugs alone or in a variety of combinations is widely adopted in the treatment of vascular disease [32, 33]. To date, numerous drugs have been studied, including heparin, paclitaxel, and the limus family. Some examples are discussed below.

19.3.1.1 Antiproliferative Agents

Paclitaxel (Taxol) is a highly lipophilic molecule derived from the Pacific yew tree *Taxus brevifolia* [34]. Taxol efficiently inhibits SMC proliferation and migration following vascular intervention, as amply demonstrated both in vitro and in vivo. The mechanism of action of Taxol is to block the depolymerization of microtubules (tubulin). Stabilization of the microtubules prevents G2/M transition in cell cycle and hence abrogates cell proliferation [32, 33, 35, 36]. Because of Taxol's hydrophobicity, it can easily pass through the hydrophobic barrier of the cell membranes, thus resulting in rapid cellular uptake and firm tissue binding, producing prolonged biological effects [37].

Sirolimus, also called rapamycin, belongs to the limus family (a total of six drugs) and is a macrolide antibiotic with potent immunosuppressive properties. Originally identified in a soil sample from Rapa Nui (Easter Island) [38], sirolimus is a natural fermentation product produced by *Streptomyces hygroscopicus*. The mode of action of sirolimus involves binding to its intracellular receptor, the FK506 binding protein (FKBP12), a member of the immunophilin protein family. By inhibiting a unique kinase called "mammalian target of rapamycin" (mTOR) that regulates the cell cycle progression, sirolimus can block cell proliferation. Sirolimus is also a strong anti-inflammatory agent without cellular toxicity at low doses [34]. It is worth mentioning that as a sirolimus-related immunosuppressant drug, **tacrolimus** also binds to the FKBP12 [34]. However, unlike sirolimus, the tacrolimus–FKBP12 complex inhibits the phosphatase calcineurin and has little or low antiproliferative and anti-migratory effects in both SMCs and ECs. Yet with its potent anti-inflammatory effects, tacrolimus may harbor promising therapeutic effects when applied in combination with other drugs, such as sirolimus [39].

Besides paclitaxel and sirolimus, other antiproliferative drugs used to prevent neointimal hyperplasia and graft-/stent-associated restenosis include **mithramycin**, an antitumor drug known to inhibit expression of genes that have G–C-rich promoters, and **sunitinib**, a small-molecule multi-target tyrosine kinase inhibitor of both PDGF receptor and VEGF receptor subtypes [40]. Red wine polyphenols, such as resveratrol (RESV) and quercetin (QUER), in clinical use for their inhibitory effects on SMC proliferation and platelet activation, have other beneficial effects, such as promoting endothelial antithrombotic and anti-inflammatory phenotype and function and interrupting key inflammatory processes in the injured vessel wall [41].

19.3.1.2 Antithrombotic Drugs

Heparin is a highly charged molecule (molecular mass 12,000–18,000 Da) and has a dose-dependent half-life of $\approx$ 1–5 h [13]. It can rapidly inhibit DNA and RNA synthesis in SMCs, limit leukocyte adhesion to damaged endothelium, restore endothelial integrity, and interact with/constrain vascular growth factors [5, 14, 42]. Partially owing to its multiple effects on diverse events accompanying vascular injury, i.e., in situ thrombosis and restenosis, heparin has in the past been regarded as the gold standard broad-spectrum inhibitor of vascular responses to injury and has been widely used for a variety of critical diseases, such as pulmonary embolism, myocardial infarction, and unstable angina [7, 14]. However, because of its relatively large molecular size and polarity, heparin passes only poorly through membranes, restricting its administration. When administered orally, it quickly degrades to inactive oligomer fragments; systemic IV administration of heparin is also questionable due to potential associated complications, such as hemorrhage, electrolyte shifts, and thrombocytopenia, thus requiring localized and sustained release [13].

Besides heparin, **aspirin** is another widely used clinical drug, which is capable of blocking platelet activation and aggregation, inhibiting SMC proliferation and endothelial activation, and protecting ECs from oxidant-induced cell damage [43].

19.3.1.3 Antibiotics

Compared to the above drugs with antiproliferative and/or antithrombotic effects, antibiotics are relatively more common due to their wide range of applications in most infection-related diseases, including those in vascular grafts. Here we only list very briefly a few antibiotics with reported application for the treatment of graft infection.

Sisomicin (SISO) is an antibiotic against *Staphylococci*, the major causative bacteria in graft infection [28]. It is also effective against *Pseudomonas aeruginosa*, and in general it shows higher antimicrobial activity than **amikacin**, an antibiotic used against gentamicin-resistant *Pseudomonas* strains [44, 45]. **Vancomycin** is another potent antibiotic, which is being tried against methicillin-resistant *S. aureus (MRSA)*. Vancomycin is not very effective when systemically administrated, thus requiring localized and sustained release [46, 47].

19.3.2 Polymer/Drug Carriers

The chemical and physical properties of a given polymer/drug carrier are of critical importance for determining the drug release profile of the drug/polymer complex.

Specifically for vascular graft applications, the mechanical property of the polymer as the base material is a critical factor. In the past decade, advances in the field of drug carriers/polymers for vascular graft diseases include the synthesis/design and modification of novel polymers or copolymers prepared via chemical grafting or physical mixing, etc. For brevity and simplicity, we will focus here on two frequently employed base polymers, PLGA and poly(ε-caprolactone) (PCL). Other biopolymers that have been designed for specific applications will be briefly introduced in the system design section.

19.3.2.1 Poly(D,L-Lactic-*co*-glycolide)

Poly(D,L-lactic-*co*-glycolide, PLGA), originally introduced as a biodegradable material with favorable mechanical properties for absorbable sutures, is a widely used polymer/drug carrier in the realm of drug delivery [7, 14, 35]. PLGA is a biomaterial with distinct advantages: PLGA copolymers are biodegradable, FDA approved, and easy to process [14, 35]; their degradation rate is tunable between a few weeks and many months, depending on the relative lactic/glycolic acid ratio and molecular weight of the polymers; and the degradation products of PLGA are natural metabolites (lactic and glycolic acid). However, as a caveat, these degradation products are acidic which can cause significant damage in the vicinity of the degrading PLGA implants. The pros and cons regarding the biodegradation and biocompatibility of PLGA microspheres as drug carriers have recently been reviewed in great detail by Anderson et al. [48].

19.3.2.2 Poly(ε-Caprolactone)

PCL is a semicrystalline aliphatic polyester well known for its slow rate of biodegradation (years), high degree of biocompatibility, and good drug permeability [23]. A number of medical and drug delivery systems composed of PCL have been approved by US Food and Drug Administration (FDA), upon promising preclinical data from extensive in vitro and in vivo biocompatibility and efficacy studies. Unlike other commonly used biodegradable polymers, such as PLGA, PCL does not produce a local acidic environment as it degrades. This, together with its beneficial mechanical properties and relative low cost, renders PCL an attractive polymer for biomedical, especially drug delivery, applications [23, 33].

19.3.2.3 Others

Besides (synthetic) polymers, natural biopolymers isolated from plasma, such as fibrinogen, or mammalian tissues, such as collagen, are also used as drug carriers in a variety of formulations, e.g., in hydrogels [49]. As an example, **fibrin glue**, a tissue adhesive commonly used for the prevention of bleeding in surgery and a pre-sealing material for prevention of blood leakage from porous vascular prostheses, has been successfully used as an effective carrier for local drug delivery [50].

19.3.3 Drug-Eluting System Design and Release Kinetics

In the past decade, research into localized and controlled drug delivery for vascular prosthetics has largely focused on using films/wraps or microspheres/nanoparticles as auxiliary vascular graft components. More recently, the incorporation of drugs into the grafts themselves has emerged, e.g., in the form of electrospun, drug-laden polymer fibers. Different from the traditional approaches for manufacturing vascular grafts (knitting, weaving, extrusion), electrospinning is a platform technology that offers the opportunity of integrating the drug delivery system during the graft manufacturing process. For drug-eluting vascular grafts, prolonged drug release is often regarded beneficial, as it increases the effectiveness/duration of residency of the therapeutics [14]. As discussed below, traditional methods like perivascular wraps and microspheres in general produce a shorter period of controlled drug release than drug-eluting electrospun fibers. In this subsection, we review a few key studies using either perivascular wraps or microspheres/nanoparticles as drug delivery systems for vascular application. We then describe in more detail the manufacture and use of drug-eluting electrospun fibers for vascular grafts along with the introduction to electrospinning technology.

19.3.3.1 Perivascular Wraps

In 1990, Edelman et al. [13] reported a local delivery system comprised of ethylene–vinyl acetate copolymer (EVAc) matrices and heparin, which was wrapped around the adventitial surface of injured arterial segments in animal models. By pouring the drug/polymer suspension into a precooled glass mold followed by freeze-drying, heparin was homogeneous dispersed within the EVAc matrix. The heparin-containing matrix was further coated with a layer of EVAc to obviate the undesirable burst release during the first 24–36 h. In vitro (<2 weeks), the resulting system showed a uniform and near zero-order release profile; with less drug loading, the release profiles were more extended. In vivo, these matrices demonstrated similar release rates as in vitro, without eliciting an adverse local tissue reaction (up to 7 months) [13]. However, as noted by the authors, EVAc is nonbiodegradable and noncompliant/nondeformable, thus not suited as an efficient perivascular drug release system.

A decade later, in 2000, the same group developed a biodegradable delivery system comprised of heparin-encapsulated PLGA microspheres, which were further embedded in a hydrophilic calcium alginate matrix [14]. The heparin-loaded PLGA microspheres were prepared by solvent extraction using a water-in-oil-in-water emulsion procedure. The microsphere–hydrogel complexes were produced by suspending heparin-loaded microspheres in alginate solution followed by freeze-drying and cross-linking via a calcium solution. This resultant system combined the tissue-like property of alginate hydrogels with the slow release properties of hydrophobic PLGA polymers and showed a controlled first-order release of heparin for over 25 days in vitro. In addition, the complex retained the bioactivity of heparin, as assessed by the inhibition of cultured bovine SMC as well as in animal models of vascular diseases [14].

A PCL-based perivascular cuff containing paclitaxel was introduced in the 2000s [33]. The drug was first blended with poly(ethylene glycol) (PEG), and then the PEG–drug complex was mixed with molten PCL to manufacture a cylindric perivascular cuff. The addition of PEG increased the water uptake by the polymer blends while decreasing the rate of paclitaxel release [51]. This perivascular cuff exhibited sustained and dose-dependent release for at least 3 weeks in vitro and showed significant reduction of intimal thickening without adverse systemic effects in a mouse model at 3 weeks [33].

19.3.3.2 Microspheres/Nanoparticles

Degradable microspheres have been widely applied as drug carriers and controlled release systems for both topical and systemic deliveries of numerous pharmaceuticals [48]. Of the many choices of polymer carriers in form of microspheres and/or nanoparticles, PLGA has been one of the most popular candidates for drug-eluting vascular grafts, due to ease of manufacturing and processing and the wide range of performance characteristics that can readily be modulated by adjusting the comonomer ratio and the molecular weight [14, 48].

In the 1990s, Yang et al. [7] and Gander et al. [52] encapsulated heparin into PLGA microspheres using a spray-drying process. Heparin release from PLGA microspheres exhibited a triphasic pattern with an initial burst for the first 24 h, a dormant period of 15–30 days, and followed by a final stage with an increased release rate lasting for an additional 10–14 days. Similar levels of reduction in SMC proliferation were observed for both native/dissolved heparin and heparin released from the PLGA microspheres [7]. Importantly, this study showed that the selection of appropriate polymer types, such as the change in the comonomer ratio in PLGA (PLGA 50:50 vs. PLGA 75:25), could significantly modulate the release profile, owing to the critical role of polymer degradation at the final stage [7].

In 2007, Lim et al. [35] reported the preparation of paclitaxel-loaded PLGA nanoparticles using an emulsion-solvent evaporation method. These drug-laden nanoparticles were transferred to the luminal surface and inner part of ePTFE grafts via micro tube pumping and spin penetration techniques. This system exhibited a controlled in vitro release profile over 21 days. Importantly, encapsulation into PLGA nanoparticles and further processing significantly reduced the initial burst release and resulted in a more sustained release profile (<30 % of cumulative release at day 21), as compared to the ePTFE grafts that had been dipped coated with paclitaxel (62 % cumulative release) [35].

Also in 2007, Westedt et al. [53] introduced paclitaxel-loaded nanoparticles (NP) (<180 nm in diameter) consisting of poly(vinyl alcohol)-graft-poly(lactide-co-glycolide) (PVA-g-PLGA) using a solvent evaporation technique, an attractive manufacturing process that is widely used for encapsulating lipophilic drugs. The brushlike PVA-g-PLGA molecules consist of a hydrophilic backbone (PVA), where the hydrophobic PLGA is grafted. The system can be fine-tuned by changing the hydrophilic/hydrophobic part of the polyesters, to modify the water uptake and swelling properties of the matrix, the polymer degradation rate, as well as the drug release kinetics. Regardless of the change in content of PLGA as well as its chain length, this grafted polymer system exhibited a biphasic release pattern, which was characterized by an initial burst release followed by a slow release period over a 22-day study. Of note, increasing the PLGA chain length led to a dramatic decrease in the burst release rates. Paclitaxel-loaded nanoparticles showed an increased anti-proliferative effect on cultured SMC as compared to the control study that used free, non-encapsulated paclitaxel [53].

Other microsphere-/nanoparticle-based delivery systems containing mixed polymer matrices have also been used to modify the bulk polymer properties and/or their drug release profile. For example, Chandy et al. [32] introduced a complex prolonged release co-matrix form, where paclitaxel-loaded polylactic acid (PLA) microspheres were encapsulated within heparin–chitosan spheres. This system exhibited a complex paclitaxel/heparin release profile with high initial release value (Taxol 15.8 %, heparin 32.7 %). Following further coating of the matrix with polyethylene glycol (PEG), this system exhibited a constant slow release profile [32].

19.3.4 Drug-Eluting Electrospun Fibers

Electrospinning is a versatile, established platform technology, which was developed in the 1930s for spinning textile fibers. "Rediscovered" in the 1990s [54], electrospinning has become the staple in tissue engineering for generating fibrous tissue scaffolds comprised of nanometer- to micrometer-sized fibers. This technique has also been used to encapsulate a plethora of pharmaceutically active compounds into fibrous scaffold [55]. For example, model proteins, such as bovine serum albumin (BSA) [56] and peptide growth factors [57, 58], have been incorporated into electrospun fibers for sustained release and tested for retained bioactivity. To date, only few reports have been published on the incorporation of clinically relevant therapeutic compounds into electrospun structures for vascular graft application, evaluating both controlled release and preservation of bioactivity. In this section, we will describe in detail these few studies as well as provide some technical background of the electrospinning platform technology. We believe that electrospinning can be used for the preparation of future drug-eluting vascular grafts with enhanced efficacy and safety.

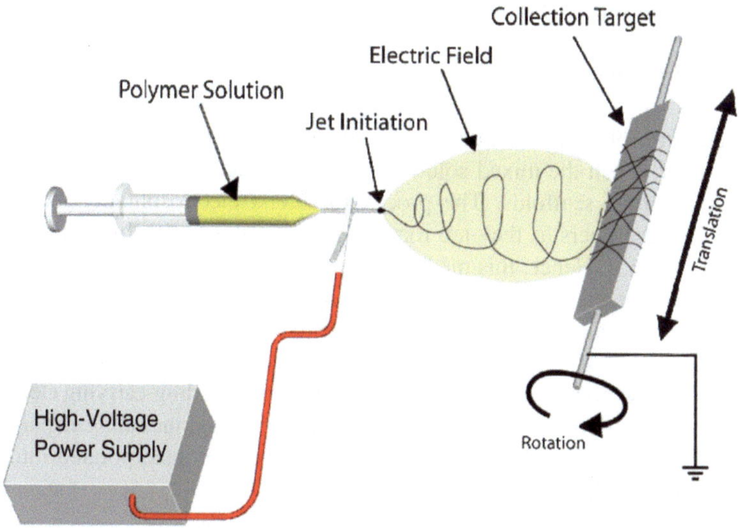

Fig. 19.2 Schematics of the setup of the electrospinning process (reproduced with permission from Barnes et al. [61])

19.3.4.1 Principles of Electrospinning

Originally introduced in the textile industry and in organic polymer science in the 1930s [59, 60], electrospinning is a rapid, efficient, and economic technique for producing nonwoven fabrics with fiber sizes in the nano-/microscale from both synthetic and natural polymers. In the basic process of electrospinning, a drop of a polymer solution hanging from a metal tube, such as a metal syringe needle, is exposed to an external electric field to form a charge imbalance between the tip of the syringe needle and a grounded target (Fig. 19.2). With the increase of the electric charge, a conical polymer jet (Taylor cone) is formed at the needle tip. Once the charge imbalance surpasses the surface tension of the polymer solution, the polymer solution is ejected from the Taylor cone. As the polymer solution travels through the air toward the target, the solvent evaporates, and the polymer fibers are deposited on the grounded target, eventually producing a nonwoven fibrous fabric/mat once sufficient polymer fibers have been accumulated on the target [61]. To produce a tubular fabric for vascular graft application, the fibers are collected on a rotating mandrel in lieu of a static flat target [2, 61].

Recently, electrospinning has reemerged as a preferred fabrication tool for generating fibrous scaffolds from a large number of both synthetic polymers, including PLGA [62], PCL [23], and natural proteins [63], as well as from mixed solutions containing both synthetic and natural polymers [4, 64]. The concentration/viscosity of the polymer solution, surface tension, applied voltage, air gap distance, and solution delivery rate are critical experimental parameters that determine the shape and size of electrospun fibers [59, 60].

19.3.4.2 Fabrication Techniques for Drug-Eluting Electrospun Fibers

Blend Electrospinning

In blend electrospinning, the drugs are mixed with the polymer solution in a common solvent, after which the mixed solution is used to electrospin the drug-containing fibers and fibrous scaffolds. The process of blend electrospinning incorporates drugs within the fibers as they are formed, rather than simply absorb them on the fiber surfaces [65]. Hence, this method is expected to provide more sustained drug release as compared to mere physical absorption. At present, a good number of diverse bioactive agents have been incorporated into electrospun fibers via this approach; they all exhibited sustained release profiles [65]. Blend electrospinning is currently the most common method of choice for generating drug-carrying electrospun fibers. A wide variety of CVD-relevant therapeutic drugs, including heparin [23], paclitaxel [16], sirolimus [2], and aspirin [43], have been incorporated into fibers via blend electrospinning.

Coaxial Electrospinning

In coaxial electrospinning, two or more solutions are coaxially and concomitantly electrospun through different feeding resources but assembled in one needle to manufacture composite fibers with a core–shell structure. As a dynamic process, coaxial electrospinning is affected by various factors, such as the delivery rates of the inner and the outer solutions and the interfacial tension and viscosity of the solutions [65]. Due to the complexity of this technique, the application of coaxially electrospun fibers as bioengineered reservoirs for the controlled delivery of bioactive agents/drugs has only been reported in the past few years [66]. It is anticipated that the core–shell structure of the fibers provided by coaxial electrospinning will allow more homogeneous drug distribution within the fibers, therefore yielding a prolonged release profile [65].

Physical Adsorption

The easiest approach to incorporating drugs into electrospun fibers is to dipcoat the fibrous constructs (mats, tubes, etc.) into appropriate organic or inorganic solvents that contain the drug in question. In this method, the drugs may cling to the fibers via electrostatic forces [65]. However, this approach is seldom used due to the putatively uncontrolled amounts of absorbed drug and release profiles. The release rates of bioactive agents adsorbed on electrospun fibers were reported to be much faster than that of the same amount of proteins/genes loaded into electrospun fibrous scaffolds using blend electrospinning [65].

Covalent Immobilization

Covalent immobilization secures/attaches the drugs and bioactive agents onto the fiber surface altered via chemical bonds and/or biophysical biochemical modifications [65, 67]. Established surface modifications of fibers include plasma treatment, graft polymerization, chemical conjugation, etc. [67–69]. This method is predominantly used to improve the surface properties of electrospun fibers to achieve sustained drug delivery. To date, a variety of biomolecules with CVD relevance, such as heparin, have been covalently attached on the surface for controlled drug delivery [67–69].

Current Progress in Drug-Eluting Electrospun Fibers/Grafts

In comparison to the numerous studies of bioactive agents releasing electrospun fibers, the few published studies focusing on drug-eluting electrospun fibers for vascular graft applications unanimously use blend electrospinning approaches. In this subsection, we paradigmatically describe a few studies, which include sirolimus-contained polyurethane (PU) fibrous mats/grafts [2], sirolimus-blended poly(L-lactide-co-caprolactone-*co*-glycolide)/phospholipid fibers [70], paclitaxel- or heparin-encapsulated PCL fibers [16, 23], and aspirin-loaded PCL grafts [43]. These data demonstrate that (blend) electrospinning is a simple and versatile technique to incorporate various drugs into polymer fibers while offering a localized and controlled drug release without significantly altering the morphology and mechanical properties of the ensuing fibrous constructs (mats, tubes, etc.). We believe that drug-eluting electrospun fibers can serve as effective drug reservoirs for the local drug delivery in vascular therapeutic applications. We surmise that with precise system design, the drug-laden electrospun fibers are promising candidates for functional vascular grafts that can exhibit long-term safety and patency.

In 2006, Luong-Van et al. [23] reported the successful incorporation of heparin into electrospun PCL fibers via blend electrospinning. Smooth, bead-free fibers were electrospun with homogeneous distribution of heparin. These heparin/PCL fibers showed a diffusion-controlled release in vitro with a cumulative release of ca. 50 % of the encapsulated heparin released after 14 days. Further in vitro studies suggest that heparin-loaded fibers did not trigger inflammatory responses in macrophages and reduced SMC proliferation, making this system a potentially promising local drug delivery vehicle for cardiovascular applications [23].

In 2009, Innocente et al. [16] introduced paclitaxel into electrospun PCL nanofibers via blend electrospinning. Generating small-diameter vascular grafts from these fibers, the authors assessed drug release profiles and graft performance both in vitro and in vivo. During a 27-day in vitro study, a triphasic release profile was observed, consisting of an initial burst release (ca. 20 % of the entrapped drug) within the first hour, followed by a sustained release over the following ca. 22 days, and a final plateau phase with a much slower release over the last 4–6 days, for a cumulative release of ca. 60 % of the entrapped drug. When implanted into the infrarenal abdominal aorta of Sprague–Dawley rats, these paclitaxel-PCL grafts exhibited significant suppression/delay of neointima formation with 100 % graft patency for up to 6 months. While the drug reduced, as expected, neointimal hyperplasia, it also significantly delayed endothelialization and cellular ingrowth. Remarkably, although loading of paclitaxel into PCL fibers affected the mechanical and morphological properties of the corresponding grafts, this did not jeopardize the grafts' suitability for implantation in a rodent model [16].

Also in 2009, Kim et al. [70] blend electrospun a solution containing poly(L-lactide-co-caprolactone-co-glycolide) (PLCG), a water-soluble amphiphilic copolymer bearing phosphorylcholine groups (PMB30W), and sirolimus and evaluated fiber morphologies and physicochemical traits as well as their in vitro release characteristics. A sustained slow release was observed during this 1-month study, with a cumulative release approaching 30 % [70]. The therapeutic efficacy of these fibers yet has to be evaluated.

Del Gaudio et al. [43] recently electrospun small-diameter aspirin-eluting PCL grafts and evaluated fiber structure and release characteristics in vitro. A dose-dependent (drug loading at 1, 5, 10 %, w/w) controlled release was observed for the first ca. 50 h. The amount of aspirin released from the fibers with the higher drug contents was effective in significantly attenuating platelet adhesion during the initial few hours of the study [43].

In our own work, we recently incorporated sirolimus into polyurethane (PU) fibers by blend electrospinning using three distinct blending methods and assessed the drug release profiles and the bioavailability of RM-containing PU fibers in the form of fibrous mats and small-diameter vascular grafts in vitro [2]. Specifically, we tested different blending methods by either dissolving RM powder in PU solution (5 %, w/v) or mixing a sirolimus solution (at 2 or 5 %, w/v) with a PU solution (5 %, w/v) for achieving final drug contents of 1, 5, 10, and 20 % (w/w), respectively.

By varying the blending methods, i.e., the total solute concentration, we were able to change the diameters of the ensuing sirolimus-eluting fibers, in a manner that seems to depend both on the solute concentration and on the method of blending. As shown in Fig. 19.3a, at 10 and 20 % RM (w/w), the RM/PU blend solutions prepared via the different blending methods produced fibers with significant different diameters [2]. A plot of the diameters of all fibers against their respective total solute concentration produced linear positive correlations ($R^2>0.94$) for samples via 2 % and powder methods (the solutions of which with varied total solute concentrations; see Fig. 19.3b) [71, 72]. In contrast to the more complex relationship between drug contents and fiber diameter, the Young's moduli of all fibrous mats generally increased with higher relative RM concentration, indicating a strengthening effect of the incorporation of RM on the resulting blend fibers (Fig. 19.3c). Of note, at a relative RM concentration of 10 or 20 % (w/w), the increase in the total solute concentration (via the different blending methods) yielded a decrease in the average Young's moduli of the ensuing fibrous mats (Fig. 19.3c). By plotting the Young's moduli of fibrous mats against the diameter of corresponding fibers (Fig. 19.3d, negative correlations), we confirmed recent results by Tan et al. [73] that with decreasing diameter, single/individual polymer fibers were stiffer and less ductile. Importantly, the average Young's moduli of RM/PU fibrous mats at 20 %

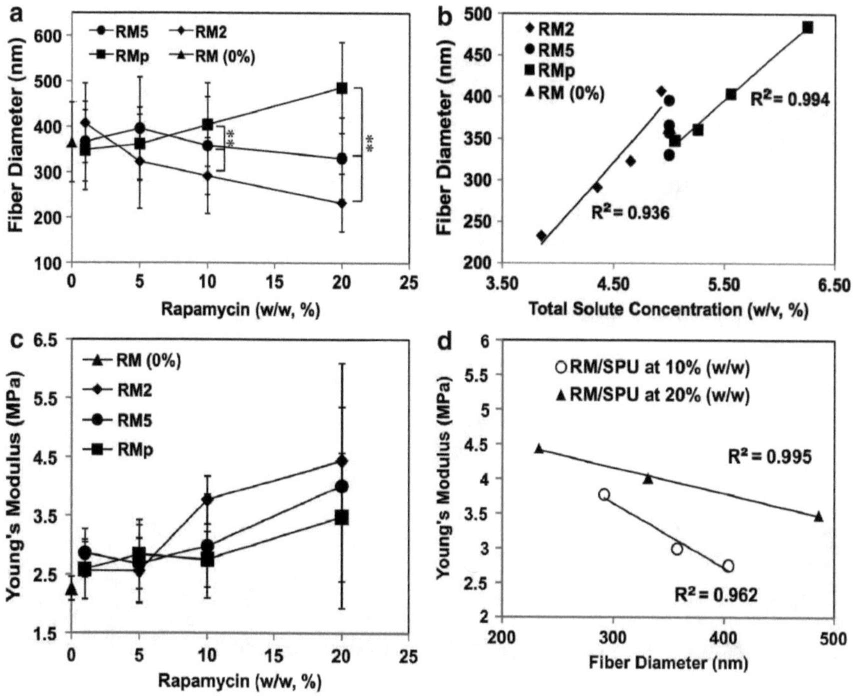

Fig. 19.3 Fiber diameter of electrospun rapamycin–polyurethane (RM–PU) fibers plotted against the relative concentration of rapamycin (w/w) (**a**); correlation of fiber diameter with the total solute concentration (w/v) of the different solutions (**b**); Young's modulus of electrospun fibrous mats plotted against the relative concentration of rapamycin (w/w) (**c**); correlation of Young's modulus of fibrous mats at a relative RM/PU ratio (w/w) at 10 or 20 % with fiber diameter (**d**). Data in (**a, c**) are represented as mean ± SD ($n=90$ in A, $n=6$ in C). **: $p<0.01$. Data in (**b, d**) represented as the mean (reproduced with permission from Han et al. [2])

RM (w/w) generated were comparable to that of natural arteries and thus relevant for the intended application of this system for vascular grafts [74].

The morphology of both fibrous mats and bilayered grafts (with a luminal spin-cast PU layer free of drug, as previously described; see [74]) was analyzed with SEM (Fig. 19.4a, b, mats; Fig. 19.4c, d, e, grafts). The purpose of this novel bilayered vascular graft design is to restrict RM-incorporated PU fibers to the outer layer of our bilayered vascular grafts [2] and with that to direct drug release only toward the *tunica media* harboring the SMCs while minimizing, if any, potentially inhibitory effects of the drug on the endothelialization of small-diameter grafts, as described by Innocente et al. [16].

In terms of functional characterization, we evaluated the in vitro drug release profiles of both mats and grafts at different sirolimus concentrations (1, 5, 10, 20 %, w/w) as well as their bioactivities. Both mats and grafts inhibited SMC proliferation in a dose-dependent fashion; at 20 % drug, both mat and graft samples "killed" essentially all cells [2]. We then examined the in vitro drug release profiles of electrospun

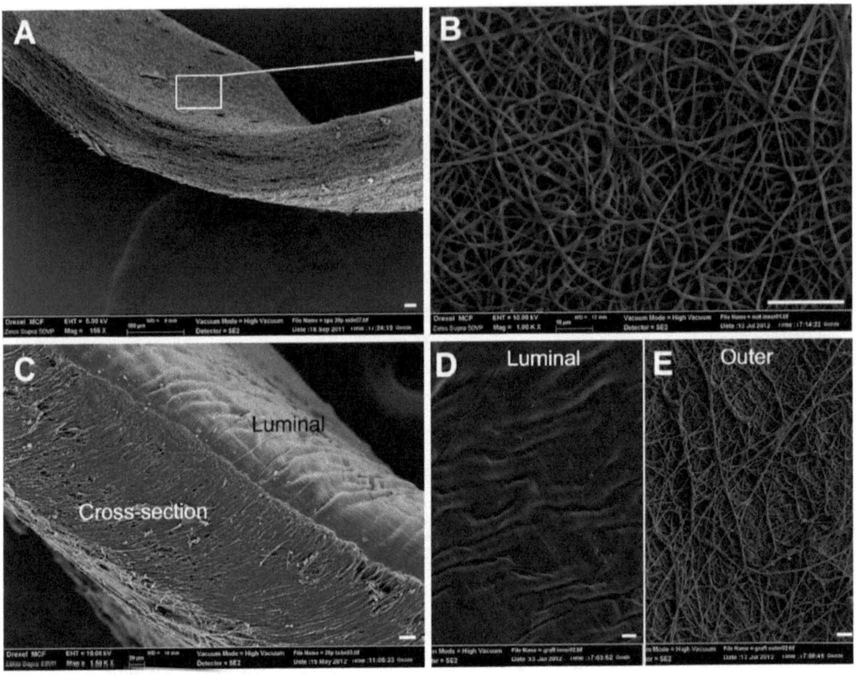

Fig. 19.4 SEM images of cross section (**a**) and topical view (**b**) of electrospun mats made of RMp (20 %) containing polyurethane fibers; cross section (**c**), luminal layer (**d**), and outer layer (**e**) of bilayered grafts made of RMp (20 %) fibers. Scale bar: 20 μm (reproduced with permission from Han et al. [2])

sirolimus-eluting fibrous mats and grafts containing 20 % drug prepared via the different blending methods for up to 77 days. All samples exhibited a relatively small initial burst release for the first week (<10 % for mats, <5 % for tubular grafts), followed by a sustained release, regardless of their form in mats or grafts (Fig. 19.5a, b). Yet the amount of drug release was different for each time point investigated: fibers produced by the 2 % method released the highest amount of RM, while those manufactured by the 5 % method released the least amount of the drug. A plot of the cumulative drug release against the square root of time (Fig. 19.5c, d, linear regression) suggested a Fickian diffusion-controlled release for all samples [23, 75]. Interestingly, some of the samples retained their competence to inhibit SMC proliferation in vitro even after 77 days of release studies, making these systems attractive candidates for potentially long-term inhibition of neointimal hyperplasia [2].

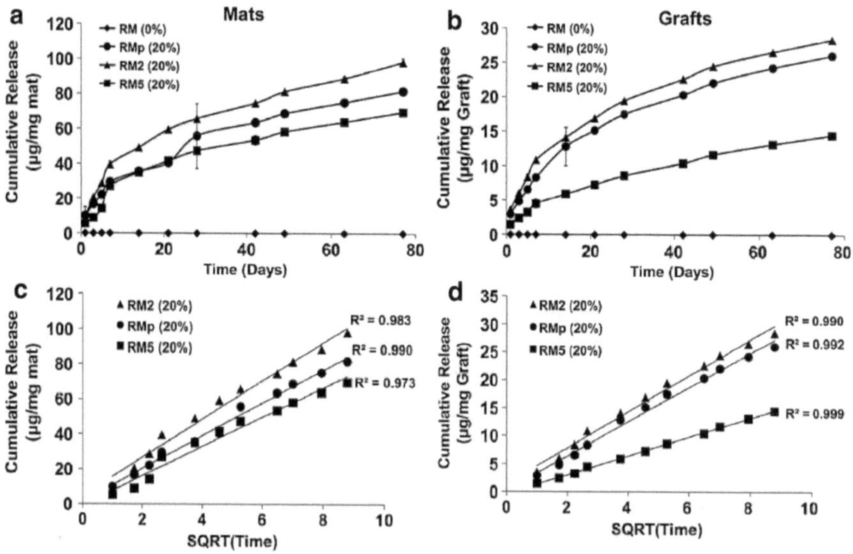

Fig. 19.5 In vitro drug release profiles of RM–PU mats (**a**) and grafts (**b**) with fibers RM2 (20 %), RM5 (20 %), or RMp (20 %) over 77-day period; data are represented as mean ± SD from triplicates. Cumulative RM release plotted against the square root of time for all fibrous mats (**c**) and grafts (**d**); data represented the mean from triplicates (reproduced with permission from Han et al. [2])

19.4 Conclusions

In this chapter, we described some of the current developments and technologies in the area of drug-eluting vascular grafts. Given the simplicity and versatility of electrospinning as platform technology, current progress in the field of drug-eluting vascular grafts suggests that this approach may be a promising step toward meeting the unmet clinical need of combining sustained localized drug release for vascular therapeutic applications with long-term efficacy and safety. This multi-faceted quest for substantial improvements of the drug-eluting vascular repair devices, grafts, and stents alike requires further innovations in the design and implementation of drugs, the polymer/drug carriers, as well as the drug delivery systems in order to provide a precise control of drug release at the correct location over a well-defined and tightly controlled time period. In addition, basic knowledge/understanding of the tissue reactions to and interactions with various vascular prosthetics is fundamental to optimizing the design and clinical practicality of these systems, thus requiring the close interdisciplinary collaborations among material scientists, tissue engineers, vascular biologists, and clinicians.

References

1. Go AS, Mozaffarian D, Roger VL et al (2013) Executive summary: heart disease and stroke statistics–2013 update: a report from the American heart association. Circulation 127(1): 143–152
2. Han J, Farah S, Domb AJ, Lelkes PI (2013) Electrospun drug-eluting fibers for vascular grafts. Pharm Res 30(7):1735–1748
3. Parang P, Arora R (2009) Coronary vein graft disease: pathogenesis and prevention. Can J Cardiol 25(2):e57–e62
4. Han J, Lazarovici P, Pomerantz C, Chen X, Wei Y, Lelkes PI (2011) Co-electrospun blends of PLGA, gelatin, and elastin as potential nonthrombogenic scaffolds for vascular tissue engineering. Biomacromolecules 12:399–408
5. Hirsh GM, Karnovsky MJ (1991) Inhibition of vein graft intimal proliferative lesions in the rat by heparin. Am J Pathol 139:581–587
6. Schwarz SM, DeBlois D, O'Brien ERM (1995) The intima: soil for atherosclerosis and restenosis. Circ Res 77:445–465
7. Yang Z, Birkenhauer P, Julmy F, Chickering D, Ranieri JP, Merkle HP, Lüscher TF, Gander B (1999) Sustained release of heparin from polymeric particles for inhibition of human vascular smooth muscle cell proliferation. J Control Release 60(2–3):269–277
8. Clowes AW, Karnowsk MJ (1977) Suppression by heparin of smooth muscle cell-proliferation in injured arteries. Nature 265:625–626
9. Porter J, Jick H (1977) Drug-related deaths among medical in-patient. JAMA 237:879
10. Serruys PM, Strauss BH, Beatt KJ, Bertrand ME, Puel J, Rickards AF, Meier B, Goy J-J, Vogt P, Kappenberger L, Sigwart U (1991) Angiographic follow-up of a self-expanding coronary artery stent. N Engl J Med 324:13–17
11. Lehmann KG, Doria RJ, Feuer JM, Hall PX, Hoang DT (1991) Paradoxical increase in restenosis rate with chronic heparin use: final results of a randomized trial. J Am Coll Cardiol 17:181A
12. Camenzind E, Kint P-P, Mario CD, Ligthart J, Van der Giessen W, Boersma E, Serruys PW (1995) Intracoronary heparin delivery in humans: acute feasibility and long-term results. Circulation 92:2463–2472
13. Edelman ER, Adams DH, Karnovsky MJ (1990) Effect of controlled adventitial heparin delivery on smooth muscle cell proliferation following endothelial injury. Proc Natl Acad Sci U S A 87(10):3773–3777
14. Edelman ER, Nathan A, Katada M, Gates J, Karnovsky MJ (2000) Perivascular graft heparin delivery using biodegradable polymer wraps. Biomaterials 21:2279–2286
15. Fattori R, Piva T (2003) Drug-eluting stents in vascular intervention. Lancet 361(9353): 247–249
16. Innocente F, Mandracchia D, Pektok E, Nottelet B, Tille JC, de Valence S, Faggian G, Mazzucco A, Kalangos A, Gurny R, Moeller M, Walpoth BH (2009) Paclitaxel-eluting biodegradable synthetic vascular prostheses: a step toward reduction of neointima formation? Circulation 120(11 Suppl):S37–S45
17. Park DW, Hong MK, Mintz GS, Lee CW, Song JM, Han KH et al (2006) Two-year follow-up of the quantitative angiographic and volumetric intravascular ultrasound analysis after non-polymeric paclitaxel-eluting stent implantation: late "catch-up" phenomenon from ASPECT Study. J Am Coll Cardiol 48(12):2432–2439
18. Loukas M, Groat C, Khangura R, Owens DG, Anderson RH (2009) The normal and abnormal anatomy of the coronary arteries. Clin Anat 22(1):114–128
19. http://nyp.org/health/cardiac-arteries.html, accessed Feb 2, 2013
20. Sarkar S, Sales KM, Hamilton G, Seifalian AM (2007) Addressing thrombogenicity in vascular graft construction. J Biomed Mater Res B Appl Biomater 82(1):100–108
21. Orr AW, Sanders JM, Bevard M, Coleman E, Sarembock IJ, Schwartz MA (2005) The subendothelial extracellular matrix modulates NF-kappaB activation by flow: a potential role in atherosclerosis. J Cell Biol 169(1):191–202

22. Stephan S, Ball SG, Williamson M, Bax DV, Lomas A, Shuttleworth CA, Kielty CM (2006) Cell-matrix biology in vascular tissue engineering. J Anat 209(4):495–502
23. Luong-Van E, Grondahl L, Chua KN, Leong KW, Nurcombe V, Cool SM (2006) Controlled release of heparin from poly(epsilon-caprolactone) electrospun fibers. Biomaterials 27: 2042–2050
24. Scott-Burden T, Vanhoutte PM (1994) Regulation of smooth muscle cell growth by endothelium-derived factors. Tex Heart Inst J 21(1):91–97
25. Beamish JA, He B, Kottke-Marchant K, Marchant RE (2010) Molecular regulation of contractile smooth muscle cell phenotype: implications for vascular tissue engineering. Tissue Eng Part B Rev 16(5):467–491
26. Tucker EI, Marzec UM, White TC, Hurst S, Rugonyi S, McCarty OJ, Gailani D, Gruber A, Hanson SR (2009) Prevention of vascular graft occlusion and thrombus-associated thrombin generation by inhibition of factor XI. Blood 113(4):936–944
27. Taite LJ, Yang P, Jun HW, West JL (2008) Nitric oxide-releasing polyurethane-PEG copolymer containing the YIGSR peptide promotes endothelialization with decreased platelet adhesion. J Biomed Mater Res B Appl Biomater 84(1):108–116
28. Osada T, Yamamura K, Yano K, Fujimoto K, Mizuno K, Sakurai T, Nabeshima T (2000) Distribution and serum concentration of sisomicin released from fibrin glue-sealed Dacron graft in the rat and human. J Biomed Mater Res 52(1):53–57
29. Huh J, Chen JC, Furman GM, Malki C, King B, Kafie F, Wilson SE (1998) Local treatment of prosthetic vascular graft infection with multivesicular liposome-encapsulated amikacin. J Surg Res 74(1):54–58
30. Zetrenne E, McIntosh BC, McRae MH, Gusberg R, Evans GR, Narayan D (2007) Prosthetic vascular graft infection: a multi-center review of surgical management. Yale J Biol Med 80(3):113–121
31. O'Brien T, Collin J (1992) Prosthetic vascular graft infection. Br J Surg 79(12):1262–1267
32. Chandy T, Rao GH, Wilson RF, Das GS (2001) Development of poly(Lactic acid)/chitosan co-matrix microspheres: controlled release of taxol-heparin for preventing restenosis. Drug deliv 8(2):77–86
33. Pires NM, van der Hoeven BL, de Vries MR, Havekes LM, van Vlijmen BJ, Hennink WE, Quax PH, Jukema JW (2005) Local perivascular delivery of anti-restenotic agents from a drug-eluting poly(epsilon-caprolactone) stent cuff. Biomaterials 26:5386–5394
34. Khan W, Farah S, Domb AJ (2012) Drug eluting stents: developments and current status. J Control Release 161:703–712
35. Lim HJ, Nam HY, Lee BH, Kim DJ, Ko JY, Park JS (2007) A novel technique for loading of paclitaxel-PLGA nanoparticles onto ePTFE vascular grafts. Biotechnol Prog 23(3): 693–697
36. Chatterjee S, Pandey A (2008) Drug eluting stents: friend or foe? A review of cellular mechanisms behind the effects of Paclitaxel and sirolimus eluting stents. Curr Drug Metab 9(6):554–566
37. Nguyen KT, Shaikh N, Wawro D, Zhang S, Schwade ND, Eberhart RC, Tang L (2004) Molecular response of vascular smooth muscle cells to paclitaxel-eluting bioresorbable stent materials. J Biomed Mater Res A 69(3):513–524
38. Marx SO, Marks AR (2001) Bench to bedside: the development of rapamycin and its application to stent restenosis. Circulation 104(8):852–855
39. Matter CM, Rozenberg I, Jaschko A, Greutert H, Kurz DJ et al (2006) Effects of tacrolimus or sirolimus on proliferation of vascular smooth muscle and endothelial cells. J Cardiovasc Pharm 48(6):286–292
40. Sanders WG, Hogrebe PC, Grainger DW, Cheung AK, Terry CM (2012) A biodegradable perivascular wrap for controlled, local and directed drug delivery. J Control Release 161(1):81–89
41. Kleinedler J, Pjescic I, Bullock KK, Khaliq A, Foley JD, Dugas TR (2012) Arterial pharmacokinetics of red wine polyphenols: implications for novel endovascular therapies targeting restenosis. J Pharm Sci 101(5):1917–1931

42. Hedin U, Daum G, Clowes AW (1998) Heparin inhibits thrombin-induced mitogen-activated protein kinase signaling in arterial smooth muscle cells. J Vasc Surg 27:512–520
43. Del Gaudio C, Ercolani E, Galloni P, Santilli F, Baiguera S, Polizzi L, Bianco A (2013) Aspirin-loaded electrospun poly(ε-caprolactone) tubular scaffolds: potential small-diameter vascular grafts for thrombosis prevention. J Mater Sci Mater Med 24(2):523–532
44. Roehrborn AA, Hansbrough JF, Gualdoni B, Kim S (1995) Lipid-based slow-release formulation of amikacin sulfate reduces foreign body-associated infections in mice. Antimicrob Agents Chemother 39:1752
45. Schassan HH (1976) Antimicrobial effectiveness of sisomicin. I: in vitro activity of sisomicin compared with gentamicin, tobramycin, amikacin and kanamycin. Infection 4(2):35–41
46. Hisashi S, Marui A, Hirose K, Nomura K, Arai Y, Chandra S, Huang Y et al (2008) Less-invasive and highly effective method for preventing methicillin-resistant Staphylococcus aureus graft infection by local sustained release of vancomycin. J Thorac Cardiovasc Surg 135(1):25–31
47. Morishima M, Akira M, Shigeki Y, Takamasa N, Naoki N, Suong-Hyu H, Tadashi I, Ryuzo S (2010) Sustained release of vancomycin from a new biodegradable glue to prevent methicillin-resistant Staphylococcus aureus graft infection. Interact Cardiovasc Thorac Surg 11(1):52–55
48. Anderson JM, Shive MS (2012) Biodegradation and biocompatibility of PLA and PLGA microspheres. Adv Drug Deliv Rev 64:72–82
49. Rajathurai T, Rizvi SI, Lin H, Angelini GD, Newby AC, Murphy GJ (2010) Periadventitial Rapamycin-eluting microbeads promote vein graft disease in long-term Pig vein-into-artery interposition grafts. Circ Cardiovasc Interv 3(2):157–165
50. Qi XR, Yan Y, Ma XD, Li TY, Nie XY, Yang BB (2007) Development of a local vasodilator delivery system using fibrin glue to prevent arterial graft from spasm. J Biomed Mater Res A 82(1):139–144
51. Winternitz CI, Jackson JK, Oktaba AM, Burt HM (1996) Development of a polymeric surgical paste formulation for taxol. Pharm Res 13(3):368–375
52. Gander B, Wehrli E, Alder R, Merkle HP (1995) Quality improvement of spray-dried, protein-loaded D, L-PLA microspheres by appropriate polymer solvent selection. J Microencapsul 12(1):83–97
53. Westedt U, Kalinowski M, Wittmar M, Merdan T, Unger F, Fuchs J, Schäller S, Bakowsky U, Kissel T (2007) Poly (vinyl alcohol)-graft-poly (lactide-co-glycolide) nanoparticles for local delivery of paclitaxel for restenosis treatment. J Control Release 119(1):41–51
54. Doshi J, Reneker DH (1995) Electrospinning process and applications of electrospun fibers. J Electrostat 35:151–160
55. Biondi M, Ungaro F, Quaglia F, Netti PA (2008) Controlled drug delivery in tissue engineering. Adv Drug Deliv Rev 60:229–242
56. Kowalczyk T, Nowicka A, Elbaum D, Kowalewski TA (2008) Electrospinning of bovine serum albumin. Optimization and the use for production of biosensors. Biomacromolecules 9:2087–2090
57. Montero RB, Vial X, Nguyen DT, Farhand S, Reardon M, Pham SM, Tsechpenakis G, Andreopoulos FM (2012) bFGF-containing electrospun gelatin scaffolds with controlled nano-architectural features for directed angiogenesis. Acta Biomater 8:1778–1791
58. Ekaputra AK, Prestwich GD, Cool SM, Hutmacher DW (2011) The three-dimensional vascularization of growth factor-releasing hybrid scaffold of poly (epsilon-caprolactone)/collagen fibers and hyaluronic acid hydrogel. Biomaterials 32:8108–8117
59. Lelkes PI, Li M, Perets A, Lin L, Han J, Woerdeman DL (2008) Electrospinning of natural proteins for tissue engineering scaffolding. In: Reis RL (ed) Handbook of natural-based polymers for biomedical applications. Woodhead, Cambridge
60. Han J (2010) Co-electrospun blends of PLGA, gelatin and elastin as nonthrombogenic scaffolds for vascular tissue engineering. Dissertation, Drexel University
61. Barnes CP, Sell SA, Boland ED, Simpson DG, Bowlin GL (2007) Designing the next generation of tissue engineering scaffolds. Adv Drug Deliv Rev 59(14):1413–1433

62. Kim K, Luu YK, Chang C, Fang D, Hsiao BS, Chu B, Hadjiargyrou M (2004) Incorporation and controlled release of a hydrophilic antibiotic using poly(lactide-co-glycolide)-based electrospun nanofibrous scaffolds. J Control Release 98(1):47–56
63. Li M, Mondrinos MJ, Gandhi MR, Ko FK, Weiss AS, Lelkes PI (2005) Electrospun protein fibers as matrices for tissue engineering. Biomaterials 26(30):5999–6008
64. Li M, Mondrinos MJ, Chen X, Gandhi MR, Ko FK, Lelkes PI (2006) Co-electrospun poly(lactide-co-glycolide), gelatin, and elastin blends for tissue engineering scaffolds. J Biomed Mater Res A 79(4):963–973
65. Ji W, Sun Y, Yang F, van den Beucken JJ, Fan M, Chen Z, Jansen JA (2011) Bioactive electrospun scaffolds delivering growth factors and genes for tissue engineering applications. Pharm Res 28(6):1259–1272
66. Chakraborty S, Liao IC, Adler A, Leong KW (2009) Electrohydrodynamics: a facile technique to fabricate drug delivery systems. Adv Drug Deliv Rev 61(12):1043–1054
67. Yoo HS, Kim TG, Park TG (2009) Surface-functionalized electrospun nanofibers for tissue engineering and drug delivery. Adv Drug Deliv Rev 61(12):1033–1042
68. Chen JP, Su CH (2011) Surface modification of electrospun PLLA nanofibers by plasma treatment and cationized gelatin immobilization for cartilage tissue engineering. Acta Biomater 7(1):234–243
69. Lee J, Yoo JJ, Atala A, Lee SJ (2012) The effect of controlled release of PDGF-BB from heparin-conjugated electrospun PCL/gelatin scaffolds on cellular bioactivity and infiltration. Biomaterials 33(28):6709–6720
70. Hyung RM II, Kim JS, Konno T, Takai M, Ishihara K (2009) Preparation of electrospun poly(L-lactide-co-caprolactone-co-glycolide)/phospholipid polymer/rapamycin blended fibers for vascular application. Curr Appl Phys 9:249–251
71. Mo XM, Xu CY, Kotaki M, Ramakrishna S (2004) Electrospun P(LLA-CL) nanofiber: a biomimetic extracellular matrix for smooth muscle cell and endothelial cell proliferation. Biomaterials 25:1883–1890
72. He SW, Li SS, Hu ZM, Yu JR, Chen L, Zhu J (2011) Effects of three parameters on the diameter of electrospun poly(ethylene oxide) nanofibers. J Nanosci Nanotechnol 11:1052–1059
73. Tan EP, Ng SY, Lim CT (2005) Tensile testing of a single ultrafine polymeric fiber. Biomaterials 26:1453–1456
74. Uttayarat P, Perets A, Li M, Pimton M, Stachelek SJ, Alferiev I, Composto RJ, Levy RJ, Lelkes PI (2010) Micropatterning of three-dimensional electrospun polyurethane vascular grafts. Acta Biomater 6:4229–4237
75. Verreck G, Chun I, Rosenblatt J, Peeters J, Dijck AV, Mensch J, Noppe M, Brewster ME (2003) Incorporation of drugs in an amorphous state into electrospun nanofibers composed of a water-insoluble, nonbiodegradable polymer. J Control Release 92:349–360

Chapter 20
Delivery Systems for Lymphatic Targeting

Indu Singh, Rajan Swami, Wahid Khan, and Ramakrishna Sistla

Abbreviations

AIDS	Acquired immunodeficiency syndrome
CNT	Carbon nanotube
DAB	Diaminobutyl
EPR	Enhanced permeation and retention
GALT	Gut-associated lymphoid tissue
HA	Hyaluronic acid
HAART	Highly active antiretroviral therapy
HER	Human epidermal growth factor receptor
HIV	Human immunodeficiency virus
i.m.	Intramuscular
i.p.	Intraperitoneal
IFN	Interferon
IgG	Immunoglobulin G
ipl	Intrapleural
LECs	Lymphatic endothelial cells
mAbs	Monoclonal antibodies

I. Singh • R. Swami • W. Khan
Department of Pharmaceutics, National Institute of Pharmaceutical Education and Research (NIPER), Hyderabad 500037, India

R. Sistla (✉)
Department of Pharmaceutics, National Institute of Pharmaceutical Education and Research (NIPER), Hyderabad 500037, India

Department of Pharmacology, Indian Institute of Chemical Technology, Hyderabad 500607, India
e-mail: sistla@iict.res.in

A.J. Domb and W. Khan (eds.), *Focal Controlled Drug Delivery*, Advances in Delivery Science and Technology, DOI 10.1007/978-1-4614-9434-8_20,

MN-MWNTs	Magnetic multiwalled nanotubes
MWNTs	Multiwalled nanotubes
NALT	Nasal-associated lymphoid tissue
PAMAM	Polyamidoamine
PEG	Polyethylene glycol
PLA	Polylactic acid
PLGA	Poly(lactic-co-glycolic acid)
PMMA	Poly(methyl methacrylate)
RES	Reticuloendothelial system
s.c.	Subcutaneous
SARS	Severe acute respiratory syndrome
SEDDS	Self-emulsifying drug delivery system
SLN	Solid lipid nanoparticles
TB	Tuberculosis
TUNEL assay	Terminal deoxynucleotidyl transferase dUTP nick end labelling assay

20.1 Introduction

20.1.1 Development of the Lymphatic Vascular System

The lymphatic system was first recognised by Gaspare Aselli in 1627, and the anatomy of the lymphatic system was almost completely characterised by the early nineteenth century. However, knowledge of the blood circulation continued to grow rapidly in the last century [1]. Two different theories are proposed which are in favour of origin of the lymphatic vessels. Firstly, centrifugal theory of embryologic origin of the lymphatics was described in the early twentieth century by Sabin and later by Lewis, postulating that lymphatic endothelial cells (LECs) are derived from the venous endothelium. Later the centripetal theory of lymphatic development was proposed by Huntington and McClure in 1910 which describes the development of the lymphatic system beginning with lymphangioblasts, mesenchymal progenitor cells, arising independently of veins. The venous connection to the lymphatic system then happens later in development [2].

The lymphatic vessels in the embryo are originated at mid-gestation and are developed after the cardiovascular system is fully established and functional [3]. A dual origin of lymphatic vessels from embryonic veins and mesenchymal lymphangioblasts is also proposed [4]. Recent studies provide strong support of the venous origin of lymphatic vessels [5–8]. The recent discovery of various molecular markers has allowed for more in-depth research of the lymphatic system and its role in health and disease. The lymphatic system has recently been elucidated as playing an active role in cancer metastasis. The knowledge of the active processes involved in lymphatic metastasis provides novel treatment targets for various malignancies.

20.1.2 Anatomy and Physiology of the Lymphatic System

The lymphatic system consists of the lymphatic vessels, lymph nodes, spleen, thymus, Peyer's patches and tonsils, which play important roles in immune surveillance and response. The lymphatic system serves as the body's second vascular system in vertebrates and functions co-dependently with the cardiovascular system [9, 10]. The lymphatic system comprises a single irreversible, open-ended transit network without a principal driving force [9]. It consists of five main types of conduits including the capillaries, collecting vessels, lymph nodes, trunks and ducts. The lymphatic system originates in the dermis with initial lymphatic vessels and blind-ended lymphatic capillaries that are nearly equivalent in size to but less abundant than regular capillaries [9, 11]. Lymphatic capillaries consist of a single layer of thin-walled, non-fenestrated lymphatic endothelial cells (LECs), alike to blood capillaries. The LECs, on the contrary to blood vessels, have poorly developed basement membrane and lack tight junctions and adherent junctions too. These very porous capillaries act as gateway for large particles, cells and interstitial fluid. Particles as large as 100 nm in diameter can extravasate into the interstitial space, get phagocytosed by macrophages and are ultimately passed on to lymph nodes [11–14]. Lymphatic capillary endothelial cells are affixed to the extracellular matrix by elastic anchoring filaments, which check vessel collapse under high interstitial pressure. These initial lymphatics, under a positive pressure gradient, distend and create an opening between loosely anchored endothelial cells letting for the entry of lymph, a protein-rich exudate from the blood capillaries [12, 15, 16]. In initial lymphatic vessels, overlying endothelial cell–cell contacts prevent fluid reflux back into the interstitial space [17, 18].

After the collection of lymph by the lymphatic capillaries, it is transported through a system of converging lymphatic vessels of progressively larger size, is filtered through lymph nodes where bacteria and particulate matter are removed and finally goes back to the blood circulation. Lymph is received from the initial capillary lymphatic by deeper collecting vessels that contain valves to maintain unidirectional flow of lymph. These collecting vessels have basement membranes and are surrounded by smooth muscle cells with intrinsic contractile activity that in combination with contraction of surrounding skeletal muscles and arterial pulsations propels the lymph to lymph nodes [19–21]. The collecting lymphatic vessels unite into lymphatic trunks, and the lymph is finally returned to the venous circulation via the thoracic duct into the left subclavian vein [22, 23]. The flow of lymph toward the circulatory system is supported by increases in interstitial pressure as well as contractions of the lymphatic vessels themselves. Roughly 25 l of lymphatic fluid enters the cardiovascular system each day [11].

20.1.3 Need for Lymphatic Targeting

The key functions of the lymphatic system are maintenance of normal tissue fluid balance, absorption of lipids and fat-soluble vitamins from the intestine and magnetism and transport of immune cells. Lymphatics transport the antigen-presenting cells as well as antigens from the interstitium of peripheral tissues to the draining lymph nodes where they initiate immune responses via B- and T-cells in the lymph nodes [9, 12, 24, 25]. Tissue fluid balance is maintained by restoring interstitial fluid to the cardiovascular system [9]. Although capillaries have very low permeability to proteins, these molecules as well as other macromolecules and bacteria accumulate in the interstitium. Due to the accumulation of these large molecules in the interstitium, significant tissue oedema would result. The lymphatic system offers the mechanism by which these large molecules re-enter the blood circulation [26]. The lymphatic system is the site of many diseases such as metastitial tuberculosis (TB), cancer and filariasis [27]. Due to the peculiar nature and anatomy of the lymphatic system, localisation of drugs in the lymphatics has been particularly difficult to achieve.

The lymphatic system has an active role in cancer metastasis. Although many cancers may be treated with surgical resection, microscopic disease may remain and lead to locoregional recurrence. Conventional systemic chemotherapy cannot prove effective for delivering drugs to the lymphatic system without dose-limiting toxicities [28]. Lymphatic system functions in the clearance of particulate matter from the interstitium following presentation to lymph nodes have created interest in developing microparticulate systems to target regional lymph nodes. Molecule's composition is important in determining uptake into the lymphatics and retention within the lymph nodes. Colloidal materials, for example, liposomes, activated carbon particles, emulsions, lipids and polymeric particulates, are highly taken up by the lymphatics; that's why nowadays these substances are emerging as potential carriers for lymphatic drug targeting [29]. The vast majority of drugs following oral administration are absorbed directly into portal blood, but a number of lipophilic molecules may get access to the systemic circulation via the lymphatic pathway [30, 31]. Intestinal lymphatic transport of lipophilic molecules is significant and presents benefits in a number of situations:

1. The total systemic bioavailability of lipophilic molecules may be increased, and the hepatic first-pass metabolism may be reduced by the intestinal lymphatic absorption [30].
2. Targeting immunomodulatory agents, cytotoxic agents and drugs used in the treatment of human immunodeficiency virus (HIV) to the lymphatic system has a potential to maximise therapeutic benefits and minimise systemic exposure [30, 32].
3. Following association of the drug with chylomicrons, intestinal lymphatic transport may alter the pharmacokinetic and the pharmacodynamic properties of the drug [32–35].

The lymphatic system also acts as the primary systemic transport pathway for B- and T-lymphocytes as well as the main route of metastatic spread of a number of solid tumours [36, 37]. Therefore, lymphatic absorption of the immunomodulatory and anticancer compounds may be more effective [38, 39]. The presence of wide amounts of HIV-susceptible immune cells in the lymphoid organs makes antiretroviral drug targeting to these sites of tremendous interest in HIV therapy. This strategy comprises once again targeting nanosystems to immune cell populations, particularly macrophages. Also evidence further suggests that lymph and lymphoid tissue, and in particular gut-associated lymphoid tissue, play a major role in the development of HIV and antivirals which target acquired immunodeficiency syndrome (AIDS) may therefore be more effective when absorbed via the intestinal lymphatics [40, 41]. Other viruses like hepatitis B [42] and morbillivirus [43] (which also replicate in gut-associated lymphoid tissue) and the closely related canine distemper virus [44] including severe acute respiratory syndrome (SARS)-associated coronavirus [45] may also spread via the lymphatic network, and the chronic persistence of hepatitis C is believed to result from uptake into systemic lymphocytes and sequestration into the lymph [46].

Targeting drugs to lymphatic system is a tough and challenging task, and it totally depends upon the intricate physiology of the lymphatic system. Targeting facilitates direct contact of drug with the specific site, decreasing the dose of the drugs and minimising the side effects caused by them. Currently, nanocarriers have encouraged the lymphatic targeting, but still there are challenges of locating drugs and bioactives to specific sites, maintaining desired action and crossing all the physiological barriers. These hurdles could be overcome by the use of modified nanosystems achieved by the surface engineering phenomena.

20.2 Targets for Lymphatic Delivery

From the growing awareness of the importance of lymph nodes in cancer prognosis, their significance for vaccine immune stimulation and the comprehension that the lymph nodes harbour HIV as well as other infectious diseases stems the development of new methods of lymph node drug delivery [47–50]. New methods of delivering drugs and other carriers to lymph nodes are currently under investigation.

20.2.1 Cancer

Lymph node dissemination is the primary cause of the spread of majority of solid cancers [51]. In regard to cancer metastasis, the status of the lymph node is a major determinant of the patient's diagnosis. The most important factor that determines the appropriate care of the patient is correct lymph node staging [52]. But patient survivals have been shown to improve by the therapeutic interventions that treat metastatic cancer in lymph nodes with either surgery or local radiation therapy [53].

20.2.2 Human Immunodeficiency Virus

Viraemia is an early indication of primary infection with HIV followed by a specific HIV immune response and a dramatic decline of virus in the plasma [54]. Long after the HIV virus can be found in the blood, HIV can be found in high levels in mononuclear cells located in lymph nodes. Viral replication in these lymph nodes has been reported to be about 10- to 100-fold higher than in the peripheral blood mononuclear cells [55]. Standard oral or intravenous drug delivery to these lymph node mononuclear cells is difficult [56]. Even if highly active antiretroviral therapy (HAART) can reduce plasma viral loads in HIV-infected patients by 90 %, active virus can still be isolated from lymph nodes even after 30 months of HAART therapy.

20.2.3 Filaria

Lymph nodes are the key element of the life cycle of several parasite organisms, including filaria. Lymphatic vessels and lymph nodes of infected patients can carry adult worms. This adult filaria obstructs the lymphatic drainage that results into swelling of extremities that are distal to the infected lymph node. These very symptoms of swollen limbs in patients with filarial disease have been termed elephantiasis. The eradication of adult worms in lymph nodes is not frequently possible, and commonly a much extended course of medical therapy is required for it to be successful [57].

20.2.4 Anthrax

New methods of curing anthrax have become a burning interest following the recent outburst of anthrax infections and deaths in the USA as a result of terrorism. In anthrax infection, endospores from Bacillus anthracis that gain access into the body are phagocytosed by macrophages and carried to regional lymph nodes where the endospores germinate inside the macrophages and become vegetative bacteria [58]. According to one literature, computed tomography of the chest was performed on eight patients infected with inhalational anthrax. Mediastinal lymphadenopathy was found in seven of the eight patients [59]. In another case report of a patient, the anthrax bacillus was shown to be rapidly sterilised within the blood stream after initiation of antibiotic therapy. However, viable anthrax bacteria were still present in postmortem mediastinal lymph node specimens [60]. Treatment and control of these diseases are hard to accomplish because of the limited access of drugs to mediastinal nodes using common pathways of drug delivery. Also, the anatomical location of mediastinal nodes represents a difficult target for external beam irradiation.

20.2.5 *Tuberculosis*

Newer methods to target antituberculosis drugs to these lymph nodes could possibly decrease the amount of time of drug therapy. TB requires lengthy treatment minimum of approximately 6 months probably because of its difficulty in delivering drugs into the tubercular lesions.

The TB infection is caused by mycobacteria that invade and grow chiefly in phagocytic cells. Lymph node TB is the most common form of extrapulmonary TB rating approximately as 38.3 %. This is frequently found to spread from the lungs to lymph nodes. In one study, total TB lymph node involvement was found as 71 % of the intrathoracic lymph nodes, 26 % of the cervical lymph nodes and 3 % of the axillary lymph nodes [61].

20.3 Approaches for Lymphatic Targeting

Targeted delivery of drugs can be achieved utilising carriers with a specified affinity to the target tissue. There are two approaches for the targeting, i.e. passive and active. In passive targeting, most of the carriers accumulate to the target site during continuous systemic circulation to deliver the drug substance, the behaviour of which depends highly upon the physicochemical characteristics of the carriers. Whereas much effort has been concentrated on active targeting, this involves delivering drugs more actively to the target site.

20.3.1 *Passive Targeting*

Passive targeting involves the transport of carriers through leaky tumour vasculature into the tumour interstitium and cells by convection or passive diffusion. Further, nanocarriers and drug then accumulate at the target site by the enhanced permeation and retention (EPR) effect [62]. The EPR effect is most prominent mainly in cancer targeting. Moreover, the EPR effect is pertinent for about all fast-growing solid tumours [63]. The EPR effect will be most positive if nanocarriers can escape immune surveillance and circulate for a long period. Very high local concentrations of drug-loaded nanocarriers can be attained at the target site, for example, about 10- to 50-fold higher than in normal tissue within 1–2 days [64].

However, there exist some limitations for passively targeting the tumour; first is the degree of tumour vascularisation and angiogenesis which is important for passive targeting of nanocarriers [65]. And, second, due to the poor lymphatic drainage in tumours, the interstitial fluid pressure increases which correlates nanocarrier size relationship with the EPR effect: larger and long-circulating nanocarriers (100 nm) are more retained in the tumour, whereas smaller molecules easily diffuse [66].

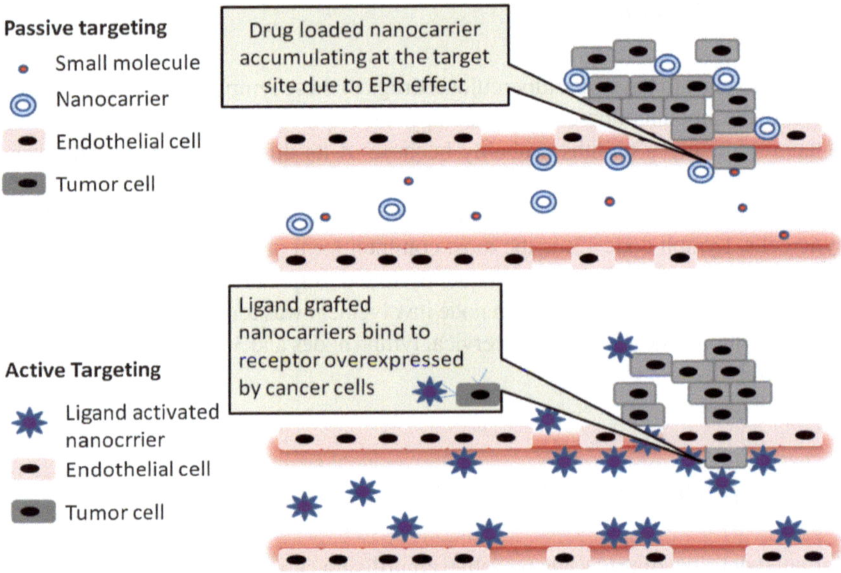

Fig. 20.1 Approaches for lymphatic targeting

20.3.2 Active Targeting

Active targeting is based upon the attachment of targeting ligands on the surface of the nanocarrier for appropriate receptor binding that are expressed at the target site. The ligand particularly binds to a receptor overexpressed in particular diseased cells or tumour vasculature and not expressed by normal cells. In addition, targeted receptors should be present uniformly on all targeted cells. Targeting ligands are either monoclonal antibodies (mAbs) and antibody fragments or non-antibody ligands (peptidic or not). These can also be termed as ligand-targeted therapeutics [67, 68]. Targeting approaches for lymphatic targeting are shown in Fig. 20.1.

20.4 Carriers for Lymphatic Targeting

Current research is focussed on two types of carriers, namely, colloidal carriers and polymeric carriers. Targeting strategies for lymphatics are shown in Fig. 20.2.

20.4.1 Colloidal Carriers

Much effort has been concentrated to achieve lymphatic targeting of drugs using colloidal carriers. The physicochemical nature of the colloid itself has been shown

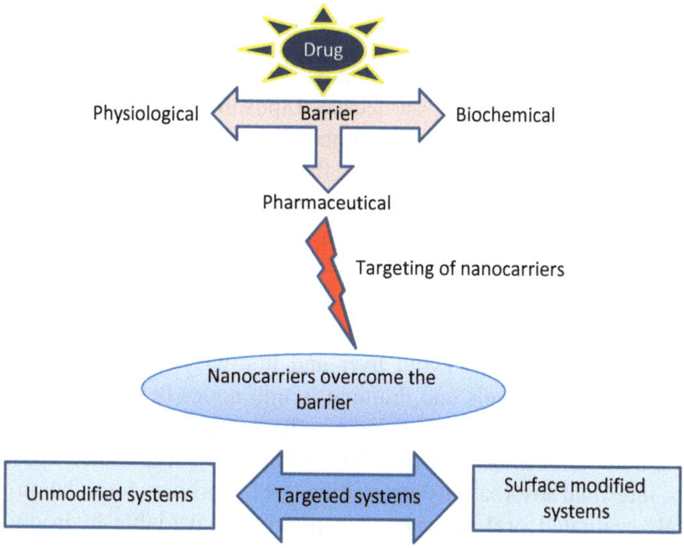

Fig. 20.2 Strategies for targeting nanocarriers to lymphatic system

to be of particular relevance, with the main considerations being size of colloid and hydrophobicity. The major purpose of lymphatic targeting is to provide an effective anticancer chemotherapy to prevent the metastasis of cancer cells by accumulating the drug in the regional lymph node.

20.4.1.1 Emulsifying Drug Delivery Systems

Emulsions are probably well-known particulate carriers with comparative histories of research and have been widely used as a carrier for lymph targeting. Hashida et al. demonstrated that injection of water-in-oil (W/O) or oil-in-water (O/W) emulsions favoured lymphatic transport of mitomycin C via the intraperitoneal and intramuscular routes and uptake into the regional lymphatics was reported in the order of O/W > W/O > aqueous solution. The nanoparticle-in-oil emulsion system, containing anti-filarial drug in gelatin nanoparticles, was studied for enhancing lymphatic targeting [69]. Pirarubicin and Lipiodol emulsion formulation was developed for treating gastric cancer and metastatic lymph nodes [70, 71]. After endoscopic injection of the pirarubicin–Lipiodol emulsion, the drug retained over 7 days at the injection site and in the regional lymph node.

Hauss et al. in their study have explored the lymphotropic potential of emulsions and self-emulsifying drug delivery systems (SEDDS). They investigated the effects of a range of lipid-based formulations on the bioavailability and lymphatic transport of ontazolast following oral administration to conscious rats and found that all the lipid formulations increased the bioavailability of ontazolast comparative to the control suspension, the SEDDS promoted more rapid absorption and maximum lymphatic transport is found with the emulsion [72, 73].

20.4.1.2 Liposomes

Lymphatic delivery of drug-encapsulated liposomal formulations has been investigated extensively in the past decade. Liposomes possess ideal features for delivering therapeutic agents to the lymph nodes which are based on their size, which prevents their direct absorption into the blood; the large amount of drugs and other therapeutic agents that liposomes can carry; and their biocompatibility. The utility of liposomes as a carrier for lymphatic delivery was first investigated by Segal et al. in 1975 [74].

Orally administered drug-incorporated liposomes enter the systemic circulation via the portal vein and intestinal lymphatics. Drugs entering the intestinal lymphatic through the intestinal lumen avoid liver and first-pass metabolism as they first migrate to lymphatic vessels and draining lymph nodes before entering systemic circulation. Lymphatic uptake of carriers via the intestinal route increases bioavailability of a number of drugs. For oral delivery of drug-encapsulated liposomal formulations, intestinal absorbability and stability are the primary formulation concerns. Ling et al. evaluated oral delivery of a poorly bioavailable hydrophilic drug, cefotaxime, in three different forms: liposomal formulation, aqueous-free drug and a physical mixture of the drug and empty liposomes [75]. The liposomal formulation of the drug turned out to exhibit a 2.7-fold increase in its oral bioavailability compared to the aqueous dosage and a 2.3-fold increase for the physical mixture. They also accounted that the liposomal formulation leads to a significant enhancement of the lymphatic localisation of the drug relative to the other two formulations. As a result, liposome systems emerged as useful carriers for poorly bioavailable hydrophilic drugs, promoting their lymphatic transport in the intestinal lymph as well as their systemic bioavailability.

Conventional liposomal formulations contain anticancer drugs incorporated in them for intravenous infusion in treating various types of cancers. Doxil, a chemotherapeutic formulation of PEGylated liposomes of doxorubicin, is widely used as first-line therapy of AIDS-related Kaposi's sarcoma, breast cancer, ovarian cancer and other solid tumours [76–80]. Liposomal delivery of anticancer drug actinomycin D via intratesticular injection has shown greater concentration of the drug in the local lymph nodes. Furthermore, a study by Hirnle et al. found liposomes as a better carrier for intralymphatically delivered drugs contrasted with bleomycin emulsions [81]. Systemic liposomal chemotherapy is preferred mainly because of its reduced side effects compared to the standard therapy and improved inhibition of the anticancer drugs from enzymatic digestion in the systemic circulation. Effective chemotherapy by pulmonary route could overcome various lacunas associated with systemic chemotherapy like serious non-targeted toxicities, poor drug penetration into the lymphatic vessels and surrounding lymph node and first-pass clearance concentrating drugs in the lungs and draining lymphatics in the case of oral delivery.

Latimer et al. developed liposomes of paclitaxel and a vitamin E analogue α-tocopheryloxy acetic acid (α-TEA) in an aerosol formulation for treating murine mammary tumours and metastases [82]. Similarly, Lawson et al. performed a comparative study for the anti-proliferative efficacy of a 9-nitro-camptothecin

(9-NC)-encapsulated dilauroylphosphatidylcholine liposomal delivery, α-TEA and a combination therapy of 9-NC and α-TEA, in a metastatic murine mammary tumour model. Liposome-encapsulated individual as well as combination treatment was delivered via an aerosol for curing metastases of lungs and of the surrounding lymph node. The animals treated with the combination therapy were found to have less proliferative cells compared to the animals treated with 9-NC alone when immunostained with Ki-67. The in vivo anticancer efficacy studies demonstrated that the combination treatment greatly hindered the tumour progression compared to each treatment alone, leading to the prolonged survival rate [83]. High levels of drugs could be targeted to lymph nodes containing TB using liposomal antituberculosis drug therapy [84].

Deep lung lymphatic drainage could also be visualised using 99mTc radioactive marker-incorporated liposomes. In addition, Botelho et al. delivered aerosolised nanoradioliposomal formulation to wild boars and observed their deep lung lymphatic network and surrounding lymph nodes [85]. Also, this technique has offered new information of the complicated structure of lymphatic network and has emerged as a new and non-invasive molecular imaging technique for the diagnosis of early dissemination of lung cancers as compared to the conventional computed tomography.

20.4.1.3 Solid Lipid Nanoparticles

Solid lipid nanoparticles (SLN) could be a good formulation strategy for incorporating drugs with poor oral bioavailability due to low solubility in GI tract or pre-systemic hepatic metabolism (first-pass effect) permitting transportation into the systemic circulation through the intestinal lymphatics. Bargoni et al. have performed various studies on absorption and distribution of SLN after duodenal administration [86–89]. In one study, ^{131}I-17-iodoheptadecanoic acid-labelled drug-free SLN were delivered into the duodenal lumen of fed rats, and transmission electron microscopy and photon correlation spectroscopy results of the lymph and blood samples verified the transmucosal transport of SLN [86].

In a later study of tobramycin-loaded SLN after duodenal administration, the improvement of drug absorption and bioavailability was ascribed mostly to the favoured transmucosal transport of SLN to the lymph compared to the blood [88]. The same group conducted a study using idarubicin-loaded SLN, administered via the duodenal route rather than intravenous route, and observed enhancement in drug bioavailability [89].

Reddy et al. prepared etoposide-loaded tripalmitin (ETPL) SLN radiolabelled with 99mTc and administered the ETPL nanoparticles subcutaneously, intraperitoneally and intravenously, to mice bearing Dalton's lymphoma tumours, and 24 h after subcutaneous administration, gamma scintigraphy and the radioactivity measurements showed that the ETPL SLN revealed a clearly higher degree of tumour uptake given via subcutaneous route (8- and 59-fold higher than that of the intraperitoneal and intravenous routes, respectively) and reduced accumulation in reticuloendothelial system organs [90].

Targeting therapies are of great potential in small cell lung cancer considering intrathoracic lymph node metastasis occurring in approximately 70 % of the limited stage patients and to nearly 80 % of the extensive stage patients [91]. Considering the case of non-small cell lung cancer, extensive rate of metastasis of lymphatics is seen in greater than 80 % of stage IV patients [92]. Videira et al. compared the biodistribution of inhaled 99mTc-D,L-hexamethylpropyleneamine oxime (HMPAO)-radiolabelled SLN with that of the free tracer administered through the same route, and gamma scintigraphic results specified that the radiolabelled SLN were primarily cleared from lungs via the lymphatics [93, 94].

20.4.1.4 Nanocapsules

Nanocapsules tend to be the most promising approach for lymphatic targeting because of their possibility of attaining distinct qualities with an easy manufacturing process. Nanocapsules coated with hydrophobic polymers could be easily captured by lymphatic cells in the body, when administered, because the hydrophobic particle is generally recognised as a foreign substance. The lymphatic targeting ability of poly(isobutylcyanoacrylate) nanocapsules encapsulating 12-(9-anthroxy) stearic acid upon intramuscular administration was evaluated and compared with three conventional colloidal carriers [69]. In vivo study in rats proved that poly(isobutylcyanoacrylate) nanocapsules retained in the right iliac regional lymph nodes in comparison with other colloidal carriers following intramuscular administration.

20.4.2 *Polymeric Carriers*

For effective targeted and sustained delivery of drugs to lymph, several polymeric particles have been designed and studied. The polymers are categorised in two types based on their origin either natural polymers like dextran, alginate, chitosan, gelatin, pullulan and hyaluronan or synthetic polymers like PLGA, PLA and PMMA.

20.4.2.1 Natural

Dextran a natural polysaccharide has been used as a carrier for a range of drug molecules due to its outstanding biocompatibility. Bhatnagar et al. synthesised cyclosporine A-loaded dextran acetate particles labelled with 99mTc. These particles gradually distributed cyclosporine A all through the lymph nodes following subcutaneous injection into the footpad of rats [95]. Dextran (average molecular weights of 10, 70 and 500 kDa)-conjugated lymphotropic delivery system of mitomycin C has been studied and it was reported that after intramuscular injection in mice, this mitomycin C–dextran conjugates retained for a longer period in regional lymph nodes for nearly 48 h while the free mitomycin was quickly cleared.

Hyaluronan, also called as hyaluronic acid, is a natural biocompatible polymer that follows lymphatic drainage from the interstitial spaces. Cai et al. demonstrated a novel intralymphatic drug delivery method synthesising a cisplatin–hyaluronic acid conjugate for breast cancer treatment. Following subcutaneous injection into the upper mammary fat pad of female rats, most of the carrier localised in the regional nodal tissue compared to the standard cisplatin formulation [96].

20.4.2.2 Synthetic

Poly(lactide-co-glycolide) as synthetic polymer that is used to prepare biodegradable nanospheres has been accounted to deliver drugs and diagnostic agents to the lymphatic system. Similarly, nanospheres coated with block copolymers of poloxamers and poloxamines with radiolabelled ^{111}In-oxine are used to trace the nanoparticles in vivo. Upon s.c. injection, the regional lymph node showed a maximum uptake of 17 % of the administered dose [97].

Dunne et al. synthesised a conjugate of block copolymer cis-diamminedichloroplatinum(II) (CDDP) and poly(ethylene oxide)-block-poly(lysine) (PEO-b-PLys) for treating lymph node metastasis. One animal treatment with 10 wt.% CDDP–polymer resulted into limited tumour growth in the draining lymph nodes and prevention of systemic metastasis [98]. Johnston and coworkers designed a biodegradable intrapleural (ipl) implant of paclitaxel consisting gelatin sponge impregnated with poly(lactide-co-glycolide) (PLGA–PTX) for targeting thoracic lymphatics. In rat model, this system exhibited lymphatic targeting capability and showed sustained drug release properties [99].

Kumanohoso et al. designed a new drug delivery system for bleomycin by loading it into a small cylinder of biodegradable polylactic acid to target lesions. This system showed significantly higher antitumour effect compared to bleomycin solution and no treatment [100]. To treat lesions, a new biodegradable colloidal particulate-based nanocarrier system was designed to target thoracic lymphatics and lymph nodes. Various nano- and microparticles of charcoal, polystyrene and poly(lactide-co-glycolide) were studied for the lymphatic distribution after intrapleural implantation in rats, and after 3 h of intrapleural injection, the lymphatic uptake was observed [101].

20.4.3 Miscellaneous Carriers

20.4.3.1 Dendrimers

Kobayashi et al. utilised dendrimer-based contrast agents for dynamic magnetic resonance lymphangiography [102]. Gadolinium (Gd)-containing dendrimers of different sizes and molecular structures (PAMAM-G8, PAMAM-G4 and DAB-G5) (PAMAM, polyamidoamine; DAB, diaminobutyl) are used as contrast agents. Size and molecular structure play a great role in distribution and pharmacokinetics of dendrimers. For example, PAMAM-G8 when injected intravenously had a comparatively long life in the circulatory system with minimum leakage out of the vessels, whereas PAMAM-G4 cleared rapidly from the systemic circulation due to rapid renal clearance but had immediate survival in lymphatic circulation. The smaller-sized DAB-G5 showed greater accumulation and retention in lymph nodes useful for lymph node imaging using MR-LG. Gadomer-17 and Gd-DTPA–dimeglumine (Magnevist) were evaluated as controls. Imaging experiments revealed that all of the reagents are able to visualise the deep lymphatic system except Gd-DTPA–dimeglumine. To visualise the lymphatic vessels and lymph nodes, PAMAM-G8 and DAB-G5 were used, respectively. While PAMAM-G4 provided good contrast of both the nodes and connecting vessels, Gadomer-17 was able to visualise lymph nodes, but not as clear as Gd-based dendrimers. Kobayashi also delivered various Gd-PAMAM (PAMAM-G2, PAMAM-G4, PAMAM-G6, PAMAM-G8) and DAB-G5 dendrimers to the sentinel lymph nodes and evaluated its visualisation with other nodes. The G6 dendrimer provided excellent opacification of sentinel lymph nodes and was able to be absorbed and retained in the lymphatic system [103].

Using a combination of MRI and fluorescence with PAMAM-G6-Gd-Cy, the sentinel nodes were more clearly observed signifying the potential of the dendrimers as platform for dual imaging. Kobayashi et al. further overcame the sensitivity limitation and depth limitations of each individual method by the simultaneous use of two modalities (radionuclide and optical imaging). Making use of PAMAM-G6 dendrimers conjugated with near-infrared (NIR) dyes and an ^{111}In radionuclide probe, multimodal nanoprobes were developed for radionuclide and multicolour optical lymphatic imaging [104, 105].

Later Kobayashi also proposed the use of quantum dots for labelling cancer cells and dendrimer-based optical agents for visualising lymphatic drainage and identifying sentinel lymph nodes [106]. Polylysine dendrimers have been best used for targeting the lymphatic system and lymph nodes.

20.4.3.2 Carbon Nanotubes

Carbon nanotubes (CNT) possess various mechanochemical properties like high surface area, mechanical strength and thermal and chemical stability which cause them to be versatile carriers for drugs, proteins, radiologicals and peptides to target tumour tissues. Hydrophilic multiwalled carbon nanotubes (MWNTs) coated with magnetic nanoparticles (MN-MWNTs) have emerged as an effective delivery system for lymphatic targeting following subcutaneous injection of these particles into the left footpad of Sprague Dawley rats; the left popliteal lymph nodes were dyed black. MN-MWNTs were favourably absorbed by lymphatic vessels following their transfer into lymph nodes and no uptake was seen in chief internal organs such as the liver, spleen, kidney, heart and lungs. Gemcitabine loaded in these particles was evaluated for its lymphatic delivery efficiency and MN-MWNTs–gemcitabine displayed the maximum concentration of gemcitabine in the lymph nodes [107]. McDevitt et al. synthesised tumour-targeting water-soluble CNT constructs by covalent attachment of monoclonal antibodies like rituximab and lintuzumab using 1,4,7,10-tetraazacyclododecane-1,4,7,10-tetraacetic acid (DOTA) as a metal ion chelator while the fluorescent probe was fluorescein. CNT-([^{111}In] DOTA) (rituximab) explicitly targeted a disseminated human lymphoma in vivo trials compared to the controls CNT-([^{111}In] DOTA) (lintuzumab) and [^{111}In]rituximab [108].

Tsuchida and coworkers evaluated the drug delivery efficiency of water-dispersed carbon nanohorns in a non-small cell lung cancer model. Polyethylene glycol (PEG)–doxorubicin conjugate bound oxidised single-wall carbon nanohorns (oxSWNHs) injected intratumourally into mice bearing human non-small cell lung cancer (NCI-H460) caused a significant retardation of tumour growth. Histological analyses showed (probably by means of interstitial lymphatic fluid transport), migration of oxSWNHs to the axillary lymph node occurred which is a major site of breast cancer metastasis near the tumour [109]. Shimada et al. described a silica particle-based lymphatic drug delivery system of bleomycin and compared its therapeutic efficacy to that of free bleomycin solution in a transplanted tumour model in animals. Silica particle-adsorbed bleomycin showed considerable inhibitory effect on tumour growth and lymph node metastasis compared to free bleomycin solution [110]. Activated carbon particles of aclarubicin are used for adsorption and sustained release into lymph nodes. Upon subcutaneous administration into the fore foot-pads of rats these particles showed significantly elevated distribution of aclarubicin to the auxiliary lymph nodes compared to aqueous solution of the drug [111]. Activated carbon particles of aclacinomycin A, adriamycin, mitomycin C and pepleomycin have also been used by another group for adsorption. Higher level of drug concentration was maintained in the new dosage form than in the solution form [112].

20.4.3.3 Antibody–Drug Conjugates

Antibody–drug conjugates enhance the cytotoxic activity of anticancer drugs by conjugating them with antibodies. Antibodies conjugated with cytostatic drugs such as calicheamicin have been used for the treatment of various lymphomas, including non-Hodgkin B-cell lymphoma (NHL), follicular lymphoma (FL) and diffuse large B-cell lymphoma (DLBCL) [113–116].

CD20 B-cell marker is expressed on the surface membrane of pre-B-lymphocytes and mature B-lymphocytes. The anti-CD20 mAb rituximab (Rituxan) is now the most potential antibody for the treatment of non-Hodgkin B-cell lymphomas (B-NHL) [117]. Rituximab-conjugated calicheamicin elevated the antitumour activity of rituximab against human B-cell lymphoma (BCL) xenografts in preclinical models [118].

CD22 is a B-lymphoid lineage-specific differentiation antigen expressed on the surface of both normal and malignant B-cells. Hence, the CD22-specific antibody could be effective in delivering chemotherapeutic drugs to malignant B-cells. Also, CD22 (Siglec-2) antibodies targeting to CD22 are suited for a Trojan horse strategy.

Thus, antibody-conjugated therapeutic agents bind to the Siglec and are carried efficiently into the cell [119]. A lot of interest has been seen in clinical progress of the conjugated anti-CD22 antibodies, especially inotuzumab ozogamicin (CMC-544) [120].

CD30 is expressed in the malignant Hodgkin and Reed–Sternberg cells of classical Hodgkin lymphoma (HL) and anaplastic large-cell lymphoma. Younes and Bartlett reported an ongoing phase I dose-escalation trial in relapsed and refractory HL patients with Seattle Genetics (SGN-35), a novel anti-CD30-antibody–monomethylauristatin E conjugate. SGN-35 was stable in the blood and released the conjugate only upon internalisation into CD30-expressing tumour cells [121]. Huang et al. constructed (anti-HER2/*neu*–IgG3–IFNα), another antibody–drug conjugate, and examined its effect on a murine B-cell lymphoma, 38C13, expressing human HER2/*neu*, and this significantly inhibited 38C13/HER2 tumour growth in vivo [122].

20.4.3.4 Hybrid Nanosystems

Hybrid systems use combination of two or more delivery forms for effective targeting. Khatri et al. prepared and investigated the in vivo efficacy of plasmid DNA-loaded chitosan nanoparticles for nasal mucosal immunisation against hepatitis B. Chitosan–DNA nanoparticles prepared by the coacervation process adhered to the nasal or gastrointestinal epithelia and are easily transported to the nasal-associated lymphoid tissue (NALT) and Peyer's patches of the gut-associated lymphoid tissue (GALT) both as IgA inductive site [123], in which chitosan–DNA might be taken in by M cell, and transported across the mucosal boundary and thereby transfect immune cells within NALT or GALT [124].

A work demonstrates targeting of three peptides containing sequences that bind to cell markers expressed in the tumour vasculature (p24-NRP-1 and p39-Flt-1) [125, 126] and tumour lymphatics (p47-LyP-1) [127] and were tested for their ability to target 3(nitrilotriacetic acid)-ditetradecylamine (NTA3-DTDA) containing liposomes to subcutaneous B16-F1 tumours. Significantly, a potential antitumour effect was seen after administration of doxorubicin-loaded PEG750 liposomes engrafted with p24-NRP-1.

Hybrid liposomes composed of L-α-dimyristoylphosphatidylcholine and polyoxyethylene (25) dodecyl ether prepared by sonication showed remarkable reduction of tumour volume in model mice of acute lymphatic leukaemia (ALL) treated intravenously with HL-25 without drugs after the subcutaneous inoculation of human ALL (MOLT-4) cells was verified in vivo. Prolonged survival (>400 %) was noted in model mice of ALL after the treatment with HL-25 without drugs [128].

In a report, LyP-1 peptide-conjugated PEGylated liposomes loaded with fluorescein or doxorubicin were prepared for targeting and treating lymphatic metastatic tumours. The in vitro cellular uptake and in vivo near-infrared fluorescence imaging results confirmed that LyP-1-modified liposome increased uptake by tumour cells and metastatic lymph nodes.

In another study, in vitro cellular uptake of PEG–PLGA nanoparticle (LyP-1-NPs) was about four times that of PEG–PLGA nanoparticles without LyP-1 (NPs).

In vivo study, about eight times lymph node uptake of LyP-1-NPs was seen in metastasis than that of NPs, indicated LyP-1-NP as a promising carrier for target-specific drug delivery to lymphatic metastatic tumours [129].

20.4.3.5 Biotherapeutics

Currently, surgery, radiation therapy and chemotherapy are the principal methods for cancer treatment. Gene therapies may act synergistically or additively with them. For example, another case demonstrated that replacement of the p53 (protein 53) gene in p53-deficient cancer cell lines enhanced the sensitivity of these cells to Ad-p53 (adenovirus-expressed protein 53) and cisplatin (CDDP) and resulted into greater tumour cell death [130]. Later, Son and Huang [131] stated that treatment of CDDP-resistant tumour cells with CDDP increased the sensitivity of these cells to transduction by DNA-carrying liposomes. Also, Chen et al. [132] described that to improve tumour killing, herpes simplex virus thymidine kinase (HSV-TK) and interleukin (IL) expression can be combined. On the whole, greater therapeutic effect can be achieved by effectively combining conventional cancer treatments and gene therapy together.

20.5 Physicochemical Aspects of Lymphatic Targeting

Mainly colloidal carriers have emerged as potential targeting agents to lymphatic system. Physicochemical properties affect the efficiency of colloid uptake into the lymphatic system [28]. These properties include size, number of particles, surface charge, molecular weight and colloid lipophilicity. Physicochemical properties are altered by adsorption of group of hydrophilic polymers like poloxamers and poloxamines to the particle surface. These properties modified the biodistribution of particles in vivo, particularly the avoidance of the reticuloendothelial system (RES) upon intravenous administration [133, 134]. In one study, it was opined that opsonisation may cause alteration of the particle surface in vivo [135].

20.5.1 Size of the Carrier

Size could be important factor in defining the behaviour of particulates after subcutaneous injection. Small particles with diameter less than a few nanometres generally exchanged through the blood capillaries, whereas larger particles of diameters up to a few tens of nanometres absorbed into the lymph capillaries. But particles over a size of few hundred nanometres remain trapped in the interstitial space for a long time [136]. Christy et al. have shown a relationship between colloid size and ease of injection site drainage using model polystyrene nanospheres after subcutaneous administration to the rat [137]. Results showed distribution of polystyrene nanospheres in the size range 30–260 nm 24 h after administration and 74–99 % of the recovered dose retained at the administration site, and as particle diameter increased, drainage became slower. It has been proposed earlier that the optimum colloid size range for lymphoscintigraphic agents is 10–50 nm [138].

Size has less importance when colloids are administered intraperitoneally (i.p.) within the nanometre size range, as drainage is only from a cavity into the initial lymphatics; hence, no diffusion is required through the interstitial space [28]. The size limit of the open junctions of the initial lymphatic wall is the only barrier to uptake from the peritoneal cavity into the lymphatics [139].

20.5.2 Concentration and Volume

More number of particles at the injection site decreases their rate of drainage, owing to increased obstruction of their diffusion through the interstitial space [139, 140]. Scientists at Nottingham University investigated this effect using polystyrene nanospheres of 60 nm. Following administration to the rat, the concentration range of nanospheres was approximately 0.05–3.0 mg/ml. Lower lymphatic uptake was seen on increasing the concentration of nanospheres in the injection volume due to slower drainage from the injection site. Injecting oily vehicles intramuscularly to the rat, the effect of injection volume has been studied. Increasing volume of sesame oil accelerated oil transport into the lymphatic system. Upon s.c. administration, volumes of aqueous polystyrene particle suspensions have been investigated in the range 50–150 μl [39].

20.5.3 Surface Charge

Surface charge studies have been done utilising liposome as colloidal carrier. The surface charge of liposomes affected their lymphatic uptake from s.c. and i.p. injection sites. Negatively charged liposomes showed faster drainage than that for positive liposomes after i.p. administration [141]. Patel et al. also indicated that liposome localisation in the lymph nodes followed a particular order negative > positive > neutral [142].

20.5.4 Molecular Weight

Macromolecule having high molecular weight has a decreased ability for exchange across blood capillaries and lymphatic drainage becomes the route of drainage from the injection site which shows a linear relationship between the molecular weight of macromolecules and the proportion of the dose absorbed by the lymphatics. For a compound to be absorbed by the lymphatics, the molecular weight should range between 1,000 and 16,000 [141, 143]. The effect of molecular weight becomes negligible when targeting carriers to the lymphatic system as the molecular weight of a colloidal carrier is generally less than 1,000 Da.

20.5.5 Lipophilicity

The most important determinant of the phagocytic response and so lymphatic uptake is the lipophilicity of a colloid [144]. Opsonins generally unite with lipophilic rather than hydrophilic surfaces; hence, the hydrophilic particles show reduced phagocytosis [145]. Hydrophobic polystyrene nanospheres adsorbed with hydrophilic block copolymers showed drastic reduction in phagocytosis prior to i.v. administration [146].

In the case of polystyrene nanospheres of 60-nm diameter, PEO chains of the poloxamers and poloxamines adsorbed onto the surface of the particle described the relationship between interstitial injection site drainage and lymph node uptake in rat [144]. Uncoated nanospheres of this diameter showed reduced drainage from the injection site with 70 % of the administered dose remaining after 24 h. The adsorption of block copolymers can enhance the drainage from the injection site such that levels remaining at the injection site may be as little as 16 % after 24 h, with very hydrophilic polymers such as poloxamine 908. Uptake of nanospheres into the regional lymph nodes may also be improved by the adsorption of block copolymers with intermediate lengths of polyoxyethylene, such as poloxamine 904. This polymer may sequester up to 40 % of the given dose by the lymph nodes after 24 h [147].

20.6 Effect of Surface Modification on Carriers

Surface modification could prove as an effective strategy for potential targeting to lymphatic system. The influence can be quoted in following ways.

20.6.1 Surface Modification with Polyethylene Glycols

Coating of a carrier with hydrophilic and sterically stabilised PEG layer can successfully enhance lymphatic absorption, reducing specific interaction of particle with the interstitial surrounding, and inhibit the formation of too large particle structure [49]. Surface modification of liposomes with PEG also does not have a significant effect on lymph node uptake. Small liposomes coated with PEG showed greatest clearance from the s.c. injection site with small 86-nm PEG-coated liposomes having <40 % remaining at the injection site at 24 h. Larger neutral and negatively charged liposomes had a clearance >60 % remaining at the initial s.c. injection site. However, this smaller amount of large liposomes that were cleared from the injection site was compensated by better retention in the lymph node [148]. Oussoren et al. reported that the amount of liposomes cleared from the injection site was somewhat greater with the PEG-coated liposomes [149]. This improved clearance did not result in improved lymph node retention because the fraction of PEG liposomes retained by the lymph node is decreased. Phillips et al. also studied the slightly improved clearance of PEG-coated liposomes from the s.c. injection site [148].

Porter and coworkers demonstrated that PEGylation of poly-L-lysine dendrimers resulted into better absorption from s.c. injection sites and stated that the extent of lymphatic transport may be improved by increasing the size of the PEGylated dendrimer complex. They estimated the lymphatic uptake and lymph node retention properties of several generation four dendrimers coated with PEG or 4-benzene sulphonate after subcutaneous administration in rats. For this surface modification study, three types of PEGs with molecular weights of 200, 570 or 2,000 Da were taken. PEG200-derived dendrimers showed rapid and complete absorption into the blood when injected subcutaneously, and only 3 % of the total given dose was found in the pooled thoracic lymph over 30 h, whereas PEG570- and PEG2000-derived dendrimers showed lesser absorption, and a higher amount was recovered in lymphatics (29 %) over 30 h. However, the benzene sulphonate-capped dendrimer was not well absorbed either in the blood or in lymph following subcutaneous injection [150].

20.6.2 Surface Modification with Ligands

Carriers capped with nonspecific human antibodies as ligands showed greater lymphatic uptake and lymph node retention compared to uncoated one at the s.c. site. Liposomes coated with the antibody, IgG, have been shown to increase lymph node localisation of liposomes to 4.5 % of the injected dose at 1 h, but this level decreased to 3 % by 24 h [151]. In a study, the liposomes containing positively charged lipids had approximately 2–3 times the lymph node localisation (up to 3.6 % of the injected dose) than liposomes containing neutral or negatively charged lipids (1.2 % of the injected dose) [149]. Attachment of mannose to the surface of a liposome increased lymph node uptake by threefold compared to control liposomes [152].

Another study demonstrated HBsAg entrapped dried liposomes with their surfaces modified with galactose. Pharmacokinetic study in rats showed that galactosylated liposomes delivered higher amounts of HBsAg to the regional lymph nodes than other ungalactosylated formulations [153].

Lectin is another ligand that can be attached to the carriers for improved targeting to intestinal lymphatics. Bovine serum albumin containing acid phosphatase model protein and polystyrene microspheres conjugated with mouse M-cell-specific Ulex europaeus lectin. Ex vivo results showed that there was favoured binding of the lectin-conjugated microspheres to the follicle-associated epithelium. Final results indicated that coupling of ligands such as lectin specific to cells of the follicle-associated epithelium can improve the targeting of encapsulated candidate antigens for delivery to the Peyer's patches of the intestine for better oral delivery [154].

20.6.3 Surface Modification with Biotin

To improve carrier retention in lymph nodes, a new method of increasing lymphatic uptake of subcutaneously injected liposome utilises the high-affinity ligands biotin and avidin. Biotin is a naturally occurring cofactor and avidin is a protein derived from eggs. Avidin and biotin are having extremely high affinity for each other. For instance, upon injection, the avidin and the biotin liposomes move into the lymphatic vessels. Biotin liposomes that migrate through the lymphatic vessels meet the avidin resulting in an aggregate that becomes trapped in the lymph nodes [155, 156]. The biotin liposome/avidin system has promising potential as therapeutic agent for delivery to lymph nodes. It can be applied not only to s.c. targeting of lymph nodes but also to intracavitary lymph node targeting [50].

Different ligands with their application in lymphatic targeting are represented in Table 20.1.

20.7 Future Trends and Conclusion

The lymphatics have the potential to play a major role in anticancer treatment as lymphatic spread is recognised to precede haematological spread in many cancers including melanoma, breast, colon, lung and prostate cancers. Currently, the focus is on the development of drug carriers that can localise chemotherapy to the lymphatic system, thus improving the treatment of localised disease while minimising the exposure of healthy organs to cytotoxic drugs. The delivery of novel carriers to lymph nodes for therapeutic purposes has much promise. Giving importance to the lymphatic route in metastasis, this delivery system may have great potential for targeted delivery of various therapeutic agents to tumours and their metastatic lymph nodes. Various delivery systems have been discussed here but colloidal carriers, especially, liposomes have been the carrier of choice to date. The purpose of this review is to provide an improved and effective lymphotropic system with a satisfactory quality for clinical use and to establish a preparation method applicable for industrial production. Surface-engineered lymphotropic systems may prove as an effective carrier for anti-HIV, anticancer and oral vaccine delivery in near future.

Table 20.1 Different ligands/chemicals used for lymphatic targeting

S. No.	Ligands/chemicals	Delivery system	Target site	Application	References
1.	Folate	Folate-PEG-CKK$_2$-DTPA carrier	Lymphatic metastasised tumour	Lymphatic metastasised tumour imaging diagnosis and targeting therapy	[157]
2.	Lectin	Nanoparticles	Metastatic spread and growth of tumour cell	Anticancer drugs	[158]
3.	Lectin	Microspheres	Delivery of antigens to gut-associated lymphoid tissue (GALT)	Intestinal delivery	[154]
4.	L-selectin	Microparticles	Active targeting of peripheral lymph nodes	Doppler ultrasonography contrast agent	[159]
5.	PEG	Dendrimers, liposome	Lymph	Vaccine delivery	[148–150]
6.	Hyaluronan	–	Lymphatic system	–	[160]
7.	Mannose	Liposome	Spleen, lymph nodes	Antiviral and anticancer drug delivery	[161]
8.	Galactose	Liposome	Lymph node	Hepatitis B	[153]
9.	Alginate/chitosan	Microparticles	Payer's patch	Tamoxifen anticancer drug	[162]
10.	Negatively charged albumins	Microparticles	Lymph nodes and lymphatic system	Antiviral drug	[163]
11.	*N*-isopropylacrylamide/(MAA)	Nanoparticles	Thoracic lymph nodes	Chemotherapeutic agent	[101]
12.	Poly(lactide-co-glycolide)	Microparticles	Thoracic lymph nodes	Chemotherapeutic agent	[101]
13.	Block copolymer of poloxamine and poloxamer	Nanospheres	Regional lymph nodes	–	[144]
14.	Lyp-1	Nanoparticles, liposomes	Targeted to lymphatic vessels and also in tumour cells within hypoxic area	Antitumour	[129]
15.	Avidin–biotin	Liposomes	Targeting to lymph node	Mediastinal lymph node targeting	[155]
16.	IgG antibody	Liposome	Targeting to lymph node	Increased lymph node retention	[145, 151]

References

1. Swartz MA (2001) The physiology of the lymphatic system. Adv Drug Deliv Rev 50:3–20
2. Huntington GS, McClure CFW (2005) The anatomy and development of the jugular lymph sacs in the domestic cat (Felis domestica). Am J Anat 10:177–312
3. Fr S (1902) On the origin of the lymphatic system from the veins and the development of the lymph hearts and thoracic duct in the pig. Am J Anat 1:367–389
4. Wilting J, Aref Y, Huang R, Tomarev SI, Schweigerer L, Christ B et al (2006) Dual origin of avian lymphatics. Dev Biol 292:165–173
5. Srinivasan RS, Dillard ME, Lagutin OV, Lin FJ, Tsai S, Tsai MJ et al (2007) Lineage tracing demonstrates the venous origin of the mammalian lymphatic vasculature. Genes Dev 21:2422–2432
6. Wigle JT, Harvey N, Detmar M, Lagutina I, Grosveld G, Gunn MD et al (2002) An essential role for Prox1 in the induction of the lymphatic endothelial cell phenotype. EMBO J 21:1505–1513
7. Wigle JT, Oliver G (1999) Prox1 function is required for the development of the murine lymphatic system. Cell 98:769–778
8. Yaniv K, Isogai S, Castranova D, Dye L, Hitomi J, Weinstein BM (2006) Live imaging of lymphatic development in the zebrafish. Nat Med 12:711–716
9. Maby-El Hajjami H, Petrova TV (2008) Developmental and pathological lymphangiogenesis: from models to human disease. Histochem Cell Biol 130:1063–1078
10. Rinderknecht M, Detmar M (2008) Tumor lymphangiogenesis and melanoma metastasis. J Cell Physiol 216:347–354
11. Heller LJ, Mohrman DE (1981) Cardiovascular physiology. McGraw-Hill
12. Cueni LN, Detmar M (2006) New insights into the molecular control of the lymphatic vascular system and its role in disease. J Invest Dermatol 126:2167–2177
13. Givoli D, Doukhovni I, Moghimi S, Rajabi-Siahboomi A (1996) Advanced colloid-based systems for efficient delivery of drugs and diagnostic agents to the lymphatic tissues. Prog Biophys Mol Biol 65:221–249
14. Leak L (1976) The structure of lymphatic capillaries in lymph formation. In: Federation Proceedings. p 1863
15. Gerli R, Solito R, Weber E, Agliano M (2000) Specific adhesion molecules bind anchoring filaments and endothelial cells in human skin initial lymphatics. Lymphology 33:148
16. Weber E, Rossi A, Solito R, Sacchi G, Agliano M, Gerli R (2002) Focal adhesion molecules expression and fibrillin deposition by lymphatic and blood vessel endothelial cells in culture. Microvasc Res 64:47–55
17. Schmid-Schönbein GW (2003) The second valve system in lymphatics. Lymphat Res Biol 1:25–31
18. Trzewik J, Mallipattu S, Artmann GM, Delano F, Schmid-Schönbein GW (2001) Evidence for a second valve system in lymphatics: endothelial microvalves. FASEB J 15:1711–1717
19. Leak L, Burke J (1968) Ultrastructural studies on the lymphatic anchoring filaments. J Cell Biol 36:129–149
20. Sharma R, Wendt JA, Rasmussen JC, Adams KE, Marshall MV, Sevick-Muraca EM (2008) New horizons for imaging lymphatic function. Ann NY Acad Sci 1131:13–36
21. von der Weid PY, Zawieja DC (2004) Lymphatic smooth muscle: the motor unit of lymph drainage. Int J Biochem Cell Biol 36:1147–1153
22. Casley-Smith J (1980) The fine structure and functioning of tissue channels and lymphatics. Lymphology 13:177
23. Moore KL, Dalley AF (1999) Clinically oriented anatomy. Lippincott Williams & Wilkins, Baltimore
24. Alitalo K, Tammela T, Petrova TV (2005) Lymphangiogenesis in development and human disease. Nature 438:946–953

25. Oliver G, Detmar M (2002) The rediscovery of the lymphatic system: old and new insights into the development and biological function of the lymphatic vasculature. Genes Dev 16:773–783
26. Mohrman ED, Heller LJ (2002) Cardiovascular Physiology. McGraw-Hill Companies Inc, USA
27. Morton DL, Eilber FR, Joseph WL, Wood WC, Trahan E, Ketcham AS (1970) Immunological factors in human sarcomas and melanomas: a rational basis for immunotherapy. Ann Surg 172:740
28. O'Hagan DT, Christy NM, Davis SS (1992) Particulates and lymphatic drug delivery. In: Charman WN, Stella VJ (eds) Lymphatic transport of drugs. CRC, Boca Raton, FL, pp 279–315
29. Charman WN, Stella VJ (1992) Lymphatic transport of drugs. CRC, Boca Raton, FL
30. Charman WN, Porter CJH (1996) Lipophilic prodrugs designed for intestinal lymphatic transport. Adv Drug Deliv Rev 19:149–169
31. Porter CJH, Charman WN (2001) Intestinal lymphatic drug transport: an update. Adv Drug Deliv Rev 50:61–80
32. Wasan KM (2002) The role of lymphatic transport in enhancing oral protein and peptide drug delivery. Drug Dev Indus Pharm 28:1047–1058
33. Humberstone AJ, Porter CJH, Edwards GA, Charman WN (1998) Association of halofantrine with postprandially derived plasma lipoproteins decreases its clearance relative to administration in the fasted state. J Pharm Sci 87:936–942
34. Shah AK, Sawchuk RJ (2006) Effect of co-administration of intralipidTM on the pharmacokinetics of cyclosporine in the rabbit. Biopharm Drug Dispos 12:457–466
35. Wasan KM, Cassidy SM (1998) Role of plasma lipoproteins in modifying the biological activity of hydrophobic drugs. J Pharm Sci 87:411–424
36. Arya M, Bott SR, Shergill IS, Ahmed HU, Williamson M, Patel HR (2006) The metastatic cascade in prostate cancer. Surg Oncol 15:117–128
37. Cense H, Van Eijck C, Tilanus H (2006) New insights in the lymphatic spread of oesophageal cancer and its implications for the extent of surgical resection. Best Pract Res Clin Gastroenterol 20:893
38. Garzon-Aburbeh A, Poupaert JH, Claesen M, Dumont P, Atassi G (1983) 1, 3-dipalmitoylglycerol ester of chlorambucil as a lymphotropic, orally administrable antineoplastic agent. J Med Chem 26:1200–1203
39. Muranishi S (1980) Lymphatic delivery of drugs and its application to cancer chemotherapy (author's transl). Yakugaku zasshi: J Pharm Soc Jpn 100:687
40. Pantaleo G, Graziosi C, Demarest JF, Cohen OJ, Vaccarezza M, Gantt K et al (1994) Role of lymphoid organs in the pathogenesis of human immunodeficiency virus (HIV) infection. Immunol Rev 140:105–130
41. Pantaleo G, Graziosi C, Fauci AS (1993) The role of lymphoid organs in the immunopathogenesis of HIV infection. AIDS 7:S19
42. Umeda M, Marusawa H, Seno H, Katsurada A, Nabeshima M, Egawa H et al (2005) Hepatitis B virus infection in lymphatic tissues in inactive hepatitis B carriers. J Hepatol 42:806–812
43. Von Messling V, Svitek N, Cattaneo R (2006) Receptor (SLAM [CD150]) recognition and the V protein sustain swift lymphocyte-based invasion of mucosal tissue and lymphatic organs by a morbillivirus. J Virol 80:6084–6092
44. Lan N, Yamaguchi R, Inomata A, Furuya Y, Uchida K, Sugano S et al (2006) Comparative analyses of canine distemper viral isolates from clinical cases of canine distemper in vaccinated dogs. Vet Microbiol 115:32–42
45. Spiegel M, Schneider K, Weber F, Weidmann M, Hufert FT (2006) Interaction of severe acute respiratory syndrome-associated coronavirus with dendritic cells. J Gen Virol 87:1953–1960
46. Kessel A, Toubi E (2007) Chronic HCV-related autoimmunity: a consequence of viral persistence and lymphotropism. Curr Med Chem 14:547–554
47. Düzgünes N, Simoes S, Slepushkin V, Pretzer E, Rossi J, De Clercq E et al (2001) Enhanced inhibition of HIV-1 replication in macrophages by antisense oligonucleotides, ribozymes and

acyclic nucleoside phosphonate analogs delivered in pH-sensitive liposomes. Nucleosides Nucleotides Nucleic Acids 20:515–523
48. Hirnle P (1997) Liposomes for drug targeting in the lymphatic system. Hybridoma 16:127–132
49. Oussoren C, Storm G (2001) Liposomes to target the lymphatics by subcutaneous administration. Adv Drug Deliv Rev 50:143–156
50. Phillips WT, Klipper R, Goins B (2000) Novel method of greatly enhanced delivery of liposomes to lymph nodes. J Pharmacol Exp Ther 295:309–313
51. Hanahan D, Weinberg RA (2000) The hallmarks of cancer. Cell 100:57–70
52. Torabi M, Aquino SL, Harisinghani MG (2004) Current concepts in lymph node imaging. J Nucl Med 45:1509–1518
53. Busby JE, Evans CP (2004) Old friends, new ways: revisiting extended lymphadenectomy and neoadjuvant chemotherapy to improve outcomes. Curr Opin Urol 14:251–257
54. Gagné JF, Désormeaux A, Perron S, Tremblay MJ, Bergeron MG (2002) Targeted delivery of indinavir to HIV-1 primary reservoirs with immunoliposomes. Biochim Biophys Acta (BBA)-Biomembranes 1558:198–210
55. Cohen OJ, Pantaleo G, Lam GK, Fauci AS (1997) Studies on lymphoid tissue from HIV-infected individuals: implications for the design of therapeutic strategies. In: Springer seminars in immunopathology. Springer, pp 305–322
56. Désormeaux A, Bergeron MG (2005) Lymphoid tissue targeting of anti-HIV drugs using liposomes. Methods Enzymol 391:330–351
57. El Setouhy M, Ramzy RMR, Ahmed ES, Kandil AM, Hussain O, Farid HA et al (2004) A randomized clinical trial comparing single-and multi-dose combination therapy with diethylcarbamazine and albendazole for treatment of bancroftian filariasis. Am J Trop Med Hyg 70:191–196
58. Dixon D, Terry C, Meselson M, Guillemin J, Hanna PC, Clínicas M (1999) Infección bacteriana por ántrax. N Engl J Med 341:815–826
59. Jernigan JA, Stephens DS, Ashford DA, Omenaca C, Topiel MS, Galbraith M et al (2001) Bioterrorism-related inhalational anthrax: the first 10 cases reported in the United States. Emerg Infect Dis 7:933
60. Barakat LA, Quentzel HL, Jernigan JA, Kirschke DL, Griffith K, Spear SM et al (2002) Fatal inhalational anthrax in a 94-year-old Connecticut woman. JAMA 287:863–868
61. Ilgazli A, Boyaci H, Basyigit İ, Yildiz F (2004) Extrapulmonary tuberculosis: clinical and epidemiologic spectrum of 636 cases. Arch Med Res 35:435–441
62. Haley B, Frenkel E (2008) Nanoparticles for drug delivery in cancer treatment. In: Urologic oncology: seminars and original investigations. Elsevier, pp 57–64
63. Maeda H, Bharate G, Daruwalla J (2009) Polymeric drugs for efficient tumor-targeted drug delivery based on EPR-effect. Eur J Pharm Biopharm 71:409
64. Iyer AK, Khaled G, Fang J, Maeda H (2006) Exploiting the enhanced permeability and retention effect for tumor targeting. Drug Discov Today 11:812–818
65. Bae YH (2009) Drug targeting and tumor heterogeneity. J Control Release 133:2
66. Pirollo KF, Chang EH (2008) Does a targeting ligand influence nanoparticle tumor localization or uptake? Trends Biotechnol 26:552–558
67. Adams GP, Schier R, McCall AM, Simmons HH, Horak EM, Alpaugh RK et al (2001) High affinity restricts the localization and tumor penetration of single-chain fv antibody molecules. Cancer Res 61:4750–4755
68. Gosk S, Moos T, Gottstein C, Bendas G (2008) VCAM-1 directed immunoliposomes selectively target tumor vasculature in vivo. Biochim Biophys Acta (BBA)-Biomembranes 1778:854–863
69. Nishioka Y, Yoshino H (2001) Lymphatic targeting with nanoparticulate system. Adv Drug Deliv Rev 47:55–64
70. Karajgi J, Vyas S (1994) A lymphotropic colloidal carrier system for diethylcarbamazine: preparation and performance evaluation. J Microencapsul 11:539–545

71. Yoshimura K, Nunomura M, Takiguchi N, Oda K, Suzuki H, Furukawa R et al (1996) Evaluation of endoscopic pirarubicin-Lipiodol emulsion injection therapy for gastric cancer. Jpn J Canc Chemother 23:1519–1522
72. Hauss DJ, Mehta SC, Radebaugh GW (1994) Targeted lymphatic transport and modified systemic distribution of CI-976, a lipophilic lipid-regulator drug, via a formulation approach. Int J Pharm 108:85–93
73. Kommuru T, Gurley B, Khan M, Reddy I (2001) Self-emulsifying drug delivery systems (SEDDS) of coenzyme Q 10: formulation development and bioavailability assessment. Int J Pharm 212:233–246
74. Segal A, Gregoriadis G, Black C (1975) Liposomes as vehicles for the local release of drugs. Clin Sci Mol Med 49:99
75. Sheue Nee Ling S, Magosso E, Abdul Karim Khan N, Hay Yuen K, Anne Barker S (2006) Enhanced oral bioavailability and intestinal lymphatic transport of a hydrophilic drug using liposomes. Drug Dev Ind Pharm 32:335–345
76. Frenkel V, Etherington A, Greene M, Quijano J, Xie J, Hunter F et al (2006) Delivery of liposomal doxorubicin (Doxil) in a breast cancer tumor model: investigation of potential enhancement by pulsed-high intensity focused ultrasound exposure. Acad Radiol 13:469–479
77. O'brien M, Wigler N, Inbar M, Rosso R, Grischke E, Santoro A et al (2004) Reduced cardiotoxicity and comparable efficacy in a phase III trial of pegylated liposomal doxorubicin HCl (CAELYX™/Doxil®) versus conventional doxorubicin for first-line treatment of metastatic breast cancer. Ann Oncol 15:440–449
78. Prescott L (1995) Doxil offers hope to KS sufferers. J Int Assoc Physicians AIDS Care 1:43
79. Soundararajan A, Bao A, Phillips WT, Perez R, Goins BA (2009) Liposomal doxorubicin (Doxil): in vitro stability, pharmacokinetics, imaging and biodistribution in a head and neck squamous cell carcinoma xenograft model. Nucl Med Biol 36:515–524
80. Tejada-Berges T, Granai C, Gordinier M, Gajewski W (2002) Caelyx/Doxil for the treatment of metastatic ovarian and breast cancer. Expert Rev Anticancer Ther 2:143–150
81. Hirnle P, Harzmann R, Wright J (1988) Patent blue V encapsulation in liposomes: potential applicability to endolympatic therapy and preoperative chromolymphography. Lymphology 21:187
82. Latimer P, Menchaca M, Snyder RM, Yu W, Gilbert BE, Sanders BG et al (2009) Aerosol delivery of liposomal formulated paclitaxel and vitamin E analog reduces murine mammary tumor burden and metastases. Exp Biol Med 234:1244–1252
83. Lawson KA, Anderson K, Snyder RM, Simmons-Menchaca M, Atkinson J, Sun LZ et al (2004) Novel vitamin E analogue and 9-nitro-camptothecin administered as liposome aerosols decrease syngeneic mouse mammary tumor burden and inhibit metastasis. Cancer Chemother Pharmacol 54:421–431
84. Bermudez LE (1994) Use of liposome preparation to treat mycobacterial infections. Immunobiology 191:578–583
85. Rabaça Roque Botelho MF, Marques T, Gomes F, Andrade Figueiredo Bairos VA, de Matos Santos Rosa M, Pedroso de Lima J (2009) Nanoradioliposomes molecularly modulated to study the lung deep lymphatic drainage. Rev Port Pneumol 15:261–293
86. Bargoni A, Cavalli R, Caputo O, Fundarò A, Gasco MR, Zara GP (1998) Solid lipid nanoparticles in lymph and plasma after duodenal administration to rats. Pharm Res 15:745–750
87. Cavalli R, Bargoni A, Podio V, Muntoni E, Zara GP, Gasco MR (2003) Duodenal administration of solid lipid nanoparticles loaded with different percentages of tobramycin. J Pharm Sci 92:1085–1094
88. Cavalli R, Zara GP, Caputo O, Bargoni A, Fundarò A, Gasco MR (2000) Transmucosal transport of tobramycin incorporated in SLN after duodenal administration to rats. Part I—a pharmacokinetic study. Pharmacol Res 42:541–545
89. Zara GP, Bargoni A, Cavalli R, Fundarò A, Vighetto D, Gasco MR (2002) Pharmacokinetics and tissue distribution of idarubicin-loaded solid lipid nanoparticles after duodenal administration to rats. J Pharm Sci 91:1324–1333

90. Harivardhan Reddy L, Sharma R, Chuttani K, Mishra A, Murthy R (2005) Influence of administration route on tumor uptake and biodistribution of etoposide loaded solid lipid nanoparticles in Dalton's lymphoma tumor bearing mice. J Control Release 105:185–198
91. Elliott JA, Osterlind K, Hirsch FR, Hansen HH (1987) Metastatic patterns in small-cell lung cancer: correlation of autopsy findings with clinical parameters in 537 patients. J Clin Oncol 5:246–254
92. Stenbygaard LE, Sørensen JB, Larsen H, Dombernowsky P (1999) Metastatic pattern in non-resectable non-small cell lung cancer. Acta Oncol 38:993–998
93. Müller RH, Mäder K, Gohla S (2000) Solid lipid nanoparticles (SLN) for controlled drug delivery–a review of the state of the art. Eur J Pharm Biopharm 50:161–177
94. Videira MA, Botelho M, Santos AC, Gouveia LF, Pedroso de Lima J, Almeida AJ (2002) Lymphatic uptake of pulmonary delivered radiolabelled solid lipid nanoparticles. J Drug Target 10:607–613
95. Bhatnagar A, Singh AK, Singh T, Sethi RS, Soni NL (1998) Inflammation imaging using Tc-99m dextran. Clin Nucl Med 23:123–124
96. Cai S, Xie Y, Bagby TR, Cohen MS, Forrest ML (2008) Intralymphatic chemotherapy using a hyaluronan–cisplatin conjugate. J Surg Res 147:247–252
97. Hawley AE, Illum L, Davis SS (1997) Lymph node localisation of biodegradable nanospheres surface modified with poloxamer and poloxamine block co-polymers. FEBS Lett 400:319–323
98. Dünne AA, Boerner HG, Kukula H, Schlaad H, Wiegand S, Werner JA et al (2007) Block copolymer carrier systems for translymphatic chemotherapy of lymph node metastases. Anticancer Res 27:3935–3940
99. Liu J, Meisner D, Kwong E, Wu XY, Johnston MR (2009) Translymphatic chemotherapy by intrapleural placement of gelatin sponge containing biodegradable Paclitaxel colloids controls lymphatic metastasis in lung cancer. Cancer Res 69:1174–1181
100. Kumanohoso T, Natsugoe S, Shimada M, Aikou T (1997) Enhancement of therapeutic efficacy of bleomycin by incorporation into biodegradable poly-d, l-lactic acid. Cancer Chemother Pharmacol 40:112–116
101. Liu J, Wong HL, Moselhy J, Bowen B, Wu XY, Johnston MR (2006) Targeting colloidal particulates to thoracic lymph nodes. Lung Cancer 51:377–386
102. Kobayashi H, Kawamoto S, Choyke PL, Sato N, Knopp MV, Star RA et al (2003) Comparison of dendrimer-based macromolecular contrast agents for dynamic micro-magnetic resonance lymphangiography. Magn Reson Med 50:758–766
103. Kobayashi H, Kawamoto S, Bernardo M, Brechbiel MW, Knopp MV, Choyke PL (2006) Delivery of gadolinium-labeled nanoparticles to the sentinel lymph node: comparison of the sentinel node visualization and estimations of intra-nodal gadolinium concentration by the magnetic resonance imaging. J Control Release 111:343–351
104. Kobayashi H, Koyama Y, Barrett T, Hama Y, Regino CAS, Shin IS et al (2007) Multimodal nanoprobes for radionuclide and five-color near-infrared optical lymphatic imaging. ACS Nano 1:258–264
105. Koyama Y, Talanov VS, Bernardo M, Hama Y, Regino CAS, Brechbiel MW et al (2007) A dendrimer-based nanosized contrast agent dual-labeled for magnetic resonance and optical fluorescence imaging to localize the sentinel lymph node in mice. J Magn Reson Imag 25:866–871
106. Kobayashi H, Ogawa M, Kosaka N, Choyke PL, Urano Y (2009) Multicolor imaging of lymphatic function with two nanomaterials: quantum dot-labeled cancer cells and dendrimer-based optical agents. Nanomedicine 4:411–419
107. Yang D, Yang F, Hu J, Long J, Wang C, Fu D et al (2009) Hydrophilic multi-walled carbon nanotubes decorated with magnetite nanoparticles as lymphatic targeted drug delivery vehicles. Chem Commun 29:4447–4449
108. McDevitt MR, Chattopadhyay D, Kappel BJ, Jaggi JS, Schiffman SR, Antczak C et al (2007) Tumor targeting with antibody-functionalized, radiolabeled carbon nanotubes. J Nucl Med 48:1180–1189

109. Murakami T, Sawada H, Tamura G, Yudasaka M, Iijima S, Tsuchida K (2008) Water-dispersed single-wall carbon nanohorns as drug carriers for local cancer chemotherapy. Nanomedicine 3:453–463
110. Shimada M, Natsugoe S, Aikou T (1995) Enhanced efficacy of Bleomycin adsorbed on silica particles against lymph node metastasis derived from a transplanted tumor. Anticancer Res 15:109
111. Hagiwara A, Takahashi T, Iwamoto A, Yoneyama C, Matsumoto S, Muranishi S (1991) Selective distribution of aclarubicin to regional lymph nodes with a new dosage form: aclarubicin adsorbed on activated carbon particles. Anticancer Drugs 2:261
112. Lucci A, Turner RR, Morton DL (1999) Carbon dye as an adjunct to isosulfan blue dye for sentinel lymph node dissection. Surgery 126:48–53
113. Advani A, Coiffier B, Czuczman MS, Dreyling M, Foran J, Gine E et al (2010) Safety, pharmacokinetics, and preliminary clinical activity of inotuzumab ozogamicin, a novel immunoconjugate for the treatment of B-cell non-Hodgkin's lymphoma: results of a phase I study. J Clin Oncol 28:2085–2093
114. Dijoseph J, Dougher M, Armellino D, Evans D, Damle N (2007) Therapeutic potential of CD22-specific antibody-targeted chemotherapy using inotuzumab ozogamicin (CMC-544) for the treatment of acute lymphoblastic leukemia. Leukemia 21:2240–2245
115. DiJoseph JF, Armellino DC, Boghaert ER, Khandke K, Dougher MM, Sridharan L et al (2004) Antibody-targeted chemotherapy with CMC-544: a CD22-targeted immunoconjugate of calicheamicin for the treatment of B-lymphoid malignancies. Blood 103:1807–1814
116. DiJoseph JF, Dougher MM, Evans DY, Zhou BB, Damle NK (2011) Preclinical anti-tumor activity of antibody-targeted chemotherapy with CMC-544 (inotuzumab ozogamicin), a CD22-specific immunoconjugate of calicheamicin, compared with non-targeted combination chemotherapy with CVP or CHOP. Cancer Chemother Pharmacol 67:741–749
117. Grillo-López AJ (2003) Rituximab (Rituxan®/MabThera®): the first decade (1993–2003). Expert Rev Anticancer Ther 3:767–779
118. DiJoseph JF, Dougher MM, Armellino DC, Kalyandrug L, Kunz A, Boghaert ER et al (2007) CD20-specific antibody-targeted chemotherapy of non-Hodgkin's B-cell lymphoma using calicheamicin-conjugated rituximab. Cancer Immunol Immunother 56:1107–1117
119. O'Reilly MK, Paulson JC (2009) Siglecs as targets for therapy in immune-cell-mediated disease. Trends Pharmacol Sci 30:240–248
120. Fayad L, Patel H, Verhoef G, Czuczman M, Foran J, Gine E et al (2006) Clinical activity of the immunoconjugate CMC-544 in B-cell malignancies: preliminary report of the expanded maximum tolerated dose (MTD) cohort of a phase 1 study. Blood 108:766a
121. Younes A, Forero-Torres A, Bartlett N, Leonard J, Rege B, Kennedy D et al (2008) Objective responses in a phase I dose-escalation study of SGN-35, a novel antibody-drug conjugate (ADC) targeting CD30, in patients with relapsed or refractory Hodgkin lymphoma. J Clin Oncol 26:8526
122. Xuan C, Steward KK, Timmerman JM, Morrison SL (2010) Targeted delivery of interferon-alpha via fusion to anti-CD20 results in potent antitumor activity against B-cell lymphoma. Blood 115:2864–2871
123. Van der Lubben I, Verhoef J, Borchard G, Junginger H (2001) Chitosan for mucosal vaccination. Adv Drug Deliv Rev 52:139–144
124. Janes K, Calvo P, Alonso M (2001) Polysaccharide colloidal particles as delivery systems for macromolecules. Adv Drug Deliv Rev 47:83–97
125. An P, Lei H, Zhang J, Song S, He L, Jin G et al (2004) Suppression of tumor growth and metastasis by a VEGFR-1 antagonizing peptide identified from a phage display library. Int J Cancer 111:165–173
126. Starzec A, Vassy R, Martin A, Lecouvey M, Di Benedetto M, Crépin M et al (2006) Antiangiogenic and antitumor activities of peptide inhibiting the vascular endothelial growth factor binding to neuropilin-1. Life Sci 79:2370
127. Laakkonen P, Porkka K, Hoffman JA, Ruoslahti E (2002) A tumor-homing peptide with a targeting specificity related to lymphatic vessels. Nat Med 8:751–755

128. Ichihara H, Ueno J, Umebayashi M, Matsumoto Y, Ueoka R (2011) Chemotherapy with hybrid liposomes for acute lymphatic leukemia leading to apoptosis in vivo. Int J Pharm 406:173–178
129. Luo G, Yu X, Jin C, Yang F, Fu D, Long J et al (2010) LyP-1-conjugated nanoparticles for targeting drug delivery to lymphatic metastatic tumors. Int J Pharm 385:150–156
130. Ramesh R, Saeki T, Templeton NS, Ji L, Stephens LC, Ito I et al (2001) Successful treatment of primary and disseminated human lung cancers by systemic delivery of tumor suppressor genes using an improved liposome vector. Mol Ther 3:337–350
131. Son K, Huang L (1994) Exposure of human ovarian carcinoma to cisplatin transiently sensitizes the tumor cells for liposome-mediated gene transfer. Proc Natl Acad Sci USA 91:12669–12672
132. Chen SH, Chen X, Wang Y, Kosai K, Finegold MJ, Rich SS et al (1995) Combination gene therapy for liver metastasis of colon carcinoma in vivo. Proc Natl Acad Sci USA 92:2577–2581
133. Davis SS, Illum L (1988) Polymeric microspheres as drug carriers. Biomaterials 9:111–115
134. Illum L, Davis S, Müller R, Mak E, West P (1987) The organ distribution and circulation time of intravenously injected colloidal carriers sterically stabilized with a blockcopolymer-poloxamine 908. Life Sci 40:367–374
135. Papahadjopoulos D (1996) Fate of liposomes in vivo: a brief introductory review. J Liposome Res 6:3–17
136. Bergqvist L, Strand S, Haftstrom L, Jonsson P (1984) The characterisation of radio colloids used for administration to the lymphatic system. In: Davis SS, Ilium L, McVic JG, Tomlinson E (eds) Microspheres and drug therapy: pharmaceutical immunological and medical aspects. Elsevier, Amsterdam, pp 263–267
137. Christy N, Moghimi S, Illum L, Davis S (1992) Effect of size on the lymphatic uptake of a model colloid system. In: Proceedings of the international symposium on controlled release of bioactive materials. pp 355–356
138. Strand S, Bergqvist L (1989) Radiolabeled colloids and macromolecules in the lymphatic system. Crit Rev Ther Drug Carrier Syst 6:211
139. Bettendorf U (1979) Electron microscopic studies on the peritoneal resorption of intraperitoneally injected latex particles via the diaphragmatic lymphatics. Lymphology 12:66
140. Winkel Kz (1972) Lymphologie mit Radionukliden. H. Hoffmann
141. Hirano K, Hunt CA, Strubbe A, MacGregor RD (1985) Lymphatic transport of liposome-encapsulated drugs following intraperitoneal administration–effect of lipid composition. Pharm Res 2:271–278
142. Patel HM, Katherine MB, Vaughan-Jones R (1984) Assessment of the potential uses of liposomes for lymphoscintigraphy and lymphatic drug delivery failure of 99-technetium marker to represent intact liposomes in lymph nodes. Biochim Biophys Acta (BBA)-General Subjects 801:76–86
143. Supersaxo A, Hein WR, Steffen H (1990) Effect of molecular weight on the lymphatic absorption of water-soluble compounds following subcutaneous administration. Pharm Res 7:167–169
144. Moghimi S, Hawley A, Christy N, Gray T, Illum L, Davis S (1994) Surface engineered nanospheres with enhanced drainage into lymphatics and uptake by macrophages of the regional lymph nodes. FEBS Lett 344:25–30
145. Patel H (1992) Serum opsonins and liposomes: their interaction and opsonophagocytosis. Crit Rev Ther Drug Carrier Syst 9:39–90
146. Florence AT, Attwood D (2011) Physicochemical principles of pharmacy. Pharmaceutical Press, London
147. Hawley A, Davis S, Illum L (1995) Targeting of colloids to lymph nodes: influence of lymphatic physiology and colloidal characteristics. Adv Drug Deliv Rev 17:129–148
148. Phillips WT, Andrews T, Liu H, Klipper R, Landry AJ, Blumhardt R et al (2001) Evaluation of [(99m) Tc] liposomes as lymphoscintigraphic agents: comparison with [(99m) Tc] sulfur colloid and [(99m) Tc] human serum albumin. Nucl Med Biol 28:435

149. Oussoren C, Storm G (1997) Lymphatic uptake and biodistribution of liposomes after subcutaneous injection: III. Influence of surface modification with poly(ethyleneglycol). Pharm Res 14:1479–1484
150. Kaminskas LM, Kota J, McLeod VM, Kelly BD, Karellas P, Porter CJH (2009) PEGylation of polylysine dendrimers improves absorption and lymphatic targeting following SC administration in rats. J Control Release 140:108–116
151. Mangat S, Patel HM (1985) Lymph node localization of non-specific antibody-coated liposomes. Life Sci 36:1917–1925
152. Wu MS, Robbins JC, Bugianesi RL, Ponpipom MM, Shen T (1981) Modified in vivo behavior of liposomes containing synthetic glycolipids. Biochim Biophys Acta (BBA)-General Subjects 674:19–29
153. Kim CK, Jeong EJ (1997) Enhanced lymph node delivery and immunogenicity of hepatitis B surface antigen entrapped in galactosylated liposomes. Int J Pharm 147:143–151
154. Akande J, Yeboah KG, Addo RT, Siddig A, Oettinger CW, D'Souza MJ (2010) Targeted delivery of antigens to the gut-associated lymphoid tissues: 2. Ex vivo evaluation of lectin-labelled albumin microspheres for targeted delivery of antigens to the M-cells of the Peyer's patches. J Microencapsul 27:325–336
155. Medina LA, Calixto SM, Klipper R, Phillips WT, Goins B (2004) Avidin/biotin-liposome system injected in the pleural space for drug delivery to mediastinal lymph nodes. J Pharm Sci 93:2595–2608
156. Medina LA, Klipper R, Phillips WT, Goins B (2004) Pharmacokinetics and biodistribution of ^{111}In avidin and 99 Tc biotin-liposomes injected in the pleural space for the targeting of mediastinal nodes. Nucl Med Biol 31:41–51
157. Gu B, Xie C, Zhu J, He W, Lu W (2010) Folate-PEG-CKK 2-DTPA, A potential carrier for lymph-metastasized tumor targeting. Pharm Res 27:933–942
158. Sinha R, Kim GJ, Nie S, Shin DM (2006) Nanotechnology in cancer therapeutics: bioconjugated nanoparticles for drug delivery. Mol Cancer Ther 5:1909–1917
159. Hauff P, Reinhardt M, Briel A, Debus N, Schirner M (2004) Molecular targeting of lymph nodes with l-selectin ligand-specific US contrast agent: a feasibility study in mice and dogs. Radiology 231:667–673
160. Drobnik J (1991) Hyaluronan in drug delivery. Adv Drug Deliv Rev 7:295–308
161. Kaur CD, Nahar M, Jain NK (2008) Lymphatic targeting of zidovudine using surface-engineered liposomes. J Drug Target 16:798–805
162. Coppi G, Iannuccelli V (2009) Alginate/chitosan microparticles for tamoxifen delivery to the lymphatic system. Int J Pharm 367:127–132
163. Swart PJ, Beljaars L, Kuipers ME, Smit C, Nieuwenhuis P, Meijer DKF (1999) Homing of negatively charged albumins to the lymphatic system: general implications for drug targeting to peripheral tissues and viral reservoirs. Biochem Pharmacol 58:1425–1435

Chapter 21
Antibiotics Delivery for Treating Bone Infections

Wahid Khan, Venu Gopala Swami Challa, Rajendra P. Pawar, Meir Nyska, Yaron S. Brin, and Abraham J. Domb

21.1 Introduction

Osteomyelitis is an infection of bone and bone marrow. It is difficult to treat and can be caused by adjacent soft tissues and joints, hematogenous seeding, or direct inoculation of microorganisms into the bone as a result of trauma or surgery. In the early 1900s 20 % of patients suffering from osteomyelitis died, and those who survived remained with significant morbidity. Today, both mortality and morbidity are lowered due to surgical treatment together with prolonged antibiotic treatment. The disease is usually caused by a single organism, but it can also be multibacterial and limited to a single portion of bone or several portions including the soft tissue surrounding the bone.

Currently, osteomyelitis treatment resolves the infective process and requires a multidisciplinary approach. Implanted devices are removed by surgery from infected tissue followed by specific antibiotic therapy. Infection risk is reduced by administrating a systemic antibiotic prophylaxis procedure. Upon systemic administration the antibiotic concentration can be irregular in bone, since blood supply to the infected area is disturbed. High local drug concentrations offering therapy are

W. Khan • V.G.S. Challa
Department of Pharmaceutics, National Institute of Pharmaceutical Education and Research (NIPER), Hyderabad 500037, India

R.P. Pawar
Department of Chemistry, Deogiri College, Aurangabad, India

M. Nyska • Y.S. Brin
Department of Orthopaedic Surgery, Meir Medical Center, Tel Aviv University, Kfar-Saba, Israel

A.J. Domb (✉)
Faculty of Medicine, Institute of Drug Research, School of Pharmacy, The Hebrew University of Jerusalem, Jerusalem 91120, Israel
e-mail: avid@ekmd.huji.ac.il

A.J. Domb and W. Khan (eds.), *Focal Controlled Drug Delivery*, Advances in Delivery Science and Technology, DOI 10.1007/978-1-4614-9434-8_21,

therefore required [1]. Cierny and Mader classified osteomyelitis according to the anatomic involvement of the infected bone [2]:

- Type 1—Medullary osteomyelitis (nidus is endosteal). No dead space management. Etiology often hematogenous or post-intramedullary rod
- Type 2—Superficial osteomyelitis. Limited to bone surface. No dead space management but needs soft tissue coverage
- Type 3—Localized osteomyelitis. Full thickness of cortex. Complex dead space management, simple osseous stabilization
- Type 4—Diffuse osteomyelitis. Circumference of cortex. Biomechanically unstable. Complex dead space and osseous management

Osteomyelitis can also be classified according to systemic and localized conditions. This classification helps to understand treatment and prognosis [2]. Others classify osteomyelitis according to the duration of illness and etiology. This classification does not implicate specific therapy [3]. Normal and healthy bone is usually highly resistant to infections.

21.2 Diagnosis

Osteomyelitis can be hematogenous or contagious to a soft tissue infection. Acute hematogenous osteomyelitis is a pediatric illness usually discussed separately. Adults usually show subacute to chronic presentation. The leading symptom is nonspecific pain around the infected focus. There is usually an absence of systemic signs and symptoms. Fever, chills, local swelling, and erythema in the proximity of the involved bone are seldom seen. A draining sinus tract may be present over the involved bone. Diagnosis should be first suspected clinically. Confirmation is done with a combination of radiologic, pathologic, and microbiologic examinations. Laboratory examination will reveal elevated erythrocyte sedimentation rate and C-reactive protein. The white blood cell count can be either normal or elevated. Radiographic abnormalities are usually seen only after 10–14 days after onset of infection. The radiographic changes can include soft tissue swelling, osteopenia, scalloping of the cortex, periosteal thickening, and elevation and loss of the trabecular architecture of the cancellous bone. Lytic changes, which are diagnostic for osteomyelitis, tend to occur later. Bone resorption, sequestration, periosteal or endosteal new bone formation, cortical irregularities, and atrophic nonunion are the main features demonstrated by plain radiographs in chronic osteomyelitis [4–6].

Nuclear bone scans are sensitive, but they can sometimes be nonspecific. Tracers most commonly in use include technetium 99m methylene diphosphonate and gallium-citrate 67. Both detect bone structural changes, but they are unable to differentiate between infections or other bone pathology [7]. Fluorodeoxyglucose positron emission tomography imaging for the diagnosis of osteomyelitis is considered to have a high degree of accuracy [8]. Indium in 111-labeled white blood cells and in vivo labeling of leucocytes by injection of radiolabelled murine monoclonal anti-granulocyte antibodies (bind to leukocyte antigens) are the standard diagnostic exams [9]. Computed tomography (CT) is useful in diagnosis, but magnetic resonance imaging (MRI) is more sensitive and is judged to be the standard in the diagnosis of osteomyelitis [7].

It is also crucial to identify the infecting organism, in order to adjust the most suitable antimicrobial treatment. The organism is identified by cultures. Samples are taken from the infection focus during operation or by needle aspiration. It is also possible to obtain a swab sample for culture by draining wounds or from sinus tracts. It is less favorable than direct intraoperative culture and needle aspiration, but may reveal valuable data, especially when the detected bacteria are *S. aureus*, *methicillin-resistant S. aureus* (*MRSA*), or *vancomycin-resistant S. aureus* [10]. It is of high importance to obtain samples for histopathologic examinations. It is noticed that in hematogenous long bone infection, infections are usually mono-microbial, and the affecting bacteria are usually *S. aureus*. In more than 50 % of cases, the pathogens are either *S. aureus* or coagulase-negative *staphylococci*. In contagious osteomyelitis the infection is usually multibacterial. MRSA has been increasingly recognized as the pathogen in long bone osteomyelitis in children [11].

21.3 Treatment with Antibiotics

The goal of treatment in adult osteomyelitis is infection eradication and return to maximum function. Treatment of acute osteomyelitis usually requires adequate debridement, drainage of pus, and prolonged course of antibiotics [12]. Antibiotics are usually prescribed intravenously for 3 weeks, followed by a 3-week oral treatment. A high parenteral dose of antibiotic is required to achieve a high concentration of therapeutic drug in the bone. It might be kept in mind that the prolonged course of treatment causes systemic antibiotic toxicity.

Local antimicrobial agent delivery is a novel therapeutic modality in chronic osteomyelitis management. It achieves high antibiotic concentrations at the infection site without systemic toxicity. The therapy requires a multidisciplinary approach to eradicate infection through debridement and appropriate antibiotic coverage. Revascularization is crucial to keep the affected area aseptic. Patients with osteomyelitis and fixation failure need eradication of the infection together with osseous union. Hence, the operative approach is sometimes multimodal, including orthopedic, plastic, and microvascular surgery. It is crucial to understand that antimicrobial therapy cannot stand as the only treatment even when given for prolonged periods, since it will not sterilize dead bone and cavities [13].

Antibiotics should be started empirically after obtaining cultures during debridement. The empirical therapy will most likely be broad spectrum, for example, glycopeptides such as vancomycin or teicoplanin often in conjunction with a broad-spectrum beta-lactam antibiotic (cephalosporins, β-lactamase inhibitor combinations, or carbapenems). The broad-spectrum combinations will usually continue until the infecting organism(s) is grown from the tissue samples and then stepped down to a narrow-spectrum regimen. Usually about 4–6 weeks are required to complete the therapy, in parallel with the time needed for debrided bone revascularization [14, 15]. The cornerstones of debridement are taking several samples for cultures, debridement of infected soft tissue and bone, and delivery of appropriate antimicrobials [16]. Antibiotic delivery can be done either as a systemic intravenous treatment or as a local treatment by antibiotic-loaded bone cement.

Fluoroquinolones are the drugs of choice used in chronic osteomyelitis treatment due to their favorable penetration into poorly vascularized sites of infection. Their bactericidal effect against all chronic osteomyelitis pathogens and the lack of serious adverse effects make them better implants [17]. Rifampin and gentamicin are commonly used to treat bone infections [18]. It should be kept in mind that quinolones have shown inhibitory effects on osteoblastic functions at high concentrations and may delay bone healing [19].

21.4 Therapeutic Drug Carriers

21.4.1 Criteria for the Fabrication of a Local Delivery System

Chronic osteomyelitis is generally caused by microbes *Staphylococcus aureus*, Group A β-hemolytic Streptococcus, and gram-negative bacteria (such as *Salmonella*, *Mycobacterium tuberculosis*, and *Pseudomonas aeruginosa*) [20]. Antimicrobial agents used in local delivery systems have the following characteristics:

- Activity against common bacterial pathogens involved in chronic osteomyelitis
- Local release at several times exceeding concentrations (ten times)
- Inability to enter into systemic circulation
- No perceivable adverse effects
- Stability at body temperature and water soluble for diffusion from the carrier
- No supra infection

Extensive research is currently being conducted on local drug delivery systems to treat osteomyelitis. An antibiotic carrier system releases the drug according to prescribed dosage. These carrier systems can be composed of both biodegradable and nonbiodegradable components (Fig. 21.1). Systems made up of biodegradable polymers have advantages over conventional poly(methylmethacrylate) (PMMA) beads and intravenous antibiotics in several ways. First, biodegradable beads provide bactericidal concentrations of antibiotics for the prolonged time needed to completely treat the particular orthopedic infection. Second, variable biodegradability from weeks to years may allow many types of infections to be treated. Third, because the biodegradable beads dissolve, there is no need for surgical removal, and fourth, because the biodegradable beads dissolve slowly, the soft tissue or bone defect will slowly fill with tissue, so there is no need for reconstruction [21, 22].

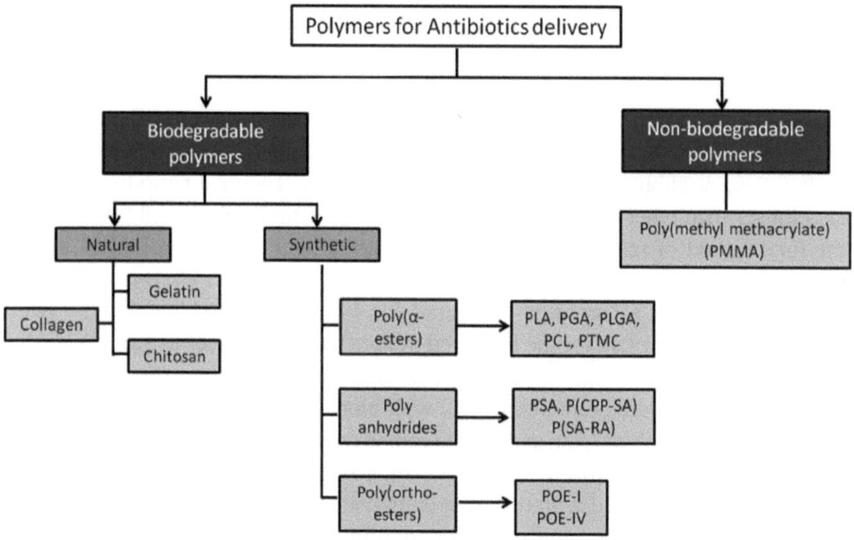

Fig. 21.1 Polymers for antibiotic delivery

21.4.2 Biodegradable Implants

Various biodegradable delivery systems have been developed and used in antibiotics local delivery for bone infection treatment [23]. Biodegradable implants provide high bactericidal concentrations for a prolonged time to complete infection eradication [24]. Biodegradation also removes unnecessary surgical implant. Initially, an implant is used to eliminate dead space and to guide repair. During the degradation phase, a secondary release of antibiotic increases antibacterial efficacy. The most common biodegradable carrier systems are discussed below.

21.4.2.1 Polylactides

Synthetic biodegradable polyester polymers are potential candidates for local drug carrier systems to avoid removal operation. Polylactide (PLA), Polyglycolide (PGA), and their copolymers, especially Poly(lactide-co-glycolide) (PLGA), are mostly widely used for controlled-release systems. They also held a central place as degradable materials for orthopedic applications in the controlled delivery of antibiotics like ampicillin, gentamicin, polymyxin B, and chloramphenicol [25]. Price et al. used a PLGA as the biodegradable carrier for gentamicin. The biocompatibility of biodegradable PLA/PGA has been well established. Tissue reaction with implanted materials is minimal; the inflammatory response is limited, which gradually diminishes as the resorbed polymer [26]. Biodegradable rods of PGA or PLGA were used in the internal fixation of a variety of fractures and osteotomies with minimal foreign-body reaction in patients [27]. Study showed that thin biodegradable implant coating of these polymers can be developed with bactericidal agents against the organisms frequently associated with osteomyelitis in cases of open fractures [28]. Among different polyesters lactic acid polymer produces sustained release of antibiotics at higher concentrations over a relatively long time span. The total period of release is also longer than other systems. When considering the molecular weight of 100 kDa, the total period of release approaches 350 days. PLAs achieve prolonged in vivo quinolone release at higher levels and produce drug release after 15 days [24].

Ramchandani et al. reported biodegradable PLGA implantable delivery system containing ciprofloxacin for the localized treatment of osteomyelitis [29]. With this, implants composed of different blends of calcium phosphates (tricalcium phosphate, TCP, and hydroxyapatite, HAP), with biodegradable PLA for local delivery of antibiotics like tobramycin and gentamicin in controlled manner, are also reported [30]. Copolymers with different polymers such as polyanhydride and polycaprolactone have been produced with different composite ratios to provide better stability, delayed decomposition, and superior elution concentrations of tobramycin, clindamycin, and vancomycin for the treatment of osteomyelitis. The local application of recombinant human bone morphogenetic protein (rhBMP-2) from collagen sponges or bio-absorbable PLA/PGA beads with osteopromotive membranes into rat mandibular defects shows a superior bone regeneration using PLA/PGA than a collagen carrier [25].

As another application, plaster of Paris implants containing vancomycin were prepared for the treatment of bone infections. The regulation of the release rate was performed by coating the carrier with a PLGA polymer composed by 10 % (w/w) PGA and 90 % (w/w) racemic poly(D,L-lactic acid). The release of the antibiotic from the biodegradable matrix was evaluated in vitro. From this investigation, it is clear that the drug elution depends on the coating depth. Plaster of Paris coated with this polymer giving a controlled release of vancomycin appears to be a promising sustained-release delivery system of antibiotics for the treatment of bone and joint infections [31].

21.4.2.2 Poly(trimethylene carbonate)

Poly(trimethylene carbonate) (PTMC) is an amorphous polymer with a glass transition temperature of 15 °C. PTMC is a biocompatible polymer that degrades enzymatically in vivo by surface erosion process. PTMC is an elastomeric material used in biomedical applications, although similarly to the majority of aliphatic polycarbonates, it becomes soft in the temperature range 40–60 °C and has rather weak mechanical properties. One of the important advantages of PTMC is its degradation products diols which are less acidic than lactic acid produced as a result of PLA hydrolytic degradation [32].

Neut et al. studied related issues regarding gentamicin-loaded PTMC, the release kinetics, and its ability to inhibit biofilm formation. Gentamicin-loaded PTMC disks show antibiotic release characteristics and biofilm inhibition characteristics [33]. Van Leeuwen et al. studied the response of surrounding tissues to developed PTMC membranes. The PTMC membranes showed minimal cellular capsule formation and showed signs of a surface erosion process. Bone tissue formed beneath the PTMC membranes comparable to that beneath the collagen membranes. The space maintaining properties of the PTMC membranes were superior to those of the collagen membrane. Newly developed PTMC membranes can be used with success as barrier membranes in critical size rat mandibular defects [34].

PTMC is also widely used in combination with calcium phosphate cements (CPC) in bone infections. One of the main disadvantages of CPC is the slow degradation rate of the material. To overcome this problem, Liao et al. studied the macroporous CPC of solid calcium phosphate salts with polymeric microspheres PLGA, gelatin, or PTMC which shown better degradation properties than CPC alone [35]. PTMC is also useful to be carried out to deliver the vancomycin, which is the antibiotic of last resort used for the treatment of clinically resistant bacteria, like MRSA [36].

21.4.2.3 Poly-ε-caprolactone

Poly-ε-caprolactone (PCL) is a biodegradable, bioresorbable polymer used for drug formulations and medical purposes (e.g., surgical sutures) and is a good candidate for drug delivery systems, as it remains active for more than a year and is regarded as nontoxic and tissue compatible [37]. The PCL is attractive due to its low cost, sustained biodegradability, and availability at low molecular weight [38]. PCL possesses appropriate rheological and viscoelastic properties to use as implants. However, is not yet widely used clinically, due to its biodegradation at a slower rate than other aliphatic polyesters [39].

Biocompatible polymer blends are the new preferred copolymers, since their preparation is less expensive and safe [40]. A porous biodegradable-guided bone regeneration membrane made of poly(L-lactide-co-ε-caprolactone) and TCP has been manufactured in order to enhance the regeneration of bone tissue cells [41]. Similarly, PCL/poloxamine blends as biocompatible and biodegradable implantable systems with tunable viscoelasticity, erosion, and ciprofloxacin-release features have been prepared. Ciprofloxacin-loaded PCL/poloxamine blends may facilitate a local treatment of infections with bone regeneration stimulation [42]. Cytocompatibility of PCL/poloxamine blends, the possible erosion rate and drug-release profile, as well as the efficacy of ciprofloxacin-loaded implantable systems against a relevant causal agent of osteomyelitis, *Staphylococcus aureus,* have been evaluated in detail. Drug-release profiles from this implantable matrice are sufficient for rapid in vitro eradication of *S. aureus*. Thus, PCL/poloxamine matrices are useful implants for the local treatment of bone infections that require antibiotic release over a long period of time [43].

21.4.2.4 Poly(sebacic-co-ricinoleic-ester-anhydride)

Polyanhydrides are a class of hydrolytically unstable surface-eroding polymers that are either aliphatic, aromatic, or a combination of the two [44, 45]. Bioerodible polyanhydride polymeric systems offer many advantages for controlled release of antibiotics than compared to regular systemic delivery in treatment of osteomyelitis [46, 47]. Biodegradable polyanhydrides, one of the polymeric carriers, have shown high and effective treatment modality. A copolymer of 1:1 sebacic acid and erucic acid dimer has been found to be useful as a potential delivery vehicle for gentamicin (Septacin®) in the treatment of osteomyelitis [48]. Laurencin et al. studied the effect of antibiotic therapy of polyanhydride copolymers of bis-carboxyphenoxypropane and sebacic acid with gentamicin in rat model [49]. Several other biodegradable polyanhydride polymers have been demonstrated to be effective antibiotic carriers in local infection treatments. Use of injectable biodegradable poly(sebacic-co-ricinoleic-ester-anhydride)s P(SA:RA) in solid tumors and osteomyelitis treatment has been explored [50–53]. According to reports, in aqueous environment, viscosity of the polymer increases, and it becomes semisolid gels. This semisolid polymeric matrix releases the antibiotic until degradation of the polymer. Efficacy studies using polyanhydride implants show an excellent effect [52]. P(SA:RA) implantation into tubular bone has resulted in minimal changes. These results are encouraging, since the gentamicin carrying soft polymers has been found to be suitable in biodegradable drug-eluting devices for internal bacterial lesions treatment. The antibiotic releases in an adequate manner kill the bacteria and prevent resistance to the antibiotics [50].

21.4.2.5 Poly(orthoesters)

Poly(orthoesters) (POEs) are amorphous hydrophobic polymers containing hydrolytically labile, acid-sensitive, backbone linkages. These were developed by ALIZA Corporation (Alzamer®) as hydrophobic surface-eroding polymers particularly for drug delivery applications. POEs are hydrophobic and bioerodible polymers that have been investigated for pharmaceutical use since the early 1970s. Among the four described generations of POE, the third (POE III) and fourth (POE IV) are promising viscous and injectable materials which have been investigated in numerous biomedical applications. POE IV is distinguishable by a highly reproducible and controlled synthesis, a higher hydrophobicity, and an excellent biocompatibility. It is currently under development for a variety of applications such as ocular delivery, periodontal disease treatment, and applications in osteomyelitis. Jasti et al. designed a new biodegradable implantable disk system based upon POE, which could release tobramycin sulfate over a 21-day period at a rate of clinical requirement for the treatment of osteomyelitis [54].

21.4.2.6 Natural Polymers

Collagen is of particular interest as a natural polymer for drug delivery since it is a major natural constituent of connective tissue and a major structural protein of any organ. Collagen has been recognized as an effective and completely biodegradable material produced from sterilized bovine tendon. It has been utilized as a carrier system for commercially available ocular application, involving pilocarpine and macrolide antibiotics. Collagen sheets impregnated with gentamicin are in regular use to treat chronic osteomyelitis [55]. Efforts in the use of collagen and collagen-synthetic polymer composites for controlled drug delivery of gentamicin and collagen sponges are widely used as bone regeneration material with rhBMP-2 [56].

Riegels-Nielsen et al. reported collagen with gentamicin for prophylaxis of post-operative infection of *S. aureus* osteomyelitis in rabbits [57].

Gelatin is derived by denaturing collagen and is hence free of antigenicity which is associated with collagen. Depending upon the conditions (acidic/alkaline) under which collagen is processed, electrically different types of gelatin are obtained with a variety of isoelectric point values ranging from 9.0 to 5.0. Khor et al. reported biodegradable gelatin sponge containing different contents of beta-tricalcium phosphate (beta-TCP) ceramic was prepared for the controlled release of vancomycin. Results showed that the composite scaffolds could achieve local therapeutic drug levels over an extended duration [58]. In another study, the use of a degradable gelatin for the combined release of therapeutic levels of both gentamicin and growth hormone was reported. The results of this study suggest that gelatin is a good DDS for the combined release of drugs for the management of acute and chronic bone and tissue infection such as osteomyelitis [59].

Alginates are naturally derived polysaccharide block copolymers composed of regions of sequential β-D-mannuronic acid monomers (M-blocks), regions of α-L-guluronic acid (G blocks), and regions of interspersed M and G units and have a structural role in giving flexibility and strength to marine plants. Commercial alginates are extracted from brown algae *Laminaria hyperborea*, *Ascophyllum nodosum*, and *Macrocystis* [60]. Yenice et al. reported biodegradable implants containing teicoplanin (a glycopeptide antibiotic) for the prevention or the treatment of bone infections using sodium alginate as the polymer material [61, 62].

Chitosan is a cationic linear polymer obtained from chitin comprising copolymers of β(1 → 4)-glucosamine and randomly located *N*-acetyl-D-glucosamine. It is the fully or partially deacetylated form of chitin. It has been proven to be biologically renewable, biodegradable, biocompatible, nonantigenic, nontoxic, and biofunctional [63]. Gentamicin-loaded chitosan bar is clinically useful for the treatment of bone infection. Different gels such as hyaluronic acid fibrin gel (Vanco-AB-FG) with bone marrow-derived mesenchymal stem cells and monoolein-water gels have been used as an alternative treatment for bone infections [64]. Biodegradable chitosan microspheres containing vancomycin hydrochloride prepared by spray drying were implanted to proximal tibia of rats with MRSA osteomyelitis. Implanted drug-loaded microspheres were found to be more effective than IM route for the treatment of experimental osteomyelitis [65].

21.4.3 Other Implants for Osteomyelitis Drug Delivery Systems

Various random copolyesters of 3-hydroxybutyrate and 3-hydroxyvalerate (PHBV) and 3-hydroxybutyrate and 4-hydroxybutyrate P(3HB–4HB) were used in the construction of biodegradable, implantable rods for the local delivery of antibiotics in chronic osteomyelitis therapy. Coating rods with the same type of polymer substantially reduced the initial burst effect observed with the uncoated rods and significantly decreased the release rate so that the release kinetics became almost zero order. Antibiotic release from coated rods was sustained for over a period of 2 weeks at a constant rate, whereas uncoated rods released their contents in less than a week [66].

The development of three-dimensional scaffolds for bone tissue engineering with controlled-release capability has been reported, for both in vitro and in vivo studies. Composites comprising polymeric matrices and added inorganic particles represent the best known systems to incorporate therapeutic drug delivery in bone tissue engineering approaches [67, 68]. Composite scaffolds are expedient alternative to osteomyelitis drug delivery systems. They have the advantages of both biodegradable polymers and bioactive ceramics for bone engineering scaffolds. This combination improves mechanical properties due to higher strength and stiffness of the inherent inorganic material. The addition of inorganic materials to bioresorbable polymers changes the polymer degradation behavior by buffering the pH of nearby solution. It prevents the autocatalytic effect of acidic end groups, resulting in polymer chain hydrolysis. Incorporation of bioactive inorganic phases in biodegradable polymer enhances water ingress owing to the internal interfaces formed between the polymer and more hydrophilic bioactive inclusions, enabling the controlled degradation kinetics of scaffolds [69].

Composite scaffolds have its distinct place for osteomyelitis drug delivery systems: calcium phosphate (CaP) has been advocated as an affective bone substitute. It maintains mechanical integrity and possesses particularly robust compressive strength after implantation. CaP is known to be immunologically nonreactive and biologically inert [70]. Calcium sulfate (CaS) has also been extensively used clinically. Biodegradable CaS exhibits an accelerated linear resorption as well as increased bone formation and is, thus, an affective bone substitute in periosteal regions [71]. Yang et al. studied about the use of a biphasic CaP/CaS composite that retains the delivery efficacy for vancomycin of PMMA [72]. Tetracycline HCl has been successfully introduced for persistent or recurrent periodontitis therapy. Demineralized bone matrix with tobramycin-impregnated calcium sulfate pellets has been effective on intramedullary *S. aureus* infection in a contaminated goat fracture model [73].

21.5 Conclusions

Osteomyelitis is attributed to several factors: antibiotic short half-life, poor blood circulation at the infected area, and systemic antibiotic toxicity usage of the required high systemic dose. All evocating osteomyelitis presents an additional challenge, because the infecting bacteria form a biofilm mode of growth, which, on devascularized surfaces, shields them from antibiotics. Widespread research is currently being conducted on local drug delivery systems to treat osteomyelitis. Much more work, however, is still needed in biodegradable and biocompatible materials. Researchers remain optimistic that many of these systems can be developed with ideal zero-order-release kinetic profiles, in vivo, over long periods of time, thus enabling widespread use in chronic osteomyelitis patients. By utilizing newer forms of sustained-release antibiotic delivery systems, it will be possible to deliver such antibiotics at constant rates over a prolonged period of time and eliminate the need for multiple dosing. It is hoped that in the future the development of new implantable systems will help reduce the cost of term release of the antibiotic and exhibit good biocompatibility gradually in in vitro studies. In summary, it appears that biodegradable polymeric materials constitute the most versatile synthetic materials useful in orthopedic devices from human health perspectives.

References

1. Schmidmaier G, Lucke M, Wildemann B, Haas NP, Raschke M (2006) Prophylaxis and treatment of implant-related infections by antibiotic-coated implants: a review. Injury 37(Suppl 2): S105–S112
2. Cierny G III, Mader JT, Penninck JJ (2003) A clinical staging system for adult osteomyelitis. Clin Orthop Relat Res (414): 7–24
3. Lew DP, Waldvogel FA (1997) Osteomyelitis. New Engl J Med 336(14):999–1007
4. Gold RH, Hawkins RA, Katz RD (1991) Bacterial osteomyelitis: findings on plain radiography, CT, MR, and scintigraphy. AJR Am J Roentgenol 157(2):365–370
5. Santiago Restrepo C, Gimenez CR, McCarthy K (2003) Imaging of osteomyelitis and musculoskeletal soft tissue infections: current concepts. Rheum Dis Clin North Am 29(1):89–109
6. Tumeh SS, Aliabadi P, Weissman BN, McNeil BJ (1987) Disease activity in osteomyelitis: role of radiography. Radiology 165(3):781–784
7. Termaat MF, Raijmakers PG, Scholten HJ, Bakker FC, Patka P, Haarman HJ (2005) The accuracy of diagnostic imaging for the assessment of chronic osteomyelitis: a systematic review and meta-analysis. J Bone Joint Surg Am 87(11):2464–2471
8. Hartmann A, Eid K, Dora C, Trentz O, von Schulthess GK, Stumpe KD (2007) Diagnostic value of 18F-FDG PET/CT in trauma patients with suspected chronic osteomyelitis. Eur J Nucl Med Mol Imaging 34(5):704–714
9. Palestro CJ, Torres MA (1997) Radionuclide imaging in orthopedic infections. Semin Nucl Med 27(4):334–345
10. Mackowiak PA, Jones SR, Smith JW (1978) Diagnostic value of sinus-tract cultures in chronic osteomyelitis. JAMA 239(26):2772–2775
11. Kaplan SL (2005) Osteomyelitis in children. Infect Dis Clin North Am 19(4):787–797, vii
12. Prasarn ML, Ahn J, Achor T, Matuszewski P, Lorich DG, Helfet DL (2009) Management of infected femoral nonunions with a single-staged protocol utilizing internal fixation. Injury 40(11):1220–1225
13. Key JA (2009) The classic: sulfonamides in the treatment of chronic osteomyelitis. 1944. Clin Orthop Relat Res 467(7):1662–1669
14. Mouzopoulos G, Kanakaris NK, Kontakis G, Obakponovwe O, Townsend R, Giannoudis PV (2011) Management of bone infections in adults: the surgeon's and microbiologist's perspectives. Injury 42(Suppl 5):S18–S23
15. Sheehy SH, Atkins BA, Bejon P, Byren I, Wyllie D, Athanasou NA, Berendt AR, McNally MA (2010) The microbiology of chronic osteomyelitis: prevalence of resistance to common empirical anti-microbial regimens. J Infect 60(5):338–343

16. Mader JT, Cripps MW, Calhoun JH (1999) Adult posttraumatic osteomyelitis of the tibia. Clin Orthop Relat Res (360): 14–21
17. Lew DP, Waldvogel FA (1995) Quinolones and osteomyelitis: state-of-the-art. Drugs 49(Suppl 2):100–111
18. Isefuku S, Joyner CJ, Simpson AH (2001) Toxic effect of rifampicin on human osteoblast-like cells. J Orthop Res 19(5):950–954
19. Holtom PD, Pavkovic SA, Bravos PD, Patzakis MJ, Shepherd LE, Frenkel B (2000) Inhibitory effects of the quinolone antibiotics trovafloxacin, ciprofloxacin, and levofloxacin on osteoblastic cells in vitro. J Orthop Res 18(5):721–727
20. Galanakis N, Giamarellou H, Moussas T, Dounis E (1997) Chronic osteomyelitis caused by multi-resistant Gram-negative bacteria: evaluation of treatment with newer quinolones after prolonged follow-up. J Antimicrob Chemother 39(2):241–246
21. Allababidi S, Shah JC (1998) Kinetics and mechanism of release from glyceryl monostearate-based implants: evaluation of release in a gel simulating in vivo implantation. J Pharm Sci 87(6):738–744
22. Wang G, Liu SJ, Ueng SW, Chan EC (2004) The release of cefazolin and gentamicin from biodegradable PLA/PGA beads. Int J Pharm 273(1–2):203–212
23. Gursel I, Korkusuz F, Turesin F, Alaeddinoglu NG, Hasirci V (2001) In vivo application of biodegradable controlled antibiotic release systems for the treatment of implant-related osteomyelitis. Biomaterials 22(1):73–80
24. Kanellakopoulou K, Giamarellos-Bourboulis EJ (2000) Carrier systems for the local delivery of antibiotics in bone infections. Drugs 59(6):1223–1232
25. Schmidmaier G, Wildemann B, Stemberger A, Haas NP, Raschke M (2001) Biodegradable poly(D, L-lactide) coating of implants for continuous release of growth factors. J Biomed Mater Res 58(4):449–455
26. Menei P, Daniel V, Montero-Menei C, Brouillard M, Pouplard-Barthelaix A, Benoit JP (1993) Biodegradation and brain tissue reaction to poly(D, L-lactide-co-glycolide) microspheres. Biomaterials 14(6):470–478
27. Bostman O, Hirvensalo E, Makinen J, Rokkanen P (1990) Foreign-body reactions to fracture fixation implants of biodegradable synthetic polymers. J Bone Joint Surg Br 72(4):592–596
28. Price JS, Tencer AF, Arm DM, Bohach GA (1996) Controlled release of antibiotics from coated orthopedic implants. J Biomed Mater Res 30(3):281–286
29. Ramchandani M, Robinson D (1998) In vitro and in vivo release of ciprofloxacin from PLGA 50:50 implants. J Control Release 54(2):167–175
30. Baro M, Sanchez E, Delgado A, Perera A, Evora C (2002) In vitro-in vivo characterization of gentamicin bone implants. J Control Release 83(3):353–364
31. Benoit MA, Mousset B, Delloye C, Bouillet R, Gillard J (1997) Antibiotic-loaded plaster of Paris implants coated with poly lactide-co-glycolide as a controlled release delivery system for the treatment of bone infections. Int Orthop 21(6):403–408
32. Zhang Z, Kuijer R, Bulstra SK, Grijpma DW, Feijen J (2006) The in vivo and in vitro degradation behavior of poly(trimethylene carbonate). Biomaterials 27(9):1741–1748
33. Neut D, Kluin OS, Crielaard BJ, van der Mei HC, Busscher HJ, Grijpma DW (2009) A biodegradable antibiotic delivery system based on poly-(trimethylene carbonate) for the treatment of osteomyelitis. Acta Orthop 80(5):514–519
34. Van Leeuwen AC, Van Kooten TG, Grijpma DW, Bos RR (2012) In vivo behaviour of a biodegradable poly(trimethylene carbonate) barrier membrane: a histological study in rats. J Mater Sci Mater Med 23(8):1951–1959
35. Liao H, Walboomers XF, Habraken WJEM, Zhang Z, Li Y, Grijpma DW, Mikos AG, Wolke JGC, Jansen JA (2011) Injectable calcium phosphate cement with PLGA, gelatin and PTMC microspheres in a rabbit femoral defect. Acta Biomater 7(4):1752–1759
36. Kluin OS, van der Mei HC, Busscher HJ, Neut D (2009) A surface-eroding antibiotic delivery system based on poly-(trimethylene carbonate). Biomaterials 30(27):4738–4742
37. Le Ray AM, Chiffoleau S, Iooss P, Grimandi G, Gouyette A, Daculsi G, Merle C (2003) Vancomycin encapsulation in biodegradable poly(epsilon-caprolactone) microparticles for

bone implantation. Influence of the formulation process on size, drug loading, in vitro release and cytocompatibility. Biomaterials 24(3):443–449
38. Kim HW, Knowles JC, Kim HE (2004) Hydroxyapatite/poly(epsilon-caprolactone) composite coatings on hydroxyapatite porous bone scaffold for drug delivery. Biomaterials 25(7–8): 1279–1287
39. Lam CX, Hutmacher DW, Schantz JT, Woodruff MA, Teoh SH (2009) Evaluation of polycaprolactone scaffold degradation for 6 months in vitro and in vivo. J Biomed Mater Res A 90(3):906–919
40. Amsden BG, Tse MY, Turner ND, Knight DK, Pang SC (2006) In vivo degradation behavior of photo-cross-linked star-poly(epsilon-caprolactone-co-D, L-lactide) elastomers. Biomacromolecules 7(1):365–372
41. Tabata Y, Yamada K, Miyamoto S, Nagata I, Kikuchi H, Aoyama I, Tamura M, Ikada Y (1998) Bone regeneration by basic fibroblast growth factor complexed with biodegradable hydrogels. Biomaterials 19(7–9):807–815
42. Castro C, Evora C, Baro M, Soriano I, Sanchez E (2005) Two-month ciprofloxacin implants for multibacterial bone infections. Eur J Pharm Biopharm 60(3):401–406
43. Puga AM, Rey-Rico A, Magarinos B, Alvarez-Lorenzo C, Concheiro A (2012) Hot melt poly-epsilon-caprolactone/poloxamine implantable matrices for sustained delivery of ciprofloxacin. Acta Biomater 8(4):1507–1518
44. Jain JP, Modi S, Domb AJ, Kumar N (2005) Role of polyanhydrides as localized drug carriers. J Control Release 103:541–563
45. Jain JP, Chitkara D, Kumar N (2008) Polyanhydrides as localized drug delivery carrier: an update. Expert Opin Drug Deliv 5(8):889–907
46. Saito N, Takaoka K (2003) New synthetic biodegradable polymers as BMP carriers for bone tissue engineering. Biomaterials 24(13):2287–2293
47. Gunatillake PA, Adhikari R (2003) Biodegradable synthetic polymers for tissue engineering. Eur Cell Mater 5:1–16, discussion 16
48. Li LC, Deng J, Stephens D (2002) Polyanhydride implant for antibiotic delivery–from the bench to the clinic. Adv Drug Deliv Rev 54(7):963–986
49. Laurencin CT, Gerhart T, Witschger P, Satcher R, Domb A, Rosenberg AE, Hanff P, Edsberg L, Hayes W, Langer R (1993) Bioerodible polyanhydrides for antibiotic drug delivery: in vivo osteomyelitis treatment in a rat model system. J Orthop Res 11(2):256–262
50. Brin YS, Nyska A, Domb AJ, Golenser J, Mizrahi B, Nyska M (2009) Biocompatibility of a polymeric implant for the treatment of osteomyelitis. J Biomater Sci Polym Ed 20(7–8):1081–1090
51. Brin YS, Golenser J, Mizrahi B, Maoz G, Domb AJ, Peddada S, Tuvia S, Nyska A, Nyska M (2008) Treatment of osteomyelitis in rats by injection of degradable polymer releasing gentamicin. J Control Release 131(2):121–127
52. Krasko MY, Golenser J, Nyska A, Nyska M, Brin YS, Domb AJ (2007) Gentamicin extended release from an injectable polymeric implant. J Control Release 117(1):90–96
53. Shikanov A, Vaisman B, Krasko MY, Nyska A, Domb AJ (2004) Poly(sebacic acid-co-ricinoleic acid) biodegradable carrier for paclitaxel: in vitro release and in vivo toxicity. J Biomed Mater Res A 69(1):47–54
54. Du J, Jasti B, Vasavada RC (1997) Controlled release of tobramycin sulfate from poly(ortho esters) implantable discs for the treatment of osteomyelitis. J Control Release 43(2–3): 223–233
55. Knaepler H (2012) Local application of gentamicin-containing collagen implant in the prophylaxis and treatment of surgical site infection in orthopaedic surgery. Int J Surg 10(Suppl 1):S15–S20
56. Geiger M, Li RH, Friess W (2003) Collagen sponges for bone regeneration with rhBMP-2. Adv Drug Deliv Rev 55(12):1613–1629
57. Riegels-Nielsen P, Espersen F, Holmich LR, Frimodt-Moller N (1995) Collagen with gentamicin for prophylaxis of postoperative infection. Staphylococcus aureus osteomyelitis studied in rabbits. Acta Orthop Scand 66(1):69–72

58. Zhou J, Fang T, Wang Y, Dong J (2012) The controlled release of vancomycin in gelatin/beta-TCP composite scaffolds. J Biomed Mater Res A 100(9):2295–2301
59. Di Silvio L, Bonfield W (1999) Biodegradable drug delivery system for the treatment of bone infection and repair. J Mater Sci Mater Med 10(10/11):653–658
60. George M, Abraham TE (2006) Polyionic hydrocolloids for the intestinal delivery of protein drugs: alginate and chitosan—a review. J Control Release 114:1–14
61. Yenice I, Calis S, Kas H, Ozalp M, Ekizoglu M, Hincal A (2002) Biodegradable implantable teicoplanin beads for the treatment of bone infections. Int J Pharm 242(1–2):271–275
62. Yenice I, Calis S, Atilla B, Kas HS, Ozalp M, Ekizoglu M, Bilgili H, Hincal AA (2003) In vitro/in vivo evaluation of the efficiency of teicoplanin-loaded biodegradable microparticles formulated for implantation to infected bone defects. J Microencapsul 20(6):705–717
63. Khor E, Lim LY (2003) Implantable applications of chitin and chitosan. Biomaterials 24:2339–2349
64. Higuchi T (1963) Mechanism of sustained-action medication. Theoretical analysis of rate of release of solid drugs dispersed in solid matrices. J Pharm Sci 52:1145–1149
65. Cevher E, Orhan Z, Mulazimoglu L, Sensoy D, Alper M, Yildiz A, Ozsoy Y (2006) Characterization of biodegradable chitosan microspheres containing vancomycin and treatment of experimental osteomyelitis caused by methicillin-resistant Staphylococcus aureus with prepared microspheres. Int J Pharm 317(2):127–135
66. Turesin F, Gursel I, Hasirci V (2001) Biodegradable polyhydroxyalkanoate implants for osteomyelitis therapy: in vitro antibiotic release. J Biomater Sci Polym Ed 12(2):195–207
67. Yates CC, Whaley D, Babu R, Zhang J, Krishna P, Beckman E, Pasculle AW, Wells A (2007) The effect of multifunctional polymer-based gels on wound healing in full thickness bacteria-contaminated mouse skin wound models. Biomaterials 28(27):3977–3986
68. Elsner JJ, Berdicevsky I, Zilberman M (2011) In vitro microbial inhibition and cellular response to novel biodegradable composite wound dressings with controlled release of antibiotics. Acta Biomater 7(1):325–336
69. Kimakhe S, Bohic S, Larrose C, Reynaud A, Pilet P, Giumelli B, Heymann D, Daculsi G (1999) Biological activities of sustained polymyxin B release from calcium phosphate biomaterial prepared by dynamic compaction: an in vitro study. J Biomed Mater Res 47(1):18–27
70. Peltier LF, Jones RH (2004) Treatment of unicameral bone cysts by curettage and packing with plaster-of-Paris pellets. 1978. Clin Orthop Relat Res (422): 145–147
71. Calhoun JH, Manring MM, Shirtliff M (2009) Osteomyelitis of the long bones. Semin Plast Surg 23(2):59–72
72. Scharer BM, Sanicola SM (2009) The in vitro elution characteristics of vancomycin from calcium phosphate-calcium sulfate beads. J Foot Ankle Surg 48(5):540–542
73. Beardmore AA, Brooks DE, Wenke JC, Thomas DB (2005) Effectiveness of local antibiotic delivery with an osteoinductive and osteoconductive bone-graft substitute. J Bone Joint Surg Am 87(1):107–112

Chapter 22
Spatiotemporal Focal Delivery of Dual Regenerating Factors for Osteochondral Defect Repair

Emil Ruvinov and Smadar Cohen

22.1 Introduction

The management of articular cartilage damage, osteochondral lesions, and frequently associated osteoarthritis (OA) of the synovial joints represents one of the most significant challenges facing the global healthcare community today. The high burden of these conditions is associated with poor intrinsic repair and regeneration ability of the articular cartilage, making an adequate treatment difficult to achieve. Osteochondral lesions, which involve both the articular cartilage and the subchondral bone, typically lead to the formation of fibrocartilage that has different biomechanical properties from the native hyaline cartilage and does not protect the subchondral bone from further degeneration. This process involves both repaired and adjacent native tissues leading to the insurgence of severe pain, joint deformity, and loss of joint motion. Thus, osteochondral defects typically require surgical procedures, in many cases requiring more than one surgical intervention. While cell-based strategies for osteochondral defect repair are already implemented in clinics, their reparative potential and long-term durability is modest, and no obvious treatment option has become apparent.

E. Ruvinov
Avram and Stella Goldstein-Goren Department of Biotechnology Engineering,
Ben-Gurion University of the Negev, Negev, Israel

S. Cohen (✉)
Avram and Stella Goldstein-Goren Department of Biotechnology Engineering,
Ben-Gurion University of the Negev, Negev, Israel

The Ilse Katz Institute for Nanoscale Science and Technology,
Ben-Gurion University of the Negev, Negev, Israel

The Regenerative Medicine and Stem Cell (RMSC) Research Center,
Ben-Gurion University of the Negev, Negev, Israel
e-mail: scohen@bgu.ac.il

A.J. Domb and W. Khan (eds.), *Focal Controlled Drug Delivery*, Advances in Delivery Science and Technology, DOI 10.1007/978-1-4614-9434-8_22,

The emergence of tissue engineering strategies may provide a novel treatment foundation for the repair and regeneration of osteochondral defects. Tissue engineering is an exciting interdisciplinary research field, combining knowledge from collaborations in life sciences and engineering, which has emerged as a potential source of future treatment strategies for a range of diseases. Conventional tissue engineering is facilitated through the use of three principal elements: scaffolds, cells, and bioactive molecules (i.e., growth factors), which are combined in a suitable biological environment to produce engineered tissue in vitro or as a regenerative medicine strategy in vivo. While the search for adequate and clinically feasible and available cell source continues, the therapeutic potential of various soluble bioactive molecules has already been realized in different pathologies, including osteochondral lesions. The use of various solid or injectable biomaterial platforms for controlled and local bioactive molecule delivery has significantly boosted the application of such substances, where now many deleterious and negative effects of the systemic high-dosage administration of potentially therapeutic growth factors and cytokines could be avoided. Moreover, these platforms could induce tissue ingrowth in situ or provide support for implanted cells. The development of systems with spatiotemporal presentation of multiple biological signals further enhances their regenerative potential of complex multiphase structures, such as the osteochondral junction.

This chapter presents an overview of the various therapeutic strategies applied for regenerating the osteochondral defect, with a special focus on the development and application of biomaterial-based acellular strategies for the spatiotemporal localized delivery of chondrogenic and osteogenic regenerative factors.

22.2 The Osteochondral Interface

22.2.1 Structure

Osteochondral tissue is a permanent interfacial structure between bone and cartilage. This junction is a gradual transition from articular cartilage (AC) to trabecular bone, in which key constituents of each tissue undergo a gradual exchange in predominance (Fig. 22.1).

Articular cartilage is a thick (e.g., 0.9 ± 0.5 mm in human phalanges and 2.4 ± 0.5 in human medial femoral condyles), avascular, aneural tissue [1]. It is a biphasic, structurally nonhomogeneous structure consisting of a solid phase of about 15–32 % and a fluid phase of about 68–85 %. The solid phase is mainly composed of collagen and proteoglycans. The fluid phase is mainly composed of water [2]. The ultrastructure of articular cartilage has a distinct zonal architecture, which is not only a developmental phenomenon but also a characteristic that is evident in adult articular cartilage, after the formation of the tidemark and the zone of calcified cartilage. Each distinct zone varies in structure and function, responds to different stimuli, and secretes different proteins. Chondrocytes isolated from each zone have unique

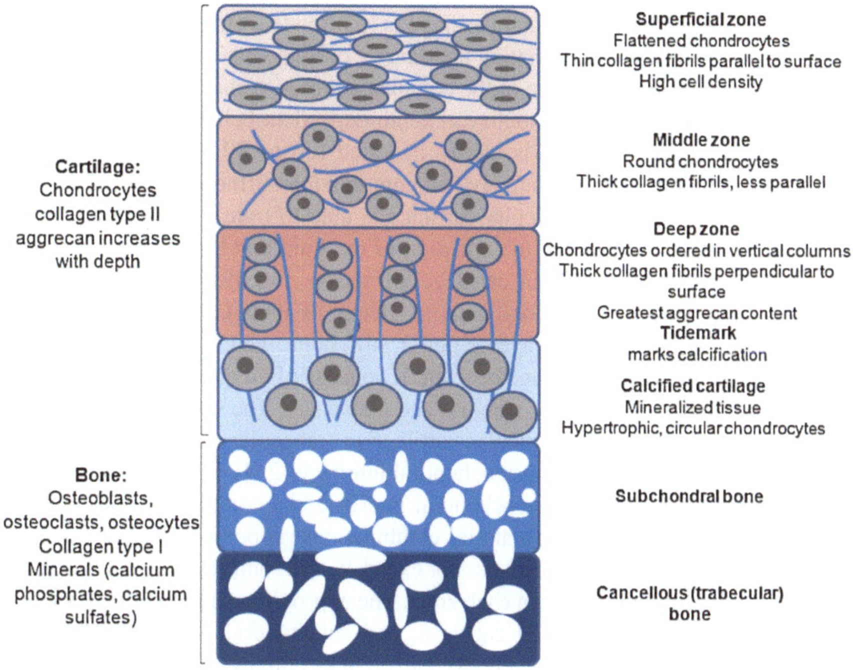

Fig. 22.1 The structure of osteochondral interface. See text for details

growth rates, gene expression profile, and levels of biosynthesis. An increase in GAG and collagen content is observed with increased depth, providing the deep zone with superior mechanical properties compared to superficial zone [3]. In each of these zones, the collagen fibrils are oriented differently.

The articular cartilage comprises four distinct zones: (1) the superficial tangential (or gliding) zone, (2) the middle (or intermediate, transitional) zone, (3) the deep (or radial) zone, and (4) the calcified zone, which is located immediately below the tidemark and above the subchondral bone. In the superficial zone, there are few small and flattened chondrocytes with low metabolic activity, and their surrounding matrix consists of collagen fibers oriented parallel to the joint surface and high water content. The middle zone, comprising 40–60 % of the cartilage weight, consists of round chondrocytes surrounded by radial bundles of thick collagen fibrils. In the deep zone, chondrocytes are frequently grouped into columns and clusters, with thick collagen bundles aligned perpendicular to articular surface. From the surface to the deep zone, the cell volume and metabolic activity increase, progressively reducing the cell density and increasing the proportion of proteoglycan relative to collagen content. Finally, the calcified zone is where the transition from soft to stiff subchondral bone occurs. The calcified (with high concentration of calcium salts) zone is responsible for firmly attaching the noncalcified cartilage to the underlying subchondral bone. Chondrocytes in this area are larger, more dispersed, and referred to as hypertrophic [1, 2, 4–7].

The main function of articular cartilage is to provide a smooth near-frictionless articulating surface while mediating the transfer of load within the joint to underlying subchondral bone. Because of the complex composite character, articular cartilage exhibits anisotropic and nonlinear, depth-dependent behavior in compression, tension, and shear. Cartilage is highly hydrated (up to 80 % of water content) composite structure with a relatively low compression stiffness (E=0.1–2.0 MPa). Despite the low stiffness, articular cartilage is able to transmit high loads (up to eight times body weight) due to exudation of the fluid and the movement of fluid through the porous structure of cartilage. Moreover, together with synovial fluid, articular cartilage provides a coefficient of friction of about 0.008 [2].

Subchondral bone is a thin layer of dense, stiff, bone, which contacts the articular cartilage and the trabecular bone beneath. It consists of an abundant matrix of collagen fibers and minerals, e.g., calcium and phosphate. The subchondral bone acts in load transition from soft articular cartilage to trabecular and cortical bone and comprises the last level in the osteochondral interface [2, 4].

In summary, the natural osteochondral interface is a complex composite structure with gradually changing material combinations/properties, volume fractions, and anisotropy at different locations. For example, the compressive modulus varies on orders of magnitude between each zone. The approximate moduli of the superficial zone, deep zone, calcified cartilage, and subchondral bone are 0.079, 2.1, 320 MPa, and 5.7 GPa, respectively [4]. Due to this complexity, the exact recreation of this interface by tissue engineering is a major challenge. However, rather simplified approaches using biphasic/bilayered systems could represent an important advancement in this direction.

22.2.2 *Pathophysiology*

The structure and function of the osteochondral tissue are frequently disrupted or lost in trauma and, more commonly, in degenerative joint diseases, such as osteoarthritis (OA). In arthritic joints, the articular cartilage is degraded (as a result of destruction of cartilage matrix), leading to a progressive loss in joint function, causing pain, loss of movement, and joint stiffness. OA is the most prevalent disorder of the musculoskeletal system, affecting about 10 % of the US population older than 30 and most of the people over the age of 65, with total direct costs estimated at $28.6 billion/year, yet current surgical treatment strategies have been largely ineffective in preventing or even impeding the development of the disease [8, 9]. Disturbingly, the strong correlation between OA and age suggests that the incidence and economic impact of the disease will continue to increase substantially unless appropriate novel treatment and management strategies can be formulated and adopted. Moreover, prior knee injuries could also lead to the development of secondary OA. The limited knowledge of the pathophysiological mechanisms responsible for the progression of OA presents an additional treatment challenge [8]. In addition to OA, which mainly affects articular cartilage, certain defects, such as those resulting from osteochondritis dissecans, may start in the subchondral bone, only secondarily affecting the overlying cartilage. Other joint pathologies involving the subchondral bone include osteonecrosis and osteochondral fractures. There has been increased awareness that the subchondral bone plays an important role even in superficial lesions limited to the articular cartilage layer, since even focal chondral defects, if left untreated, may increase in size over time and result in concomitant changes in the underlying subchondral bone plate, either overgrowth or bone loss, suggesting that conditions of articular cartilage and its supporting bone are tightly coupled and should be viewed as a connected osteochondral unit [10, 11].

Regardless of different etiology (trauma, sport injuries, or degenerative diseases), articular cartilage (partial) or osteochondral (extending into subchondral bone, full thickness) defects result in identical clinical end stage: reduced articular function, manifesting as pain and impaired movement. Importantly, endogenous repair of the two defect types follow different routes. Articular cartilage lesions are typically irreversible, due to the unique features of this tissue, including its avascular nature and consequent lack of access to a pool of potential reparative cells and humoral factors. Also, cartilage has a low cell to matrix ratio, and mature chondrocytes have a relatively low metabolic activity that limits any considerable capacity of remodeling. In cases where the lesions do not penetrate to the subchondral bone, the pluripotent progenitor cells from the bone marrow cannot be recruited and the repair will be limited. On the other hand, full-thickness (osteochondral) defects go through the subchondral bone, accessing the bone marrow cells, including mesenchymal stem cells, and also growth factors and cytokines. However, the repair response typically leads to the formation of fibrocartilage in the defect void. Fibrocartilage is a scar tissue presenting diminished resilience, reduced stiffness, and poor wear characteristics when compared to hyaline cartilage. Thus, fibrocartilage cannot withstand physiological loading and cannot guarantee to function successfully in long term [12]. Such poor substitution for articular cartilage will result in degeneration of both repaired and adjacent native tissues after long-term follow-up [9, 13].

The biomechanical alterations caused by (osteo)-chondral defects affect the articular cartilage surrounding and opposing the lesion as well as the homeostatic balance of the entire joint. Tight structural and functional connection between cartilage and bone through osteochondral interface and the importance of subchondral bone for its role in the pathogenic processes in the joint further emphasize the need for this integrity preservation. Therefore, the treatment goal for large chondral or osteochondral defects should be to restore the physiological properties of the entire osteochondral unit, aiming to achieve a more predictable repair tissue that closely resembles native articular surface and subchondral bone, and remains durable over time.

22.2.3 Current Treatments for Osteochondral Repair

The therapeutic tools aim to fill the cartilage and bone loss so as to restore joint congruence and, if possible, to induce bone formation and hyaline healing, and thus to prevent long-term degeneration. They can be classified into subchondral stimulation repair methods (also referred as marrow-stimulating procedures), most frequently resulting in a fibrous scar (microfracturing, Pridle drilling, and abrasion), reconstruction methods contributing mature cartilage to the osteochondral unit (mosaicplasty and osteochondral auto-/allografting for massive chondral defects), and regeneration through grafting of autologous chondrocytes (see Sect. 22.3) aiming for hyaline repair. Nevertheless, the follow-up biopsies are often disappointing, with hyaline cartilage frequently absent even in very costly regeneration using cell culture implantation techniques [11, 14, 15]. More recently, tissue engineering strategies (utilizing three-dimensional scaffolds or hydrogel systems) entered the field with the potential to provide a platform for effective repair and improve tissue quality and long-term effects [6, 12, 14–17].

22.2.3.1 Microfractures

These techniques rely on the stimulation of the underlying bone marrow, resulting in fibrocartilage growth. After curetting the osteochondral lesion down to subchondral bone, a tapered awl can be used to produce the microfractures, approximately 3 mm in depth and 3–5 mm apart. The microfractures result in a blood clot containing mesenchymal stem cells that then form a fibrocartilaginous tissue [2, 14, 15]. In a systematic analysis of 28 studies of over 3,000 patients, with an average follow-up duration of 41 months, the authors showed that microfracture provided effective short-term functional improvement of knee function, but there were poor long-term results. The other shortcomings revealed included poor hyaline repair, variable cartilage volume, and long-term functional deterioration [18].

22.2.3.2 Osteochondral Autograft/Allograft Transfer (Mosaicplasty)

Osteochondral autograft transfer (OAT) works by removing several plugs of hyaline cartilage and the underlying subchondral bone from an unaffected, non-weight-bearing area of the knee. These are used as autograft implants and plugged into the osteochondral defect. There are several problems with the OAT procedure, the main one being that the topography of the donor site does not match the recipient site and will therefore change the biomechanics and loading. Although very similar to OAT, osteochondral allograft transfer (OALT) does not rely on a donor site but on a cadaveric donor. OALT will theoretically have a like-for-like replacement with no donor site morbidity. It should be, however, biomechanically and topographically similar. Results have shown good to excellent outcomes in up to 80 % of cases in some reports and larger defects can be filled. Apart from having to be an open procedure, OALT carries the disadvantages of rejection, viral disease transmission, and tissue availability [2, 14, 15].

22.3 Cell Therapy for Osteochondral Repair

Achieving effective functional repair relies on reconstruction of tissue composition. Ideally, chondrocytes represent a perfect cell source to recreate damaged or lost cartilage tissue. Naturally, this possibility was further developed and subsequently applied in clinical practice. Alternatively, cartilage or bone regeneration could be achieved by controlled differentiation of mesenchymal stem cells, and their use now represents a promising alternative strategy for osteochondral repair.

22.3.1 Autologous Chondrocyte Implantation

22.3.1.1 First-Generation ACI

In 1994 Brittberg et al. described the use of Autologous Chondrocyte Implantation (ACI) in treating osteochondral defects of the human knee [19]. This was achieved with a two-stage procedure. Stage 1 involved arthroscopic biopsy of healthy AC and culture of the chondrocytes to produce between five and ten million cells over a period of 4–6 weeks. Stage 2 involved debridement of the osteochondral lesion and coverage by a periosteal flap followed by open implantation of these cells into the defect. The periosteum is sutured with fine sutures and sealed with fibrin glue to make a watertight seal. The cultured cells are injected beneath it into the defect. ACI has shown encouraging results. In 2002 Peterson et al. examined the durability of ACI grafts, showing 84 % had "good" to "excellent" results at 5–11 years [20]. One year later, they evaluated treatment of osteochondritis dissecans with ACI, revealing a 90 % successful clinical result [21].

22.3.1.2 Second-Generation ACI

Use of periosteum proved problematical as hypertrophy of the membrane producing painful clicking occurred in 25 % of patients, who required arthroscopic resection [15]. In addition, the use of periosteum patch was frequently associated with ossification, detachment, calcification, and leakage [14]. In a randomized controlled trial, Gooding et al. demonstrated the superiority of a type I/III porcine collagen membrane matrix as a cover for the graft [22]. This served as a basis for the development of second-generation ACI, a matrix-assisted chondrocyte implantation (MACI). This tissue engineering approach utilizes type I/III collagen (or other biomaterial-based) scaffold, to provide a matrix preimplantation. This eliminates the need for a periosteal patch and negative effects associated with such patching. The cells are cultured on the scaffold, which is then implanted into the defect and secured with fibrin glue. This speeds up the procedure greatly but has the potential disadvantage of a much lower number of implanted cells, up to five times fewer [15, 23].

In addition, MACI technique costs approximately three times more than the standard technique [15].

Despite thousands of treated patients and many published studies suggesting good and durable clinical results of this regenerative surgical treatment approach, there is no agreement to date about the effective superiority of one of these techniques over another, and both indications and results are still being discussed controversially [24]. Moreover, studies comparing different treatment techniques, such as microfracture, mosaicplasty, and ACI, are also inconclusive. While all treatments appear to improve clinical outcome measures compared with preoperative assessment, no technique consistently had superior results [9, 24, 25]. There is emerging evidence that the repair tissue produced in ACI is more hyaline like and more durable than that following microfracture. However, there is currently no firm evidence that ACI or MACI provides improved patient outcomes compared with either microfracture or mosaicplasty. As the collection of the data continues, the quality of trial methodology investigating osteochondral defects should be improved by adequately describing randomization procedures, including untreated control groups, use of validated outcome measures, and use of an independent investigator or outcome assessment.

22.3.2 Stem Cell-Based Strategies

Despite promising clinical results of ACI and MACI, the use of chondrocytes may be troublesome due to the requirement of large numbers of chondrocytes to fill large defect volumes and dedifferentiation of cultured chondrocytes in vitro. Moreover, the harvest of cartilage for a cartilage biopsy includes two separate operations and presents limitations associated with donor site morbidity. Stem cells represent an alternative and promising cell source for tissue regeneration in general and for osteochondral regeneration in particular [3, 26–28].

The evaluation of chondrogenic and osteogenic potential of human embryonic stem cells (hESC) as a developmentally relevant cell source is currently under investigation. Human ESCs have the potential to generate chondrocytes with phenotype closer to that of articular cartilage. However, the clinical application of these cells is much further away as differentiation protocols, and possible tissue engineering strategies require additional optimization [28]. In addition, ongoing ethical issues and immunorejection problems, together with unresolved safety concerns about the tumorigenicity of hESC, are still a major concern.

Tissue-derived mesenchymal stem cells (MSCs) represent the most practical and autologous alternative for osteochondral regeneration. Initially isolated from bone marrow, MSCs, defined as self-renewal, multipotent progenitor cells with the capacity to differentiate into several distinct mesenchymal lineages (chondrocytes, osteoblasts, adipocytes, and other cell types), represent a promising cell source for treating osteochondral defects [5, 17, 22]. More recently, it has been demonstrated that MSCs can also be reproducibly isolated from human adipose tissue, which is more abundant and easily accessible cell source. However, MSCs from adipose tissue were reported to have a reduced potential for chondrogenic differentiation, compared to bone marrow-derived cells [6, 29].

Despite promising results in vitro and in animal models, the optimal conditions to differentiate MSCs towards chondrocytes or osteoblasts still have to be clearly established and optimized. For instance, the ability of MSCs to form stable hyaline cartilage is questioned. The resulting chondrogenic cells show signs of fibrocartilage, hypertrophy, and mineralization. Hypertrophy has been associated with the development of endochondral-like cartilage and ossification in vivo after ectopic transplantation of chondro-induced cells, resulting in the loss of cartilage and its replacement by bone [29–32]. Some studies showed improved cartilage quality when MSCs were first allowed to mature (pre-differentiate) in vitro for extended period of time prior to implantation [5, 24]. Therefore, osteochondral repair by MSCs is still not sufficiently robust to allow large-scale clinical use with predictable outcomes. Moreover, the formation of osteochondral interface depends on our ability to induce and maintain osteogenesis and chondrogenesis in various regions, requiring tight and spatial control of cell differentiation, which can only be achieved by biomaterial matrix (scaffold or hydrogel)-assisted cell delivery. Spatial separation and presentation of osteo- and chondroinductive signals (see Chap. 4) within such system could guide cell differentiation into respective cell types in situ.

The applications of MSC for osteochondral repair in humans are still rare and are limited to case reports and cohort studies and not large randomized trials. A recent cohort study reported similar clinical results 2 years postoperatively for transplantation of autologous bone marrow-derived MSC as for ACI [33]. In two case reports, the patients with osteochondral patellar defects were treated with MSCs seeded within type I collagen hydrogel [34]. Fibrocartilaginous filling of the defects was found after 1 year, and both patients showed significantly improved clinical outcomes in their respective follow-ups after 1, 4, and 5 years. The same group has also used this protocol to treat another patient with a full-thickness cartilage defect in the weight-bearing area of the medial femoral condyle. The patient's clinical symptoms had improved significantly 1 year after surgery. Histologically, the defect was filled with a hyaline-like type of cartilage tissue [35]. Despite these indications, it is clear that larger clinical trials are required to evaluate the treatment potential of MSCs for osteochondral repair.

22.4 Bioactive Factors for Osteochondral Regeneration and Repair

Induction of osteochondral tissue formation refers to the capacity of many physiological stimuli to stimulate exogenous or endogenous stem cells or immature bone/cartilage cells to grow and mature, forming healthy tissue. Most of these stimuli are protein molecules (e.g., growth factors and cytokines). In osteochondral junction, numerous growth factors work in concert to regulate development, homeostasis, and repair of articular cartilage and subchondral bone. Therefore, growth factors offer promising treatments for enhanced regeneration of focal articular cartilage defects or in situations of larger full-thickness osteochondral lesions. Careful analysis and considerations, however, are required in choosing the appropriate growth factors for treatments, as most of them have pleiotropic functions, time- and location-dependent actions, and synergistic or inhibitory effects when used in combination. On the other hand, given the array and interactions of growth factors that are necessary for proper tissue development, homeostasis, and regeneration, it is unlikely that any single growth factor will lead to complete repair, but rather a combination approach will be required. Below we briefly describe major factors that are important for osteochondral tissue regeneration. Noticeably, same factors can exert their action both on chondrogenesis and osteogenesis, as these are developmentally linked pathways, further emphasizing the need for careful selection of the factors when combinations are being used, and their spatial separation, to achieve differential and effective outcome.

22.4.1 Factors Inducing Cartilage Regeneration

Uniquely, cartilage does not include blood vessels, nerve tissue, and lymphocytes. Due to lower regenerative ability, it is very difficult to recover cartilage tissue after injury. Nevertheless, it has been shown that there are a number of essential growth and differentiation factors providing regulatory effects on chondrocytes or stem cells involved in chondrocyte maturation and cartilage formation.

22.4.1.1 Transforming Growth Factor-β

Members of the highly conserved transforming growth factor-β (TGF-β) superfamily play an important role in embryonic development, tissue morphogenesis, cell proliferation, and cell differentiation. TGF-βs, activins, growth differentiation factors (GDF), bone morphogenic proteins (BMPs), and their various isoforms share similar biological activities through their homologous polypeptide structure, which only differs in the C-terminal amino acid sequences.

The growth factors from TGF-β superfamily probably belong to the most investigated biologically active substances within the field of cartilage regeneration and repair. The TGF-β family includes five members (TGF-β1–5), which are predominantly produced in bone and cartilage. Active TGF-β1, 2, and 3 are generally considered to be potent stimulators of proteoglycans (aggrecan) and of type II collagen synthesis in chondrocytes, typical to hyaline cartilage, and are able to induce the chondrogenic differentiation of MSCs in vitro. In vivo, TGF-β1 can induce the chondral differentiation of MSCs to form ectopic cartilage and was able to repair a full-thickness cartilage defect by improving chondrocyte integration into the endogenous tissue [36–38]. However, direct injection of TGF-β or of TGF-β-expressing adenoviruses resulted in side effects in the joints, such as osteophyte formation, swelling, and synovial hyperplasia, suggesting that a tightly coordinated regulation and local application of TGF-β are needed to control chondrogenesis [36, 38].

22.4.1.2 Bone Morphogenic Proteins

BMPs are homodimeric molecules and belong to TGF-β superfamily. They play crucial roles in the processes of chondrogenesis and osteogenesis in vivo [39]. Several BMPs, including BMP-2, BMP-4, BMP-6, BMP-7, BMP-13, and BMP-14, can stimulate the chondrogenic differentiation of MSCs and enhance the synthesis of collagen II and aggrecan by chondrocytes in vitro. In this regard, the effect of BMP-2 on MSCs is similar to that of TGF-β1. In vivo, BMP-2 enhanced cartilage repair in full-thickness cartilage defects [10, 29, 38]. BMP-7 is synthesized by chondrocytes and plays an important role in articular cartilage regeneration and currently appears to be one of the most potent growth factors for cartilage repair. It has significant anabolic activity by which BMP-7 protects cartilage against damage. Experiments with MSCs demonstrated that BMP-7 decreases their proliferation activity but stimulates expression of cartilaginous ECM. More recently it was shown that this effect may be significantly enhanced when BMP-7 is used in combination with TGF-β1 or TGF-β3 and in the presence of insulin-like growth factor 1 (IGF-1) [29, 38]. Although BMP-7 is also highly effective at stimulating bone repair, it does not appear to lead to osteophyte formation when administered into a joint nor does it stimulate uncontrolled fibroblast proliferation leading to joint fibrosis [38].

22.4.1.3 Insulin-Like Growth Factor-1

Within the articular cartilage, IGF-1 is the main anabolic growth factor, which is necessary for cartilage homeostasis, proteoglycan synthesis, and breakdown by the chondrocytes [29]. When added exogenously to monolayer or explant cultures of normal articular cartilage from a variety of species, IGF-1 induces a plethora of anabolic effects and decreases catabolic responses. Chondrogenic differentiation of MSCs is induced by IGF-1 but is enhanced when IGF-1 is used in combination with TGF-β1, BMP-2, or BMP-7 [38]. They have an additive effect, which leads in a significant increase of cartilage matrix synthesis [29]. In animal models, IGF-1 deficiency led to the development of articular cartilage lesions. Moreover, IGF-1 has led to enhanced repair of extensive cartilage defects and protection of the synovial membrane from chronic inflammation. Despite the diminished ability of IGF-1 to decrease catabolism in aged and OA cartilage, studies suggest that a combination of IGF-1 and BMP-7 results in greater repair potential than either growth factor alone.

These studies demonstrated that in general, BMP-7 was more potent than IGF-1 in stimulating matrix synthesis in aged and OA cells, but the greatest increase in matrix synthesis was observed after combination treatment with BMP-7 and IGF-1 [38].

22.4.1.4 Fibroblast Growth Factors

In humans, 22 members of the FGF family have been identified. Two members of the family have been investigated for their role in cartilage homeostasis, FGF-2 (or basic FGF, bFGF) and FGF-18. Basic FGF is found in relative abundance in the pericellular matrix of cartilage. On loading, FGF-2 becomes bound to cell surface receptors and activates anabolic pathways leading to decreased aggrecanase activity but no apparent change in proteoglycan content [40]. It also acts as a mitogen, inducing chondrocyte proliferation in vivo, inhibiting their terminal differentiation [37]. In MSC cultures, bFGF promotes chondrogenic differentiation [37]. However, several studies showed negative results associated with bFGF administration, such as inflammation and osteophyte formation [38]. FGF-18 appears more promising than bFGF and elicits several anabolic effects on chondrocytes [33, 41]. However, the amount of data on FGF-18 effects on cartilage repair is still insufficient to clarify its potential.

22.4.2 Factors Inducing Bone Regeneration

Bone formation is a complex process that involves a large number of hormones, cytokines, and growth factors. Several factor families and individual factors were found to be particularly important for therapeutic bone regeneration and repair [42–44].

22.4.2.1 Bone Morphogenic Proteins

BMPs are implicated in a variety of functions, mainly acting as a differentiation stimulating factor to direct endochondral ossification and chondrogenesis of mesenchymal stem cells. Of the 20 BMPs discovered, only three (BMP-2, BMP-4, and BMP-7) have been able to stimulate the osteoprogenitor differentiation into mature osteoblasts in vitro. Moreover, these three proteins have a strong efficacy for inducing de novo bone formation in ectopic and orthotopic sites, including critical size defects. BMPs have been tested in preclinical and clinical studies, showing their definite potential in osteoinduction, and have been FDA-approved for clinical use (BMP-2 and BMP-7) in open fracture of long bones, nonunions, and vertebral arthrodesis [26, 39, 42, 45, 46].

22.4.2.2 Angiogenic Factors

Vascularization for the transport of oxygen, nutrients, growth and differentiation factors, and circulating cells is essential for the formation and homeostasis of bone. The presence of a local microvascular network supports the osteogenic, chondrogenic, and mesenchymal stem cells required for bone repair. Angiogenesis is regulated by soluble molecules such as vascular endothelial growth factor (VEGF), platelet-derived growth factor (PDGF), FGF, and IGF. Bone research with angiogenic factors has primarily focused on VEGF's role in neovascularization and osteogenic recruitment, survival, and activity. VEGF delivery was found to increase blood vessel density and stimulate slight bone regeneration in rabbit and rat critical size bone defects. Recent studies have shown that the combined delivery of VEGF with osteoinductive growth factors synergistically enhances osteogenesis [42, 43].

22.4.3 Challenges in Factor Administration

The method by which a growth factor is delivered can have a significant effect on therapeutic efficacy because the dose and spatiotemporal release of such agents at the lesion site is crucial for achieving a successful outcome. Common growth factor delivery methods involve systemic administration or direct injection into the defect site close to or in direct contact to the supportive scaffold. However, as a result of the short half-life of many inductive proteins, this method requires very high doses for therapeutic effect and still may not permit the necessary concentration of the factor to be maintained for the appropriate period of time. Results from systemic administration of growth factors are often unpredictable, probably due to their short biological half-life, lack of long-term stability, tissue specificity, and potential dose-dependent carcinogenicity, toxicity, and other undesired effects (e.g., fibrosis, osteophyte formation) [16, 42].

The use of polymeric carriers in various forms (hydrogels or scaffolds) for factor delivery represents a powerful alternative to overcome the drawbacks of systemic delivery. The carrier acts as a local regulator to control doses and kinetics of released growth factors, while simultaneously protecting them from degradation, thus increasing their potential retention time at therapeutic concentration levels. The importance of the carrier is not limited to such roles. Recently, the role of carriers was extended to serving as a temporary substrate and three-dimensional matrix for cellular infiltration, in which cells can proliferate and differentiate into particular tissue types, in concert with degradation of the carrier material. The use of such carrier is indispensable for osteochondral tissue regeneration, where spatial separation of the osteoinductive and chondroinductive signals is essential.

22.5 Biomaterial-Based Delivery Systems for Osteochondral Repair and Regeneration

In this section, we review major advancements that have been made in creation of biomaterial-based acellular delivery systems for osteochondral repair and regeneration (Fig. 22.2). Our main focus will be on animal-tested systems designed for simultaneous repair/regeneration of cartilage and subchondral bone.

22.5.1 Basic Requirements and Design Criteria

The design of biomaterial-based products used for osteochondral tissue engineering and regeneration should comply with the following basic criteria:

- *Biocompatibility*—Biocompatibility refers to the ability of a material to perform as a substrate that will support the appropriate cellular activity, including the facilitation of molecular and mechanical signaling, in order to optimize tissue regeneration, without eliciting any undesirable effects in those cells or inducing any undesirable local or systemic responses in the host.

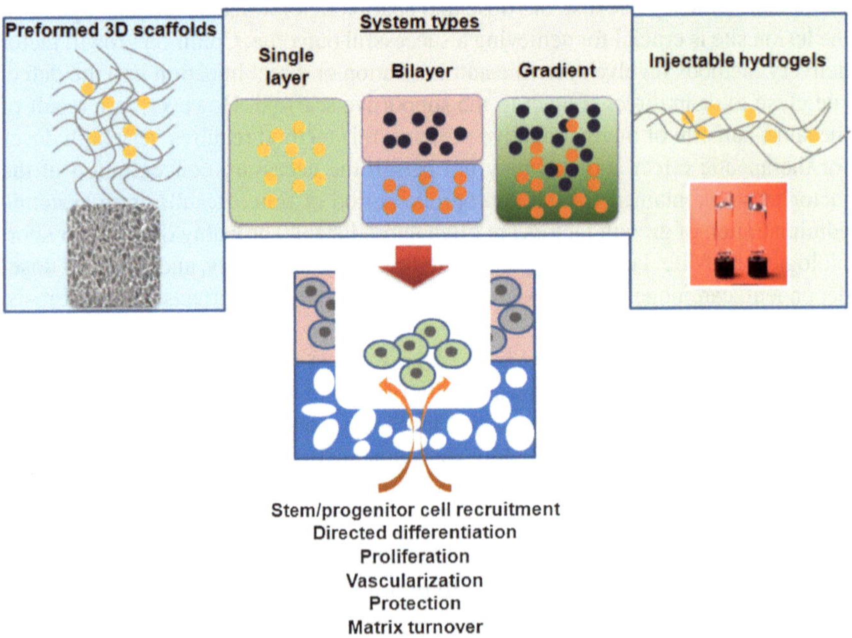

Fig. 22.2 General concepts in the design of biomaterial-based growth factor delivery systems. The biomaterials are prepared as solid 3D scaffolds or injectable hydrogels, designed as single, bilayer, or gradient systems. The delivery system should affect multiple processes that would lead to tissue repair and regeneration

- *Mechanical strength*—Scaffolds in tissue engineering should have the mechanical properties to contain and protect the seeded or recruited cells and maintain their structure under mechanical perturbations existing during cultivation and at implant site. At the same time, the scaffold mechanical properties should be compatible with the host tissue to allow its integration without interfering with the normal function of the organ. For example, matching scaffold and native cartilage compressive properties may be crucial, as stiff scaffolds shield mechanosensitive cartilage-forming cells from experiencing loading, whereas soft scaffolds may fail upon implantation. Additionally, scaffolds must possess sufficient surface and tensile properties for functioning in the high shear joint environment. Insufficiencies in these properties result in wear to opposing or adjacent cartilage due to abrasive contact with the articular surface or third-body wear from sheared off scaffold debris.
- *Biodegradation/bioresorption*—Ideally, the scaffold should disappear from the host when tissue regeneration has been accomplished and normal function was restored. Biodegradable scaffolds can do so via polymer backbone degradation (e.g., hydrolysis, enzymatic cleavage) or by dissolution of the matrix, and both should be controllable processes. It is fundamental that the products of these processes would be biocompatible and be resorbed by the body or removed from it via excretion.
- *Product fabrication*—This process should be mild using safe reagents and not affecting material properties, such as its cell recognition motifs. Cross-linking can be physical, where the polymer chains self-assemble due to electrostatic interactions, response to temperature and irradiation, or chemical, where covalent bonds are introduced between the polymer chains. Chemical cross-linking often changes the material properties (degradability, mechanical strength, and cell recognition) due to the lack of precise control over the position where the cross-link linkages are formed. In addition, chemical cross-linking often involves the use of harsh reagents, thus raising concerns about the material biocompatibility.
- *Internal morphology and surface properties*—When used as scaffolding for transplanted cells or cell ingrowth in situ, the matrix should be porous with interconnecting pore structure and pore size larger than 50 μm, to enable cell–cell interactions and construct vascularization after implantation. Surface properties and cell adhesion ability should also be carefully evaluated and matched for exact needs. For example, as adherence is generally required for cell growth and tissue integration, the specific scaffold for cartilage tissue engineering should maintain spherical chondrocyte morphology and phenotype.
- *Gradient/layered/biphasic systems*—As already mentioned, osteochondral tissue is a gradually changing interface with different cell and ECM composition. Thus, ideally, the matching engineered tissue or in situ regeneration system should mimic or recreate this complex structure. In real-world situation, layered/biphasic (for recreation of bone and cartilage) scaffolds or hydrogel systems have the potential to become functional substitutes for osteochondral defect repair. Importantly, careful matching and different specifications (i.e., composition, mechanical properties, incorporation of bioactive signals) for each phase/layer

should be taken into consideration during the system design, in order to achieve optimal long-term function and effect in vivo.

- *Bioactive factor incorporation*—For the delivery system, effective incorporation and controlled release of the factors are required. Physical entrapment of the desired agent within biomaterial construct before final production stage (gelation, scaffold preparation) is the simplest method, which, however, offers only limited control over encapsulation efficiency and release rate, governed mainly by diffusion. Chemical immobilization of bioactive molecules, either conditional/temporary or permanent, offers higher degree of control over release kinetics. However, the use of chemical processes for immobilization could raise biocompatibility concerns. An alternative strategy could rely on implementation of biological principles of specific non-covalent affinity binding, thus ensuring control over release kinetics and avoiding harsh chemical modifications (see Sect. 22.5.4). Finally, more complex "smart" systems could be designed, where the desired agents are released "on demand" by internal or external trigger.

22.5.2 Biomaterials Used in Delivery Systems

Polymers commonly used for delivery system fabrication for osteochondral repair can be categorized by their source origin (natural or synthetic) and by their chemical structure (peptides/proteins, polysaccharides, polyesters, ceramics, and others) (Table 22.1).

22.5.2.1 Natural-Based Polymers

Natural materials are biocompatible and biodegradable, and they have the ability to mimic certain aspects of native ECM, thus facilitating cell adhesion, migration, differentiation, and ECM deposition. However, natural materials have several disadvantages such as immunogenicity, difficulty in processing, and a potential risk of transmitting animal-originated pathogens. Moreover, despite the biocompatibility, these materials are mechanically weak and generally undergo rapid degradation upon implantation if not cross-linked with appropriate chemical reagents.

Collagen is a fibrous protein and the main component of ECM of mammalian tissues. About 25 types of collagen different in their chemical composition and molecular structure have been identified. Among the different collagen types, the fibrillar type I collagen is the most abundant in nature and easy to produce. The biocompatibility, biodegradability, and cell-adhesive properties of the collagen type I matrix attributed to its selection by most researchers as the candidate scaffold for tissue growth and support. Collagen can be fabricated in many forms, such as hydrogel or macroporous scaffold; its fabrication frequently requires chemical cross-linking, which may affect its biological recognition by cells, biocompatibility, and degradability. Collagen is already commercialized as injectable product; thus it has been recognized as safe material by regulatory agencies.

Table 22.1 Main biomaterials used in delivery systems for osteochondral repair and regeneration [8, 42, 47–49]

Type	Material	Biodegradability/ control over process	Cross-linking	Mechanical strength	Cell adhesion	Biocompatibility/immunogenicity
Natural—proteins	Collagen	Biodegradable; poor	Chemical	Med–low	Good, cross-linking reduce	Med immunogenicity after cross-linking
	Gelatin	Biodegradable; poor	Chemical	Low	Good	Low immunogenicity
Natural—polysaccharides	Alginate	Bioerodible matrix; good	Physical	Low	Poor	Non-immunogenic
	Hyaluronan	Biodegradable; poor	Chemical	Low	Good	Non-immunogenic
Synthetic	Polyesters (PLA, PGA, PLGA)	Biodegradable; good	Chemical	High	Poor	Med–low–no immunogenicity
	PEG	Bioerodible, m.w. dependent	Chemical	High	Poor	Non-immunogenic
	Self-assembling peptides	Biodegradable; NA	Physical	NA	Good	Non-immunogenic
Ceramics	Hydroxyapatite, other calcium phosphates, and calcium sulfates	Dissolution, osteoclastic resorption	NA	High	Poor	Non-immunogenic

PLA polylactic acid, *PGA* polyglycolic acid, *PLGA* poly(lactic-co-glycolic) acid, *PEG* polyethylene glycol, *NA* not available

Gelatin (the irreversibly hydrolyzed form of collagen) has also been used in osteochondral tissue engineering, especially in the format of a hydrogel prepared by chemical cross-linking. Gelatin is a biodegradable material, but under various conditions it can provoke an unspecific inflammatory response [50].

Alginate is an anionic polysaccharide extracted from brown algae, composed of 1→4-linked β-D-mannuronic acid (M) and α-L-guluronic acid (G). Divalent cations, such as calcium ions, interact with high affinity with the G monomer blocks to form ionic bridges between different alginate chains ("egg box" model) eventually leading to hydrogel formation. The physical cross-linking of alginate represents a significant advantage, as the use of various chemical agents for gelation is eliminated. Since there is no known mammalian enzyme which degrades the alginate backbone, it is assumed that alginate is not degradable in mammals. Yet, the calcium-cross-linked hydrogel is readily erodible with time due to exchange of calcium ions by sodium ions in physiological milieu, leading to hydrogel dissolution. The water-soluble alginate chains are excreted through the kidney if the molecular weight is below 50 kDa [51]. The long experience with alginate as cell matrix and implant indicates its biocompatibility.

Hyaluronan (HA) is a linear un-sulfated polysaccharide, composed of repeating disaccharides [(1→3)-β-N-acetyl-D-glucosamine-(1→4)-β-D-glucuronic acid]. In human body, HA is found primarily in the extracellular and pericellular matrix and is a major component of cartilage ECM; its degradation occurs by hyaluronidases. The HA has versatile biological functions, such as a lubricant material and numerous receptor-mediated roles in different cell processes. HA significantly enhances neocartilage formation from articular chondrocytes and ECM production in vitro and in in vivo studies [40]. The processing of HA as scaffolds for tissue engineering requires chemical modification of the material to achieve cross-linking, such as by photopolymerization. The un-cross-linked HA is not effective as injectable material due to its poor mechanical properties, rapid degradation, and clearance in vivo [52, 53].

22.5.2.2 Synthetic Polymers

Synthetic materials have been used extensively in tissue engineering both in vitro and in vivo due to their easy molding characteristics, relatively easy production, the ability to control dissolution and degradation, and structural and mechanical properties. Although synthetic materials used for tissue engineering applications are biocompatible, they do not have natural sites for cell adhesion, and these often need to be added. Further, their in vivo degradation by a hydrolytic reaction causes a local reduction in pH and possible inflammation response.

Poly(α-hydroxy esters) are a family of polymers that are the most widely incorporated synthetic materials in cartilage tissue engineering due to their relatively easy processability, controllable biodegradability, and already-existing FDA clinical use approval. Within this family, poly(lactic acid) (PLA), poly(glycolic acid) (PGA), and their copolymer poly(lactic-co-glycolic) acid (PLGA) have shown potential for cartilage regeneration. In addition, PLGA microspheres are widely used as drug delivery vehicles. These polymers degrade via hydrolytic de-esterification to monomeric components (lactic acid and glycine), resulting in heterogeneous bulk degradation of the polymer structure. The presence of these acidic polymer degradation by-products in vivo has been shown to elicit an inflammatory response or fibrous encapsulation in some instances [8, 16, 30, 40].

Polyethylene glycol (PEG) is a linear polyether that is used extensively in biomedical applications due to its hydrophilic and highly biocompatible properties. Even though PEG is not biodegradable, lower molecular weight (below MW 10,000) can be safely excreted by metabolism in body. The strong advantage of PEG is its ability to be cross-linked by chemical modification to conjugate acryl groups to be reactive by radical polymerization processes [40]. Further modifications to PEG, including the addition of hydrolyzable units, bioactive peptides, and others, have improved cartilage tissue growth [30].

Self-assembling peptides constitute another class of biomaterials that can be made into hydrogels and form by amino acid sequences of alternating ionic hydrophobic and uncharged hydrophilic side groups. These self-assembling peptide hydrogels form stable β-sheets of interwoven nanofibers when exposed to physiological electrolyte concentrations and pH levels. Synthetic self-assembling peptides can be modified to incorporate biologically active motifs that promote cell–matrix interactions or serve as vehicles for drug delivery [3].

22.5.2.3 Ceramics

A ceramic is a material made from an inorganic, nonmetallic material that can possess a crystalline structure. Ceramic materials, such as calcium phosphates, calcium sulfates, and bioactive glass, have been used as matrices for bone regeneration. These substances, especially the calcium phosphates, are ideal candidates for use as matrices because the inorganic component of bone is composed of the ceramic calcium hydroxyapatite. Ceramics are typified by high compressive strengths and variable degradation times, but they also have low ductility, meaning that they provide high resistance to deformation, but they tend to drastically fail because of their very brittle nature. Both calcium sulfate and crystalline hydroxyapatite have shown success as a bone-graft substitute. Ceramics such as calcium phosphate and bioactive glass are also considered biomimetic, in that they stimulate the formation, precipitation, and deposition of calcium phosphate from solution and can result in enhanced bone matrix interface strength. The calcium phosphates also have potential as drug and/or factor delivery vehicles as a result of the high binding affinities between ceramics and proteins [16, 47].

22.5.3 Types of Delivery Systems

Below we describe major systems designed for growth factor delivery for osteochondral repair and regeneration (Table 22.2, Fig. 22.2).

Table 22.2 Biomaterial-based growth factor delivery systems evaluated in animal models of osteochondral defects

		Carrier materials	Growth factor	Ref
Preformed scaffolds	Single layer	Collagen type I	BMP-2	[54]
		Gelatin	BMP-2	[55]
		Composite: collagen type I/ HAp	bFGF	[56]
		Composite: PLLA/ACP	bFGF	[41]
		Composite: PLA–PEG/HAp	BMP-2	[40]
	Bilayer/gradients	Composite: PLGA/alginate/ PLGA microspheres	BMP-2, TGF-β1	[57]
		Composite: PLGA/HAp/ PLGA microspheres	BMP-2, TGF-β1	[58]
		Composite: OPF/gelatin microspheres	IGF-1, TGF-β1	[59]
Injectable hydrogels	Single layer	Alginate	SDF-1	[60]
		Alginate	BMP-2, BMP-4	[61]
		HA	IGF-1	[62]
		HA	bFGF	[63]
		Self-assembling peptides	IGF-1, TGF-β1, dexamethasone	[64]
	Bilayer	Affinity-binding alginate (alginate-sulfate/alginate)	BMP-4, TGF- β1	[65]

PLLA poly(L-lactic acid), *HAp* hydroxyapatite, *ACP* amorphous calcium phosphate, *OPF* oligo(poly(ethylene glycol) fumarate)

22.5.3.1 Preformed Macroporous Scaffolds

Macroporous 3D scaffolds are characterized by large pore size (50–200 μm in diameter) and matrix porosity (70–90 %). The pore size in scaffolds should be at least 50 μm in diameter to enable vascularization (blood vessel penetration) after their implantation. The pore size and architecture as well as the extent of pore interconnectivity are major effectors on cell seeding, cell penetration from the host, and cell organization into a tissue. Scaffolds offer 3D ECM replacement and solid support for tissue growth, and different casting/molding methods enable fabrication of scaffolds in various sizes and shapes.

Single-Layered Scaffolds

Some authors suggested the use of single-layered scaffolds for osteochondral repair. Sellers et al. used BMP-2-containing type I collagen scaffolds for the repair of full-thickness defect in rabbits. At 24 weeks, treatment with BMP-2 greatly accelerated the formation of new subchondral bone. In addition, the defects that had been treated with BMP-2 were filled with minimally disrupted hyaline and hyaline-like cartilage, compared to empty controls or defects filled with empty scaffolds.

However, the integration of repaired tissue with adjacent cartilage was not observed. Such problem is common in cartilage repair, due to the fact that chondrocytes in normal cartilage are not involved in repair and cannot migrate [54].

Tokuhara et al. evaluated the dose-dependent effects of BMP-2 (1–40 μg) delivered by gelatin scaffolds on the repair of osteochondral defects in rabbits. At 24 weeks after surgery, there were no significant differences in cartilage repair between the groups treated with doses of 1–40 μg of BMP-2. This treatment also resulted in accelerated repair of subchondral bone that provided support for the overlying cartilage and prevented the fissuring of the cartilage as a result of decreasing biomechanical instability [55]. However, the regenerated cartilage layer was twofold thinner than its natural counterpart, suggesting that BMP-2 was unable to maintain normal cartilage thickness, probably due to accelerated endochondral ossification. Nevertheless, this study shows the potential of low protein doses for osteochondral repair, diminishing costs and possible undesired effects.

Maehara et al. developed a porous composite hydroxyapatite/collagen (HAp/Col) scaffold consisting of hydroxyapatite nanocrystals and type I atelocollagen (pepsin-solubilized type I collagen), for bFGF administration for osteochondral defect repair in femoral trochlear groove of rabbits. Twenty-four weeks after surgery, abundant bone formation was observed in the HAp/Col implanted groups as compared to the empty defect group. The group with lower bFGF concentration (10 μg/ml, compared to 100 μg/ml) displayed not only the most abundant bone regeneration but also the most satisfactory cartilage regeneration, with cartilage presenting a hyaline-like appearance, based on histological and gross examination. Subchondral bone formation may have resulted from subchondral bone repair induced by the osteoconductive property of porous HAp/Col. Moreover, a correlation between cartilage repair and subchondral bone formation observed in this study (also reported by other groups) indicates that activation of subchondral bone repair process probably enhanced the direct effects of bFGF on articular cartilage regeneration [56].

Yan and coworkers developed a bFGF-impregnated composite scaffold made of a combination of synthetic polymer (poly(L-lactic acid) (PLLA)) and bioactive bioceramics (amorphous calcium phosphate (ACP) particles). The scaffolds were implanted in the osteochondral defects in rabbits. The results demonstrated that when bFGF/ACP/PLLA was applied, most of the defects were filled with a well-established layer of cartilage tissue with abundance of cartilaginous extracellular matrix and type II collagen accumulation, compared to mainly fibrous tissue observed in animals treated with bFGF/PLLA scaffolds. In addition, results demonstrated that a continuous layer of trabecular bone was well formed below the cartilage with clusters of neobone formation and high osteoblastic activity observed inside the residual material of bFGF/ACP/PLLA. However, little bone formation was observed inside the composites, whether in bFGF/PLLA or the control group [41]. The results emphasize the beneficial effects of introduction of bioceramics on subchondral bone formation, which, in turn, could result in better regeneration of cartilage atop.

Tamai et al. used similar approach to create a composite scaffold by combining synthetic interconnected porous hydroxyapatite (IP-CHA) with BMP-2-containing poly(D,L-lactic acid)–polyethylene glycol (PLA–PEG) block copolymer and subsequent solvent evaporation. The resulting composites were implanted into osteochondral defects created in patellofemoral groove in rabbits. Six weeks after implantation, defects treated with the BMP-2/PLA–PEG/IP-CHA composite were filled with regenerated subchondral bone, which also penetrated the pores of the implant. The subchondral bone was covered with a layer of regenerated cartilage tissue of almost normal thickness. The matrix exhibited a hyaline-like cartilaginous phenotype, and the tissue was beginning to assume a columnar organization and a horizontal stratification into distinct cartilage zones, as in normal cartilage. In contrast, defects of control groups (untreated, empty composite or IP-CHA alone) were filled with hypercellular type of fibrous tissue; no hyaline cartilage was detected, despite the presence of new bone above and within the implant [40].

Bilayered or Gradient Scaffolds

To develop a sustained growth factor delivery system for osteochondral repair, Reyes et al. developed a bilayered scaffold, by overlaying bone-oriented PLGA cylinder with chondrogenic growth factor-containing (TGF-β1 or BMP-2) PLGA microspheres dispersed in alginate matrix, with subsequent freeze-drying. The scaffolds were implanted into osteochondral defects created in rabbits. The scaffolds showed a sustained in vivo release profile of the proteins for a period of 6 weeks. Twenty-four weeks after operation, all GF-treated groups showed complete repair of subchondral bone. In addition, these treatments induced marked cartilage repair and integrity, with typical features of hyaline cartilage [57].

Mohan et al. investigated the reparative potential of composite osteochondral scaffold, designed with opposing gradients of chondrogenic PLGA microspheres (encapsulating TGF-β1) and osteogenic PLGA microspheres (encapsulating BMP-2) with or without hydroxyapatite (HAp) nanoparticles. For osteoconduction and faster bone maturation, BMP-2 and HAp were co-encapsulated in the osteogenic microspheres, resulting in a gradient of both signal and material composition. The scaffolds were implanted in the osteochondral defect created in medial femoral condyle in rabbits. Twelve weeks after implantation, complete bone regeneration with a micro-architecture of the cancellous bone similar to the native tissue was observed in the TGF-β1/BMP-2/HAp gradient group. This group had the highest histological score for GAG content, cartilage thickness, bone filling, and edge integration. An even distribution of GAG content was observed in all TGF-β1/BMP-2/HAp samples and only in 50 % of HAp gradient samples. These data demonstrated that bone regeneration and maturation can be improved by inclusion of HAp in the gradient scaffolds, and cartilage regeneration may be enhanced by release of TGF-β1. Most of the joints that received the HAp and bioactive-HAp implants had higher scores for overall cartilage and bone regeneration, when compared to the other joints in the same animal that received a different implant. This clearly indicates that the HAp and TGF-β1/BMP-2/HAp implants that were investigated in this study were able to promote better osteochondral regeneration than the other groups presented in the study (empty scaffolds or scaffolds with protein gradients, but without HAp). Colocalization of osteoinductive BMP-2 and osteoconductive HAp appears to have resulted in faster stabilization of bone. This, in turn, could provide more optimal biochemically and mechanically suited environment for cartilage regeneration and tissue integrity [58]. Of note, BMP-2 in this study was associated mainly with osteoinductive function, while in previously described study it was used as a chondrogenic factor. This discrepancy only strengthens the context- and model-dependent activity of these pleiotropic factors, when rather small changes in factor localization, concentration, and release profile could significantly affect the outcome and the course of repair.

Holland et al. designed a bilayered scaffold made of oligo(poly(ethylene glycol) fumarate) (OPF) for controlled release of IGF-1 and TGF-β1. Bottom (bone-forming) layer was kept empty, while the top (cartilage-forming) layer contained separately IGF-1, TGF-β1, or their combination. IGF-1 was incorporated within gelatin microspheres, in order to provide more sustained release rate, compared to TGF-β1. Faster released TGF-β1 would act as a chemoattractant and morphogen to promote chondrogenic differentiation of progenitor cells, while IGF-1 would be required as an anabolic factor for ECM synthesis. In vitro release studies confirmed slower IGF-1 release compared to TGF-β1. In vivo, scaffolds were placed in osteochondral defects created in medial femoral condyles of rabbits. Twelve weeks postsurgery, the subchondral region in all treatment groups was similarly filled with trabecular bone. Hyaline cartilaginous regions were sometimes found in the subchondral zone, with highest incidence after TGF-β1 treatment. The neo-formed surface morphology of defects treated with scaffolds was primarily composed of fibrous tissue of variable thickness. Overall, the addition of growth factors did not improve significantly the course of repair. Nevertheless, IGF-1 delivery resulted in better surface repair, compared to TGF-β1 delivery or empty scaffolds, based on histomorphometric analysis. Surprisingly, combinatory delivery of TGF-β1 and IGF-1 did not result in widespread improvement in tissue quality, emphasizing the complex role and unpredicted interplay of these potent factors in in vivo setting [59].

22.5.3.2 Injectable Hydrogels

Hydrogel is a network of polymer chains that are water insoluble, sometimes found as a colloidal gel in which water is the dispersion medium. Hydrogels are superabsorbent (they can contain over 99 % water) natural or synthetic polymers. Hydrogels possess also a degree of flexibility very similar to natural ECM, due to their significant water content. The hydrogels can be prepared from natural and synthetic polymers by physical/ionic interactions (alginate) or via chemical cross-linking (collagen, HA, and others). Due to their resemblance to ECM texture, hydrogels are extensively being investigated as ECM replacements. Moreover, due to their aqueous nature, hydrogels can be easily applied by injection, also enabling filling of irregular voids and defects. Thus, in many cases, the clinical feasibility of hydrogels is higher than preformed solid scaffolds.

Minami and colleagues used stromal cell-derived factor-1 (SDF-1) delivery by in situ forming alginate gel for repair of osteochondral defects in the patella groove in rabbits. At 16 weeks after operation, SDF-1/alginate gel exhibited nearly normal cartilaginous structures with rich GAG matrix content and strong type II collagen staining, reconstruction of the normal subchondral bone structure, a smooth cartilage surface, and a tidemark. The neocartilagenous reparative tissue was integrated into the adjacent cartilage and bone. The compressive modulus in this group was significantly improved and reached ~81 % of that of normal cartilage. Interestingly, the alginate-only group significantly enhanced the repair of subchondral bone in the defects, compared to untreated group, with compressive modulus reaching ~63 % of that of normal cartilage, suggesting that the biomaterial has a considerable positive effect on tissue repair [60].

Marco and colleagues investigated the use of BMP-2 or BMP-4 delivered in an alginate gel for the treatment of osteochondral defects in femoral condyle in rabbits. Three months after surgery, the performed histological observations revealed subchondral bone regeneration in BMP-2 samples and moderate hyaline cartilage regeneration in BMP-4 samples, while the delivery of both proteins resulted in superior regeneration and proper location of subchondral bone and cartilage tissues. In this study, however, sole alginate injection resulted only in the formation of fibrous tissue [61].

Liu et al. used intra-articular injection of high-molecular-weight ($1.5–2 \times 10^6$ Da) HA for the delivery of IGF-1 in a rabbit model of OA in the mandibular condyle of the temporomandibular joint. Twenty-four weeks after injection, better histological repair, good cartilage preservation, and nearly normal micro-architectural properties of subchondral cancellous bone were observed in the IGF-1/HA group, compared to HA or IGF-1 alone groups, where progressive loss of cartilage occurred. As both HA and IGF-1 have roles in OA and cartilage repair, their combination seems to have a synergistic effect that results in improved outcome [62].

Itoi and coworkers also used HA for intra-articular delivery of bFGF for osteochondral defect repair of medial condyles in rabbits. Weekly administration (for 4 weeks after operation) of bFGF in HA showed significantly better restoration or articular surface than those animals treated each treatment alone. This group also showed significantly better reconstruction of subchondral bone, matrix staining, filling of the defect, and bonding of adjacent tissue. In separate set of experiment, the authors compared single bFGF delivery in HA to its delivery by gelatin microspheres (GM). Again, bFGF-HA significantly enhanced osteochondral repair, compared to bFGF delivery in GM and vehicle controls [63]. This further suggests the existence of positive synergistic effects between HA and bFGF, similar to previously described study with IGF-1.

Miller et al. used injectable self-assembling peptide hydrogel for delivery of chondrogenic factors (IGF-1, TGF-β1, and dexamethasone) for the repair of osteochondral defects in rabbits. The authors used self-assembling peptide sequence $(KLDL)_3$ (KDL) that can rapidly assemble into hydrogel when exposed to physiological pH and ionic strength, with pore size in the range of 100–500 nm. These peptides were previously shown to maintain chondrocyte phenotype and stimulate chondrogenesis of MSCs in vitro. Twelve weeks after hydrogel injection, KLD group without growth factors showed the greatest repair with significantly higher Safranin-O, collagen II immunostaining, and cumulative histology scores, compared to other groups. The addition of growth factors increased aggrecan immunostaining and subchondral bone reconstitution, but in general was not associated with beneficial or deleterious effects, with overall scores lower than in the group treated with peptides only [64]. Aside with potency of KLD peptides, this study again emphasize the need for further understanding of growth factor action in vivo, alone or in combination.

22.5.4 *Affinity-Binding Alginate Delivery System*

Bio-inspired by ECM interactions with heparin-binding proteins, our group has developed an affinity-binding alginate biomaterial to enable precise control over factor release and to allow the release of combinations of growth factors.

22.5.4.1 Alginate Sulfation

Alginate biomaterial with affinity-binding sites for heparin-binding proteins was synthesized by sulfation of the uronic acid monomers in alginate, using carbodiimide chemistry [66].

The infrared (IR) spectrum of the product alginate-sulfate confirmed the appearance of a new major peak at 1,250 cm^{-1} (assigned to S=O symmetric stretching) and a minor peak at 800 cm^{-1} (assigned to S–O–C stretching). According to nuclear magnetic resonance spectroscopy (C^{13}-NMR) spectra, the sulfate groups are added to either C-2 or C-3 or both, in an identical manner. The percentage sulfation by the Shoniger method was 8 % (wt. sulfur per wt. alginate).

22.5.4.2 Analysis of Protein Binding

Surface plasmon resonance (SPR) analysis revealed the specific and strong binding of various heparin-binding proteins to alginate-sulfate, with equilibrium binding constants at the same order of magnitude as their binding to heparin (Fig. 22.3 and Table 22.3) [66, 67]. No such interactions were recorded with pristine alginate. Thus, it appears that the binding to alginate-sulfate mimics in large the interactions of growth factors, chemokines, and cell adhesion molecules, collectively known as heparin-binding proteins. These molecules bind the proteoglycans heparin and heparan sulfate via high affinity, specific electrostatic interactions with the low- and high-sulfated sequences in these glycosaminoglycans (GAG) [50]. In this aspect, heparan sulfate GAGs play an important role in sequestering and storage of the proteins and also participate in the formation of active signaling complex with a respective cell surface receptor.

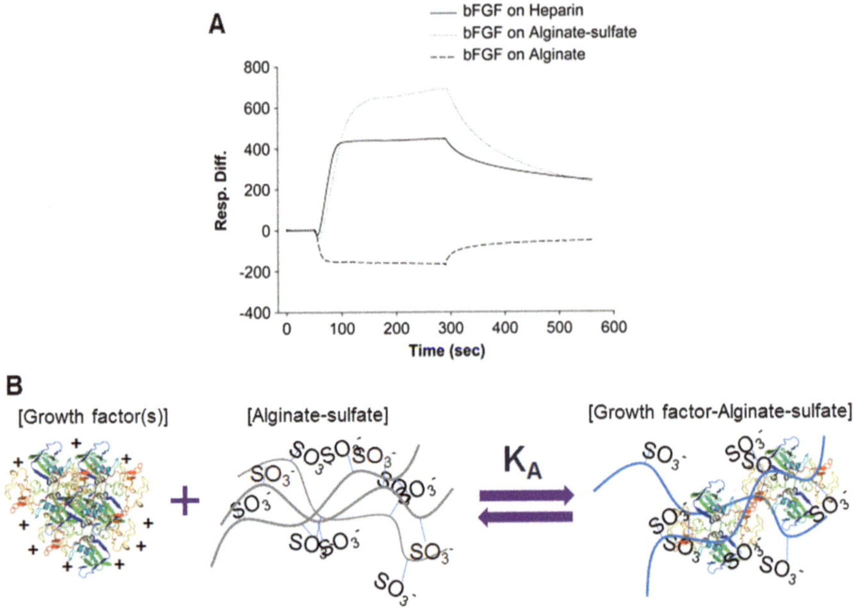

Fig. 22.3 Alginate sulfation to attain affinity-binding biomaterial. (**A**) A representative SPR sensogram of bFGF binding to unmodified alginate, heparin, and alginate-sulfate, showing strong and specific binding of bFGF to heparin and alginate-sulfate and not to unmodified alginate. (**B**) The model of reversible binding

Table 22.3 Equilibrium binding constants (K_A) calculated from the interactions of alginate-sulfate with proteins (SPR analysis) [65–67]

K_A (M^{-1})	Protein
2.80×10^7	Acidic fibroblast growth factor (aFGF)
2.57×10^6	Basic fibroblast growth factor (bFGF)
5.63×10^9	Bone morphogenic protein-4 (BMP-4)
9.93×10^6	Epidermal growth factor (EGF)
5.36×10^7	Hepatocyte growth factor (HGF)
1.01×10^8	Insulin-like growth factor-1 (IGF-1)
1.38×10^7	Interleukin-6 (IL-6)
3.53×10^7	Platelet-derived growth factor-BB (PDGF-BB)
2.06×10^8	Stromal cell-derived factor-1 (SDF-1)
6.63×10^7	Transforming growth factor- β1 (TGF-β1)
1.81×10^6	Thrombopoietin (TPO)
6.98×10^6	Vascular endothelial growth factor (VEGF)

22.5.4.3 Growth Factor Release and Bioactivity

As was already mentioned, TGF-β1 and BMP-4 are potent factors for induction of chondrogenesis and osteogenesis, respectively. Thus, these two factors were chosen as bioactive components of alginate-based system in our efforts for reconstruction of osteochondral interface in vitro and in vivo.

Protein release studies were performed on affinity binding (alginate-sulfate containing scaffolds), with proteins preloaded in the system during scaffold preparation (see below). Scaffolds without alginate-sulfate were used for control. Initial protein loading efficiency in affinity-binding scaffolds was 75 and 30 % for TGF-β1 and BMP-4, respectively. Loading efficiency in control pristine scaffolds was lower (60 and 20 %, respectively) [65, 68]. These results suggest that affinity-binding mechanism protects the protein during fabrication and ensures higher protein availability and better preservation.

Cumulative release profiles of TGF-β1 from affinity-binding scaffolds vs. from control showed a significant difference in release rates from the two scaffolds as a function of time. The profile demonstrates a sustained release of the protein for 7 days from the affinity-binding scaffolds, compared to a burst release of nearly 100 % of the entrapped protein from control scaffolds [68].

Cumulative release profiles of BMP-4 from the affinity-bound scaffold vs. from the control showed a slight but significant difference in release rates from the two scaffolds as a function of time. The profile demonstrates greater amounts of BMP-4 released from the affinity-binding scaffolds, and slightly slower release, compared to a burst release of significantly lower amounts of BMP-4 from control scaffolds (unpublished data).

To test the bioactivity of released TGF-β1 and BMP-4, monolayers of rat neonatal cardiac fibroblasts were treated with the release medium of day 1 from affinity-binding scaffolds. Protein bioactivity was assessed by their ability to enhance collagen production in these cultures and was analyzed by Sircol colorimetric assay. Collagen deposition was significantly greater in cultures exposed to TGF-β1- or BMP-4-containing medium, compared to cultures treated with medium from empty scaffolds [65, 68].

22.6 Dual Growth Factor Delivery System for Osteochondral Regeneration

The combination of alginate-sulfate with pristine alginate in one device represents a unique affinity-binding alginate biomaterial, which is capable of controlling the delivery of multiple proteins, while retaining the supporting and ECM replacing properties of the alginate. This novel material was used for recreation of osteochondral bilayered interface, in vitro (in scaffold form) or in vivo (as injectable acellular hydrogel).

22.6.1 Bilayered Scaffold for Osteochondral Tissue Engineering

Osteochondral tissue engineering requires correct and spatial differentiation of hMSCs to cartilage and bone. Thus, correct spatiotemporal presentation of chondrogenic and osteogenic inducers, in a similar fashion to their presentation by the ECM, would enhance hMSC differentiation to a respective (chondrogenic or osteogenic) lineage. For this purpose, the chondrogenic inducer TGF-β1 or the osteogenic inducer BMP-4 was affinity bound to the macroporous alginate scaffolds via specific interactions with alginate-sulfate, mimicking the specific interactions of these factors with heparan sulfate. Incorporation of protein-alginate-sulfate bioconjugates into pristine alginate before final scaffold fabrication step (by established freeze-dry technique) enabled greater growth factor loading capacity and retention in the macroporous scaffold. Moreover, this preparation strategy represents a significant advancement in product design, as ready-to-use single or multiple factor-loaded dry scaffolds could be stored and then immediately used for specific needs.

The binding and activity of attached factors (TGF-β1 or BMP-4) in the affinity-binding system, as well as their differentiation induction capabilities, were initially evaluated in separate experiments to eliminate interactions or interference between the factors. Following this, the spatial differentiation of hMSCs in the TGF-β1/BMP-4 bilayer scaffold was examined to validate regeneration of osteochondral interface (Fig. 22.4).

In the TGF-β1/affinity-bound scaffolds, the prolonged activation of TGF-β1-induced Smad-dependent (Smad2) and Smad-independent (ERK1/2) signaling pathways was found to be consistent with the appearance of differentiated cells with the typical round shape and size of matured chondrocytes with deposited collagen type II [69, 70]. By contrast, such a differentiation profile was not found when TGF-β1 was adsorbed to the alginate scaffold. The burst release from this construct, resulting in a short-term effect, as judged by the reduced activation of TGF-β1-induced signaling pathways after 14 days, may explain why chondrogenesis was incomplete. Our results are in agreement with previous studies indicating the involvement of TGF-β-induced MAP kinase signaling cascades not only in the initiation of precartilage mesenchymal condensation in early culture but also in the subsequent stages of chondrogenic differentiation and cartilage ECM production, occurring in the following days in culture up till day 21 [71]. Consistently, the affinity-binding scaffold alone (with no TGF-β1 supplementation) was insufficient to induce chondrogenesis of hMSCs [68].

In BMP-4/affinity-bound scaffolds, sustained factor release and presentation resulted in the prolonged activation of ERK1/2 signaling pathway and was consistent with the observed hMSC osteogenic differentiation, demonstrated by the increase in ALP activity and mineralized bone matrix deposition. Again, such a differentiation profile was not found when BMP-4 was directly associated with the alginate scaffold lacking alginate-sulfate (Fig. 22.4) [65].

The potential of factor/affinity-bound scaffolds to induce hMSC differentiation was further confirmed in vivo, after an ectopic transplantation of hMSC-seeded constructs in nude mice. Specifically, in the TGF-β1/affinity-bound scaffold, differentiated human chondrocytes with a deposition of specific cartilage ECM components, as collagen type II and aggrecan, were found within the implants, 3 weeks postimplantation. The cells represented different differentiation stages in the normal chondrogenesis process, from early chondroblasts up to mature chondrocytes. Importantly, the seeded hMSCs did not differentiate along the osteogenic pathway,

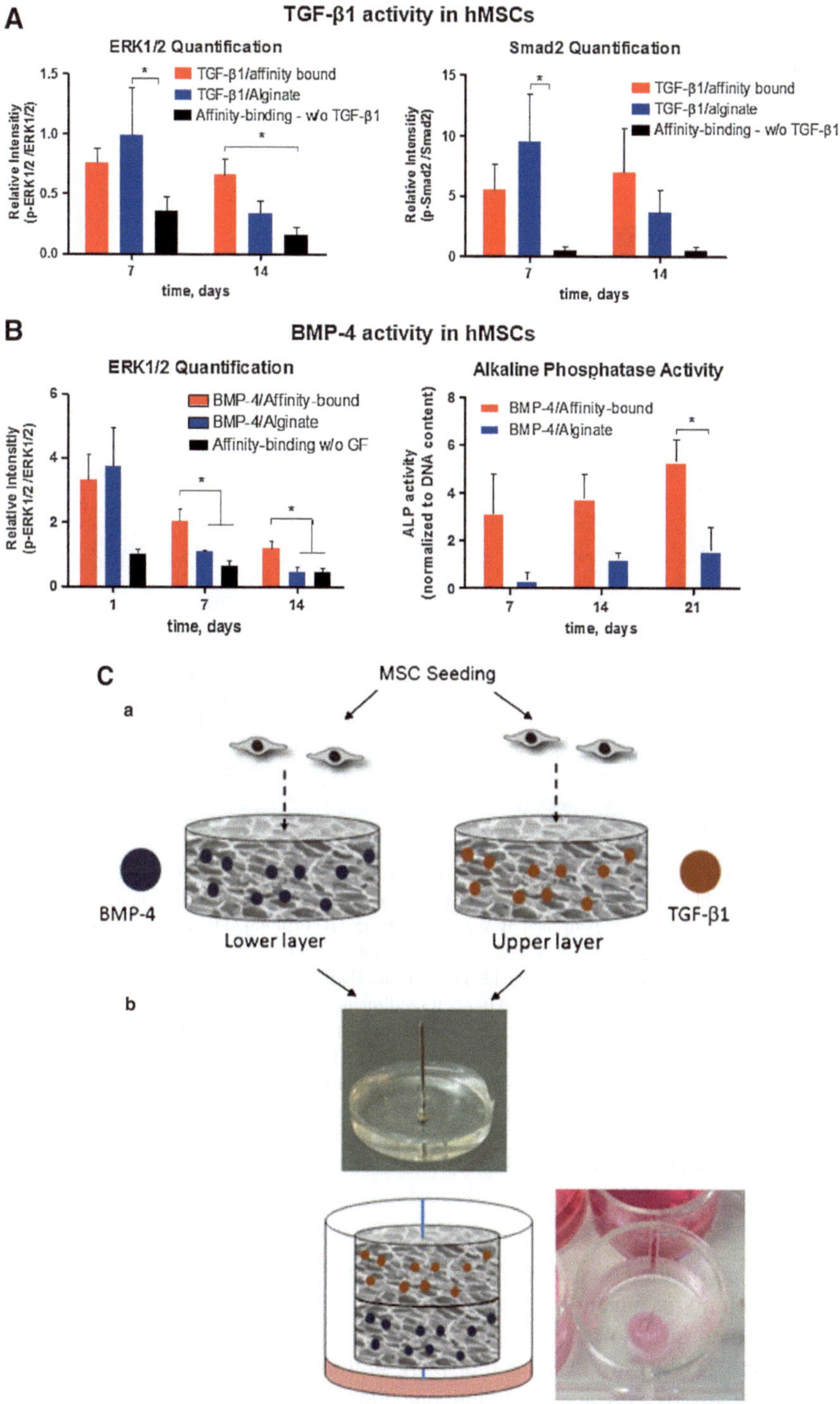

Fig. 22.4 (continued)

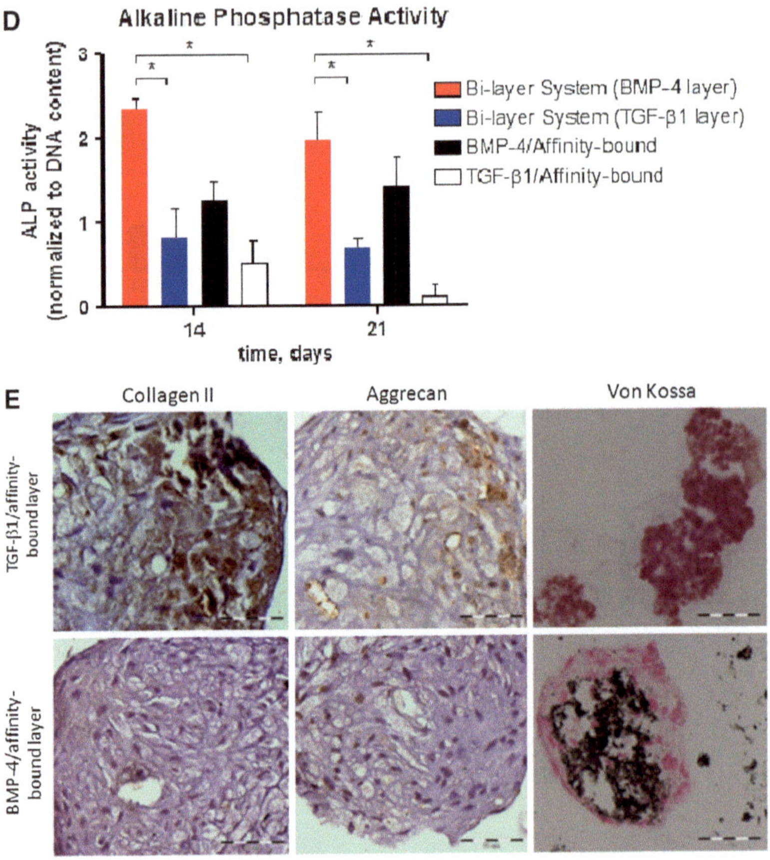

Fig. 22.4 TGF-β1 and BMP-4 activity and directed differentiation in hMSC constructs cultured within affinity-binding alginate scaffolds. (**A**) Prolonged TGF-β1-induced activation of ERK1/2 and Smad2 signaling pathways, examined by Western blotting. (**B**) Prolonged BMP-4-induced activation of ERK1/2 signaling pathway (examined by Western blotting) and ALP activity. (**C**) Construction of a bilayer affinity-binding alginate scaffold. TGF-β1 and BMP-4 were affinity bound to alginate-sulfate in macroporous alginate scaffolds. After cell seeding and a short culture of 2 days (a), the seeded scaffolds were combined together by their assembly on a stainless steel pin placed perpendicular to a supporting polydimethylsiloxane (PDMS) layer (b). (**D–E**) Osteochondral differentiation of hMSCs seeded in a bilayer system, in vitro. (**D**) ALP activity in the BMP-4 and TGF-β1 layers in a bilayer system. (**E**) Immunostaining of the different layers in the bilayer system, after 3 weeks of hMSC cultivation, for collagen type II and aggrecan—markers for cartilage ECM—and von Kossa staining for mineralized bone matrix. Bar: 50 μm. Reprinted with permission from [65, 68]

as indicated by the absence of the bone marker osteocalcin and the negative von Kossa staining for mineralized bone matrix [68].

For the construction of the bilayered in vitro system, cells were seeded separately onto TGF-β1/affinity-bound or BMP-4/affinity-bound scaffolds, and after 2 days the scaffolds were assembled into one bilayered construct, presenting TGF-β1 in one layer and BMP-4 in a second layer (Fig. 22.4). Within 3 weeks of cultivation, seeded hMSCs differentiated and formed cartilage and bone in their respective layers. In the TGF-β1 layer but not in the BMP-4 layer, type II collagen and aggrecan, typical cartilaginous ECM components, were abundant. Significantly greater levels of ALP activity and increased mineralized bone matrix deposition were observed in the BMP-4 layer. The simultaneous differentiation of hMSC in the two layers into either chondrocytes or bone pointed out to the spatial presentation of the factors and their local and prolonged activity in the bilayer system, thus implying that there was minimal cross-diffusion of the growth factors to the adjacent layers (Fig. 22.4) [65].

22.6.2 Injectable Bilayer System for Osteochondral Defect Repair and Regeneration

Osteochondral defect repair requires the simultaneous regeneration of cartilage and bone. We tested a new strategy for inducing the endogenous regeneration of the osteochondral interface in a rabbit model of osteochondral defect, by constructing an in situ formed affinity-binding alginate bilayer acellular hydrogel, designed to induce the spatial differentiation of host's migrating stem cells into chondrocytes and osteoblasts, by spatiotemporal presentation of the chondroinductive TGF-β1 and the osteoinductive BMP-4 in two distinctive hydrogel layers.

Hydrogels were prepared by mixing of TGF-β1- or BMP-4-alginate-sulfate bioconjugates with calcium-cross-linked alginate solution. Osteochondral defect, created in New Zealand rabbits in the patellofemoral groove (diameter 3 mm, depth 3 mm), was filled first with BMP-4/affinity-bound alginate solution (bottom layer, bone portion), followed by facilitated gelation with calcium chloride. Next, the top layer (cartilage portion) of TGF-β1/affinity-bound alginate solution was similarly constructed (Fig. 22.5).

Gross morphology of the explants, 2 weeks postoperation, demonstrated the integration of hydrogel with the surrounding tissues (Fig. 22.5). Four weeks postoperation, a newly formed white tissue, typical of a cartilaginous tissue, was seen to cover the defect and was well integrated with the surrounding articular cartilage (Fig. 22.5). Two weeks postoperation, microtomography (μCT) scans revealed the presence of bone tissue at the bottom and rim of the drilled hole. After 4 weeks, the formation of subchondral woven bone layer is clearly seen in the depth of the drilled hole (yellow arrow), where BMP-4 was present, while in the top soft tissue, where TGF-β1 was present, no mineralization process took place (Fig. 22.5). Since no cells were co-injected with the hydrogel, it is assumed that alginate hydrogel enabled the penetration of migrating cells from the bone marrow that were coaxed to differentiate either to chondrocytes or bone cells, depending on the layer to which they migrated, thus creating the osteochondral interface. The progression of migrated cell differentiation along the endochondral bone formation is nicely demonstrated in both layers within the defect. Specifically, the chondrocytes within the surface layer grew in terms of

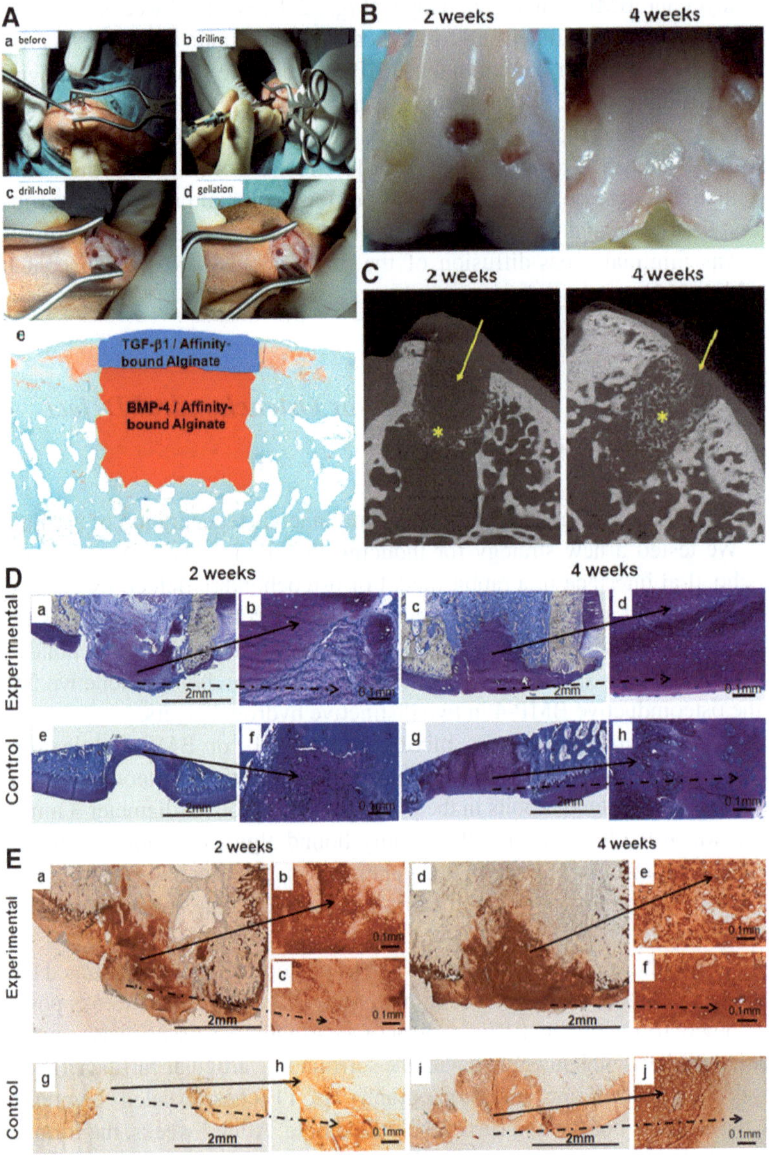

Fig. 22.5 Injectable bilayered affinity-binding alginate hydrogel for osteochondral repair and regeneration. (**A**) The surgery procedure for creating the osteochondral defect and in situ construction of the bilayer hydrogel, spatially presenting TGF-β1 in the *top layer* and BMP-4 in the *bottom layer*. (**B**) Gross morphology of the explants in a subchondral defect. (**C**) μCT for evaluation of bone formation. μCT scans of the explants revealing the formation of bone underneath the cartilage tissue. *Asterisks* denote woven bone formation; *arrows* indicate the non-mineralized tissue. (**D**) *Toluidine blue* staining for evaluation of cartilage formation (proteoglycan detection) in a subchondral defect. *Arrows* denote the area of magnification. (a, c, e, g) Bar: 2 mm; (b, d, f, h) bar: 0.1 mm. (**E**) Immunostaining for collagen type II—a specific marker of hyaline cartilage ECM. *Arrows* denote the area of magnification. (a, d, g, i) Bar: 2 mm; (b, c, e, f, h, j) bar: 0.1 mm. Reprinted with permission from [65]

cellularity and cartilage ECM deposition (proteoglycans and type II collagen) (Fig. 22.5). Four weeks after implantation, the chondrocytes were shown to be evenly distributed, isolated from each other by the cartilaginous ECM, and most importantly, this cartilage tissue was not mineralized, suggesting that the newly formed cartilage would be able to sustain the appropriate physiological load. The demonstration of endochondral differentiation progression was also observed within the deeper layer of the hole, wherein the prehypertrophic chondrocytes, observed after 2 weeks and characterized by flattened and elongated cell morphology and by deposition of proteoglycans and collagen II, were replaced by large hypertrophic cells organized in columnar structures after 4 weeks (Fig. 22.5). Moreover, the observed woven bone by μCT analysis pointed out to the progression of the endochondral bone formation within the deeper layer of the defect [65].

22.7 Conclusions and Future Aspects

The complexity of osteochondral tissue and low intrinsic regeneration of cartilage significantly limit the efficacy of currently available treatment options. As an alternative, the emergence of osteochondral tissue engineering, a hybrid of both bone and cartilage tissue engineering, may represent a feasible option for cartilage repair and associated OA prevention. In situ regeneration and repair induced by bioactive molecule-enhanced biomaterials represent an important advancement in the field of osteochondral repair. However, the current amount of preclinical data on such systems, aimed at simultaneous regeneration of bone and cartilage, is insufficient to establish a potential treatment strategy for application in patients.

The optimal design properties of biomaterial platforms are somewhat undecided with single, biphasic, and gradient structures dividing the literature. New functional composite materials, improved mechanical properties matching the target environment, innovative fabrication methods, natural tissue mimicking, and mechanoregulation models of tissue differentiation may assist to uncover the appropriate scaffold or hydrogel candidates. As was already emphasized, the use of hydrogel systems represents a major advantage for clinical applicability, as the administration of such systems could be made by minimally invasive arthroscopic surgery. In addition, the ideal delivery system design should take into consideration the complexity of bone and cartilage regeneration processes. Neo-tissue formation involves multiple growth factors and chemokines that locally expressed at a distinct time frame. The differentiated expression of several key growth factors suggests that orchestrated activities of each molecule are required in forming functional tissue. Therefore, the delivery system should not be limited to provide a single growth factor, but extended to multiple signals at an optimized ratio in a specific spatiotemporal pattern to target desirable cell types at various locations. Critically, however, more body of knowledge should be collected on the biology of each factor, with emphasis on interaction with other signals, either delivered or existing in situ, as it was repeatedly shown that apparent beneficial combinations eventfully led to only minor or even negative effects.

Several issues should also be addressed in the design of studies aimed to evaluate the potential of various systems in animal models. Although using rabbits for animal model is relatively technically convenient, the potential of therapies should be evaluated and confirmed in larger animals, where spontaneous regeneration rate is lower and cartilage thickness and weight-bearing demands are closer to humans. In addition, as many parameters in gross morphology and histology are evaluated by grading, blinded evaluation and large number of animals are required to get convincing results. Simultaneously, the use of quantitative methods (i.e., GAG content), computed tomography, mechanical-loading evaluation, and other advanced techniques could significantly upgrade the quality of the research.

In conclusion, the design of complex structured delivery carriers that possess multiple layers in a single unit in which each layer has distinct pore structure, porosity, nanoscale features for cell attachment, controlled and spatial bioactive molecule delivery and presentation profile, and chemical or mechanical characteristics may help to maintain osteochondral interface properties and induce effective tissue regeneration. By optimizing these complex parameters, the developed constructs will act as real future therapeutic alternative to damaged orthopedic tissues.

References

1. Klein TJ, Malda J, Sah RL, Hutmacher DW (2009) Tissue engineering of articular cartilage with biomimetic zones. Tissue Eng Part B Rev 15:143–157
2. Swieszkowski W, Tuan BH, Kurzydlowski KJ, Hutmacher DW (2007) Repair and regeneration of osteochondral defects in the articular joints. Biomol Eng 24:489–495
3. Chung C, Burdick JA (2008) Engineering cartilage tissue. Adv Drug Deliv Rev 60:243–262
4. Yang PJ, Temenoff JS (2009) Engineering orthopedic tissue interfaces. Tissue Eng Part B Rev 15:127–141
5. Onyekwelu I, Goldring MB, Hidaka C (2009) Chondrogenesis, joint formation, and articular cartilage regeneration. J Cell Biochem 107:383–392
6. Clouet J, Vinatier C, Merceron C, Pot-vaucel M, Maugars Y, Weiss P et al (2009) From osteoarthritis treatments to future regenerative therapies for cartilage. Drug Discov Today 14:913–925
7. Keeney M, Pandit A (2009) The osteochondral junction and its repair via bi-phasic tissue engineering scaffolds. Tissue Eng Part B Rev 15:55–73
8. O'Shea TM, Miao X (2008) Bilayered scaffolds for osteochondral tissue engineering. Tissue Eng Part B Rev 14:447–464
9. Mano JF, Reis RL (2007) Osteochondral defects: present situation and tissue engineering approaches. J Tissue Eng Regen Med 1:261–273
10. Gomoll AH, Madry H, Knutsen G, van Dijk N, Seil R, Brittberg M et al (2010) The subchondral bone in articular cartilage repair: current problems in the surgical management. Knee Surg Sports Traumatol Arthrosc 18:434–447
11. Madry H, Grun UW, Knutsen G (2011) Cartilage repair and joint preservation: medical and surgical treatment options. Dtsch Arztebl Int 108:669–677
12. Rodrigues MT, Gomes ME, Reis RL (2011) Current strategies for osteochondral regeneration: from stem cells to pre-clinical approaches. Curr Opin Biotechnol 22:726–733
13. Ahmed TA, Hincke MT (2010) Strategies for articular cartilage lesion repair and functional restoration. Tissue Eng Part B Rev 16:305–329

14. Versier G, Dubrana F (2011) Treatment of knee cartilage defect in 2010. Orthop Traumatol Surg Res 97:S140–S153
15. Perera JR, Gikas PD, Bentley G (2012) The present state of treatments for articular cartilage defects in the knee. Ann R Coll Surg Engl 94:381–387
16. Panseri S, Russo A, Cunha C, Bondi A, Di Martino A, Patella S et al (2012) Osteochondral tissue engineering approaches for articular cartilage and subchondral bone regeneration. Knee Surg Sports Traumatol Arthrosc 20:1182–1191
17. Kock L, van Donkelaar CC, Ito K (2012) Tissue engineering of functional articular cartilage: the current status. Cell Tissue Res 347:613–627
18. Mithoefer K, McAdams T, Williams RJ, Kreuz PC, Mandelbaum BR (2009) Clinical efficacy of the microfracture technique for articular cartilage repair in the knee: an evidence-based systematic analysis. Am J Sports Med 37:2053–2063
19. Brittberg M, Lindahl A, Nilsson A, Ohlsson C, Isaksson O, Peterson L (1994) Treatment of deep cartilage defects in the knee with autologous chondrocyte transplantation. N Engl J Med 331:889–895
20. Peterson L, Brittberg M, Kiviranta I, Akerlund EL, Lindahl A (2002) Autologous chondrocyte transplantation. Biomechanics and long-term durability. Am J Sports Med 30:2–12
21. Peterson L, Minas T, Brittberg M, Lindahl A (2003) Treatment of osteochondritis dissecans of the knee with autologous chondrocyte transplantation: results at two to ten years. J Bone Joint Surg Am 85-A(Suppl 2):17–24
22. Gooding CR, Bartlett W, Bentley G, Skinner JA, Carrington R, Flanagan A (2006) A prospective, randomised study comparing two techniques of autologous chondrocyte implantation for osteochondral defects in the knee: periosteum covered versus type I/III collagen covered. Knee 13:203–210
23. Kalson NS, Gikas PD, Briggs TW (2010) Current strategies for knee cartilage repair. Int J Clin Pract 64:1444–1452
24. Kon E, Filardo G, Di Martino A, Marcacci M (2012) ACI and MACI. J knee surg 25:17–22
25. Rodriguez-Merchan EC (2012) The treatment of cartilage defects in the knee joint: microfracture, mosaicplasty, and autologous chondrocyte implantation. Am J Orthop (Belle Mead NJ) 41:236–239
26. Seong JM, Kim BC, Park JH, Kwon IK, Mantalaris A, Hwang YS (2010) Stem cells in bone tissue engineering. Biomed Mater 5:062001
27. Marolt D, Knezevic M, Novakovic GV (2010) Bone tissue engineering with human stem cells. Stem Cell Res ther 1:10
28. Oldershaw RA (2012) Cell sources for the regeneration of articular cartilage: the past, the horizon and the future. Int J Exp Pathol 93:389–400
29. van Osch GJ, Brittberg M, Dennis JE, Bastiaansen-Jenniskens YM, Erben RG, Konttinen YT et al (2009) Cartilage repair: past and future–lessons for regenerative medicine. J Cell Mol Med 13:792–810
30. Grayson WL, Chao PH, Marolt D, Kaplan DL, Vunjak-Novakovic G (2008) Engineering custom-designed osteochondral tissue grafts. Trends Biotechnol 26:181–189
31. Mahmoudifar N, Doran PM (2012) Chondrogenesis and cartilage tissue engineering: the longer road to technology development. Trends Biotechnol 30:166–176
32. Freyria AM, Mallein-Gerin F (2012) Chondrocytes or adult stem cells for cartilage repair: the indisputable role of growth factors. Injury 43:259–265
33. Nejadnik H, Hui JH, Feng Choong EP, Tai BC, Lee EH (2010) Autologous bone marrow-derived mesenchymal stem cells versus autologous chondrocyte implantation: an observational cohort study. Am J Sports Med 38:1110–1116
34. Wakitani S, Mitsuoka T, Nakamura N, Toritsuka Y, Nakamura Y, Horibe S (2004) Autologous bone marrow stromal cell transplantation for repair of full-thickness articular cartilage defects in human patellae: two case reports. Cell Transplant 13:595–600
35. Kuroda R, Ishida K, Matsumoto T, Akisue T, Fujioka H, Mizuno K et al (2007) Treatment of a full-thickness articular cartilage defect in the femoral condyle of an athlete with autologous bone-marrow stromal cells. Osteoarthritis Cartilage 15:226–231

36. Vinatier C, Mrugala D, Jorgensen C, Guicheux J, Noel D (2009) Cartilage engineering: a crucial combination of cells, biomaterials and biofactors. Trends Biotechnol 27:307–314
37. Danisovic L, Varga I, Polak S (2012) Growth factors and chondrogenic differentiation of mesenchymal stem cells. Tissue Cell 44:69–73
38. Fortier LA, Barker JU, Strauss EJ, McCarrel TM, Cole BJ (2011) The role of growth factors in cartilage repair. Clin Orthop Relat Res 469:2706–2715
39. Bessa PC, Casal M, Reis RL (2008) Bone morphogenetic proteins in tissue engineering: the road from the laboratory to the clinic, part I (basic concepts). J Tissue Eng Regen Med 2: 1–13
40. Tamai N, Myoui A, Hirao M, Kaito T, Ochi T, Tanaka J et al (2005) A new biotechnology for articular cartilage repair: subchondral implantation of a composite of interconnected porous hydroxyapatite, synthetic polymer (PLA-PEG), and bone morphogenetic protein-2 (rhBMP-2). Osteoarthritis Cartilage 13:405–417
41. Huang X, Yang D, Yan W, Shi Z, Feng J, Gao Y et al (2007) Osteochondral repair using the combination of fibroblast growth factor and amorphous calcium phosphate/poly(L-lactic acid) hybrid materials. Biomaterials 28:3091–3100
42. Lee SH, Shin H (2007) Matrices and scaffolds for delivery of bioactive molecules in bone and cartilage tissue engineering. Adv Drug Deliv Rev 59:339–359
43. Vo TN, Kasper FK, Mikos AG (2012) Strategies for controlled delivery of growth factors and cells for bone regeneration. Adv Drug Deliv Rev 64:1292–1309
44. Mehta M, Schmidt-Bleek K, Duda GN, Mooney DJ (2012) Biomaterial delivery of morphogens to mimic the natural healing cascade in bone. Adv Drug Deliv Rev 64:1257–1276
45. Porter JR, Ruckh TT, Popat KC (2009) Bone tissue engineering: a review in bone biomimetics and drug delivery strategies. Biotechnol Prog 25:1539–1560
46. De Biase P, Capanna R (2005) Clinical applications of BMPs. Injury 36(Suppl 3):S43–S46
47. Khan Y, Yaszemski MJ, Mikos AG, Laurencin CT (2008) Tissue engineering of bone: material and matrix considerations. J Bone Joint Surg Am 90(Suppl 1):36–42
48. Spiller KL, Maher SA, Lowman AM (2011) Hydrogels for the repair of articular cartilage defects. Tissue Eng Part B Rev 17:281–299
49. Balakrishnan B, Banerjee R (2011) Biopolymer-based hydrogels for cartilage tissue engineering. Chem Rev 111:4453–4474
50. Akhyari P, Kamiya H, Haverich A, Karck M, Lichtenberg A (2008) Myocardial tissue engineering: the extracellular matrix. Eur J Cardiothorac Surg 34:229–241
51. Al-Shamkhani A, Duncan R (1995) Radioiodination of alginate via covalently-bound tyrosinamide allows monitoring of its fate in vivo. J Bioact Compat Polym 10:4–13
52. Prestwich GD, Kuo JW (2008) Chemically-modified HA for therapy and regenerative medicine. Curr Pharm Biotechnol 9:242–245
53. Prestwich GD (2011) Hyaluronic acid-based clinical biomaterials derived for cell and molecule delivery in regenerative medicine. J Control Release 155:193–199
54. Sellers RS, Peluso D, Morris EA (1997) The effect of recombinant human bone morphogenetic protein-2 (rhBMP-2) on the healing of full-thickness defects of articular cartilage. J Bone Joint Surg Am 79:1452–1463
55. Tokuhara Y, Wakitani S, Imai Y, Kawaguchi A, Fukunaga K, Kim M et al (2010) Repair of experimentally induced large osteochondral defects in rabbit knee with various concentrations of Escherichia coli-derived recombinant human bone morphogenetic protein-2. Int Orthop 34:761–767
56. Maehara H, Sotome S, Yoshii T, Torigoe I, Kawasaki Y, Sugata Y et al (2010) Repair of large osteochondral defects in rabbits using porous hydroxyapatite/collagen (HAp/Col) and fibroblast growth factor-2 (FGF-2). J Orthop Res 28:677–686
57. Reyes R, Delgado A, Sanchez E, Fernandez A, Hernandez A, Evora C (2012) Repair of an osteochondral defect by sustained delivery of BMP-2 or TGFbeta1 from a bilayered alginate-PLGA scaffold. J Tissue Eng Regen Med. doi:10.1002/term.1549

58. Mohan N, Dormer NH, Caldwell KL, Key VH, Berkland CJ, Detamore MS (2011) Continuous gradients of material composition and growth factors for effective regeneration of the osteochondral interface. Tissue Eng Part A 17:2845–2855
59. Holland TA, Bodde EW, Cuijpers VM, Baggett LS, Tabata Y, Mikos AG et al (2007) Degradable hydrogel scaffolds for in vivo delivery of single and dual growth factors in cartilage repair. Osteoarthritis Cartilage 15:187–197
60. Sukegawa A, Iwasaki N, Kasahara Y, Onodera T, Igarashi T, Minami A (2012) Repair of rabbit osteochondral defects by an acellular technique with an ultrapurified alginate gel containing stromal cell-derived factor-1. Tissue Eng Part A 18:934–945
61. Lopiz-Morales Y, Abarrategi A, Ramos V, Moreno-Vicente C, Lopez-Duran L, Lopez-Lacomba JL et al (2010) In vivo comparison of the effects of rhBMP-2 and rhBMP-4 in osteochondral tissue regeneration. Eur Cell Mater 20:367–378
62. Liu XW, Hu J, Man C, Zhang B, Ma YQ, Zhu SS (2011) Insulin-like growth factor-1 suspended in hyaluronan improves cartilage and subchondral cancellous bone repair in osteoarthritis of temporomandibular joint. Int J Oral Maxillofac Surg 40:184–190
63. Miyakoshi N, Kobayashi M, Nozaka K, Okada K, Shimada Y, Itoi E (2005) Effects of intraarticular administration of basic fibroblast growth factor with hyaluronic acid on osteochondral defects of the knee in rabbits. Arch Orthop Trauma Surg 125:683–692
64. Miller RE, Grodzinsky AJ, Vanderploeg EJ, Lee C, Ferris DJ, Barrett MF et al (2010) Effect of self-assembling peptide, chondrogenic factors, and bone marrow-derived stromal cells on osteochondral repair. Osteoarthritis Cartilage 18:1608–1619
65. Re'em T, Witte F, Willbold E, Ruvinov E, Cohen S (2012) Simultaneous regeneration of articular cartilage and subchondral bone induced by spatially presented TGF-beta and BMP-4 in a bilayer affinity binding system. Acta Biomater 8:3283–3293
66. Freeman I, Kedem A, Cohen S (2008) The effect of sulfation of alginate hydrogels on the specific binding and controlled release of heparin-binding proteins. Biomaterials 29:3260–3268
67. Freeman I, Cohen S (2009) The influence of the sequential delivery of angiogenic factors from affinity-binding alginate scaffolds on vascularization. Biomaterials 30:2122–2131
68. Re'em T, Kaminer-Israeli Y, Ruvinov E, Cohen S (2012) Chondrogenesis of hMSC in affinity-bound TGF-beta scaffolds. Biomaterials 33:751–761
69. Goldring MB, Tsuchimochi K, Ijiri K (2006) The control of chondrogenesis. J Cell Biochem 97:33–44
70. Petersen W, Tsokos M, Pufe T (2002) Expression of VEGF121 and VEGF165 in hypertrophic chondrocytes of the human growth plate and epiphyseal cartilage. J Anat 201:153–157
71. Tuli R, Tuli S, Nandi S, Huang X, Manner PA, Hozack WJ et al (2003) Transforming growth factor-beta-mediated chondrogenesis of human mesenchymal progenitor cells involves N-cadherin and mitogen-activated protein kinase and Wnt signaling cross-talk. J Biol Chem 278:41227–41236

Chapter 23
Polymer-Based Drug Delivery Systems for Solid Tumor Treatment

Ariella Shikanov and Abraham J. Domb

Abbreviations

ASGP	Asialoglycoprotein
CSC	Cancer stem cells
DDS	Drug delivery system
DOX	Doxorubicin
ECM	Extracellular matrix
ELP	Elastin-like polypeptides
EPR (effect)	Enhanced permeability and retention (effect)
GI	Gastrointestinal
HMDI	Hexamethylene diisocyanate
HPMA	*N*-(2-Hydroxypropyl)methacrylamide
IT	Intratumoral
LCST	Low critical solution temperature
MAbs	Monoclonal antibodies
MPEG	Methyl poly(ethylene glycol)
Mr	Relative molecular mass
Mw	Molecular Weight
NP	Nanoparticles
PCL	Poly(ε-caprolactone)
PEA	Poly(ethylene adipate)

A. Shikanov (✉)
The Department of Biomedical Engineering, University of Michigan,
1101 Beal Avenue, Ann Arbor, MI 48109, USA
e-mail: shikanov@umich.edu

A.J. Domb
Institute of Drug Research, School of Pharmacy-Faculty of Medicine, The Hebrew University of Jerusalem, Jerusalem 91120, Israel

A.J. Domb and W. Khan (eds.), *Focal Controlled Drug Delivery*, Advances in Delivery Science and Technology, DOI 10.1007/978-1-4614-9434-8_23,

PEO	Poly(ethylene oxide)
PEG	Poly(ethylene glycol)
PES	Poly(ethylene succinate)
PHA	Poly(hexamethylene adipate)
PK	Pharmacokinetics
PLGA	Poly(lactic-co-glycolic acid)
PLLA	Poly(L-lactic acid)
PNIPAM	Poly(*N*-isopropylacrylamide)
RA	Ricinoleic acid
SA	Sebacic acid
PSA	Poly(sebacic acid)
UCST	Low critical solution temperature
VEGF	Vascular endothelial growth factor

23.1 Introduction

Solid tumors account for the major cancer burden, causing millions of deaths each year [1]. Systemic chemotherapy combined with surgery or radiotherapy is the most commonly used therapeutic strategy, although considerable limitations exist. Anticancer drugs are delivered intravenously at maximum tolerable doses, which cause severe toxicities in healthy tissues or even death [2]. However, despite the adverse whole body effect, many anticancer drugs have limited distribution from blood vessels in solid tumors, which limits their effectiveness. The limited drug distribution in the solid tumor is inevitable because of the tumor architecture. Large distances between blood vessels, the dense and rigid extracellular matrix (ECM), cell–cell adhesion, high interstitial fluid pressure, lack of convection, and drug metabolism and binding all contribute to the limited drug distribution in the tumor [3] (Fig. 23.1). Thus, localized delivery of the anticancer drugs to the solid tumors holds potential advantages of improved drug delivery, because it can maintain low systemic drug levels while ensuring therapeutic levels at the tumor site [6].

Polymer-based systems have had an enormous impact on drug therapies. When a pharmaceutical agent is encapsulated in a polymer or lipid, drug safety and efficacy can be greatly improved and new therapies are possible [7]. With this approach the polymer-anticancer drug formulation can be implanted or injected in the body at the place of interest (Fig. 23.2a). Earlier formulations were composed of nondegradable polymers, such as silicone rubber used in Norplant® for delivery of contraceptives. Later on, the focus switched to the degradable polymeric systems with slow and controlled drug delivery. The degradable polymer can release the drug locally at the implantation site while fully degrading and eliminating the need for surgical removal of the empty polymer device. Solid polyanhydride matrices were used to locally deliver chemotherapeutic drug carmustine (BCNU) to treat brain tumor [9, 10], and GLIADEL® was the first polymeric system approved by FDA for clinical use in 1996. In this case, the surgeon removes as much of the tumor as possible during surgery and places up to eight small polymer–drug wafers at the

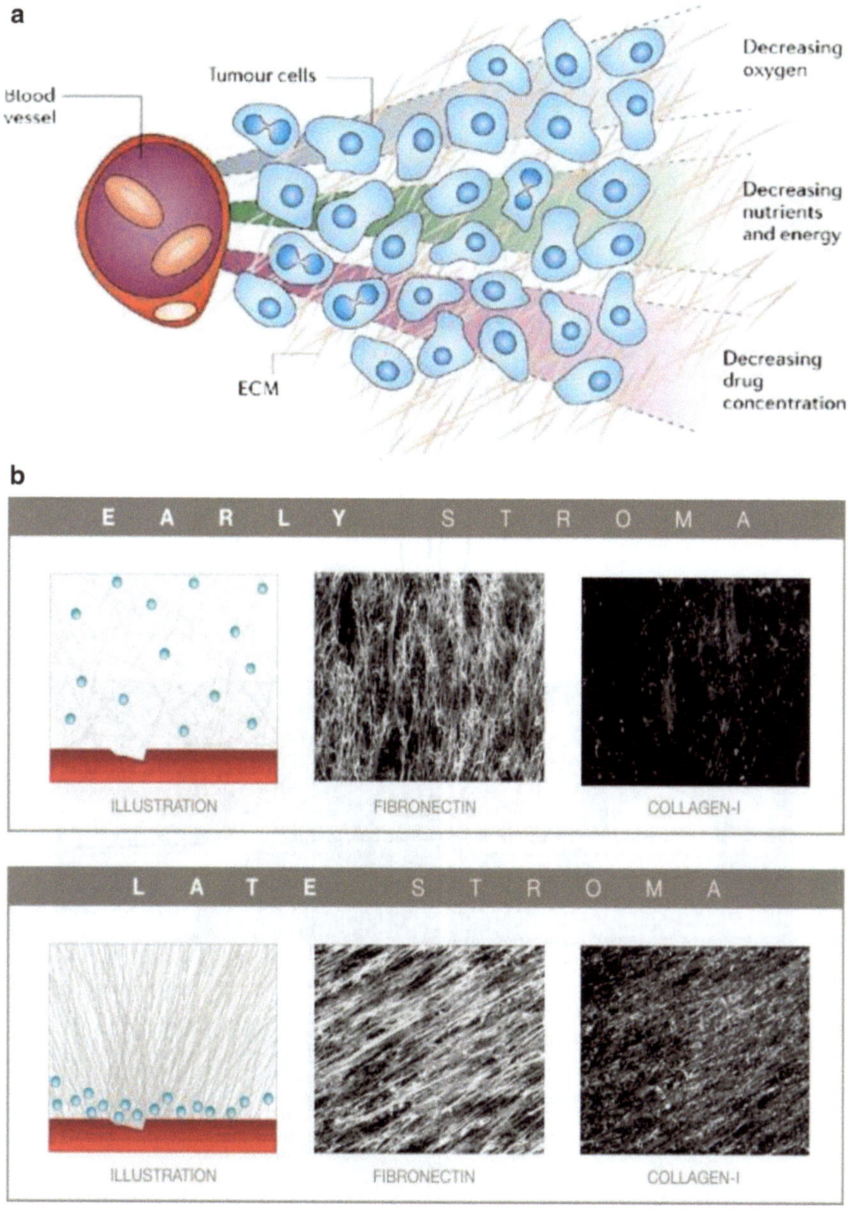

Fig. 23.1 The effect of tumor environment on drug penetration. (**a**) An illustration of drug's limited distribution from blood vessels in solid tumors. The existing gradient of oxygen, nutrients, energy, and drug concentration combined with irregular blood network limits systemically delivered anticancer drug's effectiveness. (The illustration was reproduced from Minchinton et al. [3].) (**b**) The tumor microenvironment presents with more organized and dense ECM (late stroma, *bottom*) compared to a loosely packed early tumor environment (early stroma, *top*) which results in decreased penetration of extravasated chemotherapeutic agents from circulation into the tumor microenvironment. *Left panels* are representative illustrations of damaged blood vessels and drug delivery systems or drug molecules penetrating through ECM. *Middle and right panels* depict reconstructed confocal images of indirect immunofluorescence showing ECM molecules fibronectin (*middle*) and collagen I (*right*) fibers derived from assorted fibroblasts (The illustrations are reproduced with permission from Cukierman et al. [4] and confocal images from Amatangelo et al. [5])

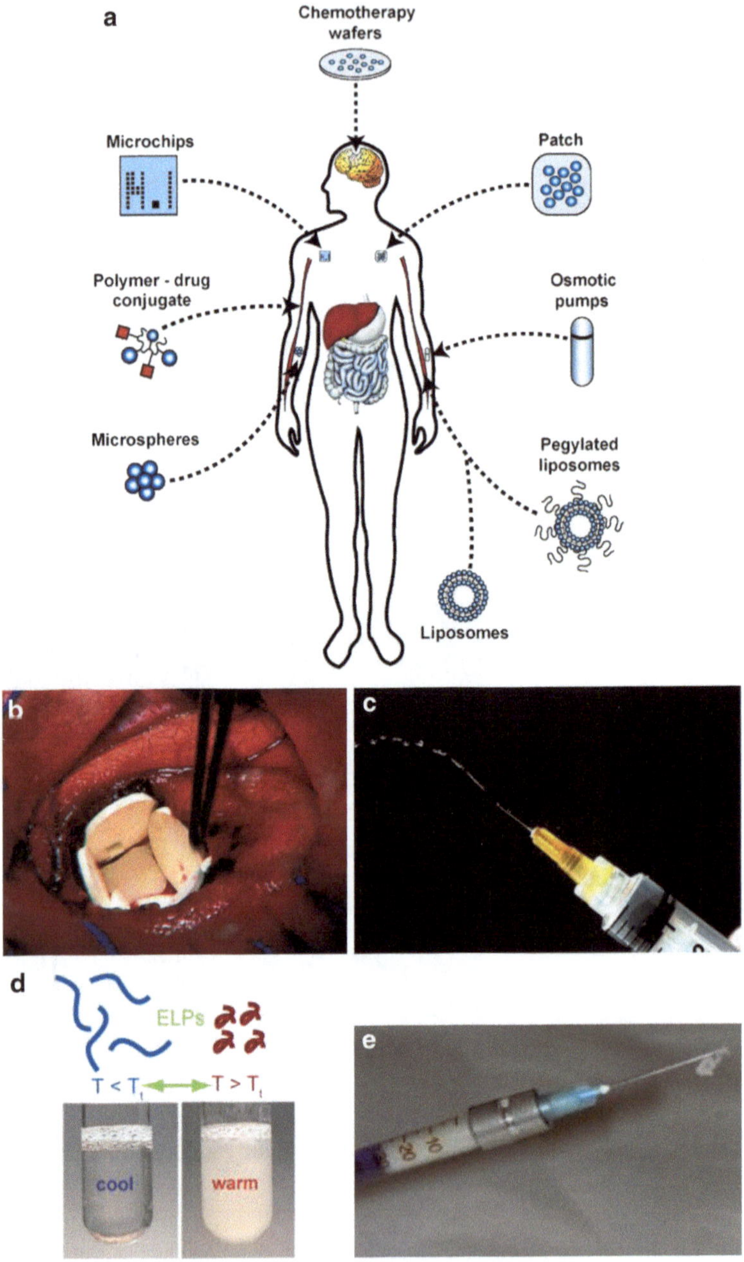

Fig. 23.2 Emerging and existing drug delivery system for treating cancer. (**a**) Examples of different delivery strategies include polymer microspheres loaded with anticancer drugs for passive or active tumor targeting, solid polymer wafers loaded with BCNU for the localized treatment of brain tumors, osmotic pumps for anti-angiogenic drug delivery, liposomes, polymer–drug moiety conjugates, and controlled release microchips. Examples of local delivery polymeric drug delivery systems that are currently used in clinic (**b**—GLIADEL®), in clinical trials (**c**—ReGel®), and in preclinical developments [(**d**) injectable elastin-like polypeptides and (**e**) injectable P(SA:RA)]. The images were reproduced from Moses et al. [7] and MacroMed [8]

tumor resection site (Fig. 23.2b). The drug is slowly released from the polymer to kill the surviving cancer cells, followed by complete degradation of the polymer. With this approach the patients' survival rate after 2 years was five times greater compared to the control group; however, the need of surgical delivery of the solid polymeric system is the main drawback with this system. To address this limitation various intratumoral (IT) liquid and pasty drug delivery systems (DDS) were developed to be directly injected into superficially accessible tumors, such as skin, breast, and cervix, or into deeper tumors, such as liver, esophageal, and pancreatic, using imaging equipment for precise placement. The main advantage of the injectable DDS is that they form solid depot system after the IT injection. Thermosensitive in situ forming gels, adhesive injectable hydrogels, elastin-like polypeptides (ELPs), and fatty acid-based polymers are some of the representative examples of the injectable DDS that will be discussed in this chapter (Fig. 23.2c–e).

Currently the main use of local anticancer DDS is as an adjuvant or neoadjuvant therapy in combination with other anticancer treatments, such as surgery, radiotherapy, and systemic chemotherapy. When the anticancer DDS serves as an adjuvant therapy after the surgical tumor removal, it is placed in the remaining tumor cavity to kill surviving cancer cells and achieve an improved local control, like the earlier mentioned GLIADEL®. Neoadjuvant DDS therapy is aimed to debulk the tumor to reduce its load before the surgery or to treat cases of inoperable tumors.

IT delivery holds an unlimited potential to treat solid tumors through incorporation of various properties in its design. For example, targeting technology can be active, passive, or physical. Passive targeting is typically accomplished through the enhanced permeability and retention effect using macromolecules for delivery. This can be used to produce greater accumulation of the drug in a tumor site as opposed to normal tissue. Active targeting is accomplished through direction of the agent to a specific body tissue by using antibodies against specific cell targets. Physical targeting can be employed to selectively treat the location or to bypass physiological barriers [6, 7]. Additionally, to address the heterogeneous population of cancer cells in some tumors or to inhibit the angiogenic potential of the tumors, multiple therapies can be incorporated and delivered in one system. This will result in better drug penetration through the tumor and improved efficiency. Treatments that substantially reduce local recurrence result in decreased mortality, firmly supporting a link between local control and overall survival in cancer patients [11].

23.2 Locally Injectable Polymeric DDS

23.2.1 Thermosensitive Sol–Gel Reversible Hydrogels

Stimuli-sensitive hydrogels are aqueous polymeric solutions, which undergo reversible volume phase transition in response to the external physical or chemical stimuli such as temperature or pH [12]. The sol phase is defined as a flowing fluid, whereas the gel phase is a non-flowing hydrogel that maintains its integrity [13]. Temperature is one of the most widely used stimuli for stimuli-sensitive hydrogels, because it is easy to control and has practical advantages both in vitro and in vivo. The solution is prepared and the drug is mixed in while the formulation is at the sol state, and the hydrogel forms in situ after the injection to the site of interest. Temperature-sensitive hydrogels undergo a sol–gel transition at a critical temperature, namely, lower critical solution temperature (LCST) or upper critical solution temperature (UCST) [14]. The LCST polymers exhibit a hydrophilic-to-hydrophobic or sol–gel transition with increasing temperatures, which is also termed a reverse thermogelation. These polymeric systems can be designed to be liquid at room temperature and form a gel upon exposure to a body temperature after injection. The UCST systems undergo a sol–gel transition with decreasing temperatures. The polymer solution, the drug incorporation, and the injection are performed at the elevated temperatures (above UCST), and the gel forms in vivo upon cooling to a body temperature [12, 14].

23.2.1.1 UCST Thermosensitive Hydrogels

Natural polymers, such as gelatin and polysaccharides, such as agarose, exhibit a sol–gel transition on lowering the temperature. The proposed mechanism for the gel formation is helix formation followed by aggregation of the helices that act as knots, i.e., the physical junctions of the gels [14]. While natural polymers remain an exciting and inspiring topic to study, they suffer from low degree of modification: hazards in using naturally derived products and low batch-to-batch reproducibility. Synthetic polymers on the other hand can be tailor-made and modified to provide desirable mechanical and chemical properties for specific applications.

Synthetic biodegradable block copolymer that exhibits UCST has been proposed for an injectable gel system by Jeong et al. [13] (Fig. 23.3a, b). The copolymers of poly(ethylene oxide) (PEO) and poly(L-lactic acid) (PLLA) were synthesized by ring opening polymerization of L-lactide, followed by coupling of the diblock units and resulting in a triblock polymer PEO–PLLA–PEO (Mr 5,000-2,040-5,000). PEO is a non-biodegradable polymer and PEO blocks of high relative molecular mass (Mr above 10,000) are unsuitable for filtration through human kidney membrane due to the large hydrodynamic radius. On the other hand, PLLA is a hydrolytically degradable polyester. Thus, to ensure biodegradability of the system, PEO units of Mr 5,000 were incorporated in the triblock polymer to allow kidney secretion of the nondegradable PEO blocks after the degradation of the polyester block. The proposed mechanism for sol–gel transition is micelle formation, which at high concentration levels is caused by the association of the micelles [13]. The sol–gel transition is a function of the polymer concentration as well as composition of the block polymer. Increasing the hydrophobicity of the block polymer by increasing the PLLA block length increases the aggregation tendency, resulting in a steepening of the gel–sol transition curve slopes and the onset of gelation at lower concentration. Thus, a delicate balance has to be kept to allow fluidity and injectability of the formulation at the temperatures that can be tolerated and not harmful for the body while having the system form a solid gel at 37 °C and not disperse from the injection site. The reported PEO–PLLA–PEO (Mr 5,000-2,040-5,000) was liquid at 45 °C,

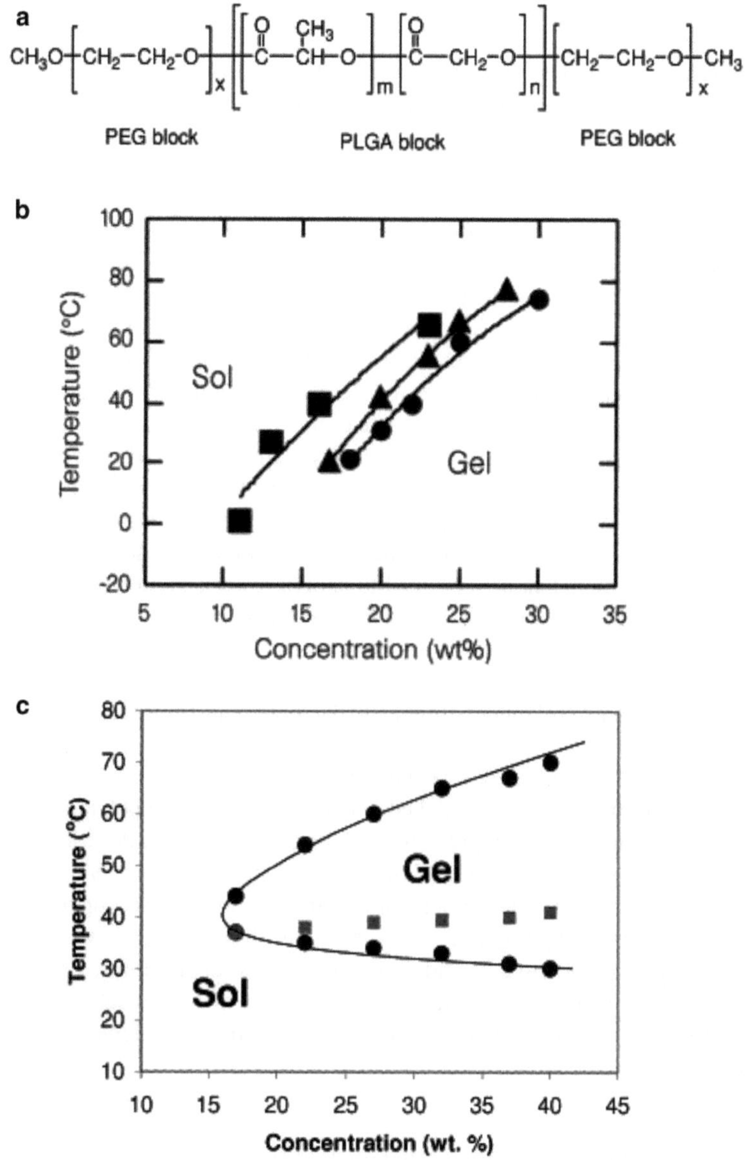

Fig. 23.3 Thermosensitive sol–gel reversible hydrogels. (**a**) The chemical structure of a PEG–PLGA–PEG (ABA) triblock copolymer. (**b**) Gel–sol transition curves of typical polymer with UCST. Here shown is a PEG–PPLA–PEG triblock copolymer (Reproduced from Jeong et al. [13]). (**c**) Phase diagram of a PEG–PLGA–PEG triblock copolymer solution exhibiting LCST (Reproduced from Jeong et al. [15])

when the drug formulation was prepared, and set a solid gel at 37 °C, after injection [13]. However, concerns about the relatively high temperature of the preparation and injection that can be harmful to some drugs and proteins, as well as painful, limit the practical applications of the UCST systems [12].

23.2.1.2 LCST Thermosensitive Hydrogels

Thermosensitive polymers that can transition from a free-flowing solution at cool temperatures (2–20 °C) to a viscous, water-insoluble hydrogel at body temperature are attractive for biomedical applications. The physiological transition point makes these polymer systems easy for fabrication, non-harmful for the encapsulated agent, and non-painful for the injection. Polymer precipitation in solution on raising the temperature often occurs in aqueous systems and results from the balance of intermolecular forces between the polymer and the solvent as well as between polymers [14]. Cellulose, a natural polymer, is not soluble in water, but by introducing hydrophilic moieties, cellulose derivatives become water soluble and exhibit reverse thermogelation. When cellulose derivatives have an optimum balance of hydrophilic and hydrophobic moieties, water becomes a poorer solvent with increasing temperature and polymer–polymer interactions become dominant at higher temperature, resulting in a gel. Methylcellulose and hydroxypropyl cellulose have a sol–gel transition temperature at 80 °C and 55 °C, respectively, making those polymers non-useful to serve as injectable in situ solidifying hydrogels [16, 17]. Advances in synthetic polymer synthesis allow adjustment of the desired qualities, and several examples of the synthetic polymers that are designed to exhibit lower and more physiologically relevant LCST are discussed in the following paragraphs.

Poly(*N*-Substituted Acrylamide)-Based Block Copolymer Hydrogels

Poly(*N*-substituted acrylamide), especially poly(*N*-isopropylacrylamide) (PNIPAM), are one class of the most widely studied temperature responsive polymers. A linear PNIPAM chain undergoes a rapid coil-to-globule (hydration-to-dehydration) transition in aqueous solution at its LCST of around 32 °C [12]. The LCST of PNIPAM can be elevated or reduced to a desirable value by incorporating *N*-isopropylacrylamide (NIPAM) with more hydrophilic or hydrophobic monomers, respectively [18], which makes it an attractive candidate for biomedical applications. However, the major drawback of poly(*N*-substituted acrylamide) is non-degradability, and only PNIPAM with limited molecular weight could be excreted by glomerular filtration without long-term accumulation in vivo.

PEO/PPO-Based Block Copolymer Hydrogels

The poly(ethylene oxide)-*b*-poly(propylene oxide)-*b*-poly(ethylene oxide) triblock copolymers (PEO–PPO–PEO), known as Pluronic® (BASF) or Poloxamer® (ICI), have surface-active properties and are widely used in pharmaceutical systems [12]. It has been established that the PEO block is dominantly hydrophilic within the temperature range from 0 to 100 °C, whereas the water solubility of the PPO block undergoes a dramatic decrease as temperature is increased above 5 °C [19]. Unimer-to-micelle transition, and further micelle aggregation as the temperature reaches the sol–gel transition point, was proposed as a gelation mechanism [19]. The Pluronic hydrogels exhibit considerable viscosity, partial rigidity, and a certain persistence time, due to the ordered micellar packing structure and intermicellar entanglements. The above properties and the gelation at physiological temperature are convenient for the incorporation of both hydrophilic and hydrophobic drugs and biomedical applications. However, there are still many drawbacks in the Pluronic hydrogel systems, such as weak mechanical strength, high permeability, non-biodegradability, and limitation of molecular weight [12].

PLGA–PEG–PLGA Triblock Copolymers

Kim and coworkers [13] have already reported that incorporation of biodegradable polyester between two PEO blocks results in degradable system, but gels as the temperature decreases (UCST), which is less favorable for biomedical applications. The next generation of thermosensitive hydrogels, which were based on PLGA–PEG–PLGA (BAB) triblock copolymers, was subsequently developed [20, 21]. The synthesis of PLGA–PEG–PLGA was simpler than that of PEG–PLGA–PEG (ABA), because the coupling procedure using HMDI could be avoided. PLGA–PEG–PLGA also exhibited a reversible sol–gel–sol transition with increasing temperature, and the phase diagram was influenced by the block length and composition as well as additives [20, 22] (Fig. 23.3c). One specific formulation of PLGA–PEG–PLGA with block lengths of 1,500-1,000-1,500 became commercially available (ReGel®), and further optimization allowed matching polymer's unique properties for the desired release rate of the active pharmaceutical agent [23]. Formulation design has demonstrated that ReGel may be used to deliver small and large hydrophobic molecules, peptides, and proteins [24–27] (Fig. 23.2c).

OncoGel™, a non-Cremophor-based formulation of paclitaxel in ReGel, was designed for local delivery of paclitaxel to solid tumors to provide targeted cytotoxicity without systemic toxicities associated with conventional treatment [28]. The in vitro release of paclitaxel from ReGel® exhibited a diffusion-controlled release profile in the initial 2 weeks, followed by a combined diffusion-degradation process for about 5 weeks. The in vivo distribution of paclitaxel was monitored after intratumoral injection of ReGel® containing [^{14}C]-radiolabeled paclitaxel. The C-14 levels in tumors decreased slowly over 6 weeks and eliminated mainly through feces and urine, with less than 0.1 % being distributed to other organs. ReGel®/paclitaxel showed higher antitumor efficacy and lower drug-related adverse effects, as compared with the maximum tolerated systemic dose of the commercially available paclitaxel formulation (Taxol®) [21]. Nonclinical studies with OncoGel demonstrated safety and improved tolerability versus systemic administration of paclitaxel, localization within and around the tumor site, safety and efficacy as a stand-alone treatment, and efficacy and tolerability in combination with other therapies. The promising results in preclinical studies led to three completed clinical studies in superficially accessible solid tumors [29] and in combination with radiation therapy (RT) in esophageal cancer [30].

Other LCST Di-, Tri-, and Multiblock Copolymer Systems

Several block copolymers composed of PEG and other biodegradable aliphatic polyesters also showed a thermoreversible sol–gel transition in the aqueous solutions. The ABA type of PEG–aliphatic polyester–PEG triblock copolymer was prepared by coupling MPEG with the polyester blocks, including poly(hexamethylene adipate) (PHA), poly(ethylene adipate) (EA), and poly(ethylene succinate) (PES), and they showed sol–gel transition as the temperature increased [31]. ABA and BAB triblock copolymers consisting of PEG and PCL also displayed a clear sol–gel transition as the temperature increased from 10 to 60 °C. Although PCL–PEG–PCL was easier to synthesize than PEG–PCL–PEG, it was unstable and turned to an opaque gel in 1 h at room temperature, which can bring inconvenience for clinical application. The gelation at room temperature was attributed to the crystallization of the polymer, which was quite different from the micellar aggregation mechanism of the thermo-induced gelation at 37 °C. The unstable behavior of the triblock copolymers led to the synthesis of PEG–PCL multiblock copolymers by terephthaloyl chloride coupling. The 20 wt.% aqueous solution of the PEG–PCL multiblock copolymer not only showed a LCST sol–gel transition but also existed as a stable transparent solution at room temperature.

PEG/PLLA and PEG-SA multiblock copolymers also showed LCST transition at 37 °C and showed controlled release of fibroblast growth factor and FITC-dextran from the hydrogels [32–34].

Poly(organophosphazenes)

Song et al. [35] designed a biodegradable thermosensitive poly(organophosphazenes) substituted with PEG and amino acid ester groups. Many different copolymers were obtained by the reaction of poly(dichlorophosphazene) with sodium salt of MPEG, followed by the reaction with an excess amount of amino acid esters. The resulting polymers exhibited LCST in the aqueous solution at the range between 25.2 and 98.5 °C, depending on the MPEG molecular weight, the amino acid ester, and the mole ratio of the two substituents. The described polymers followed the rule of the phase transition of the thermosensitive polymers that is attributed to a change in the hydrophilic–hydrophobic balance of the polymers with respect to their interaction through hydrogen bonding between the polymer and water molecules. Below the LCST, the hydrogen-bonding interactions with water predominate over the hydrophobic interaction of the polymers, which results in liquid aqueous solution. As the temperature rises, the hydrophobic interactions of the polymer increase and the polymer precipitates from the solution. For the described polymers bearing MPEG and glycine ethyl ester groups, the LCST increases with increasing MPEG content because of the increased hydrophilicity. On the other hand, increasing the hydrophobicity of the ester group with methyl, ethyl, and benzyl ester of glycine caused lower LCST of 88.5, 77.5, and 49.5 °C, respectively. The LCST was also dependent on the type of amino acids and it decreased in the order of glycine, alanine, amino-malonic acid, L-aspartic acid, and L-glutamic acid ethyl esters. The physiologically relevant LCST of 25.2, 35, and 38.5 °C were achieved at a low ratio of MPEG (0.2, 0.31, and 0.66) with L-glutamic, glycine, or aspartic acid ethyl esters [35]. The hydrolytic degradation of polyphosphazenes substituted with amino acid esters as side groups has been explained in terms of carboxylic acid-catalyzed degradation. It has been proposed that the initiation step of hydrolytic degradation is the hydrolysis of pendant ester group [36]. The generated carboxylic acid groups attack the polymer backbone, resulting in backbone cleavage [37, 38].

The biodegradable poly(organophosphazene) hydrogels were extensively used to overcome low bioavailability of highly hydrophobic anticancer drugs, such as silibinin [39], 2-methoxyestradiol [40], and paclitaxel [41]. The aqueous solution of poly(organophosphazene) enhanced the solubility of silibinin 2,000 times. Although the polymer degradation based on polymer weight loss was almost complete in 28 days, the release of silibinin in pH=7.4 was only 40 % of the loaded drug compared to 90 % at pH=6.8 after 21 days. Nonetheless, in the HT-29 human colon adenocarcinoma-xenografted mice model, the IT-injected hydrogel with silibinin exhibited a good tumor control compared to the control group [39].

Long-term theragnostic hydrogel system is another interesting application of poly(organophosphazene) polymer. Theragnosis has two major functions: the targeted therapy and the imaging diagnosis. However, achieving effective concentration in lesion sites after systemic delivery was challenging, and IV-type theragnostic agents cannot be localized within specific tumor sites over long-term periods. In a recent study [41], poly(organophosphazene) hydrogels loaded with PEGylated cobalt ferrite nanoparticles and paclitaxel were injected into H-29 subcutaneous solid tumors. The theragnostic hydrogel gradually released paclitaxel at the tumor site and simultaneously supplied MR imaging for over 3 weeks.

23.2.2 Elastin-Like Polypeptides

Another macromolecular carrier for cancer therapy—ELP—has been in development for the last decade. ELPs are a class of temperature-sensitive biopolymers based on the structural motif found in mammalian tropoelastin [42, 43]. ELPs consist of a repeated pentapeptide sequence $(VPGXG)_n$, where X is any amino acid except proline. ELPs exhibit a thermodynamic inverse phase transition in aqueous solution at a specific temperature (T_t), below which ELPs are soluble and above which ELPs become insoluble and form a gel, similar to LCST [44] (Fig. 23.2d).

ELPs exhibit many of the properties desired from a polymeric delivery system to serve in biomedical application. ELPs are composed of amino acids; they are nontoxic and biodegradable [45, 46]. They are designed and synthesized using genetic engineering, which allows precise control of their Mw and polydispersity, as well as encoding their composition at the gene level.

The tunable properties of ELPs allow using them for both systemic and local delivery using different strategies. In the first strategy, hydrophilic ELPs with a T_t much greater than a body temperature can be used as soluble, hydrophilic macromolecular carriers of conjugated drugs to advantage of the EPR effect at the tumor site for systemic delivery [47]. In a second approach ELPs can be designed to exhibit T_t between 37 and 42 °C; systemic delivery of ELPs conjugated with a drug in combination with local hyperthermia to a solid tumor leads to enhanced tumor accumulation [48]. Finally, ELPs are useful for local delivery of conjugated or mixed drugs by IT injection, as they can be designed to be soluble at room temperature and solidify at 37 °C, similar to other thermosensitive copolymers [44].

Liu et al. [49] investigated the feasibility of IT therapy using ELPs loaded with ^{125}I in 4T1 subcutaneous tumors in Balb/c nu/nu mice. They have shown that ELP_3 with T_t of 27 °C successfully formed a gel in the injection site and over 40 % of the gel was there after 24 h, decreasing to roughly 10 % after 1 week. In the efficacy studies ^{131}I-ELP_3 demonstrated inhibition of 4T1 tumor growth and prolonged survival for 1 week, compared to saline administration. To improve ELP performance, the investigators plan to redesign the ELP to prolong its tumor retention, increase the delivered dose of the drug, and allow multiple injections.

23.2.3 *Fatty Acid-Based Polymers*

Previously described thermoreversible hydrogels as well as ELPs are prepared as aqueous solutions of the polymer and the drug. As the search after the ideal injectable biodegradable polymer continues, our group was interested to explore the possibilities of designing a natural fatty acid-based polymer that can be injected as a uniform system alleviating the need for the solution preparation. New biodegradable poly(ester anhydride)s were prepared by the melt polycondensation of diacid oligomers of poly(sebacic acid) (PSA) transesterified with ricinoleic acid (RA)[50] (Fig. 23.2e). The transesterification of PSA with ricinoleic acid to form oligomers was conducted via a melt bulk reaction between a high molecular weight PSA and ricinoleic acid. Polymers with weight-average molecular weights of 2,000–60,000 and melting temperatures of 24–77 °C were obtained for P(SA:RA) containing 20–90 % (w/w) RA. The polymer properties in terms of physical state, melting point, degradation rate, and injectability can be tuned by controlling the ratio of the monomers, while ricinoleic acid contributes to the liquid properties and sebacic acid for the solid [51]. The polymer was tested as a delivery system for both hydrophilic and hydrophobic drugs [52–54]. Furthermore, the majority of the bonds in the polymer structure are ester bonds between the fatty acids. It is known that anhydride bonds are more susceptible to moisture and reactions with reactive amino or acid groups in the drugs. P(SA:RA) stability studies showed small or no effect on polymer Mw when the polymer–drug (paclitaxel and cisplatin) formulation was stored for a long period of time. The slow controlled release of paclitaxel from the formulation was dependent on the initial drug loading and mainly limited by the low aqueous solubility, when 10 % w/w paclitaxel released 15 % of the incorporated drug in 1 month compared to 30 % from a 5 % w/w paclitaxel formulation [55]. On the other hand, cisplatin release was fast and completed in 5 days [53]. Efficiency in vivo studies in heterotopic model of bladder and melanoma subcutaneous tumors in mice showed successful tumor growth inhibition and better long-term survival, compared to blank polymer or paclitaxel solution control. The formulation was also tested and showed improved efficiency and reduced toxicity in the orthotopic model of prostate cancer in rats [56]. Paclitaxel and cisplatin distribution in both the implant surrounding tumor and blood were tested in mice after IT injection. The drug released from the polymer killed the malignant cells and diffused throughout the tumor, up to 7 mm distance from the injection site, as was confirmed in the histology evaluation. Additionally, not only apoptotic cells were found in that region, but also a reduction in tumor cells density. Similar results were obtained for cisplatin distribution, and both studies showed minimal or low systemic toxicity [53, 55, 57].

23.3 Microparticles and Nanoparticles for Anticancer Drug Delivery

Particulate drug delivery offers a number of advantages. Similarly to the polymeric delivery systems injected as a bulk, particles can be injected or deposited directly at the site of action, providing a high local drug level over an extended period, while minimizing systemic toxicity. On the other hand, the particles can also be injected systemically, aimed at specific locales at which to release the encapsulated drugs, using active or passive targeting methods [58]. This feature is especially valuable to treat distant undetected metastases that have similar properties to the primary tumor and can be targeted by systemically injected particles.

During particle formation, drugs dispersed or dissolved in the solvent droplets formed by an emulsion system can partition into the surrounding external aqueous phase. The smaller the particle, the greater the proportion of the drug that will have access to the external aqueous phase. This can lead to substantial loss of payload or to a lower maximal drug loading for smaller than larger particles. In formed particles, a greater proportion of drug can leave the surface of smaller entities by diffusion. Likewise, water will penetrate smaller particles more rapidly, which will result in a "burst release" and generally in more rapid kinetics. In addition, in polymers that are degraded hydrolytically, water penetration will result in particle deterioration, which will further accelerate drug release [58, 59].

The particle size greatly affects the clearance rate. Locally injected microparticles tend to stay at the injection site, as was demonstrated by Kohane et al. [60], where 60 μm polymeric particles composed of a slowly degrading polymer injected at the sciatic nerve were still found at the injection site 8 weeks later. The difference between micro- and nanoparticles (NPs) was demonstrated by injecting NPs and 5, 25, 60, and 250 μm in diameter microparticles in the abdominal cavity. The NPs showed almost complete clearance, while the microparticles remained for at least 2 weeks [61]. When injected intravenously, large microparticles can embolize vessels with the same diameter. The exact particle size at which embolism becomes a problem is not clear, but particle aggregation, particle–blood interactions, and slurry viscosity are important issues. NPs are generally too small to cause embolic phenomena and can circulate throughout the vasculature.

The surface chemistry of the particles allows physical absorption or chemical conjugation with hydrophilic polymers, such as PEG, slowing the particles' clearance from the bloodstream by the reticuloendothelial system. For example, Park et al. [62] demonstrated that doxorubicin (DOX) cytotoxicity can be maximized and dose-limiting cardiotoxicity minimized by controlled release from PEGylated PLGA NPs, ~130 μm in diameter. They used an avidin–biotin coupling system to control PEG conjugation and showed that after intravenous injection to mice, these sterically stabilized DOX NPs showed longer $t_{1/2}$ in plasma compared to unmodified NPs or free DOX. In a subcutaneous murine B-cell lymphoma model (A20), IT-injected PEGylated NPs showed similar efficacy to regular NPs combined with reduced cardiotoxicity.

PEG–PSA-based particles are another example of biodegradable drug carriers designed for IT injection [63]. Up to 40 % etoposide was encapsulated in particles with ~1 μm in diameter. The system showed controlled drug release over 6 days and effectively suppressed subcutaneous lung tumor growth in a xenograft mouse model with 100 % survival after 31 days compared to 0 % survival in a free drug group. While the described particles are too large for intravenous delivery, they have shown efficacy not only in IT delivery but also in aerosol delivery to the lungs and good mucus penetration.

Although this chapter is focused on local delivery of anticancer drugs, one cannot avoid mentioning some of the creative ways particles were designed to actively or passively target tumors after systemic delivery. Passive targeting of systemically injected long-circulating NPs (100–600 μm in diameter) is based on the unique properties of the cancer microenvironment. A dysfunctional lymphatic drainage, which results in enhanced fluid retention in the tumor interstitium, and leaky vasculature, whose permeability is greater than in healthy tissues, result in extensive nonspecific accumulation in the tumor [59]. Stealth properties were incorporated in another example of NPs developed for passive targeting. Tamoxifen was loaded to long-circulating poly(ε-caprolactone) (PCL) NPs that were surface-modified with Pluronic® F-68 or F-108 [64]. After intravenous injection in mice bearing human breast xenograft, the long-circulating NPs induced greater drug accumulation in the tumor compared to the free drug.

Specific or active targeting of NPs has emerged as a valuable approach to target specific site of interest while avoiding the associated systemic adverse effect on healthy tissues. For example, in receptor-mediated strategy, paclitaxel targeting to liver cancer cells was investigated by using PEGylated poly(γ-benzyl-L-glutamate) particles endcapped with galactose moieties which specifically bind asialoglycoprotein (ASGP) receptors in hepatocytes. This formulation showed greater toxicity in HepG2 cells compared to a free paclitaxel [65].

Additional approaches involve monoclonal antibodies (MAbs), folate, or transferrin receptor-mediated targeting, where specific antigens or receptors on tumor cells promote NP accumulation [66–69]. Peptides that bind overexpressed integrins on the tumor cells surface allow selective targeting [70]. In conclusion, intravenously injected NPs show prolonged circulating times compared to a free drug and allow active targeting not only to the primary tumor but also to metastases and circulating tumor cells. However, the low drug loading, fast release kinetics, and relatively short cell exposure leave a lot of space of formulation design improvement.

23.4 Polymer–Drug Conjugates for Systemic and Local Anticancer Treatment

Drug conjugation to a polymer not only enhances its aqueous solubility but also changes drug pharmacokinetics in the whole organism and even at a subcellular level with the possibility to clearly enhance drug therapeutic value [71, 72]. In systemically delivered polymer–drug treatments, the "leakiness" of tumor vasculature allows selective extravasation of the conjugate in tumor tissue, and the lack of an effective lymphatic drainage subsequently promotes polymer accumulation, similarly to NPs' passive targeting. Active targeting could be also achieved by the incorporation of additional specific residues within the polymer carrier inducing a receptor-mediated endocytic uptake [73] (Fig. 23.4).

The conjugation chemistry between the polymer and the drug usually involves a covalent bond that stays stable in blood, but able to release the drug at an optimum rate at the site of interest. For example, a tetrapeptide Gly-Phe-Leu-Gly can be cleaved by lysosomal thiol-dependent proteases and release the drug in the cell. The linker chemistry used has primarily evolved to cater for the chemical functionality available in the bioactive agent to be bound (e.g., ester linkages for paclitaxel), the desired site of linker cleavage (e.g., azo bonds for degradation by bacterial azo-reductases in GI tract), and a desire to promote intracellular drug release (e.g., pH-sensitive and disulfide linkers) [75].

N-(2-Hydroxypropyl)methacrylamide (HPMA) copolymer conjugates containing DOX designed 30 years ago were the first synthetic polymer-based anticancer conjugates to enter clinical trials in 1994 [76–78], followed by clinical trials with paclitaxel, camptothecin, and platinates. Clinical use of anthracyclines and platinates is severely constrained by their acute and chronic toxicities so alteration of their PK by HPMA conjugation allowed promoting tumor targeting and limit toxicity [79]. Additionally, HPMA conjugation improved aqueous solubility and availability of paclitaxel and anthracyclines. Recent developments emerging from genomics and proteomics research have led to the incorporation of drugs directed towards inhibition of specific kinases (PI3 kinase [80]), activation of apoptosis pathways (HSP-90 chaperone inhibition [81]), or angiogenesis modulation (TNP-470 [82]). Poly(ethylene oxide-co-glycidol) (poly(EO-co-Gly)) is another example of a polymer used as a polymer–drug conjugate. Its structure resembles PEG but has multiple pendant groups in its backbone that allows drug conjugation. Poly(EO-co-Gly)-platinate showed similar antitumor activity as free cisplatin in human nasopharyngeal carcinoma xenografts in mice combined reduced toxicity [83].

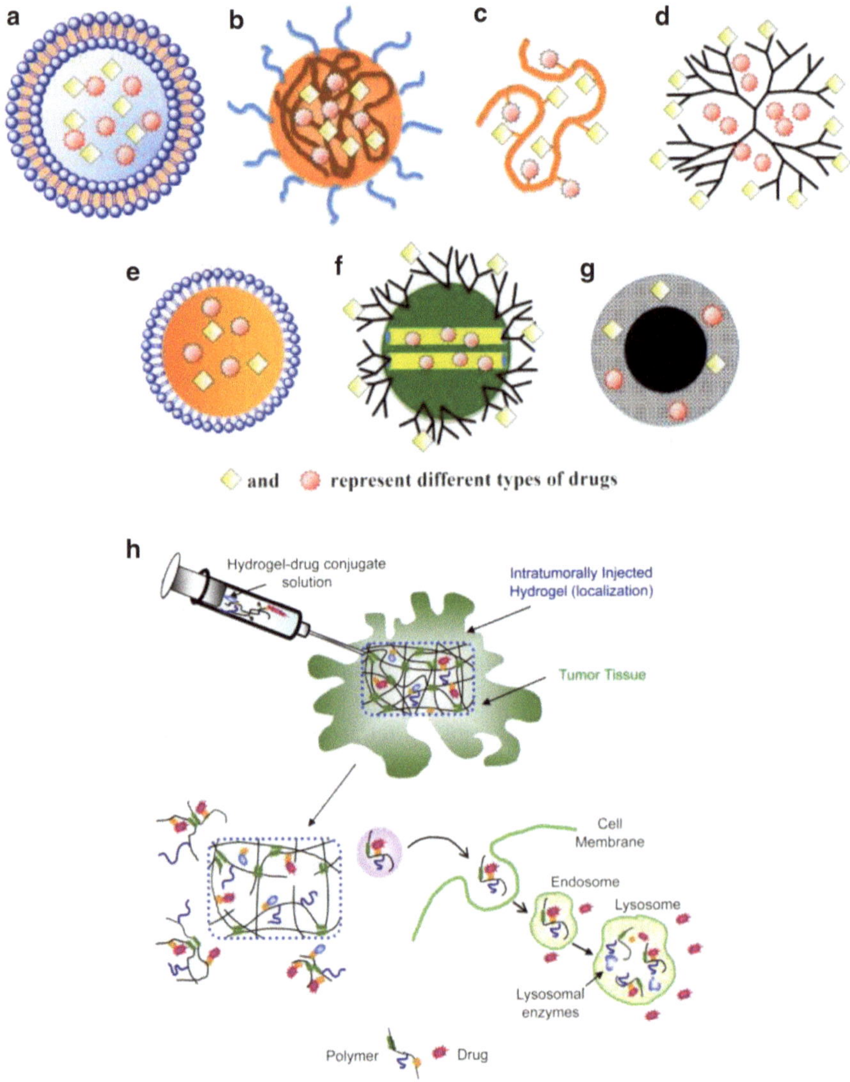

Fig. 23.4 Schematic illustration of nanoscale drug carriers used for combinatorial drug delivery aimed for passive or active targeting: (**a**) liposome, (**b**) polymeric micelle, (**c**) polymer–drug conjugate, (**d**) dendrimer, (**e**) oil nanoemulsion, (**f**) mesoporous silica nanoparticle, and (**g**) iron oxide nanoparticle (Reproduced from Hu et al. [74]). (**h**) A schematic representation for the understanding of action mechanism of poly(organophosphazene)–PTX conjugate (Reproduced from Chun et al. [38])

We have already described thermosensitive poly(organophosphazene) hydrogels as injectable DDSs with physically incorporated drugs. Chun et al. [38, 84] have recently reported the synthesis of poly(organophosphazene) hydrogels conjugated to paclitaxel or camptothecin by covalent ester linkage between the drug and the carboxylic acid-terminated polymer. The aqueous solutions of these conjugates showed a sol–gel transition at 35 and 40 °C. Release studies showed that paclitaxel was cleaved from the polymer by carboxylic acid-catalyzed degradation, which was more efficient in the acidic environment. At pH = 6.8, which corresponds with the endosome or lysosome environment, 90 % of the conjugated drug was released in 24 h, compared to a slow hydrolytic degradation of 60 % in 15 days in pH = 7.4. IT-injected polymer–drug conjugates showed tumor growth inhibition over time compared to controls or free drug combined with reduced systemic toxicity. Interestingly, free paclitaxel was more effective at day 3 compared to polymer–drug conjugate, and the researchers hypothesized that the delay in the polymer–drug group appeared because of the required time for cellular uptake and active drug release. The intracellular delivery and transfer of the drug out of the endosomal or lysosomal compartments contribute to prolonged efficacy and reduced systemic toxicity and also provide an opportunity to bypass the mechanisms of drug resistance that are reliant on membrane efflux of the free drug [38].

23.5 New Targets for Polymer-Based Combination Therapy

The complex molecular basis of cancer often means that the application of single agent therapy is insufficient for effective and sustained therapy. Furthermore, systemic administration of chemotherapeutics is associated with various and severe adverse effects that limit the dose and the number of therapies for a patient. As we mentioned earlier, polymeric DDSs can be used as adjuvant or neoadjuvant therapy, but also allow combining several drugs in one system, or combination of therapeutic drugs with imaging, targeting, or radio-sensing agents. Wu et al. [85] described the design of core–shell structured hybrid nanogels that simultaneously combine optical temperature sensing, cancer cell targeting, fluorescence imaging, and chemo-photothermal treatment in the same system. They constructed the nanogels by coating the Ag–Au bimetallic NP core with a thermoresponsive nonlinear PEG-based hydrogel as a shell and added the targeting ligands as a semi-interpenetrating hyaluronic acid chain into the surface networks of the gel shell. The Ag–Au core can emit strong visible fluorescence for cancer cell imaging, while the PEG gel provides high loading capacity for the anticancer drug.

Cancer Stem Cells

Another important cancer mechanism that can be targeted by combined therapy is the existence of cancer stem cell (CSC) in solid tumors and their contribution to tumor heterogeneity and treatment failure. From a clinical perspective these cells need to be eradicated in order to provide long-term disease-free survival. Quiescent CSCs are thought to be more resistant to chemotherapy, while targeted therapy has to target only the tumor CSCs without ablating normal stem cells. In one strategy directed at eradicating CSCs in glioblastomas, Piccirillo et al. [86] showed that bone morphogenetic proteins (BMPs) could induce differentiation of CD133+ cells, markedly attenuating their tumor-forming ability. Thus, polymer delivery system that can combine two types of chemotherapy, one against cancer cells and the other against CSCs, can be highly potent.

Tumor Angiogenesis

Angiogenesis is essential for tumor growth and metastasis; thus, controlling angiogenesis is a promising tactic in limiting cancer progression [87]. Preventing mobilization of the various bone marrow-derived cell types with anti-angiogenic drugs reduces the angiogenic response, sensitizes the tumor to therapy, and limits metastatic spread in multiple preclinical models [88, 89]. However, in both preclinical and clinical settings, resistance mechanisms limit the long-term benefit of VEGF-targeted therapy. Combination of a function-blocking antibody to placenta growth factor, which upregulation causes the resistance, improves sensitivity to anti-VEGF therapies [90]. Daenen et al. have shown that combination of chemotherapy with anti-angiogenic therapy had induced antitumor activity, compared to each treatment alone [91].

Tumor-Associated ECM

The tumor microenvironment, collectively known as stroma, is complex and is composed of connective tissue that contains an altered ECM as well as several different cell types. During tumor development, fibroblastic stroma progression, also known as stromatogenesis, results in significant changes to the surrounding mesenchyme in response to tumor growth, which are believed to promote tumorigenesis [92–95]. The outcome of these stromal modifications is the remodeling of the ECM, which can alter tumor cell responsiveness to various chemotherapeutics [96–99]. Further contributing to overall drug resistance is the low pH and hypoxic, glucose-deprived conditions that exist in the tumor environment [100–102]. Olive et al. have shown a dramatic improvement in the delivery and drug efficacy of gemcitabine in a mice model of pancreatic cancer, when coadministered with a drug known as IPI-926 [103]. IPI-926 has been shown to be involved in the depletion of tumor-associated stromal tissue via inhibition of the Hedgehog cellular signaling pathway [103, 104].

Multidrug Resistance

The use of multiple therapeutic agents in combination has become the primary strategy to treat drug-resistant cancers. Various nanoparticle platforms such as liposomes, polymeric micelles, dendrimers, and mesoporous silica particles have been used to carry broad classes of therapeutics including cytotoxic agents, chemosensitizers, siRNA, and anti-angiogenic agents that were lately reviewed [74, 105].

23.6 Conclusions and Outlook

Polymer-based anticancer DDS present a powerful and versatile platform for cancer therapy. The synthesis and design flexibility have enabled unprecedented control in delivering a wide range of therapeutics, while specifically targeting tumor cells and reducing the systemic toxicity. Various strategies focused on formulation design and engineering opened up promising options in addressing cancer complexity, such as CSCs, multidrug resistance, angiogenesis, and metastasis. The recent progress in immunology and human genomics combined with the existing polymeric and nanoparticulate platform resulted in improved targeting and brought us closer to the development of the cancer vaccine. The field of the polymer-based anticancer therapy is highly interdisciplinary and benefits from the contributions from chemists, biologists, engineers, pharmacists, and clinicians. The limitations of the preclinical tools and the inherently difficult nature of the disease resulted in a low ability to translate cancer research to clinical success. However, the latest discoveries in biomedical sciences, molecular biology, and polymer sciences, combined with improved in vitro three-dimensional culture systems and better animal models, should facilitate a transparent discovery process and consistently lead to significant patient benefit.

References

1. Visvader JE, Lindeman GJ (2008) Cancer stem cells in solid tumours: accumulating evidence and unresolved questions. Nat Rev Cancer 8(10):755–768
2. Langer R (1998) Drug delivery and targeting. Nature 392(6679):5–10
3. Minchinton AI, Tannock IF (2006) Drug penetration in solid tumours. Nat Rev Cancer 6(8):583–592
4. Cukierman E, Khan DR (2010) The benefits and challenges associated with the use of drug delivery systems in cancer therapy. Biochem Pharmacol 80(5):762–770
5. Amatangelo MD, Bassi DE, Klein-Szanto AJ, Cukierman E (2005) Stroma-derived three-dimensional matrices are necessary and sufficient to promote desmoplastic differentiation of normal fibroblasts. Am J Pathol 167(2):475–488
6. De Souza R, Zahedi P, Allen CJ, Piquette-Miller M (2010) Polymeric drug delivery systems for localized cancer chemotherapy. Drug Deliv 17(6):365–375
7. Moses MA, Brem H, Langer R (2003) Advancing the field of drug delivery: taking aim at cancer. Cancer Cell 4(5):337–341
8. Fowers KD, Baudys M, Rathi R, Shih C (2003) Thermally reversible gelling materials for safe and versatile depot delivery. Drug Dev Delivery 3(5)
9. Brem H, Ewend MG, Piantadosi S, Greenhoot J, Burger PC, Sisti M (1995) The safety of interstitial chemotherapy with BCNU-loaded polymer followed by radiation therapy in the treatment of newly diagnosed malignant gliomas: phase I trial. J Neurooncol 26(2):111–123
10. Brem H, Piantadosi S, Burger PC, Walker M, Selker R, Vick NA et al (1995) Placebo-controlled trail of safety and efficacy of intaroperative controlled delivery by biodegradable polymers of chemotherapy for recurrent gliomas. Lancet 345(8956):1008–1012
11. Punglia RS, Morrow M, Winer EP, Harris JR (2007) Current concepts: local therapy and survival in breast cancer. N Engl J Med 356(23):2399–2405

12. He CL, Kim SW, Lee DS (2008) In situ gelling stimuli-sensitive block copolymer hydrogels for drug delivery. J Control Release 127(3):189–207
13. Jeong B, Bae YH, Lee DS, Kim SW (1997) Biodegradable block copolymers as injectable drug-delivery systems. Nature 388(6645):860–862
14. Jeong B, Kim SW, Bae YH (2002) Thermosensitive sol-gel reversible hydrogels. Adv Drug Deliv Rev 54(1):37–51
15. Jeong B, Choi YK, Bae YH, Zentner G, Kim SW (1999) New biodegradable polymers for injectable drug delivery systems. J Control Release 62(1–2):109–114, Research Support, U.S. Gov't, P.H.S
16. Bekturov EA, Bimendina LA (1981) Interpolymer complexes. Adv Polymer Sci 41:99–147
17. Yoshida T, Takahashi M, Hatakeyama T, Hatakeyama H (1998) Annealing induced gelation of xanthan/water systems. Polymer 39(5):1119–1122
18. Feil H, Bae YH, Feijen J, Kim SW (1993) Effect of comonomer hydrophilicity and ionization on the lower critical solution temperature of N-isopropylacrylamide copolymers. Macromolecules 26(10):2496–2500
19. Mortensen K, Pedersen JS (1993) Structural study on the micelle formation of poly(ethylene oxide) poly(propylene oxide) poly(ethylene oxide) triblock copolymer in aqueous solution. Macromolecules 26(4):805–812
20. Lee DS, Shim MS, Kim SW, Lee H, Park I, Chang TY (2001) Novel thermoreversible gelation of biodegradable PLGA-block-PEO-block-PLGA triblock copolymers in aqueous solution. Macromol Rapid Commun 22(8):587–592
21. Zentner GM, Rathi R, Shih C, McRea JC, Seo MH, Oh H et al (2001) Biodegradable block copolymers for delivery of proteins and water-insoluble drugs. J Control Release 72(1–3):203–215
22. Shim MS, Lee HT, Shim WS, Park I, Lee H, Chang T et al (2002) Poly(D, L-lactic acid-co-glycolic acid)-b-poly(ethylene glycol)-b-poly (D, L-lactic acid-co-glycolic acid) triblock copolymer and thermoreversible phase transition in water. J Biomed Mater Res 61(2):188–196
23. Elstad NL, Fowers KD (2009) OncoGel (ReGel/paclitaxel) - clinical applications for a novel paclitaxel delivery system. Adv Drug Deliv Rev 61(10):785–794
24. Choi S, Baudys M, Kim SW (2004) Control of blood glucose by novel GLP-1 delivery using biodegradable triblock copolymer of PLGA-PEG-PLGA in type 2 diabetic rats. Pharm Res 21(5):827–831
25. Matthes K, Mino-Kenudson M, Sahani DV, Holalkere N, Fowers KD, Rathi R et al (2007) EUS-guided injection of paclitaxel (OncoGel) provides therapeutic drug concentrations in the porcine pancreas (with video). Gastrointest Endosc 65(3):448–453
26. Samlowski WE, McGregor JR, Jurek M, Baudys M, Zentner GM, Fowers KD (2006) ReGel (R) polymer-based delivery of interleukin-2 as a cancer treatment. J Immunother 29(5):524–535
27. Zhu WW, Masaki T, Bae YH, Rathi R, Cheung AK, Kern SE (2006) Development of a sustained-release system for perivascular delivery of dipyridamole. J Biomed Mater Res B Appl Biomater 77B(1):135–143
28. Rowinsky EK, Eisenhauer EA, Chaudhry V, Arbuck SG, Donehower RC (1993) Clinical toxicities encountered with paclitaxel (Taxol). Semin Oncol 20(4):1–15
29. Vukelja SJ, Anthony SP, Arseneau JC, Berman BS, Cunningham CC, Nemunaitis JJ et al (2007) Phase 1 study of escalating-dose OncoGel((R)) (ReGel((R))/paclitaxel) depot injection, a controlled-release formulation of paclitaxel, for local management of superficial solid tumor lesions. Anticancer Drugs 18(3):283–289
30. DuValla GA, Tarabar D, Seidela RH, Elstad NL, Fowers KD (2009) Phase 2: a dose-escalation study of OncoGel (ReGel/paclitaxel), a controlled-release formulation of paclitaxel, as adjunctive local therapy to external-beam radiation in patients with inoperable esophageal cancer. Anticancer Drugs 20(2):89–95

31. Song MJ, Lee DS, Ahn JH, Kim DJ, Kim SC (2004) Thermosensitive sol-gel transition behaviors of poly(ethylene oxide)/aliphatic polyester/poly (ethylene oxide) aqueous solutions. J Polym Sci A Polym Chem 42(3):772–784
32. Bae YH, Huh KM, Kim Y, Park KH (2000) Biodegradable amphiphilic multiblock copolymers and their implications for biomedical applications. J Control Release 64(1–3):3–13
33. Huh KM, Bae YH (1999) Synthesis and characterization of poly(ethylene glycol)/poly(L-lactic acid) alternating multiblock copolymers. Polymer 40(22):6147–6155
34. Lee J, Joo MK, Oh H, Sohn YS, Jeong B (2006) Injectable gel: poly(ethylene glycol)-sebacic acid polyester. Polymer 47(11):3760–3766
35. Song SC, Lee SB, Jin JI, Sohn YS (1999) A new class of biodegradable thermosensitive polymers. I. Synthesis and characterization of poly(organophosphazenes) with methoxy-poly(ethylene glycol) and amino acid esters as side groups. Macromolecules 32(7): 2188–2193
36. Kang GD, Cheon SH, Khang G, Song SC (2006) Thermosensitive poly(organophosphazene) hydrogels for a controlled drug delivery. Eur J Pharm Biopharm 63(3):340–346
37. Chen SB, Pieper R, Webster DC, Singh J (2005) Triblock copolymers: synthesis, characterization, and delivery of a model protein. Int J Pharm 288(2):207–218
38. Chun C, Lee SM, Kim SY, Yang HK, Song SC (2009) Thermosensitive poly(organophosphazene)-paclitaxel conjugate gels for antitumor applications. Biomaterials 30(12):2349–2360
39. Cho JK, Park JW, Song SC (2012) Injectable and biodegradable poly(organophosphazene) gel containing silibinin: its physicochemical properties and anticancer activity. J Pharm Sci 101(7):2382–2391
40. Cho JK, Hong KY, Park JW, Yang HK, Song SC (2011) Injectable delivery system of 2-methoxyestradiol for breast cancer therapy using biodegradable thermosensitive poly(organophosphazene) hydrogel. J Drug Target 19(4):270–280
41. Kim JI, Lee BS, Chun C, Cho JK, Kim SY, Song SC (2012) Long-term theranostic hydrogel system for solid tumors. Biomaterials 33(7):2251–2259
42. Gray WR, Sandberg LB, Foster JA (1973) Molecular model for elastin structure and function. Nature 246(5434):461–466
43. Tatham AS, Shewry PR (2000) Elastomeric proteins: biological roles, structures and mechanisms. Trends Biochem Sci 25(11):567–571
44. McDaniel JR, Callahan DJ, Chilkoti A (2010) Drug delivery to solid tumors by elastin-like polypeptides. Adv Drug Deliv Rev 62(15):1456–1467
45. Shamji MF, Betre H, Kraus VB, Chen J, Chilkoti A, Pichika R et al (2007) Development and characterization of a fusion protein between thermally responsive elastin-like polypeptide and interleukin-1 receptor antagonist - sustained release of a local antiinflammatory therapeutic. Arthritis Rheum 56(11):3650–3661
46. Urry DW, Parker TM, Reid MC, Gowda DC (1991) Biocompatibilty of the bioelastic materials, poly(GVGVP) and its gamma-irradiation cross-linked matrix - summary of generic biological test results. J Bioact Compat Polym 6(3):263–282
47. MacKay JA, Chen MN, McDaniel JR, Liu WG, Simnick AJ, Chilkoti A (2009) Self-assembling chimeric polypeptide-doxorubicin conjugate nanoparticles that abolish tumours after a single injection. Nat Mater 8(12):993–999
48. Meyer DE, Shin BC, Kong GA, Dewhirst MW, Chilkoti A (2001) Drug targeting using thermally responsive polymers and local hyperthermia. J Control Release 74(1–3):213–224
49. Liu WG, MacKay JA, Dreher MR, Chen MN, McDaniel JR, Simnick AJ et al (2010) Injectable intratumoral depot of thermally responsive polypeptide-radionuclide conjugates delays tumor progression in a mouse model. J Control Release 144(1):2–9
50. Krasko MY, Shikanov A, Ezra A, Domb AJ (2003) Poly(ester anhydride)s prepared by the insertion of ricinoleic acid into poly(sebacic acid). J Polym Sci A Polym Chem 41(8): 1059–1069

51. Shikanov A, Domb AJ (2006) Poly(sebacic acid-co-ricinoleic acid) biodegradable injectable in situ gelling polymer. Biomacromolecules 7(1):288–296
52. Shikanov A, Domb AJ, Weiniger CF (2007) Long acting local anesthetic-polymer formulation to prolong the effect of analgesia. J Control Release 117(1):97–103
53. Shikanov A, Shikanov S, Vaisman B, Golenser J, Domb AJ (2011) Cisplatin tumor biodistribution and efficacy after intratumoral injection of a biodegradable extended release implant. Chemother Res Pract 2011:175054
54. Shikanov A, Vaisman B, Krasko MY, Nyska A, Domb AJ (2004) Poly(sebacic acid-co-ricinoleic acid) biodegradable carrier for paclitaxel: in vitro release and in vivo toxicity. J Biomed Mater Res A 69A(1):47–54
55. Shikanov A, Shikanov S, Vaisman B, Golenser J, Domb AJ (2008) Paclitaxel tumor biodistribution and efficacy after intratumoral injection of a biodegradable extended release implant. Int J Pharm 358(1–2):114–120
56. Shikanov S, Shikanov A, Gofrit O, Nyska A, Corn B, Domb AJ (2009) Intratumoral delivery of paclitaxel for treatment of orthotopic prostate cancer. J Pharm Sci 98(3):1005–1014
57. Shikanov A, Vaisman B, Shikanov S, Domb AJ (2010) Efficacy of poly(sebacic acid-co-ricinoleic acid) biodegradable delivery system for intratumoral delivery of paclitaxel. J Biomed Mater Res A 92A(4):1283–1291
58. Kohane DS (2007) Microparticles and nanoparticles for drug delivery. Biotechnol Bioeng 96(2):203–209
59. Arias JL (2011) Drug targeting strategies in cancer treatment: an overview. Mini Rev Med Chem 11(1):1–17
60. Kohane DS, Lipp M, Kinney RC, Anthony DC, Louis DN, Lotan N et al (2002) Biocompatibility of lipid-protein-sugar particles containing bupivacaine in the epineurium. J Biomed Mater Res 59(3):450–459
61. Kohane DS, Tse JY, Yeo Y, Padera R, Shubina M, Langer R (2006) Biodegradable polymeric microspheres and nanospheres for drug delivery in the peritoneum. J Biomed Mater Res A 77A(2):351–361
62. Park J, Fong PM, Lu J, Russell KS, Booth CJ, Saltzman WM et al (2009) PEGylated PLGA nanoparticles for the improved delivery of doxorubicin. Nanomedicine 5(4):410–418
63. Tang BC, Fu J, Watkins DN, Hanes J (2010) Enhanced efficacy of local etoposide delivery by poly(ether-anhydride) particles against small cell lung cancer in vivo. Biomaterials 31(2):339–344
64. Shenoy DB, Amiji MA (2005) Poly(ethylene oxide)-modified poly(epsilon-caprolactone) nanoparticles for targeted delivery of tamoxifen in breast cancer. Int J Pharm 293(1–2):261–270
65. Jeong YI, Seo SJ, Park IK, Lee HC, Kang IC, Akaike T et al (2005) Cellular recognition of paclitaxel-loaded polymeric nanoparticles composed of poly(gamma-benzul L-glutamate) and poly(ethylene glycol) diblock copolymer endcapped with galactose moiety. Int J Pharm 296(1–2):151–161
66. Kos J, Obermajer N, Doljak B, Kocbek P, Kristl J (2009) Inactivation of harmful tumour-associated proteolysis by nanoparticulate system. Int J Pharm 381(2):106–112
67. Liang B, He ML, Chan CY, Chen YC, Li XP, Li Y et al (2009) The use of folate-PEG-grafted-hybranched-PEI nonviral vector for the inhibition of glioma growth in the rat. Biomaterials 30(23–24):4014–4020
68. Sahoo SK, Ma W, Labhasetwar V (2004) Efficacy of transferrin-conjugated paclitaxel-loaded nanoparticles in a murine model of prostate cancer. Int J Cancer 112(2):335–340
69. Sun B, Ranganathan B, Feng SS (2008) Multifunctional poly(D, L-lactide-co-glycolide)/montmorillonite (PLGA/MMT) nanoparticles decorated by Trastuzumab for targeted chemotherapy of breast cancer. Biomaterials 29(4):475–486
70. Hallahan D, Geng L, Qu SM, Scarfone C, Giorgio T, Donnelly E et al (2003) Integrin-mediated targeting of drug delivery to irradiated tumor blood vessels. Cancer Cell 3(1):63–74
71. Duncan R (2006) Polymer conjugates as anticancer nanomedicines. Nat Rev Cancer 6(9):688–701

72. Vicent MJ, Duncan R (2006) Polymer conjugates: nanosized medicines for treating cancer. Trends Biotechnol 24(1):39–47
73. Canal F, Sanchis J, Vicent MJ (2011) Polymer-drug conjugates as nano-sized medicines. Curr Opin Biotechnol 22(6):894–900
74. Hu CMJ, Zhang LF (2012) Nanoparticle-based combination therapy toward overcoming drug resistance in cancer. Biochem Pharmacol 83(8):1104–1111
75. Duncan R, Vicent MJ (2010) Do HPMA copolymer conjugates have a future as clinically useful nanomedicines? A critical overview of current status and future opportunities. Adv Drug Deliv Rev 62(2):272–282
76. Duncan R (2009) Development of HPMA copolymer-anticancer conjugates: clinical experience and lessons learnt. Adv Drug Deliv Rev 61(13):1131–1148
77. Nowotnik DP, Cvitkovic E (2009) ProLindac (TM) (AP5346): a review of the development of an HPMA DACH platinum polymer therapeutic. Adv Drug Deliv Rev 61(13):1214–1219
78. Rihova B (2009) Clinical experience with anthracycline antibiotics-HPMA copolymer-human immunoglobulin conjugates. Adv Drug Deliv Rev 61(13):1149–1158
79. Vicent MJ, Ringsdorf H, Duncan R (2009) Polymer therapeutics: clinical applications and challenges for development preface. Adv Drug Deliv Rev 61(13):1117–1120
80. Varticovski L, Lu ZR, Mitchell K, de Aos I, Kopecek J (2001) Water-soluble HPMA copolymer-wortmannin conjugate retains phosphoinositide 3-kinase inhibitory activity in vitro and in vivo. J Control Release 74(1–3):275–281
81. Larson N, Ray A, Malugin A, Pike DB, Ghandehari H (2010) HPMA copolymer-aminohexylgeldanamycin conjugates targeting cell surface expressed GRP78 in prostate cancer. Pharm Res 27(12):2683–2693
82. Satchi-Fainaro R, Puder M, Davies JW, Tran HT, Sampson DA, Greene AK et al (2004) Targeting angiogenesis with a conjugate of HPMA copolymer and TNP-470. Nat Med 10(3):255–261
83. Zhou P, Li ZY, Chau Y (2010) Synthesis, characterization, and in vivo evaluation of poly(ethylene oxide-co-glycidol)-platinate conjugate. Eur J Pharm Sci 41(3–4):464–472
84. Cho JK, Chun C, Kuh HJ, Song SC (2012) Injectable poly(organophosphazene)-camptothecin conjugate hydrogels: synthesis, characterization, and antitumor activities. Eur J Pharm Biopharm 81(3):582–590
85. Wu WT, Shen J, Banerjee P, Zhou SQ (2010) Core-shell hybrid nanogels for integration of optical temperature-sensing, targeted tumor cell imaging, and combined chemo-photothermal treatment. Biomaterials 31(29):7555–7566
86. Piccirillo SGM, Reynolds BA, Zanetti N, Lamorte G, Binda E, Broggi G et al (2006) Bone morphogenetic proteins inhibit the tumorigenic potential of human brain tumour-initiating cells. Nature 444(7120):761–765
87. Weis SM, Cheresh DA (2011) Tumor angiogenesis: molecular pathways and therapeutic targets. Nat Med 17(11):1359–1370
88. Shaked Y, Henke E, Roodhart JML, Mancuso P, Langenberg MHG, Colleoni M et al (2008) Rapid chemotherapy-induced acute endothelial progenitor cell mobilization: implications for antiangiogenic drugs as chemosensitizing agents. Cancer Cell 14(3):263–273
89. Shaked Y, Kerbel RS (2007) Antiangiogenic strategies on defense: on the possibility of blocking rebounds by the tumor vasculature after chemotheraphy. Cancer Res 67(15):7055–7058
90. Paez-Ribes M, Allen E, Hudock J, Takeda T, Okuyama H, Vinals F et al (2009) Antiangiogenic therapy elicits malignant progression of tumors to increased local invasion and distant metastasis. Cancer Cell 15(3):220–231
91. Daenen LG, Shaked Y, Man S, Xu P, Voest EE, Hoffman RM et al (2009) Low-dose metronomic cyclophosphamide combined with vascular disrupting therapy induces potent antitumor activity in preclinical human tumor xenograft models. Mol Cancer Ther 8(10):2872–2881
92. Beacham DA, Cukierman E (2005) Stromagenesis: the changing face of fibroblastic microenvironments during tumor progression. Semin Cancer Biol 15(5):329–341

93. Li HC, Fan XL, Houghton J (2007) Tumor microenvironment: the role of the tumor stroma in cancer. J Cell Biochem 101(4):805–815
94. Liotta LA, Kohn EC (2001) The microenvironment of the tumour-host interface. Nature 411(6835):375–379
95. Tlsty TD, Coussens LM (2006) Tumor stroma and regulation of cancer development. Annu Rev Pathol 1:119–150
96. Hodkinson PS, Mackinnon AC, Sethi T (2007) Extracellular matrix regulation of drug resistance in small-cell lung cancer. Int J Radiat Biol 83(11–12):733–741
97. Li ZW, Dalton WS (2006) Tumor microenvironment and drug resistance hematologic malignancies. Blood Rev 20(6):333–342
98. Serebriiskii I, Castello-Cros R, Lamb A, Golemis EA, Cukierman E (2008) Fibroblast-derived 3D matrix differentially regulates the growth and drug-responsiveness of human cancer cells. Matrix Biol 27(6):573–585
99. Zutter MM (2007) Integrin-mediated adhesion: tipping the balance between chemosensitivity and chemoresistance. Adv Exp Med Biol 608:87–100
100. Cosse JP, Sermeus A, Vannuvel K, Ninane N, Raes M, Michiels C (2007) Differential effects of hypoxia on etoposide-induced apoptosis according to the cancer cell lines. Mol Cancer 6:61
101. Ogiso Y, Tomida A, Tsuruo T (2002) Nuclear localization of proteasomes participates in stress-inducible resistance of solid tumor cells to topoisomerase II-directed drugs. Cancer Res 62(17):5008–5012
102. Tannock IF, Rotin D (1989) Acid pH in tumors and its potential for therapeutic exploitation. Cancer Res 49(16):4373–4384
103. Olive KP, Jacobetz MA, Davidson CJ, Gopinathan A, McIntyre D, Honess D et al (2009) Inhibition of hedgehog signaling enhances delivery of chemotherapy in a mouse model of pancreatic cancer. Science 324(5933):1457–1461
104. Tremblay MR, Lescarbeau A, Grogan MJ, Tan E, Lin G, Austad BC et al (2009) Discovery of a potent and orally active hedgehog pathway antagonist (IPI-926). J Med Chem 52(14):4400–4418
105. Oh YK, Park TG (2009) siRNA delivery systems for cancer treatment. Adv Drug Deliv Rev 61(10):850–862

Chapter 24
Topical Nanointerventions for Therapeutic and Cosmeceutical Applications

Vandana B. Patravale and Preshita P. Desai

Abbreviations

AK	Actinic keratoses
ALA	5-aminolevulinic acid
BPO	Benzoyl peroxide
HPV	Human papilloma virus
HSV	Herpes simplex virus
MNE	Magnetic nanoemulsion
MSRA	Methicillin-resistant staphylococcus aureus
NE	Nanoemulsion
NLCs	Nanostructured lipid carriers
NMSC	Non-melanoma skin cancer
NRC	Nitrosyl ruthenium complexes
PLGA	Poly-lactide-co-glycolic acid
PUVA	Psoralen and UV-A phototherapy
siRNA	Small interfering RNA
SLNs	Solid lipid nanoparticles
SNs	Silver nanoparticles
SOD-2	Superoxide dismutase 2
TEM	Transmission electron microscopy
UV	Ultraviolet
VREF	Vancomycin-resistant Enterococcus faecalis
α-IFN	Alpha interferon
γ-IFN	Gamma interferon

V.B. Patravale (✉) • P.P. Desai
Department of Pharmaceutical Sciences and Technology, Institute of Chemical Technology, Nathalal Parekh Marg, Matunga, Mumbai 400 019, India
e-mail: vbp_muict@yahoo.co.in; vb.patravale@ictmumbai.edu.in

A.J. Domb and W. Khan (eds.), *Focal Controlled Drug Delivery*, Advances in Delivery Science and Technology, DOI 10.1007/978-1-4614-9434-8_24,

Symbols

·OH	Hydroxyl radical
$AgNO_3$	Silver nitrate
C_{60}	carbon nanostructures fullerenes
IC_{50}	Half maximal inhibitory concentration
NO	Nitric oxide
ZnO	Zinc oxide
γ-Fe_2O_3	Maghemite

24.1 Introduction

The topical route is being extensively explored for possible breakthroughs in dermatological treatment modalities owing to rampant increase in skin associated clinical manifestations and the drawbacks associated with systemic drug delivery. Here nanotherapeutic approaches are establishing an edge over conventional topical formulations by ensuring better therapeutic efficacy. To appreciate this approach to topical nanodelivery systems, the most important factors which need to be understood are anatomy and barrier functions of skin.

24.1.1 Skin: An Innate Barrier

Skin, also known as integumentary system, is the largest body organ existing as external body surface that accounts for about 16 % of body weight and plays a decisive role in body homeostasis, fortification and sensory recognition. Anatomically it varies in thickness from 0.5 mm to 4 mm and comprises an upper epithelial tissue region called epidermis and an underlying connective tissue layer known as dermis along with skin appendages, viz. hair, sebaceous glands, sudoriferous glands and nails [1, 2]. Table 24.1 describes various cellular components of skin along with their key features and defence mechanisms [1–3].

Structurally, the epidermis is a stratified squamous epithelium formed of heterogeneous layers, viz. stratum basale, stratum spinosum, stratum granulosum, stratum lucidum (exclusively present in thick skin undergoing wear and tear) and stratum corneum sequentially arranged from proximal to distal end, wherein keratinocytes are present as principal cells forming the first line of defence against the entry of any external molecules. This barrier property of keratinocytes, especially in the region of outermost stratum corneum, results from synthesis of protective protein keratin, formation of close cellular network via intercellular desmosomal connections along with the fibrous proteins keratohyalin, involucrin and presence of lipid matrix. Further, sebum, a secretion from the sebaceous gland composed of triglycerides, cholesterol, proteins and inorganic salts, forms an additional defence layer. This stringent

Table 24.1 Cellular components of skin [1, 2]

Skin component	Key features	Major function and barrier mechanisms
Epidermis		
Keratinocytes (90 % of epidermal cells)	Keratin synthesis	Thermoregulation, protection against chemicals and microbes, cytokines production in response to injury
	Lamellar granules releasing hydrophobic sealant	Water repellent barrier
	Intercellular desmosomal connection	Restricts entry of foreign molecules, reflection of UV radiation
Viable epidermal cells	Synthesis of defensins and cathelicidins—pheromones/chemoattractants for APC cells	Antimicrobial function against gram-positive and gram-negative organisms fungi and viruses
Melanocytes (8 % of epidermal cells)	Melanin synthesis and melanin granule transport to keratinocytes	Skin colouration, protection against nuclear DNA damage by UV radiation, protection towards Langerhans cells
Langerhans cells	Immune response	Protection against microbes
Merkel cells	Formation of tactile disc in connection with sensory neurons	Sensation and underlined protection
Dermis		
Papillary region	Tactile receptors, areolar connective tissue	Sensation and underlined protection, elasticity
Reticular region	Connective tissue, fibroblast, collagen, elastic fibres, T-cells, vascular and lymphatic network	Elasticity, stretch, immune response, nutrient supply
Skin appendages		
Hair	Keratinised hair shaft	Protection, release of sweat and sebum
Sebaceous glands	Sebum secretion	Protection against chemicals and microbes
Sudoriferous glands	Sweat secretion	Thermoregulation, skin pH regulation, microbial resistance
Nails	Dead keratinised cell plate	Protection against trauma
Hypodermis (region below skin)		
Subcutaneous region	Areolar and adipose tissue, vascular network	Skin anchoring, thermoregulation

lipophilic set-up offers a formidable barrier against the entry of hydrophilic polar molecules, whereas it facilitates the entry of nonpolar molecules via paracellular pathways. Further, skin offers an obstruction to the entry of microbes through immunity induction generated by epidermal Langerhans cells, release of certain antimicrobial pheromone proteins like defensins, cathelicidins and acidic pH of skin (pH 5) maintained by lactic acid, amino acids and other fatty acids in sweat and sebum [2–5]. Besides this, amongst various environmental factors affecting skin, ultraviolet (UV) radiation is the most detrimental as it can cause skin damage up to an extent of malignancy. Herein skin offers a natural barricade, primarily by keratinocyte-induced light reflection restricting radiation ingress along with the more intense mechanism of formation of melanin envelope (synthesised by melanocytes present in epidermal region) around keratinocyte nucleus, arresting radiation-induced DNA damage [1–3].

Dermis, a deeper skin tissue beneath epidermis, represents a flexible layer providing protection against mechanical stress and is separated from epidermis by basal membrane. The dermis is differentiated into upper papillary region with ridges and lower reticular region which together is comprised of connective tissue, T-cells, collagen and elastic fibres that produce fibroblasts, vascular network. These are held in position by hydrophilic proteoglycan ground substance. The point to note here is that presence of vasculature makes the dermis a route for systemic absorption of drugs and being hydrophilic offers an impedance against the entry of epidermally permeated lipophilic molecules and thus affects their absorption [3, 4, 6].

As discussed earlier, though nonpolar molecules are abstained from epidermal paracellular transport pathway, they permeate skin through intracellular/follicular pathway via skin appendages wherein hair follicle forms a major pathway that comprises of keratinised hair shaft and has its root in dermal region where it provides an opening port for sebaceous and apocrine sweat glands [1, 6].

In a nutshell, the cellular components of skin together form an armoury of protective mechanisms that confine the entry of foreign molecules and ensure safety. Yet, it must be considered that the skin is not totally impervious and that it presents opportunities to deliver the desired molecules cutaneously via paracellular and/or follicular pathways, opening an entire arena of topical drug delivery systems for therapeutic and cosmeceutical application.

24.1.2 Topical Nanodelivery Approaches: Rationale and Scope

The cutaneous manifestation of skin ailments has created a remarkable impetus in the field of topical drug delivery approaches with a rationale to achieve restricted and directed delivery of drugs at the desired site of action, i.e. skin. Cutaneous delivery will not only reduce the required dose but will also avoid non-site-specific drug adverse effects associated with oral and systemic delivery. Further, the maximum protection against environmental factors, viz. radiations, chemicals, microorganisms, etc. can only be assured via topically applied protective skin coats. In addition, possible self-administration and patient compliance makes cutaneous delivery a lucrative option to treat various skin conditions topically which is well evident from large number of formulations available in the market for topical application.

Though effective, conventional gels, creams and other topical preparations often possess issues with respect to desired epidermal and dermal drug penetration and sustained release owing to barrier properties of skin as discussed in the earlier section. To counteract this, various strategies like nanoapproaches, penetration enhancers, solubilisers, drug conjugates, iontophoresis, needle-free injections, microneedles, etc. are being successfully investigated [7], amongst which nanodelivery approaches are being explored most extensively due to their multifaceted advantages, viz. noninvasiveness, enhanced skin penetration, possible modulation of site-specific and rate-controlled drug release profile, high drug loading, increased surface area, conferring stability to sensitive drugs, skin-healing potential, ease of dosage form development, etc. Recognising these benefits, topical nanoformulations have already emerged in the market. These are listed in Table 24.2.

While understanding the potential role of nanointerventions in topical treatments, selection of an appropriate nanodelivery system becomes an issue of prime importance. This selection primarily depends on various factors: physicochemical properties of drug (molecular weight, log P, etc.), desired depth of skin penetration, site of action, required drug loading, dose and drug release kinetic profile. Considering aforementioned factors and on the basis of literature reports, a probable scheme of nanodelivery system selection for topical application is depicted in Fig. 24.1.

Till date, various nanoapproaches have been investigated including lipid, polymeric, micellar, colloidal, metallic, nonmetallic and emulsion-based nanocarriers which are discussed in subsequent sections with respect to their application in ailments and cosmeceutical treatment module, which are further summarised in Table 24.3 and Fig. 24.2 gives their schematic depiction.

Looking at the available literature, amongst all lipid nanocarriers, viz. liposomes, modified liposomes, solid lipid nanoparticles (SLNs) and nanostructured lipid carriers (NLCs) have gained immense importance as they form an occlusive layer on skin and enhance permeation via interaction with lipidic component of stratum corneum. Further, liposomes in this category provide additional benefit of deep skin penetration and allow loading of both hydrophilic and lipophilic drugs [8–10, 12]. Owing to these benefits, two liposomal formulations, FUNGISOME™ Gel (drug: amphotericin B) and PSORISOME® Gel (drug: dithranol) have entered the market.

Further, to enhance drug penetration incorporation of positive charge on nanoparticles, biofunctionalisation, etc. is being investigated as cell-specific targeting approaches that ensure further reduction in drug dose and cellular drug internalisation which are especially useful for cell-specific infections [59].

Not only in dermatological conditions, but topical route is now also being explored for systemic delivery of drugs, viz. proteins, peptides, etc. which are otherwise difficult to deliver via oral route and is being established as a promising pathway for active immunisation [50, 65, 66], which is also covered within the scope of this chapter.

24.2 Antimicrobial Nanotherapeutics

Skin is most vulnerable to infections due to its incessant exposure to environmental pathogenic microorganisms that cause both dermatological and systemic complications. Common infections have fungal, viral, parasitic or bacterial origin and require meticulous treatment to ensure complete pathogenic mitigation avoiding any relapse and further seek caution owing to their contagious pattern [4, 67]. The treatment module differs with regard to type and severity of infection and is discussed in detail in subsequent sections.

Table 24.2 Topical nanoformulations: clinical trials and market insight

Trade name	Active ingredients	Nanodelivery system	Indication	Company name
FUNGISOME™ Gel	Amphotericin- B	Nano-liposome/nanosome™	Fungal infections and cutaneous leishmaniasis	Lifecare Innovations Pvt. Ltd.
PSORISOME® Gel	Dithranol	Liposomes	Psoriasis	Lifecare Innovations Pvt. Ltd.
VivaGel®	Microbicidal agent	Dendrimer	Vaginal microbicide Bacterial vaginosis	Starpharma Holdings Ltd. (Phase III clinical trials)
NB 00X	Terbinafine hydrochloride	Nanoemulsion	Herpes simplex I viral infections	Nanobio Corp. (Completed Phase II clinical trial)
Nanosomin™ Serum	Anabolisers, antioxidant vitamins, aloe vera, precious oils, aromatic essential oils	Nano-liposome/nanosome™	Anti-ageing	Elsom Research Innovative Biotechnologies
Equisomin™ Serum	Oil/water extracts of green tea, hibiscus, witch hazel bark, grape seeds, licorice root, and wild yam root	Nano-liposome/nanosome™	Hair root and scalp maintenance	Elsom Research Innovative Biotechnologies
Revitalift®	Pro-Retinol A	Nano-liposome/nanosome™	Antiwrinkle and skin firming	L'Oreal
Revitalift Double®	Pro-Retinol A and Pro-Tensium	Nano-liposome/nanosome™	Intense skin retightening	L'Oreal
Advanced Night Repair Protective	Recovery complex	Liposome	Skin repair	Estée Lauder
Zelens® Fullerene C-60 Day Cream	Fullerenes	C_{60} nanoparticles	Antioxidant	Zelens Ltd.
Trois	Natural antiarthritic components	Nanoemulsion	Arthritic pain	Venus Remedies Ltd.
Celadrin® Topical Liposome Lotion	Natural menthol and arnica	Liposome	Joint discomfort reliever	NOW Foods Ltd.
Oligo.DX cellulite reducing gel	Caffeine, TEA-hydroiodide, lotus leaf extract and ivy	Nano-liposomes/Nanosomes™	Cellulitis treatment	DS laboratories
Oxalgin® nanoGel™	Diclofenac sodium, methyl salicylate	Nanoemulsion based gel	Anti-inflammatory	Zydus Cadila Ltd.

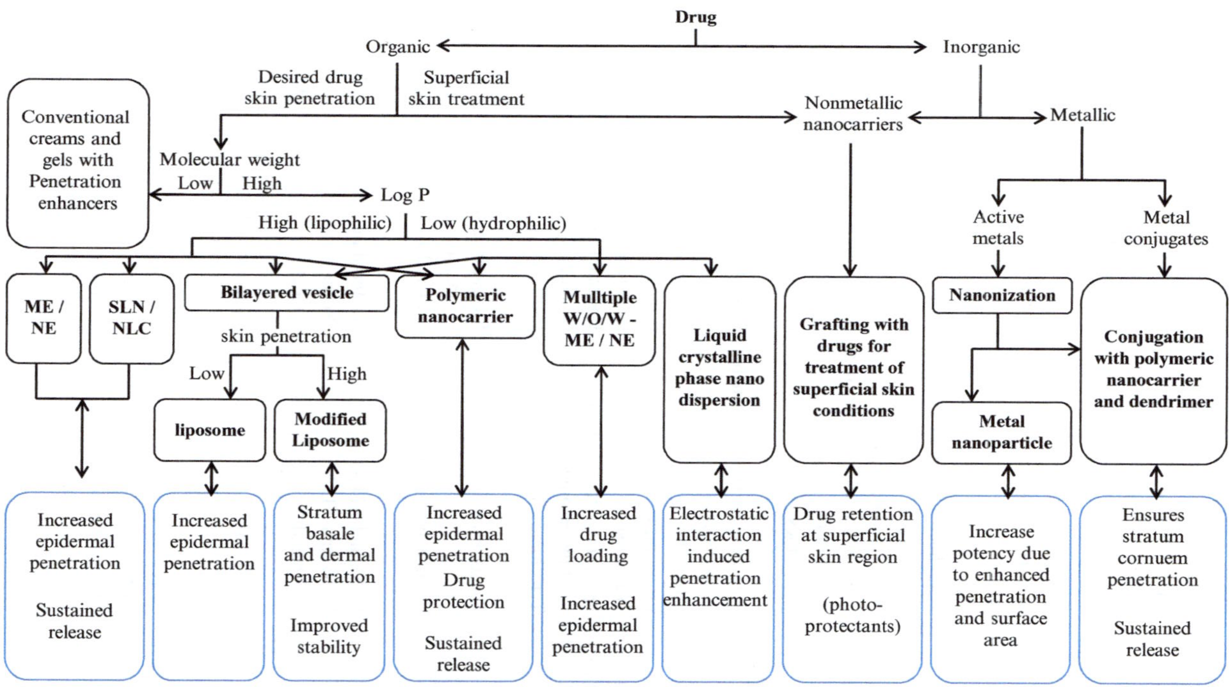

Fig. 24.1 Roadmap for topical nanodelivery system selection

Table 24.3 Topical nanoformulations: literature at glance

Drug	Nanodelivery system	Indication/use	References
Liposomes			
Fluconazole	Liposomal gel	Fungal infections majorly candidiasis	[8]
γ-IFN	Liposome dispersion	Viral infections	[9]
Lauric acid	Liposome dispersion	Acne vulgaris	[10]
Tea catechins [(−)-Epigallocatechin-3-gallate]	Liposome dispersion	Antioxidant, chemoprotective	[11]
Sodium ascorbyl phosphate	Liposome dispersion	Cutaneous photoprotection	[12]
Modified liposomal nanocarriers			
Calcipotriol	PEGylated liposomes	Psoriasis	[13]
Amphotericin B	Positively and negatively charged liposomes	Cutaneous leishmaniasis	[14]
Zinc phthalocyanine	Ultradeformable photosensitive liposomes	Cutaneous leishmaniasis	[15]
Econazole nitrate	Ethosomes	Fungal infections, majorly candidiasis	[16]
Clotrimazole	Ethosomes and ultradeformable liposomes	Fungal infections, majorly candidiasis	[17]
5-Aminolevulinic acid	Ethosome	Psoriasis	[18]
Paclitaxel	Ethosomes	Cancer	[19]
Paromomycin sulphate	Transfersomes	Cutaneous leishmaniasis	[20]
Curcuma longa extract	liposomes, ethosomes and transfersomes	UV radiation-damaged skin	[21]
Paclitaxel	TyroSomes™	Psoriasis	[22]
Niosomes			
Fluconazole	Niosomes in gel base	Fungal infections, majorly candidiasis	[8]
Tretinoin	Photostable niosomes	Antioxidant, acne vulgaris	[23]
Lipid nanoparticles			
Clotrimazole	SLNs and NLCs	Bacterial infections	[24]
Psoralen	SLNs and NLCs	Psoriasis	[25]
Nitrosyl ruthenium complex releasing NO	SLNs and NLCs	Cancer	[26]
Fluconazole	SLNs and NLCs	Fungal infections, cutaneous candidiasis	[27]
Retinoic acid	SLNs	Acne vulgaris	[28]
Sphingosine-1-phosphate	SLNs	Acne vulgaris	[29]
Resveratrol	SLNs	Cancer	[30]
Triptolide	SLNs in gel base	Anti-inflammatory	[31]
Celecoxib	NLCs in gel base	Fungal infections	[32]
Acitretin	NLCs	Psoriasis	[33]
α-IFN	Biphasic vesicles in cream base	HPV infections	[34]
Micro/nanoemulsion			
Acyclovir	W/O/W nanoemulsions	Viral infections	[35]
Flurbiprofen	Nanoemulsion	Anti-inflammatory	[36]

(continued)

Table 24.3 (continued)

Drug	Nanodelivery system	Indication/use	References
Penciclovir	Microemulsion	HSV infection	[37]
Sodium ascorbyl phosphate	Microemulsion	Antioxidant	[38]
Modified nanoemulsion			
Foscans®	Magnetic nanoemulsion	Cancer	[39]
Zinc phthalocyanine	Magnetic nanoemulsion	Cancer	[40]
camphor, menthol and methyl salicylate	Hydrogel (carbomer 940) thickened nanoemulsion	Arthritis, minor joint and muscle pain	[41]
Inclusion complexes			
Propiconazole nitrate	β-Cyclodextrin complex	Fungal infections	[42]
Polymeric nanoparticles			
Spantide II and ketoprofen	Chitosan and PLGA bilayered nanoparticles converted to nanogels	Anti-inflammatory	[43]
Chlorhexidine	Poly(ε-caprolactone) nanocapsules in gel base	Hygienic hand disinfection	[44]
Triclosan	Polymeric micelles of poloxamine	Drug-resistant bacterial infections (MRSA,VERF) and biofilm mitigation	[45]
Clarithromycin	PLGA nanoparticles for intravenous, ocular and oral and topical application	Bacterial infections	[46]
Azidothymidine triphosphate and cidofovir	Poly(*iso*butyl cyanoacrylate) nanocapsules	Viral infections	[47]
Betamethasone	Ethyl cellulose nanoparticles	Anti-inflammatory	[48]
Triclosan	Eudragit® E 100 nanoparticles	Antiacne	[49]
Hydrophilic proteins—[bovine serum albumin (67 kDa) and insulin (6 kDa)]	Chitosan-Pluronic F127 conjugated flexible nanoparticles	Protein and peptide delivery	[50]
Celecoxib	Micellar gel	Anti-inflammatory	[32]
Silver sulfadiazine	Drug dendrimer complex in cream base	Burn wound infections	[51]
Inorganic nanoparticles			
Silver	SNs, SNs in titanium implants, biofunctionalised SNs	Microbial infections with strong bactericidal effect, wound healing	[52–57]
Griseofulvin	β-Cyclodextrin inclusion complex grafted on silica nanoparticles	Superficial fungal skin infections	[58]
ZnO	ZnO micro–nanostructures with partial negative charge	HSV type 1 infcetion	[59]

(continued)

Table 24.3 (continued)

Drug	Nanodelivery system	Indication/use	References
Carbon nanostructure			
Carbon molecules with radical scavenging potential	Fullerene in gel base	Acne vulgaris	[60]
Miscellaneous			
Tretinoin	Lipopolymeric chitosan—SLNs	Antioxidant, Acne vulgaris	[61]
Ascorbyl palmitate	Self-assembled aspasomes	Antioxidant	[62]
Methotrexate	Amino acids stabilised solid in oil nanocarriers	Antirheumatic	[63]

24.2.1 Antibacterial Nanotherapeutics

Staphylococcal and *Streptococcal* bacterial infections are the most rampant bacterial skin infections. These appear on the skin as scabs, ulcerations and lesions and on the scalp as papules, abscesses and the more severe inflammatory lesions called carbuncles. These are generally treated with various classes of available antibiotics both systemically and topically.

To overcome major drawbacks of conventional antibiotic delivery like inefficient skin penetration, high-dose requirement and dose-dependent adverse effects, nanodelivery approaches are emerging as a promising solution that ensures their efficient delivery at target site. A few of them include lipidic nanocarriers of clotrimazole [24], liposomes of benzoyl peroxide (BPO) [68], ethosomes of erythromycin [69], polymeric poly (d,l-lactide-co-glycolide) (PLGA) nanoparticles of clarithromycin [46], etc.

A potent threat towards effective antibiotic therapy is emergence of antibiotic-resistant bacterial strains, viz. methicillin-resistant *Staphylococcus aureus* (MSRA), vancomycin-resistant *enterococcus faecalis* (VREF), etc. and needs major attention. To troubleshoot this, Chiappetta et al. prepared poloxamine polymeric micelles of broad-spectrum antibiotic triclosan and demonstrated their efficacy not only against aforementioned drug-resistant bacterial strains but also against eradication of *Staphylococcus epidermidis*, a hospital biofilm causative bacterium responsible for nosocomial acquired infections [45]. Further, a combinatorial approach towards the obliteration of antibiotic drug-resistant bacterial strains and the emergence of nanonisation has rejuvenated the potential use of silver as a potent antibacterial agent in arena of topical nanodelivery systems, viz. dendrimers, metallic nanoparticles, nanogels, nanolotions, etc. [51–54]. In one such study, dendrimer-conjugated silver sulfadiazine nanoparticles were prepared which possessed enhanced antibacterial activity due to increase in drug solubility at topical site and were successfully converted to a patient compliant cream base form [51]. Additionally, researchers have also developed metallic silver nanoparticles (SNs) of <100 nm size by reduction of silver nitrate ($AgNO_3$) with either chemicals (sodium citrate) or plant extracts

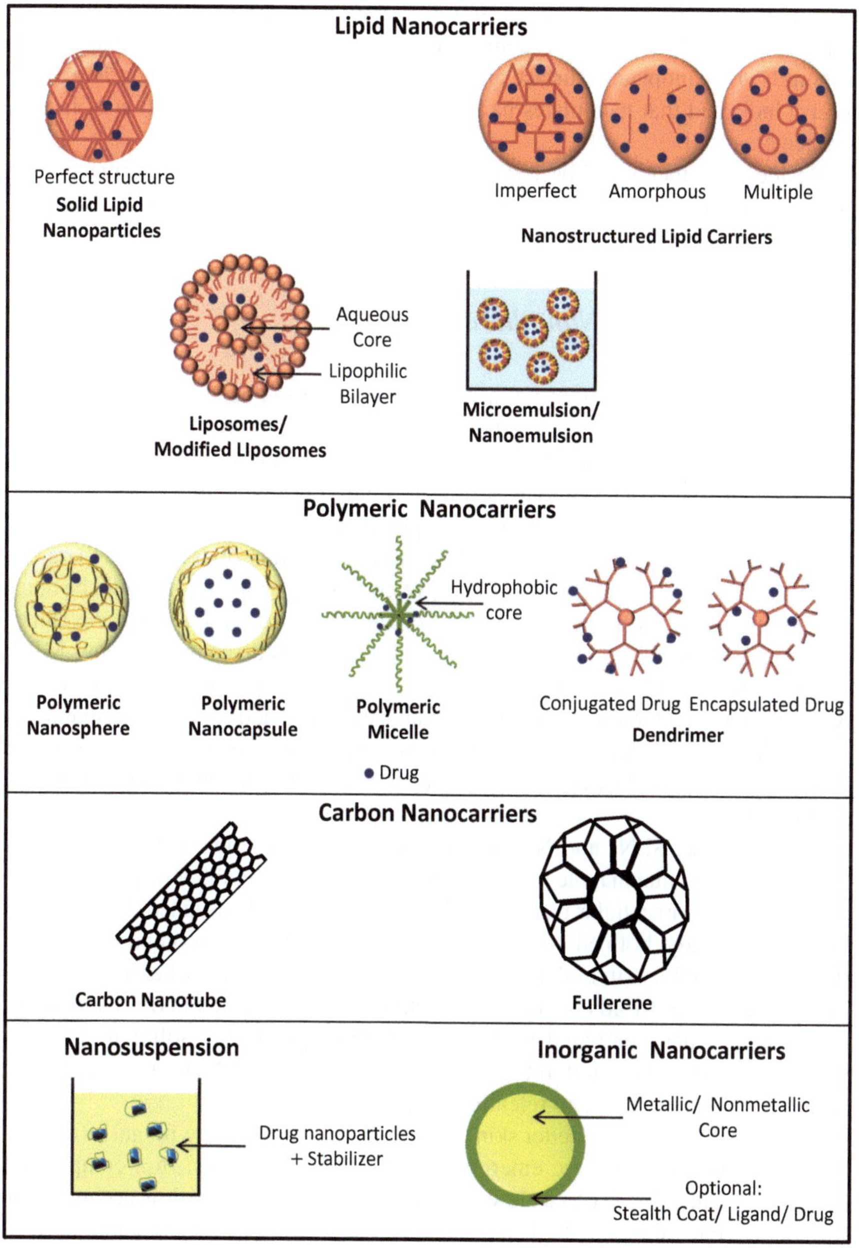

Fig. 24.2 Schematic representation of various nanodelivery systems for topical application (Modified from Desai et al. [64])

(*Artemisia nilagirica*, etc.), and these were shown to have compelling bactericidal activity against *Staphylococcus aureus, Bacillus subtilis, Escherichia coli and Proteus subtilis* [52, 54]. Further, to understand the efficacy of these nanoparticles more intricately, the effect of nanonised silver particle shape on antibacterial activity was investigated, wherein a comparative study between silver nanoparticles, nanorods and nanoplates (surface area of 17.8 m^2/g, 38.8 m^2/g, 121.1 m^2/g, respectively) indicated that nanoplates had best bactericidal activity by virtue of highest surface area [55].

Besides aforementioned bacterial infection, the most common with prime cosmeceutical concern is acne vulgaris which is discussed in detail under Sect. 24.4.

24.2.2 *Antifungal Nanotherapeutics*

The most common fungal infections are caused by candida (yeast) and dermatophytes (tinea) which appear as erythematous and scaly plaques and being superficial infections anticipate localised topical therapy. In this context, researchers are now focusing on inclusion complexes of antifungal drugs with prime focus on β-cyclodextrin inclusion complexes (e.g. propiconazole, nystatin, etc.) because of their epidermal accumulation potential, increased solubility of drug at the therapeutic site and reduced systemic side effects [42, 58]. As an evidence, it should be noted that in an antifungal assay against 56 *Candida albicans* strains, such developed propiconazole nitrate, β-cyclodextrin complexes, showed remarkably low MIC_{90} value of 0.0312 mg/L as compared to 128 mg/L in case of propiconazole alone [42]. As an advancement to this approach, with an anticipated additional epidermal depot effect, Hbaieb et al. developed a nystatin β-cyclodextrin complex and immobilised it on silica nanoparticles which are reported to accumulate in the pilosebaceous unit of skin [58].

Besides these, SLNs and NLCs of various antifungal drugs, viz. ketoconazole, clotrimazole and fluconazole [27] are reported to achieve significant epidermal skin penetration via paracellular pathways with further benefit of sustained drug delivery. In one such study, both SLNs and NLCs of fluconazole were compared for their skin retention ability (in both stratum corneum and viable skin) wherein the results indicated 5- and 3.3-fold enhancement in skin retention in case of NLCs and SLNs, respectively, when compared to the corresponding plain drug solution at end of 12 h. This can be attributed to the more flexible structure of NLCs over SLNs that ensure better penetration and retention [27].

Additionally, to ensure better skin penetration in case of severe fungal infections, liposomes and niosomes are emerging as effective approaches. In a comparative study, liposomal gel of fluconazole showed 14.2- and 3.3-fold more drug accumulation with better penetration and high antifungal activity against cutaneous candidiasis as compared to niosomal and plain drug gel [8].

The aforementioned case study undoubtedly ascribes liposomes to be an effective topical drug delivery system, but the physical instability of liposomes poses a hurdle for their effective use in topical treatments. To overcome this, scientists are now focusing on modified liposomal formulation, viz. ultradeformable liposomes with higher amount of surfactants and ethosomes which can be described as ethanolic nanovesicles comprised of phospholipids that ensure deep skin penetration by virtue of permeation-enhancing effect of ethanol and high vesicular flexibility [16, 17]. To understand their efficacy, ethosomal, liposomal and hydroethanolic gels of econazole nitrate were assessed for skin penetration using confocal laser scanning microscopy, which revealed that amongst all, only ethosomal gel penetrated up to the lowest epidermal site, i.e. stratum basale, and thus ensured its potential in treatment of deep skin fungal infections, viz. candidiasis [16].

24.2.3 Antiviral Nanotherapeutics

Amongst various viral manifestations on skin, the most prevalent are viral warts, skin lesions and chicken pox caused by human papilloma virus (HPV), herpes simplex virus (HSV) and *Varicella zoster*, respectively [67]. The conventional topical antiviral drug therapy utilises potent nucleoside analogues like acyclovir, penciclovir, cidofovir, etc. [35, 37, 47] which possess high-dose-induced skin irritation when applied as plain creams or gels due to lack of skin penetration and superficial accumulation.

As rationalised earlier, based on drug encapsulation, emulsification and enhanced permeation effect, a plethora of nanodelivery systems are being investigated for efficient antiviral therapy. To describe a few, Hillaireau et al. encapsulated azidothymidine triphosphate and cidofovir in poly(isobutyl cyanoacrylate) nanocapsules [47], whereas Yu et al. demonstrated reduced skin irritancy and significant replication inhibition of HSV-1 with penciclovir-loaded microemulsion (droplet size ~36.5 nm) as compared to the marketed cream [37]. Further, the W/O/W nanoemulsion (NE) of acyclovir (droplet size ~100 nm) was reported to be stable and interestingly in vitro skin permeation studies demonstrated 1.3-fold higher flux as compared to acyclovir aqueous solution [35].

Currently, apart from traditionally used liquid nitrogen and topical salicylic acid, an approved drug for treating HPV infections is alpha interferon (α-IFN), a19-kDa protein. Though very effective, its administration via intramuscular and intralesional injections is reported to be painful and cause systemic side effects [34, 67], and thus noninvasive delivery of α-IFN using biphasic vesicle cream becomes a promising approach, reportedly delivering α-IFN to viable epidermis up to depth of 70 μm through formation of three-dimensional cubic polymorphic phase via interaction with stratum corneum lipids that facilitate intercellular nanopathway [34]. Further, when compared, the individual components of biphasic vesicles, i.e. liposomes and submicron emulsion of gamma interferon (γ-IFN), showed negligible permeation beyond stratum corneum establishing the synergistic effect of combination approach [9].

Further, considering cell-specific infestation of HPV primarily in corneal fibroblasts, the most effective tactic is arresting cellular entry of virus that will restrict further event cascade. In one such approach, first step of binding of positively charged viral envelop with a negatively charged fibroblast cell surface was challenged by use of negatively charged filopodia-like zinc oxide (ZnO) micro–nanostructures in the presence of UV light. The developed system indicated marked virostatic behaviour by causing disruption of viral envelope glycoproteins that bind to the fibroblast [59].

24.2.4 Antiparasitic Nanotherapeutics

Parasitic skin infections show their clinical appearance in the form of skin itchiness, nodules and scales as consequence of skin infestation by small insects. The common infections include cutaneous leishmaniasis, pediculosis, scabies and onchocerciasis [4, 15, 67]. Various species of protozoal genus *Leishmania* cause cutaneous leishmaniasis and are reported to be present in amastigote and promastigote forms during their life cycle; thus, therapy anticipates protection against both forms. Clinically first-line treatment recommends use of very potent pentavalent antimonials, viz. sodium stibogluconate and meglumine antimoniate given intravenously and intramuscularly, but considering the incompliance associated with invasive therapy [70, 71], the current research is focused on topical delivery of newer drugs with more promising activity against both protozoal forms. Zinc phthalocyanine is an antileishmanial photosensitiser with 20 % anti-promastigote and antiamastigote activity which when incorporated in ultradeformable liposomes was shown to possess 100 % and 80 % activity, respectively, at the same dose and further demonstrated seven- and eightfold enhancement in amount and depth of drug penetration in stratum corneum as compared to plain liposomes. Thus this can be considered as a superior delivery system over plain drug and liposomal formulation and is claimed to be very effective at early stages of infection [15]. Paromomycin sulphate is another potent drug that is currently undergoing clinical trials and is reported to provide a significantly high fivefold drug skin retention when applied as a transferosomal formulation (particle size ~200 nm) in contrast to conventional cream [20].

Besides this, amphotericin B, a drug of choice in systemic antifungal therapy, is advised as second-line drug therapy against all forms of leishmaniasis [72]. Further to enhance its topical delivery, effect of charged liposomal amphotericin B was investigated wherein the positively charged liposomes showed higher stratum corneum transport flux of 58 ng/cm^2/h. This was attributed to their interaction with the negatively charged skin surface, and they were found to be more stable over negatively charged liposomes [14].

Shifting focus to other parasitic infections, it should be noted that their efficient management with available topical conventional formulations has restricted intensive research in this field to a greater extent.

24.3 Antiproliferative Nanotherapeutics

Hyperproliferative skin diseases commonly result from aberrant inflammatory responses and are broadly identified in the form of lesions. Amongst diverse detrimental effects, the most widespread are psoriasis and cutaneous carcinomas.

Despite intensive research over past few decades, the underlying aetiology of psoriatic disease is still not revealed completely. The known facts describe psoriasis as genetically originated pervasive inflammatory skin condition with clinical variants, viz. guttate psoriasis, sebopsoriasis and pustular form, and can worsen to psoriatic arthritis. The common forms of psoriatic lesions are seen in the form of pink to red plaques with white scales on skin showing lymphocyte and neutrophil infiltration resulting from keratinocyte hyperproliferation and immune system induction. Therapy is majorly focused on symptomatic diminution via topical psoralen and UV-A phototherapy (PUVA) along with antineoplastic drugs, viz. paclitaxel, methotrexate, etc., retinoids, vitamin D and corticosteroids. The important point to consider here is, if treated properly, psoriasis can be completely cured of its symptomatic appearance but possesses high risk of relapse due to various environmental factors like smoking, stress, alcohol and other medical conditions like HIV [22, 25, 73].

Thus to ensure efficacious treatment and to avoid non-site-specific adverse effects of antineoplastic drugs, nanodelivery approaches are being extensively investigated as promising anti-psoriatic treatment modules. One such important class of drugs is psoralen which, on photoexcitation, intercalates into a DNA adduct arresting cell proliferation. Fang et al. compared efficacy of psoralens when delivered as SLNs and in form of NLCs. Both SLNs and NLCs indicated enhanced permeation of actives in order of 8-methoxypsoralen > 5-methoxypsoralen > 4,5,8-trimethylpsoralen with zero-order kinetics as compared to respective drug suspension and lipid emulsion, wherein NLCs showed better permeation profile as compared to SLNs [25]. Besides psoralen, a prodrug, 5-aminolevulinic acid (ALA) is widely used as second-generation photosensitiser forming an active protoporphyrin IX upon photoexcitation. This is available in market as 20 % topical solution (Levulan Kerastick®) and displays drawbacks with respect to skin irritation and low skin permeation owing to hydrophilicity of the drug. To overcome these issues ALA ethosomes comprising of phosphatidylethanolamine were prepared which showed marked enhancement of about 5- and 26-fold in drug accumulation in contrast to drug solution in normal and hyperproliferating murine skin model, respectively. Further, deep skin penetration up to 80 μm was observed with ethosomes in hyperproliferative murine skin as compared to 30 μm with drug solution indicating marked efficacy of ethosomes in topical application [18].

In another interesting study, Knudsen et al. formulated lipopolymer, poly(ethylene glycol)-distearoylphosphoethanolamine-stabilised liposomes of calcipotriol, a vitamin D derivative, wherein lipopolymer used at 1 mol % concentration not only stabilised the liposome via alteration in lipid bilayer packing, forming a mushroom-like configuration, but also enhanced drug deposition in stratum corneum due to presence of head poly(ethylene glycol) molecules [13].

Amongst various polymeric nanocarriers, an appealing tyrosine-derived block copolymer poly(ethylene glycol)-β-oligo(desaminotyrosyl-tyrosine octyl ester suberate)-poly(ethylene glycol) nanospheres named TyroSpheres™ is proposed by Kilfoyle et al. for delivery of paclitaxel (particle size ~70 nm) which showed topical sustained and controlled drug delivery over a period of about 3 days with improved drug cytotoxicity proven by about 45 % reduction in the half maximal inhibitory concentration (IC_{50}) value as compared to the free drug. Further, they proved to be very apt in psoriatic treatment as TyroSpheres™ indicated preferential accumulation in epidermal sites with better control over keratinocyte proliferation and minimum systemic side effects [22].

Apart from the genetic disposition, the major advent of hyperproliferative skin disease occurs from chronic exposure to UV radiation that results in non-melanoma skin cancer (NMSC). The most common form is visualised as lesions known as actinic keratoses (AK). Microscopically these lesions show ulceration, granulomatosis papules and nodules [19].

Various conventionally used cytotoxic drugs which are investigated for their effective topical delivery to treat cutaneous carcinomas are described in Table 24.3. Apart from conventional use of nitric oxide (NO) in the treatment of hypertension and angina pectoris, recent reports have shown the apoptotic and cell destruction ability of NO at high concentration (above 10 mM), making them a reliable option in the treatment of carcinomas. To ensure release of NO only at the target site, stable transition metal complexes of NO that release NO at target site on irradiation [e.g. nitrosyl ruthenium complexes (NRC)] have been explored for its topical delivery via SLNs (particle size ~275 ± 15 nm) and NLCs (particle size ~211 ± 31 nm). Release studies in pure water revealed significantly slow release of NRC from lipid nanocarriers (about 60–70 % at end of 4 h) as compared to drug solution which was further reduced to about 22 % in case of lyophilised SLNs ensuring absence of ionic load while in contact with skin homogenates. SLNs indicated almost 100 % drug retention as compared to about 71 % in case of drug solution at the end of 1 h signifying the importance of lipid nanocarriers in topical carcinoma therapy [26].

Additionally, Rastogi et al. reported development of self-assembling triblock poly(caprolactone)-poly(ethylene glycol)-poly(caprolactone) copolymer-based polymerosomes with very low particle size of 122 ± 20 nm in the treatment of melanomas and basal cell carcinomas with ability to cross stratum corneum in <2 h with deep tissue accumulation via fluorescent labelling. Researchers have also reported their potential to replace liposomes due to their high flexibility and layered structure revealed by transmission electron microscopy (TEM) [74]. Moving further, researchers have also proposed ethosomes (described earlier in this section) for potential delivery of paclitaxel [19].

Recent research in topical anticancer treatment has emerged with a new nanodelivery system named magnetic nanoemulsion (MNE) combining the effects of magnetic nanoparticle-induced hyperthermia and photodynamic therapy for effective tumour destruction working in synergy with incorporated anticancer drug. The developed oil in water emulsion contained magnetic nanoparticle in the form of phosphate-coated maghemite (γ-Fe_2O_3) and Foscan®, a second-generation photosensitiser. In vitro skin permeation studies revealed about twofold increase in diffusional flux with fourfold improvement in drug accumulation in case of NME when compared to plain nanoemulsion [39].

24.4 Antiacne Nanotherapeutics

Global Alliance to Improve Outcomes in Acne describes acne as the most widespread chronic dermatological condition that results in development of inflammatory and/or noninflammatory lesions associated with pilosebaceous unit of skin [75, 76]. Estimated prevalence of acne is about 85 % in adolescent population out of which about 60 % of acne conditions can be cured by acute treatment, while in other cases prolonged therapy is desired [76, 77]. The pathogenic cascade in acne ramification involves 4 major steps, viz.:

1. Androgen-stimulated excessive sebum secretion (seborrhea) by the sebaceous gland
2. Follicular colonisation of anaerobic bacterium, *Propionibacterium acnes*
3. Immune response followed by release of inflammatory mediators in skin
4. Hyperproliferation of keratinocytes mimicking type IV hypersensitivity response [75–77]

Further, these lesions are reported to harbour other bacterial strains like *Pityrosporum ovale* and *Staphylococcus epidermidis* [77]. Previously, considering the bacterial origin of the disease, antibacterial agents were used as monotherapy, both orally and topically, but it resulted in development of resistant bacterial strains desiring a more effective treatment regimen. Thus, the first-line acne therapy today involves combination of antimicrobial agents with topical retinoids and an oxidising agent BPO to take care of both inflammatory and noninflammatory lesions [76]. Topical delivery of these formulations is further associated with difficulties pertaining to drug penetration in pilosebaceous unit, retention, requirement of high drug dose leading to side effects of BPO, viz. erythema, burning and scaling and of retinoids, viz. skin irritation, light sensitivity, etc. [10, 28]. Thus, considering aforementioned drawbacks and prolonged duration of therapy, nanodelivery systems are emerging as an efficient treatment modality in antiacne therapy.

Lauric acid is reported to be a strong bactericidal agent, superior to BPO. Owing to poor water solubility, liposomal formulation of lauric acid with particle size of about 120 nm was investigated for antiacne activity. It showed compelling antiacne activity proportional to loading of lauric acid, and the postulated mechanism involves fusion of liposomes with bacterial membranes, availing entry of lauric acid directly inside the bacterium [10].

Further, considering the proportional effect of drug loading on efficacy, Castro et al. formulated stable SLNs of retinoic acid with enhanced entrapment efficiency of about 94 % via ion-pair approach wherein retinoic acid was coupled with cationic stearyl amine as compared to only 13 % with retinoic acid alone. These SLNs were proven to be less skin irritant as compared to marketed cream signifying a promising solution in treatment of acne by overcoming side effects associated with retinoic acid [28].

Additionally, novel nanointerventions in antiacne therapeutics involve use of fullerenes, cage-structured spherical carbon nanostructures. Through open trial on 11 human patients, Inui et al. reported that fullerenes demonstrate antiacne activity by arresting oxidative stress-induced apoptosis of keratinocytes, neutrophil infiltration and sebum production. Further deep penetration potential in epidermis, absence of skin irritation and skin moisturising potential make them a lucrative candidate for topical treatment of acne vulgaris [60].

24.5 Anti-Ageing Nanotherapeutics

Skin ageing is an intricate biochemical process broadly classified as intrinsic and extrinsic ageing that results in both functional and aesthetic changes in the skin. Intrinsic or chronological ageing is associated with natural age-dependent genetic and hormonal changes which are identified as wrinkling, loss of skin elasticity, trivial epidermal skin atrophy, collagen bundle thickening, etc. Extrinsic ageing is mainly due to environmental factors like dietary habits, smoking, gravity, sleeping postures and radiation amongst which photo-ageing is most prevalent and is associated with photodamage, majorly caused by the UV radiation due to chronic sun exposure leading to premature ageing. This is generally exaggerated in combination with intrinsic ageing process. Starting from skin redness and swelling, this results in severe skin damage, viz. coarse wrinkling, pigmentation, hyperproliferation of keratinocytes, acanthosis, collagen bundle degradation, telangiectasia, elastosis, hypermetabolism and increased oxidative stress [78–80]. The foremost reason underlying ageing is oxidative stress resulting from the inadequate activity of enzymatic and nonenzymatic natural skin antioxidant mechanisms and increased lipoprotein oxidation with age. Thus, use of antioxidants and radical scavengers is proposed to be the most eloquent approach in anti-ageing therapeutics [79]. Antioxidants used in conventional formulations are by and large inactivated before reaching the target site and thus require multiple application at higher doses to achieve desired anti-ageing effect. In this parse, nanocarriers are emerging as an encouraging solution as they ensure stability of antioxidants with additional benefits of increase in surface area availing better scavenging activity and site-specific tissue accumulation that ensures better efficacy at lower doses. One such example is coenzyme Q10. Despite being a very potent antioxidant, issues related to lipid solubility make it a poor candidate for topical delivery. This has led to an extensive research resulting in the development of various nanodelivery systems for this drug, viz. SLNs, NLCs, self-emulsifying nanosystems, gamma cyclodextrin supramolecular complexes, PLGA polymeric nanoparticles, nanocrystals, etc. [81]. Yue et al. developed NLCs of coenzyme Q10 with particle size of about 65 nm which in comparison to simple emulsion with globule size of about 1,000 nm showed marked improvement in stratum corneum penetration and accumulation in dermis. Further, these NLCs showed 61.5 % reduction in lipid peroxidation product, malonaldehyde indicating marked reduction in oxidative stress with marked increase in activity of antioxidant enzymes [82].

Further, with a short- and long-term human volunteer study, Moddaresi et al. reported increased skin hydration potential of tocopherol acetate-loaded lipid nanocarriers when given in form of hydrophilic gel or foam in contrast to its saturated solution in silicon oil. This effect can be attributed to enhanced occlusion effect of nanodelivery system forming a lipid monolayer by virtue of nanoscale size and extensive surface packing [83]. In addition to lipid nanocarriers, use of other biocompatible and biodegradable nanodelivery systems comprising polymers, surfactants, etc. is well appreciated for topical delivery of actives. One such example is micellar lecithin organogel which is reported to form self-assembled cylindrical to rod-shaped micelles that can carry both lipophilic and hydrophilic drugs assuring deep skin penetration and is investigated for variety of applications including skin lighteners, antifungal, anticancer, vitamin, peptide, antioxidant delivery, etc. It is of prime importance from anti-ageing perspective as lecithin itself acts as protease-activated receptor 2 inhibitor asserting protection against UV-induced skin ageing and thus acts synergistically with anti-ageing bioactives [80].

Interestingly, carbon nanodelivery systems are now taking shape as promising approach in anti-ageing therapeutics. Recently, C_{60} fullerenes (initially described to have strong antiacne activity) are proven to be strong antioxidants working by accumulation at mitochondrial sites and thereby reducing mitochondrial mechanism-induced oxidative stress. Further, in hydrated form they act as strong hydroxyl (˙OH) radical scavengers thus taking care of both intrinsic and extrinsic ageing [81, 84]. Recently carbon nanotubes functionalised with gallic acid via radical grafting are reported to retain strong antioxidant and radical scavenging properties and were reported to be biocompatible. This opens a new arena of therapeutic strategies to tackle ageing via covalent coupling of actives [85].

In current scenario, metal oxide-based nanodelivery systems are gaining limelight for their anti-ageing potential. In this context, Colon et al. investigated radioprotective activity of cerium oxide nanoparticles on normal colon cells that demonstrated reduction in reactive oxygen species via reversible binding of oxygen and increase in production and activity of a natural antioxidant enzyme superoxide dismutase 2 (SOD-2) [86]. Additionally, other studies have shown neuroprotective and cardioprotective effect of these nanoparticles against oxidative stress which has established their supplementary potential towards intrinsic ageing. Besides this, pectin or poly(*N*-vinyl-2-pyrrolidone)-stabilised platinum nanoparticles are proven to combat oxidative stress via SOD-2 and catalase mimetic activity, whereas gold nanoparticles are anticipated to have anti-ageing potential via carbon-centred radical scavenging [81].

Combinatorial approach towards aforementioned strategies is reported in a form of an electrospun nanofibre mask loaded with strong antioxidants like ascorbic acid and retinoic acid in conjunction with gold nanoparticles and collagen. This multi-functional face mask ensures patient compliance and stability of antioxidants as they are moistened just before their skin application in contrast to prewetted conventionally available masks. Further the nanofibres warrant intimate skin contact resulting in high drug permeation, retention and thus better efficacy [87]. Further, l-ascorbic acid, a vitamin C derivative, has potent skin-nourishing effect, but its activity is impeded due to its high water solubility and degradation. To overcome this solid-in-oil nanosuspension of l-ascorbic acid was prepared with enhanced stability as compared to its aqueous solution which is postulated to enhance and sustain skin-nourishing effect [88].

As a precautionary measure to control photodamage and associated complications, it is advised to use sunscreens. These sunscreen agents form a chemical and/or physical barrier to prevent both UV-A and UV-B radiation-induced damage on skin. A point to note here is that initially much appreciated use of nano-zinc oxide, nano-silica and nano-titanium dioxide in sunscreens to enhance protection due to increase in surface area is currently under controversy due to reported toxicities, viz. deep skin deposition and systemic absorption of these inorganic nanomaterials and possible carcinogenicity [7, 89–91]. As a result, a company named Badger Ltd. has discontinued the use of zinc oxide nanoparticles in their marketed sunscreen formulation. Thus, nano though looks a lucrative option, it must be explored with caution and stringently examined for possible toxicities which may result from possible cellular interaction due to their nanosize [92].

24.6 Miscellaneous Nanotherapeutics

In recent years, topical nanoimmunisation is gaining avid importance for overcoming issues related with patient incompliance, possible reinfections due to lack of sterility and high cost associated with conventional invasive immunisation approaches. The mechanism proposed for topical immunisation includes presentation of antigens to Langerhans cells and dendritic cells present in the epidermis and dermis, respectively, followed by their drainage to lymphatic system where the active immune response is generated via T-cell-mediated pathways [66]. A wide range of nanodelivery systems investigated includes one novel lipid vesicular system referred to as ultradeformable archaeosomes which, in contrast to liposomes, comprise of total polar lipids derived from archaebacteria majorly glycolipids. This is reported to enhance topical adjuvancy and phagocytic uptake leading to better topical immunisation [65].

The other emerging field which is still in its infancy under preclinical trials is topical delivery of gene and small interfering RNA (siRNA) to control a variety of skin disorders [93, 94]. Their extreme hydrophilic and sensitive nature imposes a major hurdle in their effective delivery which is now being tackled by the development of novel cationic gemini surfactants-based self-assembling nanoparticles. Cationic gemini surfactants are made up of spacer linked with two cationic heads connected to hydrocarbon tails individually. So developed γ-IFN gene-loaded nanoparticles indicated better plasmid packing and lack of irritancy as compared to simple cationic lipid nanoparticles and further showed significant enhancement in gene expression as compared to untreated controlled, naked gene treatment when studied in γ-IFN-deficient mouse model [93]. Additionally, liquid crystalline phase nanodispersions comprising of monoolein with cationic polyethylenimine polymer and oleylamine lipid were investigated as nonviral vectors for delivery of siRNA wherein the nanodispersion indicated skin penetration up to viable epidermis with marked reduction in target protein expression. Besides nanodimensions, the probable underlined mechanism involves electrostatic interaction between positively charged nanodispersion with cellular negative charge which was found to be in conformity with better efficacy shown with more cationic polymeric nanodispersion (zeta potential ~31.80 ± 0.60 mV) as compared to less cationic lipid nanodispersion (zeta potential ~6.69 ± 0.60 mV) [94].

Besides aforementioned specific delivery approaches, some reported platform nanotechnologies that are gaining wide attention include bicellar systems that mimic micelles and comprised a mixture of small- and long-chain phospholipids that retain the bilayer structure as that of liposomes with smaller comparative diameter (10–50 nm) and are devoid of surfactants overcoming skin irritation associated with high surfactant content of micelles. Thus, this is reported to hold an edge over conventional liposomes and micelles as a potential carrier for variety of xenobiotics [95].

24.7 Concluding Remarks

To summarise, the chapter gives a current market and research insight into the field of topical nanodelivery systems from both therapeutic and cosmeceutical perspective. Currently, very few topical nanotechnology-based therapies are available in market amongst which liposomes are most established. Thus, in addition to extensive research going on in the field of topical nanointerventions, more attention needs to be focused on scale up and commercialisation of such promising technologies given their tremendous potential to fulfil unmet need towards efficient management of skin diseases.

References

1. Tortora GJ, Derrickson B (2006) The integumentary system. In: Roesh B (ed) Principles of anatomy and physiology, 11th edn. Wiley, NewYork, NY
2. Venus M, Waterman J, McNab I (2011) Basic physiology of the skin. Surgery 29(10): 471–474
3. Poet TS, McDougal JN (2002) Skin absorption and human risk assessment. Chem Biol Interact 140(1):19–34
4. Prabhu P, Patravale V, Joshi M (2012) Nanocarriers for effective topical delivery of anti-infectives. Curr Nanosci 8(4):491–503
5. Unnithan AR, Barakat NA, Pichiah PB, Gnanasekaran G, Nirmala R, Cha YS, Jung CH, El-Newehy M, Kim HY (2012) Wound-dressing materials with antibacterial activity from electrospun polyurethane-dextran nanofiber mats containing ciprofloxacin HCl. Carbohydr Polym 90(4):1786–1793
6. Kao J, Hall J, Helman G (1988) In vitro percutaneous absorption in mouse skin: influence of skin appendages. Toxicol Appl Pharmacol 94:93–103
7. Nabeshi H, Yoshikawa T, Matsuyama K, Nakazato Y, Matsuo K, Arimori A, Isobe M, Tochigi S, Kondoh S, Hirai T, Akase T, Yamashita T, Yamashita K, Yoshida T, Nagano K, Abe Y,

Yoshioka Y, Kamada H, Imazawa T, Itoh N, Nakagawa S, Mayumi T, Tsunoda S, Tsutsumi Y (2011) Systemic distribution, nuclear entry and cytotoxicity of amorphous nanosilica following topical application. Biomaterials 32(11):2713–2724
8. Gupta M, Goyal AK, Paliwal SR, Paliwal R, Mishra N, Vaidya B, Dube D, Jain SK, Vyas SP (2010) Development and characterization of effective topical liposomal system for localized treatment of cutaneous candidiasis. J Liposome Res 20(4):341–350
9. Du Plessis J, Egbaria K, Ramachandran C, Weiner N (1992) Topical delivery of liposomally encapsulated γ-interferon. Antiviral Res 18(3–4):259–265
10. Yang D, Pornpattananangkul D, Nakatsuji T, Chan M, Carson D, Huang CM, Zhang L (2009) The antimicrobial activity of liposomal lauric acids against propionibacterium acnes. Biomaterials 30(30):6035–6040
11. Fang JY, Hwang TL, Huanga YL, Fang CL (2006) Enhancement of the transdermal delivery of catechins by liposomes incorporating anionic surfactants and ethanol. Int J Pharm 310:131–138
12. Foco A, Gasperlin M, Kristl J (2005) Investigation of liposomes as carriers of sodium ascorbyl phosphate for cutaneous photoprotection. Int J Pharm 291:21–29
13. Knudsen N, Rønholt S, Salte RD, Jorgensen L, Thormann T, Basse LH, Hansen J, Frokjaer S, Foged C (2012) Calcipotriol delivery into the skin with PEGylated liposomes. Eur J Pharm Biopharm 81(3):532–539
14. Manosroi A, Kongkaneramit L, Manosroi J (2004) Stability and transdermal absorption of topical amphotericin B liposome formulations. Int J Pharm 270(1–2):279–286
15. Montanari J, Maidana C, Esteva MI, Salomon C, Morilla MJ, Romero EL (2010) Sunlight triggered photodynamic ultradeformable liposomes against Leishmania braziliensis are also leishmanicidal in the dark. J Control Release 147(3):368–376
16. Verma P, Pathak K (2012) Nanosized ethanolic vesicles loaded with econazole nitrate for the treatment of deep fungal infections through topical gel formulation. Nanomedicine 8(4): 489–496
17. Maheshwari RGS, Tekade RK, Sharma PA, Darwhekar G, Tyagi A, Patel RP, Jain DK (2012) Ethosomes and ultradeformable liposomes for transdermal delivery of clotrimazole: a comparative assessment. Saudi Pharm J 20(2):161–170
18. Fang YP, Huang YB, Wu PC, Tsai YH (2009) Topical delivery of 5-aminolevulinic acid-encapsulated ethosomes in a hyperproliferative skin animal model using the CLSM technique to evaluate the penetration behaviour. Eur J Pharm Biopharm 73(3):391–398
19. Paolino D, Celia C, Trapasso E, Cilurzo F, Fresta M (2012) Paclitaxel-loaded ethosomes®: potential treatment of squamous cell carcinoma, a malignant transformation of actinic keratoses. Eur J Pharm Biopharm 81(1):102–112
20. Bavarsad N, Fazly Bazzaz BS, Khamesipour A, Jaafari MR (2012) Colloidal, in vitro and in vivo anti-leishmanial properties of transfersomes containing paromomycin sulfate in susceptible BALB/c mice. Acta Trop 124(1):33–41
21. Kaur CD, Saraf S (2011) Topical vesicular formulations of Curcuma longa extract on recuperating the ultraviolet radiation-damaged skin. J Cosmet Dermatol 10(4):260–265
22. Kilfoyle BE, Sheihet L, Zhang Z, Laohoo M, Kohn J, Michniak-Kohn BB (2012) Development of paclitaxel-TyroSpheres for topical skin treatment. J Control Release 163(1):18–24. doi:10.1016/j.jconrel.2012.06.021
23. Manconi M, Valenti D, Sinico C, Lai F, Loy G, Fadda AM (2003) Niosomes as carriers for tretinoin II. Influence of vesicular incorporation on tretinoin photostability. Int J Pharm 260:261–272
24. Souto EB, Wissing SA, Barbosa CM, Müller RH (2004) Development of a controlled release formulation based on SLN and NLC for topical clotrimazole delivery. Int J Pharm 278(1):71–77
25. Fang JY, Fang CL, Liu CH, Su YH (2008) Lipid nanoparticles as vehicles for topical psoralen delivery: solid lipid nanoparticles (SLN) versus nanostructured lipid carriers (NLC). Eur J Pharm Biopharm 70(2):633–640
26. Marquele-Oliveira F, Santana DC, Taveira SF, Vermeulen DM, de Oliveira AR, da Silva RS, Lopez RF (2010) Development of nitrosyl ruthenium complex-loaded lipid carriers for topical

administration: improvement in skin stability and in nitric oxide release by visible light irradiation. J Pharm Biomed Anal 53(4):843–851
27. Gupta M, Vyas SP (2012) Development, characterization and in vivo assessment of effective lipidic nanoparticles for dermal delivery of fluconazole against cutaneous candidiasis. Chem Phys Lipids 165(4):454–461
28. Castro GA, Coelho AL, Oliveira CA, Mahecha GA, Oréfice RL, Ferreira LA (2009) Formation of ion pairing as an alternative to improve encapsulation and stability and to reduce skin irritation of retinoic acid loaded in solid lipid nanoparticles. Int J Pharm 381(1):77–83
29. Czermak P, Steinle T, Ebrahimi M, Schmidts T, Runkel F (2010) Membrane-assisted production of S1P loaded SLNs for the treatment of acne vulgaris. Desalination 250(3):1132–1135
30. Teskac K, Kristl J (2010) The evidence for solid lipid nanoparticles mediated cell uptake of resveratrol. Int J Pharm 390:61–69
31. Mei Z, Wu Q, Hu S, Li X, Yang X (2005) Triptolide loaded solid lipid nanoparticles hydrogel for topical application. Drug Dev Ind Pharm 31:161–168
32. Joshi M, Patravale V (2008) Nanostructured lipid carrier (NLC) based gel of celecoxib. Int J Pharm 346(1–2):124–132
33. Agrawal Y, Petkar KC, Sawant KK (2010) Development, evaluation and clinical studies of Acitretin loaded nanostructured lipid carriers for topical treatment of psoriasis. Int J Pharm 401(1–2):93–102
34. Foldvari M (2012) HPV infections: can they be eradicated using nanotechnology? Nanomedicine 8(2):131–135
35. Schwarz JC, Klang V, Karall S, Mahrhauser D, Resch GP, Valenta C (2012) Optimisation of multiple W/O/W nanoemulsions for dermal delivery of aciclovir. Int J Pharm 435(1): 69–75
36. Fang JY, Leu YL, Chang CC, Lin CH, Tsai YH (2004) Lipid nano/submicron emulsions as vehicles for topical flurbiprofen delivery. Drug Deliv 11(2):97–105
37. Yu A, Guo C, Zhou Y, Cao F, Zhu W, Sun M, Zhai G (2010) Skin irritation and the inhibition effect on HSV-1 in vivo of penciclovir-loaded microemulsion. Int Immunopharmacol 10(10):1305–1309
38. Spiclin P, Homar M, Zupancic-Valant A, Gasperlin M (2003) Sodium ascorbyl phosphate in topical microemulsions. Int J Pharm 256:65–73
39. Primo FL, Michieleto L, Rodrigues MAM, Macaroff PP, Morais PC, Lacava ZGM, Bentley MVLB, Tedesco AC (2007) Magnetic nanoemulsions as drug delivery system for Foscan®: skin permeation and retention in vitro assays for topical application in photodynamic therapy (PDT) of skin cancer. J Magn Magn Mater 311(1):354–357
40. Primo FL, Rodrigues MAM, Simioni AR, Andreza R, Bentley MVLB, Morais PC, Tedesco AC (2008) In vitro studies of cutaneous retention of magnetic nanoemulsion loaded with zinc phthalocyanine for synergic use in skin cancer treatment. J Magn Magn Mater 320(14): e211–e214
41. Mou D, Chen H, Du D, Mao C, Wan J, Xu H, Yang X (2008) Hydrogel-thickened nanoemulsion system for topical delivery of lipophilic drugs. Int J Pharm 353(1–2):270–276
42. Marangoci N, Mares M, Silion M, Fifere A, Varganici C, Nicolescu A, Deleanu C, Coroaba A, Pinteala M, Simionescu BC (2011) Inclusion complex of a new propiconazole derivative with β-cyclodextrin: NMR, ESI–MS and preliminary pharmacological studies. Results Pharma Sci 1(1):27–37
43. Shah PP, Desai PR, Patel AR, Singh MS (2012) Skin permeating nanogel for the cutaneous co-delivery of two anti-inflammatory drugs. Biomaterials 33(5):1607–1617
44. Nhung DT, Freydiere AM, Constant H, Falson F, Pirot F (2007) Sustained antibacterial effect of a hand rub gel incorporating chlorhexidine-loaded nanocapsules (Nanochlorex). Int J Pharm 334(1–2):166–172
45. Chiappetta DA, Degrossi J, Teves S, D'Aquino M, Bregni C, Sosnik A (2008) Triclosan-loaded poloxamine micelles for enhanced topical antibacterial activity against biofilm. Eur J Pharm Biopharm 69(2):535–545

46. Mohammadi G, Nokhodchi A, Barzegar-Jalali M, Lotfipour F, Adibkia K, Ehyaei N, Valizadeh H (2011) Physicochemical and anti-bacterial performance characterization of clarithromycin nanoparticles as colloidal drug delivery system. Colloids Surf B Biointerfaces 88(1):39–44
47. Hillaireau H, Le Doan T, Besnard M, Chacun H, Janin J, Couvreur P (2006) Encapsulation of antiviral nucleotide analogues azidothymidine-triphosphate and cidofovir in poly(iso-butylcyanoacrylate) nanocapsules. Int J Pharm 324(1):37–42
48. Abdel-Mottaleb MM, Moulari B, Beduneau A, Pellequer Y, Lamprecht A (2012) Nanoparticles enhance therapeutic outcome in inflamed skin therapy. Eur J Pharm Biopharm 82(1): 151–157
49. Domínguez-Delgado CL, Rodríguez-Cruz IM, Escobar-Chávez JJ, Calderón-Lojero IO, Quintanar-Guerrero D, Ganem A (2011) Preparation and characterization of triclosan nanoparticles intended to be used for the treatment of acne. Eur J Pharm Biopharm 79(1):102–107
50. Choi WI, Lee JH, Kim JY, Kim JC, Kim YH, Tae G (2012) Efficient skin permeation of soluble proteins via flexible and functional nano-carrier. J Control Release 157(2):272–278
51. Strydom SJ, Rose WE, Otto DP, Liebenberg W, de Villiers MM (2012) Poly(amidoamine) dendrimer-mediated synthesis and stabilization of silver sulfonamide nanoparticles with increased antibacterial activity. Nanomedicine. doi:10.1016/j.nano.2012.03.006
52. Mirzajani F, Ghassempour A, Aliahmadi A, Esmaeili MA (2011) Antibacterial effect of silver nanoparticles on Staphylococcus aureus. Res Microbiol 162(5):542–549
53. Ravindran A, Chandran P, Khan SS (2012) Biofunctionalized silver nanoparticles: advances and prospects. Colloids Surf B Biointerfaces 105:342–52. doi:10.1016/j.colsurfb.2012.07.036
54. Vijayakumar M, Priya K, Nancy FT, Noorlidah A, Ahmed ABA (2013) Biosynthesis, characterisation and anti-bacterial effect of plant-mediated silver nanoparticles using Artemisia nilagirica. Ind Crop Prod 41:235–240
55. Sadeghi B, Garmaroudi FS, Hashemi M, Nezhad HR, Nasrollahi A, Ardalan S, Ardalan S (2012) Comparison of the anti-bacterial activity on the nanosilver shapes: nanoparticles, nanorods and nanoplates. Adv Powder Technol 23(1):22–26
56. Rai M, Yadav A, Gade A (2009) Silver nanoparticles as a new generation of antimicrobials. Biotechnol Adv 27(1):76–83
57. Zhao L, Wang H, Huo K, Cui L, Zhang W, Ni H, Zhang Y, Wu Z, Chu PK (2011) Antibacterial nano-structured titania coating incorporated with silver nanoparticles. Biomaterials 32(24):5706–5716
58. Hbaieb S, Kalfat R, Chevalier Y (2012) Loading antifungal drugs onto silica particles grafted with cyclodextrins by means of inclusion complex formation at the solid surface. Int J Pharm 439(1–2):234–45. doi:10.1016/j.ijpharm.2012.09.035
59. Mishra YK, Adelung R, Röhl C, Shukla D, Spors F, Tiwari V (2011) Virostatic potential of micro-nano filopodia-like ZnO structures against herpes simplex virus-1. Antiviral Res 92(2):305–312
60. Inui S, Aoshima H, Nishiyama A, Itami S (2011) Improvement of acne vulgaris by topical fullerene application: unique impact on skin care. Nanomedicine 7(2):238–241
61. Ridolfi DM, Marcato PD, Justo GZ, Cordi L, Machado D, Durán N (2012) Chitosan-solid lipid nanoparticles as carriers for topical delivery of tretinoin. Colloids Surf B Biointerfaces 93:36–40
62. Gopinath D, Ravia D, Rao BR, Apte SS, Renuka D, Rambhaua D (2004) Ascorbyl palmitate vesicles (Aspasomes), formation, characterization and applications. Int J Pharm 271:95–113
63. Yang F, Kamiya N, Goto M (2012) Transdermal delivery of the anti-rheumatic agent methotrexate using a solid-in-oil nanocarrier. Eur J Pharm Biopharm 82(1):158–163
64. Desai PP, Date AA, Patravale VB (2012) Overcoming poor oral bioavailability using nanoparticle formulations-opportunities and limitations. Drug Discov Today Technol 9(2):e87–e95
65. Higa LH, Schilrreff P, Perez AP, Iriarte MA, Roncaglia DI, Morilla MJ, Romero EL (2012) Ultradeformable archaeosomes as new topical adjuvants. Nanomedicine 8(8):1319–1328
66. Li N, Peng LH, Chen X, Nakagawa S, Gao JQ (2011) Transcutaneous vaccines: novel advances in technology and delivery for overcoming the barriers. Vaccine 29(37):6179–6190

67. Bhagavatula M, Powell C (2011) Common superficial skin infections and infestations. Paediatr Child Health 21(3):132–136
68. Fluhr JW, Barsom O, Gehring W, Gloor M (1999) Antibacterial efficacy of benzoyl peroxide in phospholipid liposomes. A vehicle-controlled, comparative study in patients with papulo-pustular acne. Dermatology 198:273–277
69. Godin B, Touitou E, Rubinstein E, Athamna A, Athamna M (2005) A new approach for treatment of deep skin infections by an ethosomal antibiotic preparation: an in vivo study. J Antimicrob Chemother 55(6):989–994
70. Armijos RX, Weigel MM, Calvopiña M, Mancheno M, Rodriguez R (2004) Comparison of the effectiveness of two topical paromomycin treatments versus meglumine antimoniate for new world cutaneous leishmaniasis. Acta Trop 91(2):153–160
71. Frézard F, Martins PS, Bahia AP, Le Moyec L, de Melo AL, Pimenta AM, Salerno M, da Silva JB, Demicheli C (2008) Enhanced oral delivery of antimony from meglumine antimoniate/β-cyclodextrin nanoassemblies. Int J Pharm 347(1–2):102–108
72. Yardley V, Croft SL (2000) A comparison of the activities of three amphotericin B lipid formulations against experimental visceral and cutaneous leishmaniasis. Int J Antimicrob Agents 13(4):243–248
73. Mendonça CO, Burden AD (2003) Current concepts in psoriasis and its treatment. Pharmacol Ther 99(2):133–147
74. Rastogi R, Anand S, Koul V (2009) Flexible polymerosomes-an alternative vehicle for topical delivery. Colloids Surf B Biointerfaces 72(1):161–166
75. Arora MK, Yadav A, Saini V (2011) Role of hormones in acne vulgaris. Clin Biochem 44(13):1035–1040
76. Thiboutot D, Gollnick H, Bettoli V, Dréno B, Kang S, Leyden JJ, Shalita AR, Lozada VT, Berson D, Finlay A, Goh CL, Herane MI, Kaminsky A, Kubba R, Layton A, Miyachi Y, Perez M, Martin JP, Ramos-E-Silva M, See JA, Shear N, Wolf J Jr, Global Alliance to Improve Outcomes in Acne (2009) New insights into the management of acne: an update from the global alliance to improve outcomes in Acne group. J Am Acad Dermatol 60:S1–S50
77. Azimi H, Fallah-Tafti M, Khakshur AA, Abdollahi M (2012) A review of phytotherapy of acne vulgaris: perspective of new pharmacological treatments. Fitoterapia. doi:10.1016/j.fitote.2012.03.026
78. Ghersetich I, Troiano M, De Giorgi V, Lotti T (2007) Receptors in skin ageing and antiageing agents. Dermatol Clin 25(4):655–662
79. Kaur IP, Kapila M, Agrawal R (2007) Role of novel delivery systems in developing topical antioxidants as therapeutics to combat photoageing. Ageing Res Rev 6(4):271–288
80. Raut S, Singh Bhadoriya SS, Uplanchiwar V, Mishra V, Gahane A, Jain SK (2012) Lecithin organogel: a unique micellar system for the delivery of bioactive agents in the treatment of skin aging. Acta Pharm Sin B 2(1):8–15
81. Gruber J, Fong S, Chen CB, Yoong S, Pastorin G, Schaffer S, Cheah I, Halliwell B (2012) Mitochondria-targeted antioxidants and metabolic modulators as pharmacological interventions to slow ageing. Biotechnol Adv. doi:10.1016/j.biotechadv.2012.09.005
82. Yue Y, Zhoua H , Liu G, Li Y, Yan Z, Duan M (2010) The advantages of a novel CoQ10 delivery system in skin photo-protection. Int J Pharm 392 (1-2): 57–63
83. Moddaresi M, Tamburic S, Williams S, Jones SA, Zhao Y, Brown MB (2009) Effects of lipid nanocarriers on the performance of topical vehicles in vivo. J Cosmet Dermatol 8(2): 136–143
84. Andrievsky GV, Bruskov VI, Tykhomyrov AA, Gudkov SV (2009) Peculiarities of the antioxidant and radioprotective effects of hydrated C_{60} fullerene nanostructures in vitro and in vivo. Free Radic Biol Med 47(6):786–793
85. Cirillo G, Hampel S, Klingeler R, Puoci F, Iemma F, Curcio M, Parisi OI, Spizzirri UG, Picci N, Leonhardt A, Ritschel M, Büchner B (2011) Antioxidant multi-walled carbon nanotubes by free radical grafting of gallic acid: new materials for biomedical applications. J Pharm Pharmacol 63(2):179–188

86. Colon J, Hsieh N, Ferguson A, Kupelian P, Seal S, Jenkins DW, Baker CH (2010) Cerium oxide nanoparticles protect gastrointestinal epithelium from radiation-induced damage by reduction of reactive oxygen species and upregulation of superoxide dismutase 2. Nanomedicine 6(5):698–705
87. Fathi-Azarbayjani A, Qun L, Chan YW, Chan SY (2010) Novel vitamin and gold-loaded nanofiber facial mask for topical delivery. AAPS PharmSciTech 11(3):1164–1170
88. Piao H, Kamiya N, Cui F, Goto M (2011) Preparation of a solid-in-oil nanosuspension containing L-ascorbic acid as a novel long-term stable topical formulation. Int J Pharm 420(1):156–160
89. Hirai T, Yoshikawa T, Nabeshi H, Yoshida T, Akase T, Yoshioka Y, Itoh N, Tsutsumi Y (2012) Dermal absorption of amorphous nanosilica particles after topical exposure for three days. Pharmazie 67(8):742–743
90. Nohynek GJ, Lademann J, Ribaud C, Roberts MS (2007) Grey goo on the skin? Nanotechnology, cosmetic and sunscreen safety. Crit Rev Toxicol 37(3):251–277
91. Zvyagin AV, Zhao X, Gierden A, Sanchez W, Ross JA, Roberts MS (2008) Imaging of zinc oxide nanoparticle penetration in human skin in vitro and in vivo. J Biomed Opt 13(6):064031
92. Zinc oxide & nanoparticles in sunscreens. Badgerbalm. http://www.badgerbalm.com/s-33-zinc-oxide-and-nanoparticles.aspx. Accessed 31 Oct 2012
93. Badea I, Wettig S, Verrall R, Foldvari M (2007) Topical non-invasive gene delivery using gemini nanoparticles in interferon-γ-deficient mice. Eur J Pharm Biopharm 65(3):414–422
94. Vicentini FT, Depieri LV, Polizello AC, Ciampo JO, Spadaro AC, Fantini MC, Vitória Lopes Badra Bentley M (2012) Liquid crystalline phase nanodispersions enable skin delivery of siRNA. Eur J Pharm Biopharm. doi:10.1016/j.ejpb.2012.08.011
95. Rodríguez G, Barbosa-Barros L, Rubio L, Cócera M, López-Iglesias C, de la Maza A, López O (2011) Bicellar systems as modifiers of skin lipid structure. Colloids Surf B Biointerfaces 84(2):390–394

Chapter 25
Focal Drug Delivery to the Nail

Sudaxshina Murdan

25.1 The Nail Unit

25.1.1 Composition and Function

The nail unit—consisting of the nail plate, the nail bed, the nail folds, the hyponychium and the nail matrix (Fig. 25.1)—protects the delicate tips of the fingers and toes against trauma, enhances the sensation of fine touch and is used for scratching, for grooming, to relieve boredom and as a cosmetic organ and a tool [1–11]. Artefacts dating from 3000 BC show the use of the fingernail to decorate pottery (Fig. 25.2), although for optimal nail health, the use of the nail as a tool is to be discouraged. Nail art has spawned a vibrant multibillion dollar industry and has a long history; the ancient Egyptians are said to have dyed their fingernails with henna, and traces of the latter have been found on the hands of 5,000-year-old Egyptian mummies [12].

25.1.2 Nail Matrix, Bed, Hyponychium and Folds

The nail unit is anchored in place by attachment to the distal phalanx and is well perfused by blood and lymph and has a rich nerve supply. The nail matrix—also called the root of the nail—is the living tissue whose cell division gives rise to the (dead) nail plate and whose distal portion is visible (especially in the thumbs and great toes) through the transparent nail plate as a white, semilunar area,

S. Murdan (✉)
Department of Pharmaceutics, UCL School of Pharmacy,
29-39 Brunswick Square, London WC1N 1AX, UK
e-mail: s.murdan@ucl.ac.uk

A.J. Domb and W. Khan (eds.), *Focal Controlled Drug Delivery*, Advances in Delivery Science and Technology, DOI 10.1007/978-1-4614-9434-8_25,

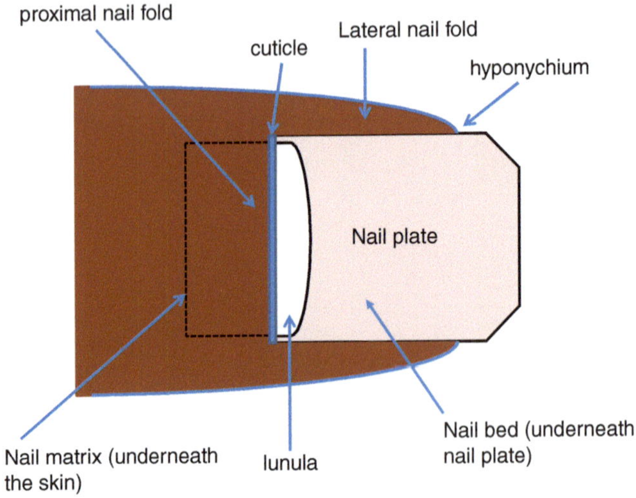

Fig. 25.1 A schematic diagram of a nail unit (and its components) at the tip of a finger/toe

Fig. 25.2 Photograph (taken at the Museum of London, London) of a pottery bowl 3000 BC. Decorative markings on the pottery were created by fingernails

called the lunula. The nail bed is a very thin epithelium underneath the nail plate, to which the nail plate is firmly held. The hyponychium is the region underneath the free edge of the nail plate where the nail plate starts to separate from the nail bed. The lateral and proximal nail folds enclose the nail plate at its lateral and proximal edges, respectively. The dorsal surface of the proximal nail fold covers part of the nail matrix and is continuous with the cuticle. The latter, also called eponychium, extends from the proximal nail fold and adheres strongly to the nail plate surface, creating a physical seal against the entry of exogenous materials.

25.1.3 Nail Plate

The nail plate—commonly called the nail—is a major focus of research due to its obvious role in disease and therapy. Its appearance is affected by diseases of the nail as well as those of more distant organs. For example, yellow nails indicate pulmonary problems. Thus, changes in the appearance of the nail plate are used to diagnose local as well as systemic diseases. In addition to its appearance, the nail plate composition and properties are being explored for disease diagnosis, e.g. dielectric properties of the nails in diabetes [13], protein content and nitration in the nails of autistic children [14]. The nail plate is the main site where medicines are applied for the treatment of nail diseases. In addition to its role in the medical field, the nail plate's composition is being used/researched for purposes as diverse as determination of binge drinking [15], a source of ancient and modern DNA [16] and to reconstruct the diets and lifestyles (e.g. whether food was stored) of the Palaeolithic era [17], as it reflects food uptake and the environment [18], during its growth.

The nail plate grows throughout life, the fingernail and toenail at, on average, 3 mm and 1.5–1 mm per month, respectively [2]. Thus, on average, fingernails grow out completely in 6 months and toenails in 12–18 months. The rate of nail plate growth rate is influenced by age (slower with age); pregnancy (faster); local and systemic disease (can increase or decrease); nutrition (slower in malnutrition); trauma/nail-biting (faster); hand dominance (faster in the dominant hand); the environment such as the weather (slower in cold climate); chemicals, e.g. drugs (can increase or decrease) [19] and references therein. The nail plate is transparent (which enables visualisation of the supporting nail bed whose capillary network imparts a pink colour to the nail plate), hard yet slightly elastic and is curved in both the longitudinal (in the direction of growth) and transverse (perpendicular to the direction of growth) directions. Its size, shape, thickness, surface ridging, curvature and mechanical properties such as flexibility vary within and among individuals, with site (finger/toe), age and other endogenous and exogenous factors such as disease states and seasons. Fingernail plates are thinner (~0.5 mm at the distal edge) than the big toe (up to 1 mm at the distal edge) and in children (~0.25 mm) [20]. In males, the fingernail plate is larger (longer and broader), thicker [20] and less curved in the transverse direction [21] compared to females.

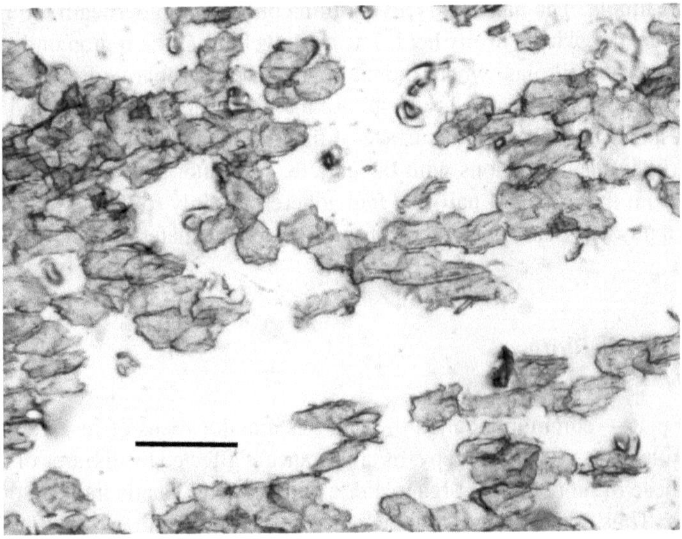

Fig. 25.3 Light micrograph of onychocytes that have been isolated by tape stripping the dorsal nail surface. Bar represents 100 μm

The nail plate may be divided into three (or two according to some researchers [22]) distinct parallel strata—dorsal, intermediate and ventral in a ratio of 3:5:2 in terms of thickness [23–25]—and consists of 80–90 layers of dead, flattened, keratinised cells and an intercellular 'cement' (thought to consist of proteins and/or mucopolysaccharides [4]) between the cells. The latter can be visualised by light microscopy following tape stripping (Fig. 25.3). On average, the dimensions of the flattened corneocytes in the upper nail layer are $34 \times 60 \times 2.2$ μm, while the cells in the lower nail layer are thicker, at $40 \times 53 \times 5.5$ μm [13]. Keratin forms the main bulk of the nail plate [26, 27]. Other constituents include water (between 5 and 30 %, [28, 29]) and lipids (<5 % w/w in adults) such as, cholesterol sulphate, ceramides, free sterols, free fatty acids, triglycerides, sterol and wax esters and squalene [27, 30]. Elements such as calcium, magnesium, sodium, potassium, iron, copper, zinc, aluminium, chlorine, selenium and fluoride can also be measured and used as biomarkers; therapeutic and recreational drugs can be measured to assess drug therapy/abuse; pollutants, e.g. arsenic, can be used to assess the environment and isotopes (e.g. 13C, 15 N) are used in archaeological samples [31–35].

A number of drug formulations are applied on the nail plate's dorsal surface. The nail plate surface properties are therefore especially important in influencing contact between the drug formulation and the nail and, subsequently, ungual (pertaining to the nail; onycheal is an alternative term) drug permeation. The dorsal nail plate surface is often ridged longitudinally, especially in older people and in certain diseases such as rheumatoid arthritis [36–38]. Scanning electron micrographs of the nail plate surfaces show dorsal surface cells' edges to overlap (Fig. 25.4) which ensures good barrier properties of the nail plate to topically applied substances. The nail plate's

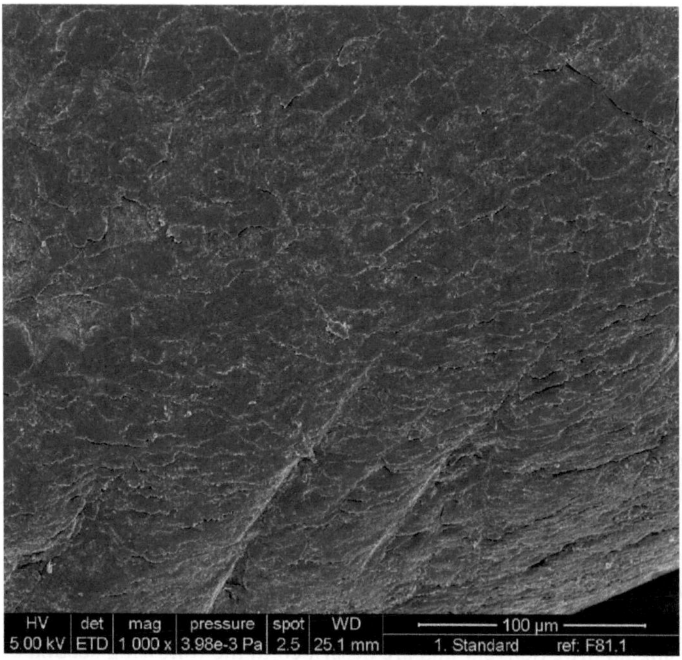

Fig. 25.4 Scanning electron microscope of the dorsal nail plate surface shows the edges of the nail plate cells overlap

surface free energy—expected to influence the adhesion of topical drug carriers as well as that of microorganisms—is approximately 34 mJ/m^2 [39], while its surface pH is around 5 [40]. The presence of a very small amount of sebum (2.1 ± 1.7 μg/cm^2) on the nail plate's surface has also been measured [11]. The presence of sebum—an oily substance produced by the skin's sebaceous glands—on the nail plate surface could be due to its spreadability which would allow its flow from the surrounding skin onto the nail plate surface and cause the sheen of nail plates. At the same time, sebum on the nail plate surface could influence the contact between the nail plate and topical formulations.

25.2 Nail Diseases

Disease can affect some or all the parts of the nail unit, for example, in some diseases, only the perionychial tissues (the nail folds) are affected, while the nail plate is normal. The nail folds can become swollen and inflamed, and pus may be present. In other diseases, the nail plate is affected—with changes in its shape, surface features, colour, mechanical properties and attachment to underlying soft tissues—without any major effects on the perionychial tissues. The nail plate may be totally or partially absent, excessively long, short, large or small, over- or under-curved, thickened and hard or excessively soft, flexible, friable, fragile and prone to breaking/splitting at the free edge. Its colour may change due to changes in the nail plate itself or in the underlying tissues which is then seen through the translucent nail plate. Its surface may be rough, excessively ridged/lined longitudinally or transversely or covered in pits (shallow depressions in the nail plate). The nail plate may also become detached from the nail bed at its distal and/or lateral edges, or even in its centre.

The symptoms mentioned above can result from a diverse range of causes, including chemicals (e.g. medicinal drugs), infections, trauma and congenital, hereditary, systemic and local diseases. Treatment of the nail disease depends on the cause, for example, antimicrobial therapy for infections; surgical removal of tumours/foreign bodies and treatment of abnormalities, such as ingrowing toenails; the wearing of rubber gloves to protect against excessive exposure of the nail to occupational hazards such as offensive chemicals and the replacement of systemic drugs which cause nail changes. Other nail symptoms, such as white spots, grow out with the growing nail plate (although new ones can reappear).

The two nail diseases which are commonly treated by focal drug application are onychomycosis and nail psoriasis. These make up the majority of nail disorders and are difficult-to-cure long-term conditions. They are thus the main targets for research on focal ungual drug delivery and are described in the following sections.

25.2.1 Onychomycosis

Onychomycosis (fungal infections of the nail plate and/or nail bed) is responsible for up to 50 % of all nail disorders [41, 42] and is a very common problem, affecting 10 % of the general population, 20 % of people aged over 60 years, up to 50 % of people aged over 70 years and up to one-third of diabetics [43]. Its incidence is on the rise, and predisposing factors include increasing age, gender (male), diabetes, immunosuppression, HIV and AIDs, peripheral vascular diseases, smoking, certain activities such as mining, sports (toes) and the wearing of tight-fitting shoes [44–52]. Most (90–95 %) of the infections are caused by dermatophytes (especially *Trichophyton rubrum and Trichophyton mentagrophytes*), the rest being caused by yeasts (*Candida albicans* mainly) and non-dermatophytes, such as *Scytalidium dimidiatum, Scytalidium hyalinum and Fusarium* species. Toenails are affected more than fingernails [53] and are also more recalcitrant to treatment.

Clinically, onychomycosis is classified depending on where the infection begins [47]:

1. Distal and lateral subungual onychomycosis (DLSO, Fig. 25.5). The infection starts at the hyponychium and the distal or lateral nail bed and progresses towards the nail matrix.
2. Superficial onychomycosis. Patches appear on the dorsal surface of the nail plate, as the latter is invaded.

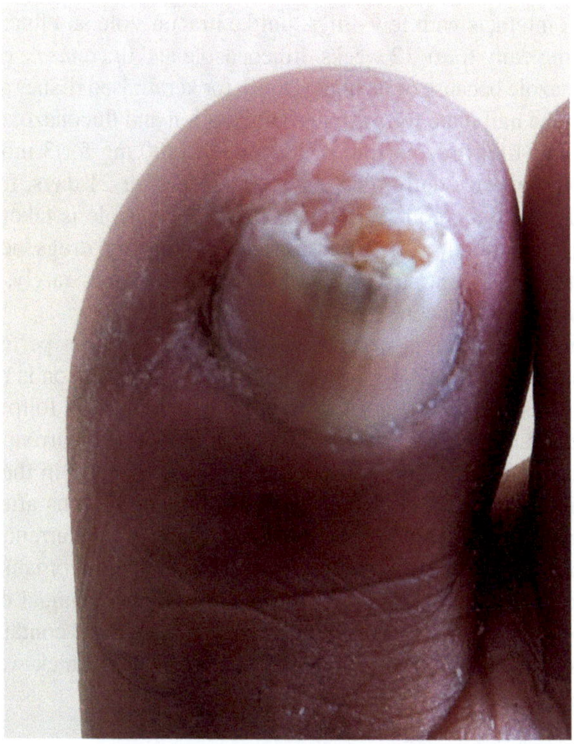

Fig. 25.5 Distal and lateral subungual onychomycosis (DLSO) in a toenail

3. Proximal white subungual onychomycosis (PWSO). The fungus penetrates via the proximal nail fold and produces a white discoloration in the area of the lunula. PWSO is rare but less so in patients with AIDS or who are otherwise immunosuppressed.
4. Endonyx onychomycosis (EO). Areas of superficial and deep fungal invasion coexist in the nail plate, which is pitted and has lamellar splits at its distal end, causing it to be friable and split.
5. Total dystrophic onychomycosis (TDO). This is the potential endpoint of all forms of onychomycosis, and the entire nail plate and bed are invaded by the fungus. The nail plate can crumble off to expose a thickened and abnormal nail bed.

25.2.1.1 Oral Drug Therapy of Onychomycosis

Onychomycosis is largely treated orally with the antifungal agents, terbinafine, itraconazole and fluconazole. Following oral administration and absorption into the systemic circulation, the drugs diffuse from the blood vessels into the nail plate via the nail bed. Terbinafine is the drug of choice due to its greater efficacy and safety profile [54]; it interacts with few drugs, unlike itraconazole and fluconazole, and is dosed at 250 mg daily for 6–12 weeks. Itraconazole and fluconazole can be dosed in pulses, itraconazole because of its high affinity for keratinised tissues and subsequent persistence in the nail plate following administration and fluconazole because of its long plasma half-life. Itraconazole is taken daily (200 mg for 3 months) or intermittently (200 mg twice daily for 7 days; repeated after 21 days, for two courses for fingernails and three courses for toenails). Fluconazole is taken once weekly (100–300 mg) for 6 months. The adverse effects of these oral drugs include headache and gastrointestinal symptoms, taste disturbance and more rarely, but seriously, liver and kidney disturbances.

Unfortunately, oral therapy of onychomycosis is far from perfect and a high recurrence rate—40 to 70 %—is found, due to relapse (infection is not completely cured and returns) or reinfection (an onychomycotic cure is followed by a new infection by the same or different organism) [43]. Infection recurrence could be due to the presence of fungal spores (which are resistant to drug) in the nail [55, 56], which germinate and produce a 'new' fungal infection months after cessation of therapy and an apparent 'cure'. Other reasons for infection recurrence include poor patient compliance; misdiagnosis; development of resistant organisms; systemic factors, e.g. diabetes and incomplete fungal kill by the antifungal drug due to its impeded movement within thickened and/or onycholytic nail containing dermatophytoma (a hyperkeratotic mass containing densely packed thick-walled dermatophyte hyphae).

25.2.1.2 Focal Drug Therapy of Onychomycosis

Focal drug therapy of onychomycosis involves the topical application of antifungal agents in diverse vehicles, such as paints, solutions, creams and lacquers. Focal therapy eliminates the adverse events and drug interaction problems associated when the drugs are administered orally and may have greater patient compliance. However, its efficacy is low [56], mainly due to the poor permeability of the nail plate, and it is generally recommended for mild and distal infection when up to two nails are affected or for superficial white onychomycosis or when systemic therapy is contraindicated, for example, in pregnancy [57]. Topical therapy is sometimes used in combination with systemic therapy as this improves cure rate [58]. Some authors suggest that in certain cases, for example, when faced with a high risk of treatment failure, the oral and topical combination approach should be considered as the first line of therapy [43]. Prior to topical drug application, as much diseased nail as possible is removed (mechanically, chemically or surgically). The nail plate can also be filed, e.g. using an abrasive to increase drug permeation into the nail.

The most convenient (and recommended) topical preparations are the nail lacquers (nail varnish) containing the antifungal agents amorolfine (Loceryl®, licensed in the UK and other countries but not the USA) and ciclopirox (Penlac®, licensed in the USA and other countries). Following application to the nail plates, the lacquers dry within a few minutes and leave a water-insoluble film, from which the drug partitions out and into the nail plate. Loceryl is applied 1–2 times weekly to filed nail plates for up to 6 months for fingernails and for 9–12 months for toenails. Penlac is applied once daily, preferably at bedtime, for up to 48 weeks. Every 7 days the existing Penlac film is removed before reapplication of the lacquer. Penlac and Loceryl are generally well tolerated, though slight burning with Loceryl and periungual erythema with Penlac have been reported (Patient Information Leaflets for Penlac and Loceryl). Success rates of topical therapy using amorolfine and ciclopirox nail lacquers have been fairly low however, being 34 and 24 % for amorolfine and ciclopirox, respectively [59].

A number of other formulations are also available, although they are not the first line of treatment, and some have been discontinued in certain countries. These include tioconazole nail solution (Trosyl®), an undecenoate solution (Monphytol® discontinued in the UK), salicylic acid paint (Phytex®), Excilor® (a pen-shaped applicator containing the active acetic acid), Fungal Free Nails™ (a mixture of natural products including essential oils) and Scholl Fungal Nail Treatment (a mixture of acid, urea, glycerine, panthenol and other ingredients). These are applied once or twice daily for long durations, in some cases up to 12 months or 'until the nail has recovered completely' (Excilor Patient Information Leaflet). They act either by directly inhibiting the fungus or creating an acidic environment in the nail which is hostile to fungal growth.

25.2.1.3 Boosted Oral Antifungal Treatment and Boosted Antifungal Topical Treatment

As mentioned above, the existence of dormant drug-resistant fungal spores within the nail plate leads to incomplete cure rates for onychomycosis. To address the fungal reservoir in the nail plate, boosted oral antifungal therapy (BOAT) and boosted antifungal topical treatment (BATT) have been developed [56, 60], where a piece of Sabouraud agar gel is secured onto the affected nail plates during systemic or topical therapy. Sabouraud medium is thought to convert dormant fungal spores into drug-susceptible fungal hyphae, thus helping fungal eradication.

25.2.1.4 Other Options for the Focal Therapy of Onychomycosis

Photodynamic therapy of onychomycosis involves application of a photosensitising agent, such as 5-aminolevulinic acid, to the nail plate, followed by irradiation with visible light, to generate reactive oxygen species which then kill fungal cells [61]. Drug-free therapies have also been tried. For example, a drug-free polymer film applied to the nail plate was found to have some success at treating onychomycosis. The film occludes the nail plate and thereby alters the latter's microenvironment, which is said to be conducive to cure [62]. Application of laser light to onychomycotic nails has been found to exert beneficial effects, although the mechanisms of action remain(s) to be elucidated [63].

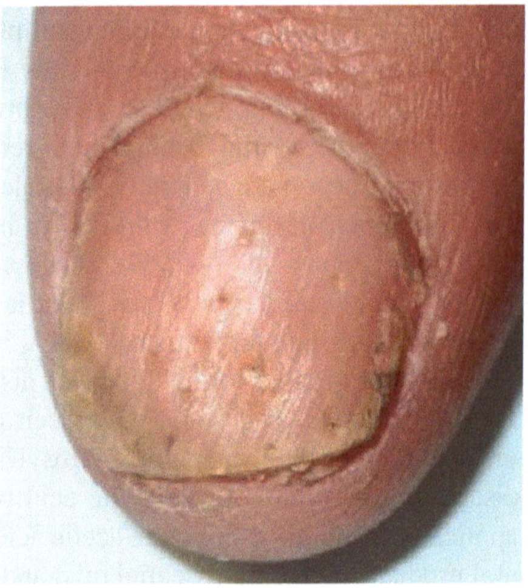

Fig. 25.6 Nail pitting—a symptom of nail psoriasis. Reproduced from Jiaravuthisan et al. [68], with kind permission from Elsevier

25.2.2 Nail Psoriasis

It is thought that 80–90 % of patients with skin psoriasis (which affects 1 and 3 % of most populations [64]) will also suffer from nail psoriasis over their lifetime [65]. At the same time, 1–5 % of patients with nail psoriasis do not present any overt cutaneous disease [66]. Psoriasis can affect the nail matrix, nail bed and nail folds. The location of psoriasis dictates the symptoms of the nail disease, which include white spots, pitting, transverse furrows, salmon-coloured 'oil drop' discolouration of the nail plate, onycholysis, subungual hyperkeratosis, splinter haemorrhages, paronychia, nail fragility, crumbling and nail loss (Fig. 25.6).

Nail psoriasis is very difficult to treat [67] and has been treated by a range of oral and focal—rather than one standardised—options as certain symptoms respond better to certain therapies, to accommodate patient preference, and due to the lack of a clear optimal treatment [68, 69]. An expert panel recently concluded that systemic therapy would most likely be required to treat patients suffering from both skin and nail psoriasis and recommended methotrexate and the TNF antagonist infliximab as the first and second lines of systemic therapy, respectively [67]. Other drugs that have been used systemically include retinoids, cyclosporine and nimesulide, although they are not recommended for the treatment of nail psoriasis in the absence of skin psoriasis [67]. Focal options include intralesional, topical and radiation therapies and are discussed below.

25.2.2.1 Intralesional Drug Therapy of Nail Psoriasis

Injection of long-acting corticosteroids (most commonly triamcinolone acetonide) into the nail folds (most commonly the proximal ones) is effective in treating psoriasis lesions of the nail matrix. The injections are repeated (e.g. monthly, weekly) using a fine gauge needle, and a local anaesthetic may be used to ease the pain of the injection. As well as being painful, these injections are time consuming, are unpopular with patients and can cause subungual hematomas, reversible atrophy at the injection site and potential atrophy of the underlying bone [68, 69].

25.2.2.2 Topical Drug Therapy of Nail Psoriasis

Topical application of drugs is much more patient-friendly than injections in the nail folds, and different drugs have been evaluated. Glucocorticoids (e.g. betamethasone, clobetasol, fluocinolone acetonide, triamcinolone acetonide) and vitamin D3 analogs (calcipotriol, tacalcitol, calcitriol) are the most popular drugs, although anthralin, 5-fluorouracil, tazarotene and cyclosporine have also shown promise [68]. A recent expert panel recommended that a corticosteroid be used as the first-line topical therapy, as the efficacy of other drugs currently used for the management of skin psoriasis (such as tazarotene, 5-fluorouracil, calcipotriol) is limited when nail disease is being treated, due to their poor permeation into the nail following topical application [67].

Topical drug formulations are applied onto the nail plate and/or nail folds and/or nail bed (or as close to it as possible), depending, to some extent, on the location of the psoriatic lesion. For example, if onycholysis is present, the nail plate is trimmed to the point where it separates from the nail bed prior to application. Sometimes, the nail plate is chemically removed (following the topical application of urea formulations), and the medicine is applied onto the nail bed. To enhance the drug movement into the nail unit, patients may be advised to cover the topical formulations with an occlusive dressing, such as by wearing plastic gloves.

25.2.2.3 Other Focal Therapies for the Treatment of Nail Psoriasis

Phototherapy (UVB) and photochemotherapy (oral or topical psoralen plus UVA) has shown benefits for certain, but not all types of, nail dystrophies. Superficial radiotherapy, Grenz rays, electron beam therapy and pulsed dye laser applications have also demonstrated variable response rates.

25.3 Topical Ungual Formulations

Topical therapy of nail diseases is the ideal mode of treatment, as it is non-invasive, painless and patient-friendly. A range of topical formulations, such as lotions, oily solutions, paints, lacquers, creams, ointments and gels, have been applied to the nail plate for the treatment of nail diseases [70]. Many of these formulations were originally developed for application to the skin, and there have been relatively fewer topical preparations developed specifically for the treatment of nail diseases. This could be due to the former perception that nail diseases did not provide a large enough market for drug development and/or that nail diseases were mainly a cosmetic issue. Fortunately, the situation is now improved with new nail-specific products (e.g. Loceryl and Penlac nail lacquers) launched fairly recently, more products in the pipeline and much research on topical ungual drug delivery [19, 71–74]. An ideal ungual formulation should be cosmetically acceptable, be easy to apply (and easy to remove when removal is desired), have a long residence time on the nail plate such that frequent application is not required and, most importantly, release the drug which can then permeate into the nail. Topical ungual formulations are described below.

25.3.1 Solutions, Lotions and Paints

Solutions (formulation where all the components, i.e. drug, excipients and solvent, are molecularly dispersed) have been investigated for the treatment of nail diseases. Examples include a solution of cyclosporine in maize oil for nail psoriasis, a solution of vitamin E in dimethyl sulfoxide for the treatment of yellow nail syndrome and terbinafine solution for onychomycosis. A lotion is a liquid for external use, which may be a solution, a suspension or an emulsion. Belanyx® lotion—a solution composed of propylene glycol and urea in water—was developed to serve as a carrier for 5-fluorouracil and has shown efficacy in the treatment of nail psoriasis. Paints are concentrated aqueous or alcoholic antimicrobial solutions which are formulated with a volatile vehicle. Upon application, the vehicle evaporates leaving a film of active substance on the nail plate. Tioconazole nail solution (Trosyl) contains the antifungals tioconazole and undecylenic acid and ethyl acetate as the solvent. Following application, the solution takes 10–15 min to dry, leaving a transparent, slightly greasy film on the nail plate. Salicylic acid paint (Phytex) contains tannic acid, boric acid, salicylic acid and methyl salicylate as the active agents and industrial methylated spirit and ethyl acetate as the solvents. The borotannic complex possesses fungistatic properties. When dissolved, the active complex ionises producing a low pH (about 2.0) which is fungicidal. After application the solvent evaporates to leave a clear film over the infected area. Monphytol® paint is a solution containing the actives methyl undecenoate, propyl undecenoate, salicylic acid, methyl salicylate, propyl salicylate and clorobutanol. Due to the easy removal of these liquids upon washing, they have to be applied once (or more) daily.

25.3.2 Creams, Ointments and Gels

These semi-solid preparations, formulated for skin conditions, have been applied to the nail plate, nail bed and nail folds, mainly for the treatment of nail psoriasis.

Examples include tazarotene gel, calcipotriol cream/ointment, anthralin in petroleum and betamethasone ointment. Tazarotene gel has also been evaluated for the treatment of brittle nails. Gels are also being researched for onychomycosis. These formulations are however easily wiped/washed off the nail during routine activities and have to be administered frequently. Patients are sometimes advised to apply an occlusive dressing or wear a pair of plastic gloves following drug application, to prevent loss of the formulation.

25.3.3 Nail Lacquers

Nail lacquers can be used to treat both onychomycosis and nail psoriasis, although commercially available pharmaceutical lacquers are only available for onychomycosis. They seem to be the most popular drug carriers for the focal treatment of nail diseases and are thus discussed at greater length. A lacquer is defined as a 'type of surface coating that consists of a film forming system dissolved in a volatile organic solvent phase' [75, 76]. In typical cosmetic nail lacquers, the film-forming phase is composed of a primary film former (nitrocellulose), a secondary film former (also called 'modifying resin' which improves the film's properties such as adhesion to the nail plate, water resistance and gloss), plasticiser(s) (which improve(s) the flexibility of the film) and miscellaneous additives (e.g. silicones, micronised waxes, which make the film surface smooth and slippery). The volatile organic solvent phase consists of active solvent(s) (which dissolve(s) the film-forming phase), coupling solvent(s) (which improve(s) the solvency power of the active solvents) and sometimes diluent(s) (volatile material(s) added to reduce bulk cost). In addition to the film-forming and volatile phases, pigmented cosmetic lacquers also contain pigments, suspending agents and pearlescent materials [76].

In contrast to cosmetic nail lacquers, the commercially available pharmaceutical nail lacquers Loceryl and Penlac contain far fewer chemicals, and most obviously do not contain nitrocellulose as the film former. Despite its many advantages (low toxicity, compatibility with many resins and plasticisers, low cost, manufacture from renewable resources), nitrocellulose has some serious drawbacks such as its explosivity, flammability and sensitivity to degradation by sunlight and heat [76], which may have precluded its use in the pharmaceutical lacquers. Loceryl is composed of the solvents butyl acetate, ethyl acetate and ethanol, the film former methacrylic acid copolymer and the plasticiser triacetin, in addition to the antifungal amorolfine at 5 %. Penlac® contains the solvents ethyl acetate and isopropyl alcohol and the polymer butyl monoester of poly(methyl vinyl ether/maleic acid) in addition to the antifungal ciclopirox at 8 %.

Like cosmetic nail lacquers, drug-containing nail lacquers must be chemically and physically stable, the different components must be compatible, and the viscosity of the lacquer must allow it to flow freely into all the edges and grooves of the nail for ease of application. Once applied, the lacquer must dry quickly (within 3–5 min) and form an even film. The latter must adhere well to nail plates and must not come off during daily activities, but must be able to be removed cleanly with nail varnish remover and the film must be well tolerated. Drug-containing lacquers must also be colourless and nonglossy to be acceptable to male patients, and most importantly, the drug must be released from the film so that it can penetrate into the nail [77, 78].

Both Loceryl and Penlac are clear colourless liquids which give rise to clear colourless films following application to the nail plate and rapid (within a few minutes) evaporation of the solvents. Both films are water insoluble, although the film formed by Loceryl was found to have longer staying power on the nail plate than that of Penlac [79], which might explain the application regimen (1–2× weekly for Loceryl versus daily for Penlac). The lacquer films have higher drug concentrations than the original formulations (25 % for Loceryl from the original 5 % and 35 % for Penlac from the original 8 %) due to the loss of the volatile phase from the lacquers [80, 81] and form a drug reservoir on the nail plate, from which drug can permeate into the nail plate along the thermodynamic activity gradient. Film formation on the nail plate reduces transonychial water loss (TOWL, loss of water from the body into the atmosphere via the nail plate) [82, 83], which is likely to cause increased hydration of the upper nail layers [83], which could enhance ungual drug diffusion [81], drug diffusivity being known to increase with nail plate hydration [84]. Ciclopirox and amorolfine have been reported to penetrate into nails following just one application of the nail lacquer, while repeated applications result in drug concentrations above the inhibitory and fungicidal concentrations of most pathogens in all nail layers [80, 85]. However, as expected from the inherently low nail plate permeability, systemic absorption of the antifungals from the nail lacquers is negligible [80, 86].

Other antifungal lacquers in development include EcoNail™—a formulation containing econazole and 2-*n*-nonyl-1,3-dioxolane. The latter is also known as SEPA (soft enhancement of percutaneous absorption) and is thought to act at the lacquer/nail interface to enhance the ungual drug permeation [87, 88]. Terbinafine HCl has also been formulated into a nail lacquer, with dodecyl-2-*N*,*N*-dimethylaminopropionate hydrochloride (DDAIP HCl, trade name NexACT-88) as an ungual permeation enhancer [89–92]. Clotrimazole lacquer, comprising cellulose derivative(s) as film former(s), dibutyl phthalate as plasticizer, acetone and ethanol as solvents, and urea is also being developed [93]. A lacquer based on a hydroalcoholic solution of hydroxypropyl chitosan (HPCH) polymer has been formulated with the antifungal ciclopirox, amorolfine and octopirox. Interestingly, HPCH polymer and the film formed by this lacquer are water soluble; the film thus is easily removed by washing with water, without the need for organic solvents. While easy removal could be considered an advantage, it does necessitate daily application of the lacquer at bedtime when avoidance to water exposure over a relatively long period can reasonably be expected. Other lacquers in investigation include bilayered and phase-separating lacquers [94, 95].

In addition to their role in onychomycosis, pharmaceutical nail lacquers are also being investigated for the topical treatment of nail psoriasis, for chemical nail avulsion and to improve nail quality. A nail lacquer containing 8 % clobetasol-17-propionate was found to improve onycholysis, pitting and salmon patches in psoriatic nails as well as nail pain, without causing any local or systemic adverse effects [96, 97]. A drug-free water-soluble nail varnish containing HPCH, horsetail extract (*Equisetum arvense*) and methylsulphonylmethane was also reported to improve the appearance of psoriatic nails [98]. Apart from having a beneficial effect on psoriatic nails, the HPCH lacquer improved the quality of brittle nails, reducing nail roughness and fragility and increasing nail strength. Nail lacquers containing urea at 10 % have also been tested for their ability to increase nail plate quality [99], while lacquers containing 40 % urea have been used to chemically avulse diseased nail plates [100].

Despite the increased activity in the research and development of nail lacquers as evidenced by the number of patents and lacquers in clinical trials, the literature pertaining to the systematic investigation of lacquer formulation is limited. Obviously, the formulation of the nail lacquer is critical for optimal drug delivery to the nail unit. The nature and concentrations of the solvents, film-forming polymer, resins, plasticisers and drug affect the lacquer's properties such as the liquid's flow and drying rate, the film's quality in terms of adhesion to the nail plate, brittleness, flexibility, resistance to washing and to other mechanical barriers to its residence time, and the drug release from the film. The hydrosoluble HPCH-based lacquer was reported to be a better vehicle for ciclopirox and amorolfine, compared to the commercially available preparations of these two antifungal agents [101]. In contrast, Hui et al. reported similar ungual drug permeation from three different lacquer formulations of AN2690 (a new boron-containing antifungal agent) where the films formed were water insoluble and resistant to removal, water soluble and easily washed off or water insoluble but removable by peeling or scratching, using the polymers poly(vinyl methyl ether-alt-maleic acid monobutyl ester), poly(2-hydroxyethyl methacrylate) and poly(vinyl acetate), respectively [102]. Reports on the influence of the solvent on ungual permeation have also been conflicting; Franz reported greater ungual flux of amorolfine from a methylene chloride lacquer compared to an ethanol lacquer, while Polak showed a lack of influence of solvent nature (ethanol or methylene chloride) [85, 103]. Nalamothu and Schwartz observed that higher concentrations of drug (clotrimazole), film-forming polymer poly(methyl vinyl ether-co-maleic anhydride) and tailing solvent (propylene glycol) resulted in increased adhesion of the lacquer film [104]. Much more systematic research is needed in this field to assist the rational design of pharmaceutical nail lacquers.

25.3.4 Patches

There has been very little research into drug-containing patches for application to the nail plate for the treatment of nail diseases. Like nail lacquers, patches have the benefits of 'containing' the drug so that the latter is not lost from the nail during routine activities and of a long residence at the site of action, if desired. The patch can also be designed to be occlusive, which would reduce transonychial water loss and increase nail plate hydration, which would in turn favour drug flux into the nail, drug diffusivity being known to increase with nail plate hydration. A drug-in-adhesive patch containing the antifungal sertaconazole was developed and tested in volunteers. Following application for 6 weeks (the patch being replaced at weekly intervals), drug concentrations above the minimum inhibitory concentrations (MICs) were found in the nail plate, and as expected from the low nail plate permeability, no systemic absorption was observed [105]. Patches have also been used as vehicles (for 5-aminolevulinic acid) in investigations on the photodynamic therapy of onychomycosis [61]. Myoung and Choi showed that the nature of the pressure-sensitive adhesive and of the vehicles influenced drug (ciclopirox) permeation across hoof membranes (used as model for nail plate) [106].

25.3.5 *Films*

Drug-containing films consisting of hydroxypropyl cellulose (HPC) and/or polyethylene oxide (PEO), prepared by hot-melt extrusion, have also been investigated as topical drug delivery vehicles for onychomycosis. Repka et al. characterised the films in terms of their surface appearance, mechanical properties, adhesion to human nail, the state of the drug (crystalline or amorphous) within the films, drug release and in vitro ungual drug permeation. The films' properties were influenced by their composition, and ungual drug permeation was enhanced when films were applied to nail plates whose dorsal surface had been etched with phosphoric acid, etching resulting in increased surface roughness, hence area, and increased adhesion of the film to the nail plate [107–110].

25.4 Improving Focal Drug Delivery for the Treatment of Nail Diseases

The current focal therapies for nail diseases are not ideal. Injection of steroids into the nail folds—the main treatment for nail psoriasis—is painful, time consuming and can cause subungual hematomas, as well as atrophy at the injection site and of the underlying tissues, as mentioned earlier. On the other hand, topical therapies for both onychomycosis and nail psoriasis have a low success rate and need to be used for long durations [59]. Topical drug treatment of onychomycosis and nail psoriasis has advantages over the respective mainstay oral and intralesional therapies, such as elimination of adverse effects and drug interactions. Following the topical application of a drug formulation onto the nail plate, the drug is expected to partition out of the formulation into the nail plate, diffuse through the latter and, finally, partition into the nail bed. Drug transport into the nail plate is thus expected to be influenced by the physico-chemical properties of the drug molecule (e.g. size, shape, charge, hydrophobicity), its interaction with nail keratin, the formulation characteristics (e.g. nature of vehicle, pH, drug concentration) and the properties of the nail (e.g. hydration, disease state) [19]. In order to improve the focal therapy of nail diseases, a number of different strategies are needed/currently being researched and are discussed below.

25.4.1 Elimination of the Needle in Intralesional Therapy of Nail Psoriasis

Elimination of the needle to deliver drugs inside the nail unit may eliminate/reduce the incidence of injection-related pain, hematomas and atrophy. The needle could be replaced by other modalities such as iontophoresis, ultrasound and laser which are known to enhance the permeability of the skin and could therefore enable drug delivery to the nail unit via the nail folds. Iontophoresis has already been shown to increase the amount of drug delivered to the matrix when it is applied to the proximal nail fold [111].

25.4.2 Increased Knowledge of the Properties of Diseased Nails

The diseased nail is obviously very different to the healthy nail described above. Onychomycotic nails are often thicker and more crumbly. While the appearance of diseased nails has been described and is used in diagnosis, other properties such as the mechanical strength, permeability, constituents and intercellular contact are little known. We recently reported that the disulphide content of diseased nails was similar to that of healthy nails [112]. Work is underway to further understand the different properties of diseased nails in order to inform the treatment of nail diseases.

25.4.3 Development of Improved Ungual Drug Formulations

The low success rate of the commercially available topical onychomycotic preparations could be due to suboptimal formulation. The active agents in the older preparations were not sufficiently potent against the fungal organisms. The newer formulations, such as the lacquers containing amorolfine or ciclopirox, have short residence on the nail plate following topical application [79, 113]. Amorolfine lacquer was almost completely lost within 2 days [113]. This lacquer is supposed to be applied once or twice a week, which means that under the prescribed application regimen, there would be a number of drug-free days every week, and as a result insufficient drug would be delivered to the nail. It is also possible that insufficient drug partitions out of topical preparations and into the nail unit. While it is claimed that drug concentrations above the MICs of the causative fungal organisms are achieved in the nail plate following application of many topical onychomycotic preparations, the MICs used are those that have been determined in standard laboratory assays and may be well below the MICs needed in the nail plate to cure a fungal infection. Research is underway to develop improved formulations as described in Sect. 25.3.

25.4.4 Development of Drugs with Optimal Properties for Ungual Permeation After Topical Application

Very few drugs have been specifically designed for ungual uptake, especially after topical application. Synthesis of new active chemical entities taking into account of the existing literature on ungual drug delivery is one way of improving the therapy of nail diseases. Such drug molecule must be potent, as very little of the applied drug normally permeates into the nail, due to the nail plate's low permeability. In addition, it is probably preferable for topical therapy that the drug does not bind to keratin and is thus in a free form to act, for example, kill fungus in onychomycotic nails; drug binding to keratin has been shown to have a deleterious effect on its antifungal efficacy [114]. The drug molecule must be small—molecular weight has been found to be the most important parameter which influences its permeation into and through the nail plate [27, 28]. The smaller the molecule, the greater is its permeation. The molecule would also be uncharged (unless it is delivered by iontophoresis); ionic species have ~10-fold lower permeability than their unionised counterparts due to a small increase in their apparent molecular size upon their hydration [28]. Permeant hydrophilicity/lipophilicity is also thought to be important, although investigations into the influence of permeants' lipophilicity on their permeability have been hampered by the fact that increasing lipophilicity in a series is also accompanied by increasing permeant molecular weight.

25.4.5 Enhancement of Ungual Drug Permeation Following Topical Application

The poor permeability of the nail plate to topically applied drugs limits the success of topical therapy. For example, van Hoogdalem calculated the total drug uptake into the nail to be <0.2 % of the applied dose after twice daily application for 6 weeks [7]. For effective topical therapy, ungual drug permeation must be enhanced. As mentioned above, one way of doing this is to optimise the drug formulation so that drug partitioning out of the vehicle and into the nail plate is enhanced. Other ways of increasing ungual drug permeation involve the use of physical and chemical enhancers which disrupt the nail plate to a small extent, reducing its barrier properties or, which encourage drug permeation into intact nail plate [74]. The physical techniques include filing the dorsal nail surface, either manually using an abrasive or using electrical equipments such as dental drills; etching the dorsal nail surface using an acidic preparation, such as tartaric acid solution or a phosphoric acid gel and creating holes in the nail surface using lasers, electricity or low-frequency ultrasound and iontophoresis. Except for the latter, the nail is exposed to the physical techniques prior to drug application. In contrast, chemical ungual enhancers can be included in the formulations, making drug administration a one-step process. The chemical enhancers disrupt the chemical and physical (disulphide, peptide, hydrogen and polar) bonds responsible for the stability of nail plate keratin. Of these, the disulphide bond has received the most attention, due to its critical role in protein stability. Cleavage of the disulphide bond has mainly been achieved by its reduction, though oxidation has also been attempted. Only a few ungual chemical enhancers have been identified; these include thiols [-SH containing compounds such as *N*-acetylcysteine, mercaptoethanol, thioglycolic acid, *N*-(2-mercaptopropionyl) glycine (MPG)], sulphites (e.g. sodium sulphite), salts, hydrogen peroxide, urea (acts synergistically with other enhancers) and enzymes (e.g. keratinase, papain). More research is needed to identify and characterise additional enhancers.

25.5 Concluding Remarks

Nail diseases—whether localised to the nail or related to systemic disorders—are not trivial and need to be treated effectively to improve the health and well-being of sufferers. Focal therapy of localised nail diseases is ideal as it avoids systemic adverse reactions and drug interactions. However, current focal therapies are not sufficiently effective, and more research and drug development is needed in this field to produce new topical medicines and improve the treatment of nail diseases.

References

1. Dawber RPR, de Berker DAR, Baran R (2001) Science of the nail apparatus. In: Baran R, Dawber RPR, De Berker DAR, Haneke E, Tosti A (eds) Baran and Dawber's diseases of the nails and their management. Blackwell, Oxford, pp 1–47
2. Fleckman P (2005) Structure and function of the nail unit. In: Scher R, Daniel CR III (eds) Nails diagnosis therapy surgery. Elsevier Saunders, Shanghai, pp 13–25
3. de Berker DAR, Andre J, Baran R (2007) Nail biology and nail science. Int J Cosmet Sci 29(4):241–275
4. de Berker D, Forslind B (2004) The structure and properties of nails and periungual tissues. In: Forslind B, Linberg M, Norlén L (eds) Skin, hair, and nails, structure and function. Marcel Dekker, New York, NY, pp 409–464
5. Zaias N (1990) The nail in health and disease, 2nd edn. Appleton & Lange, Norwalk, CT
6. Barron JN (1970) The structure and function of the skin of the hand. Hand 2(2):93–96
7. Dawber RPR, Baran R (1984) Structure, embryology, comparative anatomy and physiology of the nail. In: Baran R, Dawber RPR (eds) Diseases of the nails and their management. Blackwell, Oxford

8. Chapman RE (1986) Hair, wool, quill, nail, claw, hoof and horn. In: Bereiter-Hahn J, Matoltsy AG, Richards KS (eds) Biology of the Integument, Vol 2, Vertebrates. Springer: New York, NY
9. Gonzalez-Serva A (1997) Structure and function. In: Scher RK, Daniel CR (eds) Nails: therapy, diagnosis, surgery. WB Saunders, Philadelphia, PA, pp 12–31
10. Williams TI, Rose R, Chisholm S (2007) What is the function of nail biting: an analog assessment study. Behav Res Ther 45(5):989–995
11. Murdan S (2012) The nail: anatomy, physiology, diseases, and treatment. In: Murthy SN, Maibach H (eds) Topical nail products and ungual drug delivery. CRC, Boca Raton, FL, pp 1–36
12. Roome L (1998) Mehndi: the timeless art of henna painting. St Martin's, New York, NY
13. Marzec E, Olszewski J (2009) Molecular interactions in human nail plate analysed by dielectric spectroscopy. Colloids Surf B Biointerfaces 69(1):91–94
14. Priya MDL, Geetha A (2011) A biochemical study on the level of proteins and their percentage of nitration in the hair and nail of autistic children. Clin Chim Acta 412(11–12): 1036–1042
15. Morini L et al (2012) Determination of ethyl glucuronide in nails by liquid chromatography tandem mass spectrometry as a potential new biomarker for chronic alcohol abuse and binge drinking behavior. Anal Bioanal Chem 402(5):1865–1870
16. Bengtsson CF et al (2012) DNA from keratinous tissue. Part I: hair and nail. Ann Anat 194(1):17–25
17. Williams JS, Katzenberg MA (2012) Seasonal fluctuations in diet and death during the late horizon: a stable isotopic analysis of hair and nail from the central coast of Peru. J Archaeol Sci 39(1):41–57
18. Oyoo-Okoth E et al (2012) Element profiles in hair and nails of children reflect the uptake from food and the environment. Environ Toxicol Chem / SETAC 31(7):1461–1469
19. Murdan S (2002) Drug delivery to the nail following topical application. Int J Pharm 236(1–2):1–26
20. Hamilton JB, Terada H, Mestler GE (1955) Studies of growth throughout the lifespan in Japanese: growth and size of nails and their relationship to age, sex, heredity, and other factors. J Gerontol 10(4):401–415
21. Murdan S (2010) Transverse fingernail curvature: a quantitative evaluation, and an exploration into the influence of gender, age, handedness, height and hand size. Presentation at UK Pharm Sci 2010, September 2010
22. Forslind B, Thyresson N (1975) Structure of normal nail-scanning electron-microscope study. Archiv Fur Dermatologische Forschung 251(3):199–204
23. Lewis BL (1954) Microscopic studies of fetal and mature nail and surrounding soft tissue. AMA Arch Derm Syphilol 70(6):733–747
24. Jarrett A, Spearman RI (1966) Histochemistry of human nail. Arch Dermatol 94(5):652–657
25. Sowa MG et al (1995) Infrared spectroscopic investigation of in-vivo and ex-vivo human nails. Vib Spectros 10(1):49–56
26. Rice RH et al (2010) Proteomic analysis of human nail plate. J Proteome Res 9(12): 6752–6758
27. Gniadecka M et al (1998) Structure of water, proteins, and lipids in intact human skin, hair, and nail. J Investig Dermatol 110(4):393–398
28. Stern DK et al (2007) Water content and other aspects of brittle versus normal fingernails. J Am Acad Dermatol 57(1):31–36
29. Egawa M, Ozaki Y, Takahashi M (2006) In vivo measurement of water content of the fingernail and its seasonal change. Skin Res Technol 12(2):126–132
30. Helmdach M et al (2000) Age and sex variation in lipid composition of human fingernail plates. Skin Pharmacol Appl Skin Physiol 13(2):111–119
31. He K (2011) Trace elements in nails as biomarkers in clinical research. Eur J Clin Invest 41(1):98–102
32. Vellar OD (1970) Composition of human nail substance. Am J Clin Nutr 23(10):1272–1274

33. Sirota L et al (1988) X-ray - microanalysis of the fingernails in term and preterm infants. Pediatr Dermatol 5(3):184–186
34. Wilhelm M et al (2005) Comparison of arsenic levels fingernails with urinary as species as biomarkers of arsenic exposure in residents living close to a coal-burning power plant in Prievidza District, Slovakia. J Expo Anal Environ Epidemiol 15(1):89–98
35. Nardoto GB et al (2006) Geographical patterns of human diet derived from stable-isotope analysis of fingernails. Am J Phys Anthropol 131(1):137–146
36. Lewis BL, Montgomery H (1955) The senile nail. J Investig Dermatol 24(1):11–18
37. Tosti A, Piraccini BM (2007) Biology of nails and nail disorders. In: Wolff K et al (eds) Fitzpatrick's dermatology in general medicine. McGraw Hill, New York, NY, pp 778–794
38. Michel C et al (1997) Nail abnormalities in rheumatoid arthritis. Br J Dermatol 137(6): 958–962
39. Murdan S et al (2012) In vivo measurement of the surface energy of human fingernail plates. Int J Cosmet Sci 34(3):257–262
40. Murdan S, Milcovich G, Goriparthi GS (2011) An assessment of the human nail plate pH. Skin Pharmacol Physiol 24(4):175–181
41. Achten G, Wanet-Rouard J (1978) Onychomycoses in the laboratory. Mykosen Suppl 1:125–127
42. Pardo-Castello V (1960) Diseases of the nail, 3rd edn. Charles C Thomas, Springfield, IL
43. Thomas J et al (2010) Toenail onychomycosis: an important global disease burden. J Clin Pharm Ther 35(5):497–519
44. Roberts DT (1999) Onychomycosis: current treatment and future challenges. Br J Dermatol 141:1–4
45. Chabasse D, Baran R, De Chauvin MF (2000) Onychomycosis I: epidemiology and etiology. J Mycol Med 10(4):177–190
46. Pierard G (2001) Onychomycosis and other superficial fungal infections of the foot in the elderly: a Pan-European survey. Dermatology 202(3):220–224
47. Hay RJ, Baran R, Haneke E (2001) Fungal (onychomycosis) and other infections involving the nail apparatus. In: Baran R, Dawber RPR, De Berker DAR, Haneke E, Tosti A (eds) Baran and Dawber's diseases of the nails and their management. Blackwell, Oxford, pp 129–171
48. Caputo R et al (2001) Prevalence of superficial fungal infections among sports-active individuals: results from the Achilles survey, a review of the literature. J Eur Acad Dermatol Venereol 15:312–316
49. Daniel CR (1991) The diagnosis of nail fungal infection. Arch Dermatol 127(10):1566–1567
50. Cohen PR, Scher RK (1994) Topical and surgical treatment of onychomycosis. J Am Acad Dermatol 31(3):S74–S77
51. Scher RK (1996) Onychomycosis: a significant medical disorder. J Am Acad Dermatol 35(3):S2–S5
52. Gupta AK, Shear NH (1997) Onychomycosis – going for cure. Can Fam Physician 43:299–305
53. Midgley G, Moore MK, Cook JC (1994) Mycology of nail disorders. J Am Acad Dermatol 31:S68–S74
54. Sigurgeirsson B et al (2002) Long-term effectiveness of treatment with terbinafine vs itraconazole in onychomycosis: a 5-year blinded prospective follow-up study. Arch Dermatol 138(3):353–357
55. Baran R, Hay RJ, Garduno JI (2008) Review of antifungal therapy, part II: treatment rationale, including specific patient populations. J Dermatolog Treat 19(3):168–175
56. Pierard GE, Pierard-Franchimont C, Arrese JE (2000) The boosted antifungal topical treatment (BATT) for onychomycosis. Med Mycol 38(5):391–392
57. BNF (ed) (2008) British national formulary, 56 edn. BMJ Group & RPS, London
58. Baran R et al (2000) A randomized trial of amorolfine 5% solution nail lacquer combined with oral terbinafine compared with terbinafine alone in the treatment of dermatophytic toenail onychomycoses affecting the matrix region. Br J Dermatol 142(6):1177–1183
59. Halmy K (2005) Experience with nail lacquers containing amorolfine 5% and ciclopirox 8% in patients with onychomycosis. J Am Acad Dermatol 52(3):P126

60. Pierard G, Pierard-Franchimont C, Arrese JE (2000) The boosted oral antifungal treatment for onychomycosis beyond the regular itraconazole pulse dosing regimen. Dermatology 200(2):185–187
61. Donnelly RF et al (2005) Bioadhesive patch-based delivery of 5-aminolevulinic acid to the nail for photodynamic therapy of onychomycosis. J Control Release 103(2):381–392
62. Nasir A et al (2011) Clinical evaluation of safety and efficacy of a New topical treatment for onychomycosis. J Drugs Dermatol 10(10):1186–1191
63. Carney C et al (2011) How is laser effective in treatment of onychomycosis? J Am Acad Dermatol 64(2):AB103
64. Schofield OMV, Hunter JAA (1999) Diseases of the skin. In: Haslett C, Chilvers ER, Hunter JAA, Boon NA (eds) Davidson's principles and practice of medicine. Churchill Livingstone, London, p 900
65. Samman P (1978) The nails in disease, 3rd edn. Heinemann, London
66. Del Rosso JQ, Basuk PJ, Scher RK, Ricci AR (1997) Dermatologic diseases of the nail unit. In: Scher RK, Daniel CR (eds) Nails: therapy, diagnosis, surgery. WB Saunders, Philadelphia, PA, pp 172–200
67. Langley RG et al (2012) Recommendations for the treatment of nail psoriasis in patients with moderate to severe psoriasis: a dermatology expert group consensus. J Eur Acad Dermatol Venereol 26(3):373–381
68. Jiaravuthisan MM et al (2007) Psoriasis of the nail: anatomy, pathology, clinical presentation, and a review of the literature on therapy. J Am Acad Dermatol 57(1):1–27
69. Edwards F, de Berker D (2009) Nail psoriasis: clinical presentation and best practice recommendations. Drugs 69(17):2351–2361
70. Shivakumar HN et al (2012) Topical nail formulations. In: Murthy SN, Maibach H (eds) Topical nail products and ungual drug delivery. CRC, Boca Raton, FL, pp 61–86
71. Gupchup GV, Zatz JL (1999) Structural characteristics and permeability properties of the human nail: a review. J Cosmet Sci 50(6):363–385
72. Sun Y, Liu J-C, Wang JCT (1999) Nail penetration. Focus on topical delivery of antifungal drugs for onychomycosis treatment. In: Bronaugh RL, Maibach HI (eds) Percutaneous absorption. Drugs cosmetics mechanisms methodology. Marcel Dekker, New York, pp 759–778
73. Elkeeb R et al (2010) Transungual drug delivery: current status. Int J Pharm 384(1–2):1–8
74. Murdan S (2008) Enhancing the nail permeability of topically applied drugs. Expert Opin Drug Deliv 5(11):1267–1282
75. Hochberg S (1979) Kirk-Othmer encyclopedia of chemical technology, vol 6, 3rd edn. Wiley, New York, NY
76. Patil A, Sandewicz R (1999) Nail lacquer technology. Society of Cosmetic Chemists: Revlon Research Centre, New Jersey, NJ
77. Pittrof F et al (1992) Loceryl(R) nail lacquer - realization of a new galenical approach to onychomycosis therapy. Clin Exp Dermatol 17:26–28
78. Mitsui T (1997) Make up cosmetics, in new cosmetic science. Elsevier, Amsterdam, pp 370–405
79. Sidou F, Soto P (2004) A randomized comparison of nail surface remanence of three nail lacquers, containing amorolfine 5%, ciclopirox 8% or tioconazole 28%, in healthy volunteers. Int J Tissue React 26(1–2):17–24
80. Bohn M, Kraemer KT (2000) Dermatopharmacology of ciclopirox nail lacquer topical solution 8% in the treatment of onychomycosis. J Am Acad Dermatol 43(4):S57–S69
81. Marty JP (1995) Amorolfine nail lacquer: a novel formulation. J Eur Acad Dermatol Venereol 4(Suppl 1):S17–S21
82. Murdan S, Hinsu D, Guimier M (2008) A few aspects of transonychial water loss (TOWL): inter-individual, and intra-individual inter-finger, inter-hand and inter-day variabilities, and the influence of nail plate hydration, filing and varnish. Eur J Pharm Biopharm 70(2):684–689
83. Spruit D (1972) Effect of nail polish on the hydration of the fingernail. Amer Cosmet Perf 87:57–58

84. Gunt HB, Kasting GB (2007) Effect of hydration on the permeation of ketoconazole through human nail plate in vitro. Eur J Pharm Sci 32(4–5):254–260
85. Polak A (1993) Kinetics of amorolfine in human nails. Mycoses 36(3–4):101–103
86. Reinel D (1992) Topical treatment of onychomycosis with amorolfine 5 percent nail lacquer - comparative efficacy and tolerability of once and twice weekly use. Dermatology 184:21–24
87. Chan TC et al (2005) EcoNail (TM) (5% econazole plus 18% SEPA (R)) nail lacquer in patients with severe onychomycosis: safety, local tolerability, and systemic exposure. J Investig Dermatol 125(1):926
88. Hui XY et al (2003) Enhanced econazole penetration into human nail by 2-n-nonyl-1,3-dioxolane. J Pharm Sci 92(1):142–148
89. Hossain M et al (2005) Efficacy of terbinafine hydrochloride nail lacquer formulations in a guinea pig model of Trichophyton mentagrophytes dermatophytosis. J Am Acad Dermatol 52(3):P8
90. Gupta A et al (2005) Evaluation of terbinafine HCL nail lacquer in a phase 2 study in onychomycosis patients: clinical efficacy assessment by computerized planimetry. J Am Acad Dermatol 52(3):P126
91. Long L, Pfister W, Ghannoum M (2006) In vitro evaluation of terbinafine HCl nail lacquer in a bovine hoof penetration bioassay. J Am Acad Dermatol 54(3):AB147
92. Ghannoum MA, Long L, Pfister WR (2009) Determination of the efficacy of terbinafine hydrochloride nail solution in the topical treatment of dermatophytosis in a guinea pig model. Mycoses 52(1):35–43
93. Wohlrab W, Wellner K (1994) Nail lacquer for the treatment of onychomycosis. Roehm Pharma GmbH, Weiterstadt DE http://www.freepatentsonline.com/5346692.html, last accessed 21 Jun 2010
94. Shivakumar HN et al (2010) Bilayered nail lacquer of terbinafine hydrochloride for treatment of onychomycosis. J Pharm Sci 99(10):4267–4276
95. Makvana R, Brown MB, McAuley WJ (2010) Development of a novel phase separating nail lacquer for the treatment on onychomycosis. J Pharm Pharm 62(6):802–803
96. Baran R, Tosti A (1999) Topical treatment of nail psoriasis with a new corticoid-containing nail lacquer formulation. J Dermatol Treat 10(3):201–204
97. Regana MS et al (2005) Treatment of nail psoriasis with 8% clobetasol nail lacquer: positive experience in 10 patients. J Eur Acad Dermatol Venereol 19(5):573–577
98. Cantoresi F et al (2009) Improvement of psoriatic onychodystrophy by a water-soluble nail lacquer based on hydroxypropyl-chitosan. J Am Acad Dermatol 60(3):AB163
99. Kruger N et al (2006) Effect of urea nail lacquer on nail quality. Clinical evaluation and biophysical measurements. Hautarzt 57(12):1089–1093
100. Baran R, Tosti A (2002) Chemical avulsion with urea nail lacquer. J Dermatolog Treat 13(4):161–164
101. Monti D et al (2010) Hydrosoluble medicated nail lacquers: in vitro drug permeation and corresponding antimycotic activity. Br J Dermatol 162(2):311–317
102. Hui X et al (2007) In vitro penetration of a novel oxaborole antifungal (AN2690) into the human nail plate. J Pharm Sci 96:2622–2631
103. Franz TJ (1992) Absorption of amorolfine through human nail. Dermatology 184:18–20
104. Nalamothu V, Schwartz J (2004) Influence of formulation variables on the adhesion characteristics of clotrimazole nail lacquers. Available from http://www.aapsj.org/abstracts/AM_2004/AAPS2004-000732.PDF Cited 9 Jun 2010
105. Susilo R et al (2006) Nail penetration of sertaconazole with a sertaconazole-containing nail patch formulation. Am J Clin Dermatol 7(4):259–262
106. Myoung Y, Choi HK (2003) Permeation of ciclopirox across porcine hoof membrane: effect of pressure sensitive adhesives and vehicles. Eur J Pharm Sci 20(3):319–325
107. Repka MA, Prodduturi S, Stodghill SP (2003) Production and characterization of hot-melt extruded films containing clotrimazole. Drug Dev Ind Pharm 29(7):757–765

108. Repka MA, Mididoddi PK, Stodghill SP (2004) Influence of human nail etching for the assessment of topical onychomycosis therapies. Int J Pharm 282(1–2):95–106
109. Mididoddi PK, Prodduturi S, Repka MA (2006) Influence of tartaric acid on the bioadhesion and mechanical properties of hot-melt extruded hydroxypropyl cellulose films for the human nail. Drug Dev Ind Pharm 32(9):1059–1066
110. Mididoddi PK, Repka MA (2007) Characterization of hot-melt extruded drug delivery systems for onychomycosis. Eur J Pharm Biopharm 66(1):95–105
111. Manda P et al (2012) Iontophoresis across the proximal nail fold to target drugs to the nail matrix. J Pharm Sci 101(7):2392–2397
112. Baraldi A et al (2012) A study on the structural integrity of healthy and diseased nails: the retention of disulphide bonds in perspectives in percutaneous penetration. La Grande Motte, France
113. Murdan S et al (2011) Does the strength of adhesion between an ungual topical film and the nail plate influence its in vivo residence time? UKPharmSci, Nottingham
114. Tatsumi Y et al (2002) Therapeutic efficacy of topically applied KP-103 against experimental tinea unguium in guinea pigs in comparison with amorolfine and terbinafine. Antimicrob Agents Chemother 46(12):3797–3801

Chapter 26
Drug Delivery to Wounds, Burns, and Diabetes-Related Ulcers

Sonam Jain, Abraham J. Domb, and Neeraj Kumar

26.1 Introduction

Skin is the largest organ of the body covering 1.7 m^2. It is a bilayered construct with a stratified epidermis and a thick layer of dermal connective tissue. Epidermis provides a barrier to microbial invasion and prevents loss of moisture or fluid, as well as plays a vital role in other physiological functions, such as protection from ultraviolet light, thermoregulation, sensation, and appearance. The lower, thicker dermal layer consists of collagen fibers and glycosaminoglycans (GAG) [1, 2]. The dermis provides mechanical functions of strength and elasticity, and it supplies nutrition and immune defense to both the skin layers by means of its vasculature [3]. Beneath the dermis lies the hypodermis which is a loose connective layer comprising primarily of adipose tissue.

The skin's continuity and integrity can be compromised as a result of trauma arising from physical or thermal damage or an underlying pathological condition resulting in wounds. The problem of large skin losses occurs most often in wounds, diabetic ulcers, and burns. Wounds extend to different layers of skin and thus result in variable thickness of injury. First-degree burns or superficial partial-thickness wounds cause epidermal destruction only, which heal spontaneously and require no skin replacement. Second-degree wounds involve partial damage to dermis and skin appendages and heal spontaneously by epithelialization from the deeper appendages but with the likelihood of scar formation and wound contracture [4]. Third-degree or full-thickness wounds involve complete destruction of the epidermis and dermis.

S. Jain • N. Kumar (✉)
Department of Pharmaceutics, National Institute of Pharmaceutical Education & Research (NIPER), S.A.S. Nagar, Mohali 160062, Punjab, India
e-mail: neeraj@niper.ac.in

A.J. Domb
Faculty of Medicine, School of Pharmacy, Institute of Drug Research, The Hebrew University, Jerusalem 91120, Israel

A.J. Domb and W. Khan (eds.), *Focal Controlled Drug Delivery*, Advances in Delivery Science and Technology, DOI 10.1007/978-1-4614-9434-8_26,

The unassisted wound healing process in these cases does not result in the formation of normal dermis. A full-thickness wound is life-threatening, not only due to the local loss of barrier functions but also due to a systemic physiological response of fever and hypercatabolic metabolism. Examples of injuries or diseases that can result in large, full-thickness wounds include burns and other trauma (e.g., diabetes ulcers), benign skin tumor surgery, cancer surgery, reconstructive surgery, and congenital cutaneous anomalies like the removal of giant hairy nevus [5]. At this scale of burn/wound, a skin dressing and drug delivery is needed to heal the wound rapidly and aesthetically [6].

This chapter discussed different wound management dressings and advanced technologies for achieving improved healing for wounds, burns, and diabetes-related ulcers. Several dressings have been used for delivering drugs to acute and chronic wounds. These dressings could be classified in number of ways depending on the type of material used for healing [7], type of wound, and physical state of dressings [8]. Active ingredients (antimicrobials, growth factors (GFs), or other supplements) could be incorporated into medicated dressing which will be useful in wound healing either directly or indirectly [9]. Apart from incorporating drugs in the dressings, several pharmacological agents have been given systemically route and/or locally to inhibit the microbial growth or to improve the wound healing process. These agents include antibiotics [10], gene therapy and cytokines [11], GFs [12], stem cells [13], and other drugs like antioxidants [14], antihistamines [15], phenytoin [16], antisense gel [17], and traditional drugs [18] and will be discussed in the later part of the chapter along with the pathogenesis of wounds, burns, and diabetes ulcers.

26.2 Skin Anatomy and Pathophysiology of Wounds, Burns, and Diabetes-Related Ulcers

Skin is the largest organ of the body, both in terms of outer surface area and in mass. Skin is amassed of three layers: dermis, epidermis, and hypodermis (Fig. 26.1). These layers protect the body from external mechanical and thermal injury like burns and wounds. Epidermis is a thin and outermost layer of the skin which acts as a protection barrier against external factors. Dermis lies below the epidermis, covers major component of skin, and supports it structurally and nutritionally. Appendages such as hairs and glands are derived from the epidermis, but they project deep into the dermal layer. The epidermal layer is 0.1–0.2 mm in depth and consists of a self-renewing population of keratinocytes programmed to be replaced continuously for a lifetime from basal cells located in the basement membrane (BM) zone [1].

Skin acts as a barrier to the external environment and protects the body from harmful chemicals and microbial invasion, maintains moisture balance, and ensures thermoregulation [2]. Approximately 1 in 100 individuals suffers from wounds requiring medical intervention. The skin barrier function is achieved by layers of differentiated keratinocytes fusing together in a tightly integrated epidermal sheet. This natural barrier might get wounded as a result of trauma, including surgery, or

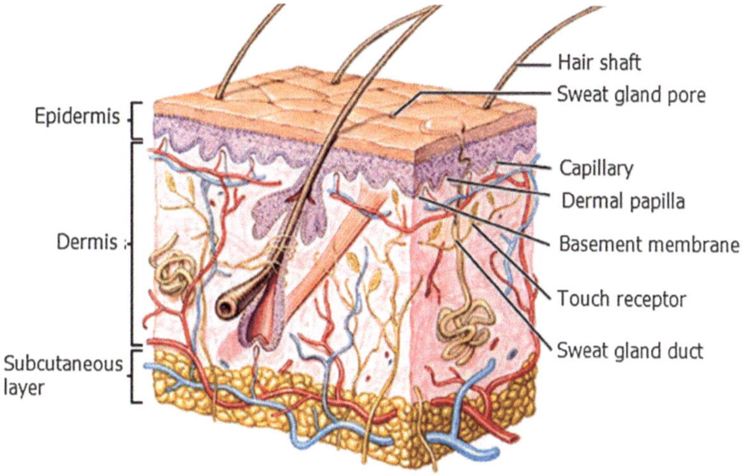

Fig. 26.1 Schematic representation of human skin

it can become damaged secondary to an underlying pathology, such as reduced venous or arterial circulation, thus causing wounds. Superficial wounds are capable of self-healing and cause no impairment; however, full-thickness wounds and third-degree burns (extending to deeper layers of the skin) are a significant threat to the patient because of fluid and protein loss and the potential for bacterial invasion. For the development of effective therapies, it is important to understand the pathophysiology of acute, chronic, and nonhealing wounds. Acute wound healing involves three phases: inflammation, proliferation (new tissue formation), and remodeling/contraction. Disruption in normal process leads to chronic nonhealing wounds [14]. In chronic wounds phenotypes of some cells are altered [19], as keratinocytes are not able to migrate on wound edges properly and thus cannot be closed completely [20, 21]. This could be because of the downregulation of cell signaling pathways that encourage cell migration after tissue injury.

Tissue injury elicits immediate activation of intrinsic and extrinsic coagulation pathways which is associated with inflammation. Inflammation is an important event of injury which contributes in tissue repair and regeneration in wound healing. During wound healing process, local vasodilatation, blood, extravasation fluid in extravascular cavity, and blocking drainage of lymphatic fluid can produce fundamental symptoms of inflammation, including redness, swelling, and heat. The first phase of wound healing inflammation to injury provides necessary framework to forming new protection barrier. In proliferative phase major events involve introduction of permeability barrier (reepithelialization), provide proper blood supply (angiogenesis), and support the injured dermal tissues (fibroplasias) which leads to fibrin clot fabrication. The third one which is remodeling phase comprises matrix deposition, and it occurs throughout the wound healing process. In this phase fibrin clot in proliferation phase is replaced by granulation tissues which is ample of type III

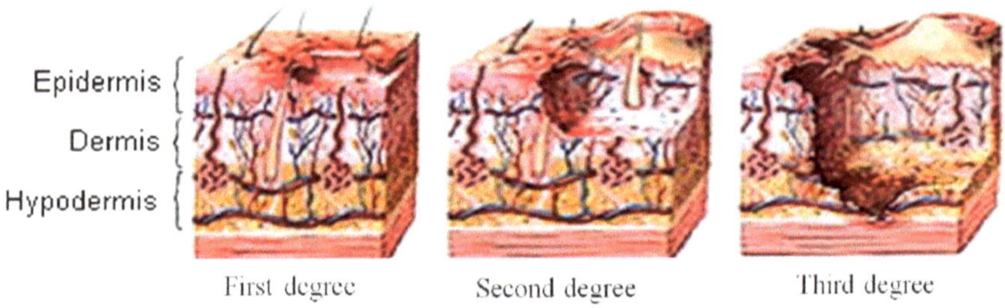

Fig. 26.2 Different types of skin wounds

collagen and matured blood vessels later on substituted by collagenous scar which is predominant of type I collagen and less mature blood vessel. As in the case of diabetes, burns, infection, poor nutrition, or other pathological disease, disruption of normal healing process or excessive fibrosis (hypertrophic scars and keloids) may lead to altered structure and loss of function. Acute wounds follow all phases of normal wound healing, but chronic wounds do not comply with orderly process. All wounds may extend to different skin layers and could be responsible for first-degree, second-degree, or third-degree wound (Fig. 26.2).

In chronic wounds like diabetic ulcers, fibroblasts depicted a decreased response to TGF-β1 and other GFs as well as diminished expression of the TGF-β receptor and vitiated signal transduction [22]. Patients suffering from diabetis show slow wound healing because of impairment in immune response with lesser enrolment of inflammatory cells at damaged site due to reduced chemotactic effects, leading to the slowing down of healing and higher risk of bacterial infection [23–25]. When an inflammatory response is instituted for longer period of time, it shifts to an aggravation of inflammation and proteolysis [26]. Hyperglycemia for prolonged time responsible for glycation of proteins and commotion of cell responses, which extends to improper fibrosis process and tissue repair [27–29]. Chronic diabetic ulcers may involve some specific processes like prevention of quick entrance and accumulation of leucocytes in the ulcers, which, consequently, fail to accomplish normal healing. Another reason of chronic ulcer formation is the concentration of the high-molecular-weight hyaluronic acid in the pericellular matrix of fibroblast from diabetic patients.

Burn is a type of acute wound and progressively occurring injury to the skin precipitated due to thermal shock and leads to invasion of microorganism and antigen and generation of free radicals. Pathophysiology of burns starts with increasing the body surface temperature, as higher temperature produces local responses, and dissipation of heat involves vasodilatation and proceeds many cascade events. As the temperature increases it causes increase in local inflammatory mediator's forms leading to systemic organ injury. Free radicals are also involved in cell damage in case of thermal shock in connection with tissue ischemia and hypoxia.

Skin is the largest organ of the body and serves as a protective barrier against foreign materials. It might be affected by various injuries which can cause loss of skin integrity (Fig. 26.3). Acute wounds, including surgical incisions, generally

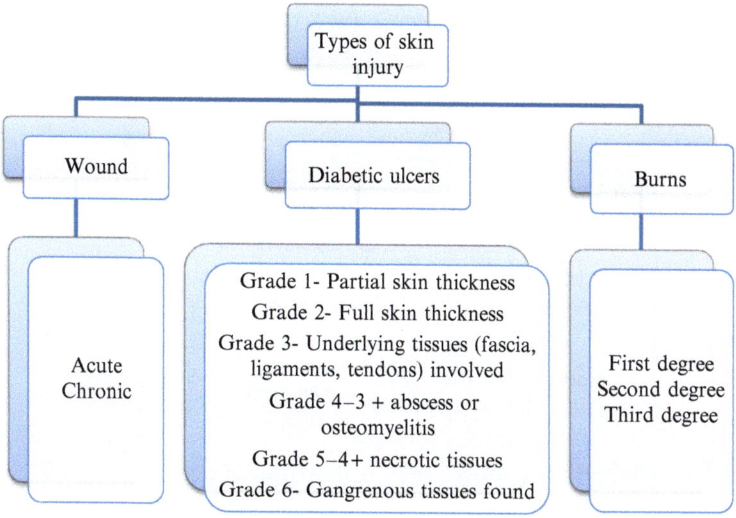

Fig. 26.3 Different types of skin injuries

communicate through the phases of inflammation, proliferation, and remodeling. But in case of chronic wounds, it exhibits delayed healing 12 weeks after the initial shock often due to prolonged pathological infection and inflammation [30]. Diabetic foot ulcers have been classified in different categories and grade on the basis of size, area, and depth of ulcers. Wagner-Meggitt classification is a practical idea of possibility of amputation and prediction for healing [31–33].

Burns are classified based on depth and the surface area of the skin that is affected. First-degree burns involve only the epidermal layer but result in pain and erythema and generally heal within several days without any scarring. Second-degree burns involve the intact epidermis and some portion of dermis. It is further classified in superficial partial-thickness or deep partial-thickness burns, based on the depth of injury to the dermis. Superficial partial-thickness burns usually heal within 2 weeks with minimal scarring, while deep partial-thickness burns may require up to 3 weeks for healing with significant scarring. Full-thickness or third-degree burns involve the entire epidermis and dermis with destruction of nerves and maximal scarring.

26.3 Drug Delivery to Wound, Burns, and Diabetes-Related Ulcers

26.3.1 Dressings

For effective treatment of skin condition, the primary choice is wound dressing which could be either medicated or non-medicated. These dressings should posses some ideal properties like removal of excess exudates, maintaining moist environment,

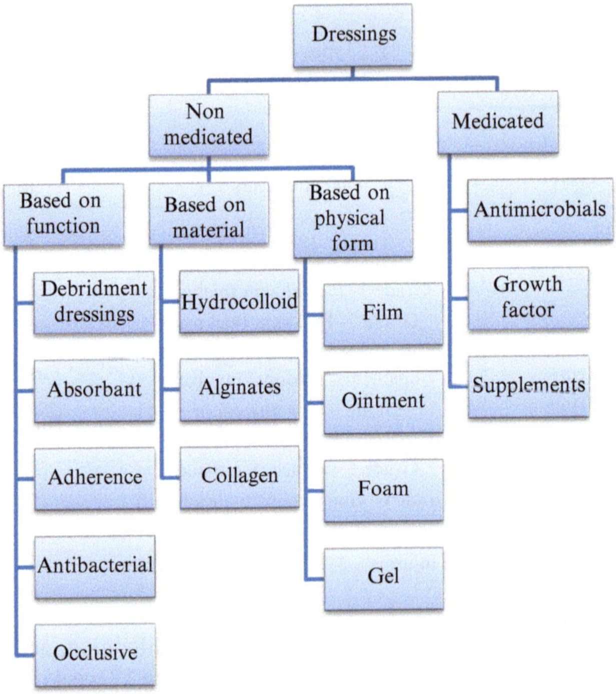

Fig. 26.4 Dressings for wound healing

no trauma on removal, protecting against bacteria, no allergies, providing thermal insulation, relieving pain, and cost-effective. Dressings can be classified in different ways based on their function (debridement, antibacterial, occlusive, absorbent, adherence) [34], type of material (e.g., hydrocolloid, alginate, collagen) [7], and physical form (ointment, film, foam, gel) [8]. Dressings are further classified into primary, secondary, and island dressings [35] and also into medicated and non-medicated dressings (Fig. 26.4). Non-medicated dressings were initially prepared from traditional treatment of plant source, animal fat, and honey and recently from synthetic scaffold. To fasten the healing process and avoid infection at wound site, dressing has been medicated with antimicrobials, GFs, and supplements. The choice of dressing is usually determined by the type and pathological condition of wound. The aspects that need consideration are wound location, depth, amount of eschar or slough present, amount of exudate, condition of the wound margins, presence of infection, need for adhesiveness, and conformability of the dressing. Periodic evaluation of wound is needed since wound environment changes constantly during treatment [36].

26.3.1.1 Non-medicated Dressing

Non-medicated dressings have essential characteristics to hold and make moist surroundings around the wound to facilitate wound healing. These dressings are mainly classified on the basis of material from which they are developed such as hydrocolloids, alginates, and hydrogels and generally occur in the form of gels, thin films, and foam sheets.

Hydrocolloid Dressings

Hydrocolloid term describes the products obtained from colloidal (gel-forming agents) materials combined with other materials such as elastomers and adhesives. Hydrocolloid dressings are biodegradable and adhere to the skin, so external taping is not required. They mostly occur in the form of thin films and sheets or in combination with other material such as alginates. These are helpful in wound healing because they adhere to both dry and moist wound sites [37]. These dressings can be used for light to moderately exuding wounds such as minor burns, pressure sores, and traumatic injuries and also in leg ulcers [38]. A randomized trial showed that the hydrocolloid attains faster healing and is a less painful dressing [39]. These dressings are not useful for infected wounds that require a certain amount of oxygen to heal rapidly. Examples of hydrocolloid dressings include Granuflex™ and AQUACEL® (ConvaTec, Hounslow, UK), Comfeel™ (Coloplast, Peterborough, UK), and Tegasorb™ (3M Health Care, Loughborough, UK).

Alginate Dressings

Alginate dressings are developed from the calcium and sodium salts of alginic acid, a polysaccharide-containing mannuronic and guluronic acid units. These occur either in the form of freeze-dried porous sheets (foams) or as flexible fibers. These dressings form a protective film of gel over wounds by exchanging calcium ions present in alginates with those present in exudates and blood, resulting in the formation of polymer constructs [40]. Calcium alginate dressings are ideal material as scaffold for tissue engineering [41, 42]. These dressings are useful for moderate to heavily exuding wounds, but it cannot be used for dry wounds and those covered with necrotic tissue because it requires moisture to function properly. Some examples of alginate dressings are Sorbsan™ (Maersk, Suffolk, UK), Kaltostat™ (ConvaTec), and Tegagen™ (3M Health Care). Comfeel Plus™ is a hydrocolloid/alginate combination dressing.

Hydrogel Dressings

Hydrogels are insoluble, swellable hydrophilic materials constituted from synthetic polymers such as poly(methacrylates) and polyvinylpyrrolidone. These can be given either as an amorphous gel or as an elastic, solid sheet or film. These dressings contain significant amounts of water (70–90 %) and thus cannot take up much exudate and are suitable only for light to moderately exuding wounds. Additionally, they clean dry, sloughy, or necrotic wounds by rehydrating dead tissues and enhance self-digesting debridement [43]. Examples of hydrogel-based dressings are NU-GEL® (Johnson & Johnson, Ascot, UK), Purilon™ (Coloplast), Opsite™ (Smith & Nephew, Hull, UK), Cutifilm™ (B.D.F. Medical, Milton Keynes, UK), BIOCLUSIVE® (Johnson & Johnson), and Tegaderm™ (3M Health Care).

26.3.1.2 Medicated Dressing

Pharmaceutical agents can be incorporated into dressings to overcome the disadvantages of topical agents as already discussed in the previous section and provide better and faster healing than non-medicated dressings [44]. Drugs incorporated in dressings could act as cleansing or debriding agent for removing necrotic tissue, antimicrobials which cures infection or growth agents (factors) to assist in tissue regeneration.

Antimicrobials

Antimicrobials and antibiotics are basically applied for the purpose of preventing infection at the affected site, e.g., for treating infections in case of diabetic foot ulcers [45, 46]. These can also be used in surgical and accident wounds where chances of infection are very high due to the higher trauma. Antibiotics and antimicrobials like minocycline-incorporated dressings have shown better healing in severe burn wounds [47].

Some other Antimicrobial incorporated dressings that delivered drugs for wound healing include tetracycline delivery via freeze-dried fibrin discs [48] and ofloxacin delivery via lactic acid-based dressings to inhibit the *Staphylococcus aureus* and *P. aeruginosa* in split-thickness wounds [49]. Antimicrobial-incorporated silicone gel sheets treated superficial burns by promoting epithelialization [50]. Topical delivery of antibiotics to the wounds through dressings can provide tissue compatibility, more bacterial sensitivity, and less interference with wound healing [51].

In systemic delivery there is a chance of drug toxicity like accumulation and organ toxicity of aminoglycosides in the kidneys and ears [52, 53]. So the use of dressing reduces the dose of antibiotics which further leads to lowering systemic toxicity. Along with its local delivery from dressings can also overwhelm the ineffective systemic antibiotic therapy problem resulting from poor blood circulation at the appendages in diabetic foot ulcers. Cutisorb™ is a dialkyl carbamoyl chloride antibiotic-loaded dressing used for delivery to wounds. A wide variety of semiocclusive dressing formats, such as foams (Contreet®, Coloplast), hydrocolloids (Urgotul SSD, Urgo), alginates (Silvercel®, Johnson & Johnson), and Hydrofiber® (Aquacel, ConvaTec), are commercially available. For instance, Acticoat® is a 3-ply gauze dressing made of an absorbent rayon polyester core, with upper and lower layers of nanocrystalline silver-coated high-density polyethylene mesh [54].

Growth Factors

Though antibacterial agents are helpful for the treatment of infections and wound healing, they do not take an active physiological part in the wound healing process. Growth factors (GFs) promotes wound healing and involves in cell division, migration, differentiation, protein expression, and enzyme production. These growth factors show their healing activity by the angiogenesis induction and cellular proliferation, which affects both the production and the degradation of the extracellular matrix (ECM), and is also important in cell inflammation and fibroblast activity [55]. Hence, all phases such as inflammatory, proliferation, and migratory of wound healing are affected by GFs which involve in normal wound healing [56]. A number of GFs have been reported which are helpful in wound healing including epidermal growth factor (EGF), platelet-derived growth factor (PDGF), fibroblast growth factor (FGF), transforming growth factor (TGF-β1), insulin-like growth factor (IGF-1), human growth hormone, and granulocyte-macrophage colony-stimulating factor (GM-CSF) [12, 57].

Growth factors (GFs) were delivered through medicated dressings which include hydrogel dressings for delivering TGF-β1 [58, 59], collagen film dressings for delivering PDGF [60], alginate dressings for delivering human growth hormone [61] and endothelial growth factor [62], and polyurethane and collagen film dressings for delivering EGF [63]. Collagen-hyaluronic acid matrix, containing tobramycin, basic FGF, and PDGF, significantly increased wound healing in comparison to dressings which comprise only the antibiotic [64].

Supplements

In wound healing process another groups of active compound which can be used are vitamins and mineral supplements [65], including vitamins A, C, and E as well as zinc and copper. Vitamin A is helpful in epithelial cell differentiation, collagen synthesis, and bone tissue development [66] and also involved in facilitating normal physiological wound healing [67]. Vitamin C plays an important role in the synthesis of collagen and other components of skin and other connective tissues. In addition, vitamin C helps in ameliorating immune system especially during infection. Vitamin A and C incorporated collagen sheet dressings were used in burn wound treatment. Vitamin E has antioxidant and anti-inflammatory activity; along with this, it also promotes angiogenesis and reduces scarring, resulting in accelerating wound healing [68]. Vitamin E has also been used in hypertrophic and keloid scar treatment in combination with silicone gel sheets. Zinc-containing topical formulation reduced healing times in deep second-degree burn wounds significantly [69] and also enhanced reepithelialization in diabetic foot ulcers. This healing time reduction was further improved when basic FGF and EGF were used in combination with the zinc preparation. Copper-based dressings were shown to improve wound contraction and wound healing at a faster rate than zinc-based dressings.

Table 26.1 Ideal properties of topical antimicrobial for wound healing [126]

Specific antibacterial activity
Rapid bactericidal activity
Sustained action at infected site
Activity in the presence of body fluids and proteins in wound exudate
Should not produce resistant
No systemic side effects
No allergic reactions
Acceptable cosmetic and aesthetic qualities
Cost-effective

26.3.2 Antibiotics

The moist, warm, and nutritious environment provided by the exposed surface after wounding along with compromised immune function and inadequate wound perfusion creates an ideal milieu for the growth of bacteria and microbial colonization. Clinical studies have revealed that 86 % of the ulcers contain more than one bacterial species though without any clinical signs of infection [70]. Most predominant species found in infected wounds include *Staphylococcus aureus*, *Pseudomonas aeruginosa*, *Escherichia coli*, *Enterobacter cloacae*, *Klebsiella* species, *Streptococcus* species, *Enterococcus* species, and *Proteus* species [71]. If a wound becomes infected, the inflammatory phase during healing becomes chronic, suppressing the next phase, that is, the regenerative phase. Moreover, the enzymes and toxins expressed by the colonizing microorganisms affect the structural integrity of healing tissue along with the surrounding skin as well as the wound dressing [72]. In such a situation, wound healing is compromised, the patient suffers increased trauma, treatment cost rises, and general wound management practices become more complicated and critical [73]. Thus, it is critical for proper healing to prevent infections in the healing wound bed.

To suppress the infection during wound healing, antibiotics are delivered systemically (e.g., silver sulfadiazine, fusidic acid, and metronidazole) or topically (such as chlorhexidine and povidone-iodine). Systemic delivery of antibiotics entails poor penetration into ischemic and necrotic wounded tissue and can cause systemic toxicity with associated renal and liver complications due to larger amount of drugs required to achieve therapeutic concentrations at the infection site and needs longer duration of treatment [74, 75]. Alternatively, local delivery of antibiotics by either topical administration or using a delivery device may enable the maintenance of a high local antibiotic concentration for an extended duration of time without causing systemic toxicity. These topical antimicrobials should have some ideal properties, which are described in Table 26.1. Some wounds are infected because they have pussy secretions or some of the fundamental expressions of inflammation that have determined the host response to tissue damage caused by infective microorganisms [76]. In most of the wounds, bacteria remain inside the wound cavity where they induce chronic inflammation that retards healing and also produce resistant to antimicrobial therapy. Some wound infections can heal without antimicrobial therapy, but in case of immunocompromised patients, it can progress to involve deeper tissues and leads to systemic pathogenesis. These inflammatory processes are mostly mediated by toxins and metabolic wastes produced by microorganisms, but it also depends on the sensitivity of host towards infection [77].

A variety of antibiotic products are available for treating chronic wounds [10]. Antibiotics are chemicals produced either naturally (by a microorganism) or synthetically that in dilute solution inhibit or kill other microorganisms. Topical antimicrobial therapy is not commonly used for uninfected wound healing; however, it can be used for critically infected wounds which may be either chronic or acute. Antimicrobials have also been reported for burn wounds in which blood vessels to the skin are frequently ruined and help in reducing sepsis as well as infection [78]. Topical agents can also be used prior to skin grafting for reducing bacterial burden or for reducing odor linked with nonhealing and necrotic wounds.

26.3.3 Gene Therapy and Cytokines

Full-thickness wound and thermal injury are a major trauma characterized by high cardiac output, increased oxygen consumption, and protein and fat wasting. Increased hypermetabolic state compromises the immune system and decreases wound healing. Moreover, it causes tissue damage by membrane destabilization and energy depletion at the cellular level, resulting in tissue necrosis. Skin trauma could be prevented by blocking immediate triggering of the inflammatory cascades that result in prolonged metabolic imbalances. Gene therapy and cytokines have been reported as a predicting approach for the treatment of skin diseases, wound healing [79, 80], and diabetic pathologies [81, 82].

Therapeutic approach to promote wound healing would be to block the immediate activation of the inflammatory reactions, and second approach for therapy would be to enhance wound healing factors. Another approach for wound healing is the effect of "positive" growth hormones, and cytokines may be raised and that of "negative" factors suppressed through molecular or genetic manipulation [11]. Cytokines such as TGF-β and HB-EGF (heparin-binding epidermal growth factor-like growth factor) used for the local treatment of wounds was found to be ineffective due to lack of enzymes and proteases locally present in the wounds and also because of lack of adequate receptors [83]. Gene delivery to the wounds is important phenomenon for overexpressing wound healing-promoting factors for a definite time period. Local application of adenoviral constructs has shown wound healing, especially in processes of reepithelialization and vascularization.

Gene therapy is rising as an effective therapeutic approach to promote wound healing including after thermal injury [84]. Gene therapy to the skin is dependent, however, on a number of factors and has the following aims. Firstly, suitable therapeutic gene should be selected, and it should not be restricted by the size of the gene. Secondly, therapeutic genes should be expressed at application site for desirable time. Eventually, cells should take up the therapeutic gene through particular delivery vehicle [85].

Gene therapy has been classified into two categories of viral and nonviral gene delivery. Viruses that include retroviruses, adenoviruses, and adeno-associated viruses are the most commonly used natural vehicles for gene delivery. Some nonviral vector such as plasmid DNA have also been used because they have potential to target the genes at the affected site without recombination with virus so due to this site injury can also be prevented which is due to repeated exposure of virus. Direct plasmid application, lipofection, and receptor-mediated delivery vectors are mostly used nonviral vectors for gene delivery. Electroporation and laborious microinjection of DNA are also other nonviral techniques [86].

Viral vectors [84], naked nonencapsulated DNA or plasmid DNA constructs alone, without viral genes and nonviral liposomal cDNA genes as stable complexes have shown faster wound healing [83, 87]. Gene has delivered through subcutaneous injection with the use of microseeding, microfabricated needles, and puncture-mediated DNA transfer at the wound margin [88–90] and nucleic acid vaccine to promote wound healing [90].

26.3.4 Growth Factors

For wound healing, tissue engineering approach could be adopted to create an environment to mimic the natural "healing cascade" for induction of regeneration and to hasten tissue regrowth capability that is generally termed regenerative medicine. In wound healing cascades, GFs are known to play a primal role in information transfer between a wide range of cells and their ECM. GFs regulate most aspect of fibroblast, endothelial cell, and keratinocyte function, including cell proliferation, migration, and synthesis of ECM in wound healing process [91]. GFs and ECM molecules could be incorporated into biomaterials for regulating cellular proliferation, migration, differentiation, adhesion, and gene expression [92].

Various GFs have shown to accelerate cell proliferation in vitro and to promote wound healing in animal models. TGF-β might be helpful in wound repair as it promotes cell proliferation, differentiation, and matrix production. TGF-β has shown accelerated cutaneous wound repair in animal models when administered either topically or systemically [93, 94]. Similarly, PDGF promoted reepithelialization and revascularization in diabetic wounds [95, 96]. PDGF is released from cells which involved in the initiation of wound healing such as platelets, macrophages, endothelial cells, and fibroblasts [97]. PDGF helps in wound healing by inducing granulation tissue formation [95] and collagen gene expression [98].

HB-EGF was applied topically through slow-release cholesterol lecithin pellets shown to accelerate wound healing in murine burn models [99]. EGF alleviates epidermal cell regeneration and helps in dermal wound healing through stimulation of proliferation and migration of keratinocytes. It also promotes the formation of granulation tissue and induces fibroblast motility [100]. Endothelial cell growth factor (ECGF) is responsible for promoting angiogenesis in rats when applied via controlled release formulation [101], and this may have important significance for wound healing.

Basic fibroblast growth factor (bFGF) has been used in impaired wound healing which is angiogenic in several animals [102] and accelerates wound repair in full-thickness skin wounds [103]. The topical application of GFs (EGF, bFGF, HB-EGF, TGF-β, PDGF) is being a useful treatment [104, 105], and combinations of these factors are being the most efficacious for wound healing [106, 107].

26.3.5 Stem Cells

Wound healing involves a well-organized complex biological and molecular event of cell migration and proliferation; along with this, it also requires ECM deposition, angiogenesis, and remodeling [108]. Stem cells, due to their differentiation ability into various tissue types and self-renewal capacity, may create such a complex structure. They can be classified on the basis of the source, and thus, stem cells can be considered as either embryonic or adult in nature [13]. Stem cells have been isolated from a variety of sources, such as bone marrow, peripheral blood, umbilical cord blood, adipose tissue, and skin/hair follicles (epithelial cells), to regulate healing response of acute and chronic wounds.

26.3.5.1 Bone Marrow-Derived Mesenchymal Stem Cells (MSCs)

Mesenchymal stem cells were first isolated from bone marrow, and these are generally defined as plastic-adherent cells with a fibroblast-like morphology and have in vitro multipotent differentiation capacity [109]. MSC-based therapies target wound repair by stimulating cellular responses to injury and promote regeneration rather than scar formation. Murine excisional wound treated with bone marrow stem cells in diabetic mouse model by topical delivery of single treatment with 7.5×10^5 cells showed enhanced epithelialization, granulation tissue formation, and angiogenesis [110]. MSC treatment affects all phases of wound healing including inflammation, epithelialization, granulation tissue formation, and tissue remodeling [111].

26.3.5.2 Umbilical Cord-Derived Stem Cells

Stem cells from umbilical cord blood were able to differentiate into epithelial cells under in vitro conditions, so it could also be a good source for large skin defect treatment [112]. It was found that the amniotic membrane of the umbilical cord to be an extremely rich source of stem cells for burn healing and the isolated cells called as cord lining stem cells was divided into cord lining epithelial cells and cord lining mesenchymal cells. These cells have been employed to treat partial-thickness and full-thickness burns as well as chronic diabetic wounds [113].

26.3.5.3 Adipose-Derived Stem Cells

Adipose-derived stem cells encouraged human dermal fibroblast proliferation through direct cell-to-cell contact and by secretory-induced paracrine activation, which is important in wound healing by enhancing reepithelialization and regeneration [114].

26.3.5.4 Epithelial Stem Cells

Epithelial stem cells are a type of adult somatic stem cells as they are dormant but self-renew and differentiate into at least one type of daughter cells [115]. When these cells activate, it enrolls a number of proliferating cells and plays an important function in wound healing. Stem cells extracted from the human bulge region demonstrate hair follicle differentiation and form epidermal and sebaceous cells in vitro [116]. The role of Epithelial stem cell is not only limited for epidermis regeneration in wound healing, but also in the center of wounds, these cells can also assume morphogenetic plasticity, and along with wound dermis, it can employ in an embryonic-like process of hair follicle neogenesis [115].

26.3.6 Other Drugs

26.3.6.1 Antioxidants

Reactive oxygen species (ROS) at low levels are required for the defense against infecting pathogens and also essential mediators of intracellular signaling. However, excessive amounts of ROS are deleterious due to their high reactivity [117]. Free radicals or ROS are generated in wound and burn trauma, which are responsible for promoting injury; most deleterious ROS are superoxide anion and hydrogen peroxide that overcome the scavenging capacity of endogenous enzymes.

Antioxidants which have been used to antagonize ROS include (1) vitamin E, the lipid-soluble chain-breaking antioxidant; (2) vitamin C, which, in addition to its antioxidant effect, helps to recycle vitamin E; (3) zinc sulfate, which has acute and chronic antioxidant effects; (4) allopurinol, the well-known xanthine oxidase (XO) inhibitor, which might be beneficial for burn healing; (5) melatonin, the pineal gland product that has potent antioxidant effects through its direct scavenging effect and stimulation of antioxidant enzymes; and (6) *N*-acetylcysteine, which is rapidly metabolized to cysteine, which is a direct precursor in the synthesis of intracellular glutathione (GSH), the natural antioxidant. The use of these antioxidants in the treatment of burns and wounds may have improving effects [118]. Polyphenolic drugs such as quercetin-incorporated collagen matrix [119] and curcumin have shown wound healing by modulating collagen and decreasing ROSs [120].

26.3.6.2 Nutritions

For wound healing and wound decontamination processes, metal ion-ligand interactions are essential, and these metal elements are required at desirable concentration along with other nutritions for repair and regeneration. When wound occurs, some trace metal elements are essential to support the healing processes, and deficiency of fundamental supply of these elements may lead to slow or nonhealing. A more direct route to producing nutritional components at their optimum levels is that of total parenteral nutrition (TPN) [14].

Essential elements of TPN are calcium, chromium, copper, fluorine, iodine, iron, manganese, molybdenum, phosphorus, selenium, and zinc. Other important components of TPN fluids which, with the exception of specially selected drugs, are required by healing wounds are (1) energy sources such as amino acids, glucose, soya protein, phospholipids, and glycerol, (2) nitrogen source such as amino acids, (3) minerals such as electrolytes, (4) trace elements, (5) impurities, (6) vitamins, and (7) water and drugs such as anticoagulants [14].

26.3.6.3 Antihistamines and Gabapentin

Itching is associated with wound healing, and it may be defined as the sensation that induces the urge to scratch [121]. The H_1 receptor antagonists such as diphenhydramine, hydroxyzine, and cetirizine and selective H_2 receptor antagonist such as cimetidine and gabapentin are the most commonly used agents for treatment of itching associated with wound healing. Although, gabapentin is basically an antiepileptic drug used in neuralgias, it has also been used in the treatment of postburn pruritus [15].

26.3.6.4 Phenytoin

Phenytoin has dose-dependent hyperplasia adverse effect due to collagen proliferation, which could be useful for wound healing. Phenytoin has been used in the treatment of superficial burn wound [16], diabetic ulcer [122], and variety of chronic nonhealing wound ulcers [123].

26.3.6.5 Connexin-Targeted Antisense Gel

The gap junction protein Cx43, dominantly found in epidermis, plays an important role in acute wound healing at different stages. The Cx43 is presented in the wound edge keratinocytes of human chronic wounds [124]. Unusual expression of Cx43 in wound-edge keratinocytes was shown to cause improper wound healing in diabetic rats, and by using antisense gel, its expression was targeted to restored normal healing [125]. Antisense gel accelerated the downregulation of Cx43, and it has the effect of speeding the migration of keratinocytes and fibroblasts, which results in closing the wound and forming the granulation tissue [17].

26.3.6.6 Traditional Drugs

Many Ayurvedic herbal plants have a very important role in the process of wound healing. Traditional drugs are more potent wound healers because they promote the repair mechanisms in the natural way. Various medicinal plants have been used for wound healing [18].

26.4 Conclusions

Skin is the largest organ in the body which might be associated with a lot of injuries like wound, burns, and diabetic ulcers. Various drugs in dressings as well as topically without dressings for wound healing have been discussed here. Dressings were used as matrices for skin replacement and delivery of therapeutic agents like antibiotics, vitamins, and mineral supplements for the management of skin injuries. Apart from dressings different classes of drugs were also delivered topically to the skin. Antibiotics/antimicrobials were delivered topically to reduce bioburden on wounds and prevented from infection. Many cytokine proteins and genes were delivered for promoting the expression of wound healing GFs, and a number of GFs were also applied directly to the affected site. Stem cells were used for improving skin repair and regeneration, and antioxidant enhanced wound healing by scavenging free radicals. Other drugs like antihistamines for preventing itch, phenytoin for accelerating collagen formation, antisense gel for the downregulation of connexin gap junction proteins, and medicinal drugs for wound healing were also considered in this chapter.

References

1. Fuchs E (2007) Scratching the surface of skin development. Nature 445:834–842
2. MacNeil S (2008) Biomaterials for tissue engineering of skin. Mater Today 11:26–35
3. Ehrenreich M, Ruszczak Z (2006) Update on tissue-engineered biological dressings. Tissue Eng 12:2407–2424
4. Schulz JT, Tompkins RG, Burke JF (2000) Artificial skin. Annu Rev Med 51:231–244
5. Williams WG, Phillips LG (1996) Pathophysiology of the burn wound. In: Hernndon DN (ed) Total burn care. Harcourt Health Sciences, Philadelphia, PA, pp 63–69
6. Boyce ST (1996) Cultured skin substitutes: a review. Tissue Eng 2:255–266
7. Queen D, Orsted H, Sanada H, Sussman G (2004) A dressing history. Int Wound J 1:59–77
8. Falabella A (2006) Debridement and wound bed preparation. Dermatol Ther 19:317–325
9. Boateng J, Matthews K, Stevens H, Eccleston G (2008) Wound healing dressings and drug delivery systems: a review. J Pharm Sci 97:2892–2923

10. Lipsky BA, Hoey C (2009) Topical antimicrobial therapy for treating chronic wounds. Clin Infect Dis 49:1541–1549
11. Atiyeh B, Gunn S, Hayek S (2005) State of the art in burn treatment. World J Surg 29:131–148
12. Steenfos H (1994) Growth factors and wound healing. Scand J Plast Reconstr Surg 28:95–105
13. Peng L, Fung K, Leung P, Gao J (2011) Genetically manipulated adult stem cells for wound healing. Drug Discov Today 16:957–966
14. Williams D (1996) Wounds: decontamination and healing. Coord Chem Rev 151:161–174
15. Zachariah J, Rao A, Prabha R, Gupta A, Paul M, Lamba S (2012) Post burn pruritus-a review of current treatment options. Burns 38:621–629
16. Meena K, Mohan A, Sharath B, Somayaji S, Bairy K (2011) Effect of topical phenytoin on burn wound healing in rats Indian. J Exp Biol 49:56–59
17. Becker D, Thrasivoulou C, Phillips A (2012) Connexins in wound healing; perspectives in diabetic patients. Biochim Biophys Acta 1818:2068–2075
18. Kasarla R, Elumalai A, Eswaraiah MC, Ravi P, Naresh V (2012) An annual review on wound-healing medicinal plants. J Nat Prod Plant Resour 2:182–185
19. Loot M, Kenter S, Au F (2002) Fibroblasts derived from chronic diabetic ulcers differ in their response to stimulation with EGF, IGF-I, bFGF and PDGF-AB compared to controls. Eur J Cell Biol 81:153–160
20. Stojadinovic O, Brem H, Vouthounis C (2005) Molecular pathogenesis of chronic wounds: the role of beta-catenin and c-myc in the inhibition of epithelialization and wound healing. Am J Pathol 167:59–69
21. Waikel R, Kawachi Y, Waikel P, Wang X, Roop D (2001) Deregulated expression of c-Myc depletes epidermal stem cells. Nat Genet 28:165–168
22. Kim B, Kim H, Park S (2003) Fibroblasts from chronic wounds show altered TGF-beta–signaling and decreased TGF-beta type II receptor expression. J Cell Physiol 195:331–336
23. Goova M, Kislinger T, Qu W, Lu Y, Bucciarelli L (2001) Blockade of receptor for advanced glycation end products restores effective wound healing in diabetic mice. Am J Pathol 159:513–515
24. Bennett S, Grifflth G, Schor A, Leese G, Schor S (2003) Growth factors in the treatment of diabetic foot ulcers. Br J Surg 90:133–146
25. Mast B, Schultz G (1996) Interactions of cytokines, growth factors, and proteases in acute and chronic wounds. Wound Repair Regen 4:411–420
26. Lobmann R, Ambrosch A, Schultz G, Waldmann K, Schiweek S, Lehnert H (2002) Expression of matrix-metalloproteinases and their inhibitors in the wounds of diabetic and non-diabetic patients. Diabetologia 45:1011–1018
27. Brigstock D (2003) The CCN family: a new stimulus package. J Endocrinol 178:169–175
28. Wetzler C, Kampfer H, Stallmeyer B, Pfeilschifte J, Frank S (2000) Large and sustained induction of chemokines during impaired wound healing in the genetically diabetic mouse: prolonged persistence of neutrophils and macrophages during the late phase of repair. J Invest Dermatol 115:245–253
29. Portero O, Pamplona R, Bellmunt M, Ruiz M, Prat J, Salvayre R (2002) Advanced glycation end product precursors impair epidermal growth factor receptor signalling. Diabetes 51:1535–1542
30. Alistair Y (2011) The physiology of wound healing. Surgery 29:475–479
31. Treece K, MacFarlane R, Pound N, Game F, Jeffcoate W (2004) Validation of a system of foot ulcer classification in diabetes mellitus. Diabet Med 21:987–991
32. Lawrence S, Wraight P, Campbell D, Colman P (2004) Assessment and management of inpatients with acute diabetes related foot complications. Intern Med J 34:229–233
33. Margolis K, Taylor L, Hoffstad O, Berline J (2003) Diabetic neuropathic foot ulcer: the association of wound size, wound duration and would grade. Diabetes Care 25:1835–1839
34. Purner S, Babu M (2000) Collagen based dressings—a review. Burns 26:54–62
35. Rijswijk L (2006) Ingredient-based wound dressing classification: a paradigm shift that is passé and in need of replacement. J Wound Care 15:11–14
36. Millington J, Norris T (2000) Effective treatment strategies for diabetic foot wounds. J Fam Pract 49:40–48

37. Heenan A (1998) Hydrocolloids: frequently asked questions. World Wide Wounds 1:1–17
38. Koksal C, Bozkurt A (2003) Combination of hydrocolloid dressing and medical compression stocking versus Unna's boot for the treatment of venous leg ulcers. Swiss Med Wkly 133:364–368
39. Cadier M, Clarke J (1996) Dermasorb versus Jelonet in patients with burns skin graft donor sites. J Burn Care Rehabil 17:246–251
40. Thomas S (2000) Alginate dressings in surgery and wound management—Part 1. J Wound Care 9:56–60
41. Kuo C, Ma P (2001) Ionically crosslinked alginate hydrogels as scaffolds for tissue engineering: part 1. Structure, gelation rate and mechanical properties. Biomaterials 22:511–521
42. Wang L, Shelton R, Cooper P, Lawson M, Triffitt J (2003) Evaluation of sodium alginate for bone marrow cell tissue engineering. Biomaterials 24:3475–3481
43. Morgan D (1999) Wound management products in the drug tariff. Pharm J 263:820–825
44. Heenan A (1998) Dressings on the drug tariff. Worldwide Wounds, 4th edn. Surgical Material Testing Laboratory, Wales
45. O'Meara S, Callum N, Majid M, Sheldon T (2000) Systematic reviews of wound care management (3) antimicrobial agents for chronic wounds (4) diabetic foot ulceration. Health Technol Assess 4:1–237
46. Nelson E, O'Meara S, Craig D, Iglesias C, Golder S, Dalton J et al (2006) A series of systematic reviews to inform a decision analysis for sampling treating infected diabetic foot ulcers. Health Technol Assess 10:1–221
47. Aoyagi S, Onishi H, Machida Y (2007) Novel chitosan wound dressing loaded with minocycline for the treatment of severe burn wounds. Int J Pharm 330:138–145
48. Kumar T, Bai M, Krishnan L (2004) A freeze-dried fibrin disc as a biodegradable drug release matrix. Biologicals 32:49–55
49. Sawada Y, Tadashi O, Masazumi K, Kazunobu S, Koichi O, Sasaki J (1994) An evaluation of a new lactic acid polymer drug delivery system: a preliminary report. Br J Plast Surg 47:158–161
50. Sawada Y, Ara M, Yotsuyanagi T, Sone K (1990) Treatment of dermal depth burn wounds with an antimicrobial agent-releasing silicone gel sheet. Burns 16:347–352
51. Doillon C, Silver F (1986) Collagen-based wound dressing: effect of hyaluronic acid and fibronectin on wound healing. Biomaterials 7:3–8
52. Chu H, Xiong H, Zhou X, Han F, Wu Z, Zhang P et al (2006) Aminoglycoside ototoxicity in three murine strains and effects on NKCC1 of stria vascularis. Chin Med J 119:980–985
53. Patrick B, Rivey M, Allington D (2006) Acute renal failure associated with vancomycin and tobramycin-laden cement in total hip arthroplasty. Ann Pharmacother 40:2037–2042
54. Fraser JF, Bodman J, Sturgess R, Faoagali J, Kimble RM (2004) An in vitro study of the antimicrobial efficacy of a 1 % silver sulphadiazine and 0.2 % chlorhexidine digluconate cream, 1 % silver sulphadiazine cream and a silver coated dressing. Burns 30:35–41
55. Komarcevic A (2000) The modern approach to wound treatment. Med Pregl 53:363–368
56. Dijke P, Iwata K (1989) Growth factors for wound healing. Biotechnology 7:793–798
57. Greenhalgh D (1996) The role of growth factors in wound healing. J Trauma Inj Infect Crit Care 41:159–167
58. Puolakkainen P, Twardzik D, Ranchalis J, Pankey S, Reed M, Gombotz W (1995) The enhancement in wound healing by transforming growth factor-b1 (TGF-b1) depends on the topical delivery system. J Surg Res 58:321–329
59. Defail A, Edington H, Matthews S, Lee W, Marra K (2006) Controlled release of bioactive doxorubicin from microspheres embedded within gelatin scaffolds. J Biomed Mater Res 79:954–962
60. Koempel J, Gibson S, O'Grady K, Toriumi D (1998) The effect of platelet-derived growth factor on tracheal wound healing. Int J Pediatr Otorhinolaryngol 146:1–8
61. Maeda M, Kadota K, Kajihara M, Sano A, Fujioka K (2001) Sustained release of human growth hormone (hGH) from collagen film and evaluation of effect on wound healing in mice. J Control Release 77:261–272

62. Gu F, Amsden B, Neufeld R (2004) Sustained delivery of vascular endothelial growth factor with alginate beads. J Control Release 96:463–472
63. Grzybowski J, Oldak E, Antos-Bielska M, Janiak M, Pojda Z (1999) New cytokine dressings. I. Kinetics of the in vitro rhG-CSF, rhGM-CSF, and rhEGF release from the dressings. Int J Pharm 184:173–178
64. Park S, Kim J, Suh H (2004) Evaluation of antibiotic loaded collagen-hyaluronic acid matrix as a skin substitute. Biomaterials 25:3689–3698
65. Wallace E (1994) Feeding the wound: nutrition and wound healing. Br J Nurs 3:662–667
66. Flanigan K (1997) Nutritional aspects of wound healing. Adv Wound Care 10:48–52
67. Ehrlich H, Tarver H, Hun T (1973) Effects of vitamin A and glucocorticoids upon inflammation and collagen synthesis. Ann Surg 177:222–227
68. Porto R, Lucio D, Souza T, Pereira S, Fernandes G (2002) Effects of a vitamin pool (vitamins A, E, and C) on the tissue necrosis process: experimental study on rats. Aesthetic Plast Surg 26:197–202
69. Yang H, Chai J, Guo ZR (2001) Effect of improved topical agents on healing time of deep second-degree burn wound. Zhongguo Xiu Fu Chong Jian Wai Ke Za Zhi 15:3162–3164
70. Hansson C, Hoborn J, Moller A, Swanbeck G (1995) The microbial flora in venous leg ulcers without clinical signs of infection: repeated culture using a validated standardised microbiological technique. Acta Derm Venereol 75:24–30
71. Mertz PM, Ovington LG (1993) Wound healing microbiology. Dermatol Clin 11:739–747
72. Yates CC, Whaley D, Babu R, Zhang J, Krishna P, Beckman E et al (2007) The effect of multifunctional polymer-based gels on wound healing in full thickness bacteria-contaminated mouse skin wound models. Biomaterials 28:3977–3986
73. Kumar MS, Kirubanandan S, Sripriya R, Sehgal PK (2008) Triphala promotes healing of infected full-thickness dermal wound. J Surg Res 144:94–101
74. Shanmugasundaram N, Uma TS, Ramyaa Lakshmi TS, Babu M (2009) Efficiency of controlled topical delivery of silver sulfadiazine in infected burn wound. J Biomed Mater Res A 89:472–482
75. Teo EY, Ong SY, Chong MS, Zhang Z, Lu J, Moochhala S et al (2011) Polycaprolactone-based fused deposition modeled mesh for delivery of antibacterial agents to infected wounds. Biomaterials 32:279–287
76. Cutting K, White R (2005) Criteria for identifying wound infection—revisited. Ostomy Wound Manage 51:28–34
77. Drosou A, Falabella A, Kirsner R (2003) Antiseptics on wounds: an area of controversy. Wounds 15:2–11
78. Neely A, Gardner J, Durkee P (2009) Are topical antimicrobials effective against bacteria that are highly resistant to systemic antibiotics? J Burn Care Res 30:19–29
79. Meng X (2002) Keratinocyte gene therapy: cytokine gene expression in local keratinocytes and in circulation by introducing cytokine genes into skin. Exp Dermatol 11:456–461
80. Steinstraesser L (2001) Feasibility of biolistic gene therapy in burns. Shock 15:272–277
81. Yao F, Eriksson E (2000) Gene therapy in wound repair and regeneration. Wound Repair Regen 8:443–451
82. Eming S, Morgan J, Berger A (1997) Gene therapy for tissue repair: approaches and prospects. Br J Plast Surg 50:491–500
83. Jeschke M, Richter G, Hofstadter F (2002) Non-viral liposomal keratinocyte growth factor (KGF) cDNA gene transfer improves dermal and epidermal regeneration through stimulation of epithelial and mesenchymal factors. Gene Ther 9:1065–1074
84. Galeano M, Deodato B, Altavilla D (2003) Effect of recombinant adeno-associated virus vector-mediated vascular endothelial growth factor gene transfer on wound healing after burn injury. Crit Care Med 31:1017–1025
85. Braddock M, Campbell C, Zuder D (1999) Current therapies for wound healing: electrical stimulation, biological therapeutics, and the potential for gene therapy. Int J Dermatol 38:808–817
86. Eming SA, Krieg T, Davidson JM (2007) Gene therapy and wound healing. Clin Dermatol 25:79–92

87. Jeschke M, Barrow R, Hawkins H (1996) IGF-I gene transfer in thermally injured rats. Gene Ther 6:1015–1020
88. Henry S, McAllister D, Allen M (1998) Microfabricated micro-needles: a novel approach to transdermal drug delivery. J Pharm Sci 87:922–925
89. Ciernik I, Krayenbuhl B, Carbone D (1996) Puncture-mediated gene transfer to the skin. Hum Gene Ther 7:893–899
90. Jia C, Chen B, Wang D (1999) Effects of PCMV4-hTGF beta 1 as nucleic acid vaccine on II* burn wound healing and postburn scarring in rats. Zhonghua Zheng Xing Shao Shang Wai Ke Za Zhi 15:170–172
91. McLennan S, McGill M, Twigg S, Yue D (2007) Improving wound-healing outcomes in diabetic foot ulcers. Expert Rev Endocrinol Metabol 2:205–213
92. Chen F, Zhang M, Wu Z (2010) Toward delivery of multiple growth factors in tissue engineering. Biomaterials 31:6279–6308
93. Sporn M, Roberts A (1993) A major advance in the use of growth factors to enhance wound healing. J Clin Invest 92:2565–2566
94. Beck L, Deguzman L, Lee W (1991) TGFβ accelerates wound healing: reversal of steroid-impaired healing in rats and rabbits. Growth Factors 5:295–304
95. Lepisto J, Laato M, Niinikoski J (1992) Effects of homodimeric isoforms of platelet-derived growth factor (PDGF AA and PDGF BB) on wound healing in rats. J Surg Res 53:596–601
96. Steed D (1995) Clinical evaluation of recombinant human platelet-derived growth factor for the treatment of lower extremity diabetic ulcers. J Vasc Surg 21:71–81
97. Ross R, Raines E, Bowen-Pope DF (1986) The biology of platelet derived growth factor. Cell 46:155–169
98. Lepisto J, Peltonen J, Vaha-Kreula M (1996) Selective modulation of collagen gene expression by different isoforms of platelet-derived growth factor in experimental wound healing. Cell Tissue Res 286:449–455
99. Cribbs R, Luquette M, Besner E (1998) Acceleration of partial-thickness burn wound healing with topical application of heparin-binding EGF-like growth factor (HB-EGF). J Burn Care Rehabil 19:95–101
100. Hardwicke J, Schmaljohann D, Boyce D, Thomas D (2008) Epidermal growth factor therapy and wound healing-past, present and future perspectives. Surgeon 6:172–177
101. Ko C, Dixit V, Shaw W, Gitnick G (1997) Extensive in vivo angiogenesis from the controlled release of endothelial cell growth factor: implications for cell transplantation and wound healing. J Control Release 44:209–214
102. Folkman J, Klagsbrun M (1987) Angiogenic factors. Science 235:442–447
103. Klingbeil C, Cesar LB, Fiddes JC (1991) Clinical and experimental approaches to dermal and epidermal repair: normal and chronic wounds. In: Barbul A, Caldwell M, Eaglstein W, Hunt T, Marshall D, Pines E, Skover G (eds) Basic fibroblast growth factor accelerates tissue repair in models of impaired wound healing. Wiley, New York, NY, pp 443–458
104. Pierce G, Mustoe T, Lingelbach J, Masakowski V, Gramates P, Deuel T (1989) Transforming growth factor beta reverses the glucocorticoid-induced wound-healing deficit in rats: possible regulation in macrophages by platelet derived growth factor. Proc Natl Acad Sci U S A 86(7):2229–2237
105. Pierce G, Mustoe T, Lingelbach J, Masakowski V, Griffin G, Senior R et al (1989) Platelet-derived growth factor and transforming growth factor-beta enhance tissue repair activities by unique mechanisms. J Cell Biol 109:429–440
106. Greenhalgh D, Sprugel K, Murray M, Ross R (1990) PDGF and FGF stimulate wound healing in the genetically diabetic mouse. Am J Pathol 136:1235–1246
107. Mustoe T, Pierce G, Morishima C, Deuel T (1991) Growth factor-induced acceleration of tissue repair through direct and inductive activities in a rabbit dermal ulcer model. J Clin Invest 87:694–703
108. Wu Y, Chen L, Scott P, Tredget E (2007) Mesenchymal stem cells enhance wound healing through differentiation and angiogenesis. Stem Cells 25:2648–2659

109. Pittenger M, McKay A, Beck S, Jaiswal R, Douglas R, Mosca J et al (1999) Multilineage potential of adult human mesenchymal stem cells. Science 284:143–147
110. Javazon E, Keswani S, Badillo A, Crombleholme T, Zoltick P, Radu A et al (2007) Enhanced epithelial gap closure and increased angiogenesis in wounds of diabetic mice treated with adult murine bone marrow stromal progenitor cells. Wound Repair Regen 15:350–359
111. Anne M, Nicole S (2010) Mesenchymal stem cells: paracrine signaling and differentiation during cutaneous wound repair. Exp Cell Res 316:2213–2219
112. Kamolz L, Kolbus A, Wick N, Mazal P, Eisenbock B, Burjak S (2006) Cultured human epithelium: human umbilical cord blood stem cells differentiate into keratinocytes under in vitro conditions. Burns 32:16–19
113. Branski L, Gauglitz G, Herndon D, Jeschke M (2009) A review of gene and stem cell therapy in cutaneous wound healing. Burns 35:171–180
114. Kim W, Park B, Sung J, Yang J, Park S, Kwak S (2007) Wound healing effect of adipose-derived stem cells: a critical role of secretory factors on human dermal fibroblasts. J Dermatol Sci 48:15–24
115. Plikusa M, Gaya D, Treffeisena E, Wanga A, Supapannacharta R, Cotsarelisa G (2012) Epithelial stem cells and implications for wound repair. Semin Cell Dev Biol 23:946–953
116. Roh C, Lyle S (2006) Cutaneous stem cells and wound healing. Pediatr Res 59:100–103
117. Matthias S, Sabine W (2008) Oxidative stress in normal and impaired wound repair. Pharmacol Res 58:165–171
118. Al-Jawad F, Sahib A, Al-Kaisy A (2008) Role of antioxidants in the treatment of burn lesions. Ann Burns Fire Disasters 186–191
119. Gomathi K, Gopinath D, Ahmed MR, Jayakumar R (2003) Quercetin incorporated collagen matrices for dermal wound healing processes in rat. Biomaterials 24:2767–2772
120. Panchatcharam M, Miriyala S, Gayathri V, Suguna L (2006) Curcumin improves wound healing by modulating collagen and decreasing reactive oxygen species. Mol Cell Biochem 290:87–96
121. Brooks J, Malic C, Judkins K (2008) Scratching the surface—managing the itch associated with burns: a review of current knowledge. Burns 34(2008):2751–2760
122. Albsoul A, Younes N, Badran D (2006) Topical phenytoin ointment increases autograft acceptance in rats. Saudi Med J 27:962–971
123. Pendse A, Sharma A, Sodani A, Hada S (1993) Topical phenytoin in wound healing. Int J Dermatol 32:214–223
124. Brandner J, Houdek P, Husing B, Kaiser C, Moll I (2004) Connexins 26, 30, and 43: differences among spontaneous, chronic, and accelerated human wound healing. J Invest Dermatol 122:1310–1320
125. Wang C, Green C, Duft B, Becker D (2007) Targeting connexin43 expression accelerates the rate of skin and diabetic wound repair. J Biotechnol 131:S64–S64
126. Patel P, Vasquez S, Granick M, Rhee S (2008) Topical antimicrobials in pediatric burn wound management. J Craniofac Surg 19:913–922

Chapter 27
Vaginal Drug Delivery

Emily A. Krogstad, Michael J. Rathbone, and Kim A. Woodrow

Abbreviations

ARV	Antiretroviral
CAP	Cellulose acetate phthalate
CTAB	Cetyltrimethylammonium bromide
CVF	Cervicovaginal fluid
EFV	Efavirenz
FTC	Emtricitabine
HAART	Highly active antiretroviral therapy
HIV	Human immunodeficiency virus
HSV-2	Herpes simplex virus 2
IRV	Intravaginal ring
nanoART	Nano-antiretroviral therapy
NRTI	Nucleotide reverse transcriptase inhibitor
PBMC	Peripheral blood mononuclear cell
PCL	Poly(caprolactone)
PDLLA	Poly(D,L)-lactic acid
PEG	Poly(ethylene glycol)
PEO	Poly(ethylene oxide)
PLA	Poly(lactic acid)
PLGA	Poly(lactic-co-glycolic acid)

E.A. Krogstad • K.A. Woodrow (✉)
Department of Bioengineering, University of Washington,
William H. Foege Building N410D, 3720 15th Ave NE, Seattle, WA 98195, USA
e-mail: emkrog@uw.edu; woodrow@uw.edu

M.J. Rathbone
School of Pharmacy, International Medical University, Kuala Lumpur, Malaysia
e-mail: michael_rathbone@imu.edu.my

A.J. Domb and W. Khan (eds.), *Focal Controlled Drug Delivery*, Advances in Delivery Science and Technology, DOI 10.1007/978-1-4614-9434-8_27,

PLLA	Poly(L-lactic acid)
PPI	Poly(propyleneimine)
PVA	Poly(vinyl alcohol)
SHIV	Simian-human immunodeficiency virus
SIV	Simian immunodeficiency virus
SLS	Sodium lauryl sulfate
SQV	Saquinavir
STI	Sexually transmitted infection
TDF	Tenofovir disoproxil fumarate
TFV	Tenofovir
VFS	Vaginal fluid simulant
VP5k	Vitamin E/5 kDa PEG

27.1 Gross Anatomy and Physiology of the Vagina

The usefulness of the vagina as an administration site for local or systemic drug delivery has been known for many years; however, the successful development of a reliable vaginal delivery system for humans is fraught with challenges. However, topical delivery to the vaginal mucosa has many salient applications for preventing human immunodeficiency virus (HIV) and other sexually transmitted infections (STIs), mucosal vaccines, treatment of various STIs, and maintenance of reproductive health. From a drug delivery viewpoint, the vagina is characterized by a constantly changing environment. It is a complex organ with multiple functions and physicochemical, physiological, histological, and anatomical parameters that change with age, pregnancy, and sexual excitement, all of which play a role in efficiency and potential duration of vaginal drug delivery. Thus, the design of any effective intravaginal drug delivery system must take into account the anatomy and physiology of the vagina.

27.1.1 Gross Anatomy

The human vagina may be described as a slightly s-shaped highly expandable, fibromuscular tube ~7–10 cm in length [1]. The vagina extends from the lower part of the uterine cervix to the external part of the vulva known as the labia minor [2]. It can be divided into four areas relative to the cervix, these being the posterior fornix, which is capacious; the anterior fornix, which is shallow; and two lateral fornices. The anterior and posterior walls of the vagina come together to form two ridges of folds causing it to assume an H-shape appearance in cross section [3]. The vaginal wall is composed of three layers: the epithelial layer, the muscular coat, and the tunica adventitia [3]. The epithelium consists of a lamina propria and an epithelial cell layer which, at different stages of the menstrual cycle, changes in thickness by ~200–300 μm. Therefore, the menstrual cycle may affect drug permeation through the vaginal mucosa. The muscular coat consists of smooth and elastic fibers in a spiral arrangement which are covered by the tunica adventitia (a loose connective tissue). These features allow stretching of the epithelium. At its surface, the vagina has transverse muscular folds called "rugae." The surface area of the vagina is increased by these folds and by microridges covering the epithelial cell surface which provides the vagina with a relatively large surface area [4]. Characterization of the vaginal mucosa on a cellular level was performed and compiled in a complete textbook on the human vagina edited by Hafez and Evans in 1978 [5]. This book provides an excellent reference for the fundamentals, and its individual chapters have been extensively referenced in publications involving the human vagina.

27.1.2 Physiology of the Vagina

27.1.2.1 Blood Supply

The vagina receives its blood supply from the uterine and pudendal arteries which arise from the internal iliac arteries. Blood leaving the vagina enters the peripheral circulation through an abundant venous plexus surrounding the vagina and eventually empties directly into the internal iliac veins, thus bypassing the liver metabolism [1, 4, 6]. This immediate access to the systemic circulation also circumvents the gastrointestinal tract, thereby eliminating the problems of degradation and gastrointestinal side effects that oral delivery systems must sometimes contend with [4].

27.1.2.2 Vaginal Fluid

Origin and Contents

Even though the epithelia have no secretory glands, it is usually covered with a thin film of vaginal fluid. The composition of the fluid is complex and changes with age, stage of menstrual cycle, and health condition [1, 6]. This vaginal fluid arises from cervical secretions and exudate from the blood vessels, and its presence in the vagina is dependent upon adequate blood flow [7]. This fluid also contains secretions from the endometrium and fallopian tubes, desquamated vaginal epithelial cells, and leukocytes [8]. The chemical composition of the fluid is reported to include carbohydrates, amino acids, aliphatic acids, proteins and immunoglobulins [9], several enzymes, enzyme inhibitors, alcohols, hydroxy ketones, and aromatic compounds [4].

Mucus

The chemical and physical nature of the vaginal fluids can significantly affect drug adsorption [1, 10], sometimes positively, for example, thick cervical mucus may assist in the bioadhesion of microparticles, and sometimes negatively, for example, the thick cervical mucus may also act as a diffusional barrier to drug diffusion. The main physical properties of the vaginal mucus that may affect absorption are its rheological parameters. These include viscosity, flow elasticity, spinnbarkeit, thixotropy, and tack [4] which are dependent upon the amount, composition, and physical characteristics of the cervical mucus which in turn change with the menstrual cycle [4]. For example, there is an increase in fibrosity, pH, and mucin content and a decrease in the viscosity, cellularity, and albumin concentration at the time of ovulation due to an increase in the amount of cervical secretions at this time of the menstrual cycle. Overall, mucus composition may influence drug delivery [10].

pH

The pH value of the vaginal fluids depends upon the location within the vagina (the pH is lowest near the anterior fornix and highest near the cervix [11]) as well as upon the menstrual cycle (pH values are reported to be at their lowest during ovulation and pre-menstruation [1]). The basal pH value of vaginal fluid in healthy mature women ranges between pH 4 and 5. The acidity is produced mainly by the commensal microorganism *Lactobacillus acidophilus*. The carbohydrate glucagon is released into the lumen from exfoliated epithelial cells arising from the superficial layers of the vaginal wall, and *Lactobacillus acidophilus* converts the glucagon into lactic acid [6]. The cervical mucus varies in pH from 7 to 9 during different days of the cycle, reaching its most basic during ovulation. The acidic nature of the vaginal fluid is important since it offers natural resistance to the colonization of bacteria that cause pus, and changes in pH can influence the release profile from an intravaginal drug delivery system through alteration of the extent of ionization of the drug in the vaginal fluids or its solubility in the vaginal fluid. It should also be taken into consideration that any drug delivery system administered to the vagina must take into account the presence of these lactic acid producing bacteria and not disturb their processes which are essential in maintaining a healthy vaginal environment [1].

Volume

The volume (weight) of vaginal discharge can be variable. For example, the weight of discharge measured in women of reproductive age was observed to be 3–4 g per 4 h, while menopausal women produced 50 % less [6]. When cervical caps were fitted to women, only 2.7 g/24 h secretion was measured. The variability in fluid volume can have a variable effect upon vaginal drug delivery since drugs have to be in solution in order to become absorbed. A large volume of fluid can be advantageous for adsorption of drugs which need to dissolve before their subsequent absorption, but the fluid will gradually flow out of the vagina and thus may remove released drug with it as it leaves the vagina. Alternatively, copious amounts of fluid may reduce drug delivery by increasing drug diffusional distances.

Enzymes

The vagina plays host to a variety of enzymes present in its epithelial and basal cell layers [6] which provides a potential enzymatic barrier toward the intravaginal delivery of peptides and proteins. An excellent review by Lee describes the various enzymatic barriers to peptides and proteins present in the vagina [12]. The activity of vaginal enzymes varies with the menstrual cycle and declines with the onset of menopause.

Microflora

The vaginal environment contains a mix of anaerobic microbes, mainly *Lactobacillus*, *Bacteroides*, and *Staphylococcus* species. It is known that some medications can have a direct influence on the vaginal microflora, for example, administration of antibiotics may result in the colonization of the vagina by opportunistic pathogens.

27.1.3 Drug Candidates and Absorption

27.1.3.1 Physicochemical Properties of Drug

Drug absorption across the vaginal mucosa is influenced by the physicochemical properties of the drug [13]. These properties include its molecular size, molecular weight, partition coefficient (lipophilicity), extent of ionization, and chemical nature [6].

The list of drugs that are absorbed across the vaginal mucosa includes peptides and proteins [4], several local or topically active agents such as spermicides and antifungals [1], steroids, prostaglandins, antimicrobials, proteins, and nonoxynol-9 [2].

27.1.3.2 Absorption Mechanisms

The pathways for drug diffusion across vaginal epithelium are essentially similar to other mucosa of the body [1, 6] and can occur via three primary routes: diffusion

Table 27.1 Key considerations when designing an intravaginal drug delivery system

Design consideration	Comment
Whether the delivery system is for local or systemic delivery	Utilize traditional dosage forms such as creams or gels. Enhance the delivery system through the use of excipients that promote increased intravaginal residence time
Whether the goal of treatment is site specific or if the drug needs to be distributed rapidly throughout the vaginal space	Site-specific delivery requires a self-locating system, typically a mucoadhesive formulation, although an intravaginal ring, due to its elastomeric nature, will remain located high in the vaginal space
	For rapid distribution throughout the vaginal cavity, semisolid or fast-dissolving solid systems are appropriate. However, remember that the rheological properties of the formulation will be critical determinants of their ability to spread and speed of spreadability
The desired drug release profile	Different formulations may be required dependent upon whether an immediate or modified (sustained or controlled release) drug release profile is required for optimal treatment
For systemic drug delivery via the intravaginal route	Considerations to take into account include the physicochemical and potency of the drug. These include partition coefficient, molecular weight and size, and epithelial permeability
The cultural acceptability of the vagina as a route for drug delivery	Between and within different cultures, the personal bias for the application of delivery systems that can be self-inserted and removed, or considerations relating to leakage, will vary
Commercial considerations	From an industrial perspective, the cost-benefit in relation to the commercial value of the active component(s) and the likely benefits in relation to the disease state must be considered

through the transcellular route in the presence of a concentration gradient, diffusion between cells (intercellular or paracellular transport), or vesicular or receptor-mediated transport [4]. These absorption routes can be affected by many physiological factors which are in turn influenced by the stage of the menstrual cycle. For example, vaginal epithelial thickness and porosity change during the menstrual cycle.

27.1.4 Vaginal Drug Delivery Systems

There are a wide range of systems that can be used to deliver drugs to the vagina for local or systemic activity. These include traditional pharmaceutical formulations such as creams, pessaries, foams, gels, tablets, and particulate systems; more advanced delivery systems that attempt to account for the conditions prevailing in the vagina which may incorporate the use of one or more mucoadhesive polymeric components; and specifically designed intravaginal delivery systems that generally involve solid polymeric systems, usually either elastomers or hydrogels [14]. Some key considerations when designing an intravaginal drug delivery system were described by Woolfson et al. [14]. These are shown in Table 27.1.

27.2 Rational Design of Vaginal Drug Delivery Systems

The rational design of vaginal drug delivery systems is directed toward the goal of delivering target concentrations of one or more agents to tissues and cells that will elicit desirable pharmaceutical responses for a specified duration. For the vaginal route of drug delivery, agents may be of various compositions that act locally or systemically to achieve therapeutic or prophylactic outcomes. Therapeutic or prophylactic opportunities for vaginal drug delivery include HIV/AIDS, cancer, and a multitude of other indications (STIs, contraception, vaccines). The development of topical microbicides for HIV prevention chronicles both the history of development and innovation in vaginal drug delivery systems as well as lessons learned in the rational design of effective products. Herein we examine the design constraints associated with the development of topical microbicides that are based on product structure-function relationships known to correlate with protection against sexual HIV transmission.

27.2.1 Protecting the Frontlines: Tissue and Cellular Targets for HIV

Studies from high-dose vaginal challenge models using nonhuman primates support the conclusion that prevention strategies should target early events that occur during the first week of HIV-1 infection [15]. The critical events following mucosal exposure to high-dose simian immunodeficiency virus (SIV)/simian-human immunodeficiency virus (SHIV) indicate that the virus can cross the mucosal epithelial barrier within hours to infect susceptible target cells that are spatially dispersed within the epithelium and stroma of the vagina, ectocervix, and endocervix [16–18]. Therefore, successful delivery strategies for highly active antiretroviral (ARV) microbicides targeting HIV-1 must be highly retentive, penetrate mucosal epithelial barriers, and make bioavailable a combination of ARV agents at concentrations that inhibit the expansion of locally infected cells and preempt systemic infection.

The cervicovaginal tract provides natural barriers to infection that can be additionally fortified by topical prevention strategies but only if these strategies are able to localize protective concentrations of drugs to the tissues and cells targeted by the virus. Therefore, an effective intravaginal drug delivery system for HIV-1 prevention must (1) penetrate mucus and tissue barriers before being diluted and cleared by vaginal secretions (cervical vaginal mucus is produced at a rate of 2–8 mL/day and cleared every 6–17 h [19]); (2) deliver and target ARV agents to free virions and infected cells that are dispersed in genital secretions, the epithelium and submucosa; and (3) respond to variations in vaginal pH and microflora that occur during intercourse, menstrual cycle, and infection with bacterial and viral pathogens. In addition, microbicides products must be safe, easy to use, and cost-effective.

27.2.2 Pharmacological Considerations: The Right Drug at the Right Place

Pharmacological agents must be maintained at effective concentrations in tissue and cellular compartments to inhibit various mechanisms of HIV transmission at exposure sites within the vaginal mucosa. Current ARV drugs being evaluated for oral or topical pre-exposure prophylaxis show varying levels of drug concentration within tissues (vaginal, rectal) and biological fluids (blood plasma, cervicovaginal fluids (CVFs)) that depend on the route of delivery, on the physicochemical properties of the drug, and on drug pharmacokinetics. For example, tenofovir (TFV) is a water-soluble nucleotide reverse transcriptase inhibitor (NRTI) that achieves 100-fold lower concentration in CVF and genital tissue when administered orally compared to vaginal dosing [20, 21]. In contrast, emtricitabine (FTC), an NRTI with tenfold higher aqueous solubility than TFV, shows higher concentrations in CVF than in blood plasma after oral administration [21]. The disoproxil diester prodrug of tenofovir (TDF), which is more lipophilic and cell-permeable than the dianionic TFV, results in up to 1,000-fold higher intracellular concentration of the phosphorylated active metabolite than the parent drug when delivered orally [22]. Finally, the intracellular half-lives of specific pharmacological agents will dictate the required dosing frequency to achieve protective efficacy. Therefore, the correlation between cellular pharmacology and clinical therapeutic or prophylactic outcome cannot be generalized but rather should be determined specifically for each drug.

To date, the majority of clinical HIV microbicide trials have used gel-based systems for intravaginal drug delivery. The CAPRISA 004 clinical trial showed that significant protection from HIV infection was associated with TFV concentration of >1,000 ng/mL in CVF, which is tenfold higher than can be achieved from oral TFV delivery [23]. Although the before-and-after sex dosing regimen of the TFV gel in the CAPRISA 004 trial was sufficient to be protective, low user adherence resulted in an overall reduction in HIV acquisition of only 39 % [24]. Therefore, design of vaginal drug delivery systems must also account for aesthetic aspects of the dosage form that support user compliance in addition to facilitating the pharmacological properties of the delivered drug. Aesthetic properties that effect user compliance include organoleptic properties such as color, odor, taste, and texture of the dosage form. User preferences can vary widely based on cultural reasons, but it is generally accepted that dosage forms be colorless and smooth and have no odor or taste so as to allow for discreteness. In addition, user requirements that involve dosage frequency, use of an applicator, potential for lubrication, or messiness of the formulation will also be decisive factors in the success or failure of the dosage form. The ability to control all these attributes individually and in combination while maintaining effective drug activity *in vivo* is at the core of rational design of vaginal drug delivery systems.

27.2.3 *Design Constraints: Physical and Biological Evaluation of Vaginal Delivery Systems*

Studies show that inadequate coating of vaginal rugae by intravaginal delivery systems is associated with incomplete protection against vaginal challenge with STIs [25]. For HIV prevention, protecting the cervix is equally as important to protecting the vaginal epithelium since the first cellular targets for HIV have been observed in the endocervical epithelium [26]. Microbicides that coat *both* the cervix and the vagina will be the most effective in preventing sexual HIV transmission. Vaginal delivery systems are also subject to dilution by genital secretions during application and coitus that cause erosion and reduce retention of the protective coating layer [27]. Therefore, products that resist erosion and elimination and are instead rapidly absorbed by the mucosa should significantly enhance drug bioavailability. Finally, we propose that delivery of a combination of agents with different mechanisms of actions will have the greatest likelihood of enhancing potency and decreasing toxicity [28–30]. However, co-formulating agents with different physicochemical properties while preserving mechanical properties that maintain the structure or sustain the aesthetic qualities of the vaginal drug delivery system have been challenging [30].

Despite the prevalence of gel-based systems in human clinical trials of microbicides, there are several limitations with this dosage form [29, 31–33]. Inadequate retention and spreading may contribute to low tissue and cellular concentrations of active agents that have likely contributed to the outcome that inhibitors that act with nanomolar potency are required at millimolar concentrations for consistent protection against HIV-1 transmission in nonhuman primate models using a high-dose viral challenge [34–40]. Gel-based formulations also limit the ability to deliver multiple ARV compounds that have desirable synergistic properties but are chemically incompatible [41]. As a consequence, the next-generation delivery strategies must enable a more tailored approach for addressing the unique challenges associated with delivery of different chemical classes of compounds while preserving physical attributes that support user adherence.

27.2.3.1 Physical, Rheological, and Retentive Properties of Vaginal Delivery Systems

Vaginal drug delivery systems are comprised of the specific pharmaceutical agents to achieve the therapeutic or prophylactic outcomes and may also include various excipients in the formulations. Excipients may include binders, diluents, preservatives, antioxidants, and other agents that affect drug stability as well as user acceptability. Current vaginal dosage forms can be classified as solids (films, rings), semisolids (gels, ointments), or liquids (foams, douche) and have different requirements for

Table 27.2 Comparison of current dosage forms for vaginal drug delivery

Strategy	Examples	Advantages	Disadvantages
Solid	Vaginal ring	Ease of application	Mechanical requirements of vaginal rings limit drug selection
	Film	Capacity for sustained release	Must be hydrated for drug release
	Sponge	Long shelf life	
	Tablet	Minimal packaging	
	Capsule	Controllable geometry	
	Suppository		
	Diaphragm		
Semisolid	Gel	High user acceptability in sub-Saharan Africa	Limited retention time
	Cream	Can be designed for bioadhesive and rheological properties	Messiness, leakage
	Ointment	Large area of initial coverage	Limited chemical compatibility with drugs with poor aqueous solubility
Liquid	Foam	Can be designed for bioadhesive and rheological properties	Very limited retention time
	Spray	Large area of initial coverage	Messiness, leakage
	Douche		Limited chemical compatibility with drugs with poor aqueous solubility

excipients. The drug-excipient interactions are designed to maintain the drug stability during manufacturing, transport, and storage while also preserving and facilitating the intended biological activity upon deployment. The physical attributes of a dosage form include its dimensions (shape), volume (size), and rheological and bioadhesive properties. These attributes are guided by specific design specifications that are important for deployment and biological function of the vaginal delivery system. Advantages and disadvantages of solid, semisolid, and liquid dosage forms are compared in Table 27.2.

Only solid dosage forms allow for control of the geometry and shape of the vaginal delivery system and are important considerations for administration and user compliance as well as manufacturing and packaging. Among the solid dosage forms, intravaginal rings (IVRs) and quick-dissolving films are the lead technologies, which also include tablets, capsules, and suppositories. The physical attributes of IVRs and films are constrained by the materials and their fabrication process. Poly(ethylene-co-vinyl acetate), silicone, and polyurethane polymers are the primary materials used in the fabrication of IVRs. Balancing the mechanical flexibility and strength of an IVR are needed to permit insertion and retention within the vaginal cavity without causing tissue abrasion [43]. Differences in the material mechanical stiffness impact the requirement for the cross-sectional diameter of IVRs to avoid expulsion. Unlike IVRs, quick-dissolving vaginal films have greater potential to design for physical dimensions since these dosage forms are typically fabricated via solvent-casting processes [44]. However, all vaginal films have a quadrilateral geometry (squares, rectangles) to maximize the surface area for rapid dissolution and are typically smaller than 5 cm^2. The average thickness (<0.3 mm) and mass (~300–400 mg) of the final vaginal film dimensions will also depend on the requirement for the amount of drug loading. Vaginal films are advantageous due to their ease of application, minimum packaging requirements, and reduced messiness compared to conventional vaginal gels. Overall, there has been limited opportunity to design the geometry of existing vaginal drug delivery systems. This limitation may be due to the particular fabrication and manufacturing processes used for some solid dosage forms. The greater challenge, however, is likely the inability to design independently the physical attributes of the dosage form without consequently affecting their biological function.

In contrast to solid dosage forms, semisolid and liquid vaginal delivery systems do not have strict physical dimensions. The lead semisolid and liquid vaginal dosage forms include vaginal gels and foams, respectively. However, liquid dosage forms show low efficacy associated with poor user adherence due to inconveniences such as leakage, messiness, and poor retention. Therefore, these dosage forms are generally not a development priority for the field. An attractive feature of semisolid and liquid vaginal delivery systems, as well as hydrated vaginal films, is the ability to design and control for rheological and bioadhesive properties. The rheological properties of a dosage form are important during manufacturing (packaging, storage, transport) as well as for their end use (application, spreading, and retention). Viscosity is a standard rheological measurement for gels and films, which have reported specific viscosity values in the range of 80 and 0.01 Pa·s, respectively [33, 45]. These viscosity values may be appropriate initial targets to inform design specifications that are conducive for spreading and retention under both steady and dynamic shear stresses experienced within the vaginal cavity. Delivery systems that are designed to be bioadhesive can promote retention of dosage forms that are rapidly cleared after application due to short residence times in the vagina. Materials that promote bioadhesion include acrylic acid-based polymers, high-molecular-weight poly(ethylene glycol) (PEG), and natural polymers such as chitosan and carrageenan [46]. Target values for bioadhesion would allow for formulations to be maintained in the vaginal cavity for 24 h [42]. Several methods to measure bioadhesion have been developed but no single method has been standardized. The most accepted method is to use tensile testing to measure the force required to separate the dosage form adhered to mucosal tissue.

Collectively, if designed correctly, physical and rheological properties of the vaginal delivery systems will promote user adherence and biological function. A critical criterion for biological function is to deliver therapeutic and protective concentrations of the drugs to tissues and cells. Programming drug release kinetics is a primary goal of many vaginal delivery systems.

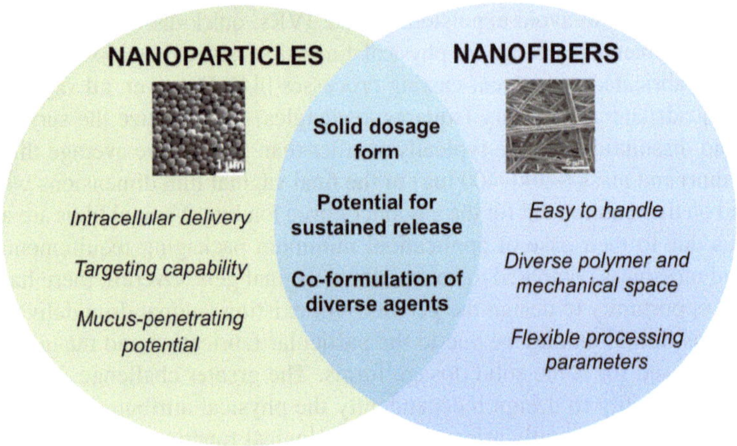

Fig. 27.1 Rationale for novel nanomaterial strategies for vaginal drug delivery. Nanoparticles and nanofibers represent two nanomaterial drug delivery platforms that offer individual and combined advantages for vaginal drug delivery

27.2.3.2 Engineering Vaginal Delivery Systems for Controlled Release

For topical products that are used against HIV prevention, most vaginal drug delivery systems are designed to provide bolus delivery combined with sustained drug release profiles that can achieve rapid onset as well as prolonged drug activity. One advantage of IVRs is that they allow for sustained release of agents up to several months. Zero-order release kinetics up to 90 days has been reported for a number of different drug compounds. Released drug acts locally where the dissolution in vaginal fluid will dictate the drug biodistribution and coating of cervicovaginal surfaces. Unlike IVRs, quick-dissolving vaginal films are hydrated upon insertion by vaginal fluid. Hydration and dissolution of films have been reported to be in the range of 4–20 min [47, 48]. The resulting drug release from quick-dissolving vaginal films is typically complete after 20 min and shows linear release kinetics over this time span.

The goals of programming drug release in vaginal drug delivery systems are to deliver from the dosage form effective concentrations of the active agents into tissues and cells. As discussed above, the physicochemical and pharmacological properties of the drug will have important consequences related to tissue absorption, cell permeation, and biological activity. Vaginal drug delivery systems have the potential to achieve specific delivery kinetics of the drug that is aligned with the drug pharmacokinetics. However, many novel delivery formulations are also focused on altering the pharmacokinetics of the parent compound so as to enhance potency, stability, or decrease cytotoxicity. Nanomaterials offer one such strategy to control over both delivery kinetics and drug pharmacokinetics (Fig. 27.1). The remainder of this chapter will focus on nanoparticles and nanofibers as next-generation strategies for vaginal drug delivery.

27.3 Nanomaterial Strategies for Vaginal Delivery

27.3.1 Nanocarriers for Drug Delivery: Overview

Nanocarriers have been widely investigated for many drug delivery applications due to their potential for intracellular delivery, sustained release, formulation of a diverse range of agents, and targeting capabilities. Nanocarriers are identified by their size of 10–1,000 nm and include carriers such as nanoparticles, liposomes, micelles, dendrimers, nanotubes, and nanolipogels. They have been studied as delivery vehicles to the cardiovascular, pulmonary, and central nervous systems for applications such as cancer therapy, hormone delivery, and vaccination, and many have advanced to clinical trials [49]. Many extensive reviews on nanocarriers for drug delivery have been published [50–52], so here we will provide only an overview of this topic.

One advantage that nanocarriers facilitate compared to other delivery systems is their ability to deliver contents intracellularly, which may allow for greater concentrations of drug to reach their targets. Submicron-sized particles have been shown to have greater cell uptake and better submucosal penetration than microparticles [51]. Nanocarriers have been documented to enter cells via endocytic pathways, pinocytosis, receptor-mediated transport, and facilitated transport [53]. The small size of nanocarriers that enables intracellular delivery is an attractive feature of this system, particularly for agents with intracellular sites of action.

Nanocarriers also provide a strategy to sustain the release of agents over time, decreasing the need for frequent administration. Poly(lactic-co-glycolic acid) (PLGA) particles have been shown to escape the endosomal compartment and remain in the cytoplasm for extended periods of time [54]. Animal studies using PLGA nanoparticles for gene delivery have shown sustained levels of tissue gene expression for at least 4 weeks after administration [55]. Additional studies have shown that nanoparticles can extend the period of drug delivery from 3 days to 2 weeks or longer compared with free drug in solution [54, 56].

Besides providing intracellular delivery and sustained release of agents, nanocarriers enable versatility in the types of agents that can be delivered. Nanocarriers can be made from polymers with varying rates of degradation, chemical, and physical properties. They have been used to deliver a diverse range of agents including hydrophilic drugs [57], hydrophobic drugs [56], DNA [55], siRNA [58, 59], and proteins [47]. For many of these agents, encapsulation within nanocarriers provides protection from degradation due to surrounding environment. This is particularly true for drugs and proteins that are susceptible to enzymatic or hydrolytic degradation [60].

Both synthetic and natural materials have been used to create nanocarriers [61]. PLGA has been one of the most widely used synthetic polymers for nanoparticles, due to its controllable degradation rate and biocompatibility, and it is already being used in humans for sutures, bone screws, and contraceptive implants [51]. Other synthetic materials such as poly(lactic acid) (PLA)[62], poly(caprolactone) (PCL) [63], and poly(acrylates)[64] have also been used as materials for nanoparticles.

Synthetic materials offer the advantage of providing greater control over the molecular weight distribution and side chain identity compared to natural materials. Some natural polymers that have been used for fabricating nanocarriers include chitosan, alginate, collagen, and gelatin [65]. Natural polymers often possess reactive sites that can be cross-linked or modified with ligands, and they are generally cytocompatible, although some concerns have been raised about their immunogenicity [61].

Several fabrication methods have been used to synthesize nanoparticles, including single and double emulsion techniques [66], nanoprecipitation [67], spray drying, and electrospraying [68]. Concerns have been raised with the ability to scale up emulsion and nanoprecipitation techniques, and spray drying can result in reduced product yields [66]. After fabrication, particles are generally characterized for their size and zeta potential using dynamic light scattering, morphology with scanning electron microscopy, and drug loading and release using chromatographic methods. Various assays are used to confirm that agents retain their activity after encapsulation into particles and are biocompatible, depending on the application.

Despite all the potential advantages that nanocarriers offer for enhanced delivery, several barriers exist that have prevented nanocarriers from advancing toward clinical application. First, a trade-off exists between enhanced uptake and controlled release. While the small size of nanocarriers enhances tissue penetration and intracellular uptake, this comes at a cost of a less controllable release profile due to their large surface area. Smaller nanoparticles generally display significant burst release, while larger particles display more controlled release but have reduced uptake [53]. Second, depending on surface properties, nanoparticles can have poor colloidal stability and problems with aggregation [69, 70]. Third, low drug encapsulation efficiencies have been observed for some agents [71], likely due to drug partitioning into the aqueous phase during formulation. Finally, nanocarriers have been shown to either stimulate or suppress immune responses, and methods to characterize potential adverse immunologic responses are not yet well established [72]. Immunoglobulins and complement proteins C3, C4, and C5 are capable of binding to the surface of some nanoparticles, making them detectable by the mononuclear phagocytic system and preventing them from reaching their targets [73]. New approaches are needed to address these concerns and more extensively evaluate the immune response to nanoparticles. For instance, modifying particle surface chemistry to prevent aggregation may enhance colloidal stability and reduce interactions with the immune system, and higher encapsulation efficiencies may be obtained using electrospraying technology for particle synthesis.

27.3.2 Nanocarriers for Vaginal Drug Delivery

Due to their ability to be rationally designed for specific environments, nanocarriers are a promising platform to overcome barriers associated with mucosal delivery routes including to the vaginal and cervical epithelium. Nanocarriers are being developed that can penetrate the mucus layer [74, 75] and respond to pH changes within the vagina [76, 77] for optimal delivery. Local administration of nanocarriers to the vagina also allows for the bypass of hepatic first-pass metabolism, higher local drug concentrations at the site of infection, and reduced side effects compared to systemic delivery. This is especially beneficial for diseases that are transmitted sexually like HIV, in which the initial cellular targets of viral infection are the immune cells of the vagina and cervix. Additionally, the cervicovaginal tract offers a large area for nanocarrier delivery because the surface is comprised of numerous folds called rugae. Several helpful reviews on nanoparticles for mucosal drug delivery have been published [44, 53, 78, 79]. Here we will focus on nanomaterials engineering strategies that have been used to overcome specific barriers related to vaginal drug delivery (Fig. 27.2).

27.3.2.1 Overcoming the Mucus Barrier

Although local delivery offers many advantages, one particularly challenging aspect of vaginal delivery is that of the mucus barrier produced by mucus-secreting cervical cells. Cervicovaginal mucus is a cross-linked viscoelastic hydrogel composed primarily of water with ~2–5 % mucin by weight [79]. The composition, pH, and rheological properties are highly variable, depending on the menstrual cycle and health and reproductive status of the woman. The presence of semen (pH 7–8) also influences mucus properties [80]. The pore size of cervicovaginal mucus has been estimated to be 340 nm on average, with a range of 50–1,800 nm [81]. Notably, CVF is a discontinuous layer, with some areas along the cervicovaginal tract having no mucus and some areas having a mucus layer that is several millimeters thick [82]. Because mucus is shear thinning, it is able to maintain an unstirred layer of mucus called the adherent layer that lies directly underneath the more rapidly cleared luminal layer [79, 83]. The mucus barrier must be crossed in order for nanoparticles to reach the underlying target cells, but its variable properties make it challenging to characterize.

Nanoparticles are being designed to overcome this barrier by controlling the surface properties and size of particles to modulate their interaction with the mucus layer [53, 79, 80, 84]. Particle size in the range of 200–500 nm has been found to be optimal for diffusion through mucus [79]. Large particles (>1,000 nm) are too large to penetrate porous mucus network, but small particles (100 nm) can become physically entrapped within the tortuous polymer network [82]. The hydrophobicity of the particle surface also appears to greatly impact particle diffusivity. PEGylation of particles has been shown to increase transport in mucus [79, 85, 86]. Further, surface charge has been shown to greatly influence particle diffusion through mucus. Negatively charged particles are able to move more quickly through mucus than positively charged particles, likely due to electrostatic repulsion between particles and negatively charged functional groups on mucin [82]. The effect of electrostatics on transport through mucus has been demonstrated with HIV virions themselves. The production of lactic acid in cervical mucus has been shown to neutralize the negative surface charge of the HIV virion and significantly slow its diffusion

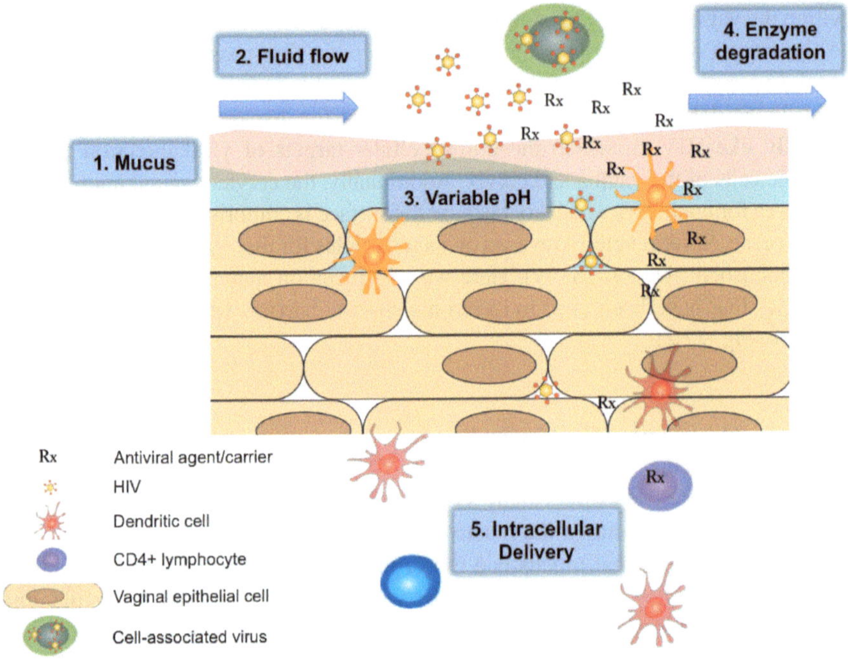

Barrier	1. Mucus	2. Fluid flow	3. Variable pH	4. Enzyme degradation	5. Intracellular delivery
Strategy	Mucus-penetrating particles	Nanoparticles that interact with mucus Delivery systems for sustained release	pH-responsive drug delivery	Protection of contents within particles	Targeted nanoparticles Engineering particle size, shape, surface charge for intracellullar uptake
References	[84, 80, 79, 74]	[74, 79, 82, 84, 71]	[76,77, 140]	[47, 59]	[113, 106, 116, 119]

Fig. 27.2 Strategies to overcome barriers in vaginal drug delivery. Intravaginal delivery of drugs presents several unique barriers depicted above: (1) cervicovaginal mucus that can trap drugs and nanocarriers, (2) fluid flow out of the body due to normal discharge of vaginal fluid, (3) variable vaginal pH arising from presence of semen or stage of menstrual cycle, (4) potential for drug degradation due to vaginal enzymes and reactive oxygen species, and (5) the vaginal epithelium layer, which provides a barrier to drugs intended for intracellular delivery to underlying cell populations. In designing vaginal drug delivery systems such as topical microbicides, much can be learned from how HIV itself has evolved to penetrate these barriers. Examples of strategies and drug delivery systems that have been engineered to overcome these barriers are listed above

compared to cervicovaginal mucus at pH 6–7, in which HIV can maintain its negative charge [87].

Mucoadhesive particles and mucoevasive particles have proposed as two strategies to overcome the mucosal barrier. Mucoadhesive particles are typically fabricated with a positively charged surface that will electrostatically bind to the negatively charged carboxyl and sulfate groups on mucin [53]. These particles are designed to remain in the mucus, slowly releasing their contents that can then diffuse through the mucus layer into underlying cells. Chitosan has been used as a material for mucoadhesive particles for nasal siRNA delivery [58] and vaginal drug delivery [71] because of its cationic nature resulting from protonated amine groups. In addition to surface charge, another determinant of mucoadhesion is size. Meng et al. studied the effect of particle size on mucoadhesion and encapsulation efficiency of the microbicidal drug candidate TFV [71]. They found that 188 nm particles demonstrated a twofold increase in mucoadhesion to porcine vaginal tissue compared with 900 nm particles. However, the smaller particles only resulted in 5 % encapsulation efficiency compared with 20 % for the larger particles, so the authors concluded that the larger-sized particles were most optimal for delivery of TFV. One must consider the balance between size and diffusivity in designing mucoadhesive particles. While smaller particles have increased surface area that leads to greater mucoadhesion, they may also have increased diffusivity in mucus compared to larger particles.

A contrasting approach to mucoadhesive particles is to create mucus-penetrating particles that are able to reach the underlying cells in the submucosa by penetrating the mucus layer. Since particle retention is also limited by the periodic sloughing of the mucus layer, particles that are trapped within this layer may be expelled before reaching their cellular targets [53]. As such, several groups have designed the surface properties of nanoparticles so that they can better penetrate the negatively charged mucus layer [74, 79, 82, 85]. One of the more successful strategies has been to mimic viral strategies for mucus penetration by creating hydrophilic, net neutrally charged surface to minimize adhesion-causing hydrophobic and electrostatic interactions [53, 79].

The density of PEG surface coating and molecular weight of PEG were found to greatly influence transport, as demonstrated by Wang et al. [86]. A highly dense surface coating of PEG was found to result in faster transport due to minimized hydrophobic interactions with mucin. Low-molecular-weight PEG was also found to enhance transport, with 2 kDa and 5 kDa PEG resulting in much faster transport than 10 kDa PEG. The authors suggested that high-molecular-weight PEG may result in entanglements with mucin polymer chains, preventing diffusion [86]. Using this strategy, Lai et al. modified the surface of polystyrene nanoparticles with a high density of low-molecular-weight (2 kDa) PEG to create a net neutrally charged, hydrophilic surface [79]. They found transport rates in human mucus were enhanced by three orders of magnitude for 200 and 500 nm particles. Mert et al. have also shown that a dense coating of low-molecular-weight PEG is optimal for mucus penetration [85]. They report that vitamin E/5 kDa PEG (VP5k)-coated PLGA nanoparticles better penetrate human cervicovaginal mucus than poly(vinyl alcohol) (PVA) or 1 kDa PEG-coated particles and that VP5k allowed for controlled release of paclitaxel over 4 days. They attributed this difference to the denser PEG coating and more neutral surface charge for 5 kDa PEG-coated particles compared with the 1 kDA PEG particles.

Conferring a negative surface charge to nanoparticles also appears to enhance their ability to penetrate mucus. Das Neves et al. have studied the influence of surface charge on rate of transport in simulated vaginal fluid-containing mucin adjusted to pH 4.2 and 7.0 [82]. They evaluated dapivirine-loaded PCL particles with three different surface coatings: poloxamer 338 NF (PEO), sodium lauryl sulfate (SLS), and cetyltrimethylammonium bromide (CTAB). They found that all PCL particles were mildly mucoadhesive on their own, likely due to hydrophobic interactions between the PCL polymer and mucin. However, they reported that nanoparticles with negatively charged surfaces (modified with PEO or SLS) were able to move through mucus simulant by subdiffusive transport at rates that may be appropriate for microbicide delivery. In contrast, transport of nanoparticles with positively charged surfaces (CTAB modified) was impaired, likely from electrostatic interactions with mucin. This work highlights the importance of considering surface charge in engineering nanoparticles with mucoevasive properties.

There is emerging evidence that conventional (i.e., uncoated) polymeric nanoparticles may be unable to penetrate cervicovaginal mucus to reach the underlying target cells. For example, Yu et al. report that PLGA particles were slowed by 12,000-fold in human cervicovaginal mucus compared to water, whereas particles made from the diblock copolymer PLGA-PEG were slowed only eightfold in human cervicovaginal mucus compared to water [75]. Cu et al. have studied the distribution of PEG-modified (mucus-penetrating), avidin-modified (mucoadhesive), and unmodified PLGA nanoparticles following intravaginal administration in mice [84]. They found that that surface properties significantly impact their ability to penetrate tissues. Approximately five times greater vaginal retention was observed for both the mucus-penetrating PEG particles and the mucoadhesive avidin-modified particles over 24 h compared to unmodified PLGA particles. However, significantly more PEG-modified PLGA particles remained in tissue extracts up at to 6 h following administration, leading the authors to conclude that the mucus-penetrating modification was most effective for intravaginal delivery. Ensign et al. developed mucus-penetrating 5 kDa PEG-conjugated polystyrene particles loaded with acyclovir that were able to prevent herpes simplex virus 2 (HSV-2) in mice [74]. The acyclovir mucus-penetrating particles protected 53 % of mice against HSV-2 challenge compared to 16 % of mice receiving soluble drug. In addition, they found over a nearly threefold improvement in tissue coverage (from 30 to 87 %) and longer tissue retention provided by mucus-penetrating particles compared with conventional (carboxyl-modified polystyrene) particles.

While the improved mucus penetration and tissue retention attained by mucus-penetrating particles represents an important advance in vaginal drug delivery, some challenges remain. Particularly, the effect of modifying surface properties for mucus penetration on intracellular uptake has yet to be determined. Particles with a net neutral charge may be optimal for penetrating mucus, but there may be a trade-off with lower levels of intracellular uptake. Size, material composition, and surface chemistry have been shown to affect cell uptake [88–90]. Specifically, PEG modification has been shown to reduce phagocytosis in macrophages [88, 91]. Studies that pair mucus penetration capabilities of nanoparticles with their intracellular uptake, such as those performed by Das Neves et al., would provide valuable insight into this potential setback [63, 82]. Balance between the ability of nanoparticles to penetrate the mucus layer and deliver their payload intracellularly may best be achieved by evaluating particle transport, intracellular uptake, and cytotoxicity in parallel for all formulations.

27.3.2.2 Overcoming Variations in Vaginal pH

Another challenge unique to vaginal drug delivery is the varying pH present in the reproductive tract. Normal physiologic pH in the vagina ranges from 4 to 6 but can change with the presence of bacterial infections, changes in the menstrual cycle, or presence of semen [53, 92]. Lactobacilli convert glycogen to lactic acid and are the primary microbe responsible for maintaining vaginal pH in premenopausal women [53]. Once particles penetrate the vaginal epithelium, they would be subject to increasing pH in the subepithelial layer and intracellularly.

Instead of seeing vaginal pH fluctuation as a barrier, several groups have used this variation as a stimulus for "smart" drug delivery from nanoparticles. The normally acidic pH of the vagina is altered during sexual intercourse, since the pH of semen is known to be basic (7–8) with a high buffering capacity [93]. Polymers that are soluble in alkaline solutions but insoluble in acidic solutions could serve as stimuli-responsive materials to release drug contents upon an increase in pH. Such polymers could be used to trigger release of agents that block HIV action when the virus enters the body via semen. Zhang et al. have created Eudragit® S-100/PLGA pH-responsive particles for semen-triggered vaginal delivery of TFV [76]. Eudragit® S-100 is a pH-responsive copolymer made from methacrylic acid-methyl methacrylate (1:2) that is soluble in an alkaline environment. They show a fourfold increase in drug release rate from these particles over 72 h in the presence of a semen fluid simulant (pH 7.6) compared with vaginal fluid simulant (VFS) (pH 4.2). Yoo et al. also investigated Eudragit® S-100 as a material to create pH-responsive particles for mucosal delivery [77]. They show that particles loaded with hydrophilic or hydrophobic model drugs release <40 % of contents in vitro at pH 4.0, compared with >90 % of contents at pH 7.4 over 6 h. While this system has yet to be evaluated in vivo, initial results suggest that pH-triggered release from Eudragit® particles may be a feasible method for stimuli-responsive drug delivery to the vagina.

27.3.2.3 Vaginal Delivery of Antiretroviral Agents, Proteins, and Nucleic Acids

First generation microbicides for HIV chemoprophylaxis were based on materials with nonspecific action against HIV and included surfactants, polyanions, and dendrimers [94]. Disappointingly, these products failed to show efficacy in clinical trials, with some even increasing susceptibility to HIV. The first microbicide to show clinical efficacy was a microbicide gel containing 1 % TFV, resulting in an overall 39 % reduction in HIV infections among South African women [24]. The current paradigm for designing microbicides has shifted to focus on agents with specific mechanisms of actions against HIV [95]. While delivering agents with specific against HIV may prove to be more effective than nonspecific methods, formulating them into vaginal products so that they are active and bioavailable has been challenging.

Nanoparticles offer several advantages for the development of HIV microbicides in particular, including overcoming challenges of drug stability and solubility, co-delivery of diverse agents, and providing sustained release. First, many ARVs have low aqueous solubility, making them hard to administer in their free forms. Nanoparticles have been used to encapsulate and release dozens of hydrophobic drugs, including many ARVs [53, 96]. They also offer a means to protect their contents from premature degradation by vaginal pH and enzymes. Nanoparticles have been made that encapsulate ARVs including zidovudine [91, 97], efavirenz (EFV) [56], saquinavir [98], lopinavir [56], ritonavir [56], TFV [71, 76], TFV disoproxil fumarate [76], dapivirine [63], and indinavir [99]. Some of these particles were designed for systemic delivery for HIV therapy, and some were proposed for topical application for HIV prevention. Several helpful reviews have previously been published on ARV-loaded nanoparticles [96, 100, 101].

One paradigm for engineering next-generation microbicides is that of highly active antiretroviral therapy (HAART). HAART has been used successfully to treat people infected by HIV through the coadministration of multiple ARV drugs, surmounting the drug-resistant strains that come from a rapidly mutating virus. However, co-formulating antiviral drugs of diverse physicochemical properties into a single product has posed a challenge for topical microbicide development. Polymeric nanoparticles are a platform that allows for individual formulation of ARVs that can subsequently be delivered simultaneously.

Chaowanachan et al. have investigated the potential for ARV nanoparticles to provide drug synergy [102]. They formulated PLGA nanoparticles loaded with EFV or saquinavir (SQV) and evaluated their antiviral activity and synergy with free TFV in TZM-bl cells. EFV nanoparticles resulted an ~50-fold decrease in IC50 value compared to free EFV, in addition to strong synergism when delivered with TFV. SQV nanoparticles resulted in a ~2-fold reduction in IC50 and also showed synergistic activity in combination with TFV. Even though intracellular delivery was not measured in this study, the authors expect this is the mechanism by which nanoparticles produced increased activity compared to free drug. This work demonstrates the potential for nanoparticles not only to facilitate greater delivery than unformulated ARVs but also to allow for drug–drug interactions that may not be possible just by delivering free drug combinations.

In addition to the delivery of hydrophobic ARV drugs, nanocarriers have also been used as a vehicle for the vaginal delivery of biologic agents, including peptides and nucleic acids. These agents would likely be quickly degraded by the acidic vaginal pH or enzymes in their free form, so nanocarriers may prove especially helpful for their delivery.

PSC-RANTES, a chemokine analogue that blocks CCR5 expression, is of particular interest for vaginal drug delivery due to its picomolar potency and its demonstrated in vivo protection against HIV-1 when topically applied in rhesus macaques [36]. However, for in vivo efficacy, doses were required which were orders of magnitude higher than concentrations necessary for in vitro efficacy. Hypothesizing that nanoparticles could overcome problems with free protein delivery such as poor submucosal tissue penetration or premature protein degradation, Ham et al. have created PLGA nanoparticles loaded with PSC-RANTES [47]. Indeed, they found a fivefold increase in tissue uptake over 4 h for PSC-RANTES encapsulated in nanoparticles versus unformulated PSC-RANTES in an ex vivo cervical tissue model. Nanoparticles also provided enhanced tissue penetration of the peptide and localization at basal layers of the epithelium. They also report that encapsulation into nanoparticles did not affect the in vitro anti-HIV activity of PSC-RANTES. Approximately 70 % of the peptide was released in vitro over 30 days, with release being affected by the pH of release media and the L (lactide):G (glycolide) ratio of PLGA used for nanoparticle formulation. Decreased PSC-RANTES release was observed with increased pH (4.6–7.4) and with increasing L:G ratio (50:50–85:15). These results provide evidence that PLGA nanoparticles can be used to encapsulate and sustain release of a peptide relevant to vaginal drug delivery and further that the L:G ratio of PLGA may be used to control release kinetics.

PLGA nanoparticles have also been investigated as a strategy for the vaginal delivery of nucleic acids. Woodrow et al. loaded PLGA nanoparticles with siRNA targeted against the gene encoding enhanced green fluorescent protein and studied gene expression after topical intravaginal application to transgenic mice [59]. In vitro release studies showed that particles were able to provide sustained, linear release of siRNA over 30 days, with about 50 % of total encapsulated siRNA released. Sustained gene silencing was observed throughout the mouse reproductive tract for at least 14 days after a single topical application of nanoparticles, and histological analysis revealed that particles were able to penetrate deeply into the epithelial tissue. The authors concluded that PLGA nanoparticles provide effective and sustained release of siRNA and are less inflammatory than siRNA lipoplexes and thus demonstrate an expanded application of PLGA nanoparticles to mucosal surfaces. This work has been expanded on by Eszterhas et al. who hypothesized that by lowering expression of CD4 and CCR5 receptors in tissue, infection could be reduced because HIV would be less able to enter cells [103]. They delivered CD4- and CCR5-specific siRNA in INTERFERin® (Genesee Scientific) nanoparticles to human cervical explants and monitored gene expression and HIV-1 infection over 5 days. Explants exposed to nanoparticles were found to have lower levels of CD4 and CCR5 transcripts, as well as reduced HIV-1 reverse transcripts. They also observed increased production of IFN-α, a potent antiviral cytokine, which may strengthen antiviral activity. These examples demonstrate that nanoparticles can be used to effectively deliver siRNA intravaginally, which may have applications for topical microbicides for HIV prevention.

27.3.2.4 Drug Targeting and Intracellular Vaginal Delivery

Besides the challenge of formulating multiple drugs for simultaneous delivery, user adherence has been a substantial challenge in clinical trials of vaginal microbicides for HIV prevention. Low adherence to vaginal gels requiring daily administration has been suggested as a primary reason for the lack of efficacy observed in FEM-PREP and VOICE trials [104, 105]. The development of a long-acting microbicide that requires less frequent administration provides a means to increase adherence. As discussed previously, nanoparticles have been shown to provide sustained release of agents for 2–4 weeks after a single administration. As such, nanoparticles may offer a means to create a long-acting microbicide by allowing for sustained intracellular delivery and targeting of agents with activity against HIV.

Research done by Destache et al. has provided insight into the persistence time and intracellular release profiles of ARV-loaded PLGA nanoparticles in peripheral blood mononuclear cells (PBMCs) [56]. They synthesized EFV-, lopinavir-, and ritonavir-loaded nanoparticles and measured intracellular drug levels over 28 days. Free drugs administered to PBMCs reached a peak concentration at 8 h and were eliminated within 48 h. In contrast, ARV-loaded nanoparticles reached peak drug concentrations at 24–96 h, and drug levels persisted at >0.9 μg for the full 28 days of the study. This study provides strong evidence for the ability of PLGA nanoparticles to provide sustained release of ARVs and potential application for a long-acting microbicide.

Das Neves et al. have done work to investigate the relationships between surface modification of nanoparticles, cellular uptake, and antiviral activity using six cell types relevant to vaginal HIV transmission [63]. They formulated dapivirine-loaded PCL nanoparticles with three different surface modifiers (CTAB, PEO, or SLS) to create particles with either positively or negatively charged surfaces. Nanoparticles with a positive surface charge (CTAB) resulted in higher drug concentrations in VK2/E6E7 vaginal epithelia cells and HeLa cervical cells compared to negatively charged particles (SLS, PEO). They found that all nanoparticles resulted in enhanced intracellular drug levels in phagocytic cells and similar or improved activity against HIV compared to free drug, indicating a passive targeting mechanism of nanoparticles. Due to the relatively high levels of cytotoxicity observed with CTAB-modified particles, they concluded that particles with a negative surface charge (PEO or SLS modified) were better candidates for vaginal delivery of dapivirine. This study is consistent with other research that has shown that for some drugs, ARV nanoparticles result in enhanced antiviral activity when compared to free ARVs [102].

Another strategy for the delivery of ARVs has been termed "nanoART" (nano-antiretroviral therapy), which refers to crystalline drug broken into nano-sized pieces using techniques like wet-milling. NanoART has been proposed as a strategy to overcome pharmacokinetic limitations of ARVs by providing a more long-acting formulation with better biodistribution. Nowacek et al. have investigated several wet-milled ARVs for their physicochemical properties, intracellular delivery, and ability to prevent HIV-1replication in monocyte-derived macrophages [106]. They found that properties such as particle size, surface charge, and shape influence cell uptake and ARV efficacy. Specifically, a strong correlation of 0.92 was found between intracellular drug levels and protection against HIV for EFV and atazanavir. These results suggest that intracellular drug delivery is a key component of establishing high levels of efficacy. Roy et al. evaluated nanoART in HIV-1-infected human peripheral blood lymphocyte-reconstituted mice for the combination delivery of ARVs as a long-acting formulation [107]. They found decreased viral replication and higher CD4+ T-cell populations for mice receiving weekly subcutaneous injections of nanoART compared to orally administered conventional ARVs. While nanoART has shown promise as a strategy for systemic delivery of ARVs for HIV therapy, little work has been done yet to evaluate nanoART for topical application for HIV prevention.

A distinctive feature of nanoparticles compared with other delivery systems is their ability to be targeted to specific cell types through passive or active targeting. Passive targeting is based on physicochemical properties like hydrophobicity and size that lead to preferential uptake, whereas active targeting involves the use of targeting ligands or molecules. Targeted nanoparticles have been widely investigated as a strategy for cancer therapy, as many types of tumors are known to overexpress specific receptors [108]. Nanoparticles have been targeted to tumor surfaces using molecules such as monoclonal antibodies, aptamers, oligopeptides, and folic acid [109]. Targeted nanoparticles may also have valuable applications in vaginal drug delivery, particularly for HIV treatment and prevention. However, unlike in some types of tumor targeting, a surface marker that is unique to all cells infected with HIV has not been identified [110]. HIV is known to infect only certain cell types, including CD4+ T cells, CD4+ monocytes/macrophages, dendritic cells, follicular dendritic cells, some fibroblasts, and microglial cells [110]. Consequently, the HIV receptor CD4, coreceptors CCR5 and CXCR4, and macrophage or dendritic cell receptors may serve as potential targets. Gunaseelan et al. have published an informative review on targeting strategies for HIV infection [110].

Much research on HIV targeting has been aimed at developing a drug delivery system to eradicate virus from already infected persons through targeting reservoir cell populations that are latently infected with HIV. In particular, many strategies have focused on targeting the CD4+ lymphocyte [111] or macrophage reservoirs [112, 113]. Macrophages constitute a major HIV reservoir that harbors the virus in its latent form and are resistant to the cytotoxic effects of HIV, preventing its complete eradication from the body. Though CD4+ lymphocytes represent the largest HIV reservoir by number, the dynamics of HIV replication in macrophages indicate that this reservoir may be of even greater importance. Unlike CD4+ lymphocytes, which are rapidly killed by HIV, macrophages have been shown to continually produce high levels of HIV for at least 60 days after virus challenge [112].

Tuftsin is a tetrapeptide (Thr-Lys-Pro-Arg) that has been shown to naturally activate macrophages [113]. It binds specifically to macrophages, monocytes, and polymorphonuclear leukocytes. Dutta et al. have designed poly(propyleneimine) (PPI) nanoparticle dendrimers conjugated with tuftsin to deliver EFV to macrophages [113]. They showed a 34.5-fold increased cellular uptake for the dendrimers conjugated with tuftsin compared with free drug, as well as significantly higher cellular uptake of tuftsin-conjugated PPI nanoparticles versus PPI nanoparticles. Interestingly, they also observed increased uptake of tuftsin-conjugated PPI nanoparticles in HIV-infected macrophages versus uninfected macrophages, which they attribute to an increased activation state of macrophages infected with HIV. Other strategies for targeting macrophages include molecules that target the formyl peptide receptor 1, mannose receptor, or the Fc receptor [110].

While macrophages may represent a primary target for targeting HIV reservoirs for therapy, other cell types may be more appropriate targets for HIV prevention strategies that are intravaginally administered. Perhaps the most relevant targets for prevention are the cells considered to be the initial sites of HIV infection, including CD4+ T cells and Langerhans cells. HIV infection is thought be established through the transport of free HIV virions or cell-associated virus (primarily in macrophages) through the vaginal or cervical epithelium to underlying CD4+ cells [114]. Targeting strategies that would prevent this transport of free or cell-associated virus are potential approaches for designing novel microbicides.

In addition to cells expressing CXCR4 or CCR5, dendritic cells also play an important role in establishing productive HIV infection. Dendritic cells can uniquely bind to HIV without the use of these receptors by means of interactions between DC-SIGN and viral gp120 [115]. Langerhans cells can mediate trans-infection by transporting virions across the cervicovaginal epithelium to susceptible cells and migrating to T-cell-rich lymph nodes [115]. Targeting dendritic cells may therefore be a useful strategy for creating a more effective HIV microbicide. For example, Penadés et al. have developed mannosylated gold nanoparticles that interfere with DC-SIGN as a potential microbicide [116]. They report that these gold particles are able to inhibit DC-SIGN-mediated trans-infection of human T cells. By mimicking HIV in its cluster presentation of oligomannosides, this group was able to utilize the high surface area of nanoparticles, allowing for maximal interaction with gp120. Other targeting strategies specific to dendritic cells include targeting the C-type lectins present on dendritic cells, the CD205 receptor, or Langerin found on intraepithelial Langerhans cells [117].

Given the variety of cell types that can be infected by HIV, nanoparticles targeted to just one cell type may not be sufficient to prevent infection or completely eradicate virus from all reservoir sites. However, as the biology of HIV transmission becomes better understood, nanoparticles offer a strategy to specifically target early stage viral-host interactions. It is also conceivable that multiple types of nanoparticles that target different cell populations could be delivered simultaneously. Nanoparticles have also been made that directly target the HIV itself instead of indirectly targeting cells that could be infected by HIV. Examples include silver nanoparticles that have been found to specifically interfere with HIV entry [118] and polystyrene particles with concanavalin-A on their surface that can capture viral gp120 [119].

27.3.2.5 Future Challenges and Directions for Nanoparticles for Vaginal Delivery

Beyond the advances made in overcoming mucus barrier and variations in vaginal pH, several challenges remain for successful nanoparticle-mediated vaginal drug delivery. First, there is a need for studies that investigate how long nanoparticles can sustain the delivery of drugs to the vagina and surrounding tissue. As discussed previously, much work has demonstrated the potential of nanoparticles to sustain drug release when delivered via other routes of administration. Cellular uptake of dapivirine-loaded nanoparticles and intracellular dapivirine levels has been monitored over 6–10 h in six cell types relevant to vaginal drug delivery [63]. Similar studies are needed that compare differences in intracellular drug concentrations from delivering nanoparticle-formulated drug versus free drug, evaluated in vaginal cell lines but over time scales extending to several days or weeks. Discerning whether these drug levels actually correspond to protective concentrations that could prevent diseases such as HIV-1 would be of great relevance to this field.

Even if nanoparticles are shown to sustain intracellular drug levels and provide a long-acting dosage form, a practical method of administration of nanoparticles to the vaginal mucosa is still lacking. Several in vivo studies have administered nanoparticles intravaginally to mice in aqueous suspensions [74, 84], but it is unclear how this would translate to human use. A method of vaginal delivery for nanoparticles is needed that allows for a long product shelf life and enhanced retention time to provide better coverage of the vaginal tissue. One potential solution may be combination dosage forms, such as nanoparticles that release from a solid-state dosage form such as a vaginal ring, diaphragm, or electrospun nanofiber mesh. Further work that evaluates shelf stability, vaginal retention, and biodistribution that result from various dosage forms would serve to more quickly advance this work to clinical applications. Although there are many complex barriers that have prevented effective vaginal drug delivery in the past, nanocarriers represent a versatile delivery vector that may be able to overcome them.

27.3.3 Nanofibers for Drug Delivery: Overview

Electrospinning technology has been an area of tremendous growth in the last decade [120], as it offers great flexibility to engineer platforms that can be tailored to specific applications. Electrospinning is a simple process in which a polymer solution is pumped through a needle and an electric field is applied across the positively charged needle and grounded collector. Static electrical charges are induced on the molecules of the polymer solution, which then repel each other [121]. The polymer liquid is stretched into fibers upon the force of the electric field overcoming the surface tension of the solution. As the fibers stretch toward the collector and the solvent evaporates, they undergo whipping instability and form a mat of fibers on the collector. We refer the reader to several reviews on the topic of electrospun fibers for a more detailed explanation [120, 122, 123].

One of the attractive features of electrospinning is its versatility in terms of material selection, control over processing parameters, and product geometries (Fig. 27.3). Over 100 polymers of both hydrophilic and hydrophobic nature have been electrospun as of 2007 [123]. Polymers can be blended prior to electrospinning to adjust properties like degradation rate and mechanical behavior [124]. Properties including fiber diameter, pore size, spatial deposition, and alignment can be controlled through modifying parameters like flow rate, applied voltage, and polymer concentration [120, 122, 125, 126]. Electrospun fibers can also be engineered to mimic the

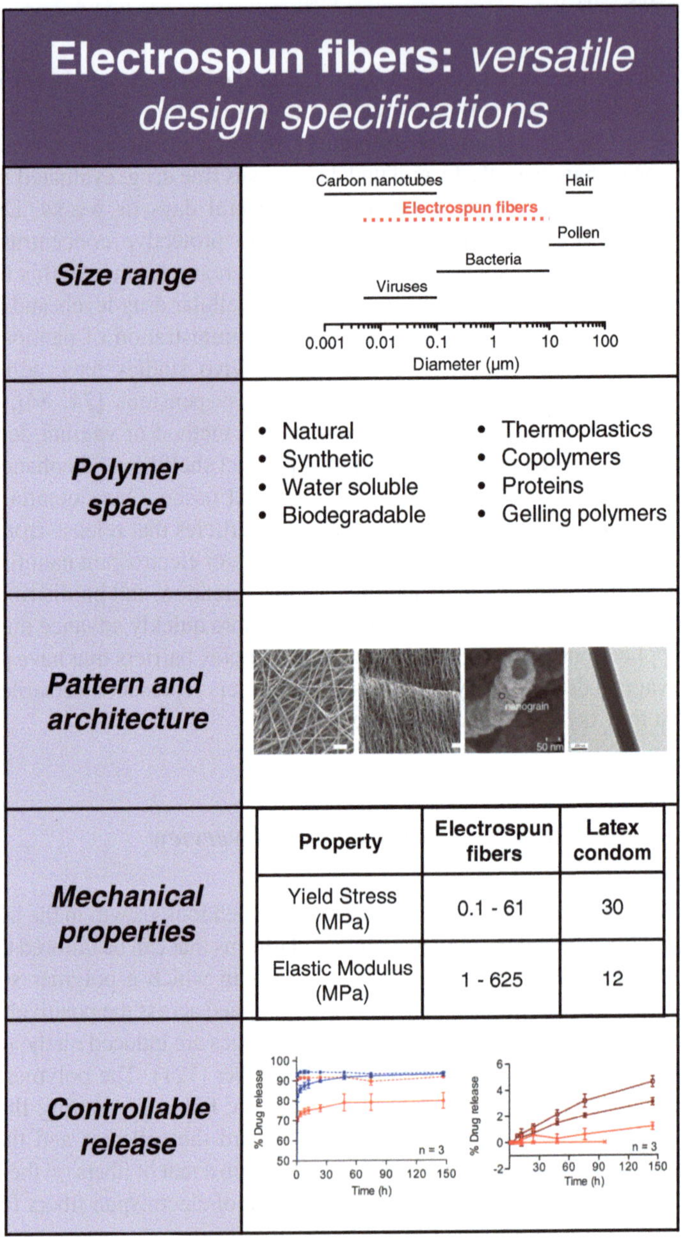

Fig. 27.3 Electrospun fibers enable multiple design specifications within a single device. One attractive feature of electrospun fibers is that their versatility may allow for multiple design requirements to be achieved in a single product for vaginal drug delivery. Fibers can be electrospun with controllable fiber diameters over a wide size range in comparison to microorganisms and carbon nanotubes (figure adapted from Greiner and Wendorff [199]). A diverse polymer space compatible with electrospinning allows for selection of materials that are most appropriate for the application. Various patterns including aligned and randomly oriented fibers are possible. Fiber architectures like the hollow fiber and core-shell fiber displayed above can also be obtained (reprinted with permission from Xia et al. [200]). Such control over polymer selection and patterning allows for fibers to be spun with versatile mechanical properties comparable with a latex condom [201]. Electrospinning also allows for a wide range of release kinetics, such as the burst release and sustained release of maraviroc from PLLA/PEO and PDLLA/PLLA fibers [141]

structure of the native extracellular matrix, a cross-linked porous network of glycosaminoglycans with collagen fibers ranging from 500 nm to 15 μm in diameter [127]. As such, they have been investigated for applications including 3-D tissue engineering scaffolds [121, 128, 129] and medical device coatings to reduce the foreign body reaction [130]. Geometries of the fiber mat (e.g., aligned mats, hollow tubes, yarn) and the degree of fiber alignment can be controlled by changing the nozzle configuration or type of collector [122]. Another advantage of electrospinning is that it is economical, allowing for high encapsulation efficiencies and relatively easy scale-up [121].

Electrospun fibers have recently been investigated for their potential as drug delivery systems. Fibers offer a large surface area-to-volume ratio that is amenable to quick drug delivery and have the capacity for high drug loading [131]. In addition, fibers can be engineered to achieve desired release properties by selecting materials with appropriate degradation rates, controlling electrospinning process parameters, and varying the drug binding mechanism. Fibers have been electrospun to deliver drugs such as antibiotics, anticancer drugs, proteins, and DNA [122]. Methods for incorporating drugs into fibers include adding drug to the polymer solution prior to electrospinning, or coating fibers with drug after electrospinning. Some have also utilized hydrogen bonding, hydrophobic, or electrostatic interactions between drug and polymer to complex drugs with fibers [120, 132]. Drug release from fibers is governed by three mechanisms: (1) desorption of drugs from fibers surface, (2) solid-state diffusion of drug through fibers, and (3) polymer degradation of fibers in vivo [120]. Several strategies have been employed to control drug release from fibers, including embedding drug within fibers versus coating drugs on the outside of fibers, modulating polymer crystallinity, changing fiber diameter, and creating a rate-controlling outer fiber shell using coaxial or emulsion spinning [120]. One of the key strategies has been using coaxial electrospinning to create "core-shell" fibers. Coaxial spinning allows for two different polymers to be electrospun into a single fiber, with one material forming an outer shell and another forming an inner core. This technique can protect biologic agents, nucleic acids, and even cells from organic solvents and the effects of the electric field, in addition to providing another way to modulate release [133, 134].

27.3.4 Considerations in Engineering Nanofibers for Vaginal Drug Delivery

Although electrospun fibers have been widely investigated for many drug delivery applications including transdermal [135], oral [136, 137], ocular [138], and abdominal delivery [139], they are just beginning to be explored for vaginal drug delivery. There are currently only two publications to our knowledge in which electrospun fibers have been studied as a drug delivery platform to the vaginal mucosa [140, 141].

Huang et al. have electrospun cellulose acetate phthalate (CAP) fibers that dissolve quickly upon increase in pH as potential semen-triggered microbicides [140]. CAP is an especially interesting material for HIV prevention given its ability to induce conformational changes in HIV glycoproteins and interfere with entry. To enhance the inherent antiviral activity of the CAP fibers, the fibers were loaded with two reverse transcriptase inhibitors (TMC 125 or TFV disoproxil fumarate). Unloaded CAP fibers were found to neutralize 50 % of HIV at 0.05 mg/mL CAP, and drug-loaded CAP fibers were found to achieve complete neutralization at 0.5 μg TFV disoproxil fumarate/mL. Electrospun fibers displayed low toxicity toward vaginal epithelial cells (<1.8 mg/mL CAP) as well as three strains of *Lactobacillus* (<0.1 mg/mL CAP). Fibers were observed to completely dissolve in <20 s when added to mixtures of semen and VFS, in contrast to their insolubility in VFS alone. The authors expect that these fibers would remain intact upon vaginal insertion, and then locally dissolve and release ARVs upon exposure to semen.

Ball et al. report electrospinning fiber meshes from blends of PLLA and PEO loaded with ARV and contraceptive agents for the prevention of HIV-1 infection and unwanted pregnancy [141]. They electrospun fibers incorporating inhibitors of viral reverse transcriptase and CCR5 binding that potently inhibited HIV infection in vitro. The PLLA and PEO blends demonstrated rapid burst release of >70 % of ARV payload within 1 h in VFS. By blending the amorphous isomer poly(D,L)-lactic acid (PDLLA) with semicrystalline PLLA, they were able to prevent the burst release and obtained small (5 %) but sustained release of maraviroc over 144 h. They further demonstrated that fiber meshes act as a both a chemical and physical barrier to sperm. The fibers could act to chemically inhibit sperm by releasing glycerol monolaurate to impair sperm motility and physically block sperm from penetrating the tortuous fibrous mesh. Fiber meshes were electrospun into a cylindrical geometry intended to provide physical coverage of both the vaginal epithelium and cervix. Nanofibers were expected to provide enhanced coverage of the mucosal tissue and vaginal rugae, supported by the extent of coverage observed when fluorescent fiber meshes were applied to mice. The authors envision that fiber meshes could be inserted simply with a tampon applicator, rendering this platform as discreet, female controlled, and reversible. They project that the application of drug-eluting fibers for vaginal delivery can extend beyond use as a microbicide and contraceptive to other applications such as mucosal vaccine delivery, STI treatment, and rectal microbicides. Figure 27.4 summarizes potential applications of nanofibers for vaginal drug delivery.

While there are no other examples of electrospun fibers for vaginal drug delivery in literature from which to draw direct conclusions, much can be learned from related work in transdermal drug delivery, wound dressings, and tissue engineering. We will now discuss aspects of electrospinning fibers for drug delivery using examples from literature and comment on their implications for vaginal drug delivery.

27.3.4.1 Engineering Nanofibers with Controllable Release Profiles

Electrospun fibers have been engineered to provide long-term sustained release of drugs over several months, ultrafast burst release within seconds, and asynchronous release of multiple agents. All three of these release profiles may be desirable for

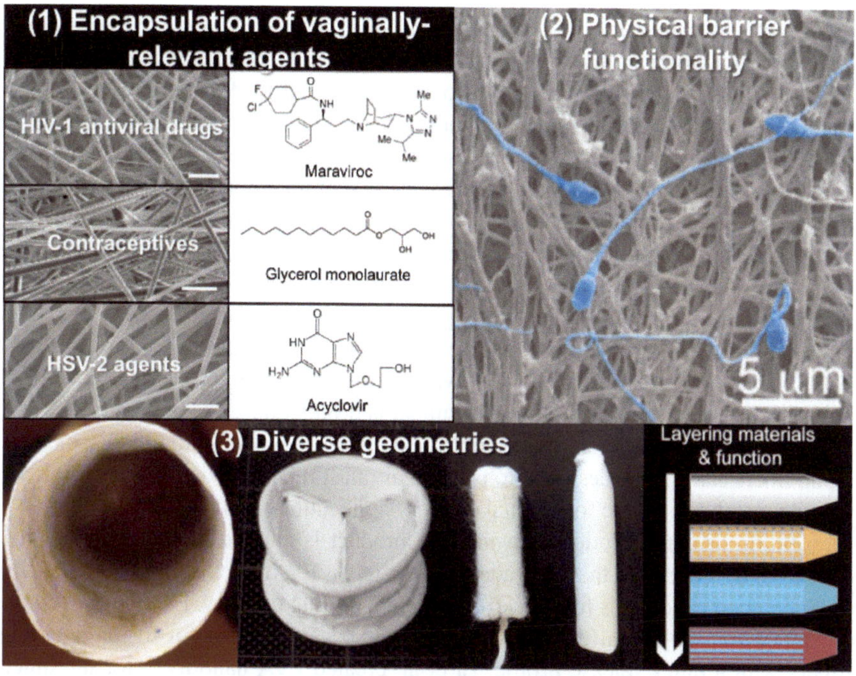

Fig. 27.4 Potential applications of nanofibers for vaginal drug delivery. Nanofibers are relevant for vaginal drug delivery with applications in HIV prevention, contraception, and HSV-2 prevention, among other reproductive health prevention and therapy applications. (1) Electrospun fibers have been shown to encapsulate agents relevant to vaginal delivery, including maraviroc, glycerol monolaurate, and acyclovir (shown *above*) [141]. (2) The tortuous network of an electrospun fiber mesh has been shown to physically block sperm penetration [141]. (3) Fibers have been fabricated into diverse geometries varying from simple tubular structures [141] to complex structures like an artificial heart valve (reprinted with permission from Szentivanyi et al. [196]). It is conceivable that fibers could be electrospun into vaginal dosage forms similar to films, rings, diaphragms, or coatings for tampon applicators

various applications for delivery to the lower female reproductive tract. For example, a quick-dissolving fiber platform may be most appropriate for a pericoitally administered microbicide so that the drug is able to reach target cell populations to provide protection before the virus. However, for a fiber-incorporating vaginal ring or diaphragm intended to remain in place for weeks to months, fibers that can provide long-term, sustained drug release with zero-order kinetics would be preferable. One can envision many types of products being created from electrospun fibers for vaginal delivery that facilitate release on time scales of <5 min to several months.

Many studies have been done showing that controlled release is possible from nanofibers [142–145]. Strategies to control drug release from fibers include coating them after electrospinning to prevent burst release [146, 147], creating a core-shell fiber structure using emulsion or coaxial electrospinning [148–150], cross-linking the polymer to slow release [151], and introducing air into fibers with an air displacement technique [143]. Simply modifying the choice or ratio of the polymer/drug combination can also be used as a means to control release. Jannesari et al. demonstrated that by lowering ciprofloxacin HCl drug loading from 10 to 5 % in PVA/poly(vinyl acetate) fibers, the burst release effect was reduced from ~75 to ~25 % of total drug content, with sustained release observed for 80 days [152]. Xie et al. developed PLGA nanofibers as chemotherapy implants and demonstrated in vitro release of drug for 60 days [153]. In settings where adherence to a product that must be frequently administered is a challenge, such as sub-Saharan Africa, a vaginal product that provides predictable, sustained release of active agents over a longer time is desirable. Furthermore, given the high rate of clearance in the vagina and presence of enzymes and reactive oxygen species that can degrade drugs [4, 154], the potential of fibers to slowly release drug while protecting unreleased contents from the vaginal microenvironment is promising.

Besides their ability to provide controlled release over weeks to months, electrospun fibers can also be designed to provide ultrafast release of drugs. Such a release profile may be desirable for vaginal delivery applications such as for pericoital HIV prevention or contraception in which the product is applied just prior to or after intercourse. Polymers that degrade quickly (e.g., PVA, PEO, alginate, chitosan) may be good candidates for quick release vaginal drug delivery. Examples of fibers designed for sublingual or oral delivery are especially insightful for engineering fibers with a quick release profile. Li et al. created PVA nanofibers for the quick release of caffeine and riboflavin that dissolved in <5 s and displayed burst release of 100 % of caffeine and 40 % of riboflavin within 60 s [136]. Macri et al. created fibers for the topical delivery of a hydrophilic peptide with controllable release kinetics [155]. By changing polymer composition, they were able to change the dominate release mechanism from polymer erosion to diffusion and thus obtain release over 9 h to 4 days. Similar strategies may be useful for engineering fibers as quick-dissolving platforms for vaginal drug delivery.

Another attractive feature of electrospun fibers is the ability to obtain asynchronous release profiles, in which the timing, order, and duration of the release of multiple drugs can be controlled. By layering or simultaneously electrospinning different fiber types, customized release profiles for multiple agents can be obtained within a single device. Okuda et al. show time-programmed release of two model dyes from tetra-layered poly(L-lactide-co-ε-caprolactone) fibers incorporating "barrier" and "basement" fiber layers to modulate release [156]. By changing the thickness and fiber diameters of the mesh layers, they were able to extend the release-suppressing period for one dye from 30 min to 1 h. Combination therapies have improved clinical outcomes for many diseases, including the advent of HAART for the treatment of HIV [157]. Applying the paradigm of HAART to HIV prevention has been considered as a promising strategy in the microbicide development field, but strategies to deliver multiple drugs simultaneously from one product are lacking. Electrospun fibers may offer a drug delivery platform that allows not only for the co-formulation of multiple drugs but also for asynchronous release of those drugs. Few existing drug delivery platforms can offer the same versatility in release profiles as electrospun fibers.

27.3.4.2 Fibers Enable Encapsulation and Simultaneous Delivery of Diverse Agents Including Drugs, Biologics, Cells, and Nanoparticles

Many diverse agents have been proposed for vaginal drug delivery, including drugs with a wide range of physicochemical properties, peptides, nucleic acids, antibodies, and nanocarriers. These agents are often difficult to co-formulate into a single device and sometimes even challenging to individually formulate into carriers by themselves. Additionally, due to the acidic environment and degrading enzymes present in the vagina [4], many of these agents degrade before reaching their targets. Since some polymers can be electrospun in water instead of organic solvents, fibers offer a platform to protect biologic agents from harsh formulation conditions, as well as from conditions present in the vagina. The versatility of materials that can be used in electrospinning can allow for individual fiber types that can be customized to encapsulate specific agents. Multiple fiber types could then be layered or simultaneously electrospun using multi-jet configurations, enabling the creation of a single product capable of simultaneously delivering multiple diverse agents. This would have profound implications for multipurpose prevention technologies, which are products that act against multiple indications, such as preventing HIV-1 infection and unwanted pregnancy. In addition, fibers may be a suitable platform for the delivery of nanoparticles, which are of great relevance for vaginal delivery but lack a practical method for uniform intravaginal distribution.

The materials selection and processing requirements for vaginal rings impose constraints on what pharmaceutical agents can be loaded [43], and hydrophobic drugs have limited solubility in vaginal gels [158]. In contrast, numerous studies show that electrospun fibers can incorporate drugs with a wide range of solubility [135, 141, 159, 160]. Electrospun fibers have been used as a means to increase bioavailability of poorly water-soluble drugs by solid dispersion. Verreck et al. created fibers loaded with a poorly water-soluble drug, itraconazole, and confirmed with differential scanning calorimetry the lack of a melting endotherm for the drug, suggesting a solid drug dispersion within fibers [161]. Yu et al. have shown similar improvement of delivery of ketoprofen from drug-loaded fibers via solid dispersion [162]. This strategy is of particular interest for vaginal drug delivery, since many ARVs of interest for topical HIV prevention have low aqueous solubility that limits their formulation in products like vaginal gels [163]. Fibers offer potential to increase bioavailability of such drugs for intravaginal administration.

Biologic agents including peptides, antibodies, enzymes, bacteria, and nucleic acids have been proposed for vaginal delivery. However, designing vehicles that can both encapsulate and deliver biologics to the vaginal mucosa without altering their activity remains a challenge. Electrospinning has been used to encapsulate a wide range of biologics, including proteins [164, 165], viruses [166], DNA [167], siRNA [168], bacteria [169, 170], and cells [171]. Maretschek et al. demonstrated tunable release of cytochrome C, a hydrophilic protein, from PLLA nanofibers by blending in a more hydrophilic polymer (PEG) to increase release [172]. Growth factors including human β-nerve growth hormone and basic fibroblast growth factor have also been embedded within [165, 173] or immobilized onto the surface of nanofibers [174] for tissue engineering applications. Two groups utilized coaxial spinning to protect adenovirus for gene delivery or plasmid DNA within the core of fibers during electrospinning [166, 167]. They demonstrated that these agents retained bioactivity when released from fibers in vitro. Another group reported on the controlled release of siRNA from PCL fibers for 28 days, with a silencing efficiency of 61–81 % similar to conventional siRNA transfection [168]. Although many of the examples discussed here has been targeted toward cancer therapy or tissue engineering applications, nanofibers may serve as useful drug delivery systems for antibodies against HIV-1 or sperm, bacteria such as *Lactobacilli* for the treatment of bacterial vaginosis, antigens for mucosal vaccines, or siRNA or DNA for vaginally administered gene therapy.

One aspect of nanofibers that may prove especially valuable for vaginal drug delivery is their potential to deliver nanoparticles, as a method suitable for the vaginal delivery of nanoparticles is currently lacking. Delivering nanoparticles in a conventional vaginal gel raises concerns about nanoparticle stability in an aqueous environment, long-term storage, and suboptimal gel retention. Given the breadth of work being done in designing nanoparticles for vaginal drug delivery as discussed previously, a platform suitable for uniform delivery of nanoparticles that provides long-term stability is greatly needed.

Several groups have shown that nanoparticles can be successfully incorporated into and released from nanofibers [68, 175–177]. Release profiles vary based on the drug/material combination. Composites made from coumarin-6 PLGA nanoparticles and PVA/PEO fibers showed 100 % release within 24 h [175]. In contrast, a zero-order release of fibroblast growth factor from heparin-based nanoparticle/chitosan fiber composites was observed over 30 days [176]. Besides a wide range in release profiles, nanofibers can also serve to protect agents from the surrounding environment. Chen et al. observed sustained release of intact siRNA for 50 days from chitosan nanoparticles embedded within PLGA fibers [177]. Nanoparticle/nanofiber composites may offer not only a suitable platform for the vaginal delivery of nanoparticles but also another method for controlling release profiles and enabling intracellular delivery from a single product.

27.3.4.3 Versatility in Engineering Bulk Properties of Fibers: Functionalized Surfaces, Mechanical Properties, and Stimuli Responsive

Along with versatility in the types of agents that can be encapsulated in electrospun fibers, they offer versatility in bulk properties that can be engineered, including modifying surface chemistry, mechanical properties, and stimuli-responsive drug release. Such features are attractive for vaginal drug delivery, potentially allowing for products to be engineered with mucoadhesive coatings to confer greater retention, mechanical properties appropriate for vaginal insertion, and “smart” drug release mechanisms that are triggered by vaginal physiologic cues.

The surfaces of electrospun fibers have been chemically functionalized with drugs, enzymes, cytokines, bioactive ligands, and polysaccharides to produce desired functions [178]. Strategies that have been used for surface modification are well described in the reviews by Yoo et al. [178] and Agarwal et al. [179] and include plasma treatment, surface grafting, co-electrospinning, and a wet chemical method. Of particular interest to vaginal drug delivery would be engineering the surface of fibers to be mucoadhesive. Poor vaginal retention has been cited as a problem for vaginal gel and other microbicide formulations, which can wash out with vaginal fluid flow or can leak from normal movement or due to gravity [44, 115, 180]. Thus, engineering the surface of a nanofiber-based drug delivery system to be mucoadhesive may alleviate some of these problems. Puttipaiboon et al. electrospun gastro-mucoadhesive fibers out of a blend of chitosan, PEO, and the corn-derived biopolymer zein [181]. They selected chitosan for its mucoadhesive properties and zein for its resistance to gastric fluid degradation and found that electrospun fiber films had adhesive strength similar to other gastro-mucoadhesive materials. By selecting materials for specific surface properties or modifying the fiber mesh surface post-electrospinning, fibers offer potential for a more efficient, longer-lasting formulation for vaginal drug delivery.

Depending on the method of insertion and the desired time scale of release, vaginal products have differing mechanical requirements. Vaginal rings must be flexible enough to be inserted without damaging the tissue but stiff enough so they do not slip or become expelled during normal use [182]. Diaphragms require a low compression relaxation value so they can form-fit closely around the cervix [183]. A relatively high tensile strength of vaginal films has been found to be important for optimal insertion and handling [184]. Nanofibers have previously been engineered to have a wide range in mechanical properties. Baji et al. have published an informative review on the effects of electrospinning parameters on fiber morphology, microstructure, and tensile properties [185]. For example, mechanical properties such as elastic modulus and strength have been found to increase with decreasing fiber diameter [185]. Additives like carbon nanotubes have been used to increase polymer chain orientation within fibers and thus strengthen their mechanical properties [186]. Another way to modulate mechanical properties is by increasing fiber alignment with a rotating mandrel collector [187, 188]. The ability to tune mechanical properties may offer a unique advantage for electrospun fibers to be tailored to the mechanical requirements of a variety of vaginal devices.

With the advances in "smart" polymers for drug delivery, there are opportunities to create nanofibers with stimuli-triggered release. Huang et al. has published a review on stimuli-responsive fibers, documenting reports of different types of fibers that change in response to pH, temperature, photo/optical, magnetic field, humidity, ethanol, glucose, or proteins [125]. One potential physiologic cue relevant to vaginal delivery is change in pH, as the presence of semen (pH 7–8) during intercourse raises the normal vaginal pH (pH 4–6) [140]. Other potential stimuli for vaginal delivery are the presence of reactive oxygen species, response to a vaginal enzyme, or an increase in temperature upon fiber insertion that would cause a transition in material properties and subsequent release of active agents.

27.3.4.4 Diverse Geometries in Three-Dimensional Nanofiber Structure Can Be Obtained

Another unique feature of nanofibers is their ability to be fabricated into a variety of geometries. Not only can the intrinsic geometry of individual fibers be controlled (i.e., fiber morphology, porosity, diameter) but also the structural geometry of the bulk fiber mesh. Much work has been done in designing three-dimensional scaffolds from nanofibers for tissue engineering [189–192]. Using techniques such as mixed electrospinning or multilayered electrospinning, one can obtain spatial control over the material composition of three-dimensional fiber structures [190]. McLure et al. have created a three-layered blood vessel scaffold from PCL, elastin, and collagen blends with appropriate mechanical properties for each layer [192]. Teo et al. created tubular electrospun scaffolds with fibers aligned to a 54.7° angle (optimal for blood vessel engineering) by using a collector with angled sharp metal blades [193]. Layering or co-electrospinning drug-loaded fibers would allow for the creation of composite materials that could serve as multipurpose prevention technologies to simultaneously address needs for HIV prevention and contraception.

Even more complex geometries of electrospun fibers have been obtained, such as a three-dimensional heart valve scaffold [194, 195]. These groups used convex or mildly concave collectors to obtain scaffolds, but more concave surfaces could be coated using air-blowing-assisted electrospinning [196]. Given preliminary results showing that electrospun fibers serve as an effective barrier to sperm [141], their ability to be configured into complex geometries may allow for their application to contraceptive devices such as vaginal diaphragms, rings, or sponges.

27.3.4.5 Considerations in Creating Products for Global Settings

Since an estimated 95 % of the 36 million people infected with HIV live in developing countries [197], it is important to consider what types of products would be most realistic for low-resource settings. Microbicides or contraceptive products that are available in a solid-state dosage form, have a long shelf life, and can be scaled-up in a cost-effective manner are highly desirable. Electrospun fibers represent a solid-state dosage form that may provide enhanced stability compared with semisolid dosage forms like vaginal gels. Kayaci et al. encapsulated vanillin within cyclodextrin inclusion complexes in PVA nanofibers and found prolonged shelf life and high temperature stability of the meshes [198]. Additionally, PVA nanofibers were found to enhance the shelf life of lipase enzyme, even at elevated temperature and humidity [164]. Similar enhancements in stability for ARVs or contraceptive agents would be valuable for vaginal drug delivery applications. Several manufacturers have also established cost-effective methods for large-scale productions of electrospun fibers [121]. Electrospun fibers represent a solid-state dosage form that would be easy to store and transport, result in minimal waste, and be scaled-up for mass production. Thus, fibers may offer advantages for creating economic products for low-resource settings.

27.3.4.6 Future Challenges and Directions in Nanofibers for Vaginal Drug Delivery

While the versatility of nanofibers offers many potential benefits for vaginal drug delivery, several challenges remain. One important limitation of nanofibers for vaginal drug delivery is that of limited penetration depth. Unlike nanoparticles, which are able to diffuse through tissue and be targeted for intracellular delivery, nanofibers do not offer opportunities for such targeting. Nanofibers are better suited to deliver agents that at locally within the penetration depth of the unformulated drug. On a basic science level, the fundamental physics of electrospinning are still not fully understood [122]. Continued research in this area may allow for even more control over fiber morphology and drug distribution, which will be key for reproducible and predictable pharmacokinetics. Release profiles that occur through both diffusion and degradation mechanisms can be difficult to predict, so fibers made from biodegradable materials must be carefully engineered so that the drug contents do not burst release at toxic levels.

Further, limited in vivo testing of nanofibers for drug delivery has been done, especially for vaginal delivery. Biocompatibility, inflammatory, and immune responses of nanofibers administered intravaginally must be well characterized before fiber-based vaginal products advance to clinical trials. Additionally, although conceptually fibers could be formed into diverse geometries, work remains to be done to evaluate what forms are practical for vaginal administration. Characterization of the mechanics, retention, and dissolution of various configurations will be necessary for advancing fibers for use in vaginal delivery. Safety and efficacy of fibers for vaginal delivery will need to be thoroughly evaluated before preclinical research can be translated to clinical applications. Despite these challenges, electrospun fibers offer much potential as a vaginal delivery platform for products beyond HIV microbicides and contraception to other reproductive health applications like STI treatment, yeast infections, surgical meshes, and mucosal vaccines.

References

1. Brannon-Peppas L (1993) Novel vaginal drug release applications. Adv Drug Deliv Rev 11:169–177
2. Benziger DP, Edelson J (1983) Absorption from the vagina. Drug Metab Rev 14:137–168
3. Platzner W, Poisel S (1978) Functional anatomy of the human vagina. In: Hafez ES, Evans TN (eds) Human reproductive medicine: the human vagina. North Holland, New York, NY, pp 39–54
4. Richardson JL, Illum L (1992) (D) Routes of delivery: case studies: (8) the vaginal route of peptide and protein drug delivery. Adv Drug Deliv Rev 8:341–366
5. Hafez ESE, Evans TN (1978) Human reproductive medicine: the human vagina. Elsevier North-Holland, New York, NY. ISBN 072040648X
6. Knuth K, Amiji M, Robinson JR (1993) Hydrogel delivery systems for vaginal and oral applications: formulation and biological considerations. Adv Drug Deliv Rev 11:137–167

7. Burgos MH, Roig de Vargas-Linaires CE (1978) Ultrastructure of the vaginal mucosa. In: Hafez ES, Evans TN (eds) Human reproductive medicine: the human vagina. North Holland, New York, NY, pp 63–93
8. Paavonen J (1982) Physiology and ecology of the vagina. Scand J Infect Dis Suppl 40:31–35
9. Wagner G, Levin RJ (1978) Vaginal fluid. In: Hafez ES, Evans TN (eds) Human reproductive medicine: the human vagina. North Holland, New York, NY, pp 121–137
10. Katz DF, Dunmire EN (1993) Cervical mucus: problems and opportunities for drug delivery via the vagina and cervix. Adv Drug Deliv Rev 11:385–401
11. Kistner RW (1978) Physiology of the vagina. In: Hafez ES, Evans TN (eds) Human reproductive medicine: the human vagina. North Holland, New York, NY, pp 109–120
12. Lee VHL, Yamamoto A (1989) Penetration and enzymatic barriers to peptide and protein absorption. Adv Drug Deliv Rev 4:171–207
13. Chien YW (1982) Controlled administration of estrus-synchronizing agents in livestock. In: Chien YW (ed) Novel drug delivery systems. Marcel Dekker, New York, NY, pp 51–95
14. Woolfson D, Malcolm K (2008) Intravaginal drug delivery technologies. In: Modified-release drug delivery technology, 2nd edition. Rathbone MJ, Roberts K, Lane ME, Hadgraft J (Eds.). Informa, New York, NY, pp 481–498
15. Haase AT (2010) Targeting early infection to prevent HIV-1 mucosal transmission. Nature 464:217–223
16. Ma Z, Lü FX, Torten M, Miller CJ (2001) The number and distribution of immune cells in the cervicovaginal mucosa remain constant throughout the menstrual cycle of rhesus macaques. Clin Immunol 100:240–249
17. Zhang Z-Q, Wietgrefe SW, Li Q, Shore MD, Duan L, Reilly C, Lifson JD, Haase AT (2004) Roles of substrate availability and infection of resting and activated CD4+ T cells in transmission and acute simian immunodeficiency virus infection. Proc Natl Acad Sci U S A 101:5640–5645
18. Pudney J, Quayle AJ, Anderson DJ (2005) Immunological microenvironments in the human vagina and cervix: mediators of cellular immunity are concentrated in the cervical transformation zone1. Biol Reprod 73:1253–1263
19. Saltzman WM (2001) Drug delivery: engineering principles for drug therapy: engineering principles for drug therapy. Oxford University Press, New York, NY
20. Schwartz JL, Rountree W, Kashuba ADM, Brache V, Creinin MD, Poindexter A, Kearney BP (2011) A multi-compartment, single and multiple dose pharmacokinetic study of the vaginal candidate microbicide 1 % tenofovir gel. PLoS One 6:e25974
21. Patterson KB, Prince HA, Kraft E, Jenkins AJ, Shaheen NJ, Rooney JF, Cohen MS, Kashuba ADM (2011) Penetration of Tenofovir and emtricitabine in mucosal tissues: implications for prevention of HIV-1 transmission. Sci Transl Med 3:112re4
22. Durand-Gasselin L, Van Rompay KKA, Vela JE, Henne IN, Lee WA, Rhodes GR, Ray AS (2009) Nucleotide analogue prodrug tenofovir disoproxil enhances lymphoid cell loading following oral administration in monkeys. Mol Pharm 6:1145–1151
23. Karim SSA, Kashuba AD, Werner L, Karim QA (2011) Drug concentrations after topical and oral antiretroviral pre-exposure prophylaxis: implications for HIV prevention in women. Lancet 378:279–281
24. Karim Q, Karim SSA, Frohlich JA, Grobler AC, Baxter C, Mansoor LE (2010) Effectiveness and safety of tenofovir gel, an antiretroviral microbicide, for the prevention of HIV infection in women. Science 329:1168
25. Achilles SL, Shete PB, Whaley KJ, Moench TR, Cone RA (2002) Microbicide efficacy and toxicity tests in a mouse model for vaginal transmission of Chlamydia trachomatis. Sex Transm Dis 29:655–664
26. Zhang Z-Q, Schuler T, Zupancic M et al (1999) Sexual transmission and propagation of SIV and HIV in resting and activated CD4+ T cells. Science 286:1353–1357

27. Geonnotti AR, Peters JJ, Katz DF (2005) Erosion of microbicide formulation coating layers: effects of contact and shearing with vaginal fluid or semen. J Pharm Sci 94:1705–1712
28. Doncel GF (2006) Exploiting common targets in human fertilization and HIV infection: development of novel contraceptive microbicides. Hum Reprod Update 12:103–117
29. Klasse PJ, Shattock R, Moore JP (2008) Antiretroviral drug-based microbicides to prevent HIV-1 sexual transmission. Annu Rev Med 59:455–471
30. Friend DR, Doncel GF (2010) Combining prevention of HIV-1, other sexually transmitted infections and unintended pregnancies: development of dual-protection technologies. Antiviral Res 88(Suppl 1):S47–S54
31. Pope M, Haase AT (2003) Transmission, acute HIV-1 infection and the quest for strategies to prevent infection. Nat Med 9:847–852
32. Ndesendo V, Pillay V, Choonara Y, Buchmann E, Bayever D, Meyer L (2008) A review of current intravaginal drug delivery approaches employed for the prophylaxis of HIV/AIDS and prevention of sexually transmitted infections. AAPS PharmSciTech 9:505–520
33. Andrews GP, Donnelly L, Jones DS, Curran RM, Morrow RJ, Woolfson AD, Malcolm RK (2009) Characterization of the rheological, mucoadhesive, and drug release properties of highly structured Gel platforms for intravaginal drug delivery. Biomacromolecules 10:2427–2435
34. Dyer JR, Kazembe P, Vernazza PL et al (1998) High levels of human immunodeficiency virus type 1 in blood and semen of seropositive men in sub-Saharan Africa. J Infect Dis 177:1742–1746
35. Tsai C-C, Emau P, Jiang Y, Tian B, Morton WR, Gustafson KR, Boyd MR (2003) Cyanovirin-N gel as a topical microbicide prevents rectal transmission of SHIV89.6P in macaques. AIDS Res Hum Retroviruses 19:535–541
36. Lederman MM, Veazey RS, Offord R et al (2004) Prevention of vaginal SHIV transmission in rhesus macaques through inhibition of CCR5. Science 306:485–487
37. Tsai C-C, Emau P, Jiang Y, Agy MB, Shattock RJ, Schmidt A, Morton WR, Gustafson KR, Boyd MR (2004) Cyanovirin-N inhibits AIDS virus infections in vaginal transmission models. AIDS Res Hum Retroviruses 20:11–18
38. Boadi T, Schneider E, Chung S, Tsai L, Gettie A, Ratterree M, Blanchard J, Neurath AR, Cheng-Mayer C (2005) Cellulose acetate 1,2-benzenedicarboxylate protects against challenge with pathogenic X4 and R5 simian/human immunodeficiency virus. AIDS 19:1587–1594
39. Wang Y, Abel K, Lantz K, Krieg AM, McChesney MB, Miller CJ (2005) The Toll-like receptor 7 (TLR7) agonist, imiquimod, and the TLR9 agonist, CpG ODN, induce antiviral cytokines and chemokines but do not prevent vaginal transmission of simian immunodeficiency virus when applied intravaginally to rhesus macaques. J Virol 79:14355–14370
40. Klasse PJ, Shattock RJ, Moore JP (2006) Which topical microbicides for blocking HIV-1 transmission will work in the real world? PLoS Med 3:e351
41. Herrera C, Cranage M, McGowan I, Anton P, Shattock RJ (2009) Reverse transcriptase inhibitors as potential colorectal microbicides. Antimicrob Agents Chemother 53:1797–1807
42. Garg S, Tambwekar KR, Vermani K, Kandarapu R, Garg A, Waller DP, Zaneveld LJD (2003) Development pharmaceutics of microbicide formulations. Part II: formulation, evaluation, and challenges. AIDS Patient Care STDS 17:377–399
43. Malcolm RK, Edwards K-L, Kiser P, Romano J, Smith TJ (2010) Advances in microbicide vaginal rings. Antiviral Res 88(Suppl 1):S30–S39
44. Rohan LC, Sassi AB (2009) Vaginal drug delivery systems for HIV prevention. AAPS J 11:78–87
45. Dobaria N, Badhan A, Mashru R (2009) A novel itraconazole bioadhesive film for vaginal delivery: design, optimization, and physicodynamic characterization. AAPS PharmSciTech 10:951–959
46. Asane GS, Nirmal SA, Rasal KB, Naik AA, Mahadik MS, Rao YM (2008) Polymers for mucoadhesive drug delivery system: a current status. Drug Dev Ind Pharm 34:1246–1266

47. Ham A, Cost M, Sassi A, Dezzutti C, Rohan L (2009) Targeted delivery of PSC-RANTES for HIV-1 prevention using biodegradable nanoparticles. Pharm Res 26:502–511
48. Akil A, Parniak M, Dezzutti C, Moncla B, Cost M, Li M, Rohan L (2011) Development and characterization of a vaginal film containing dapivirine, a non-nucleoside reverse transcriptase inhibitor (NNRTI), for prevention of HIV-1 sexual transmission. Drug Deliv Transl Res 1:209–222
49. Zhang L, Gu F, Chan J, Wang A, Langer R, Farokhzad O (2007) Nanoparticles in medicine: therapeutic applications and developments. Clin Pharmacol Ther 83:761–769
50. Rawat M, Singh D, Saraf S, Saraf S (2006) Nanocarriers: promising vehicle for bioactive drugs. Biol Pharm Bull 29:1790–1798
51. Panyam J, Labhasetwar V (2003) Biodegradable nanoparticles for drug and gene delivery to cells and tissue. Adv Drug Deliv Rev 55:329–347
52. Ganta S, Devalapally H, Shahiwala A, Amiji M (2008) A review of stimuli-responsive nanocarriers for drug and gene delivery. J Control Release 126:187–204
53. Mallipeddi R, Rohan LC (2010) Nanoparticle-based vaginal drug delivery systems for HIV prevention. Expert Opin Drug Deliv 7:37–48
54. Panyam J, Zhou W-Z, Prabha S, Sahoo SK, Labhasetwar V (2002) Rapid endo-lysosomal escape of poly(dl-lactide-co-glycolide) nanoparticles: implications for drug and gene delivery. FASEB J 16:1217–1226
55. Cohen H, Levy RJ, Gao J, Fishbein I, Kousaev V, Sosnowski S, Slomkowski S, Golomb G (2000) Sustained delivery and expression of DNA encapsulated in polymeric nanoparticles. Gene Ther 7:1896–1905
56. Destache CJ, Belgum T, Christensen K, Shibata A, Sharma A, Dash A (2009) Combination antiretroviral drugs in PLGA nanoparticle for HIV-1. BMC Infect Dis 9:198
57. Alukda D, Sturgis T, Youan B-BC (2011) Formulation of tenofovir-loaded functionalized solid lipid nanoparticles intended for HIV prevention. J Pharm Sci 100:3345–3356
58. Liu X, Howard KA, Dong M, Andersen M, Rahbek UL, Johnsen MG, Hansen OC, Besenbacher F, Kjems J (2007) The influence of polymeric properties on chitosan/siRNA nanoparticle formulation and gene silencing. Biomaterials 28:1280–1288
59. Woodrow KA, Cu Y, Booth CJ, Saucier-Sawyer JK, Wood MJ, Saltzman WM (2009) Intravaginal gene silencing using biodegradable polymer nanoparticles densely loaded with small-interfering RNA. Nat Mater 8:526–533
60. Perera G, Greindl M, Palmberger TF, Bernkop-Schnürch A (2009) Insulin-loaded poly(acrylic acid)-cysteine nanoparticles: stability studies towards digestive enzymes of the intestine. Drug Deliv 16:254–260
61. Chakravarthi S, Robinson D, De S (2007) Nanoparticles prepared using natural and synthetic polymers. In: Thassu D (ed) Nanoparticulate drug delivery systems. Informa Healthcare, New York, NY, pp 51–60
62. Xiong XY, Tam KC, Gan LH (2005) Release kinetics of hydrophobic and hydrophilic model drugs from pluronic F127/poly(lactic acid) nanoparticles. J Control Release 103:73–82
63. Das Neves J, Michiels J, Ariën K, Vanham G, Amiji M, Bahia M, Sarmento B (2012) Polymeric nanoparticles affect the intracellular delivery, antiretroviral activity and cytotoxicity of the microbicide drug candidate dapivirine. Pharm Res 29:1468–1484
64. Turos E, Reddy GSK, Greenhalgh K, Ramaraju P, Abeylath SC, Jang S, Dickey S, Lim DV (2007) Penicillin-bound polyacrylate nanoparticles: restoring the activity of β-lactam antibiotics against MRSA. Bioorg Med Chem Lett 17:3468–3472
65. Dang JM, Leong KW (2006) Natural polymers for gene delivery and tissue engineering. Adv Drug Deliv Rev 58:487–499
66. Jain RA (2000) The manufacturing techniques of various drug loaded biodegradable poly(lactide-co-glycolide) (PLGA) devices. Biomaterials 21:2475–2490
67. Bilati U, Allémann E, Doelker E (2005) Development of a nanoprecipitation method intended for the entrapment of hydrophilic drugs into nanoparticles. Eur J Pharm Sci 24:67–75
68. Bock N, Dargaville TR, Woodruff MA (2012) Electrospraying of polymers with therapeutic molecules: state of the art. Prog Polym Sci 37:1510–1551

69. Avgoustakis K, Beletsi A, Panagi Z, Klepetsanis P, Livaniou E, Evangelatos G, Ithakissios DS (2003) Effect of copolymer composition on the physicochemical characteristics, in vitro stability, and biodistribution of PLGA–mPEG nanoparticles. Int J Pharm 259:115–127
70. Santander-Ortega MJ, Csaba N, Alonso MJ, Ortega-Vinuesa JL, Bastos-González D (2007) Stability and physicochemical characteristics of PLGA, PLGA:poloxamer and PLGA:poloxamine blend nanoparticles: a comparative study. Colloids Surf A Physicochem Eng Asp 296:132–140
71. Meng J, Sturgis TF, Youan B-BC (2011) Engineering tenofovir loaded chitosan nanoparticles to maximize microbicide mucoadhesion. Eur J Pharm Sci 44:57–67
72. Dobrovolskaia MA, McNeil SE (2007) Immunological properties of engineered nanomaterials. Nat Nanotechnol 2:469–478
73. Owens DE III, Peppas NA (2006) Opsonization, biodistribution, and pharmacokinetics of polymeric nanoparticles. Int J Pharm 307:93–102
74. Ensign LM, Tang BC, Wang Y-Y, Tse TA, Hoen T, Cone R, Hanes J (2012) Mucus-penetrating nanoparticles for vaginal drug delivery protect against herpes simplex virus. Sci Transl Med 4:138ra79
75. Yu T, Wang Y-Y, Yang M et al (2012) Biodegradable mucus-penetrating nanoparticles composed of diblock copolymers of polyethylene glycol and poly(lactic-co-glycolic acid). Drug Deliv Transl Res 2:124–128
76. Zhang T, Sturgis TF, Youan B-BC (2011) pH-responsive nanoparticles releasing tenofovir intended for the prevention of HIV transmission. Eur J Pharm Biopharm 79:526–536
77. Yoo J-W, Giri N, Lee CH (2011) pH-sensitive eudragit nanoparticles for mucosal drug delivery. Int J Pharm 403:262–267
78. Takeuchi H, Yamamoto H, Kawashima Y (2001) Mucoadhesive nanoparticulate systems for peptide drug delivery. Adv Drug Deliv Rev 47:39–54
79. Lai SK, Wang Y-Y, Hanes J (2009) Mucus-penetrating nanoparticles for drug and gene delivery to mucosal tissues. Adv Drug Deliv Rev 61:158–171
80. Das Neves J, Amiji M, Sarmento B (2011) Mucoadhesive nanosystems for vaginal microbicide development: friend or foe? Wiley Interdiscip Rev Nanomed Nanobiotechnol 3:389–399
81. Lai SK, Wang Y-Y, Hida K, Cone R, Hanes J (2009) Nanoparticles reveal that human cervicovaginal mucus is riddled with pores larger than viruses. Proc Natl Acad Sci 107:598–603
82. Das Neves J, Rocha CMR, Gonçalves MP, Carrier RL, Amiji M, Bahia MF, Sarmento B (2012) Interactions of microbicide nanoparticles with a simulated vaginal fluid. Mol Pharm 9:3347–3356
83. Cone RA (2009) Barrier properties of mucus. Adv Drug Deliv Rev 61:75–85
84. Cu Y, Booth CJ, Saltzman WM (2011) In vivo distribution of surface-modified PLGA nanoparticles following intravaginal delivery. J Control Release 156:258–264
85. Mert O, Lai SK, Ensign L, Yang M, Wang Y-Y, Wood J, Hanes J (2012) A poly(ethylene glycol)-based surfactant for formulation of drug-loaded mucus penetrating particles. J Control Release 157:455–460
86. Wang Y-Y, Lai SK, Suk JS, Pace A, Cone R, Hanes J (2008) Addressing the PEG mucoadhesivity paradox to engineer nanoparticles that "slip" through the human mucus barrier. Angew Chem Int Ed 47:9726–9729
87. Lai SK, Hida K, Shukair S, Wang Y-Y, Figueiredo A, Cone R, Hope TJ, Hanes J (2009) Human immunodeficiency virus type 1 is trapped by acidic but not by neutralized human cervicovaginal mucus. J Virol 83:11196–11200
88. Alexis F, Pridgen E, Molnar LK, Farokhzad OC (2008) Factors affecting the clearance and biodistribution of polymeric nanoparticles. Mol Pharm 5:505–515
89. Zahr AS, Davis CA, Pishko MV (2006) Macrophage uptake of core–shell nanoparticles surface modified with poly(ethylene glycol). Langmuir 22:8178–8185
90. Schäfer V, von Briesen H, Andreesen R, Steffan A-M, Royer C, Tröster S, Kreuter J, Rübsamen-Waigmann H (1992) Phagocytosis of nanoparticles by human immunodeficiency virus (HIV)-infected macrophages: a possibility for antiviral drug targeting. Pharm Res 9:541–546

91. Mainardes RM, Gremião MPD, Brunetti IL, Da Fonseca LM, Khalil NM (2009) Zidovudine-loaded PLA and PLA–PEG blend nanoparticles: influence of polymer type on phagocytic uptake by polymorphonuclear cells. J Pharm Sci 98:257–267
92. Ravel J, Gajer P, Abdo Z et al (2010) Vaginal microbiome of reproductive-age women. Proc Natl Acad Sci 108:4680–4687
93. Owen DH, Katz DF (2005) A review of the physical and chemical properties of human semen and the formulation of a semen simulant. J Androl 26:459–469
94. Nuttall J (2010) Microbicides in the prevention of HIV infection: current status and future directions. Drugs 70:1231–1243
95. Whaley KJ, Hanes J, Shattock R, Cone RA, Friend DR (2010) Novel approaches to vaginal delivery and safety of microbicides: biopharmaceuticals, nanoparticles, and vaccines. Antiviral Res 88(Suppl 1):S55–S66
96. Lembo D, Cavalli R (2010) Nanoparticulate delivery systems for antiviral drugs. Antivir Chem Chemother 21:53–70
97. Bender A, Schäfer V, Steffan AM, Royer C, Kreuter J, Rübsamen-Waigmann H, Von Briesen H (1994) Inhibition of HIV in vitro by antiviral drug-targeting using nanoparticles. Res Virol 145:215–220
98. Shah LK, Amiji MM (2006) Intracellular delivery of saquinavir in biodegradable polymeric nanoparticles for HIV/AIDS. Pharm Res 23:2638–2645
99. Dou H, Morehead J, Destache CJ et al (2007) Laboratory investigations for the morphologic, pharmacokinetic, and anti-retroviral properties of indinavir nanoparticles in human monocyte-derived macrophages. Virology 358:148–158
100. Hillaireau H, Le Doan T, Couvreur P (2006) Polymer-based nanoparticles for the delivery of nucleoside analogues. J Nanosci Nanotechnol 6:2608–2617
101. Sharma P, Garg S (2010) Pure drug and polymer based nanotechnologies for the improved solubility, stability, bioavailability and targeting of anti-HIV drugs. Adv Drug Deliv Rev 62:491–502
102. Chaowanachan T, Krogstad E, Ball C, Woodrow KA (2013) Drug synergy of tenofovir and nanoparticle-based antiretrovirals for HIV prophylaxis. PLoS One 8(4):e61416
103. Eszterhas SK, Ilonzo NO, Crozier JE, Celaj S, Howell AL (2011) Nanoparticles containing siRNA to silence CD4 and CCR5 reduce expression of these receptors and inhibit HIV-1 infection in human female reproductive tract tissue explants. Infect Dis Rep. doi:10.4081/idr.2011.e11
104. Van Damme L, Corneli A, Ahmed K et al (2012) Preexposure prophylaxis for HIV infection among African women. N Engl J Med 367:411–422
105. Microbicide Trials Network (2013) Daily HIV prevention approaches didn't work for African women in the VOICE Study, Atlanta
106. Nowacek AS, Balkundi S, McMillan J, Roy U, Martinez-Skinner A, Mosley RL, Kanmogne G, Kabanov AV, Bronich T, Gendelman HE (2011) Analyses of nanoformulated antiretroviral drug charge, size, shape and content for uptake, drug release and antiviral activities in human monocyte-derived macrophages. J Control Release 150:204–211
107. Roy U, McMillan J, Alnouti Y et al (2012) Pharmacodynamic and antiretroviral activities of combination nanoformulated antiretrovirals in HIV-1-infected human peripheral blood lymphocyte-reconstituted mice. J Infect Dis. doi:10.1093/infdis/jis395
108. Brannon-Peppas L, Blanchette JO (2012) Nanoparticle and targeted systems for cancer therapy. Adv Drug Deliv Rev 64:206–212
109. Gu FX, Karnik R, Wang AZ, Alexis F, Levy-Nissenbaum E, Hong S, Langer RS, Farokhzad OC (2007) Targeted nanoparticles for cancer therapy. Nano Today 2:14–21
110. Gunaseelan S, Gunaseelan K, Deshmukh M, Zhang X, Sinko PJ (2010) Surface modifications of nanocarriers for effective intracellular delivery of anti-HIV drugs. Adv Drug Deliv Rev 62:518–531
111. Dahl V, Josefsson L, Palmer S (2010) HIV reservoirs, latency, and reactivation: prospects for eradication. Antiviral Res 85:286–294
112. Aquaro S, Caliò R, Balzarini J, Bellocchi MC, Garaci E, Perno CF (2002) Macrophages and HIV infection: therapeutical approaches toward this strategic virus reservoir. Antiviral Res 55:209–225

113. Dutta T, Garg M, Jain NK (2008) Targeting of efavirenz loaded tuftsin conjugated poly(propyleneimine) dendrimers to HIV infected macrophages in vitro. Eur J Pharm Sci 34:181–189
114. Collins KB, Patterson BK, Naus GJ, Landers DV, Gupta P (2000) Development of an in vitro organ culture model to study transmission of HIV-1 in the female genital tract. Nat Med 6:475
115. Das Neves J, Amiji MM, Bahia MF, Sarmento B (2010) Nanotechnology-based systems for the treatment and prevention of HIV/AIDS. Adv Drug Deliv Rev 62:458–477
116. Arnáiz B, Martínez-Ávila O, Falcon-Perez JM, Penadés S (2012) Cellular uptake of gold nanoparticles bearing HIV gp120 oligomannosides. Bioconjug Chem 23:814–825
117. Woodrow KA, Bennett KM, Lo DD (2012) Mucosal vaccine design and delivery. Annu Rev Biomed Eng 14:17–46
118. Lara HH, Ayala-Nuñez NV, Ixtepan-Turrent L, Rodriguez-Padilla C (2010) Mode of antiviral action of silver nanoparticles against HIV-1. J Nanobiotechnol 8:1
119. Hayakawa T, Kawamura M, Okamoto M, Baba M, Niikawa T, Takehara S, Serizawa T, Akashi M (1998) Concanavalin A-immobilized polystyrene nanospheres capture HIV-1 virions and gp120: potential approach towards prevention of viral transmission. J Med Virol 56:327–331
120. Leung V, Ko F (2011) Biomedical applications of nanofibers. Polym Adv Technol 22:350–365
121. Teo W-E, Inai R, Ramakrishna S (2011) Technological advances in electrospinning of nanofibers. Sci Technol Adv Mater 12:013002
122. Sill TJ, Von Recum HA (2008) Electrospinning: applications in drug delivery and tissue engineering. Biomaterials 29:1989–2006
123. Liang D, Hsiao BS, Chu B (2007) Functional electrospun nanofibrous scaffolds for biomedical applications. Adv Drug Deliv Rev 59:1392–1412
124. Gunn J, Zhang M (2010) Polyblend nanofibers for biomedical applications: perspectives and challenges. Trends Biotechnol 28:189–197
125. Huang C, Soenen SJ, Rejman J, Lucas B, Braeckmans K, Demeester J, De Smedt SC (2011) Stimuli-responsive electrospun fibers and their applications. Chem Soc Rev 40:2417
126. Fridrikh S, Yu J, Brenner M, Rutledge G (2003) Controlling the fiber diameter during electrospinning. Phys Rev Lett. doi:10.1103/PhysRevLett.90.144502
127. Zhang K, Qian Y, Wang H, Fan L, Huang C, Yin A, Mo X (2010) Genipin-crosslinked silk fibroin/hydroxybutyl chitosan nanofibrous scaffolds for tissue-engineering application. J Biomed Mater Res A 95A:870–881
128. Teo W-E, He W, Ramakrishna S (2006) Electrospun scaffold tailored for tissue-specific extracellular matrix. Biotechnol J 1:918–929
129. Venugopal J, Ramakrishna S (2005) Applications of polymer nanofibers in biomedicine and biotechnology. Appl Biochem Biotechnol 125:147–158
130. Cao H, Mchugh K, Chew SY, Anderson JM (2010) The topographical effect of electrospun nanofibrous scaffolds on the in vivo and in vitro foreign body reaction. J Biomed Mater Res A 93(3):1151–1159
131. Yu D-G (2009) Electrospun nanofiber-based drug delivery systems. Health 01:67–75
132. Patel SK, Lavasanifar A, Choi P (2009) Roles of nonpolar and polar intermolecular interactions in the improvement of the drug loading capacity of PEO-b-PCL with increasing PCL content for two hydrophobic cucurbitacin drugs. Biomacromolecules 10:2584–2591
133. Heunis TDJ, Dicks LMT (2010) Nanofibers offer alternative ways to the treatment of skin infections. J Biomed Biotechnol 2010:1–11
134. Tiwari SK, Tzezana R, Zussman E, Venkatraman SS (2010) Optimizing partition-controlled drug release from electrospun core–shell fibers. Int J Pharm 392:209–217
135. Taepaiboon P, Rungsardthong U, Supaphol P (2007) Vitamin-loaded electrospun cellulose acetate nanofiber mats as transdermal and dermal therapeutic agents of vitamin A acid and vitamin E. Eur J Pharm Biopharm 67:387–397

136. Li X, Kanjwal MA, Lin L, Chronakis IS (2013) Electrospun polyvinyl-alcohol nanofibers as oral fast-dissolving delivery system of caffeine and riboflavin. Colloids Surf B Biointerfaces 103:182–188
137. Ignatious F, Sun L, Lee C-P, Baldoni J (2010) Electrospun nanofibers in oral drug delivery. Pharm Res 27:576–588
138. Bernards DA, Bhisitkul RB, Wynn P, Steedman MR, Lee O-T, Wong F, Thoongsuwan S, Desai TA (2013) Ocular biocompatibility and structural integrity of micro- and nanostructured poly(caprolactone) films. J Ocul Pharmacol Ther 29(2):249–57
139. Meinel AJ, Germershaus O, Luhmann T, Merkle HP, Meinel L (2012) Electrospun matrices for localized drug delivery: current technologies and selected biomedical applications. Eur J Pharm Biopharm 81:1–13
140. Huang C, Soenen SJ, Van Gulck E et al (2012) Electrospun cellulose acetate phthalate fibers for semen induced anti-HIV vaginal drug delivery. Biomaterials 33:962–969
141. Ball C, Krogstad E, Chaowanachan T, Woodrow KA (2012) Drug-eluting fibers for HIV-1 inhibition and contraception. PLoS One 7:e49792
142. Cui W, Li X, Zhu X, Yu G, Zhou S, Weng J (2006) Investigation of drug release and matrix degradation of electrospun poly(dl-lactide) fibers with paracetanol inoculation. Biomacromolecules 7:1623–1629
143. Yohe ST, Colson YL, Grinstaff MW (2012) Superhydrophobic materials for tunable drug release: using displacement of air to control delivery rates. J Am Chem Soc 134:2016–2019
144. Zhang YZ, Wang X, Feng Y, Li J, Lim CT, Ramakrishna S (2006) Coaxial electrospinning of (fluorescein isothiocyanate-conjugated bovine serum albumin)-encapsulated poly(ε-caprolactone) nanofibers for sustained release. Biomacromolecules 7:1049–1057
145. Zamani M, Morshed M, Varshosaz J, Jannesari M (2010) Controlled release of metronidazole benzoate from poly ε-caprolactone electrospun nanofibers for periodontal diseases. Eur J Pharm Biopharm 75:179–185
146. Chunder A, Sarkar S, Yu Y, Zhai L (2007) Fabrication of ultrathin polyelectrolyte fibers and their controlled release properties. Colloids Surf B Biointerfaces 58:172–179
147. Zeng J, Aigner A, Czubayko F, Kissel T, Wendorff JH, Greiner A (2005) Poly(vinyl alcohol) nanofibers by electrospinning as a protein delivery system and the retardation of enzyme release by additional polymer coatings. Biomacromolecules 6:1484–1488
148. Yang Y, Li X, Qi M, Zhou S, Weng J (2008) Release pattern and structural integrity of lysozyme encapsulated in core–sheath structured poly(dl-lactide) ultrafine fibers prepared by emulsion electrospinning. Eur J Pharm Biopharm 69:106–116
149. Yu D-G, Zhu B-W, Yang W, Li Q (2011) Solid dispersions in the form of electrospun core-sheath nanofibers. Int J Nanomedicine 6:3271–3280
150. Yarin AL (2011) Coaxial electrospinning and emulsion electrospinning of core-shell fibers. Polym Adv Technol 22:310–317
151. Kenawy E-R, Abdel-Hay FI, El-Newehy MH, Wnek GE (2007) Controlled release of ketoprofen from electrospun poly(vinyl alcohol) nanofibers. Mater Sci Eng A 459:390–396
152. Jannesari M, Varshosaz J, Morshed M, Zamani M (2011) Composite poly(vinyl alcohol)/poly(vinyl acetate) electrospun nanofibrous mats as a novel wound dressing matrix for controlled release of drugs. Int J Nanomedicine 6:993–1003
153. Xie J, Wang C-H (2006) Electrospun micro- and nanofibers for sustained delivery of paclitaxel to treat C6 glioma in vitro. Pharm Res 23:1817–1826
154. Mehta S, Verstraelen H, Peremans K, Villeirs G, Vermeire S, De Vos F, Mehuys E, Remon JP, Vervaet C (2012) Vaginal distribution and retention of a multiparticulate drug delivery system, assessed by gamma scintigraphy and magnetic resonance imaging. Int J Pharm 426:44–53
155. Macri LK, Sheihet L, Singer AJ, Kohn J, Clark RAF (2012) Ultrafast and fast bioerodible electrospun fiber mats for topical delivery of a hydrophilic peptide. J Control Release 161:813–820
156. Okuda T, Tominaga K, Kidoaki S (2010) Time-programmed dual release formulation by multilayered drug-loaded nanofiber meshes. J Control Release 143:258–264

157. Palella FJ, Delaney KM, Moorman AC, Loveless MO, Fuhrer J, Satten GA, Aschman DJ, Holmberg SD (1998) Declining morbidity and mortality among patients with advanced human immunodeficiency virus infection. N Engl J Med 338:853–860
158. Veazey RS, Ketas TJ, Dufour J, Moroney-Rasmussen T, Green LC, Klasse PJ, Moore JP (2010) Protection of rhesus macaques from vaginal infection by vaginally delivered maraviroc, an inhibitor of HIV-1 entry via the CCR5 co-receptor. J Infect Dis 202:739–744
159. Prabaharan M, Jayakumar R, Nair SV (2011) Electrospun nanofibrous scaffolds-current status and prospects in drug delivery. In: Jayakumar R, Nair S (eds) Biomedical applications of polymeric nanofibers. Springer, Berlin, pp 241–262
160. Xu X, Chen X, Wang Z, Jing X (2009) Ultrafine PEG–PLA fibers loaded with both paclitaxel and doxorubicin hydrochloride and their in vitro cytotoxicity. Eur J Pharm Biopharm 72:18–25
161. Verreck G, Chun I, Peeters J, Rosenblatt J, Brewster ME (2003) Preparation and characterization of nanofibers containing amorphous drug dispersions generated by electrostatic spinning. Pharm Res 20:810–817
162. Yu D-G, Branford-White C, Shen X-X, Zhang X-F, Zhu L-M (2010) Solid dispersions of ketoprofen in drug-loaded electrospun nanofibers. J Dispers Sci Technol 31:902–908
163. Ham AS, Rohan LC, Boczar A, Yang L, Buckheit KW, Buckheit RW (2012) Vaginal film drug delivery of the pyrimidinedione IQP-0528 for the prevention of HIV infection. Pharm Res 29:1897–1907
164. Wang Y, Hsieh Y-L (2008) Immobilization of lipase enzyme in polyvinyl alcohol (PVA) nanofibrous membranes. J Membr Sci 309:73–81
165. Choi JS, Choi SH, Yoo HS (2011) Coaxial electrospun nanofibers for treatment of diabetic ulcers with binary release of multiple growth factors. J Mater Chem 21:5258
166. Liao I, Chen S, Liu J, Leong K (2009) Sustained viral gene delivery through core-shell fibers. J Control Release 139:48–55
167. Saraf A, Baggett LS, Raphael RM, Kasper FK, Mikos AG (2010) Regulated non-viral gene delivery from coaxial electrospun fiber mesh scaffolds. J Control Release 143:95–103
168. Cao H, Jiang X, Chai C, Chew SY (2010) RNA interference by nanofiber-based siRNA delivery system. J Control Release 144:203–212
169. Salalha W, Kuhn J, Dror Y, Zussman E (2006) Encapsulation of bacteria and viruses in electrospun nanofibres. Nanotechnology 17:4675–4681
170. Gensheimer M, Becker M, Brandis-Heep A, Wendorff JH, Thauer RK, Greiner A (2007) Novel biohybrid materials by electrospinning: nanofibers of poly(ethylene oxide) and living bacteria. Adv Mater 19:2480–2482
171. Townsend-Nicholson A, Jayasinghe SN (2006) Cell electrospinning: a unique biotechnique for encapsulating living organisms for generating active biological microthreads/scaffolds. Biomacromolecules 7:3364–3369
172. Maretschek S, Greiner A, Kissel T (2008) Electrospun biodegradable nanofiber nonwovens for controlled release of proteins. J Control Release 127:180–187
173. Chew SY, Wen J, Yim EK, Leong KW (2005) Sustained release of proteins from electrospun biodegradable fibers. Biomacromolecules 6(4):2017–2024
174. Patel S, Kurpinski K, Quigley R, Gao H, Hsiao BS, Poo M-M, Li S (2007) Bioactive nanofibers: synergistic effects of nanotopography and chemical signaling on cell guidance. Nano Lett 7:2122–2128
175. Beck-Broichsitter M, Thieme M, Nguyen J, Schmehl T, Gessler T, Seeger W, Agarwal S, Greiner A, Kissel T (2010) Novel "nano in nano" composites for sustained drug delivery: biodegradable nanoparticles encapsulated into nanofiber non-wovens. Macromol Biosci 10:1527–1535
176. Zomer Volpato F, Almodóvar J, Erickson K, Popat KC, Migliaresi C, Kipper MJ (2012) Preservation of FGF-2 bioactivity using heparin-based nanoparticles, and their delivery from electrospun chitosan fibers. Acta Biomater 8:1551–1559
177. Chen M, Gao S, Dong M, Song J, Yang C, Howard KA, Kjems J, Besenbacher F (2012) Chitosan/siRNA nanoparticles encapsulated in PLGA nanofibers for siRNA delivery. ACS Nano 6:4835–4844

178. Yoo HS, Kim TG, Park TG (2009) Surface-functionalized electrospun nanofibers for tissue engineering and drug delivery. Adv Drug Deliv Rev 61:1033–1042
179. Agarwal S, Wendorff JH, Greiner A (2010) Chemistry on electrospun polymeric nanofibers: merely routine chemistry or a real challenge? Macromol Rapid Commun 31:1317–1331
180. El-Sadr WM, Mayer KH, Maslankowski L et al (2006) Safety and acceptability of cellulose sulfate as a vaginal microbicide in HIV-infected women. AIDS 20:1109–1116
181. Puttipaiboon N, Yoovidhya T, Wongsasulak S (2012) Fabrication and gastro-mucoadhesive property of Zein-PEO-Chitosan blend ultrafine fibrous films. Proc NanoThailand 2012:13–16
182. Johnson TJ, Gupta KM, Fabian J, Albright TH, Kiser PF (2010) Segmented polyurethane intravaginal rings for the sustained combined delivery of antiretroviral agents dapivirine and tenofovir. Eur J Pharm Sci 39:203–212
183. Major I, Boyd P, Kilbourne-Brook M, Saxon G, Cohen J (2013) Malcolm RK A modified SILCS contraceptive diaphragm for long-term controlled release of the HIV microbicide dapivirine. Contraception 88(1):58–66. doi:10.1016/j.contraception.2012.10.018
184. Yoo J-W, Dharmala K, Lee CH (2006) The physicodynamic properties of mucoadhesive polymeric films developed as female controlled drug delivery system. Int J Pharm 309:139–145
185. Baji A, Mai Y-W, Wong S-C, Abtahi M, Chen P (2010) Electrospinning of polymer nanofibers: effects on oriented morphology, structures and tensile properties. Compos Sci Technol 70:703–718
186. Ayutsede J, Gandhi M, Sukigara S, Ye H, Hsu C, Gogotsi Y, Ko F (2006) Carbon nanotube reinforced Bombyx mori silk nanofibers by the electrospinning process. Biomacromolecules 7:208–214
187. Katta P, Alessandro M, Ramsier RD, Chase GG (2004) Continuous electrospinning of aligned polymer nanofibers onto a wire drum collector. Nano Lett 4:2215–2218
188. Li W-J, Mauck RL, Cooper JA, Yuan X, Tuan RS (2007) Engineering controllable anisotropy in electrospun biodegradable nanofibrous scaffolds for musculoskeletal tissue engineering. J Biomech 40:1686–1693
189. Wang S, Zhang Y, Yin G, Wang H, Dong Z (2009) Electrospun polylactide/silk fibroin–gelatin composite tubular scaffolds for small-diameter tissue engineering blood vessels. J Appl Polym Sci 113:2675–2682
190. Kidoaki S, Kwon IK, Matsuda T (2005) Mesoscopic spatial designs of nano- and microfiber meshes for tissue-engineering matrix and scaffold based on newly devised multilayering and mixing electrospinning techniques. Biomaterials 26:37–46
191. Pham QP, Sharma U, Mikos AG (2006) Electrospun poly(ε-caprolactone) microfiber and multilayer nanofiber/microfiber scaffolds: characterization of scaffolds and measurement of cellular infiltration. Biomacromolecules 7:2796–2805
192. McClure MJ, Sell SA, Simpson DG, Walpoth BH, Bowlin GL (2010) A three-layered electrospun matrix to mimic native arterial architecture using polycaprolactone, elastin, and collagen: a preliminary study. Acta Biomater 6:2422–2433
193. Teo WE, Kotaki M, Mo XM, Ramakrishna S (2005) Porous tubular structures with controlled fibre orientation using a modified electrospinning method. Nanotechnology 16:918–924
194. Van Lieshout MI, Vaz CM, Rutten MCM, Peters GWM, Baaijens FPT (2006) Electrospinning versus knitting: two scaffolds for tissue engineering of the aortic valve. J Biomater Sci Polym Ed 17:77–89
195. Del Gaudio C, Grigioni M, Bianco A, De Angelis G (2008) Electrospun bioresorbable heart valve scaffold for tissue engineering. Int J Artif Organs 31:68–75
196. Szentivanyi A, Chakradeo T, Zernetsch H, Glasmacher B (2011) Electrospun cellular microenvironments: understanding controlled release and scaffold structure. Adv Drug Deliv Rev 63:209–220
197. Food and Agricultural Organization of the United Nations AIDS—a threat to rural Africa. http://www.fao.org/FOCUS/E/aids/aids1-e.htm Accessed 16 Jan 2013

198. Kayaci F, Uyar T (2012) Encapsulation of vanillin/cyclodextrin inclusion complex in electrospun polyvinyl alcohol (PVA) nanowebs: prolonged shelf-life and high temperature stability of vanillin. Food Chem 133:641–649
199. Greiner A, Wendorff JH (2007) Electrospinning: a fascinating method for the preparation of ultrathin fibers. Angew Chem Int Ed 46:5670–5703
200. Xia X, Dong XJ, Wei QF, Cai YB, Lu KY (2011) Formation mechanism of porous hollow SnO_2 nanofibers prepared by one-step electrospinning. Express Polym Lett 6:169–176
201. Mauck RL, Baker BM, Nerurkar NL, Burdick JA, Li W-J, Tuan RS, Elliott DM (2009) Engineering on the straight and narrow: the mechanics of nanofibrous assemblies for fiber-reinforced tissue regeneration. Tissue Eng Part B Rev 15:171–193

Chapter 28
Prolonged Duration Local Anesthesia

J. Brian McAlvin and Daniel S. Kohane

28.1 Introduction

Local anesthetics produce reversible blockade of conduction of nerve impulses. Since the introduction of cocaine as a topical anesthetic into clinical medicine in 1884, a broad variety of synthetic local anesthetic agents have been developed for analgesia following surgery or for management of acute and chronic pain. However, their durations of action are generally limited to several hours, which may be insufficient for prolonged acute pain (e.g., perioperative pain) and certainly inadequate for neuropathic and other chronic types of pain. The exact nerve block duration that is optimal is very context dependent; there are situations where prolongation by a few hours or by a factor of 2–3 may be all that is needed, while there are others where durations of days to weeks or longer might be best.

J.B. McAlvin
Department of Medicine, Medicine Critical Care Program, Boston Children's Hospital, Harvard Medical School, Boston, MA, USA

Laboratory for Biomaterials and Drug Delivery, Department of Anesthesiology, Division of Critical Care Medicine, Boston Children's Hospital, Harvard Medical School Boston, Boston, MA, USA

Division of Health Sciences and Technology, Harvard-Massachusetts Institute of Technology, Cambridge, MA, USA
e-mail: brian.mcalvin@childrens.harvard.edu

D.S. Kohane (✉)
Laboratory for Biomaterials and Drug Delivery, Department of Anesthesiology, Division of Critical Care Medicine, Boston Children's Hospital, Harvard Medical School Boston, Boston, MA, USA

Division of Health Sciences and Technology, Harvard-Massachusetts Institute of Technology, Cambridge, MA, USA
e-mail: daniel.kohane@childrens.harvard.edu

A.J. Domb and W. Khan (eds.), *Focal Controlled Drug Delivery*, Advances in Delivery Science and Technology, DOI 10.1007/978-1-4614-9434-8_28,

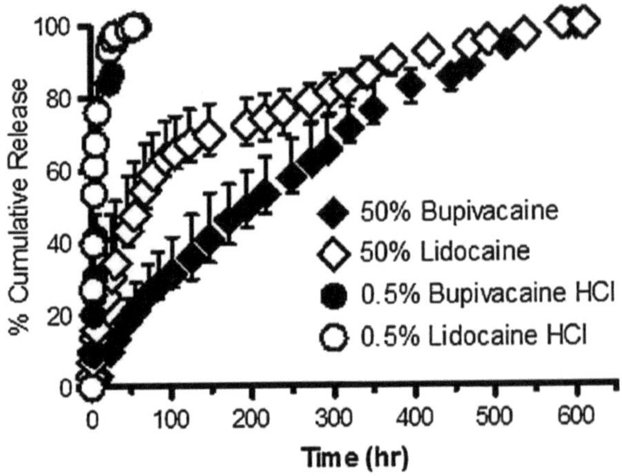

Fig. 28.1 Cumulative release of bupivacaine or lidocaine eluted from 50 % (w/w) drug-loaded PLGA microspheres compared to the release of unencapsulated 0.5 % (w/v) bupivacaine or lidocaine. Controlled release systems contain very large doses of drug (potentially toxic) but safely deliver them over a long period of time. Data are medians with 25th and 75th percentiles. Reproduction from [9] with permission from Lippincott Williams & Wilkins (2013)

Efforts to provide prolonged duration local anesthesia (PDLA) by modification of the physicochemical properties of local anesthetics have been successful, but only by relatively small multiples; their effect continues to be measured in hours [1]. Simply increasing the dose is unlikely to result in very large increases in duration, and cardiovascular and central nervous system toxicity may be limiting [2, 3]. Clinicians have attempted to extend the duration of action through the addition of adjuncts that increase the duration of effect via different mechanisms of action [4]. The same is true of experimental drug combinations, such as the co-injection of tetrodotoxin or saxitoxin with more "conventional" local anesthetics [5, 6] (intended here to mean amino-amide and amino-ester compounds in current clinical practice). Local or regional anesthesia lasting days can be achieved by indwelling catheters (e.g., epidural), but these may require inpatient hospitalization. If placed percutaneously, such devices present a risk of infection and require that the patient be tethered to a drug reservoir (syringe pump); if not, then the device would require implantation.

To address these shortcomings—but that of inadequate duration in particular—the development of prolonged duration local anesthetics has been the subject of extensive research for several decades. The exact properties desired of PDLA formulations may vary depending on the specific clinical context and perhaps anatomical location. Obviously, they should be effective and safe; this may necessitate a reliable and consistent rate of drug elution while minimizing the initial, uncontrolled release of large amounts of payload (burst release) [3], which could give rise to a rapid elevation of local and systemic drug levels and increase toxicity [2, 3, 7, 8]. A key concept is that although PDLA controlled release systems often contain drug dosages which could be lethal without controlled release [2, 3], they create a local concentration of drug high enough to block nerves without systemic toxicity by slowly releasing drug over a long period (Fig. 28.1). Ideally, tissue reaction should be benign (more on that topic below); this is an oft-neglected aspect of the evaluation of a formulation. If the device is to be biodegradable (i.e., nonrefillable), the degradation products should be nontoxic and should be cleared at a rate that is reasonable in relation to the release kinetics of the encapsulated drug so that residue is not present for prolonged periods after nerve blockade resolves. Injectability is likely to be desirable, to avoid the need for implantation. In general, the nerve blockade should be fully reversible, i.e., nerve function should return to normal when the formulation wears off. The duration of block achieved by a given dose of a formulation should be fairly reproducible, i.e., have a small coefficient of variation. It is also important that the relative durations and intensities of sensory and motor blockade be appropriate; a formulation which resulted in extended immobility of a body part without analgesia might not be desirable. Other important attributes that are common to most biomedical devices include cost-effectiveness, relative simplicity, ease of use, stability, and a reasonable shelf life. Arguably, patentability is also an important consideration, and there are downstream issues that may come into play (e.g., regulatory, whether a third-party payer will pay for it).

This chapter will review some representative examples of controlled release systems that have been developed for PDLA, selected from a very extensive literature. A brief discussion of topical local anesthetics will follow. We note that comparison of the pros and cons of specific formulations is very difficult due to the marked heterogeneity in experimental design, definitions of nerve blockade, reporting of tissue reaction, dosing regimens, and even *in vitro* methodologies of device characterization.

28.2 Physiochemical Properties of Conventional Local Anesthetics

The chemical properties of conventional local anesthetics are major determinants of their biologic effect as well as of the manufacturing technique and pharmacokinetics of the controlled release systems into which they are incorporated. In general conventional local anesthetic agents share a common chemical structure consisting of a lipophilic aromatic ring and a tertiary amine group [10] connected by an ester or an amide linkage. Esters include cocaine, procaine, chloroprocaine, tetracaine, and benzocaine. Amides include lidocaine, bupivacaine, mepivacaine, etidocaine, prilocaine, ropivacaine, and articaine.

Local anesthetics may also be differentiated by the direction in which they deflect polarized light and are labeled as either S-(−)-levorotatory or R-(+)-dextrorotatory [10]. Despite the similarity of their physicochemical properties, the clinical effect of individual enantiomers can be dramatically different. For example, with epidural anesthesia, S(−)-bupivacaine (levobupivacaine) produces longer sensory blockade than racemic bupivacaine [11] and has significantly less cardiovascular and central nervous system toxicity [10]. S(−)-bupivacaine and S(−)-ropivacaine (ropivacaine) are examples of single-isomer local anesthetic preparations.

Most conventional local anesthetics are weak bases and have pK_a values (7.6–8.9) that are greater than physiological pH (approximately 7.4) [10]. In general, they exist as either charged cations (from protonation of the tertiary amine) or neutral bases according to the Henderson-Hasselbalch equation: pH = pK_a + log [neutral base]/[charged cation] [10]. Therefore, it is the ambient pH that determines the proportion of each species present in a given solution such that charged cations (water soluble) are the dominant species in low pH solutions and neutral bases (lipophilic) dominate at high pH. Conventional local anesthetics are therefore amphiphilic.

These chemical considerations can have a marked impact on the manufacturing process for PDLA formulations. For example, in emulsion-based systems charged/hydrophilic species selectively partition into the aqueous phase(s), whereas relatively hydrophobic ones partition into the lipophilic environments, such as droplets of polymer in organic solvents. Similar considerations apply for liposomal and other systems.

The physiochemical properties of local anesthetics determine their potency, speed of onset, intensity, and duration of anesthetic action. After injection, usually in a mildly acidic solution so that the drug is in the cationic state, local anesthetics must transit across sequential biologic barriers to their intended site of action [10, 13]. These barriers include the fibrous and cellular components of the epineurium, perineurium, endoneurium, and the neuronal lipoprotein membrane, known as the plasmalemma. In order to penetrate into nerve cells, the drug must partition into the plasmalemma as the neutral unionized form [10, 13]. Once inside the cell, the lower intracellular pH induces the ionized cationic form, which then associates with the D4-S6 domain of the α-subunit of voltage-gated sodium channels situated within the plasmalemma, blocking the inward flux of sodium ions [10].

In general, conventional local anesthetics with high octanol/buffer partition coefficients (i.e., highly hydrophobic drugs) possess greater anesthetic potency [10, 13]. Similarly, the speed of onset of block is determined by the proportion of neutral base relative to cationic charged species and correlates with the pK_a of the drug [2]. The closer the pK_a is to physiological pH (7.4), the faster the onset of action.

28.3 Nonconventional Local Anesthetic Agents and Adjuvants

The bulk of this review will concern itself with conventional local anesthetics, in large part because those are what have been employed in the majority of the PDLA literature. However, it bears noting that other types of compounds have been used to extend the duration of PDLA formulations and/or to mitigate local tissue reaction. The incorporation of the glucocorticoid receptor agonist dexamethasone into polymeric microspheres containing bupivacaine extended their duration of block from 6 to 12 h to several days [15]. This prolongation was also seen with co-encapsulation of bupivacaine and dexamethasone in lipid-protein-sugar particles [16]. The co-encapsulation of site 1 sodium channel blockers (compounds that act at a different site on the sodium channel than conventional local anesthetics) such as tetrodotoxin and saxitoxin extended the duration of action of bupivacaine-containing

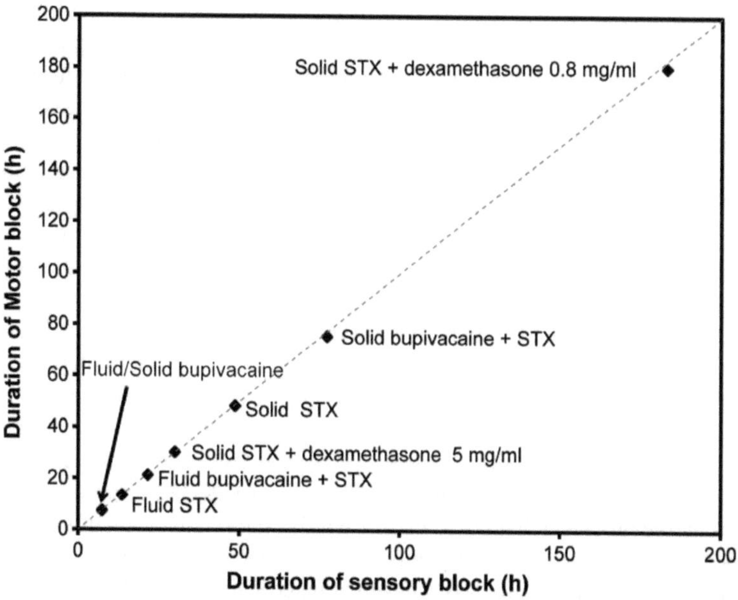

Fig. 28.2 Following sciatic nerve injection with 75 mg of liposomes loaded with bupivacaine, STX, and/or dexamethasone, the duration of sensory and motor blockade was tested. "Fluid" liposomes are comprised of low-phase transition (T_c) lipids, whereas "solid" liposomes are prepared with high-T_c lipids. Solid liposomes prolonged the duration of nerve blockade more than did fluid liposomes. At lower concentrations dexamethasone greatly extended the duration of nerve blockade. At higher concentration dexamethasone altered the stability of the liposome, resulting in "leakage" and rapid release. Reproduction from [18] with permission from National Academy of Sciences, USA (2009)

polymeric microspheres [17] and liposomes [18] severalfold. Dexamethasone had a similar effect on saxitoxin liposomes [18], prolonging block duration from 2 days to 1 week with very little variability (Fig. 28.2). The triple combination of bupivacaine, tetrodotoxin, and dexamethasone encapsulated in polymeric microspheres resulted in block durations (median 9 days) that were greater than those from compounds encapsulated singly or in pairs [17]. That formulation had three drawbacks: the therapeutic index was very narrow because of the burst release of tetrodotoxin, perhaps due to difficulties in incorporating that relatively toxic hydrophilic small molecule within a hydrophobic matrix; wide variability in duration of block due to a range of factors including needle clogging from the use of large (60 μm) particles; and relatively active tissue reaction. All three problems were markedly improved by the use of liposomes [11, 18].

The interaction between site 1 sodium channel blockers and conventional local anesthetics [5, 6, 19] extends to chemically similar compounds [19, 20] and is just one example of interactions that result in prolonged nerve blockade even without the use of sustained-release systems. Vanilloid receptor agonists such as capsaicin, the pungent ingredient in chili peppers, resulted in synergistic prolongation of both nociceptive and motor blockade when coadministered with tetrodotoxin (TTX) [21]. Interestingly, this synergistic combination was sensory selective at certain concentrations. Of note, high concentrations of capsaicin were neurotoxic. Epinephrine [19, 22], phenylephrine [22], and clonidine [22] prolonged block duration by tetrodotoxin and reduced its toxicity probably in large part by local vasoconstriction (by slowing drug release from the site of injection) [19, 22, 23]. (Bupivacaine also reduced the toxicity of TTX, presumably by vasoconstriction [19].) Very high concentrations of phentolamine, propranolol, and yohimbine [23] (much higher than the concentrations required for specific receptor blockade) also greatly prolonged the duration of block from TTX. It is possible that they acted as chemical permeation enhancers (CPEs), enhancing penetration of tetrodotoxin to the nerve surface.

In fact, established CPEs such as those used in transdermal drug delivery also increased the duration of nerve blockade from hydrophilic tetrodotoxin but were not effective with amphiphilic molecules such as bupivacaine [24], which presumably do not have difficulty reaching their receptors. CPEs probably act by enhancing penetration to the intended site of action. CPEs have also been used in combination with the quaternary lidocaine derivatives QX-314 or QX-222 to produce sensory-selective nerve blockade that lasted up to 7 h [25]. The relative lack of motor blockade may have been due to drug entrapment in the myelin sheaths of well-myelinated Aα motor fibers, whereas the less well-myelinated pain fibers (Aδ fibers or small, unmyelinated C fibers) were not protected. In rabbits, CPEs have also been shown to improve transmeningeal passage of ropivacaine into the intrathecal space, producing an intrathecal ropivacaine concentration 1.6 times greater than administration without a CPE [26].

Combinations of local anesthetics with adjuvants that do not belong to the class of local anesthetics have been employed in clinical practice. Epinephrine is commonly used with lidocaine or bupivacaine to extend the duration of action [4]. The use of opioids, alpha-2 agonists, and *N*-methyl-D-aspartate receptor antagonists has also been described [4]. In all cases, however, the effects have been relatively small.

28.4 Local Tissue Reaction and Systemic Toxicity

The specific drug employed can also have marked effects on tissue reaction and systemic toxicity. Tissue reaction can limit the usefulness of local anesthetics. For example, conventional local anesthetics are intrinsically myotoxic [7, 27–29] and neurotoxic [18, 30, 31], especially when delivered for extended periods [7, 28]. Both sequelae have been demonstrated for amino-amide and amino-ester local anesthetics [18, 32–36]. The degree of toxicity is agent specific [33, 35]: procaine and tetracaine produce the least and chloroprocaine and bupivacaine the most severe muscle injury [37]. When compared in rats, single injections of lidocaine produced milder muscle damage than did bupivacaine [35]. However, differences in myotoxic potential do not necessarily persist when these agents are administered as controlled release preparations; lidocaine and bupivacaine microspheres were equally myotoxic [9]. Multiple large doses [38, 39] and continuous catheter infusions [29, 40] can produce severe myopathy and myonecrosis that is clinically relevant, and the risk of muscle injury is related to the duration of exposure [29, 40]. Neurotoxicity generally occurs at higher concentrations than those that cause myotoxicity [18, 30, 31, 41]. Tricyclic antidepressants used as local anesthetics produced severe tissue injury characterized by ischemic necrosis and myotoxicity; there was also evidence of nerve injury [42, 43]. Similarly, QX-314, which could produce very long nerve blocks in animals [44], caused far worse local tissue injury than did conventional agents [20].

In contrast, site 1 sodium channel blockers, such as tetrodotoxin (TTX), saxitoxin (STX), neosaxitoxin, and decarbamoyl saxitoxin, do not cause myotoxicity [45] or neurotoxicity [46]. However, these extremely potent local anesthetics can cause systemic toxicity when given as free drugs [19, 47] or if their release is not controlled properly. Conventional local anesthetics may also cause systemic toxicity from inadvertent intravascular or intrathecal injection, or after administration of very large doses, and primarily affect the central nervous system (CNS) and—at higher concentrations—the cardiovascular system [2, 10]. As mentioned above, stereospecificity influences the local anesthetic binding affinity for sodium channels and as a result influences the severity of systemic toxicity. For example, (+)-(R)-enantiomers bind cardiac sodium channels with greater affinity than (−)-(S)-enantiomers [10]. The (−)-(S)-enantiomeric agents levobupivacaine and ropivacaine produce less CNS and cardiovascular toxicity than racemic local anesthetic preparations, allowing higher plasma concentrations before signs of systemic toxicity occur [10]. Systemic toxicity tends to be largely mitigated by controlled release [2, 3, 14].

Considerations regarding adverse tissue reaction extend to controlled release formulations as well. In vivo, the presence of delivery vehicles themselves enhances local anesthetic myotoxicity [7] and can cause inflammatory responses at the nerve that may considerably outlast the duration of blockade [7, 9, 17, 28]. Reporting of local tissue injury from conventional local anesthetic controlled release formulations has been spotty and variable, in both animal and human studies. In the authors' experience, muscle injury has been a ubiquitous finding in a wide range of extended-release bupivacaine formulations independently of the delivery vehicle [7, 9, 27, 28, 48] or co-encapsulated agent [17, 18, 45, 49], and it is sometimes severe. Tissue reaction has, on at least one occasion, proven fatal to the commercial prospects of a formulation en route to market [50].

28.5 Specific Drug Delivery Vehicles

28.5.1 Particulates

A variety of micro- and nanoparticulates have been employed in drug delivery and in PDLA. There are generalizations that can be made regarding particulate drug delivery systems, but it is important to realize that those generalizations do not necessarily hold for all formulations or across classes/types of particles. Particle size in particular has numerous effects [14], in many cases because the surface area-to-volume ratio of smaller particles is greater than that of larger ones. Therefore, all other things being equal, a larger particle is likely to have higher drug loading, release drug more slowly, and degrade more slowly. Larger particles are less likely to move from the site where they are deposited [51], and tissue reaction will be different [28]. Although nanoparticulate formulations may successfully provide PDLA [52, 53], the rationale for microparticles is perhaps more compelling for this application as will be discussed below. For the purposes of this review, the terms "microparticle" and "nanoparticle" refer to particles where the principal dimensions are usefully measured in micrometers and nanometers, respectively.

28.5.2 Liposomes

Phospholipids in aqueous systems can form closed bilayered structures known as liposomes [54]. The basic structures of liposomes are either small or large unilamellar, multilamellar, or multivesicular membranes consisting of concentric lipid bilayers with an aqueous phase inside and between the lipid layers [55–57]. Lipid-soluble drugs partition into the lipid bilayers, whereas aqueous drugs partition into the internal aqueous compartments [54]. Due to the amphiphilic nature of local anesthetic agents, they may partition into the aqueous compartment either as charged species or as the hydrophobic freebase within the lipid bilayer [55] (Fig. 28.3a). Early formulations incorporated either the neutral freebase into the lipid bilayer [58] or the hydrophilic charged species into the aqueous compartments [59]. Encapsulation efficiency and drug loading were found to be greater for vesicles with drug loaded into the aqueous core compared to those with freebase incorporated into the lipid bilayer [60]. "Remote" drug loading, a process that utilizes a transmembrane pH [61, 62] or chemical [63–65] gradient to drive charged (hydrophilic) species into the aqueous compartments of the liposome, increases encapsulation efficiency and drug loading.

Liposomes can suffer from a lack of stability (i.e., their shelf life may be short) due to oxidation and hydrolysis during storage, which result in leaking of encapsulated anesthetic [2, 66, 67]. One solution to this problem was to prepare liposomes with high-transition temperature (T_c) lipids. The resulting "solid" vesicles demonstrated greater stability during storage [57] and produced longer durations of nerve blockade compared to "fluid" liposomes [18] that utilized low-T_c lipids, presumably by slowing the rate of drug release. To further enhance stability, "dehydration-rehydration vesicles" (DRVs) were prepared with either high- or low-T_c lipids ("solid" or "fluid," respectively), remotely loaded with an ammonium sulfate gradient to encapsulate charged species, and freeze-dried until immediately prior to use [64]. Solid vesicle formulations resulted in greater encapsulation efficiency compared to fluid vesicles. Independent of lipid composition, dehydration conferred a significantly longer shelf life (i.e., DRVs were more stable).

Tissue reaction to liposomes is generally considered to be benign [55–57, 66]. However, compounds and metabolites formed as part of production or metabolism

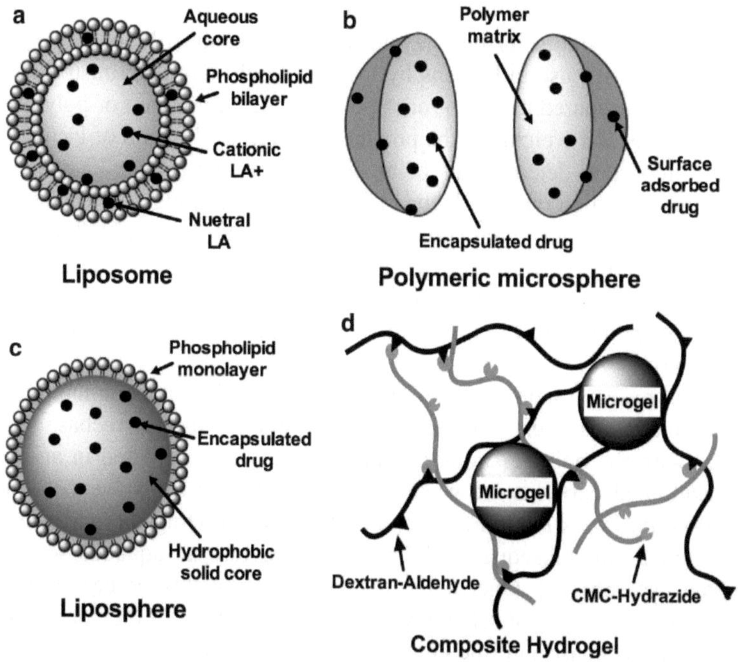

Fig. 28.3 Schematic representations of (**a**) a liposome loaded with an amphiphilic local anesthetic drug partitioned into either the internal aqueous space as a charged cation (LA+) or the hydrophobic lipid bilayer as a neutral base (LA); (**b**) a polymeric biodegradable microsphere with drug encapsulated within the polymer matrix and adsorbed onto the surface; (**c**) a liposphere comprised of a solid hydrophobic core matrix impregnated with local anesthetic neutral base; (**d**) a composite hydrogel containing entrapped microgels containing local anesthetics. The hydrogel network is comprised of carboxymethylcellulose (CMC) and dextran cross-linked by aldehyde-hydrazide bonds

of some liposomes have been shown to be neurotoxic [66]. Hydrolysis and oxidation products of lecithin, such as lysophosphatidylcholine, fatty acid free radicals, and peroxides, are cytotoxic [68] and neurotoxic [69] and may cause demyelination. Despite these reports of toxicity, other liposomal formulations have been used for PDLA with no evidence of neurotoxicity from in vitro cytotoxicity studies, neurobehavioral experiments, high-resolution microscopy, and analyses of changes in gene expression in dorsal root ganglia [18]. Presently, the only commercially available PDLA formulation for clinical use is liposomal bupivacaine (DepoFoam, Exparel) [70–73], which has completed four phase three clinical trials in humans.

28.5.3 Formulations Composed of Synthetic Hydrophobic Polymers

A variety of polymers have been used to form hydrophobic matrices that contain the local anesthetic freebase (Fig. 28.3b). Thermoplastic aliphatic poly(esters) like poly(lactic acid), poly(glycolic acid), and especially poly(lactic-co-glycolic acid) (PLGA) have been studied extensively due to their biocompatibility, biodegradability, and approval for use in humans by the Food and Drug Administration of the USA. They are prepared as homo- or copolymers of poly(lactic) or poly(glycolic) acids. The molecular weights of the polymers as well as the ratio of lactide to glycolide are important determinants of biodegradation and release kinetics. PLGA formulations with high ratios of lactide to glycolide and/or high molecular weights degrade more slowly and release drug for longer periods of time [74, 75]. Although the majority of polymeric controlled release systems reported to date have utilized PLGA, other polymer preparations have been used with success, including poly(ε-caprolactone) (microspheres [76] and nanospheres [53]), polycarbonates [77], and PLA [53].

Emulsion-based systems (e.g., oil/water emulsification) are commonly used to make polymeric particles. Typically, local anesthetics in the freebase form are dissolved with the polymer in an organic solvent and emulsified in an aqueous buffer. Hardened particles are produced by eliminating the solvent through an evaporation and/or extraction process and then lyophilizing the particles into a powder, which can subsequently be reconstituted into an aqueous solution for immediate use [56, 78]. One of the disadvantages of this process is the poor encapsulation efficiency of relatively hydrophilic species, as they partition from the dispersed oil phase into the continuous aqueous phase in which the nascent particles are being formed [78]. For example, lidocaine (low octanol/buffer partition coefficient) [79–81] required greater molar concentrations in the starting materials to attain loading that was comparable to bupivacaine (high octanol/buffer partition coefficient) [7, 9, 16, 17, 28, 51]. Synthesis by spray drying provided an alternative method of preparation that was faster and more suitable for scale-up and significantly improved encapsulation efficiency compared to traditional oil/water emulsification method [76, 82].

Both oil/water emulsification and spray-drying methods produced relatively polydisperse particle populations with a broad range of particle diameters [7, 9, 16, 17, 28, 51, 76, 79–82]. The polydispersity and batch-to-batch variation in both the size and morphology of the resulting particles may produce undesirable variation in the rate of microparticle degradation, the stability of the drug, and the kinetics of drug release [83]. Monodisperse PLGA microspheres were synthesized using a microfluidic flow-focusing device, which resulted in a significantly lower initial burst release and slower overall drug release compared to the polydisperse formulations made by conventional emulsion-based methods [83].

PLGA has also been used to manufacture absorbable local anesthetic matrices for implantation into surgical incision sites for controlled release of various local anesthetics [84–87]. Such systems produced analgesia ranging from hours [84, 85] to several days following surgical incision [86]. Bupivacaine-loaded PLGA sutures manufactured by electrospinning conferred incisional analgesia to rats for 7 days in a skin wound model [12]. Other surgically implantable, non-PLGA polymeric formulations have produced similar anesthetic effect [88, 89]. Neither implantable anesthetic matrices nor analgesic sutures impaired wound healing, suggesting suitability for perioperative analgesia and the potential to mitigate the need for standard postoperative analgesics.

In general, tissue reaction to polymeric delivery systems is benign and is characterized by generic acute and chronic inflammation [28]. As we have noted above, that tissue reaction may not be benign once local anesthetics are incorporated [50]. A drawback of many of these systems is that residual excipient materials persist in the tissues long after the drug payload has been delivered [9, 28, 76].

28.5.4 Lipid-Protein-Sugar Particles

Delivery vehicles that are characterized by short tissue dwell times relative to the duration of drug release may improve biocompatibility when compared to delivery systems with extended tissue dwell times. Lipid-protein-sugar particles composed of dipalmitoylphosphatidylcholine (DPPC), albumin, and lactose are absorbed relatively quickly and were used to encapsulate bupivacaine. When compared to conventional PLGA microspheres (which generate particle residues that last for prolonged periods) with 50 % (w/w) drug loading for rat sciatic nerve blockade [16], tissue reaction 4 days after injection was similar for both particle types (acute inflammation with predominance of neutrophils). However, at 2 weeks animals in the LPSP group had mild, patchy inflammation with macrophages and lymphocytes and no particles visualized, while animals injected with PLGA microspheres still had significant polymeric residue surrounded by a foreign-body giant cell reaction that persisted for at least 8 weeks [28]. This difference in tissue reaction was attributed to the fact that the LPSPs were absorbed much more rapidly, although differences in composition of matter may have played a role as well.

28.5.5 Lipospheres

Lipospheres were developed to address the high production costs and inherent instability of liposomes [90]. They are microparticles that consist of a solid, hydrophobic, triglyceride or fatty acid core impregnated with a hydrophobic drug, such as bupivacaine. The core is surrounded by a phospholipid monolayer [90–92] (Fig. 28.3c). Early versions of the formulation containing bupivacaine were dispersions formulated with natural phospholipids such as egg or soy phosphatidylcholine (PC) and produced sensory blockade in the rat for 24–72 h depending on the method used to measure analgesia [91]. Because these formulations were found to solidify at room temperature within days of manufacture, they were limited to surgical implantation rather than injection [88, 89]. The problem of rapid gelification was overcome by adding synthetic phospholipids (such as dimyristoyl PC, dipalmitoyl PC, distearoyl PC, Precirol, and Miglyol) or a cellulosic polymer (carboxymethylcellulose) to improve the physical stability of the lipospheres [92, 93].

28.5.6 Hydrogels

Hydrogels are three-dimensional, cross-linked networks of water-soluble polymers that form a hydrophilic microstructure whose pore size is readily tunable by manipulation of the cross-linking chemistry and density [94, 95]. They can also be chemically modified to exhibit tunable affinities for target drugs. Hydrogels typically possess favorable biocompatibility and produce only mild inflammation [94, 95].

One example of a hydrogel used for PDLA is hyaluronic acid (HA), a naturally occurring mucopolysaccharide which was used because (a) it can be made into a viscous aqueous solution; (b) the negative charge of the molecule can bind conventional local anesthetics, which are all cationic when charged; and (c) it has excellent biocompatibility [27]. Unmodified HA has variable effects on the duration of block from co-dissolved local anesthetics (including no effect at all). In one attempt to remedy this, HA was modified to aldehyde and hydrazide forms that would cross-link upon mixing [96] at the time of injection, forming a gel in situ, trapping local anesthetics in the vicinity of the nerve for extended release [27]. The resulting cross-linked hydrogel doubled the duration of nerve blockade compared to the same concentrations of free bupivacaine. These HA formulations did not achieve durations of block such as those seen with some polymeric or other microsphere preparations. However, the duration of blockade was adequate for many postoperative and/or dental situations without entailing the sequelae of microparticle injection (e.g., PLGA microspheres persisting at the injection site for weeks [9, 28]). Tissue reaction to cross-linked HA-based injections was restricted to a well-circumscribed area at the site of injection and consisted of interstitial chronic inflammation (primarily lymphocytes and macrophages). Muscle injury was comparable to that from free bupivacaine at the same concentration. The principal limitation of cross-linked HA formulations was the need for celerity by the practitioner during injection, in order to avoid needle blockage from hydrogel gelation; the use of a double-barreled syringe addressed that problem.

Synthetic hydrogel solutions have also been used for PDLA. Poloxamer 407 is a synthetic hydrogel that demonstrates reverse-phase thermal gelation (liquid phase at cold temperatures, gel at body temperature). When prepared as a 25 % (w/w) solution with 2 % (w/v) lidocaine HCL and administered at the rat sciatic nerve [97] or injected epidurally in pigs [98], sensory nerve blockade resulted for up to 210 min. Compared to free lidocaine, the lidocaine gel resulted in less systemic absorption (lower peak serum concentration) and a longer time to peak serum concentration [98]. Cellulose additives (hydroxypropyl methylcellulose, carboxymethylcellulose, dextran) were found to increase the rate of release of lidocaine from the poloxamer gel while paradoxically reducing lidocaine flux across porcine dura mater [97]. The authors hypothesized that the cellulose molecules formed a diffusion-reducing barrier with the relatively porous dura mater, thereby reducing the rate of lidocaine penetration across the membrane.

28.5.7 Cyclodextrins

Cyclodextrins (CDs) have been used to enhance aqueous solubility of poorly water-soluble drugs, while reducing drug toxicity and prolonging the duration of anesthetic effect [57]. Cyclodextrins are cyclic oligosaccharides comprised of 6–8 sugar subunits that form a hydrophobic internal cavity into which inclusion complexes with conventional local anesthetic agents can be formed [99, 100]. Naturally occurring cyclodextrins are poorly soluble in water and organic solvents, limiting their use in pharmaceutical formulations. However, the introduction of alkyl [101] or sugar side chains [99, 100] confers improved aqueous solubility, thereby facilitating the development of aqueous injectable solutions. An aqueous solution of maltosyl-β-CD complexed with levobupivacaine more than doubled the duration of both intrathecal [99] and sciatic nerve [99, 100] blockade. Similar results were achieved by complexation of ropivacaine with hydroxypropyl-β-CD [101].

Although the duration of nerve blockade for cyclodextrin-complexed local anesthetics was modest, tissue reaction was favorable [57]. Hydroxypropyl-β-CD formulations of bupivacaine and ropivacaine produced less local anesthetic myotoxicity on histologic analysis than did free drug injections [102]. However, unmodified cyclodextrins, such as β-CD, produce metabolites that are nephrotoxic when administered parenterally [57, 99, 100] and may produce hemolysis [99, 100]. Side chain functionalization may alter the metabolic fate and mitigate the potential for toxicity [57].

The entrapment of hydrophobic drugs that have been rendered hydrophilic by cyclodextrin complexation into the aqueous core of liposomes has been demonstrated, giving rise to drug-in-cyclodextrin-in-liposome systems [103]. Liposomes can be "double-loaded" with the cyclodextrin-complexed drug in the aqueous core and the hydrophobic plain drug in the lipophilic shell [74, 103]. In a rabbit model of corneal analgesia, the use of this double-loading technique resulted in faster onset, greater intensity, and longer duration of analgesia [74].

28.5.8 Injectable Liquid Polymers

Extensive investigation of poly(fatty ester-anhydrides) to dissolve bupivacaine within a hydrophobic matrix for controlled release has been performed. An early formulation comprised of poly(sebacic-co-ricinoleic acid) with 10 % (w/v) bupivacaine produced sensory nerve blockade lasting 30 h [75]. However, the anhydride bonds readily underwent hydrolysis, leading to rapid drug release. In an effort to produce a more hydrophobic polymer that would result in slower drug release from the hydrophobic matrix, polyester-poly(lactic acid-co-castor oil) was prepared with 10 % (w/v) bupivacaine [8]. The formulation extended the duration of nerve blockade to 48 h but was limited by burst release that produced systemic toxicity. Interestingly, the burst release was eliminated, and the duration of nerve blockade extended to 96 h by increasing the bupivacaine loading to 15 % (w/v) [104]. The paradoxical reduction in burst release from increased drug loading was attributed to increased formulation density and hydrophobicity, resulting in reduced water penetration into the drug-polymer matrix. The tissue reaction to these formulations was characterized by moderate inflammation of the muscle and nerve at 3 days, with one of five mice found to have necrosis of fat surrounding the nerve [104]. The injected polymer was present after 2 weeks and was accompanied by mild inflammation that persisted for 3 months.

The polymers that are used for these preparations (without drug) demonstrated benign tissue reaction [105]. The main disadvantage was the prolonged time the polymer carriers persisted at the injection site, far beyond the duration of effect of the local anesthetic agent [2, 104, 105].

28.5.9 Injectable Lipid Matrix

Lipid mixtures comprised of medium-chain triglycerides (MCTs) provide a physically stable liquid drug delivery system that is easy to inject and can be precisely tuned to contain large quantities of local anesthetics [106]. Such formulations can be used to produce ultra-long nerve blockade lasting for weeks. For example, a 1:1 eutectic mixture of lidocaine and prilocaine was dissolved into an MCT carrier fluid matrix. The eutectic mixture was freely soluble in the MCT solution, and the concentration of local anesthetic was easily tunable from 2 to 80 % (v/v) [106]. The duration of sensory nerve blockade was 14 days for animals injected with 80 % (v/v) drug in MCT. However, "moderate" to "marked" neurotoxicity was observed for formulations greater than 60 % (v/v) loading and appeared as axonal swelling and neuronal degeneration accompanied by myelin degeneration. The degree of neurotoxicity was directly correlated with drug loading. The MCT carrier alone did not produce any adverse tissue reaction, suggesting that the unfavorable tissue biocompatibility was attributable to the presence of local anesthetics [107]. All formulations were physically stable, with a shelf life of 1 year.

28.5.10 Hybrid Formulations

Controlled release systems may be combined to form formulations with properties that differ from the individual delivery vehicles. Polymeric microparticulate formulations are capable of high drug loading and prolonged drug release, but their clinical applicability can be hindered by a tendency to aggregate during storage, leading to needle clogging at the time of injection; they also tend to cause local inflammation. Liposomes can possess good biocompatibility, short tissue dwell times, and favorable injectability, but they can be limited by a lack of physical stability that results in considerable leakage during storage. In situ gelling systems offer favorable injectability, but are limited by their modest duration of nerve blockade. Composite formulations can mitigate the limitations of the individual systems and augment their desirable properties.

Systems have been developed incorporating microspheres within in situ-forming hydrogels. One such preparation, comprised of PLGA microparticles (31 % [w/w] lidocaine loading) suspended in the thermosensitive poloxamer 407 gel, yielded sciatic nerve blockade in rats lasting 8.5 h. Bupivacaine-loaded poly(ε-caprolactone) microspheres (50–90 % [w/w] drug loading) suspended in a hyaluronic acid-Pluronic F-127 (a proprietary name of poloxamer 407) matrix achieved drug release in vitro for up to 42 days [108].

Nanoparticles have also been employed in composite systems. Lidocaine-loaded poly(ε-caprolactone)-poly(ethylene glycol)-poly(ε-caprolactone) nanoparticles were prepared with 40 % (w/w) drug loading and mixed into a Pluronic F-127 gel [109]. The duration of analgesia following subcutaneous injection in rats was only 6 h. Histologic analysis was not performed.

Microscale bupivacaine-loaded multivesicular liposomes, called Bupisomes, were prepared by remote loading and then entrapped within a calcium alginate cross-linked hydrogel (Bupigel) to form stable, injectable 3–5 mm beads that became a homogenous dispersion when injected [110]. Minimal bupivacaine leakage from the Bupigel system was observed after 2 years of storage at 4 °C. Subcutaneous injection of Bupigel in mice produced 35 h of analgesia. Tissue reaction was not described.

A formulation consisting of soft composite injectable hydrogels containing entrapped copolymer microgels was developed as a delivery system for bupivacaine [95]. Anionically functionalized copolymer microgels composed of *N*-isopropylacrylamide and acrylic acid were entrapped within an in situ gelling carboxymethylcellulose-based hydrogel (Fig. 28.3d). By controlling the hydrogel cross-link density, degree of swelling, and affinity for cationic bupivacaine of both the hydrogel and microgel phases, the investigators were able to engineer composites that achieved more prolonged drug release kinetics than could be achieved by the same hydrogels or microgels alone. Bupivacaine release was sustained for up to 60 days in vitro, whereas the constituent hydrogel or microgel alone resulted in drug release for less than 1 week.

28.5.11 Calcium Phosphate Apatite Matrix

Calcium phosphate materials are widely used in human surgeries for bone repair because their chemical composition, which is similar to that of bone, fosters the resorption/substitution process of bone healing [111]. They have also been used as drug delivery systems to bone [112, 113]. Recently, a bioresorbable, osteoconductive (i.e., facilitates reparative natural bone growth) synthetic matrix comprised of calcium-deficient apatites (CDA) was used in rats as a carrier for bupivacaine for provision of analgesia to the periosteum following bone harvest [111]. CDA granules containing 1, 4, or 16 % (w/w) bupivacaine were prepared by isostatic compaction [111], whereby a powder comprised of bupivacaine adsorbed to CDA was subjected to high pressure to produce granules. All 3 formulations demonstrated similar in vitro release kinetics, with approximately 80 % of drug released within 5 h and 100 % by 24 h. Postoperative analgesic threshold was assessed by the von Frey monofilament test. By their definition of analgesic effect, nerve blockade lasted for up to 72 h in a dose-dependent fashion.

28.6 Topical Local Anesthesia

Many clinical procedures including needle insertions (vascular access, immunizations, regional nerve block, bone marrow aspiration) and dermatological procedures (skin biopsy, laser treatment) are painful [67]. Topical local anesthetic formulations bear brief mention here, given their adoption into broad clinical use as a method to reduce pain associated with procedures.

The impervious nature of skin is the greatest barrier to successful transdermal drug delivery [114]. Enhanced local anesthetic penetration of skin has been a major focus of research for more than 60 years [57]. A variety of topical local anesthetic delivery systems have been developed, intended to reduce the necessary concentration of local anesthetic, increase skin drug permeability, prolong local drug dwell time by reducing clearance, and/or limit local toxicity.

Bioadhesive patches with incorporated eutectic anesthetic mixtures have been studied extensively. One commercially available system in broad clinical use is a eutectic mixture of 2.5 % (w/v) lidocaine and 2.5 % (w/v) prilocaine in oil and water emulsion (EMLA™) [57]. Onset of anesthesia with EMLA is slow (up to 60 min) [57, 115]. To hasten the onset of analgesia, a lidocaine/tetracaine (1:1 mass ratio) medicated plaster (Synera™ in the USA, Rapydan™ in Europe) was developed with an integrated heating element designed to warm the skin by up to 5 °C upon contact with air, thereby enhancing drug flux. Analgesia was achieved within 10 min of application [115].

Skin transit may also be augmented through the use of chemical permeation enhancers. Hydroxypropyl methylcellulose (HPMC) impregnated with mepivacaine has been used as a bioadhesive gel and evaluated in rats with the tail-flick analgesiometer. The addition of a CPE (polyoxyethylene 2-oleyl ether) and a vasoconstrictor (tetrahydrozoline) were found to enhance local anesthetic action compared to HPMC alone [116], presumably by augmenting skin transit (by the CPE) and by maintaining the drug at the site of placement (by the vasoconstrictor). Liposomal delivery systems, prepared as liquid (emulsions), semisolid (gels, creams, ointments), or solid formulations, have also been reported for transdermal local anesthetic delivery [57].

Local anesthetics encapsulated in microemulsions comprised of permeation enhancers have also been shown to produce rapid transdermal penetration and effect [117, 118]. However, the surfactants and used to create the microemulsions may produce hemolysis and local tissue toxicity [117].

28.7 PDLA in Humans

Controlled release PDLA formulations have been used in humans for brachial plexus blockade [119], epidural injection [120, 121], subcutaneous infiltration [63, 122, 123], and intercostal nerve blockade [124]. Yet, despite at least 2 decades of published work, they have not been widely adopted clinically. A major limitation to broader clinical use has been adverse tissue reaction [50]. That tissue injury is a crucial consideration for PDLA systems is seen in the example of a sustained-release bupivacaine-dexamethasone formulation, where inflammation and nerve and muscle injury in preclinical animal studies and clinical human trials led to withdrawal of its Investigational New Drug application (IND#53,441) [50]. Few PDLA formulations are commercially available. One example is a liposomal bupivacaine preparation (DepoFoam bupivacaine, Exparel). In animal studies, mild granulomatous inflammation was reported to be the major histopathologic consequence [125, 129, 130]. To date Exparel has undergone 4 clinical trials. Anesthetic effect was measured for perianal infiltration in patients undergoing hemorrhoidectomy [70], peri-metatarsal infiltration for bunionectomy [71], surgical site infiltration for breast augmentation [73], and wound infiltration for knee arthroplasty [72]. Patients from the hemorrhoidectomy and bunionectomy cohorts who received Exparel remained opioid-free for longer and required less cumulative opioid dosing compared to saline placebo. In studies where the saline placebo was replaced with conventional bupivacaine preparations, Exparel failed to produce statistically significant differences in primary efficacy measures [72, 73].

28.8 Summary

The large number of reported PDLA formulations confirms the intense interest in the development of effective and safe PDLA. Nonetheless, since the introduction of dibucaine-loaded PLA microspheres more than 3 decades ago [126], only one formulation has become commercially available [70–73], and its recommended uses are limited to surgical incision sites (i.e., not recommended for epidural, intrathecal, intra-articular, or regional nerve blockade). Ultra-long duration anesthesia is achievable by a variety of formulations (perhaps even lasting for months [107]), but there are concerns regarding the association of PDLA with adverse tissue reaction. When a PDLA formulation is used in anatomic locations with relatively low blood flow, or in patients with poor peripheral circulation, the potential for significant tissue injury might be further increased, as these conditions might lead to the maintenance of high local tissue concentrations. Consequently, placement of PDLA formulations near nerves and major muscles should be approached with circumspection. Furthermore, surveillance for these complications will be necessary as these agents are adopted into broader clinical use. These considerations are not unique to any single formulation, but apply to all PDLA formulations that employ conventional local anesthetics.

PDLA is also being investigated for uses beyond the simple ablation of pain. For example, it was possible to delay by 1 month the onset of neuropathic pain after spared-nerve injury of the rat sciatic nerve by delivering three sequential injections of dexamethasone-saxitoxin-loaded liposomes at the sciatic nerve that provided 18 days of nerve blockade [11].

New or nonconventional local anesthetic agents may also enhance PDLA formulations. For example, tetrodotoxin, saxitoxin, or other site 1 sodium channel blockers may provide solutions to the problem of local tissue toxicity. As noted above, their narrow therapeutic indices indicate a need for caution in developing formulations with these compounds. Their use in humans to treat pain is already being explored [83, 127].

Future studies in this developing field will likely continue to be hampered by a lack of standardization of many important metrics relating to effectiveness or safety. More importantly, the relative durations of sensory and motor blockade and measures of systemic and local toxicity are frequently completely overlooked. These deficiencies make it impossible to evaluate crucial performance criteria and make comparisons between systems very difficult.

The broader field of drug delivery is growing rapidly and is incorporating many scientific disciplines, which could potentially bring great benefits to PDLA. For example, the development of remotely triggerable drug delivery systems [128] holds the promise of devices that allow the patient or doctor to determine exactly when pain relief begins and ends and how intense the analgesia is (with attendant motor deficits), i.e., on-demand drug delivery. One can further envision—in a future where precise physiological or chemical correlates of pain were known and had been quantitated—entirely self-contained closed-loop systems that would be able to sense the presence of pain and release drugs to treat that pain.

References

1. Wang GK, Quan C, Vladimirov M, Mok WM, Thalhammer JG (1995) Quaternary ammonium derivative of lidocaine as a long-acting local anesthetic. Anesthesiology 83(6):1293–1301
2. Weiniger CF, Golovanevski M, Sokolsky-Papkov M, Domb AJ (2010) Review of prolonged local anesthetic action. Expert Opin Drug Deliv 7(6):737–752. doi:10.1517/17425241003767383
3. Weiniger CF, Golovanevski L, Domb AJ, Ickowicz D (2012) Extended release formulations for local anaesthetic agents. Anaesthesia 67(8):906–916. doi:10.1111/j.1365-2044.2012.07168.x
4. Wiles MD, Nathanson MH (2010) Local anaesthetics and adjuvants–future developments. Anaesthesia 65(Suppl 1):22–37. doi:10.1111/j.1365-2044.2009.06201.x
5. Adams HJ, Blair MR Jr, Takman BH (1976) The local anesthetic activity of saxitoxin alone and with vasoconstrictor and local anesthetic agents. Arch Int Pharmacodyn Ther 224(2):275–282
6. Adams HJ, Blair MR Jr, Takman BH (1976) The local anesthetic activity of tetrodotoxin alone and in combination with vasoconstrictors and local anesthetics. Anesth Analg 55(4):568–573
7. Padera R, Bellas E, Tse JY, Hao D, Kohane DS (2008) Local myotoxicity from sustained release of bupivacaine from microparticles. Anesthesiology 108(5):921–928. doi:10.1097/ALN.0b013e31816c8a48

8. Sokolsky-Papkov M, Golovanevski L, Domb AJ, Weiniger CF (2009) Prolonged local anesthetic action through slow release from poly (lactic acid co castor oil). Pharm Res 26(1):32–39. doi:10.1007/s11095-008-9699-8
9. McAlvin JB, Reznor G, Shankarappa SA, Stefanescu CF, Kohane DS (2013) Local toxicity from local anesthetic polymeric microparticles. Anesth Analg 116(4):794–803
10. McLure HA, Rubin AP (2005) Review of local anaesthetic agents. Minerva Anestesiol 71(3):59–74
11. Shankarappa SA, Tsui JH, Kim KN, Reznor G, Dohlman JC, Langer R, Kohane DS (2012) Prolonged nerve blockade delays the onset of neuropathic pain. Proc Natl Acad Sci U S A 109(43):17555–17560. doi:10.1073/pnas.1214634109
12. Weldon CB, Tsui JH, Shankarappa SA, Nguyen VT, Ma M, Anderson DG, Kohane DS (2012) Electrospun drug-eluting sutures for local anesthesia. J Control Release 161(3):903–909. doi:10.1016/j.jconrel.2012.05.021
13. Strichartz GR, Sanchez V, Arthur GR, Chafetz R, Martin D (1990) Fundamental properties of local anesthetics. II. Measured octanol:buffer partition coefficients and pKa values of clinically used drugs. Anesth Analg 71(2):158–170
14. Kohane DS (2007) Microparticles and nanoparticles for drug delivery. Biotechnol Bioeng 96(2):203–209. doi:10.1002/bit.21301
15. Castillo J, Curley J, Hotz J, Uezono M, Tigner J, Chasin M, Wilder R, Langer R, Berde C (1996) Glucocorticoids prolong rat sciatic nerve blockade in vivo from bupivacaine microspheres. Anesthesiology 85(5):1157–1166
16. Kohane DS, Lipp M, Kinney RC, Lotan N, Langer R (2000) Sciatic nerve blockade with lipid-protein-sugar particles containing bupivacaine. Pharm Res 17(10):1243–1249
17. Kohane DS, Smith SE, Louis DN, Colombo G, Ghoroghchian P, Hunfeld NG, Berde CB, Langer R (2003) Prolonged duration local anesthesia from tetrodotoxin-enhanced local anesthetic microspheres. Pain 104(1–2):415–421
18. Epstein-Barash H, Shichor I, Kwon AH, Hall S, Lawlor MW, Langer R, Kohane DS (2009) Prolonged duration local anesthesia with minimal toxicity. Proc Natl Acad Sci U S A 106(17):7125–7130. doi:10.1073/pnas.0900598106
19. Kohane DS, Yieh J, Lu NT, Langer R, Strichartz GR, Berde CB (1998) A re-examination of tetrodotoxin for prolonged duration local anesthesia. Anesthesiology 89(1):119–131
20. Shankarappa SA, Sagie I, Tsui JH, Chiang HH, Stefanescu C, Zurakowski D, Kohane DS (2012) Duration and local toxicity of sciatic nerve blockade with coinjected site 1 sodium-channel blockers and quaternary lidocaine derivatives. Reg Anesth Pain Med 37(5):483–489. doi:10.1097/AAP.0b013e31826125b3
21. Kohane DS, Kuang Y, Lu NT, Langer R, Strichartz GR, Berde CB (1999) Vanilloid receptor agonists potentiate the in vivo local anesthetic activity of percutaneously injected site 1 sodium channel blockers. Anesthesiology 90(2):524–534
22. Kohane DS, Lu NT, Cairns BE, Berde CB (2001) Effects of adrenergic agonists and antagonists on tetrodotoxin-induced nerve block. Reg Anesth Pain Med 26(3):239–245. doi:10.1053/rapm.2001.23215
23. Kohane DS, Lu NT, Crosa GA, Kuang Y, Berde CB (2001) High concentrations of adrenergic antagonists prolong sciatic nerve blockade by tetrodotoxin. Acta Anaesthesiol Scand 45(7):899–905
24. Simons EJ, Bellas E, Lawlor MW, Kohane DS (2009) Effect of chemical permeation enhancers on nerve blockade. Mol Pharm 6(1):265–273. doi:10.1021/mp800167a
25. Sagie I, Kohane DS (2010) Prolonged sensory-selective nerve blockade. Proc Natl Acad Sci U S A 107(8):3740–3745. doi:10.1073/pnas.0911542107
26. Brandhonneur N, Dollo G, Ratajczak-Enselme M, Deniau AL, Chevanne F, Estebe JP, Legrand A, Le Corre P (2011) Ex vivo and in vivo diffusion of ropivacaine through spinal meninges: influence of absorption enhancers. Int J Pharm 404(1–2):36–41. doi:10.1016/j.ijpharm.2010.10.049
27. Jia X, Colombo G, Padera R, Langer R, Kohane DS (2004) Prolongation of sciatic nerve blockade by in situ cross-linked hyaluronic acid. Biomaterials 25(19):4797–4804

28. Kohane DS, Lipp M, Kinney RC, Anthony DC, Louis DN, Lotan N, Langer R (2002) Biocompatibility of lipid-protein-sugar particles containing bupivacaine in the epineurium. J Biomed Mater Res 59(3):450–459
29. Pere P, Watanabe H, Pitkanen M, Wahlstrom T, Rosenberg PH (1993) Local myotoxicity of bupivacaine in rabbits after continuous supraclavicular brachial plexus block. Reg Anesth 18(5):304–307
30. Zimmer C, Piepenbrink K, Riest G, Peters J (2007) Cardiotoxic and neurotoxic effects after accidental intravascular bupivacaine administration. Therapy with lidocaine propofol and lipid emulsion. Anaesthesist 56(5):449–453. doi:10.1007/s00101-007-1147-3
31. Yamashita A, Matsumoto M, Matsumoto S, Itoh M, Kawai K, Sakabe T (2003) A comparison of the neurotoxic effects on the spinal cord of tetracaine, lidocaine, bupivacaine, and ropivacaine administered intrathecally in rabbits. Anesth Analg 97(2):512–519, table of contents
32. Benoit PW, Belt WD (1970) Destruction and regeneration of skeletal muscle after treatment with a local anaesthetic, bupivacaine (Marcaine). J Anat 107(Pt 3):547–556
33. Benoit PW, Belt WD (1972) Some effects of local anesthetic agents on skeletal muscle. Exp Neurol 34(2):264–278
34. Yagiela JA, Benoit PW, Buoncristiani RD, Peters MP, Fort NF (1981) Comparison of myotoxic effects of lidocaine with epinephrine in rats and humans. Anesth Analg 60(7):471–480
35. Foster AH, Carlson BM (1980) Myotoxicity of local anesthetics and regeneration of the damaged muscle fibers. Anesth Analg 59(10):727–736
36. Brun A (1959) Effect of procaine, carbocain and xylocaine on cutaneous muscle in rabbits and mice. Acta Anaesthesiol Scand 3(2):59–73
37. Zink W, Graf BM (2004) Local anesthetic myotoxicity. Reg Anesth Pain Med 29(4):333–340
38. Nouette-Gaulain K, Bringuier S, Canal-Raffin M, Bernard N, Lopez S, Dadure C, Masson F, Mercier J, Sztark F, Rossignol R, Capdevila X (2010) Time course of mitochondrial metabolism alterations to repeated injections of bupivacaine in rat muscle. Can J Anaesth 57(9):836–842. doi:10.1007/s12630-010-9347-8
39. Hogan Q, Dotson R, Erickson S, Kettler R, Hogan K (1994) Local anesthetic myotoxicity: a case and review. Anesthesiology 80(4):942–947
40. Neuburger M, Breitbarth J, Reisig F, Lang D, Buttner J (2006) Complications and adverse events in continuous peripheral regional anesthesia Results of investigations on 3,491 catheters. Anaesthesist 55(1):33–40. doi:10.1007/s00101-005-0920-4
41. Selander D (1993) Neurotoxicity of local anesthetics: animal data. Reg Anesth 18(6 Suppl):461–468
42. Barnet CS, Louis DN, Kohane DS (2005) Tissue injury from tricyclic antidepressants used as local anesthetics. Anesth Analg 101(6):1838–1843. doi:10.1213/01.ANE.0000184129.50312.C1
43. Estebe JP, Myers RR (2004) Amitriptyline neurotoxicity: dose-related pathology after topical application to rat sciatic nerve. Anesthesiology 100(6):1519–1525
44. Lim TK, Macleod BA, Ries CR, Schwarz SK (2007) The quaternary lidocaine derivative, QX-314, produces long-lasting local anesthesia in animal models in vivo. Anesthesiology 107(2):305–311. doi:10.1097/01.anes.0000270758.77314.b4
45. Padera RF, Tse JY, Bellas E, Kohane DS (2006) Tetrodotoxin for prolonged local anesthesia with minimal myotoxicity. Muscle Nerve 34(6):747–753. doi:10.1002/mus.20618
46. Sakura S, Bollen AW, Ciriales R, Drasner K (1995) Local anesthetic neurotoxicity does not result from blockade of voltage-gated sodium channels. Anesth Analg 81(2):338–346
47. Kohane DS, Lu NT, Gokgol-Kline AC, Shubina M, Kuang Y, Hall S, Strichartz GR, Berde CB (2000) The local anesthetic properties and toxicity of saxitonin homologues for rat sciatic nerve block in vivo. Reg Anesth Pain Med 25(1):52–59
48. Colombo G, Langer R, Kohane DS (2004) Effect of excipient composition on the biocompatibility of bupivacaine-containing microparticles at the sciatic nerve. J Biomed Mater Res A 68(4):651–659. doi:10.1002/jbm.a.20074
49. Colombo G, Padera R, Langer R, Kohane DS (2005) Prolonged duration local anesthesia with lipid-protein-sugar particles containing bupivacaine and dexamethasone. J Biomed Mater Res A 75(2):458–464. doi:10.1002/jbm.a.30443

50. Kohane DS, Langer R (2010) Biocompatibility and drug delivery systems. Chem Sci 1(4):441–446. doi:10.1039/c0sc00203h
51. Kohane DS, Tse JY, Yeo Y, Padera R, Shubina M, Langer R (2006) Biodegradable polymeric microspheres and nanospheres for drug delivery in the peritoneum. J Biomed Mater Res A 77(2):351–361. doi:10.1002/jbm.a.30654
52. Leng F, Wan J, Liu W, Tao B, Chen X (2012) Prolongation of epidural analgesia using solid lipid nanoparticles as drug carrier for lidocaine. Reg Anesth Pain Med 37(2):159–165. doi:10.1097/AAP.0b013e31823fc058
53. De Melo NF, De Araujo DR, Grillo R, Moraes CM, De Matos AP, de Paula E, Rosa AH, Fraceto LF (2012) Benzocaine-loaded polymeric nanocapsules: study of the anesthetic activities. J Pharm Sci 101(3):1157–1165. doi:10.1002/jps.22829
54. Lian T, Ho RJ (2001) Trends and developments in liposome drug delivery systems. J Pharm Sci 90(6):667–680
55. Torchilin VP (2005) Recent advances with liposomes as pharmaceutical carriers. Nat Rev Drug Discov 4(2):145–160. doi:10.1038/nrd1632
56. Rose JS, Neal JM, Kopacz DJ (2005) Extended-duration analgesia: update on microspheres and liposomes. Reg Anesth Pain Med 30(3):275–285
57. de Paula E, Cereda CM, Tofoli GR, Franz-Montan M, Fraceto LF, de Araujo DR (2010) Drug delivery systems for local anesthetics. Recent Pat Drug Deliv Formul 4(1):23–34
58. Boogaerts J, Declercq A, Lafont N, Benameur H, Akodad EM, Dupont JC, Legros FJ (1993) Toxicity of bupivacaine encapsulated into liposomes and injected intravenously: comparison with plain solutions. Anesth Analg 76(3):553–555
59. Mashimo T, Uchida I, Pak M, Shibata A, Nishimura S, Inagaki Y, Yoshiya I (1992) Prolongation of canine epidural anesthesia by liposome encapsulation of lidocaine. Anesth Analg 74(6):827–834
60. Mura P, Maestrelli F, Gonzalez-Rodriguez ML, Michelacci I, Ghelardini C, Rabasco AM (2007) Development, characterization and in vivo evaluation of benzocaine-loaded liposomes. Eur J Pharm Biopharm 67(1):86–95. doi:10.1016/j.ejpb.2007.01.020
61. Mowat JJ, Mok MJ, MacLeod BA, Madden TD (1996) Liposomal bupivacaine. Extended duration nerve blockade using large unilamellar vesicles that exhibit a proton gradient. Anesthesiology 85(3):635–643
62. Madden TD, Harrigan PR, Tai LC, Bally MB, Mayer LD, Redelmeier TE, Loughrey HC, Tilcock CP, Reinish LW, Cullis PR (1990) The accumulation of drugs within large unilamellar vesicles exhibiting a proton gradient: a survey. Chem Phys Lipids 53(1):37–46
63. Grant GJ, Barenholz Y, Bolotin EM, Bansinath M, Turndorf H, Piskoun B, Davidson EM (2004) A novel liposomal bupivacaine formulation to produce ultralong-acting analgesia. Anesthesiology 101(1):133–137
64. Grant GJ, Barenholz Y, Piskoun B, Bansinath M, Turndorf H, Bolotin EM (2001) DRV liposomal bupivacaine: preparation, characterization, and in vivo evaluation in mice. Pharm Res 18(3):336–343
65. Grant GJ, Piskoun B, Bansinath M (2003) Analgesic duration and kinetics of liposomal bupivacaine after subcutaneous injection in mice. Clin Exp Pharmacol Physiol 30(12):966–968
66. Grant SA (2002) The Holy Grail: long-acting local anaesthetics and liposomes. Best Pract Res Clin Anaesthesiol 16(2):345–352
67. Shipton EA (2012) New formulations of local anaesthetics-part I. Anesthesiol Res Pract 2012:546409. doi:10.1155/2012/546409
68. Smith KJ, McDonald WI (1980) Spontaneous and mechanically evoked activity due to central demyelinating lesion. Nature 286(5769):154–155
69. Hall SM, Gregson NA (1971) The in vivo and ultrastructural effects of injection of lysophosphatidyl choline into myelinated peripheral nerve fibres of the adult mouse. J Cell Sci 9(3):769–789
70. Gorfine SR, Onel E, Patou G, Krivokapic ZV (2011) Bupivacaine extended-release liposome injection for prolonged postsurgical analgesia in patients undergoing hemorrhoidectomy: a multicenter, randomized, double-blind, placebo-controlled trial. Dis Colon Rectum 54(12):1552–1559. doi:10.1097/DCR.0b013e318232d4c1

71. Golf M, Daniels SE, Onel E (2011) A phase 3, randomized, placebo-controlled trial of DepoFoam(R) bupivacaine (extended-release bupivacaine local analgesic) in bunionectomy. Adv Ther 28(9):776–788. doi:10.1007/s12325-011-0052-y
72. Bramlett K, Onel E, Viscusi ER, Jones K (2012) A randomized, double-blind, dose-ranging study comparing wound infiltration of DepoFoam bupivacaine, an extended-release liposomal bupivacaine, to bupivacaine HCl for postsurgical analgesia in total knee arthroplasty. Knee 19(5):530–536. doi:10.1016/j.knee.2011.12.004
73. Smoot JD, Bergese SD, Onel E, Williams HT, Hedden W (2012) The efficacy and safety of DepoFoam bupivacaine in patients undergoing bilateral, cosmetic, submuscular augmentation mammaplasty: a randomized, double-blind, active-control study. Aesthet Surg J 32(1):69–76. doi:10.1177/1090820X11430831
74. Bragagni M, Maestrelli F, Mennini N, Ghelardini C, Mura P (2010) Liposomal formulations of prilocaine: effect of complexation with hydroxypropyl-ss-cyclodextrin on drug anesthetic efficacy. J Liposome Res 20(4):315–322. doi:10.3109/08982100903544169
75. Shikanov A, Domb AJ, Weiniger CF (2007) Long acting local anesthetic-polymer formulation to prolong the effect of analgesia. J Control Release 117(1):97–103. doi:10.1016/j.jconrel.2006.10.014
76. Blanco MD, Bernardo MV, Sastre RL, Olmo R, Muniz E, Teijon JM (2003) Preparation of bupivacaine-loaded poly(epsilon-caprolactone) microspheres by spray drying: drug release studies and biocompatibility. Eur J Pharm Biopharm 55(2):229–236
77. Kojima T, Nakano M, Juni K, Inoue S, Yoshida Y (1985) Preparation and evaluation in vitro and in vivo of polycarbonate microspheres containing dibucaine. Chem Pharm Bull (Tokyo) 33(11):5119–5125
78. Jain RA (2000) The manufacturing techniques of various drug loaded biodegradable poly(lactide-co-glycolide) (PLGA) devices. Biomaterials 21(23):2475–2490
79. Chen PC, Kohane DS, Park YJ, Bartlett RH, Langer R, Yang VC (2004) Injectable microparticle-gel system for prolonged and localized lidocaine release. II. In vivo anesthetic effects. J Biomed Mater Res A 70(3):459–466. doi:10.1002/jbm.a.30101
80. Chen PC, Park YJ, Chang LC, Kohane DS, Bartlett RH, Langer R, Yang VC (2004) Injectable microparticle-gel system for prolonged and localized lidocaine release. I. In vitro characterization. J Biomed Mater Res A 70(3):412–419. doi:10.1002/jbm.a.30086
81. Horie RT, Sakamoto T, Nakagawa T, Tabata Y, Okamura N, Tomiyama N, Tachibana M, Ito J (2010) Sustained delivery of lidocaine into the cochlea using poly lactic/glycolic acid microparticles. Laryngoscope 120(2):377–383. doi:10.1002/lary.20713
82. Le Corre P, Estebe JP, Clement R, Du Plessis L, Chevanne F, Ecoffey C, Le Verge R (2002) Spray-dryed bupivacaine-loaded microspheres: in vitro evaluation and biopharmaceutics of bupivacaine following brachial plexus administration in sheep. Int J Pharm 238(1–2):191–203
83. Xu Q, Hashimoto M, Dang TT, Hoare T, Kohane DS, Whitesides GM, Langer R, Anderson DG (2009) Preparation of monodisperse biodegradable polymer microparticles using a microfluidic flow-focusing device for controlled drug delivery. Small 5(13):1575–1581. doi:10.1002/smll.200801855
84. Wang CF, Pancaro C, Gerner P, Strichartz G (2011) Prolonged suppression of postincisional pain by a slow-release formulation of lidocaine. Anesthesiology 114(1):135–149. doi:10.1097/ALN.0b013e3182001996
85. Gerner P, Wang CF, Lee BS, Suzuki S, Degirolami U, Gandhi A, Knaack D, Strichartz G (2010) The relationship between functional sciatic nerve block duration and the rate of release of lidocaine from a controlled-release matrix. Anesth Analg 111(1):221–229. doi:10.1213/ANE.0b013e3181dd2690
86. Tobe M, Obata H, Suto T, Yokoo H, Nakazato Y, Tabata Y, Saito S (2010) Long-term effect of sciatic nerve block with slow-release lidocaine in a rat model of postoperative pain. Anesthesiology 112(6):1473–1481. doi:10.1097/ALN.0b013e3181d4f66f
87. Blanco MD, Bernardo MV, Gomez C, Muniz E, Teijon JM (1999) Bupivacaine-loaded comatrix formed by albumin microspheres included in a poly(lactide-co-glycolide) film: in vivo biocompatibility and drug release studies. Biomaterials 20(20):1919–1924

88. Masters DB, Berde CB, Dutta S, Turek T, Langer R (1993) Sustained local anesthetic release from bioerodible polymer matrices: a potential method for prolonged regional anesthesia. Pharm Res 10(10):1527–1532
89. Masters DB, Berde CB, Dutta SK, Griggs CT, Hu D, Kupsky W, Langer R (1993) Prolonged regional nerve blockade by controlled release of local anesthetic from a biodegradable polymer matrix. Anesthesiology 79(2):340–346
90. Hersh EV, Maniar M, Green M, Cooper SA (1992) Anesthetic activity of the lipospheres bupivacaine delivery system in the rat. Anesth Prog 39(6):197–200
91. Masters DB, Domb AJ (1998) Liposphere local anesthetic timed-release for perineural site application. Pharm Res 15(7):1038–1045
92. Toongsuwan S, Li LC, Erickson BK, Chang HC (2004) Formulation and characterization of bupivacaine lipospheres. Int J Pharm 280(1–2):57–65. doi:10.1016/j.ijpharm.2004.04.020
93. Pietkiewicz J, Sznitowska M, Placzek M (2006) The expulsion of lipophilic drugs from the cores of solid lipid microspheres in diluted suspensions and in concentrates. Int J Pharm 310(1–2):64–71. doi:10.1016/j.ijpharm.2005.11.038
94. Yeo Y, Kohane DS (2008) Polymers in the prevention of peritoneal adhesions. Eur J Pharm Biopharm 68(1):57–66. doi:10.1016/j.ejpb.2007.03.027
95. Sivakumaran D, Maitland D, Hoare T (2011) Injectable microgel-hydrogel composites for prolonged small-molecule drug delivery. Biomacromolecules 12(11):4112–4120. doi:10.1021/bm201170h
96. Bulpitt P, Aeschlimann D (1999) New strategy for chemical modification of hyaluronic acid: preparation of functionalized derivatives and their use in the formation of novel biocompatible hydrogels. J Biomed Mater Res 47(2):152–169
97. Paavola A, Yliruusi J, Kajimoto Y, Kalso E, Wahlstrom T, Rosenberg P (1995) Controlled release of lidocaine from injectable gels and efficacy in rat sciatic nerve block. Pharm Res 12(12):1997–2002
98. Paavola A, Tarkkila P, Xu M, Wahlstrom T, Yliruusi J, Rosenberg P (1998) Controlled release gel of ibuprofen and lidocaine in epidural use–analgesia and systemic absorption in pigs. Pharm Res 15(3):482–487. doi:10.1023/A:1011992702604
99. Karashima K, Taniguchi M, Nakamura T, Takasaki M, Matsuo K, Irikura M, Irie T (2007) Prolongation of intrathecal and sciatic nerve blocks using a complex of levobupivacaine with maltosyl-beta-cyclodextrin in rats. Anesth Analg 104(5):1121–1128. doi:10.1213/01.ane.0000260309.15034.52, tables of contents
100. Suzuki R, Arai YC, Hamayasu K, Fujita K, Hara K, Yamaguchi T, Sasaguri S (2009) Complex of branched cyclodextrin and lidocaine prolonged the duration of peripheral nerve block. J Anesth 23(2):295–297. doi:10.1007/s00540-008-0720-5
101. de Araujo DR, Tsuneda SS, Cereda CM, Del GFCF, Prete PS, Fernandes SA, Yokaichiya F, Franco MK, Mazzaro I, Fraceto LF, de FABA, de Paula E (2008) Development and pharmacological evaluation of ropivacaine-2-hydroxypropyl-beta-cyclodextrin inclusion complex. Eur J Pharm Sci 33(1):60–71. doi:10.1016/j.ejps.2007.09.010
102. Cereda CM, Tofoli GR, Maturana LG, Pierucci A, Nunes LA, Franz-Montan M, de Oliveira AL, Arana S, de Araujo DR, de Paula E (2012) Local neurotoxicity and myotoxicity evaluation of cyclodextrin complexes of bupivacaine and ropivacaine. Anesth Analg 115(5):1234–1241. doi:10.1213/ANE.0b013e318266f3d9
103. Maestrelli F, Gonzalez-Rodriguez ML, Rabasco AM, Ghelardini C, Mura P (2010) New "drug-in cyclodextrin-in deformable liposomes" formulations to improve the therapeutic efficacy of local anaesthetics. Int J Pharm 395(1–2):222–231. doi:10.1016/j.ijpharm.2010.05.046
104. Sokolsky-Papkov M, Golovanevski L, Domb AJ, Weiniger CF (2010) Poly(DL:lactic acid-castor oil) 3:7-bupivacaine formulation: reducing burst effect prolongs efficacy in vivo. J Pharm Sci 99(6):2732–2738. doi:10.1002/jps.22025
105. Vaisman B, Motiei M, Nyska A, Domb AJ (2010) Biocompatibility and safety evaluation of a ricinoleic acid-based poly(ester-anhydride) copolymer after implantation in rats. J Biomed Mater Res A 92(2):419–431. doi:10.1002/jbm.a.32342

106. Soderberg L, Dyhre H, Roth B, Bjorkman S (2006) Ultralong peripheral nerve block by lidocaine:prilocaine 1:1 mixture in a lipid depot formulation: comparison of in vitro, in vivo, and effect kinetics. Anesthesiology 104(1):110–121
107. Dyhre H, Soderberg L, Bjorkman S, Carlsson C (2006) Local anesthetics in lipid-depot formulations–neurotoxicity in relation to duration of effect in a rat model. Reg Anesth Pain Med 31(5):401–408. doi:10.1016/j.rapm.2006.05.008
108. Lee JW, Lim TH, Park JB (2010) Intradiscal drug delivery system for the treatment of low back pain. J Biomed Mater Res A 92(1):378–385. doi:10.1002/jbm.a.32377
109. Yin QQ, Wu L, Gou ML, Qian ZY, Zhang WS, Liu J (2009) Long-lasting infiltration anaesthesia by lidocaine-loaded biodegradable nanoparticles in hydrogel in rats. Acta Anaesthesiol Scand 53(9):1207–1213. doi:10.1111/j.1399-6576.2009.02030.x
110. Cohen R, Kanaan H, Grant GJ, Barenholz Y (2012) Prolonged analgesia from Bupisome and Bupigel formulations: from design and fabrication to improved stability. J Control Release 160(2):346–352. doi:10.1016/j.jconrel.2011.12.030
111. Verron E, Gauthier O, Janvier P, Le Guen H, Holopherne D, Cavagna R, Bouler JM (2010) Analgesic properties of calcium phosphate apatite loaded with bupivacaine on postoperative pain. J Biomed Mater Res B Appl Biomater 94(1):89–96. doi:10.1002/jbm.b.31628
112. Radin S, Campbell JT, Ducheyne P, Cuckler JM (1997) Calcium phosphate ceramic coatings as carriers of vancomycin. Biomaterials 18(11):777–782
113. Hamanishi C, Kitamoto K, Tanaka S, Otsuka M, Doi Y, Kitahashi T (1996) A self-setting TTCP-DCPD apatite cement for release of vancomycin. J Biomed Mater Res 33(3):139–143. doi:10.1002/(SICI)1097-4636(199623)33:3<139::AID-JBM3>3.0.CO;2-R
114. Tiwary AK, Sapra B, Jain S (2007) Innovations in transdermal drug delivery: formulations and techniques. Recent Pat Drug Deliv Formul 1(1):23–36
115. Sawyer J, Febbraro S, Masud S, Ashburn MA, Campbell JC (2009) Heated lidocaine/tetracaine patch (Synera, Rapydan) compared with lidocaine/prilocaine cream (EMLA) for topical anaesthesia before vascular access. Br J Anaesth 102(2):210–215. doi:10.1093/bja/aen364
116. Cho CW, Choi JS, Shin SC (2011) Enhanced local anesthetic action of mepivacaine from the bioadhesive gels. Pak J Pharm Sci 24(1):87–93
117. He CX, He ZG, Gao JQ (2010) Microemulsions as drug delivery systems to improve the solubility and the bioavailability of poorly water-soluble drugs. Expert Opin Drug Deliv 7(4):445–460. doi:10.1517/17425241003596337
118. Yuan JS, Yip A, Nguyen N, Chu J, Wen XY, Acosta EJ (2010) Effect of surfactant concentration on transdermal lidocaine delivery with linker microemulsions. Int J Pharm 392(1–2): 274–284. doi:10.1016/j.ijpharm.2010.03.051
119. Lafont ND, Boogaerts JG, Legros FJ (1994) Use of liposome-associated bupivacaine for the management of a chronic pain syndrome. Anesth Analg 79(4):818
120. Lafont ND, Legros FJ, Boogaerts JG (1996) Use of liposome-associated bupivacaine in a cancer pain syndrome. Anaesthesia 51(6):578–579
121. Boogaerts JG, Lafont ND, Declercq AG, Luo HC, Gravet ET, Bianchi JA, Legros FJ (1994) Epidural administration of liposome-associated bupivacaine for the management of postsurgical pain: a first study. J Clin Anesth 6(4):315–320
122. Holte K, Werner MU, Lacouture PG, Kehlet H (2002) Dexamethasone prolongs local analgesia after subcutaneous infiltration of bupivacaine microcapsules in human volunteers. Anesthesiology 96(6):1331–1335
123. Pedersen JL, Lilleso J, Hammer NA, Werner MU, Holte K, Lacouture PG, Kehlet H (2004) Bupivacaine in microcapsules prolongs analgesia after subcutaneous infiltration in humans: a dose-findingstudy.AnesthAnalg99(3):912–918.doi:10.1213/01.ANE.0000143791.77608.0B, table of contents
124. Kopacz DJ, Lacouture PG, Wu D, Nandy P, Swanton R, Landau C (2003) The dose response and effects of dexamethasone on bupivacaine microcapsules for intercostal blockade (T9 to T11) in healthy volunteers. Anesth Analg 96(2):576–582, table of contents
125. Richard BM, Ott LR, Haan D, Brubaker AN, Cole PI, Nelson KG, Ross PE, Rebelatto MC, Newton PE (2011) The safety and tolerability evaluation of DepoFoam bupivacaine

(bupivacaine extended-release liposome injection) administered by incision wound infiltration in rabbits and dogs. Expert Opin Investig Drugs 20(10):1327–1341. doi:10.1517/13543784.2011.611499

126. Wakiyama N, Juni K, Nakano M (1982) Preparation and evaluation in vitro and in vivo of polylactic acid microspheres containing dibucaine. Chem Pharm Bull (Tokyo) 30(10): 3719–3727
127. Rodriguez-Navarro AJ, Lagos M, Figueroa C, Garcia C, Recabal P, Silva P, Iglesias V, Lagos N (2009) Potentiation of local anesthetic activity of neosaxitoxin with bupivacaine or epinephrine: development of a long-acting pain blocker. Neurotox Res 16(4):408–415. doi:10.1007/s12640-009-9092-3
128. Timko BP, Dvir T, Kohane DS (2010) Remotely triggerable drug delivery systems. Adv Mater 22(44):4925–4943. doi:10.1002/adma.201002072
129. Richard BM, Rickert DE, Newton PE, Ott LR, Haan D, Brubaker AN, Cole PI, Ross PE, Rebelatto MC, Nelson KG (2011) Safety evaluation of exparel (DepoFoam bupivacaine) administered by repeated subcutaneous injection in rabbits and dogs: Species comparison. J drug deliv 2011:467429. doi:10.115/2011/467429
130. Richard BM, Newton P, Ott LR, Haan D, Brubaker AN, Cole PI, Ross PE, Rebelatto MC, Nelson KG (2012) The safety of Exparel® (bupivacaine liposomal injectable suspension) administered by peripheral nerve block in rabbits and dogs. J Drug deliv 2012:962101. doi:10.1155/2012/962101

About the Editors

Abraham J. Domb is a Professor for Medicinal Chemistry and Biopolymers at the Faculty of Medicine of the Hebrew University, Jerusalem, Israel. He earned Bachelor's degrees in Chemistry, Pharmacy and Law studies and Ph.D. degree in Chemistry from Hebrew University. He did his postdoctoral training at MIT and Harvard University, USA, and was R&D Manager at Nova Pharm. Co., Baltimore, USA, during 1988–1992. During 2007–2012 he headed the Division of Identification and Forensic Sciences of the Israel Police. His primary research interests are in biopolymers, controlled drug delivery, cancer therapy, nanoparticulate systems, and forensic sciences.

Wahid Khan obtained his Master's and Ph.D. in Pharmaceutics at the National Institute of Pharmaceutical and Educational Research (NIPER), Mohali, India, and worked with Prof. Abraham J. Domb at the Hebrew University of Jerusalem, Israel, for his postdoctoral research. Currently, he is working as an Assistant Professor in the Department of Pharmaceutics, NIPER, Hyderabad, India. He is having experience of working in areas of drug delivery, drug targeting, nanomedicine, and biodesign of implantable medical devices. Dr. Khan has also won various national and international awards and fellowships including Israel government prestigious PBC fellowships for outstanding postdoctoral researcher and Valazzi-Pikovsky and Lady Davis fellowship award.

A.J. Domb and W. Khan (eds.), *Focal Controlled Drug Delivery*, Advances in Delivery Science and Technology, DOI 10.1007/978-1-4614-9434-8,

Index

A.J. Domb and W. Khan (eds.), *Focal Controlled Drug Delivery*, Advances in Delivery Science and Technology, DOI 10.1007/978-1-4614-9434-8,

B

C

P

Q

R

MIX
Papier aus verantwortungsvollen Quellen
Paper from responsible sources
FSC® C105338
FSC
www.fsc.org

If you have any concerns about our products,
you can contact us on
ProductSafety@springernature.com

In case Publisher is established outside the EU,
the EU authorized representative is:
Springer Nature Customer Service Center GmbH
Europaplatz 3, 69115 Heidelberg, Germany

Printed by Libri Plureos GmbH
in Hamburg, Germany